NORTH AMERICAN FOREST HISTORY

NORTH AMERICAN FOREST HISTORY:

A Guide to Archives and Manuscripts in the United States and Canada

COMPILED BY

Richard C. Davis

PUBLISHED UNDER CONTRACT WITH
THE FOREST HISTORY SOCIETY, INC.

CLIO BOOKS
Santa Barbara, California

Library of Congress Cataloging in Publication Data

Davis, Richard C 1939
North American forest history.

Bibliography: p.
Includes index.
1. Forest and forestry—United States—History—Sources—Bibliography. 2. Forest and forestry—Canada—History—Sources—Bibliography. 3. Archives—United States—Inventions, calendars, etc. 4. Archives—Canada—Inventories, calendars, etc. 5. Forests and forestry—Library resources—United States—Directories. 6. Forests and forestry—Library resources—Canada—Directories. I. Forest History Society. II. Title.
Z5991. D33 [SD143] 016.6349'0973 76-46460

ISBN 0-87436-237-7

Manufactured in the United States of America.

American Bibliographical Center — Clio Press
2040 Alameda Padre Serra
Santa Barbara, California

European Bibliographical Center — Clio Press
Woodside House, Hinksey Hill
Oxford OX1 5BE, England

Published under contract with The Forest History Society, Inc.

To the archivists, librarians, and others who collaborated in this effort.

Acknowledgement

The publisher acknowledges with thanks the contributions of the compiler and the Forest History Society for their supervision of the entire editorial process.

CONTENTS

Canada

NORTH AMERICAN FOREST HISTORY

INTRODUCTION

This guide describes manuscripts and archival records relating to the history of the lumber and forest products industries, forestry, conservation politics, and national and state parks movements, wilderness preservation, and associated topics. Materials merely descriptive of forests are not necessarily included, nor are records relating to the use and marketing of manufactured forest products, although many of the entries herein do concern such topics. This guide describes materials located within the United States and Canada. Time limits imposed at the outset of the project did not permit inclusion of material in British and other European repositories which relate to North American forests.

In 1956 the Forest History Foundation, a precursor of the Forest History Society, released in processed form *Forest History Sources of the United States and Canada,* compiled by Clodaugh M. Neiderheiser. That work contained descriptions of 972 groups of manuscripts and records held by 108 repositories. The contrast between these figures and the 3830 groups held by 358 repositories reporting today suggests a tremendously increased interest in forest and conservation history by historians and those responsible for the safekeeping of historical records.

The collections described in this guide consist largely of textual records, such as holograph and typed correspondence, reports, memoranda, diaries, journals, accounts, minutes, notes, memoirs, speeches, contracts and bills. Printed or processed materials are mentioned when contained in manuscript collections or archives, but collections consisting entirely of such materials are generally excluded. On the other hand, photocopies of manuscript or archival materials are included, along with a statement concerning the location of the originals, if known.

Master's theses written from a historical perspective and a few undergraduate theses are listed here when reported by repositories, but no technical forestry theses are included. Doctoral dissertations are excluded, as most of those relating to forest history may be found in the companion to this volume, Ronald J. Fahl's *North American Forest and Conservation History: A Bibliography*.

No great effort has been made to pursue pictorial material, but this guide includes collections of photographs and motion picture films whenever repositories have reported them. However, a more thorough search has been made for oral history transcripts and tape recordings, and a large number are listed herein. Others of interest may be found in *Oral History Collections,* compiled and edited by Alan M. Meckler and Ruth McMullin, New York and London: R.R. Bowker Company, 1975, 344 pp.

The institutions represented in this guide include libraries, archives, historical societies, and museums. Except in a few cases where private industrial archives have been established and opened to scholarly use on a regular basis, there are no entries for materials remaining in the custody of the originating agencies. Historical investigators should be aware, however, that both public and private organizations often welcome use of their records by serious researchers, and that many national park and national forest headquarters maintain "historical files" of materials relating to local history. Since this places them in the category of manuscript repositories, the holdings of a few such offices are listed.

Compilation of this work began with the examination of finding aids issued by individual repositories, as well as a number of national, regional, and subject guides covering American and Canadian institutions. At the end of this volume is a bibliography of those guides which provided useful information. The bibliography does not include registers of single manuscript collections, but such aids are cited in the appropriate descriptive entries in the body of the guide. Only a few of the bibliographies in the monographic literature of forest history have been consulted in search of additional collections, and many of the works listed in *North American Forest and Conservation History: A Bibliography* may list materials not described here.

Letters of inquiry enclosing descriptions of the previously reported materials were mailed to 398 institutions credited in those finding aids with forest history holdings. The repositories were asked to verify the accuracy of this information; to indicate any of the listed collections not now in their possession; to supply, when possible, more detailed information on the relevance of their collections to forest history; and to send descriptions of any additional pertinent accessions.

The first few repositories to be queried were asked to use questionnaires modeled on those employed by the *National Union Catalog of Manuscript Collections.* It quickly became apparent that the use of such questionnaires placed an impractical burden on repository staffs, and prompt and comprehensive replies could more easily be obtained by inviting repositories to send information in any convenient format. The subsequent resolution of discrepancies and the frequent need for more detailed information necessitated several hundred follow-up letters.

One hundred and twelve repositories did not respond to the initial letter of inquiry, and even two or three reminders failed to elicit reply from forty-nine institutions. Those thought to hold only scattered items of marginal research value were dropped from further consideration. Eighteen nonrespondents believed to have significant collections are listed in this guide, with the sources of information on their holdings indicated. In order that the user may be aware that the descriptions of holdings of these institutions have not been verified, such entries are marked with an asterisk (*) next to the repository name. Also marked by asterisks are the names of six repositories with holdings described only from published reports received too late for mail contact with the holding institutions to have been practical.

Two hundred and ninety-nine of the queried institutions reported forest history holdings, while fifty replied that they did not now hold, or never did hold, such material. Reports usually took the form of annotations jotted on the descriptions submitted with the initial inquiry, accompanied by photocopies of catalog cards or other brief descriptions originally compiled for some other purpose covering new accessions. Often, copies of in-house registers or processed finding aids were provided. In a few cases, curators compiled descriptions specifically for this project.

Nine hundred and fifty-six additional manuscript and archival institutions received form letters inquiring about their holdings. In this group were all state and provincial archives and state historical societies not previously contacted. Twenty-three institutions reported holdings in response to this circular. Although no reply had been requested from institutions without materials in the field, sixty-four courteously advised that they had nothing to report.

All major historical and archival journals received statements concerning the project, and at least eleven published this item in their news sections. Five forestry and forest-industries journals carried similar announcements. In response to these notices, three institutions reported holdings.

During the course of the project, the compiler of the guide visited twelve repositories. Time limitations largely restricted his work to the examination of in-house finding aids. Either volunteer assistants or Forest History Society employees provided information on the holdings of several other institutions.

Almost all the reports incorporated in the body of this guide were received between December, 1973, and May, 1975. Editorial work and indexing were completed early in September, 1975. Later in the same month, the compiler inserted in the manuscript of the guide newly received information concerning five additional repositories and thirteen collections.

The information received from these various sources has been edited to focus the description on the relevance of the material to forest history. Thus the user should not assume that the entries represent comprehensive descriptions of the records or of manuscript collections.

The body of this guide reflects some unevenness in reporting of information. Units of entry for archival materials may be either record groups or subgroups, but whenever possible subgroups are kept together under record group headings. On the other hand, diverse materials, either of minor significance or about which little information was made available, sometimes are described in collective entries. Such variations in description should cause no confusion to moderately experienced researchers.

The space required to relate the known importance of the described material to the subject of forest history determines the length and detail of any particular entry. Large homogeneous collections are described in fewer words than smaller collections composed of diverse types of records or relating to a variety of topics. Thus, unprocessed collections, whatever their size, are represented by very brief entries. The emphases of the descriptions are on information concerning the records, rather than information about the person or organization responsible for their creation.

When there are known to be restrictions on the research use of materials, that fact has been noted. In most cases, it has been left to the prospective researcher to determine, through correspondence with the repositories, the exact nature of the restrictions.

Unless specified otherwise, measurements of the volume of records should be assumed to be given in linear feet.

The index refers the user to entry numbers, not page numbers. Index citations are to the descriptive entries, not to the records themselves. Index headings include all proper names mentioned in the entries. Specific geographical locations are listed under state or provincial headings. Topical headings are divided into geographical and chronological groupings. Regional subheads, such as "New England," "Pacific Northwest," and "Maritime Provinces," are used for materials that cannot be attributed to specific states or provinces. Cross-references are provided between these subheadings and states and provinces within the regions. The terms "United States" and "Canada" are used only for entries which cannot be attributed to a specific state, province, or region, and for references to the national governments of the two countries. References to localities in other countries may be found indexed under the name of the particular nation involved.

In the selection of subject headings, no attempt has been made to establish a thesaurus of mutually exclusive terms. Synonyms have been examined, however, and compared with the usage suggested by *Forest Terminology: A Glossary of Technical Terms Used in Forestry,* 3rd ed., Washington: Society of American Foresters, 1958; L.G. Sorden, *Lumberjack Lingo,* Spring Green, Wisconsin: Wisconsin House, Inc., 1969; and Walter F. McCulloch, *Woods Words: A Comprehensive Dictionary of Loggers Terms,* Portland: Oregon Historical Society and the Champoeg Press, 1958; as well as standard dictionaries. Cross-references are provided wherever appropriate.

A major grant to the Forest History Society from the National Endowment for the Humanities supported the compilation of this guide and the companion bibliography by Ronald J. Fahl. The following donors generously contributed supplemental funding: Laird Norton Foundation, John M. Musser, Weyerhaeuser Foundation, Inc., Garrett Eddy, Gene C. Brewer, Susan L. Flader, Harold A. Miller, Nils B. Hult, James D. Bronson, J. Paul Neils, Joseph E. McCaffrey, John H. Hauberg, Edmund G. Hayes, George E. Lamb, The Langdale Company, George S. Kephart, Robert H. Noyes, and Hans Schneider; also John H. Hinman, James W. Craig, E.R.Titcomb, F.P. Keen, Connwood, Inc., Philip H. Jones, G.E. Karlen, Selwyn J. Sharp, Hardy L. Shirley, Clare W. Hendee, Carl G. Krueger, Rolf B. Jorgensen, Gene S. Bergoffen, and H.E. Ruark.

Completion of this project in a two-year period would have been impossible without the fine cooperation and assistance of hundreds of curators, archivists, and manuscript librarians who generously provided information on their holdings. More than a few made the supply of information for this guide one of their own major projects.

Many Forest History Society staff members and others worked on the project from time to time. Elwood R. Maunder, Executive Director, and George A. Garratt, President, marshalled the necessary financial resources. Roberta M. Barker handled the details of administration. Kathryn Fahl provided research assistance in Santa Cruz, while several others gathered information in the holdings of a number of repositories. Arlene Towne typed most of the letters of inquiry. Pamela O'Neal typed the manuscript and assisted in indexing. Karen Burman proofread the manuscript. Other indexing assistants included Katrina Kuizenga and Pamela Mathis. Technical advice during the formative states of the project came from several experienced archivists, most frequently Harold T. Pinkett. Above all others, however, Harold K. Steen, Associate Director of the Forest History Society for Research and Library Services, and Ronald J. Fahl, Bibliographer, daily provided useful criticism and encouragement.

The Forest History Society hopes to produce occasional supplements to this work, and welcomes notification of any errors or omissions. Information on new accessions will be reported in the Archival News column of the *Journal of Forest History.*

UNITED STATES

UNITED STATES

Alabama

AUBURN

Auburn University, Department of Archives

Archivist, Ralph Brown Draughon Library, South College Street 36830 Telephone: (205) 826-4465

1 **Auburn University, Extension Service,** Records, ca. 150-200 boxes. Unprocessed.

2 **Auburn University, Forestry Department,** Records, 1947-1956, ca. 1000 items. Primarily the papers of Wilbur DeVall, head of the Forestry Department after 1951. Includes pamphlets, memoranda, conference material, correspondence with forestry related industries, campus personnel, budget material, public relations material, summer camp information. There is also scattered correspondence of Terrill D. Stevens, the first department head after Forestry was separated from the Department of Horticulture. Register.

3 **Duncan, Luther Noble,** Papers, 1945-1946, ca. 50 items. Topics include Auburn University School of Agriculture, Forestry Department; development of a degree course in forestry in Alabama; forestry research in Alabama; status of forestry education in the South. May be used only with permission from the President's Office, Auburn University.

4 **Draughon, Ralph B.,** Papers, 1947-1965, ca. 300 items. Correspondence with other officers of Auburn University and with federal and state agencies. Includes material on the Auburn University Forestry Department. May be used only with permission of the President's Office, Auburn University.

5 **Auburn University. Historical Collection.** Includes "History and Contributions of the Forestry Program of the Alabama Polytechnic Institute" (June, 1947); correspondence of historical value to the department; history and growth of operations within the state. 75 pp.

MONTGOMERY

State of Alabama, Department of Archives and History

Director, 624 Washington Ave. 36104

6 **Alabama. Forestry Commission,** Records. Among the historic records of the Forestry Commission are materials of the Civilian Conservation Corps, including timber survey data sheets and notes (1937-1946), and applications and correspondence (1937-1946).

7 **Smith, W.T., Lumber Company** (Chapman, Alabama), Records, ca. 1925-1950. Records relating to purchase of tracts of timberland, and statistical data. The collection may include material as early as 1900. Unprocessed.

UNIVERSITY

University of Alabama, Library

P.O. Box 5 35486

8 **Gravner, William Charles** (1883-1955), Family Papers, 1803-1963, ca. 5018 items. Includes records of the Sterling Lumber and Supply Company, Birmingham, Ala. (1940-1953, 1955-1956). Unpublished finding aid.

9 **Jemison, Robert** (1802-1871), Papers, 1797-1898, 2214 items. Lawyer, businessman, politician, and planter. The collection contains correspondence, 15 letter books, and other papers, including material concerning Jemison's interest in a sawmill and other businesses. Unpublished guide.

10 **Shelby Iron Company** (Shelby, Ala.), Records, 1862-1923, ca. 494,000 items. Contains minutes of directors and stockholders; manufacturing recorde, including charcoal reports, time books, payrolls in each department, etc.; furnace record books; and other records. Time books and payroll records include details of employment in the charcoal workings. 5 books record the receipt of charcoal shipments by rail from various sources. Unpublished guide.

Alaska

FAIRBANKS

University of Alaska, Elmer E. Rasmuson Library

University Archivist and Curator of Manuscripts, 99701

11 **Bartlett, Edward Lewis** (1904-1968), Papers. U.S. Delegate from Alaska, 1944-59; U.S.Senator, 1959-68; served on House Committee on Public Lands and Committee on Agriculture. There are scattered materials relating to forests in his private papers (1944-1959), and material on the U.S. Forest Service in the Executive and Independent Agencies File (1959-68). Unprocessed.

12 **Davis, Edby,** Diaries, 1914-1965. Fairbanks, Alaska, vicinity. Scattered entries on logging for firewood. Guide.

13 **Dimond, Anthony Joseph** (1881-1953), Papers. U.S. Delegate from Alaska, 1933-1945; served on the Public Lands

Committee; also Agriculture committee. There are two file folders relating to H.R. 1548, 75th Congress, 1st session, a tax on wood pulp.

14 **Farnsworth, Charles Stewart** (1862-1955), Papers, 1899-1943, 622 items. Army officer. Correspondence, monthly reports, progress reports, articles relating to his service in Alaska; scattered mention of forests and firefighting. Unpublished description and index.

15 **Gruening, Ernest** (b. 1887), Papers. Governor of Alaska Territory; U.S. Senator, 1959-69; served on the Senate Committee on Agriculture and Forestry. Papers relating to forests are in the Bill File, General File, and Rampart Dam File. Unprocessed.

16 **Houck, Jonas B.,** Papers, 1898. Isolated references to logging can be found in the correspondence. Guide.

17 **Mears, Frederick C.,** Papers, 1909-1938. Engineering files relating to the construction of the Alaska Railroad. Materials relating to forests may be found under timber cutting provisions, railroad ties, etc. Guide.

18 **Pilcher, George M.** (b. 1864), Diaries, 1898-1933, 21 volumes. Miner, woodcutter. Diaries kept by Pilcher relating to his life along the Yukon and Innoko Rivers, with scattered references to woodcutting. Guide.

19 **Rivers, Ralph Julian** (b. 1903), Papers. U.S. Representative from Alaska, 1959-1967; served on the Interior and Insular Affairs Committee and the Outdoor Recreation Resources Review Commission. There are materials dealing with forests in the Legislative and Bill files; also files on the U.S. Forest Service, and the National Park Service. Other material pertaining to forests is scattered through the general files on the Bureau of Land Management. A legal case file includes materials on lumber and the lumber industry. The collection is described in: *Ralph J. Rivers, U.S. Representative from Alaska, 1959-1966: An Inventory of his Congressional Papers,* compiled by Paul McCarthy, College: University of Alaska, 1971, 122 pp.

20 **Wickersham, James** (1857-1939), Diaries, microfilm. Alaska Territorial Judge and U.S. Delegate, 1909-17, 1919, 1921, 1931-33; served on the Public Lands Committee. Mention of forests.

KETCHIKAN

Tongass Historical Society Inc., Museum

Director, Box 674 99901

21 **Ketchikan Spruce Mills** (Ketchikan, Alaska), Records, 425 cubic feet. Unprocessed.

Arizona

FLAGSTAFF

Norther Arizona University, Library

Special Collections Librarian—Archivist, P.O. Box 6022 86001

22 **American Forestry Association,** Miscellaneous Records, 1958. Records of meeting at Tucson, Arizona (1958). Correspondence and notes pertaining to W.L. Swager, *et al., A Study of the Cooperative Forest-fire Control Problems,* prepared by the Battelle Memorial Institute of Columbus, Ohio (1958). Material collected by Jay H. Price.

23 **Applequist, Martin B.,** Papers, 1957, 1966, 2 inches. Professor of Forestry, Nothern Arizona University. Correspondence, papers relating to the Reineke core holder (1957); thinning in Coconino National Forest; pinyon dating experiments at the Fort Valley Experimental Forest summer camp workshop (1966).

24 **Arizona Cattle Growers Association,** Records, 1922-1956, 5 cartons. Includes transcripts of meeting between the Association and the U.S. Forest Service (June 19, 1944), Prescott, Arizona, concerning grazing on forest lands.

25 **Arizona Lumber and Timber Company** (Flagstaff, Ariz.), Records, 1882-1947, ca. 368 feet. Includes also records of the Ayer Lumber Company and the Greenlaw Lumber Company. There are correspondence, financial records, and annual reports. Among the important correspondents are Edward Ayer, Denis M. Riordan, Michael J. Riordan, Tim A. Riordan, Joseph C. Dolan. In the Northern Arizona Pioneers Historical Society Collection.

26 **Arizona State College Forestry Club** (Flagstaff, Ariz.), Records, 1962-1966, 2 inches. Newsletters, correspondence, workshop-conference records.

27 **Arizona Water Resources Committee,** Records, 1956-1960, 2 feet. Correspondence, maps, pamphlets, newsletters, clippings, published and unpublished reports pertaining to watershed and forest management for the Salt River Project and the Central Arizona Project, Arizona. Also correspondence and other materials relating to Jay H, Price's critique of "A Ten Year Program for the Iowa State Conservation Commission," prepared by the Wildlife Management Institute (1958).

28 **Buggelin, Martin** (1867-1939), Papers, 2 inches. Maps of Tusayan National Forest, Arizona (1910), annotated to show proposed Grand Canyon National Park boundaries; correspondence relating to U.S. Congress, Senate Bill No. 3362 (1938), to revise the Grand Canyon National Park boundary and abolish the Grand Canyon National Monument, and H.R. 20721 (1917), to extend the time limit of timber cutting rights of the Saginaw and Manistee Lumber Company in Arizona; and correspondence between Buggelin and U.S. Senator Carl Hayden concerning establishment of the Grand Canyon National Park (1917).

29 **Collingwood, George Harris** (1890-1958), Correspondence, 1910-1915, 5 inches. U.S. Forest Service ranger, Apache National Forest, Arizona. Copies of letters to his family and fiancee describing his life as a ranger. Originals are held by the Forest History Society, Santa Cruz, California. Typed description, 18 pp.

30 **Horsehoe Land and Livestock Company** (Flagstaff, Ariz.), Records, 6½ feet. Agreements, bills, maps, correspondence pertaining to lease of Tonto National Forest land for grazing (1930-1950). John Hennessey was owner of the firm.

31 **Northern Arizona University Photograph Collection.** Approximately 150 forest-related photographs depicting forests, foresters, loggers, logging, and logging railroads in Arizona and California. Places represented include the Coconino, Kaibab, Apache, Crook, Sitgreaves, and Coronado (Huachuca) National Forests. Organizations include the Arizona Lumber and Timber Company, Flagstaff; Ayer Lumber Company, Flagstaff; Ed Dorado Lumber Company, Ed Dorado, California; Ponderosa Pine Management Conference, Fort Valley Experimental Station, Flagstaff (Sept. 20, 1944). Persons include George Harris Collingwood, Jim Owen, Clyde Leavitt, Bill Donovan, Jim Sizer, and Jesse T. Fears, all U.S. Forest Service personnel in Arizona.

32 **Pearson, Mary S.** "Edward Bert Perrin, 1838-1932: A Southern Entrepreneur in the American West." Northern Arizona University: M.A. thesis, 1968.

33 **Perrin, Edward Bert** (1838-1932), Papers, 1 foot. Williams, Arizona. Estimates of timber on Perrin's land, Tehama, California (1911); correspondence (1874-1916), including material on the sale of timber land in Guerrero, Mexico, and the transfer of private land to the San Francisco Mountains Forest Reserve, Arizona (1901).

34 **Price, Jay H.,** Papers, ca. 1950s, 2 cartons. Maps, pamphlets, and correspondence pertaining to the establishment of the International Peace Memorial Forest in Quetico Provincial Park, ONtario, and Superior National Forest, Minnesota. Also a manuscript by Samuel T. Dana, "Wildland Ownership in California—History—Present Patterns and Problems," with correspondence and papers relating to Price's critique of that manuscript.

35 **Saginaw and Manistee Lumber Company** (Williams and Flagstaff, Ariz.), Records, 1902-1954, 21½ feet. Arizona operations of a Michigan corporation. The papers include journals, ledgers, other financial records, inventory, sales, shipment, and time records (1900-1920); correspondence, annual reports, land and timber records, tax reports, other papers (1919-1946); correspondence and miscellaneous papers (1900-1915); also contracts, agreements, deeds, maps, blue prints. There is a draft inventory for documents of the period 1900-1920.

36 **Society of American Foresters, Southwest Section,** Records, 1948-1958, 1 foot. Albuquerque, New Mexico. Includes correspondence.

37 **Soil Conservation Society, Arizona Chapter,** Records, 1956-1957, 2 inches. Phoenix, Ariz. Correspondence, photographs, negatives, other records.

GRAND CANYON

Grand Canyon National Park, South Rim Visitor Center

Curator/Librarian 86023. The collections are available to researchers by appointment.

38 **Grand Canyon National Park Study Collections,** 1871-1955. Includes unsorted notes and other papers of Francois Emile Matthes on the topographical mapping of the Grand Canyon (1902-1903); packets of unsorted letters (1890-1910) on mining and the hotel business in the area prioir to establishment of the park; "A Townsite Plan for Grand Canyon National Monument"; "Drift Fence and Other Improvements, Tusayan National Forest," by W.R. Mattoon, Forest Supervisor (July 18, 1910), typed, carbon, with photos and maps; "Grand Canyon Working Plan: Uses, Information, Recreational Development," by Don. P. Johnston and Aldo Leopold (1916, revised, 1917), typed; also printed materials; and journals and reports relating to the exploration of the Colorado River.

TEMPE

Arizona State University, University Library, Special Collections

Head, Special Collections 85281

39 **Hayden, Carl Trumbull** (b. 1877), Papers, 1912-1968, 720 boxes. U.S. Representative, 1912-27, and Senator, 1927-69, from Arizona; served on the House Public Lands and Irrigation of Arid Lands committees, and the Senate Committee on Interior and Insular Affairs. The papers consist mainly of correspondence between Hayden, government officials, and constituents; proposed legislation; bills; memoranda; and other documents. Over seventy boxes are entirely concerned with agriculture, reclamation, forests, and parks. Topics include the report on timber resources (1946); utilization of forest products on Indian reservations (1950); highway funds for forest lands (1937); national forest mining restrictions (1950); snow surveys (1951). There are photographs of balloon logging in Oregon (1964); and Bernalillo watershed projects, Cibola National Forest, New Mexico (1955). Many items under the general heading of reclamation also concern forestry. Partially unprocessed. Guide for part of the collection.

TUCSON

Arizona Historical Society

Director, 949 East Second Street 85719. Holdings are described in *Documents of Southwestern History: A Guide to Manuscript Collections of the Arizona Historical Society,* compiled by Charles C. Colley, Tucson: Arizona Historical Society, 1972, 233 pp.

40 **Barnes, Will C.** (1858-1936), Papaers, 1879-1940, 21 feet. Cattleman and forester, Arizona. The collection contains letters (2,000 items, 1900-40); scrapbooks (25 vols.); diaries (1879-1936); manuscripts and notes by Barnes; photographs. Subjects include ranching, wildlife, national forests and parks, and other topics.

41 **Cosulich, Bernice** (1897-1956), Papaers, 1937-1947, 6 feet. Journalist. The papers include manuscripts on Arizona subjects such as the Grand Canyon, National Park Service, etc. Mostly notes for her column and articles. Little primary source material.

42 **Gosney, E.S.,** Manuscript, 1898-1909, 86 pp. A history of the first ten years of the Arizona Wool Grovers' Association, 1899-1909, compiled by the association's first president, E.S. Gosney (1931).

43 **Kelley, Lon** (1873-1943), Papers, 1910-1938, 1½ inches. Kelley worked for logging and mining companies around Tombstone after 1880. The papers include a diary (1880-1937), material concerning mining claims, and autobiographical material. Unprocessed.

44 **Kneipp, Leon Frederick** (1880-1966), Papers, 1890-1961, 3½ feet (400 items). U.S. Forest Service official, 1905-1946. The papers contain correspondence; mimeographed bulletins; pamphlets; minutes of the American Forestry Association and other organizations (1926-1955); personal correspondence; and awards given upon Kneipp's retirement (1946); other papers relating to agriculture and the Forest Service; and photographs.

45 **Pack, Charles Lathrop** (1857-1937), Papers, 1908-1919, 2 inches. Pack was a lumberman-economist active in national conservation movements. The collection includes letters from William Howard Taft, Henry L. Stimson, Woodrow Wilson, Luther Burbank, and others to C.L. Pack relating to his service to national conservation and to his writings. There is also a letter from Theodore Roosevelt appointing Pack to the Commission on the Conservation of National Resources (1908). Unprocessed.

46 **Pearce, Joseph Harrison** (b. 1873), Papers, 1943-1950, 3 inches. U.S. Forest Service official, Black Mesa Forest Reserve, Arizona. The collection includes manuscripts by Pearce, including "Line Rider," an autobiography concentrating upon the development of national forestry and the sheep and cattle industries; also related newspaper clippings and publications by Pearce in *Improvement Era.*

47 **Roberts, Paul H.,** Papers, 1897-1931, 2 inches (315 pp.). U.S. Forest Service official, Southwestern Regio n, and super-

visor, Sitgreaves National Forest (1912-31). The papers include typed minutes of the Arizona Wool Grower's Association (1897-1923), edited with historical introduction by Roberts.

48 **Van Valkenburgh, Richard Fowler** (1904-1957), Papers, 1880-1946, 4 feet. Van Valkenburgh made an economic study of the Navajo Indian ½ Reservation, 1935-1941, for the Bureau of Indian Affairs, and was employed by the Navajo tribe to head the Department of Land Uses and Surveys, 1952-1957. The papers include research notes, typescripts, documents, and copies of publications by Van Valkenburgh. There is material on tree ring dating.

49 **Winn, Frederick** (1880-1945), Papers, 1907-1955, 7 feet. U.S. Forest Service official, Arizona and New Mexico, 1907-1942; supervisor, Coronado National Forest, 1925-1942; collector of material for a history of U.S. national forests. The collection includes diaires and notebooks (67 items, 1907-1942); personal correspondence; Forest Service publications and guide books; maps; material on pioneer graves within the Coronado National Forest; notes on wildlife management, conservation, ranch life, Arizona Rangers, Civilian Conservation Corps camps, folklore, historical sites, etc.

University of Arizona, University Library, Special Collections

Manuscript Librarian 85721. Theses are listed in *Checklist of Theses and Dissertations*, Tucson: University of Arizona Library, 1964—date.

50 **Cameron, Ralph Henry** (1863-1953), Papers, 1899-1915, 1 volume and 5 boxes. U.S. Delegate from Arizona, 1909-12; U.S. Senator, 1921-27. Correspondence and papers relate chiefly to Cameron's career in Congress. There are proclamations, petitions, and some correspondence regarding game reserves in national forests. There are also letters and legal papers relating to various interests in the Grand Canyon National Park, such as the Bright Angel Trail, mining claims, and a proposed power project; also registers (4 vols.) for tourist facilities operated by Cameron. Inventory.

51 **Douglass, Andrew Ellicott** (1867-1962), Papers, 1792-1965, 90 feet. Astronomer; founder of the science of dendrochronology. Includes correspondence, personal papers, financial records, typescripts of speeches and writings, astronomical and tree ring records, including records of Douglass' work with various observatories and the Laboratory of Tree Ring Research, University of Arizona, Tucson. Approximately 60 boxes contain tree ring material.

52 **Lauver, Mary E.** "A History of the Use and Management of the Forested Lands of Arizona, 1862-1936," Tucson: University of Arizona, M.A. thesis, 1938.

53 **Pollock, Thomas E.** (b. 1868), Papers, 1900-1941, 65 boxes. Arizona sheepman and cattleman. The papers include, among diverse business records, correspondence and permits relating to land and cattle companies in northern Arizona, and their use of U.S. forest lands. Detailed list of contents.

54 **Udall, Stewart Lee** (b. 1920), Papers, U.S. Representative from Arizona, 1955-1961; member, committee on Interior and Insular Affairs; U.S. Secretary of the Interior, 1961-1969. The collection contains materials relating to his congressional career and all his cabinet years. It includes working drafts of his book, *The Quiet Crisis* (1963). Unprocessed.

55 **Verkamp, Margaret M.** "History of Grand Canyon National Park," Tucson: University of Arizona, M.A. thesis, 1940.

Arkansas

FAYETTEVILLE

University of Arkansas, University Library, Special Collections

Curator, Special Collections 72701 Telephone: (501)575-4101

Manuscript Collections

56 **McRae, Thomas Chipman** (1851-1929), Collection, ca. 2 feet (1571 items). Governor of Arkansas; U.S. Representative, 1885-1903 (chairman, House Committee on Public Lands). A sizable portion of the 1½ feet of correspondence pertains to McRae's management of yellow pine timber lands in Nevada County and in other parts of south Arkansas. There is virtually no correspondence for 1887-1901, most of McRae's congressional career. Finding aid, typed, 28 pp.

57 **Spaulding, Dudley J.** (b. 1834), Papers, 1879-1899, 1 reel microfilm. Letters, legal papers, land plats, and other documents pertaining to Spaulding's lumbering activities in Arkansas and in other states. Originals are in the Manuscript Division of the Wisconsin State Historical Society, LaCrosse Area Research Center.

58 **Twiford, Ormie** (1869-1913), Diary, January 1-December 1, 1901. Transcript of a diary of a timber cutter working for various logging contractors in the vicinity of Mena, Arkansas.

Theses and Research Papers

59 **Cheatham, Andrew Reynolds.** "Utilization of Land in a Cut-Over District of Arkansas," M.S. thesis, 1930.

60 **Corliss, Colby Curry.** "A History of the Timber Industry in Ashley, Bradley, and Drew Counties, Arkansas," M.A. thesis, 1953.

61 **Dodson, J.M.** "The Demand for Arkansas Lumber," Seminar thesis, 1937.

62 **Grobmyer, John R., Jr.** "The Marketing of Arkansas Forest Products," Seminar thesis, 1938.

63 **Leslie, James W.** "The Arkansas Lumber Industry," M.S. thesis, 1938.

64 **Leslie, James W.** "Some Economic Aspects of the Arkansas Lumber Industry up to 1935," Seminar thesis, 1935.

65 **Vaughter, Ray B.** "Arkansas Forests," Seminar thesis, 1937.

California

ARCATA

California State University, Humboldt Campus, Library

66 A large collection of records (dating from ca. 1890) of the Northern Redwood Company, a firm absorbed into the Simpson Timber Company, are in storage. There are also account books of several firms that acted as suppliers to local lumber companies, and a large photograph collection documenting the lumber industry in Humboldt County, California.

BERKELEY

University of California, The Bancroft Library

Head, Manuscripts Division 94720. A portion of the holdings is described in *A Guide to the Manuscript Collections of the Bancroft Library, Vol. I: Pacific and Western States (Except Cali-*

fornia), edited by Dale L. Morgan and George P. Hammond, Berkeley and Los Angeles: University of California Press, 1963, 379 pp. A more general description of the holdings of the Bancroft and other libraries on the Berkeley campus is *Guide to Special Collections, University of California, Berkeley, Library,* compiled and edited by Audrey E. Phillips, Metuchen, N.J.: The Scarecrow Press, Inc., 1973, 151 pp.

67 **Abrams, William Penn** (1820-1851?), Diary, 1849-1851, 1 volume and 2 folders in portfolio. Refers to the location of a steam sawmill near Oregon City, Oregon.

68 **Albright, Horace Marden** (b. 1890), Interview, 1962, 858 l. Copy of the transcript of the tape-recorded interviews conducted 1960 by William T. Ingersoll for Oral History Research Office, Columbia University. Photographs inserted. Restricted.

69 **Allen, Benjamin Shannon** (1883-1963), Manuscript, ca. 160 l. "They Did Not Quit: A Study in Dynamic Capitalism." The California redwood lumber strike of 1946-48, its influence on the passage of the Taft-Hartley Act, and its effect on the redwood lumber industry.

70 **Allen, Benjamin Shannon** (1883-1963), Notes, 1869-1953, 4 folders in portfolio. Notes on the Pacific Lumber Company and lumbering in Humboldt County, California. Included are questionnaires and answers obtained by Allen from company employees and residents of Scotia and vicinity; interview with A.P. Alexanderson (August 23, 1944); history of the Pacific Lumber Company, as told by George Douglas to Derby Bendorf; notes by Hugh Bower.

71 **Allen, Benjamin Shannon** (1883-1963), Papers, 1940-1965, 1 box. Press agent and publisher. Correspondence, memoranda, clippings, and papers, relating to the redwood strike of 1945-1947, Allen's work with the Redwood Region Conservation Council and the California Redwood Association, and his interest in tree farms. Includes letters and reprints by Emanuel Fritz and others.

72 **Allen, George B.** (b. 1825), Statements, 1886?, 4 l. Mining, lumbering, and ranching in Colorado after 1858, most recently in Jefferson County.

73 **Allen Family**, Papers, 1853-1923, Microfilm, 1 reel. Diaries, accounts, and miscellaneous papers of Oliver Allen and his son, Charles Denison Allen, relating to early life in Marin County, California, construction of sawmills at Bolinas, the Pioneer Paper Mill at Daniel's Creek. Originals are in private possession.

74 **Alvarado, Juan Bautista**, Manuscript, 1876, 5 volumes. "Historia de California," Vol. 5, pp. 5-7, relates to Stephen Smith's grant of land from the governor, and the erection of a steam sawmill in 1843.

75 **Anderson, W.A.**, Dictation, 1887. 1 l. Concerns his career as sawmill operator and as auditor of Chehalis, Washington.

76 **Arcata & Mad River Rail Road Company**, Correspondence and Papers, 1883-1927, 1 box, 1 carton, and 4 volumes. Correspondence with other railroads and with Joseph and Anton Korbel, who controlled the railroad; accounts; freight register (1883-1885); journal (1883-1887); letterpress copy book containing cash reports, bills, freight tariffs, etc. (1888-1890); and train orders (Sept.-Dec. 1927).

77 **Arizona Miscellany**, 9 items. Includes U.S. General Land Office, lieu selection blanks issued to Atchison, Topeka, & Santa Fe Railroad Company for areas in Grand Canyon Forest Reserve (1904, 4 items).

78 **Austin, Lloyd**, Correspondence and Papers, ca. 1926-1930, 2 folders. Relates to his work on breeding pines, as director in the Forest Products Laboratory. Includes correspondence with the Burbank Experiment Farms at Santa Rosa, notes and reports on visits and conferences on forest genetics. Unprocessed.

79 **Bale Family**, Papers, ca. 1841-1899, 2 boxes. Some accounts and legal papers of Dr. Edward T. Bale and some papers relating in part to the establishment of his flour and sawmills in Napa County, California (1841-1849). Report and Key to arrangement.

80 **Banning, Phineas** (1830-1885), Biographical sketch based on a dictation, 1883, 15 l. "Settlement of Wilmington." Mentions the first steam sawmill in southern California and the lumber trade at Wilmington, California.

81 **Basset, Alden** (b. 1845), Statement, 1886, 2 pp. Lumbering and mining in Colorado after 1860.

82 **Bates, Bernar**, Correspondence and Papers, ca. 1948-1969, 2 cartons. Miscellaneous material, including printed items, photographs, clippings, in large part relating to forestry, especially to redwoods. Mr. Bates was public relations consultant for Pacific Lumber Co. Unprocessed.

83 **Beebee, William L.** (1829-1899), Dictation, n.d., 3 l. Recorded for H.H. Bancroft by R.H. Collier. Beebee describes his residence at San Luis Obispo from 1847, public offices held, and lumber business. Included are newspaper clippings about his lumber business.

84 **Beith, James** (b. 1832?), Diaries and Journals, 1862-1888, 8 vols. Record of his life and activities, mainly in Arcata, California. Includes notes on his participation in logging and other activities.

85 **Blair, Ramona**, Paper, n.d., 9 l. "El Molino, First Commercial Sawmill in California. Santa Rosa. California." Xerox copy of a paper written for the dedication of the sawmill site as a Historical Landmark.

86 **Bliss, Duane L.** (1833-1907), Manuscript, 1887, 9 l. "Data Concerning the Virginia & Truckee Railroad, and Those Who Planned and Carried Out that Work." Contains material concerning the Carson & Tahoe Lumber & Fluming Company.

87 **Bowles, R.R.** (b. 1854), Reminiscences, 1886, 2 l. "Lumbering in Colorado." Contractor, builder, and lumberman at Aspen after 1880.

88 **Boynton, Joseph S.**, Statement, 1878?, 10 l. "Statement of a Pioneer of 1849." Mentions a sawmill and launching of a ship in 1849.

89 **Branham, Isaac** (1803-1887), Reminiscences, ca. 1885, 2 folders in portfolio. "Lumbering in 1847." Describes his sawmills and other property in Santa Clara County, California.

90 **Briggs, Robert** (b. 1836), Biographical Sketch, ca. 1887, 3 folders in portfolio. Includes accounts of lumbering, milling, and mining in Santa Clara, Alameda, Amador, and Calaveras Counties, California.

91 **Brown, Alan K.**, Manuscript, 1961, 40 l. "Lumbering in the Pulgas Redwoods, 1777-1853." The history of the lumber industry near Woodside, California.

92 **Brown, Charles** (1814-1883), Manuscript, 1878, 28 l. "Early Events in California." Includes notes on lumbering, ca. 1830s.

93 **Bryan, Thomas J.** (b. 1838), Statement, 1885, 8 l. Lumbering and cattle raising in Montana after 1881.

94 **Buhne, Hans Henry** (1882-1894), Papers, 1865-1897, microfilm (944 pp.). Selected items from originals in family possession, Eureka, California, including accounts, legal papers, etc., concerning shipping, lumber, railroad and mining interests in the Humboldt Bay area, of Buhne and D.R. Jones, associates of John Kentfield & Co., San Francisco.

95 **Burgess, Sherwood Denn,** Manuscript, ca. 1962, 7 l. "Loggers, Lumber Haulers, Merchants, Sea Captains and Ships Engaged in the Lumber Trade Before 1848 in the Monterey-Santa Cruz Area." Originally prepared as an appendix for the article "Lumbering in Hispanic California," *California Historical Society Quarterly*, vol. 41, September, 1962, 237-48.

96 **Burke, John** (b. 1842), Dictation, 1886, 3 l. Mentions lumbering and ranching near Cañon City, Colorado.

97 **Burris, Davis** (1824-1904), Dictation and Biographical Sketch, 1887, 1 portfolio. Dictation, recorded for H.H. Bancroft, concerns mining and lumbering in California to 1851, among other topics (5 l.); also Biographical Sketch based upon dictation, prepared for *Chronicles of the Builders of the Commonwealth.*

98 **Caldor Lumber Company,** Records, ca. 1954-1956, 1 carton. Correspondence, particularly in relation to promoting Philippine mahogany; material on log purchases and specifications; mailing lists; publicity and pictures, etc. Unprocessed.

99 **California. Department of Natural Resources, Division of Beaches and Parks,** Records, ca. 1929-1958, 1 box. A few items pertaining to administration of the state parks from the files of Newton B. Drury.

100 **California. Department of Natural Resources,** Reports to the Governor's Council, 1950-1952, 1 box. Mimeographed copies of monthly statements on activities of the Department, covering conservation education, beaches and parks, forestry, mines, oil and gas, and soil conservation.

101 **California Forest Protective Association,** Articles of Incorporation, 1909, 4 l. Proposed organization paper, signed by C.X. Wendling, with an explanatory note attached.

102 **California Redwood Association,** Records, ca. 1924-1960, 3 cartons. Material on American Lumber Standards, reports and publications of the National Lumber Manufacturers Association, reports and publications of the California Redwood Association. Unprocessed.

103 **California. State Board of Forestry,** Records, 1885-1889, 1 box. Includes correspondence and accounts, relating in part to illegal timber cutting, forest fires, the production of a survey map of forest lands, and plans for instituting an Arbor Day in California. Unprocessed.

104 **California State Parks Council,** Records, ca. 1925-1929, 3 boxes, 1 carton. Relates to the campaign for a comprehensive state park program, including copies of letters by Newton B. Drury and staff, incoming correspondence, interoffice memoranda, minutes of meetings, press releases, articles prepared for newspapers and magazines, speeches and statements, lists of speakers and speaking engagements, clippings, and reference materials, particularly relating to parks in other states. Report and Key to arrangement.

105 **Caspar Lumber Company,** Records, ca. 1907-1945, 10 cartons, 31 vols. Relates to lumbering in California.

106 **Chaffee, R.R.,** Collection, ca. 1909-1954, 18 cartons, 15 vols. and 1 package of maps. Business records of several lumber companies in California and the Pacific Northwest. Companies represented include Wolf Creek Timber Co., California Coast Lumber Co., Trask Timber Co., Silver Fork Lumber Co., Soper-Wheeler Co., Nelson Lumber Co., Rogue River Timber Co., N.P. Wheeler Land Co., Portville Timber Co.

107 **Chambers, A.J.** (b. 1835), Dictation, 1885?, 3 l. Farmer and lumberman in Texas sometime after 1853.

108 **Clar, C. Raymond,** Manuscript, n.d., 6 l. "The Structure of California State Government in Respect to Forestry," typescript, with organization chart for the Department of Natural Resources, April, 1957.

109 **Cleveland, Abner Coburn** (1839-1903), Manuscript, 1888, 6 l. "Data Regarding Progress and Growth of Nevada." Mentions lumbering and ranching in Washoe and White Pine Counties, Nevada, after 1863.

110 **Colby, William Edward** (1875-1964), Correspondence and Papers, 1892-1964, 2 boxes, 3 cartons. Letters, manuscripts of his writings, notes, reprints and copies of articles, diaries, newspaper clippings, and photographs, relating to the Sierra Club, John Muir, conservation (including material on the recession of Yosemite and the Hetch Hetchy controversy), mining law, and Colby's other interests. Also includes typed transcripts of letters written by John Muir, mainly to his daughter Helen. Report and Key to the arrangement.

111 **Colby, William Edward** (1875-1964), Interview, Feb. 27-28, 1961, ca. 91 l. Typed transcript of interview conducted by Hal Roth. Some of the material was used in Roth's book, *Pathway in the Sky.*

112 **Cole, Harry W.,** Correspondence and Papers, ca. 1924-1961, 22 cartons. Includes much material (reports and statements) for the Little River Redwood Company and the Hammond Lumber Company as well as records relating to their consolidation as Hammond & Little River Redwood Co.; papers relating to the California Redwood Association, some of them concerning the Lagunitas property; price schedules, memoranda, maps; files on area and ownership of forest land in California; timber cruise reports; tax and litigation files. Unprocessed.

113 **Collins, John** (b. 1834), Dictation, 188-?, 2 l. Typescript. Concerns Collins' sawmill and hotel business, Port Gamble, Washington, 1857-1867.

114 **Conservation Associates,** Records, 3 cartons. Records relating to redwoods. Unprocessed.

115 **Contra Costa Hills Fire Protection Committee,** Collection, 1921-1926, 1 vol. Letters, clippings, reports and pamphlets relating to the formation and work of the Committee and to the Berkeley fire of September 17, 1923. Assembled from the files of Professors Woodbridge Metcalf and Emanuel Fritz by the latter. Includes copies of their letters, and replies from various Berkeley, Oakland, and Piedmont officials, California State Board of Forestry, University of California, and Charles Keeler, Director of the Berkeley Chamber of Commerce and the Secretary of the Committee.

116 **Courteney, Albert A.** (b. 1854), Statement, 1887, 3 l. Relates to lumber business in Southern California from 1883.

117 **Dana, Samuel Trask** (b. 1883), Interviews, 1965, 98 l. "The Development of Forestry." A composite of selections from separate interviews conducted by Amelia R. Fry of the Regional Oral History Office, Bancroft Library, and Elwood R. Maunder, the Forest History Society, in preparation for the edited version published as "The Dana Years" in *American Forests,* Nov. and Dec., 1966.

118 **Denman, William** (1872-1959), Papers, ca. 1900-1959, 42 boxes, 32 cartons, and 21 volumes. Correspondence; re-

ports; speeches and publications; legal and financial papers; briefs, arguments and other legal documents; copies of judicial opinions; scrapbooks; clippings. Includes material on Denman's association with the Coos Bay Lumber Co. Report and Key to the arrangement.

119 **Dolbeer & Carson Lumber Company** (Eureka, California), Records, ca. 1884-1941, 54 volumes. Correspondence and accounts, including records of the Bucksport & Elk River Railroad, Humboldt Northern Railway Company, and William Carson Estate Company. Key to the arrangement.

120 **Dollar Family,** Collection, ca. 1872-1967, 78 cartons, 72 vols. Personal papers, including memoirs, and other writings of Robert Dollar; photographs; scrapbooks; clippings; etc.; with some material relating to other members of the family; also business records of the Robert Dollar Company and its many divisions, including Dollar Steamship Lines, Dollar Portland Lumber Company, Egmont Timber Company, etc. Report and Key to the arrangement.

121 **Dolloff, John W.** (b. 1852), Biography, 1886, 2 l. Sawmill operator in Gilpin County, Colorado, after 1878; lumber business and general store at Berthoud.

122 **Drury, Aubrey** (1891-1959), Manuscript, 1953, 95 l. "The Livermore Family: Pioneers in California." Typescript concerning Horatio G. Livermore and his sons, with information about lumbering on the American River.

123 **Eldridge, Edward,** Recollections, 1880, 39 pp. Contains references to a sawmill near Bellingham, Washington.

124 **Elk River Mill and Lumber Company,** Records, 1884-1933, 74 vols. One of the early sawmill operations in Humboldt County at Falk on the Elk River. Included are letterbooks, cashbooks, journals, ledgers, timebooks for the mill and for wood and lumber sales, invoices, receipts, etc. Unprocessed.

125 **Ellsworth, Rodney S.,** Photographs of redwoods, 1 box, 1 carton. A small amount of related correspondence and papers, mainly with and concerning the Sierra Club and the Save-the-Redwoods League (ca. 1928-1950), is also included. Unprocessed.

126 **Farquhar, Francis Peloubet** (1887-1974), Correspondence and Papers. Includes correspondence with Horace Albright, Newton Drury, Stephen T. Mather, and other conservationists. List of contents.

127 **Flanigan, D.J.,** Manuscript, ca. 1888, 5 l. "California Lumber." A history of Flanigan's association with T.F. Bosnan in the lumber business in Humboldt County after 1871, including an analysis of the San Francisco market.

128 **Flowers, Jacob** (b. 1827), Dictation, 1886, 2 l. Lumber business at Bellvue, Colorado, from 1871.

129 **Forest History Society,** Oral History Interviews, 1953-1956, 1 portfolio. Typed transcripts of incomplete interviews with members of the Caspar Lumber Company, concerning lumbering operations in California. Subjects include Casmir J. Wood, J.W. Lilly, and William McCarthy.

130 **Forest History Society,** Oral History Interviews, 1953-1960, 8 items. Typed transcripts of interviews with various persons in the lumber industry, especially the Pacific Northwest and the Redwood Region of California. Key to contents.

131 **Forest History Society,** Oral History Interviews, 1971-1972, microfiche, 8 sheets. 2 interviews conducted by the Forest History Society, Santa Cruz, California, including J.P Kinney, "The Office of Indian Affairs: A Career in Forestry" (1971, 3 sheets), and J. Herbert Stone, "A Regional Forester's View of Multiple Use" (1972, 5 sheets).

132 **Fritz, Emanuel** (b. 1886), Correspondence and papers, ca. 1924-1958, 1 box, 6 cartons. Principally his records relating to redwoods, with correspondence and report from Save-the-Redwoods League; material from the California Redwood Association; redwood forestry codes and circulars; files on forest fires, fire protection, and the Berkeley Fire of 1923, with materials relating to wood shingles. Unprocessed.

133 **Fritz, Emanuel** (b. 1886), Correspondence and papers—additions, ca. 1919-1954, 28 boxes, 33 cartons. Correspondence, subject files and technical files relating to his career in the School of Forestry, University of California, Berkeley, and to his work with the Society of American Foresters, his interest in redwoods, his work for various lumber companies and other consultation work relating to forestry. Unprocessed.

134 **Galloway, John Debo** (1869-1943), 1939, Microfilm, 77 exposures. "Memorandum on Some of the Engineering Works of the Comstock at Virginia City and Gold Hill, Nevada." Includes geological and historical background of the Comstock region, including the wood and lumber industry, with photographic illustrations. Original owned by the Nevada Bureau of Mines.

135 **Gay, Theresa,** Interviews, 1958, 1 portfolio. Interviews with inhabitants of Trinity County, California, relating to local history, including lumbering.

136 **Gilbert, J.J.,** Manuscript, 1878, 13 l. "Logging and Railroad Building on Puget Sound." Includes information on early lumbering in the Puget Sound area of Washington.

137 **Gilfrey, Henry H.,** Address, 1876, 3 pp. "History and Resources of Oregon." Delivered August 4, 1876, at the Centennial Exposition, Philadelphia, Pennsylvania.

138 **Grant, Mortimer N.** (b. 1851), Dictation, 1885, 6 pp. Mentions surveys of Union Pacific Railroad lands and other natural resources, including timber, in Wyoming Territory after 1869.

139 **Great Britain. Board of Trade, Archive of the Companies Registration Office,** 1844-1951, 542 reels of microfilm. Negative and positive, filmed at Bush House, London. Originals are in various repositories in England. Selected documents concerning the organization and activities of limited companies interested chiefly in the North American West. Includes records of the A.S. Agency and Trading Company (1918-25, 38 exposures), a firm organized to engage in agency businesses in relation to agriculture, mining, and forestry, and registered to do business in California; Alberni Timber Company (1912-27, 76 exposures), loggers and general contractors in British Columbia; Anglo-United States Timber Company (1919-31, 82 exposures), organized to undertake a lumber business in the United States, and agents for Jay-Tarbell Lumber Co. of Chicago; British and American Freehold Land and Timber Company (1880-88, 19 exposures), organized to deal in timber, minerals, and manufactured goods in Arkansas; British Columbia and Vancouver Island Spar Lumber and Sawmill Company (1864-82, 14 exposures), organized to open lumber mills in British Columbia; British Columbia Gold Trust (1897-1904, 53 exposures), organized to acquire mines, timber, and real estate holdings in British Columbia; British Columbia Lumber and Development Trust (1910-14, 27 exposures), organized to engage in finance, lumbering, land development and manufacturing in British Columbia; Canada-North West Coal and Lumber Syndicate (1889-1928, 193 exposures), organized to acquire coal and timber lands in Canada; Canadian Lumber Company (1919-50, 115 exposures), to acquire timber lands in Canada; Canadian Pacific Timber Company (1906-09, 74 exposures), to engage in the lumber business in Canada; Cana-

dian Timber Company [British Columbia] (1911-15, 30 exposures), to acquire timber lands in British Columbia; Canadian Timber Importers (1925-37, 46 exposures), to import timber to Britain from Canada; Canadian Timber Investment Company (1911-33, 118 exposures), to acquire timber in British Columbia; Dominion of Canada Free Hold Estate and Timber Company (1881-1906, 36 exposures), to acquire lands and mines in Quebec; Fort George Syndicate (1910-19, 47 exposures), to acquire timber lands in Canada; Fort Steele Development Syndicate (1898-1931, 222 exposures), to acquire mines and timber in British Columbia; Fraser River Gold Dredging Company (1901-07, 118 exposures), to acquire timber and mines in British Columbia; Graham Island (British Columbia) Coal and Timber Syndicate (1911-17, 113 exposures); North American Land and Timber Company (1882-1920, 572 exposures), to purchase lands and mines; North West Timber Company of Canada (1883-91, 41 exposures), to acquire timber lands and mill properties in Canada; Northern Exploration Company of British Columbia (1898-1910, 52 exposures), to acquire mineral, timber, and land resources in British Columbia; Oregon Land and Timber Company (1882-1889, 35 exposures), to acquire timber lands in Oregon; Pacific Coast Timber Lands (1910-14, 32 exposures), organized to undertake banking or capital investment; Pacific Timber Company (1915-34, 189 exposures), organized to acquire timber lands, and to sell lumber, pulp, and power in Canada; Plymouth Consolidated Gold Mines (1914-25, 443 exposures), formed to purchase mines, timber lands in California; Purcell Mining Company (1898-1929, 42 exposures), organized as the Selkirk Mother Lode Copper Mines, Ltd., to acquire mines and timber in British Columbia, and renamed in 1901; Timber Lands of British Columbia (1910-20, 62 exposures), organized to invest in lands, timber, and mineral resources in British Columbia; Vancouver Island Timber Syndicate (1911-50, 281 exposures), organized to purchase and exploit timber resources and to mill and market lumber; War Eagle Gold Mining Company (1896-98, 15 exposures), organized to acquire lands and timber in West Kootenay district, British Columbia; West Canadian Corporation (1900-04, 26 exposures), to carry on lumber and provision businesses in Calgary, Alberta; and the Western Canada Timber Company (1907-35, 376 exposures), organized to acquire timber properties in Canada and to purchase the Canadian Pacific Timber Company, Ltd. Shelf list and filming lists available. The records of each firm are cataloged individually in the manuscript card catalog.

140 **Halderman, Daniel,** Statement, 1887, 2 l. Lumbering and homestead entry in Colorado after 1860.

141 **Henderson, Amos** (b. 1852), Statement, 1886, 1 l. Lumbering, contracting, and railroad grading after 1871 in Colorado.

142 **Higby, William** (1813-1887), Correspondence, 1837-1856, 36 folders in portfolio. Includes some letters concerning his sawmill near Mokelumne Hill, California (1854-1855).

143 **Hobbs, Caleb Seechomb,** Biography, n.d., 7 l. Mainly describes his box making and lumber business in San Francisco from 1854. Includes a biographical sketch, 4 l.

144 **Holmes Eureka Lumber Company,** Records, 1940-1958, 1 box. Includes standard moulding book (1940); material for brochure for the Holmes Eureka Building Service; and reports, mainly financial, for the Redwood Export Company, California Redwood Association, and California Association of Timber Truckers. Unprocessed.

145 **Horner Family,** History and photographs, 1 box. Photographs (many taken by Walter Horner) including some depicting lumbering at Usal and other places in Mendocino County, California.

146 **Hughes, John T.** (b. 1836), Statement, 1886, 3 l. Lumber business since 1865 at Roaring Creek, Denver, Pueblo, and Trinidad, Colorado.

147 **Hughes, Josiah** (b. 1841), Statement, 1886, 4 l. Lumber merchant, Pueblo, Colorado, after 1871.

148 **Hyman, Frank J.,** Reminiscences, 4 l. "The Making." Xerox copy of typescript relating to Hyman's work for the Little Valley Lumber Company, cutting wood and making ties.

149 **Innes, Charles** (b. 1844), Statement, 1887, 2 l. Contractor and builder at Trinidad, Colorado, from 1874; discusses types of lumber used.

150 **Innes, William** (b. 1844), Statement, 1887, 2 l. Lumber business at Laramie, Wyoming, 1874, and later in El Paso County, Colorado.

151 **Johnson, Mary,** Notes on California history and Biography, ca. 1939-1941, 4 folders in portfolio. Contains a sketch of the Z.B. Heywood family, pioneers in the California lumber trade, who established yards in Berkeley.

152 **Johnson, Robert Underwood** (1853-1937), Papers, 1889-1924, 13 boxes (1243 folders). Editor. Correspondence, printed reports, circulars, copies of laws, and other papers, chiefly concerning preservation of the Yosemite area, including the campaign against the Hetch Hetchy reservoir, and material relating to Sequoia National Park, General Grant Grove, and to other national parks and forest reserves. Among the correspondents whose letters include conservation topics are Frederick Law Olmsted [b. 1870] (9 letters, 1913-14); Newton B. Drury (1 letter, 1923); William E. Colby (13 letters, 1907-1914); Robert Underwood Johnson (169 letters, 1891-1914); John W. Noble (n.d., and 1904); John Muir (146 letters, 1889-1914); Austin Foster Hawes (3 letters, 1913); Edward T. Parsons (67 letters, 1909-1914). Guide and key to arrangement.

153 **Johnson, S.S.** (b. 1844), Dictation, 1887, 6 l. Describes his work as a lumber foreman in National City, California, after 1882.

154 **Jones, Henry** (b. 1807), Dictation, 1887, 8 pp. Mentions sawmill interests in Texas after 1837.

155 **Jones, Herbert Coffin** (1880-1970), Manuscript, 1955, 4 l. "History of Acquisition of the Big Basin as State Park." Prepared as President, Sempervirens Club of California. Accompanied by a copy of a speech before the California Assembly (February 18, 1901) by D.M. Delmas in behalf of an appropriation by the State Legislature for purchase of the area. Also included is a related album of pictures.

156 **Jordan, Arthur L.** (b. 1876), Interview, 1970, ca. 22 l. Photocopy of transcript of tape-recorded interview conducted by Sally Bush. Includes recollections of Sierra Club outings with John Muir and William E. Colby.

157 **Joy, T.B., & Co.,** Records, 1854?-1934, 24 vols. Daybooks, journals, ledgers, miscellaneous correspondence and accounts for lumbering operations near Bodega, California; records of the Sebastopol Lumber Yard; notes by Emanuel Fritz concerning the property; maps; and a speech given by Howard McCaughey (1951) relating to the history of T.B. Joy & Co., and containing information about the Joy family. List of contents.

158 **Kaweah Colony,** Collection, 1 portfolio and 2 reels microfilm. Transcripts of originals in the Sequoia National Park Library, including a letter and autobiographical sketch by C.F. Keller, and letters and articles by James John Martin.

The microfilm includes reports of special agents to the General Land Office Commissioner (1889-1891; originals in the National Archives) and a list of members and report and letters from James John Martin and C.F. Keller (originals at Sequoia National Park, Three Rivers, Calif.).

159 **Kentfield, John, and Company,** Records, 1853-1923, ca. 174,000 pieces (184 vols., 129 boxes, 20 cartons and portfolio). Correspondence, accounts, and papers of a lumbering and shipping firm founded by John Kentfield, H.H. Buhne, and D.R. Jones. Some Kentfield family correspondence and papers are included. Guide and Key to the arrangement.

160 **Kingwell, William Ira** (b. 1878), Autobiography, 1957-1958, 58 l. Xerox copy of original in private possession. Recollections concerning his experiences in Sonoma County, logging for T.B. Joy & Co., and others.

161 **Kizer, W.B.** (b. 1825), Dictation, 1887, 5 l. Discusses the lumber business and timber resources in Texarkana, Texas, vicinity, after 1876.

162 **Koch, William C.E.** (b. 1857), Statement, 1886, 4 l. "Life and Adventures in Colorado." Discusses lumbering and other business activities in Colorado, 1880s.

163 **Koogle, W.C.** (b. 1849), Dictation, 1885, 4 l. Discusses lumbering in Texas after 1876.

164 **Kraebel, Charles J.,** Correspondence and papers, ca. 1916-1956, 15 cartons. U.S. Forester. Subject files, photographs, and related printed materials, relating to California, U.S. National Parks and Hawaii. Unprocessed.

165 **Lake, William B.** (1833-ca. 1914), 1852-1899, 7 vols. Records of Lake's association with his father, Jefferson Lake, in a sawmill near Sacramento, California, in the 1850s.

166 **Le Conte Family,** Papers, 1858-1949, 3 boxes, 2 cartons, and 1 portfolio. Correspondence, diaries, lecture notes, and miscellaneous papers of various members of the Le Conte family, including Joseph, Joseph Nisbet, and Caroline Eaton Le Conte, relating chiefly to trips in the Sierra Nevada in California, 1871-1946. and to activities with the Sierra Club. Report and Key to arrangement.

167 **Lewis, Ruth (Krandis),** Correspondence, Feb. 1942, 1 portfolio. Correspondence with the U.S. District Court, Southern District of California, and the Tulare County Board of Trade concerning the Kaweah Cooperative Colony.

168 **Lowdermilk, Walter Clay** (b. 1888), Papers, 1912-1969, 8 cartons, 1 portfolio, 1 tube. Includes correspondence, with letters from Carl Hayden, Hubert G. Schenck, Clinton P. Anderson, and others; personalia; diaries and field notes (1912-1969); etc.

169 **McReavy, John** (b. 1842), Dictation, 1888?, 26 pp. McReavy was a rancher and logger in Oregon.

170 **Madera Sugar Pine Lumber Company** (Madera, California), Photographs, 1 carton. Pictures of the company's fluming operations. Unprocessed.

171 **Mann, Samuel Stillman** (1819-1888), Letter to Mrs. Frances Fuller Victor, 1880, 4 pp. Mentions shipbuilding, lumbering, and other activities around Coos Bay, Oregon. Accompanied by a typed copy.

172 **Manson, Marsden** (1850-1931), Papers, 1887-1927, 2 cartons. City engineer of San Francisco, California. Correspondence, papers, reports, manuscripts, and miscellaneous printed material pertaining chiefly to Manson's office as city engineer (1908-1912), and concerning the city's water supply, particularly the Hetch Hetchy and Lake Eleanor Reservoirs in Yosemite National Park. Report and partial analytics in the library.

173 **Marshall, Robert Bradford** (1867-1949), Papers, 1898-1949, 23 boxes, 4 scrapbooks, 1 portfolio, 1 volume. Correspondence, notes, manuscripts of his writings, speeches, memoranda, clippings and scrapbooks, mainly relating to the Marshall plan for water development, conservation, Hetch Hetchy Valley, roads, Yosemite National Park, other parks; also family and personal papers. Key to arrangement.

174 **Martin, James John** (1845-1938?), Papers, 1889-1937, 2 volumes, 1 box, and 1 carton. Includes, among other subjects, materials relating to the Kaweah Colony, with correspondence, clippings, and a manuscript history of the colony. Key to the arrangement.

175 **Matthews, Gordon Frazer** (b. 1877), Manuscript, 1959?, 87 l. "Shipbuilding History of the Matthews Family." Prepared from typescript. Relates experiences in wooden ship construction, mainly at Eureka, California, and Hoquiam, Washington. Includes vessels from the E.K. Wood Lumber and the Pacific Lumber Companies, E.T. Kruse, and S.S. Freeman.

176 **Maxwell, James P.** (b. 1839), Biographical sketch, 1885, 2 l. Lumberman after 1862, Boulder, Colorado.

177 **Mendocino Lumber Company,** Records, ca. 1907-1922, 1 carton. Orders (1907); orders and correspondence relating to orders (1921-1922); manifests; stock sheets; hotel board time books, etc. Unprocessed.

178 **Merriam, John Campbell** (1869-1945), Papers, ca. 1904-1932, 15 boxes and 1 carton. Paleontology and historical geology professor, University of California; president of Carnegie Institution, Washington. Correspondence, manuscripts, and reprints of articles, and miscellaneous papers, relating to Merriam's career and to his interest in conservation. Report and Key to the arrangement.

179 **Merriam, John Campbell** (1869-1945), Papers, 6 boxes and 5 cartons. Material relating to the Save-the-Redwoods League. Unprocessed.

180 **Merrimac Land and Lumber Company,** Report, ca. 1910, 6 l. Typescript copy. Unsigned appraisal of assets of the company, located in Butte and Plumas Counties, California. Contains timber estimates, a proposal for a railroad, note on income and expenses, and a proposal that cutover lands be sold for agricultural purposes.

181 **Metcalf, Woodbridge,** Correspondence and papers, ca. 1950-1972, 24 cartons. Relate to his career in forestry, and as professor in the School of Forestry, University of California, Berkeley, and to his interest in eucalyptus. Included are slides, writings, notes, diaries, subject files, etc. Unprocessed.

182 **Mirov, Nicholas Tiho** (b. 1893), Correspondence and papers, ca. 1928-1968, 2 cartons. Correspondence, reprints, etc., relating in large part to his studies on pines and turpentine, and to the production of heptane, with California Timber Products Co. and the Institute of Forest Genetics. Unprocessed.

183 **Montgomery, A.W.** (b. 1840), Statement, 1886, 2 l. Lumber business at Central City, Colorado.

184 **Moore, Alexander,** Manuscript, 1878, 9 l. "A Pioneer of '47." Written and signed for Moore by an amanuensis. Mentions a lumber mill in California ca. late 1840s.

185 **Morris, John M.** (b. 1835), Diary and autobiography, 1885-1906, microfilm (14 vols.). Originals in private possession. The autobiography describes work in sawmills and logging camps in California. Restricted.

186 **Morse, Eldridge** (b. 1847), Notebooks, ca. 1880, 2 vols. "Notes on the History and Resources of Washington Territory." 2 volumes concern logging and mining in the Puget Sound vicinity. Key to arrangement.

187 **Mount Diablo Museum Advisory Committee,** Records, ca. 1935-1939, 11 cartons. Correspondence, minutes of committee meetings, administrative reports, photos, reference materials used in the preparation of reports, and copies of reports prepared by the staff on the history of the redwoods and Mt. Diablo, for projects sponsored by the California State Division of Parks under the California State Emergency Relief Administration and the Work Projects Administration, for the purpose of preparing graphic educational material for the California Redwoods Park and the Mount Diablo State Museum. Inventory.

188 **Muir, John** (1838-1914), Papers, ca. 1860-1914, 2 boxes and 1 portfolio. Naturalist, explorer and author. Correspondence, notes and drawings of one of Muir's inventions, copy of a report (1875) by Muir on the geology and botany of Mount Shasta, notes by Katherine Putnam Hooker relating mainly to Muir, magazine articles about him, and poems by Helen Muir. Much of the correspondence concerns Muir's daily activities, writings, and travels, and his interest in conservation and Yosemite National Park. Correspondents include J.D. Butler, William E. Colby, Katherine Putnam Hooker, Marian O. Hooker, S.M. Ilsley, R.U. Johnson, Charlotte Kellogg, J.B. McChesney, Muir's daughter, Helen, and other members of the family.

189 **Munns, Edward N.,** Manuscript, 1915, 75 l. "Resumé of Eucalyptus Planting on the Angeles National Forest." California.

190 **Nevada Miscellany,** 1862-1877, 12 items. Includes a statement of Sherwood & Bro., Lumber Merchants, Pioche, Nevada (January, 1874) and bill (October 3, 1873).

191 **Olive, S.C.** (b. 1833), Dictation, 6 l. Olive came to Bastrop County, Texas, in 1849. He discusses milling and lumbering, among other interests.

192 **Pacific Coast Redwood Company,** Records, 1903-1929, 6 vols. Bylaws (1907); minute book, including articles of incorporation (1903-1929); journal (1906-1929); stock ledger and journal (1907-1929); ledger (1906-1929); and miscellaneous loose material removed from preceding volumes.

193 **Pacific Lumber Company,** Records, ca. 1868-1949, 3 cartons. Includes early business records such as letterpress copies, vouchers, receipts, bills and orders (ca. 1868-1870); history of the company; price lists; catalogs; miscellaneous printed materials, mainly on redwoods. Unprocessed.

194 **Pardee, George Cooper** (1857-1941), Papers, 1890-1941, 58 volumes, 117 boxes, 11 cartons, and 1 portfolio. Correspondence, speeches, reports, articles, and bound volumes of newspaper clippings, relating to Pardee's career as governor of California, 1903-1907; his work as chairman of the California Conservation Commission, 1911-1915; chairman of the State Board of Forestry, 1919-1923; and other conservation, irrigation, and reclamation activities. Report and Key to arrangement.

195 **Parsons Family,** Papers, ca. 1880-1953, 2 volumes, 9 boxes, 5 cartons, and 1 portfolio. Correspondence, manuscripts, notes, photographs, newspaper clippings, and other papers of Marion R. Parsons, relating to her interest in mountaineering, her writings, and other matters; and correspondence, diaries, accounts, and other papers of Edward T. Parsons, relating to his interest in mountaineering and conservation, and his business interests. Key to arrangement.

196 **Patterson, Arthur H.** (b. 1844), Statement, 1885, 2 l. Lumberman of Larimer County, Colorado, after 1866.

197 **Pfeuffer, George** (b. 1830), Biography, 1886?, 2 l. The lumber business in Texas after 1845.

198 **Pierce, John B.** (b. 1827), Dictation, 1885, 4 pp. Contains information on lumbering in Siskiyou County, California, before 1862.

199 **Prendergast, Jeffry Joseph** (1875-1962), Diaries, 1931-1962, 32 vols. Association with the Bear Valley Mutual Water Company in San Bernardino County; interest in water conservation, construction and maintenance of dams, canals, and irrigation projects; flood control; water levels; agriculture; forest conservation.

200 **Rabbeson, Antonio B.** (b. 1824), Manuscript, 1878, 31 l. "Growth of Towns. Olympia, Tumwater, Portland, and San Francisco." Mentions the beginnings of logging on Puget Sound.

201 **Rand, George** (b. 1837), Statement, 1886, 2 l. "Agriculture in Colorado." Lumbering in Jefferson and Boulder Counties, Colorado, since 1864.

202 **Redwood Empire Association,** Records, ca. 1930-1960, 137 transfer cases. Includes correspondence, reports, minutes, accounts, material on legislation, on the Shoreline Highway Association. Much of the material appears to relate to highways and to publicity. Unprocessed.

203 **Redwood Manufacturers Company,** Records, ca. 1930-1954, 54 boxes, 6 vols. Correspondence, payrolls, accounts receivable, ledgers, price lists, etc. Unprocessed.

204 **Regional History Project, University of California Library, Santa Cruz,** Oral History Interviews. Typed transcripts of tape-recorded interviews. Relating to forest history are interviews conducted by Elizabeth S. Calciano with Michael Bergazzi (b. 1886) on redwood lumbering in Santa Cruz County, California (1964, 207 l.), and with Albretto Stoodley (b. 1873) on the Loma Prieta Lumber Company, redwood lumbering, and the fruit box industry in Santa Cruz County.

205 **Regional Oral History Office,** Oral History Collection. Transcripts of interviews completed by the Regional Oral History Office, Room 486, Bancroft Library. Among the items relating to forestry, parks, and recreation are interviews with Horace M. Albright (b. 1890) and Newton B. Drury (b. 1889) on conservation, chiefly in national parks (1962, 53 pp.); Rexford Black (b. 1894) on private and state forestry in California (1968, 159 pp.); Harold C. Bryant (1886-1968) and Newton B. Drury on development of the naturalist program in the U.S. National Park Service (1964, 49 pp.); Ralph Works Chaney (1890-1971) on his work with the Save-the-Redwoods League and the National Parks Advisory Board (1960, 277 pp.); Henry Clepper (b. 1901) on the Society of American Foresters (1968, 36 pp.); John D. Coffman (1882-1973) on forestry in the U.S. National Park Service (1973, 126 pp.); William E. Colby (1875-1964) on John Muir, the Hetch Hetchy water project, and the California State Park Commission (1954, 145 pp.); Richard Colgan (b. 1891) on private forestry in the California pine region (1968, 50 pp.); Gladys Austin, Jack Carpenter, William G. Cumming, Alfred R. Liddicoet, Nicholas T. Mirov, and Frances I. Righter on the Eddy Tree Breeding Station [Placerville, California] (1974); Samuel T. Dana (b. 1883) on federal forest policies and forestry education (1967, 98 pp.); Newton B. Drury on the U.S. National Park Service and California redwood state parks (1972, 2 vols.,

772 pp.) ; Herbert Evison and Newton B. Drury on the National Park Service and the Civilian Conservation Corps (1963, 143 pp.) ; Francis P. Farquhar (b. 1887) on the Sierra Club and National Park Service (1958, 34 pp.) ; Enoch Percy French (1882-1970) and Newton B. Drury on preservation of the California redwoods (1963, 86 pp.) ; Emanuel Fritz (b. 1886) on forestry education and consultant work (1972, 336 pp.) ; Tom Gill (1891-1972) on U.S. contributions to foreign forestry (1969, 75 pp.) ; Christopher Granger (1885-1967) on U.S. Forest Service policies (1965, 131 pp.) ; Ansel F. Hall on Yosemite National Park (1958, 34 pp.) ; R. Clifford Hall (b. 1885) on the Fairchild Forest Taxation Study, 1926-1935 (1967, 113 pp.) ; John W. Keller on Gifford Pinchot (1974) ; George B. Hartzog (b. 1920) on the U.S. National Park Service (1973, 92 pp.) ; Fred E. Hornaday (b. 1900) on the American Forestry Association (1968, 20 pp.) ; Leon F. Kneipp on U.S. Forest Service land planning and acquisition (1975) ; Leo Isaac (1892-1970) on Douglas-fir research in the Pacific Northwest (1967, 152 pp.) ; Evan W. Kelley on the U.S. Forest Service Guayule Rubber Project (1974) ; Joseph Russell Knowland (1873-1966) and Newton B. Drury on the California State Park Commission (1965, 120 pp.) ; I.E. Kotok on research in the U.S. Forest Service and state forestry (1975) ; Myron E. Krueger (b. 1890) on logging technology in northern California (1968, 27 pp.) ; Walter C. Lowdermilk (b. 1888) on his experiences in the U.S. Forest Service and in international forestry (1969, 704 pp. [2 vols.]) ; Walter H. Lund (b. 1902) on U.S. Forest Service timber management in the Pacific Northwest (1967, 83 pp.) ; Walter McCulloch (1905-1973) on forestry education (1968, 216 pp.); Woodbridge Metcalf (1888-1972) on extension forestry in California (1969, 138 pp.) ; personal acquaintances of John Muir on their recollections of Muir (1971, 106 pp.) ; Thornton T. Munger (b. 1883) on forest management research in the Pacific Northwest (1967, 145 pp.) ; Earl S. Pierce (b. 1886) on the U.S. Forest Service timber salvage program following the 1938 New England hurricane (1968, 52 pp.) ; Kenneth B. Pomeroy on the American Forestry Association (1968, 21 pp.) ; Paul H. Roberts on the Prairie States Forestry Project (1974) ; Arthur C. Ringland (b. 1882) on his career with the U.S. Forest Service and in international forestry (1970, 538 pp.) ; William R. Schofield (1894-1963) on forestry legislation in California (1968, 159 pp.) ; Harold B. Shepard on the U.S. Forest Service Forest Insurance Study, 1929-1939 (1967, 6 pp.) ; Stuart Bevier Show (1886-1963) on National Forests in California (1965, 215 pp.) ; John H. Sieker on recreation and wilderness policy in the U.S. Forest Service (1968, 49 pp.) ; Lloyd Swift (b. 1904) on wildlife management in the U.S. Forest Service (1968, 29 pp.). Interviewers included Amelia R. Fry, Edna Tartaul Daniel, Corinne L. Gilb, Willa K. Baum, Elwood R. Maunder, Francis P. Farquhar, Fern S. Ingersoll, Malca Chall, Evelyn Bonnie Fairburn, John Jencks and Ernest Lowe of radio station KPFA, and others. Restrictions.

206 **Renton, William** (1818-1891), Dictation and Biographical Material, 1887-1890, 3 items. Renton's role in lumbering on Puget Sound, especially the Port Blakely Mill Company, from 1852-1853. Prepared for Bancroft's *Chronicles of the Builders.*

207 **Rich, Frederic R.** (b. 1851), Diaries, 1873-1878, 1880-1887, 15 vols. Concerns, in part, work at various sawmills in Butte County, California, including the Sierra Lumber Company.

208 **Riley, F.C.,** Manuscript, 1961, 22 l. "Opportunities for Operations of Pulp Mills in Humboldt County, California." Photocopy of a typescript report.

209 **Rockport Lumber Company** (Rockport, California), Records, ca. 1925-1968, 19 cartons. Correspondence, reports, papers concerning timber cruises, timber and land purchases and sales, inventories, financial statements, and files relating to cooperation with the American Forest Products Industries, California Forest Protective Association, Redwood Region Conservation Council, and the Society of American Foresters. Includes records (1925-1928) of the Finkbine-Guild Lumber Company of Rockport, California, and Jackson, Mississippi, and notes by B.Z. Agrons on the Mendocino, California, area and the logging industry. Closed until May 28, 1980.

210 **Roder, Henry,** Dictation, 1878, 48 l. "Bellingham Bay and the San Juan Island Difficulty." Describes lumber business on Puget Sound following the San Francisco fire of 1852, and beginnings of cities in Puget Sound area.

211 **Rowell, Chester Harvey** (1867-1948), Papers, 1887-1946, 27 boxes, and 10 cartons. Correspondence (1893-1946 [chiefly 1900-1931]) ; editorials, *Fresno Republican* (1898-1920), and other writings, pamphlets and clippings. Relating to California and national politics and various causes in which Rowell was interested. Among the correspondents whose letters relate to conservation matters are Milton T. U'Ren (6 letters, 1911-28) ; Overton Price (1 letter, 1910) ; Gifford Pinchot (17 letters) ; T.H. Goodspeed (1 letter, 1920) ; William E. Colby (2 letters, 1917, 1928) ; Hu Maxwell (5 letters, 1913) ; Virginia Ferguson (2 letters, 1928-29). Organizations represented include the National Conservation Association, Save-the-Redwoods League, and Sierra Club. Key to arrangement.

212 **Russell, Warren T.,** Manuscript, 1947?, 248 l. "Chispas by an Old Miner." A collection of stories and notes on gold mining, which includes a chapter on logging operations.

213 **Save-the-Redwoods League,** Interviews, 2 items. Interviews conducted by Susan R. Schrepfer with Tom Grieg (1970, 48 pp.) concerning acquisition of land for California state parks; and with Enoch Percival French (1969, 30 pp.) concerning experiences as a timber cruiser in the redwoods. Photocopies of typed transcripts.

214 **Save-the-Redwoods League,** Records, ca. 1919-1944, 10 cartons. Correspondence and papers, particularly files of John C. Merriam; publicity material; memoranda; material on National Conference on State Parks; files on forestry and the lumber industry; redwood groves. Unprocessed.

215 **Save-the-Redwoods League Semicentennial,** Proceedings, June 15-16, 1968, 41 l. Corrected typed transcript of tape-recorded proceedings at dedication of redwood groves, Pepperwood Flat, honoring Newton B. Drury, Thomas A. Greig, and Norton R. Cowden, and at a banquet at Eureka Inn, Humboldt County, California. Comments by Ralph W. Chaney, Horace M. Albright, Newton B. Drury, and others. With this is a photocopy of final corrected transcript.

216 **Sayward, William T.,** Reminiscences, 1882?, 36 pp. Brief notes of Sayward's lumber interests in California and the Northwest.

217 **Scammon, J.L.** (b. 1822), Statement, 1887, 2 pp. Mentions lumbering at Coloma, California.

218 **Scotland. Treasury, Archive of the Companies Registration Office,** 1856-1951, 120 reels microfilm. Negative and positive microfilm, filed at the Office of the King's Remembrancer, Edinburgh. Selected documents concerning the organization and activities of limited companies interested chiefly in the North American West. Includes records of the American Lumber Company (1882-88, 53 exposures), organized to acquire and deal in timber lands in Ontario; Athabasca Saw-

mills, Ltd. (1916-48, 136 exposures), organized to acquire a lumber and sawmilling business in Alberta; British Columbia Lumbering & Timber Company (1880-1901, 103 exposures), to acquire timber properties in Canada, Michigan, and Wisconsin; British Columbia Pulp and Paper Mills (1898-1903, 29 exposures), to manufacture and deal in paper and wood pulp, British Columbia; California Redwood Company (1883-1890, 105 exposures), to acquire timberlands and mills in Humboldt County, California, from J. Russ & Company, and the lines of the Humboldt Logging Railway, assets sold later to Edinburgh & San Francisco Redwood Company, Ltd.; Edinburgh & San Francisco Redwood Company, Ltd. (1885-1906, 242 exposures); Humboldt Redwood Company (1885-1909, 136 exposures), organized to acquire timber lands in Humboldt County; Little River Forests Syndicate (1911-19, 38 exposures), organized to invest in timber and mining enterprises in Newfoundland; Quebec Timber Company (1881-98, 80 exposures), lumbering in the United States and Canada; St. John Sulphite Pulp Company (1897-1906), wood pulp and paper making in New Brunswick; Scottish Canadian Timber Preserving Company (1910-12, 28 exposures), organized to acquire Canadian rights on a patent for preserving and fireproofing timber; Scottish Carolina Timber & Land Company (1884-98, 86 exposures), organized to acquire lands in North Carolina; and the Woodruff Land & Timber Company (1893-98, 44 exposures), organized to acquire timber in Woodruff and Cross Counties, Arkansas. Indexed. Each group is cataloged separately in the manuscripts card catalog.

219 **Show, Stuart Bevier** (1886-1963), Papers, 1921-1954, 4 boxes. Forester. Notes, manuscripts of Show's writings, relating mainly to the U.S. Forest Service and to fire control in California, and other papers relating to Show's career with the U.S. Forest Service. Some of the papers were written in cooperation with R.W. Ayers.

220 **Sierra Club,** Records, ca. 100 cartons. Included are papers relating to the U.S. Forest Service, redwoods, lumber and logging. Unprocessed. Further accessions are anticipated.

221 **Sierra Nevada Wood & Lumber Company,** Records, 1899-1900, 1 box. Wood records for the Hobart Lumber Mills at Overton, California. Unprocessed.

222 **Skellenger, W.B.,** Letter, November 2, 1885, 7 pp. Letter to H.H. Bancroft containing reminiscences on lumbering, chiefly in Placer County, California; erection of a steam sawmill near Coloma, California, in 1849; and wages and prices.

223 **Slaughter, John** (b. 1809), Dictation, 1884, 7 l. "Life in Colorado and Wyoming." In handwriting of H.H. Bancroft. Slaughter came to Colorado, 1861; Wyoming, 1867, in the lumber business. Discusses lumbering practices.

224 **Smith, Relmond,** Reminiscences, August 29, 1971, 3 l. Xerox copy. Relates to lumbering in Mendocino, California, at the turn of the century. Includes a Xerox copy of a typed transcript.

225 **Sonoma Lumber Company,** Records, ca. 1877-1884, 6 volumes. The records consist of minutes of meetings of directors and stock holders, ledgers, stock books, bylaws, cashbooks, and other papers.

226 **Sopris, Richard** (b. 1813), Statement, 1884, 16 l. "Settlement of Denver." Carpenter and joiner; came to Colorado in 1859; discusses early lumbering and other activities.

227 **Spink, Henry Makinson.** "The Economic Development of the Pacific Coast of North America." Master's Thesis (Geography), Liverpool, England, 1921, 2 vols. (ca. 121 pp. + maps). Includes materials on lumber and lumbering.

228 **Stanwood, Richard Goss,** Journals and Letterbooks, 1852-1884, 5 volumes. Describes employment in lumber business, Marysville, California, 1850s.

229 **Starr, Walter Augustus** (b. 1877), Manuscript, n.d., 2 l. "From Yosemite to Kings River Canyon." Notes on his trip as Sierra Club member in 1896, and on the construction of trails in this area. With this is condensed biography and family history, 4 l.

230 **Steel, James** (b. 1834), Dictations, notes, and drafts of biographical sketch, ca. 1889, 5 items. Discussion of lumberingand other business enterprises, Portland, Oregon, after 1862. Prepared for Bancroft's *Chronicles of the Builders.*

231 **Sudworth, George Bishop** (1862-1927), Manuscript, 1899, ca. 75 l. "General Notes on the Distribution of Trees inCalifornia." Typescript. Prepared while Sudworth worked for the U.S. Forest Service. Relates to Sequoia trees, Sierra Nevada forests.

232 **Union Lumber Company** (Fort Bragg, California), Records, ca. 1887-1939, 1245 vols., 58 boxes. Includes correspondence; cashbooks; ledgers, journals; payrolls; receipts; time-books, etc. Also includes miscellaneous correspondence and records for the Mendocino Lumber Co., Glen Blair Redwood Co., Acme Lumber Co., Fort Bragg Lumber Co., Fort Bragg Electric Co., and the National Steamship Co. Unprocessed.

233 **U.S. National Park Service,** Interviews with agency officials, Sept.-Dec. 1962, 1 box. Photocopies of typed transcripts of tape-recorded interviews of officials, active, and retired, associated primarily with the Western Region Office, and with western national parks (Yellowstone, Yosemite, Sequoia, Kings Canyon, Death Valley, etc.). Conducted by Herbert Evison. Interviewees include Thomas J. Allen, Harold C. Bryant, Blanton Clement, John D. Coffman, John M. Davis, Edward D. Freeland, Lemuel A. Garrison, Theodore R. Goodwin, Aubrey F. Houston, Granville B. Liles, Herbert Maier, George L. Mauger, Lawrence C. Merriam, Keith Neilson, Hilmer Oehlman, Carl P. Russell, John A. Rutter, Max Walliser, Volney J. Westley, John B. Wosky. Included also is an interview with Herman H. Hoss of the Western Conference of National Park Concessioners.

234 **Van Dyke, Walter** (b. 1823), Recollections, 1878, 28 l. "Recollections on matters concerned with early years of California and Oregon, 1849-63." Dictation recorded for H.H. Bancroft. Mentions timber resources of the Humboldt Bay area, California.

235 **Washington Miscellany,** Collection, 10 items. Mostly letters to H.H. Bancroft and John S. Hittell from various persons (1878-1884). Includes J.M. Rhorb, Seattle, 1884, 3 l., remarks on the Blakely, Washington, Sawmill.

236 **Washington Miscellany,** Collection, 12 items. Includes Puget Mill Company letter to Mannee McMucken, Port Gamble, Washington (September 10, 1881), asking for tracings of certain townships.

237 **Weed Lumber Company,** Report, 1911, 1 vol. Transcript (carbon) of report for fiscal year ending December 31, 1911. With some comparative reports (1907-1911).

238 **Welch, James W.** (b. 1842), Dictations, 1887, 1 portfolio. Information concerning lumbering and other business activities in Astoria, Oregon.

239 **Wilde, Willard H.,** Manuscript, n.d., 168 l. "Chronology of the Pacific Lumber Company, 1869 to 1945." Typescript. Includes biographical sketches of men associated with the company, and bibliography.

240 **Wiley, Hugh** (b. 1884), Correspondence and papers, ca. 1925-1963, 1 portfolio, 1 box. Contains material relating to his management of the Hazard Mine in Placer County, California, and its related lumbering business.

241 **Williams, Henry F.** (1828-1911), Recollections, 1878, 16 l. "Recollections on Early Days in California by the Pioneer of 1849." Dictation recorded by H.H. Bancroft. Concerns activities as carpenter and builder in San Francisco, with brief mention of lumber supplies and of the building of the city, 1849.

242 **Williams, Mary Floyd** (1886-1959), Papers, 45 items, in portfolio. Information relating to her father, Edward C. Williams (1820-1913), including his account of the first redwood operations in California.

243 **Wilson, B.F.** (b. 1832), Dictation, 188-?, 1 l. Lumbering in Shasta and Lassen Counties, California.

244 **Withrow, Chase** (b. 1839), Statement, 1885, 2 l. "Central City, Colorado, in 1860." Withrow was a lumberman, along with other activities.

245 **Wood, E.K., Lumber Company,** Records, ca. 1906-1957, 93 cartons. Correspondence and records from the Company's offices at Reedsport, Oregon, Southern California and San Francisco, dealing with the company's mill and lumber yard operations. Include journals, ledgers, cash registers, sales records, deeds, appraisals of property, tax records, corporation minutes; correspondence files for various offices, etc. Unprocessed.

246 **Works, John Downey** (1847-1928), Papers, ca. 1910-1917, 11 boxes. U.S. Senator from California, 1911-1917; member of the Committee on Public Lands. The papers include correspondence, printed copies of Works' speeches, clippings, reports, bills, and government documents, relating mainly to Works' career in the U.S. Senate and his stand on such issues as Hetch Hetchy legislation. Report and Key to arrangement.

247 **Yeaton, John Gardener,** Journal, 1852-1853, 63 pp. Mentions work in sawmills in California, among other activities.

248 **Yerington, Henry Marvin** (1828-1910), Correspondence and Papers, 1864-1950, 45 volumes, 1 box and 4 cartons. Includes material relating to the Carson & Tahoe Lumber and Fluming Company, El Dorado Wood and Flume Company, Sierra Nevada Wood and Lumber Company, Walker Lake Wood and Lumber Company, among other business records.

249 **Yesler, Henry Leiter** (1810-1892), Dictation, 1878, 20 pp. "Settlement of Washington Territory." Largely in the handwriting of H.H. Bancroft. Describes the establishment of Yesler's lumber mill in Seattle, Washington. "Incidents in the life of Hon. H.L. Yesler," with a note concerning J.D. Lowman, Yester's nephew, pasted in.

University of California, Forestry Library

Librarian, 260 Mulford Hall 94720. The following is a partial list of the manuscript and mimeographed unpublished materials held by the Forestry Library. They are cataloged and shelved along with other holdings.

250 **Barrett, Louis A.** "A record of forest and field fires in California from the days of the early explorers to the creation of the forest reserves," San Francisco: 1935, ca. 171 l., typewritten.

251 **Barrett, Louis A.** "History of National Forest Land Problems," San Francisco: 1929?, ca. 8 l. Paper read at the 1929 Annual Meeting of the California Section, Society of American Foresters, San Francisco.

252 **Bauer, Patricia McCollum.** "History of Lumbering and Tanning in Sonoma County, California, since 1812," Berkeley: University of California, M.A. thesis, 1951, 189 pp. Microfilm of the original in the University Library.

253 **Ellsworth, Rodney Sydes.** "Discovery of the Big Trees of California, 1833-1852," Berkeley: University of California, M.A. thesis, 1933, 146 pp.

254 **Knowles, Constance Darrow.** "A history of lumbering in the Truckee Basin from 1856 to 1936," Berkeley: U.S. Forest Service, California Forest and Range Experiment Station, 1942, 54 pp., typescript. Derived from the Bibliography of Early California Forestry.

255 **Loughman, Michael Lawrence.** "National Parks, Wilderness Areas, and Recreation in the Southern Sierra Nevada. California: An Historical Geography," Berkeley: University of California, M.A. thesis (Geography), 1967, 193 pp.

256 **Rettie, Dwight Fay.** "National Forest Timber Sale Policy: A Case Study of the Disposal of Federally-Owned Natural Resources Severable from Land," Berkeley: University of California, M.A. thesis (Political Science), 1955, ca. 171 l.

257 **Schlappi, Elizabeth Roe.** "Saving the Redwoods," Berkeley: University of California, M.A. thesis (History), 1959, ca. 152 pp. Concerns the Save-the-Redwoods League.

258 **Stanford, Everett Russell.** "A Short History of California Lumbering, including a Descriptive Bibliography of Material on Lumbering and Forestry in California," Berkeley: University of California, M.S. thesis, 1924, 92 l.

259 **Sudworth, George B.,** Notebooks, 1899-1904, 14 volumes. Notebooks kept by a U.S. Forest Service botanist while he was collecting material on the trees of California. The notes, which are intended to identify photographs he took, give detailed descriptions of vegetation and topography, and are interspersed with diary entries. A typescript of selected extracts from these notebooks was recently donated by Elbert Little, U.S. Forest Service, Washington, D.C.

260 **U.S. Forest Service, California Forest and Range Experiment Station.** "Bibliography of Early California Forestry," 69 volumes. The bibliography is divided into six sections as follows: (1) Forest fires up to 1930; (2) lumbering operations to 1920; (3) other forest utilization operations to 1920; (4) records of lumber and other forest products shipments and/or consumption to 1920; (5) records of the amount of standing timber of specific areas to 1910; (6) vegetation of specific areas to 1870. The entries have been compiled from newspapers, diaries, books, periodicals, documents, and maps. They include full bibliographic information and quotations or abstracts of the material.

261 **Vale, Thomas Randolph.** "The Redwood National Park: A Conservation Controversy," Berkeley: University of California, M.A. thesis (Geography), 1968, 189 l.

262 **Warren, Judith Ann.** "The Lumber Industry in the Plumas-Lassen Area of the Northern Sierra-Southern Cascades in California: An Historical Geography," Berkeley: University of California, M.A. thesis (Geography), 1971, 160 l.

263 **Wattenburger, Ralph Thomas.** "The Redwood Lumbering Industry on the Northern California Coast, 1850-1900," Berkeley: University of California, M.A. thesis, 1931, 87 l.

264 **Wilson, R.C.,** Interview, 1935-1936, 27 l. Typescript of interviews with early California and Nevada settlers on sawmills and lumbering in the Reno-Tahoe area. Produced for the U.S. Forest Service, California Forest and Range Experiment Station, Berkeley.

University of California, Water Resources Center Archives

Librarian, Room 40, North Gate Hall 94720

265 Manuscript holdings include papers (1906-62, ca. 2 feet) of Max J. Bartell (1879-1968); papers (1905-40, ca. 9 feet) of John Debo Galloway (1869-1943); papers (1910-50, ca. 13 feet) of Walter LeRoy Huber (1883-1960); papers (1882-1942, ca. 34 feet) of Joseph Barlow Lippincott (1864-1942); papers (1908-58, ca. 4 feet) of Thomas Herbert Means (1875-1965); and papers of others involved in water engineeringwork. The records are largely reports, documents, photographs, and maps, but include some correspondence, relating to irrigation, water supply, reservoirs, hydroelectric power, and flood control in the western U.S. and Canada. Each of the collections named above includes material relating to the Hetch Hetchy Valley project in Yosemite National Park, California. There are also numerous files pertaining to permits for reservoirs, conduits, and stream flow investigations on national forests. The Huber papers contain a file of correspondence (chiefly 1910-19) concerning establishment of the Devil's Postpile National Monument, California. The collection is described in the *Dictionary Catalog of the Water Resources Center Archives, University of California, Berkeley,* 5 vols., Boston: G.K. Hall & Co., 1970; and supplements, 1971, 1972, 1973. There are also calendars for the Huber and Lippincott papers in *Archives Series Reports,* No. 12 (1962) and 21 (1970).

CHICO

California State University, Learning Activities Resource Center, University Library

95929

266 **Lassen Lumber and Box Company** (Susanville, California), Miscellaneous Papers, 1918-1951, 3 vols. Included are articles of incorporation, certificate of increase of capital stock, certificate of winding up and dissolution (1 vol.); transcript on appeal in the Supreme Court of the State of California (1 vol.); photograph album (1 vol.).

267 **Taylor, Will L.,** Photographs, 30 items (in portfolio). The Sierra Flume and Lumber Company of California, with descriptive material prepared by W.H. Hutchinson (1956, 9 l.).

Theses

268 **Brouilette, Joseph F.** "A History of Stirling City, Butte County, California," Chico: Chico State College, M.A. thesis, 1959, 168 pp.

269 **Spence, Vernon Charles.** "A History of the Redwood Lumber Industry in Sonoma County," Chico: Chico State College, M.A. thesis, 1962, 96 pp.

269a **Weed, Abner Edward, Jr.** "Weed: The Evolution of A Company Town," Chico: California State University, M.A. thesis (Geography), 1974. Relates to a mill town after 1900, the Long-Bell Lumber Company, and Abner Weed.

270 **Wolf, Virginia Sue.** "The Burney Basin, Shasta County: Human Utilization of an Intermontaine Habitat," Chico: Chico State College, 1969, ci + 81 pp.

DAVIS

University of California, the University Library, F. Hal Higgins Library of Agricultural Technology

Head, Department of Special Collections 95616

271 The Higgins Library is a collection of materials relating to the development of agricultural implements, containing, in general, advertising literature, tearsheets, photographs, parts books, manuals, correspondence, and some manuscript materials. Material on the application of traction equipment to problems of logging technology since ca. 1900 includes ca. 3 linear feet of vertical file folders and 2 linear feet of photographs, chiefly of Caterpillar and Holt machines. Card catalog.

FRESNO

California State University, Fresno Campus, Library

Head, Department of Special Collections, Shaw Avenue at Cedar 93740

272 **Johnson, Donald R.** "West Coast Lumbermen and Federal Forest Policies, 1901-1909," Fresno State College: M.A. thesis, 1966, vi + 122 l.

273 **Luehe, Frederick William.** "A Historical Study of the Clemmie Gill School of Science and Conservation," Fresno State College: M.A. thesis, 1969, xiii + 271 + 42 l.

274 **Moore, Carol L. Peterson.** "The Wilderness Preservation Movement 1920 to 1964," Fresno State College: M.A. thesis, 1965, v + 126 l.

275 **Pidgeon, Harry,** Collection, 1913-1925, 630 items. Glass negatives and contact prints of the camp and logging activities at Sugar Pine, near Yosemite National Park, California. There is one set of prints at the Yosemite National Park Museum. Prints may be obtained by interlibrary loan from California State University, Fresno.

Fresno County Historical Society

Executive Director, 7160 West Kearney Blvd. 93706
Telephone: (209) 264-8317

276 Newspaper clippings from the *Fresno Republican* relating to national parks in Fresno and surrounding counties (1919-31, 70 items) and forest fires (1923-30, 37 items); monthly time book of the Sanger Lumber Co.; history of efforts to preserve the Mountain Home tract of redwoods through the correspondence of A.H. Drew and articles in the *Fresno Republican* (1942-45); recollections of Frank Dusy and early days in the Sierra by J.W. Humphreys (manuscript letter and transcription); brief manuscript descriptions of sites in the Sierra Nevada, giving explanations for place names (102 items); photographs of scenes in the Sierra Nevada, including logging activities (ca. 1852-1935, 342 items).

GOLETA

Los Padres National Forest

Supervisor's Office, 42 Aero Camino Street 93017
Telephone: (209) 264-8317

276a Historical files include diaries and other journals and notebooks of Forest Service, U.S. Interior Department, and U.S. Geological Survey personnel, letterpress books, maps, township plats, atlases, newspaper clippings, work plans, narrative and statistical reports, photographs, ledgers, and other records (1858-1970) relating to the area of the Los Padres (formerly Santa Barbara) National Forest. Subjects include fires, land ownership, grazing, insect control, land classification, timber management, water rights, recreation, forest boundaries, use regulations, personnel, roads and trails, historical sites, rainfall, accidents, forest lieu selections, mining, tree nurseries, eucalyptus planting, and other aspects of forest administration and expenditures.

LAGUNA NIGUEL

Federal Archives and Records Center

Chief, Archives Branch, 24000 Avila Road 90377 Telephone: (714) 831-4220. Records that have not been accessioned by the Archives Branch remain under the jurisdiction of the agency of origin. Holdings are described in *A Guide to Research Records,* Federal Records Center, Los Angeles, National Archives and Records Service, 1967, 21 1, processed.

Archives Branch

277 **Land Management Bureau,** Records, 1, 530 cubic feet (Record Group 49). Records of the U.S. Surveyor General, Phoenix, Arizona, include letters received (1860-1925, 93 vols.), containing information on timber lands; letters received from the Commissioner of the General Land Office (1870-1925, 67 vols.), including information relative to forest reserves; survey register (ca. 1880?, 1 vol.), recording the progress of surveys on forest and timber and other reserves; township survey sheets (1873-ca. 1896, 2 feet), containing remarks on the quality of timber; chronology of homestead entry surveys in national forests (1909-49, 2 vols.); and field notes of homestead entry surveys made on national forests (1909-1923, 7 inches), including some surveys made for railroad rights-of-way. Records of the Office of the Register and Receiver, Prescott, Arizona, include a register of timber culture entries (1883-1891, 1 vol.); records of the Register and Receiver, Tucson, include a register of timber culture entries (1889-1891, 1 vol.), and a register of final timber culture certificates (1884-1903, 1 vol.); records of the Register and Receiver, Phoenix, contain files of the Grazing Division relating to the Civilian Conservation Corps (1939-43, 7 feet). In the records of the Office of the Register and Receiver, Los Angeles, California, are applications for timber culture lands (1881-ca. 1892, 5 feet); a register of timber culture entries (1875-1892, 2 vols.); a register of timber culture receipts (1875-1892, 2 vols.); a register of final timber culture receipts (1886-1905, 1 vol.); a register of final timber culture certificates (1886-1895, 1 vol.); and a register of lieu selections on forest reserves (1898-1908, 1 vol.); and approved lists of lands selected by and granted to the State of California for use in the State Park System (1871-1921, 1933-38, 1½ feet). There is an unpublished preliminary inventory (1966, 83 pp.).

Federal Records Center

278 **Forest Service,** Records, 1892-1974, 2,537 cubic feet (Record Group 95). Contains operating and historical files, including correspondence, permits, mineral rights, and leases, and other records of the U.S. Forest Service field offices in southern California and Arizona, including the Los Padres, Angeles, Inyo, and San Bernardino national forests in California (Region 5); the Tonto, Apache, Kaibab, Coronado, Sitgreaves, Prescott, and Coconino national forests in Arizona (Region 3); and the Riverside, California, Fire Laboratory, and other small offices and experiment stations.

279 **Indian Affairs Bureau,** Records, 507 cubic feet (Record Group 75). Records of the Area Office, Phoenix, Arizona, include records of the Forestry and Grazing Section (1932-49).

280 **National Park Service,** Records, 175 cubic feet (Record Group 79). Correspondence files of the Tonto National Monument, Arizona (1935-61); Grand Canyon National Park (1924-56); Tumacacori National Monument (1932-54); and the Lake Mead National Recreation Area (1926-59).

281 **Public Roads Bureau,** Records, 225 cubic feet (Record Group 30). Records of the Phoenix, Arizona, office contain forest highway and national park project files (1932-53).

LOS ANGELES

University of California, Department of Special Collections, University Research Library

Historical Manuscripts Librarian 90024. Some of the holdings are described in the *Guide to the Special Collections in the Library of the University of California at Los Angeles,* U.C.L.A. Occasional Paper Number 7, Los Angeles: University of California Library, 1958, 76 pp., processed. Holdings include copies of the oral histories produced by the Regional Oral History Office, University of California, Berkeley, which are described in the section of the present Guide containing entries for the Bancroft Library.

282 **Albright, Horace Marden** (b. 1890), Papers, 1910s-1970, over 50 feet. Contains material on national parks, especially Yellowstone, and Albright's national conservation interests. Included are personal papers, records as Director of the National Park Service, manuscripts, clippings, pamphlets, and other printed materials. Unprocessed. Closed until processed.

283 **California Labor Camps,** Collection, ca. 1932-1935, ca. 100 items. Mostly carbon copies of originals in the California State Archives. The collection relates to a state-operated system of camps that was a forerunner of the Civilian Conservation Corps. Includes memos, letters, reports, and inventory lists made by camp directors and sent to Merritt B. Pratt, California State Forester; and Pratt's summary reports to Governor James Rolph. The material is arranged by camp name. A list of the camps is available.

284 **Clements, George Pidgeon** (1867-1958), Papers, 1925-1945, ca. 40 feet. Agricultural economist and physician. The collection includes correspondence, other papers, and printed material concerning Clements' career as an agricultural economist. Includes material relating to southern California agriculture, forestry, soil and water conservation, and other matters. Also includes papers concerning his work with the Los Angeles Chamber of Commerce. Partially arranged. Unpublished inventory.

285 **Dawson's Book Shop** (Los Angeles), Records, ca. 1905-1960s, 19 feet (ca. 2,000 items). Correspondence, accounts, clippings, and photographs. Primarily the records of an antiquarian bookshop founded by Ernest M. Dawson in 1905, but also contains personal correspondence of Dawson and members of his family, including scattered items relating to the Sierra Club and various California state park and U.S. national park issues (ca. 1920s-1940s). Unprocessed. A rough description of this collection has been published in Mary Patricia Dixon, *Ernest Dawson* (Los Angeles: Dawson's Book Shop, 1967), pp. 47-48.

286 **Fultz, Francis Marion** (b. 1857), Papers, ca. 30 feet. Educator. Forester, vocational department, Los Angeles City Schools, 1925-32. The collection includes printed material, including copies of books and articles by Fultz and others; brochures advertising Fultz's lectures; radio scripts; manuscripts of Fultz's writings; photographs; poems; a few letters. Many of the manuscripts relate to Fultz's interests in reforestation, recreation, and botany, particularly of the Angeles and San Bernardino and other national forests in California. Scat-

tered material relates to Sierra Club outings and to the Conservation Association of Southern California. Unprocessed.

287 **Kerchoff, William George** (1856-1929), Letters, 1900. 2 items to Kerchoff from Thomas R. Bard and George C. Perkins regarding the California forest reserves.

288 **Knight, William Henry** (1835-1925), Correspondence, 1860-1919, 205 pieces. Letters concern publishing and lectures; some relate to issues of conservation and politics in California. A list of correspondents is available.

289 **Littlefield Family,** Correspondence, 1850-1867, 91 letters. Correspondence of Lambert Littlefield and members of his family, dealing primarily with mining, but including some material on lumbering in California.

290 **Swing, Philip David** (1883-1963), Papers, ca. 1890-1963, ca. 65 feet. U.S. Representative from California, 1921-1933. Attorney. The collection relates primarily to water resources in Southern California with complimentary and supplementary material on conservation. Congressional files include correspondence, clippings, printed materials, bills, etc., including material on legislation relating to Colorado River development, interstate water agreements, Mexican Water Treaty; also material relating to legislation sponsored by Swing and proposed legislation concerning parks and forests, including a proposed national monument for the preservation of palm trees on the Agua Caliente Indian Reservation, California; proposed transfer of lands from the San Bernardino National Forest to the state of California for state park purposes; the regulation of mining claims on recreational lands in the San Bernardino National Forest; control of forest fires; the Laguna Mountains forest road, Cleveland National Forest; etc. There is also correspondence (ca. 1938-39) relating to the proposed Kings Canyon National Park. Legal files relate largely to irrigation and water law. Partly processed. Partial container lists.

291 **Willamette Steam Mill Lumber and Manufacturing Company** (Los Angeles), Journals and Cash Book, 1888-1897, 3 vols. The journals include accounts for the company's main office in Los Angeles and the Fullerton yard. The cash book is for the yard at Pasadena.

University of California, William Andrews Clark Memorial Library

Librarian, 2520 Cimarron Street at West Adams 90018

292 **Montana Papers,** 1833-1934, 549 items. Material, chiefly collected by Charles N. Kessler of Helena, Montana, relating to the history of Montana and the Northwest. Included are letters of Walter L. Brown, Charles L. Camp, C.W. Cook, Henry W. Elliott, D.E. Folsom, A.H. Holmes, and John D. Sherman concerning explorations within the area of Yellowstone National Park in 1869, 1871, and later; and also concerning published accounts of and manuscripts about these explorations and about William Henry Jackson's photography. There are also letters (ca. 1916-22) concerning Indian place names of the Glacier National Park region, motion picture making in Glacier (1919), and the travels of Raphael Pumpelly in the Glacier region, 1882-83; the authors include Joseph Culbertson, Raphael Pumpelly, and James Willard Schultz. The collection is described in *Calendar of the Montana Papers in the William Andrews Clark Memorial Library (University of California at Los Angeles),* Los Angeles: Southern California Historical Records Survey Project, 1942.

REDLANDS

University of Redlands, George and Verda Armacost Library

Special Collections Supervisor 92373 Telephone: (714) 793-2121

293 A few student papers are of interest, including Helen Pruitt Beattie, "Lumbering in the San Bernardino Mountains" (1934, 45 pp.); Ann Ford Farran, "Biota of the Oak Glen Region," which includes a map showing location of species (ca. 1959, 44 pp.); Lloyd S. Parratt, "Some Ecological Studies of the Effects of Chaparral Fire on Small Rodent Populations in Southern California" (1972).

SACRAMENTO

California State Archives

Chief of Archives, 1020 O Street, Room 200 95814 Telephone: (916) 445-4293

294 Forestry records in the California State Archives document the period 1885-date as created by the following agencies: State Board of Forestry (1885-1927); Division of Forestry, Department of Natural Resources (1927-1961); Division of Forestry, Department of Conservation (1961-date). Records prior to 1905, the date of appointment of the first State Forester, are limited to Board of Forestry minutes (1885-92) and very fragmentary Board correspondence. Records from 1905 to date are more complete and reflect the development of the Division of Forestry from virtually a one-man operation to an agency which now employs in excess of 5800 people. Records series of significant research value include biographies of State Foresters and rangers (1905-62); correspondence of Board members George H. Rhodes (1914-24) and W.S. Rosecrans (1944-57); State Forester M.B. Pratt personal and general correspondence (1919-46); State Forester correspondence with Board of Forestry members, including George C. Pardee (1928-41); Gifford Pinchot correspondence with Pratt and Pardee (1904-44); origins and development of State Nursery Program (1916-); State Forestry, Civilian Conservation Corps and State Emergency Relief Administration Labor Camp Programs records include files on individual camps and detail work projects and overall operations (1931-42); fire control programs, including implementation of Weeks Law of 1911, Compulsory Fire Patrol Law of 1923, and the Clarke-McNary Act of 1924; photographic materials on Forestry personnel and activities. Records for the past 30 years document ongoing programs from those listed above and also cover such subjects as the development of agency planning, policy, and programs. Subject areas include Conservation Camp Program, Forest Management, Fire Control, Pest Control, Reforestation, Law Enforcement (Forest Practices Act), Range and Watershed Management. Other agency records which contain forestry or forestry related materials include: State Relief Administration; Division of Highways, Department of Public Works; Division of Beaches and Parks, Department of Natural Resources; Departments of Natural Resources and Conservation—Directors Administrative files; Governor's Office records for the Administrations of Earl Warren (1943-53), Goodwin J. Knight (1953-58), and Edmund G. Brown (1959-66); and miscellaneous agency records as the California Redwood Park Commission (1901-27) and Yosemite Valley and Mariposa Big Tree Commission (1872-1906). There are also materials pertaining to forestry and state park legislation among the committee records of the legislature, which contain transcripts of hearings and, occasionally, working papers. These records are described in *Records of the California Legislature: California State Archives*

Inventory No. 2, compiled by David L. Snyder, Sacramento: California State Archives, 1971.

California State Library

California Section, P.O. Box 2037, Library-Courts Building 95809 Telephone: (916) 445-2585

295 **Arents, H[iram] & Company** (Sacramento), Records, 1853, 2 volumes. Daybook and ledger of a lumberyard.

296 **California. Division of Forestry,** Lantern slides, ca. 1910-1927, ca. 270 items. Includes pictures of various tree species, fire damage to forests, fire fighters, and fire fighting equipment.

297 **Clark, David, & Company,** Account book, 1857-1858, 1 volume. Contains lumber and shingle accounts of a company in Mariposa County, California.

298 **Conaway, J[ames] C., & Brother** (Grass Valley, California), Records, 1851-1909, 11 volumes. Account books (1886-1891, 1895-1901), journals (1856-58, 1894-1905), ledgers (1858-1861, 1894-1897), cashbooks (1851-1857, 1879-1881, 1894-1895) of a company which dealt in lumber, doors, sashes, and mouldings.

299 **Harbour White & Company,** Account book, 1853-1856, 1 volume. Account book of a lumber mill in Mariposa County, California.

300 **Houghton, James Franklin** (1827-1903), Business records, 1853-1855, 58 items. Civil engineer, businessman, and surveyor general of California. Chiefly correspondence relating to the California lumber firm of Pine and Houghton. Most of the letters were written by Houghton to his brother C.B. Houghton, who was also in the lumber business. Index.

301 **Leech, C[harles],** Records, 1858-1963, 2 volumes. Accounts of a Grass Valley, Nevada County, California, dealer for scantling, shingles, siding, fencing, shakes, and flooring.

302 **Lepetit, Louis,** Account books, 1859-1872, 4 volumes. Account books of sawmills in El Dorado County, California. Sites include Diamond Springs and Baltic Mills.

303 **Rider, William Brown,** Photograph album, ca. 1900-1920, 1 vol. Photograph album of a ranger with the U.S. Forest Service in California.

304 **Sellers, Charles Henry,** Photographs, 1910, 1 volume. Contains about 100 photographs on the cultivation of several species of eucalyptus as forest trees. Some of the photographs were printed in Sellers, *Eucalyptus, Its History, Growth, and Utilization,* Sacramento, 1910.

305 **Union Lumber Company** (Marysville, California), Collection, ca. 1869-1874, ca. 100 items. Correspondence, receipts, etc.

306 **U.S. Forest Service,** Reports, 1903-1907, 1 reel microfilm. Microfilm of typed reports on California forests. Originals are in the National Archives, Washington, D.C.

307 **U.S. Forest Service, California Region,** "A sawmill history of the Sierra National Forest, 1852-1940," 1 volume, 60 photos. Photostat of original held by U.S. Forest Service.

308 **Winchester, Jonas** (1810-1887), Papers, 1829-1884, ca. 240 items. Correspondence, pictures, and miscellaneous items. Includes letters from Winchester to his family in New York describing his lumber mill operations in Grass Valley, California.

SAN BERNARDINO

San Bernardino County Free Library

104 West 4th Street 92401 Telephone: (714) 383-1734

309 There are a number of typescripts relating to the history of San Bernardino County shelved in a locked case. Items related to the forest history of the county include: "Mormon Road to forests on San Bernardino Mountain Range," prepared under the direction of George William Beattie (1932, 127 pp., includes 4 photos and 1 map); "Historic Crossing Points in the San Bernardino Mountains," by George William Beattie (n.d.); "History of Lake Arrowhead Country," comp. by Mrs. C.E. LaFuze (8 pp.; includes clippings); material on the early history of Seely Flats, San Bernardino mountains, by David Randolph Seely (2 pp.); diary of a family mill enterprise in the San Bernardino mountains, 1872-1891, by Joseph B. Tyler (ed. by Mrs. C.E. LaFuze, 1955-56, 133 pp.; includes photos and map).

SAN BRUNO

Federal Archives and Records Center

Chief, Archives Branch, 1000 Commodore Drive 94066. Records that have not been accessioned into the Archives Branch remain under the jurisdiction of the agency of origin. Some of the holdings are described in *Opportunities for Research in Federal Records for California, Nevada, and the Pacific Ocean Area: Special List of Research Records,* San Francisco: Regional Archives Branch, Federal Records Center, National Archives and Records Service, General Services Administration, 1970, 16 pp. National Archives microfilm publications available for use in this and other centers and through inter-institutional loan are described in *National Archives Microfilm Resources in the Archives Branch, Federal Archives and Records Center,* San Bruno: General Services Administration, 1973, ca. 21 pp.

Regional Archives Branch

310 **District Courts of the United States,** Records, 14,909 cubic feet (Record Group 21). Records of the U.S. District Court, Northern District of California, include, among the criminal indictments and case files (1867-1912, 58 feet), cases pertaining to the cutting and removal of timber on federal lands; commissioners' docket files, record of proceedings, trial of petty offenses (1893-1960, 8 feet) contain minor proceedings before the U.S. Commissioner for Lassen National Park, California (1946-1950). Unpublished preliminary inventory (1964, 40 pp.).

311 **Land Management Bureau,** Records (Record Group 49). Microfilm holdings include the journal and report of James Leander Cathcart and James Hutton, agents appointed by the U.S. Secretary of the Navy to survey timber resources between the Mermentau River, Louisiana, and the Mobile River, Alabama (1818-1819, 1 reel, #M-8). Original is in the National Archives.

Federal Records Center

312 **Economic Research Service,** Records, 16 cubic feet (Record Group 354). Records of the San Francisco office, including reports on forest land ownership (1941-58) and Santa Cruz County land-use study (1941).

313 **Entomology and Plant Quarantine Bureau,** Records, 4 cubic feet (Record Group 7). Records relating to blister rust.

314 **Forest Service,** Records, 6431 cubic feet (Record Group 95). Records of the Region 5 office, San Francisco (1910-1972, ca. 2400 cubic feet), include correspondence files on trespass, insect control, primitive areas, fire control, etc.; Civilian Conservation Corps camp records (1911-49); wildlife

management maps; road and trail development maps; fire atlases; aerial maps; reports on tree planting, roads, trails, water resources, land acquisition, range planning, fire prevention and control, etc.; aerial photographs of watersheds; diaries of fire control personnel; timber sales records; erosion control records; historical manuscripts relating to forest conditions, water supply, vegetation, inspections, etc., on various national forests and the California region. Records of the Pacific Southwest Forest and Range Experiment Station, Berkeley, California (1,400 cubic feet), include maps of land ownership and vegetation (1940-50); records of forest management (1868-1960); research programs (1908-26); fire suppression records (1911-45); silviculture research reports (1927-55); forest pathology records (1947); research program and flood report correspondence (1915-72); redwood region maps (1913-50); insect research records (1919-53); lumber census atlases; watershed management research correspondence and reports (1904-55); and technical records and reports (1908-48). Records of National Forest Supervisors' offices in California and Nevada (1920-72, 1950 cubic feet) include correspondence for the El Dorado, Klamath, Lassen, Mendocino, Modoc, Plumas, Sequoia, Shasta-Trinity, Sierra, Six Rivers, Stanislaus, Tahoe, and Toiyabe National Forests.

315 **Land Management Bureau,** Records, 3,244 cubic feet (Record Group 49). Included are records of Bureau of Land Management offices in Northern California and Nevada. Records of the Office of the U.S. Surveyor, San Francisco, contain photostats of official field notes of surveys of the Yosemite grant (1883) and Yosemite National Park (1902 and 1905). In the records of the Bureau Regional Office, San Francisco, Branch of Land Classification and Planning, are general files of the chief field examiner, with material on the San Gabriel Canyon Project (1929-32) and the activities of the Forest Protection Board (1927-33); miscellaneous investigations (1900-46) with material on timber trespass and on 50 percent donation purchases for national park lands; index to land cases (1902-16) including timber trespass; index to Oregon land cases including material on the Oregon Lumber Company (1890-1902). In the records of the California State Office, Sacramento, Division of Engineering, Branch of Cadastral Surveys, are field notes of national forest area surveys relating to the Santa Barbara National Forest (chiefly 1904-06) and the San Jacinto Forest Reserve (chiefly 1904-05); supplemental plat survey case files (1954-63) including correspondence relating to resurveys of national forest areas; national forest homestead entry survey case files (1908-28, 1 foot) for California; Forest Service exchange survey case files (1929-34, 75 cubic inches) including three California cases. California State Office, Division of Land and Mineral Program Management and Land Office Records contain unpatented land entry serial case files including timber sales (1916-67, 4 feet), and Forest Service surface management (ca. 1946-66); reclamation and power reservation contest case files, including material on San Gabriel damsite (1926-29), Joshua National Monument and Angeles National Forest (1939-45), Isabella Dam (1949-52), Pine Flat reservoir (1949-52), Folsom reservoir (1950-52), Trinity project (1955-67); National Park reservation case files including material on Kings Canyon National Park; also timber sales case files (1960-66, 2 feet); and reforestation contracts (1960-65, 1 foot). Records of the Susanville District Office include a timber trespass docket (1903-06, 1 vol.) containing entries for Wisconsin and California; and timber sales case files (1945-64, 3 feet). Records of the Ukiah District Office include timber sales case files (1951-59, 3 feet). In the records of the U.S. Land Office, Carson City, Nevada, are a record of fees received for timber culture entries and final timber culture proofs (1886-1908, 1 vol.); an account of fees received under the Timber Lands Act [sic] of June 3, 1878; record of patents delivered (1884-1909) including patents for forest reserve lieu lands; record of applications for entry (1906-08, 1 vol.) including entries under the timber and stone act; register of timber culture entries (1876-88, 1 vol.); abstract of timber and stone and timber culture entries (1906-07, 1 vol.); and a register of forest reserve lieu selections (1898-1908, 1 vol.). Records of the Nevada Land Office include general correspondence (1935-66) containing timber trespass case files; forest homestead entry survey plats (1911-21, 3 feet, and 1910-31, 2 feet); unpatented land entry survey case files (1881-1968), including surface management investigations by the Bureau of Land Management and the U.S. Forest Service (1958-65, 2 feet). These records are described in a draft preliminary inventory (1970, 60 pp., processed).

316 **National Park Service,** Records, 1,263 cubic feet (Record Group 79). Records, primarily from the San Francisco Regional Office, date from about 1920 and contain recreation studies and reports; land acquisition files; correspondence files on forestry, fires, pest control, interpretative activities, park development and maintenance, the Civilian Conservation Corps, and individual parks; fire atlases; reservoir projects; annual and special reports of individual parks, including park histories.

317 **Public Roads Bureau,** Records, 1917-1959, 1,814 cubic feet (Record Group 30). Contains forest highway and national park-national monument project files, including photographs.

SAN FRANCISCO

California Historical Society

Manuscript Librarian, 2090 Jackson Street 94109 Telephone: (415)567-1848. The following is an incomplete listing of collections cited as relating to timber, etc., in the Society's card catalog.

318 **Chipman, William W.,** Papers, 1850-1929. The papers contain occasional information on redwood lumbering in Alameda County, California, after 1852.

319 **Cowden, James S.,** Papers. Diary (p. 42) includes a short description of the lumber industry at Yreka, California, 1853.

320 **Dring, David,** Papers. Includes a receipt (1847) for 162,750 feet of lumber sent on the bark *Janet,* from the Columbia River to San Francisco.

321 **Freeman, Leslie J.,** Manuscript, 1 item. "Tehama Counties [sic] First Sawmill." n.d., 3 pp.

322 **Galloway, John D.,** Papers, 1939. Memoranda on some of the engineering works of the Comstock at Virginia City and Gold Hill, Nevada. 103 pp. Includes photographs, chart, map.

323 **Hays, John Coffey,** Papers. The papers contain testimony (November 17, 1857) in a case concerning redwood lumbering in Alameda County. Howe's sawmill (1854) and Whipple's steam sawmill, both in San Francisco, are mentioned in the papers.

324 **Hotchkiss, Edward,** Correspondence, 1850-1851. The correspondence, largely to Hotchkiss' family, contains occasional reference to his interest in a San Francisco sawmill.

325 **Jones, David R., and Company,** Daybooks, 1856-1862, 2 volumes. The daybooks contain information on a sawmill operated in San Francisco by Jones and John Franklin and on a subsidiary redwood sawmill operated by Jones and John Kentfield in Humboldt County, California (1861-62).

326 **Kellogg, Walter Y.,** Papers. Correspondence, accounts, etc., of Wilson & Brother and California Door Company.

327 **Leidesdorff, William Alexander,** Correspondence, 1810-1848, 3 items. The correspondence (1847-48) between Leidesdorff and Stephen Smith concerns shipments of lumber and wheat from Bodega, Sonoma County, California, to San Francisco.

328 **Martin, J.O.,** Papers. Includes a description of the lumber industry in Humboldt County, California, January 26, 1853.

329 **Miscellaneous.** Papers of James King of William (1 folder) include an agreement relating to the lumber industry in San Francisco (1850); papers of Samuel Merritt (1 folder) include an agreement relating to the shipping of lumber from Mendocino to San Francisco (1856).

330 **O'Brien, William Shoney** (1826-1878), Estate Records, 1878, 2 boxes. Includes material relating to various firms active in the Comstock, including the Pacific Wood, Lumber & Flume Company.

331 **Odall, R.P.,** Papers. Includes reference to lumbering in Santa Cruz County, California, January 4, 1852.

332 **Seward, Thomas T.,** Papers. Included is a description of lumber valued at $8450 taken from one redwood tree, August, 1849.

333 **Smith, Stephen S.,** Account book, 1850-1851, 1 volume. San Francisco lumber dealer.

334 **Towle Brothers Company** (Towle, California), Records, 1861-1909, 400 pieces. The records of a lumbering firm. The collection also includes papers of the Pioneer Pulp Company and the Capital Box Factory.

335 **Willcox, Robert N.,** Papers. Willcox operated a lumber mill in El Dorado County, California, 1853-56.

SAN MARINO

Huntington Library, Art Gallery and Botanical Gardens

Manuscript Department, 1151 Oxford Road 91108

336 **Carr, Jeanne C. (Smith),** Papers, 1842-1903, 192 items. Correspondence, articles, and other papers of Jeanne Carr and her husband, Ezra S. Carr. The papers relate to botany, to early California, and to Carr's work in the universities of Wisconsin and California. Several pieces relate to forests and horticulture. Summary report and catalog cards in the library.

337 **Fall, Albert Bacon** (1861-1944), Papers, 1887-1941, ca. 55,000 items. U.S. Senator from New Mexico, 1913-21; U.S. Secretary of the Interior, 1921-23. Personal and political correspondence and other papers relating to natural resources, Indian lands, and other Interior Department affairs. Many items relate to national forests in New Mexico (1912-22), various national parks, and the lumber business, especially in New Mexico; there is also a file on grazing fees (1916-40). Finding aid in the library. The University of New Mexico holds microfilm copies (42 reels).

338 **Lukens, Theodore Parker** (1848-1918), Papers, 1869-1942, ca. 3,600 items. Correspondence and papers of Lukens, conservationist and agent of the U.S. Bureau of Forestry in southern California. The papers contain material on reforestation; field diaries (ca. 1900-09) with observations on conditions in forest reserves; letters from John Muir (ca. 80 items) relating to forest reserves, parks, reforestation, the Hetch Hetchy Valley controversy, personal matters, etc. Other correspondents include Gifford Pinchot. Inventory.

339 **Muir, John** (1838-1914), Collection. Letters and literary manuscripts. This is a composite description for Muir material in various collections in the library, in part represented by their own entries in the present work. Muir letters to Charles A. Keeler (28 items) are personal or literary in subject matter. Muir letters to Theodore P. Lukens (80 items) relate to conservation and personal matters. Muir letters (1902-55, 86 pieces) include 22 signed letters from John Muir, chiefly to Helen and Enos Mills. Muir Family letters (1860-1906, 155 items) and Xerox copies of letters to Daniel Muir, chiefly from other family members, but including 26 letters from John Muir (1860-91). The originals are owned by John C. Bell. There are restrictions on copying.

340 **Nevada Railroads Collection,** 1862-1950, ca. 9,000 items. Correspondence, business records, and other papers of various Nevada railroad companies; also material on the Nevada mining and lumber industries. Report.

341 **Thoreau, Henry David** (1817-1862), Papers, 1836-1876. Author. Correspondence and literary manuscripts. Correspondents include John and Sophia Thoreau. Mainly of a literary nature.

342 **Thrall, William H.,** Papers, ca. 1873-1962, ca. 3500 items. Editor of *Trails* magazine, published by the Mountain League of Southern California. Correspondence, notes, articles, maps, photos, and other papers relating to the history and development of the San Gabriel Mountains and to the water supply, flood control, and sewage disposal of Los Angeles County, California.

SAN MATEO

San Mateo County Historical Association and County Historical Museum

College of San Mateo Campus, 1700 West Hillside Boulevard 94402

343 The holdings include a number of manuscripts relatingto the forest history of San Mateo County, California, among them "Early Lumber Industry in San Mateo County," a paper read by Mrs. A.S. Kalenborn before the association, October 19, 1944; and a transcript of a memoir on problems faced by early sawmill owners in redwood timber, particularly in southwestern San Mateo County, probably written by John Pope Davenport.

SANTA BARBARA

University of California, Santa Barbara, Library

Manuscript Curator, Manuscripts Division, Department of Special Collections 93106. Some of the holdings are described in "A Checklist of Manuscripts," University Library, Department of Special Collections, 1970.

344 **Chase, Pearl** (b. 1888), Papers, 1918-1971. The Chase Papers are designated as part of the Community Development and Conservation Collection. Included are records of the California Conservation Council from 1934, an organization devoted to conservation education, containing correspondence with state and federal resource agencies, conservation organizations, and individuals throughout California and the United States. U.S. Forest Service correspondents (1938-68) include C.E. Fox and W.I. Hutchinson, Directors of Information and Education, California Region. National Park Service correspondents include Newton Drury, Frank A. Kittredge, and Aubrey Neasham. Conservation organizations represented in the correspondence include ca. 60 groups represented by at

least one file folder, among them the Sierra Club, National Conference on State Parks, Friends of the Earth, Audubon Society, Wilderness Society, Conservation Education Association, and the Save-the-Redwoods League. Additions to the collection are anticipated. Inventory in process (1974).

345 **Peattie, Donald Culross** (1898-1964), Papers, 1921-1964, ca. 60 feet (15,343 items). The collection contains ca. 13,000 letters; manuscripts of all of Peattie's articles and most of his books, also photographs, clippings, and pamphlets. Included are the complete notes and manuscripts for *A Natural History of Trees of Eastern and Central North America* (1950), and *A Natural History of Western Trees* (1953). Cataloged on cards.

SANTA CRUZ

Forest History Society, Inc.

Associate Director for Library and Research Services, P.O. Box 1581 95061. 733 River Street. Telephone: (408)426-3770

346 **American Forest Institute,** Records, 1935, 1940-1967, 65 feet. Formerly called the American Forest Products Industries, the American Forest Institute is an educational organization designed to supplement the lobbying activities of the National Forest Products Association. The records contain minutes, mailings, memoranda, and other materials of the Board of Trustees; general correspondence; reading files, press releases, mailings; and other materials of various divisions and committees; reading files of Charles Alton Gillett (b. 1904), managing director (1947-67); publicity materials and reports of the National 4-H Forestry Program; also photographs, AFI publications, phonograph records, motion picture films, and miscellaneous educational and public relations material. Typed container list, 12 pp. Additional accessions are expected.

347 **American Forestry Association,** Records, 1875-1971, 78 feet. Correspondence, articles, manuscript surveys and reports on forestry, newspaper clippings, and published material relating to the association. Included are studies and investigations (61 boxes), awards (21 boxes), directors' minutes (8 boxes), directors' correspondence (16 boxes), annual meetings (6 boxes), annual reports (12 boxes), elections (6 boxes), miscellaneous (25 boxes), Redwood National Park (13 boxes), miscellaneous administration (25 boxes). There are few materials dating before 1900; the bulk of the collection dates from the 1930s-1960s. Among the prominent correspondents are William B. Greeley, Ovid Butler, Gifford Pinchot, and Samuel T. Dana. Typed list of contents, 62 pp. Additional accessions are expected.

348 **Boerker, Richard H.D.,** Papers, 1890s-1957, 1 carton (4 letter cases and 5 vols.). The letter cases contain drafts of Boerker's writings on silviculture, national forests, and conservation, along with related correspondence with publishers and others; also some miscellaneous correspondence, clippings, etc. There are also ledger books (4 vols.) and one loose leaf notebook containing drafts of writings.

348a **California State Council of Lumber and Sawmill Workers, AFL-CIO,** Records, ca. 1945-1973, ca. 25 feet. Files of the San Francisco office include U.S. Department of Labor case files and arbitration cases (ca. 2 feet); pension fund records (ca. 1 foot); incomplete subject files (ca. 5 feet) relating to insurance, labor-management relations in the lumber industry, referendums, resolutions, strike benefits, etc.; industrial accident cases (ca. 4 feet); convention proceedings; bulletins; subject files of the economic representative (1951-1973, 7 feet); miscellaneous correspondence and other records. Inventory.

349 **Clark, Lawrence S.,** Papers, 1944-1971, 3 boxes. Material relating to the U.S. Office of Price Administration control of the lumber industry including regulations, minutes of the Hardwood Distribution Yards Advisory Committee, financial reports, etc. (1942-46) ; material relating to the Wood Ply Research Foundation, including balance sheets, correspondence, lists of members, reports, news releases, etc. (ca. 1941-47) ; miscellaneous correspondence and reports relating to lumber grading and standards, and the wholesale lumber industry. Typed list of contents, 4 pp.

350 **Cliff, Edward Parley** (b. 1909), Papers, 1931-1972, 7 vols. Forester; Chief, U.S. Forest Service, 1962-72. Duplicates of the principal collection of Cliff's papers, committed for deposit at Utah State University, Logan, Utah. Included are Xerox copies and typescripts of speeches, articles, statements, press releases, and photographs documenting Cliff's career in the U.S. Forest Service. In loose leaf binders.

351 **Collingwood, George Harris** (1890-1958), Papers, 1889-1890, 1909-1958, 4 feet. Includes professional and family papers, diaries, forestry course notes, photographs, clippings, mementos, manuscripts, drafts of articles, correspondence, relating to Collingwood's work as ranger on the Apache National Forest, Arizona (1914-15) ; U.S. Bureau of Aircraft Production; National Lumber Manufacturers' Association (1940-46) ; Society of American Foresters; American Forestry Association (1928-40) ; Cornell University Forestry Extension (1916-23) ; U.S. Legislative Reference Service (1952-58) ; and the U.S. Forest Service; Letters of Collingwood to Jean Cummings (later Mrs. Jean Cummings Collingwood) describing life on the Apache National Forest were published as "Sincerely Yours, Harris," edited by Joseph A. Miller and Judith C. Rudnicki, *Forest History,* XII, January, 1969, pp. 10-28. Typed container list, 4 l.

352 **Drake, George L.,** Papers, ca. 1935-51, 1 folder. Material relating to the Pacific Logging Congress (1935, 1951), speech at the 50th Anniversary Intermountain Logging Conference; miscellaneous.

353 **Forest History Society Photograph Collection,** ca. 22 linear feet. Black and white paper prints originating with the U.S. Forest Service, the Weyerhaeuser Timber Company, and the American Forest Institute, and documenting the work of these organizations in a broad range of activities; also photographs collected for editorial purposes of the *Journal of Forest History.* The bulk of the collection is post-1940, but there are a few portraits of prominent lumbermen and foresters dating from the late 19th century. Some negatives are included.

354 **Gill, Thomas Harvey** (1891-1972), Papers, 1921-1972, 4 feet. Forester with the U.S. Forest Service, 1915-25, Charles Lathrop Pack Forestry Foundation, 1926-60; and the Food and Agriculture Organization; a founder of the International Society of Tropical Foresters. The collection includes personal and professional correspondence; published and draft articles; photographs; printed materials, including books; excerpts from travel diaries (1924-29) ; directors' minutes of the Charles Lathrop Pack Forest Education Board (1930-40). Among the prominent correspondents are Ralph S. Hosmer, Adalbert Ebner, Randolph G. Pack, Arthur N. Pack, the Tropical Plant Research Foundation, Gifford Pinchot, Carl Alwin Schenck, Ferdinand A. Silcox, and others. Typed container list, 7 pp.

355 **International Concatenated Order of Hoo-Hoo, Inc.,** Records, 1 box and 53 rolls of microfilm. Correspondence of Lawrence S. Clark concerning Hoo-Hoo meetings, conventions,

membership, budget, etc. (1939-72). Microfilm includes the *Hoo-Hoo* magazine (now *Log and Tally*) (7 rolls); applications for degrees (43 rolls); applications for membership (3 rolls). The organization was originally called the Fraternal Order of Lumbermen.

356 **Jewett, George F.,** Papers, 1908-1952, 4 boxes. Correspondence (1936-41); materials relating to training program, engineering studies, annual reports, public relations materials, trial balances, audits, annual statements, minutes, and other records of Potlatch Forests and associated companies, chiefly relating to the lumber industry in Idaho. Among the firms represented are the Red Collar Line, Inc.; Yakima Valley Lumber Company; Washington, Idaho, Montana Railway Co.; St. Joe Boom Co.; Potlatch Yards; Potlatch Forests; Clearwater Timber Co.; Wood Briquettes, Inc.; the Bonners Ferry Lumber Co.; and the Northwest Paper Company.

357 **Kneipp, Leon Frederick** (1880-1966), Papers, 1915-1966, 2 boxes. U.S. Forest Service official. The collection contains largely printed matter, but there is a small amount of miscellaneous correspondence, and diaries and notebooks (1915-46, ca. 37 items).

358 **Lacey, James D., Company,** Records, 1898-1958, 30 feet. Records of firm engaged in timberland acquisition and management in the United States, Canada, Mexico, Central and South America, and to a lesser extent in the Philippines and other parts of the world. Included are correspondence between the Lacey Company and many lumber and forest products corporations and private timberland purchasers; personal and household papers of J.W. McCurdy, which contain business correspondence of the Lacey Company after 1935 (14 boxes). Also ledgers, journals, and other financial records of the company (1898-1958, 6 cartons). Typed container list, 11 pp.

359 **Laughead, William B.,** Papers, 1915-1959, 2 boxes. Correspondence, articles, drawings, sketches, clippings, printed materials, much of it relating to Paul Bunyan stories; photographs of the manufacture of California white pine and sugar pine lumber by the Red River Lumber Company.

360 **Miscellaneous Manuscripts,** Collection, ca. 1880-ca. 1970, ca. 20 feet (8 vertical file drawers). Correspondence, drafts of articles, speeches, reports, handbooks, clippings, photograph albums, memoirs, surveyors' notebooks, miscellaneous booklets and other printed items, relating to the lumber industry and to forestry and conservation in the United States, Canada, and the world, including some editorial material relating to the *Journal of Forest History,* but largely small lots of material collected from various sources; arranged alphabetically by author or subject.

361 **National Community Christmas Tree,** Records, 1923-54, 1 box. Brief history of the national community Christmas tree (1923-41); correspondence (1931-45); programs, guest lists, reports on lighting and location, invitations, photographs, clippings.

362 **National Forest Products Association,** Records, 1928-1971, 71 feet. Records of an organization, formerly called the National Lumber Manufacturers Association, concerned with lumber grade standardization, trade extension and promotion, economic and statistical surveys, and reporting on lumber industry conditions. Included are correspondence and reference files on federal forest legislation and regulation of forestry practices; taxation; soil and water resources conservation; American Forest Congresses; Forest Industries Council meetings; and many other forestry, lumber trade, and conservation associations, and lumber companies; national parks; the Civilian Conservation Corps; related activities of the U.S. Departments of the Interior, Agriculture, and Commerce; the U.S. Forest Service, particularly the Timber Resources Review (1952); the Society of American Foresters; activities of state forestry agencies; forestry schools; forest fires and fire fighting; the Hoover Commission on Government Reorganization; state forestry legislation; forest land acquisition; the Timber Conservation Board (1931-33); lumber industry surveys; insect and disease problems; lumber supply and the regulation of the lumber industry during World War II and the Korean War periods; price controls by the Economic Stabilization Agency; Indian reservations; mining laws; national forest recreation values; National Lumber Manufacturers Inter-Insurance Exchange; Lumber and Timber Products War Committee; also articles and addresses; statistics on lumber production, shipments, orders, prices; NLMA and NFPA annual reports, press releases, publications, and other printed matter; obituaries; photographs; etc. Typed container list, 47 pp. Additional accessions are expected.

363 **Natural Resources Council of America,** Records, 1946-1971, 4 feet. These records are primarily the files of Clinton Raymond Gutermuth (b. 1900), a founder, secretary (1946-57), and chairman (1959-61) of the Natural Resources Council. Included are correspondence, typed and processed reports, memoranda, printed and processed newsletters, lists of officers and members (1947-65), bylaws, articles of incorporation, typed and mimeographed minutes (1946-71), special conferences, correspondence, meeting notices, ballots, dues notices, general correspondence, form letters, mailing lists, correspondence and memoranda relating to the organization of the council and to conservation problems, committee files, correspondence and agreements relating to publications, photographs of members (1948-63). Typed list of contents, 4 pp.

364 **North American Wholesale Lumber Association, Inc.,** Records, 1897-1967, 9 feet. Records of the national trade association in the United States of manufacturers, manufacturing wholesalers, and the wholesalers in lumber, formed by the merger of the National Wholesale Lumber Dealers Association and the American Wholesale Lumber Association in 1923 and known as the National-American Wholesale Lumber Association, Inc., until 1972 when the present name was adopted. The records contain typed carbons, mimeographed, and other processed and printed material distributed to members and officers of the association, including reports of annual meetings (1955-63), programs (1924-26, 1951, 1960), annual reports (1924-53), directors' minutes and circulars (1942-67), bulletins (1933-65), executive committee minutes and circulars (1956-65), bylaws, and publications of the New York and Portland, Oregon, offices. Also included are bulletins (1948-60, incomplete) of the Southern Wholesale Lumber Association; annual reports (1897-1923), programs, certificate of incorporation, constitution and bylaws (1913) of the National Wholesale Lumber Dealers Association; and the secretary's weekly report of the Pacific Coast Shippers' Association (1921-23). Additional accessions are anticipated.

365 **Price, Jay H.,** Diaries, 1916, 1920-1954, 57 vols. Field diaries of a U.S. Forest Service official who served in Region 5 (California) as logging engineer, assistant regional forester for fire control, and associate regional forester (1930s); as associate director of the Prairie States Forestry Project (Lincoln, Nebraska); and in Region 2 (Rocky Mountain Region); as Regional forester, Region 9 (Eastern); and in other Forest Service offices.

366 **Schenck, Carl Alwin** (1868-1955), Papers, ca. 1893-1955, 1 box and 1 vol. Booklets, writings of Schenck, manu-

scripts, photographs, miscellaneous items, relating to Schenck, the Biltmore Forest and the Biltmore Forest School, Schenck's explorations in Minnesota (1896), the Cradle of Forestry in America (1964); and correspondence of Schenck with Elwood R. Maunder (1952-55). Includes a draft of the original version of *The Biltmore Story.*

367 **Schofield, William R.,** Papers, ca. 1920-1972, 3 boxes, 4 cartons, and 7 cardboard filing drawers. Secretary-Manager of the California Forest Protective Association. The collection primarily consists of pamphlets and other printed and processed material. Only 3 boxes contain a significant proportion of correspondence, including material on the California Forest Protective Association (1964-72); small lots of letters and reports on a proposed extension of state parks in Humboldt County, California, and opposition by the redwood lumber industry (1959); the proposed Redwoods National Park (1966); miscellaneous personal correspondence; timber taxation in California (1930s-60s); censure of S. Rexford Black by the Society of American Foresters (1930s); draft timber survey of California (1931); proposed legislation relating to wages of lumber workers (1957); State Board of Equalization reports relating to taxes (1930s); statistical data on land holdings of the Hammond Lumber Company and other lumber firms in Humboldt County (1913, 1927, n.d.); etc.

368 **Shepard, William C.,** Papers, 1948-1958, 1 box. Miscellaneous correspondence and other material relating to the Connecticut Forest and Park Association, Connecticut forests, forestry, and forest utilization.

369 **Society of American Foresters,** Records, 1903-1966, ca. 96 feet. Records of the professional organization of foresters in the United States. Included are correspondence and reference files concerning society administration, elections, and membership; forestry education; publication of the *Journal of Forestry;* timber and forestry conferences; wildlife; society committees, especially the Committee on Natural Areas; surveys of industrial forestry; ethics; survey of state forestry administration; internal controversies; accreditation of forestry schools; employment in the forestry profession; the Civilian Conservation Corps; recreation; state parks; the U.S. Forest Service. Individuals prominently represented include H.H. Chapman, Raphael Zon, Christopher Mabley Granger, and various society officials. Also included are papers of Samuel T. Dana (19 boxes), dealing mainly with forestry education. Typed container list, 34 pp. Additional accessions are anticipated. Inventory.

370 **Weaver, Harold,** Notebooks, 1930s-1966, 2 cartons (49 vols.). Reports, largely consisting of 8"x10" black and white photographic prints, but also containing narrative material, resulting from studies of fire as an ecological factor on timberland; also contains related correspondence. Most of the sites studied are on Indian reservations in Washington, Oregon, Montana, Arizona and New Mexico; also national forests in the Pacific Northwest, the Tillamook Burn forest (1966), etc. In loose leaf binders.

Oral History Transcripts

371 **Forest History Society,** Oral history interviews relating to forestry and the lumber industry in the United States, 1955-1973, ca. 10 items. Unpublished transcripts of oral history interviews, with magnetic tapes. Some transcripts are in unedited or rough draft form. Subjects of interviews chiefly conducted by Forest History Society staff members and others, and relating primarily to forestry and the lumber industry in the United States, include Wilson Compton (1890-1967), largely on the National Lumber Manufacturers Association (1965, 155 pp.); Samuel Trask Dana (b. 1883) on Forestry education, U.S. Forest Service work, and evaluations of various foresters (1960, 87 pp.; 1964, 98 pp.); Earl H. Frothingham, including an account of preparations for the defense of the U.S. Forest Service office in Washington, D.C., during World War I (1968, 27 pp.); Mrs. William B. Greeley, on William B. Greeley (1960, 21 pp.); William Joseph Griffin, on the timber economy of the Adirondack region, New York (1958, 26 pp.); Ralph S. Hosmer (1957, 73 pp.); Royal S. Kellogg, on the Biltmore Forest School, U.S. Forest Service research, and the Newsprint Service Bureau (1955, 54 pp.); Jay P Kinney, on forestry in the Bureau of Indian Affairs (1960, 120 pp.); Frank K. Mitchell, Sr., on forestry in Connecticut (1968, 29 pp.); A.B. Recknagel, on experiences in the U.S. Forest Service in the Southwest U.S. during the early 20th century, on contributions to conservation of Gifford Vinchot and others, and on the St. Regis Paper Company (1958, 46 pp.); T.R. Williams, on mahogany and imported woods (1961, 80 pp.); and George H. Wirt, on forestry and forestry education in Pennsylvania (1959, 62 pp.). Among topics discussed in these interviews are the National Recovery Administration Lumber Code, Gifford Pinchot, Raphael Zon, Earle Clapp; the Clarke-McNary Act; McSweeney-McNary Act; the U.S. Forest Service, Northeastern Forest Experiment Station; other foresters; Weyerhaeuser leadership in the lumber industry; the Forest Products Laboratory, Madison, Wisconsin; trade associations; lumber imports; etc. Interviewers include Elwood R. Maunder, Amelia R. Fry, Charles D. Bonsted, Bruce C. Harding. Restricted.

372 **Forest History Society,** Oral history interviews relating to forestry and the lumber industry in California, 1953-1967, ca. 12 items. Subjects of unpublished interviews conducted by Forest History Society staff members and relating primarily to forestry and the lumber industry in California include Clarence W. Broback, on the work of the Union Lumber Company in the redwood region (1953, 9 pp.); Silas B. Carr, on lumbering near Gualala (1953, 2 pp.); Harry W. Cole, on the Hammond Lumber Company (1953, 4 pp.); Charles G. Dunwoody on fforest protection, lobbying in Washington, D.C., for the California Chamber of Commerce, and on the U.S. Department of the Interior under Secretary Harold Ickes (1967, 58 pp.); Frank Fraser, on logging techniques and equipment (1953, 4 pp.); Jack Furlong, on saws (1953, 6 pp.); Herman Heitman, on logging techniques (1953, 11 pp.); Caspar Hexberg, on shipping for the Union Lumber Company (1953, 6 pp.); Howard A. Libbey, on the Arcata Lumber Company (1953, 6 pp.); N.A. McCallum (1953, 5 pp.); Mrs. Elsie Miller, on the Hammond Lumber Company (1953, 7 pp.); E.J. Stewart, on logging equipment (1953, 8 pp.). Interviewers include John Larson, Elwood R. Maunder, and others. Restricted.

373 **Forest History Society,** Oral history interviews relating to forestry and the lumber industry in the Lake States and the Upper Mississippi Valley, 1952-1964, 14 items. Subjects of interviews conducted by Forest History Society staff members and relating primarily to forestry and the lumber industry in the Lake States and the Upper Mississippi Valley include Clifford Ahlgren, on the Quetico-Superior Research Center, Ely, Minnesota, and on wilderness policy (1964, 47 pp.); John G. Ballord (b. 1870), on Minnesota lumbering (1962); James Beatty (1861-1957), on Minnesota lumbering (1952, 19 pp.); Paul Caplazi (1967-1957), on lumbering in the St. Croix Valley (1953, 4 pp.); Mrs. Maud Carlgren and Mrs. Hope Garlick Mineau, primarily on logging in the St. Croix Valley (1955, 27 pp.); Leonard Costley (b. 1887), on life in northern Minnesota logging camps and Paul Bunyan stories (1957, 20

pp.) ; Walter Ernest Dexter, on lumbering in Mennesota, in Nova Scotia in the 1870s, and on the J. Neils Lumber Company mill, Libby, Montana (1953, 10 pp.) ; George W. Dulany, Jr., on the Empire Lumber Company, Eau Claire, Wisconsin, and comments on various lumbermen (1956, 37 pp.) ; Herbert F. Foster, on various lumber firms (1956, 26 pp.) ; Julius Joel, on life in Michigan logging camps in the early 20th century, and on the Finns and Swedes (1953, 6 pp.) ; William B. Laughead, on the lumber industry in Minnesota and California, the Red River Lumber Company and Paul Bunyan (1957, 106 pp.) ; James Arthur Matheiu, on rafting and logging (1957, 10 pp.) ; Mrs. Maggie Orr O'Neill, on lumber camp life in Wisconsin in the 1880s (1955, 18 pp.) ; L.J. Olson, on the J. Neils Lumber Company operations in Minnesota and Montana (1953, 4 pp.) ; Hugo Schlenk, on the Cloquet, Minnesota, forest fire of 1918, on the C.N. Nelson Company, and on Stillwater and Cloquet lumbermen. Interviewers include Elwood R. Maunder, Willaim Trygg, Helen McCann White, Burce C. Harding, John Larson, Clodaugh Neiderheiser, and others. Restricted.

374 **Forest History Society,** Oral History interviews relating to forestry and the lumber industry in the Pacific Northwest, 1953-1973, ca. 25 items. sSubjects of interviews conducted by Forest History Society staff members and relating primarily to forestry and the lumber industry in the Pacific Northwest include Emmit Aston (1958, 41 pp.) ; P.B. Anderson on the cedar shingle industry of British Columbia (1958, 41 pp.) ; Don Bruce (1966, 68 pp.) ; W.W. Clark (1954, 10 pp.) ; Charles S. Cowan, on fire protection (1956, 55 pp.) ; George Drake (1958, 32 pp. ; 1967 and 1968, 49 pp.) ; O.D. Fisher (1966, 127 pp.) ; Francis Frink (1958, 13 pp.) ; Rudolph L. Fromme, on his work in the U.S. Forest Service ; William D. Hagenstein (q960, 126 pp.) ; Edmund Hayes (1967, 64 pp.) ; Ed Heacox, on the Weyerhaeuser Timber Company (1967, 60 pp.) ; Donald Mackenzie, on logging activities of the Anaconda Copper Mining Company, Montana (1957, 24 pp.) ; David Townsend Mason (1883-1973), on sustained-yield forestry (1973, 1054 pp.) ; Stuart Moir (1958, 18 pp.) ; A.W. Moltke (1960, 57 pp.) ; L.T. Murray, Sr. (1957, 58 pp.) ; George Neils, on sustained-yield practices of the J. Neils Lumber Company in Washington State (1953, 9 pp.) ; Walter Neils, on the J. Neils Lumber Company (1953, 9 pp.) ; Walter Russell, on the lumber industry in Idaho (1953, 12 pp.) ; H.C. Shellworth (1963, 200 pp.) ; Orrin W. Sinclair, on lumbering in Minnesota and Washington (1954, 10 pp.) ; Paul R. Smith (1970, 96 pp.) ; James Stevens, on Paul Bunyan myths (1957, 78 pp.) ; and Arch Whisnant on the publication of the *Timberman* (1958, 40 pp. ; 1969, 40 pp.). Among topics frequently mentioned by the interviewees are labor relations, the cedar shingle industry, fires, timber cruising, sawmills, the Weyerhaeuser Timber Company, logging equipment and methods, fire protection, logging congresses, the Western Forestry and Conservation Association, West Coast Lumberman's Association, tree farming, wilderness, wood products, William B. Greeley, World War I, federal and state cooperation, the Industrial Workers of the World, lumber marketing, Paul Bunyan stories, legislation, Indian timber, shingles and shakes, and log transportation. Interveiwers included Elwood R. Maunder, John W. Larson, and George T. Morgan. Restricted.

375 **Forest History Society,** Oral history interviews relating to forestry and the lumber industry in the Southern United States, 1953-1964, ca. 13 items. Subjects of interviews conducted by Forest History Society staff members and relating primarily to forestry and the lumber industry in the Southern United States include Wendell R. Becton, on the Civilian Conservation Corps (1958, 19 pp.) ; Clinton Hux Coulter, State Forester of Florida (1958, 21 pp.) ; W.J. damtoff, on the pulp and paper industry (1959, 66 pp.) ; E.L. Demmon (1958-59, 75 pp.) ; D.D. Devereaux (1954, 15 pp.) ; Inman F. Eldredge (1959, 99 pp.) ; Marc Leonard Fleishel (1960, 56 pp.) ; David Kenley, on logging in Texas (1954, 5 pp.) ; Clarence F. Korstian, on his career in the U.S. Forest Service, on the Duke University School of Forestry, and on forestry education in the South (1959, 74 pp.) ; Joseph E. McCaffrey, on the pulp and paper industry (1964, 270 pp.) ; John W. McClure, on the hardwood industry (1953, 62 pp.) ; Elis Olsson, on the pulp and paper industry in Virginia (1959, 28 pp.) ; Reuben B. Robertson and E.L. Demmon, on theChapmion Paper and Fiber Company (1959, 32 pp.) ; and Clyde Thompson (1954, 9 pp.) and Robert Weeks (1953, 4 pp.), on the Southern Pine LUmber Company, Diboll, Texas. Topics discussed include fire protection; reforestation; Austin Cary; the Biltmore Forest School; the Pinchot-Ballinger Affair; U.S. Forest Service experiment stations; southern attitudes toward forest fires; forestry congresses; the pulp and paper industry; Negro labor; taxation; turpentine induestry; trade associations in the lumber industry; labor organizations; logging techniques; national forests in Southern states, in the Southwest and in Utah; World War I; the Florida State Forest Service; opposition of the Champion Paper and Fiber Company to the formation of the Great Smokies National Park; federal-private cooperation in the lumber industry. Interviewers include Elwood R. Maunder and John Larson. Restricted.

376 **Miscellaneous Oral History Interviews,** 1955-1964, 5 items. Miscellaneous oral history transcripts include an interview with Frederick K. Weyerhaeuser, conducted by Fred Kohlmeyer of the Oral History Research Office, Columbia University (1956, 167 pp.) ; with Jack Fish, on lumbering in Arizona and California, conducted by Peter C. Gaffney and James M. Boyd (1961) ; with J.C. Rutledge, on logging railroads in the Pacific Northwest, conducted by Kramer Adams (1960, 54 pp.) ; with H.C. Shellworth, on lumbering in Idaho and the Boise-Payette Lumber Company, conducted by Ralph W. Hidy (1955, 69 pp.) and by Michael P. Malone (1964, 13 pp.). Restricted.

377 **Ogle, Charles E.,** Oral History interviews, 1966-1967, 11 items. Transcripts of interviews conducted by Charles E. Ogle relating to Oregon timber tax law. Subjects include Clarence A. Barton (1966, 12 pp.) ; Bruce N. Cowan (1966, 10 pp.) ; Wallace B. Eubanks (1966, 15 pp.) ; Paul F. Liniger (1966, 10 pp.) ; Harry J. Loggan (1967, 15 pp.) ; Charles H. Mack (1966, 12 pp.) ; David T. Mason (1966, 11 pp.) ; Robert Oslund (1966, 10 pp.) ; Earl Plummer (1966, 2 pp.) ; Leland W. Swarverud (1966, 14 pp.) ; Richard L. Uhlman (1966, 8 pp.). Restricted.

378 **Regional Oral History Office-Forest History Society,** Oral history interviews, 1965-1968, 10 items. Transcripts of unpublished interviews conducted by staff members of the Regional Oral History Office, Bancroft Library, under the auspices of the Forest History Society. Subjects include S. Rexford Black, on private and state forestry in California (1966, 183 pp.) ; C. Raymond Clar, on California State Forestry (1966, 65 (1966, 65 pp.) ; Richard A. Colgan, on forestry in the California pine region (1966, ca. 50 pp.) ; Leo A. Isaac, on Douglas-fir research in the Pacific Northwest region of the U.S. Forest Service (1967, 152 pp.) ; Myron E. Krueger, on forestry and technology in Northern California (1965, ca. 27 pp.) ; Walter H. Lund, on U.S. Forest Service timber management in the Pacific Northwest (1967, 81 pp.) ; Walter F. McCulloch,, on forestry education at Oregon State University (1967, 135 pp.) ; Woodbridge Metcalf, on University of California forestry

extension work (1968, 138 pp.); Thornton Munger, on U.S. Forest Service research in the Pacific Northwest (1966, 245 pp.); and William R. Schofield, on lobbying activities for the California Forest Protective Association (1966, 175 pp.). Interviewers included Amelia R. Fry and Bonnie Fairburn. Restricted.

379 **Regional Oral History Office,** Oral History interviews, 1966-1974, 6 items. Transcripts of unpublished interviews conducted by staff members of the Regional Oral History Office of the Bancroft Library, University of California, Berkeley. Subjects include Henry Clepper, Kenneth B. Pomeroy, and Fred Hornaday (1968, 81 pp.); Charles G. Dunwoody (1966, 81 pp.); Tom Gill (1969, 75 pp.); R. Clifford Hall and Harold B. Shepard (1967, 121 pp.); Morris Kleiner (1974, 141 pp.); Earl S. Peirce (1968, 52 pp.). Interviewers include Amelia R. Fry, Malca Chall, and Fern Ingersoll. Restricted.

380 **White, Roy R.,** Oral History interviews, 1956-1959, 7 items. Transcripts of interviews conducted by Roy R. White on Austin Cary and Southern forestry. Subjects include Charles A. Cary (1956, 7 pp.); E. Worth Hadley (1959, 9 pp.); Frank Heyward, Jr. (1959, 12 pp.); James H. Jones (1959, 8 pp.); Herbert L. Kayton (1959, 7 pp.); Brooks Lambert and Ed Leigh McMillan (1959, 8 pp.); and G.P. Shingler (1959, 10 pp.). Restricted.

University of California, Santa Cruz, Library

Head, Special Collections 95064 Telephone: (408) 429-2547

380a **Santa Cruz History Collection.** Includes typed transcripts of oral history interviews, master's theses, photographs, pamphlets, ephemera, and a bibliography of Santa Cruz County forestry. Major emphasis is on materials relating to the Santa Cruz Mountains. Among those interviewed by Elizabeth S. Calciano are Michael Bergazzi (b. 1886) on redwood lumbering (1964, 207 l.) and Albretto Stoodley (b. 1873) on the Loma Prieta Lumber Company, redwood lumbering, and the fruit box industry.

STANFORD

Hoover Institution on War, Revolution and Peace

Archivist, Hoover Institution, Stanford University Campus 94305 Telephone: (415) 321-2300

381 **Lowdermilk, Walter Clay** (b. 1888), Collection, ca. 1922-1961, 6 boxes. Papers and printed materials on forestry, water, soil, etc., pertaining to China, Japan, and the Pacific Islands.

382 **Wilbur, Ray Lyman** (1875-1949), Papers, 1917-1949, ca. 80 feet. U.S. Secretary of the Interior, 1929-33. The papers relating to Wilbur's career as Interior Secretary comprise 31 boxes; subject headings for these files include conservation, national parks, U.S. Forest Service, etc. Inventory.

Stanford University Libraries, Department of Special Collections

Manuscripts Librarian 94305. The holdings are listed in *Cataloged Manuscripts, Department of Special Collections, Manuscripts Division, Stanford University Libraries,* 1973, mimeographed, 39 l.

383 **California and Western Manuscripts,** Collection, 1849-1906, 3½ feet (ca. 300 items). Small collections and miscellaneous single items pertaining to various periods, persons, and phases of California history. Relating to the Point Lobos State Reserve are letters from the Save-the-Redwoods League and reports on conditions (1 folder); also reports of the Advisory Committee to the State Park Commission (1936, 1 vol.). Register.

384 **Deitrick, James** (b. 1864), Papers, 1900-1918, ca. 475 items. Correspondence, extracts from a diary, contracts, legal papers, financial papers, memoranda, notes and drafts, sketches, maps, and other papers, relating to business affairs in Russia, China, Nicaragua, Mexico, and the United States, including railroad construction, development of mining and agricultural lands and resources. Correspondence, arranged chronologically, includes scattered references to the Nicaragua Timber and Fruit Lands Co., use of redwood railroad ties in California, timberland near Albion, California (all ca. 1913-15). Register.

385 **De Voto, Bernard Augustine** (1897-1955), Papers, 1918-1955, 51 feet (ca. 42,400 items). Correspondence, typescripts and proofs of DeVoto's writings, speeches, lectures, and other papers. Outgoing letters (6 boxes) are arranged chronologically; incoming letters (27 boxes) are alphabetical by name of correspondent. Topics of articles and drafts of articles by DeVoto include conservation, national park and national forest policies and problems, fire fighting, public lands, etc. Research material in notebooks relating to conservation relates to the Advisory Committee on Conservation, the Barrett Committee; the Citizens Committee on Natural Resources; the U.S. Congress; Dinosaur National Monument; firefighting, the Forest Service; national forest recreation and timber policies; grazing and forests; the U.S. Grazing Service; Hells Canyon; the "land Grab" issue; Massachusetts forests; mining claims on public lands; the National Parks Advisory Board; public lands; public power; conservation issues in Adlai Stevenson's presidential campaign (1952); tourists in national parks; wilderness; the Wichita National Wildlife Refuge; national forests in Idaho, Utah, Vermont; and many other conservation issues of the 1940s and 1950s. There are photographs of Grand Teton National Park, Glacier National Park, Boise National Forest, reseeding and grazing on national forests, forest fires in Idaho, Dinosaur National Monument, etc. There is also correspondence and photos (1 box) relating to the DeVoto Memorial Plaque and Grove, Clearwater National Forest, Idaho; and maps (1 case) of national forests in Idaho and Montana, annotated to show routes of the Lewis and Clark expeditions (1804-06). Restricted; any photocopying or use for publication of materials in the collection must be approved by Mrs. DeVoto. Register.

386 **Ely, Northcutt,** Papers, 1903-1965, 30 feet (ca. 2500 items -87 boxes•). Material, largely 1920-50, used by Ely in his position as legal representative in disputes over water projects and in his position as executive assistant to the Secretary of the Interior during the Hoover administration. Ely's working files relate to California flood control projects and the California Central Valley Project. Subject files of material collected by him include material on the California Water Project, the Colorado River controversy, Boulder Canyon Project, Colorado River Storage Project, the Colorado Compact, Glen Canyon, the Upper Colorado River Storage Project, etc. Restricted; permission of Ely is required for access. Register.

387 **Hill, Andrew P.,** Photographs, 1900, 1 vol. Photographs (ca. 25 items) taken by Hill showing activities of the Sempervirens Club in the area of Big Basin State Park, Santa Cruz County, California. Shown are views of the club exploring party, camp life, redwood trees, destruction caused by logging operations. Two photos are panoramic views of the vicinity. Also included are related clippings (1901, 1950, 1954, 3 items). Uncataloged.

388 **Hopkins, Timothy** (1859-1937), Collection, 1816-1942, 33 feet (3,300 items). Largely relates to the Central Pacific Railroad. Additional papers (1887-1936, ca. 6¼ feet) are mostly working materials including reports, histories, statistics, briefs, maps, diagrams, etc., for applications to hearings before the California Railroad Commission and the Interstate Commerce Commission, including information on a number of California logging and other railroads. Included are are files on the Minarets and Western Railway Co. (1 folder); Great Northern & Western Pacific Railroads (2 folders); Southern Pacific Company, including an agreement with the Red River Lumber Co. (1929). There are aerial photographs (no date) of the Great Northern-Western Pacific, showing timberlands, barren lands, grazing lands, etc., in northeastern and central eastern California; also reconnaissance maps of the same territory. Photographs relating to the Southern Pacific Company show routes of the N.C. & O. & Fernley Lassen Branch, Long Bell Lumber Co. railroad, McCloud River Lumber Co. railroad, Lassen Lumber & Box Co. railroad, and the Fruit Growers Supply Co. logging railroad. Register.

389 **Jones, Herbert Coffin** (b. 1881), Papers, 1911-1954, 43 feet (ca. 40,000 items). California State Senator from San Jose, 1913-34. Correspondence, documentsn pamphlets, clippings, and photographs, pertaining to Jones' work in the legislature and including a considerable amount of material on legislative actions. Within the files for each legislative session, correspondence is generally arranged alphabetically, and reference material, sometimes also including correspondence, is arranged by subject. Included are files concerning Big Basin State Park (1917-25), bird and arbor conservation (1915), conservation (1913), fish and game (1913-33), forest fires (1915), forestry (1913-21), forests (1923), Hetch Hetchy Valley and San Francisco water supply (1917-34), Reclamation, Recreational Enquiry Committee (1913-16), Save-the-Redwoods League 1919-21), Redwood State Park (1915), state parks and State Park Commission (1915-33), the John Muir Trail (1915). Register, indexed.

390 **Needham, James Carson** (1864-1942), Papers, 1893-1936, 5 feet (ca. 9,000 items). U.S. Representative from California, 1899-1913; member of the House Public Lands Committee. Correspondence, legal and financial papers, speeches, notes, clippings, and photos (largely 1898-1908), relating to Needham's congressional career, and business and personal affairs. The congressional correspondence, arranged alphabetically and chronologically, reflects Needham's interest in conservation, forestry, and irrigation. Register.

390a **Ralston, William C.,** Papers, 1870s-1880s, ca. 3,000 items. The collection largely relates to banking in San Francisco and mining in California and Nevada, but includes some material on the lumber trade between the Pacific Coast and the Orient. Copies of originals owned by the Bank of California. Restricted.

391 **Taylorn Frank J.,** Papers, 1922-1967, ca. 24,000 items. The collection consists of articles and books written by Taylor, manuscripts of his writings, and correspondence with publishers. Some writings pertain to conservation and to national parks, including items written in collaboration with Horace Albright. Register.

392 **Younger, Jesse Arthur** (1893-1967), Papers, 1954-1964 12½ feet (75,000 items). U.S. Representative from California, 1953-67. Miscellaneous papers relating to Younger's life and duties in the House of Representatives, including material on contacts with constituents, committee activity, elections, House rules and procedures, public and private bills, and public opinion. Topics, chiefly relating to California, include flood control, lumber (1963-66), national parks and forests (1963-66), public lands, Sierra Club, water resources, wildlife conservation. Register.

Stanford University, Graduate School of Business, J. Hugh Jackson Library of Business

Director 94305

393 **Alaska Commercial Company** (San Francisco), Records, 1868-1940, 18 feet (ca. 11,400 items). Documents and correspondence on the Yukon Saw Mill Company (Dawson, Yukon Territory) include several pages of descriptions of timber berths, maps, correspondence, and timber licenses (1906-17, 3 folders). There is also material on the Hammond Lumber Company (Dutch Harbor, Alaska) (1929, 1 folder). Inventory.

394 **Pacific Improvement Company,** Records, 1869-1931, 180 feet. Records of a holding company and many of its subsidiaries. Among the companies represented are the Pacific Lumber Transportation Company, with journals (1908-12, 2 boxes), ledgers (1909-16, 2 boxes), and cashbooks (1909-12, 1 box). Inventory.

STOCKTON

University of the Pacific, Pacific Center for Western Historical Studies, Stuart Library of Western Americana

Curator of Manuscripts 95204

395 **California. State Board of Forestry,** Records, 1885-1889, 27 l. Xerox copy of manuscript minutes of Board meetings (missing leaves 1-2, 4, 6, 10, 16, 18, 20).

396 **Emparan, Madie Brown,** Papers, ca. 1941-1970, 4 items. Conservationist, historian. Papers include notes on John McLaren, letter from Linnie Marsh Wolfe, minutes of the meeting of the John Muir Association, March 27, 1941. Typescript in part.

397 **Muir, John** (1838-1914), Papers, 1856-1914, 30 feet. Naturalist, explorer, writer, conservationist. Includes related papers collected by William Bade and Linnie Marsh Wolfe as well as correspondence, journals, writings, and pictorial works of Muir. Among correspondents are Charles N. Eliot, Hamlin Garland, Asa Gray, E.H. Harriman, R.U. Johnson, Gifford Pinchot, and Theodore Roosevelt. On loan from the children of Wanda Muir Hanna. Index and inventories.

398 **Stanford, Ernest Elwood,** Papers, ca. 1913-1965, ca. 8 feet (19 boxes). Biologist, educator, author. Includes research notes on redwood distribution and hay fever, textbook manuscripts, typescript drafts of books for children, reprints and carbons of published articles, and over 100 magazines containing articles by the author.

THREE RIVERS

Sequoia and Kings Canyon National Parks, Ash Mountain Headquarters

Chief Park Interpreter 93271

399 **Historical Collection.** Materials relating to Kaweah Cooperative Colony Company, Limited, include copies of the *Kaweah Commonwealth* and other printed items, time checks, meal tickets, applications for membership, and photographs; and correspondence, including letters (1926-47) regarding the history of the colony, letters from and information about several of the colonists, memoranda, testimony in the Kaweah case, Los Angeles courts (1894), photostats of timber claim, con-

tract, ledger, receipts; copies of newspapers articles. There are also correspondence files, designated as historical, relating to Sequoia and Kings Canyons National Parks; topics include dedications of roads and sites, anniversary programs, army occupation (1891-1913), Kaweah Colony, pioneer settlers, first automobile in Sequoia National Park, John Muir Centennial, general history, and the proposed John Muir National Park. There are also several thousand historical photographs of the region, papers of Francois Matthes, material on tree mensuration, old pamphlets and maps relating to the park, reports of superintendents (1891-1953), and archeological materials from Indian sites.

YOSEMITE NATIONAL PARK

Yosemite Research Library

Curator, P.O. Box 577 95389

400 Correspondence, manuscripts, and other materials relating the human and natural history of the Sierra Nevada. There are letters and reports written during the U.S. Army administration of Yosemite National Park (1891-1914); letters, diaries, biographical sketches of persons who contributed to the history of Yosemite Valley, including James D. Savage, Galen Clark, James Conway, James Hutchings, and John Muir; forestry manuscripts, including reports by Alfred J. Bellue on the timber in Cherry Creek and the Sequoia groves (1930, 4 items, including illustrations), correspondence of William Gianella concerning prescribed burning (1950-71), report of R.N. McIntyre on the human impact on the Sequoia in Yosemite (1954, illustrated), survey of the Mariposa Grove by E.C. Smith (1942); general correspondence of the National Park Service concerning the Mariposa Grove (1910-1939), and trees (1927-53, 6 vols.); study of Sequoia reproduction by Wulff, Lyons, and Dudley (1930, illustrated); and material of the Yosemite Sugar Pine Company concerning logging activities adjoining the park (1935-38), cruise data, and land data.

Colorado

BOULDER

University of Colorado Libraries, Western Historical Collections

Curator 80302

401 **Cockerell, Theodore Dru Allison** (1866-1948), Papers, 1885-1949, 28 feet (ca. 11,500 items). Publications, writings, pictures, and correspondence of Cockerell, professor of zoology at the University of Colorado, 1904-1934. Includes material on forests. Guide.

402 **Darley Family** Papers. Genealogical material, correspondence, personal effects, and photographs of George Marshall Darley, his three sons, George S., Ward, and William M., and their families. Includes material on forests. Guide.

403 **Detwiler, Sam B.,** Papers, 1912-1959, 11 boxes. Head of the Forestry Division, U.S. Soil Conservation Service. Much of The collection relates to investigations of errosion-resistant plants. Guide, 2 pp.

404 **Holt, George Hubbard** (1852-1924), Papers, 1875-1880, 1 foot (55 items). Mine manager in Leadville, sawmill operator, Crested Butte, Colorado. Business correspondence and records. Typed list of contents, 3 pp.

405 **Miscellaneous** Manuscripts. Included are a letter of A.G. Wallihan, U.S. Surveyor, to S.H. Babcock of the Denver, Northwestern and Pacific Railway, evaluating the agricultural potentials of Moffat County, Colorado, including information on forests (January 10, 1913); transcript of hearings held by the U.S. Forest Service on the Uncompahgre Primitive Area, Grand Junction, Colorado (November 15-16, 1971, 252 pp.); and other single items.

DENVER

Division of State Archives and Public Records

Director and State Archivist, Colorado State Archives Building, 1530 Sherman 80203 Telephone: (303) 892-2055

406 Primarily published documents of Colorado Department of Natural Resources; papers of the governors; records of the Colorado State Bureau of Corporations; papers of Hubert Work (1860-1942), U.S. Secretary of the Interior, 1923-1928. There are no finding aids.

Denver Public Library, the Conservation Library

Conservation Specialist, 1357 Broadway 80203 Telephone: 573-5152, extension 254 or 262

407 **American Association for Conservation Information,** Papers, ca. 1961-1962, 11 cartons. Correspondence, yearbooks, material for *The Balance Wheel*, material for awards program.

408 **American Forestry Association,** Records, 2 boxes. Financial records (1949, 1955-62), letters, programs (1969), miscellaneous.

409 **Association for the Protection of the Adirondacks,** Papers, 2 boxes. Bylaws, minutes (1939-48), correspondence files, material relating to conservation.

410 **Bruce, Robert Keady.** "History of the Medicine Bow National Forest, 1902-1910," Laramie: University of Wyoming, M.A. thesis, 1959.

411 **Carhart, Arthur Hawthorne** (b. 1892), ca. 50 boxes. Recreation engineer, U.S. Forest Service, 1919-22; later with Colorado Game and Fish Department; author; conservationist. Manuscripts, articles, letters, photographs, diaries, miscellaneous. Topics include conservation, water pollution, national parks, forest industries, writings of Carhart, outdoor recreation, and personal material. Unarranged.

412 **Colorado Open Space Council, Inc.** Denver), Miscellaneous records, 1965-73, ca. 2 boxes. Formerly the Colorado Open Space Coordinating Council. Records include material relating to wilderness workshops sponsored by the Council; materials relating to various proposals for wilderness areas in Colorado National Forests; membership lists; location of highways in Colorado scenic areas; state parks in Colorado; various related organizations, such as Colorado Open Space Conference, minutes (1965-66); Colorado Open Space Foundation, reports and recommendations (1966-69); Citizens Committee for the Outdoor Recreation Resources Review Commission Report (1965). Calendar on 3x5 cards.

413 **Cliff, Edward Parley** (b. 1909), 2 boxes. Major speeches, articles, and statements while chief of the U.S. Forest Service 1962-72. Copies of materials available in bound sets at the Forest History Society, Santa Cruz, California, Utah State University, Logan, Utah, and elsewhere.

414 **Edge, Mrs. Charles Noel (Mabel Rosalie) (1877-1962), Papers, 1929-1955, ca. 1,200 items. Correspondence, pamphlets, photographs, miscellaneous. Mostly relating to wildlife protection; Hawk Mountain Sanctuary, Pennsylvania; internal affairs of the National Audubon Society and the**

Emergency Conservation Committee; also preservation of Calaveras Sequoia Grove (ca. 1953-54); Dinosaur National Monument (ca. 1953-54); Olympic National Park (1952-53); and the proposed Kings Canyon National Park (ca. 1938-39). Calendar (processed), ca. 100 pp.

415 **Evison, S. Herbert,** Papers. Materials relating to the US.S National Park Service, including correspondence files relating to the reorganization of the National Park Service; national park lands; the National Park Service and its relation to other federal agencies; proposals for wilderness areas in national parks; agreements between the National Park Service and the Bureau of Public Roads for road surveys, construction, and maintenance within parks; photographs; etc.

416 **Farb, Peter**, Collection, ca. 8 boxes (?). Manuscripts of published and unpublished writings of Farb, notes and research materials, and copies of published articles and books. Includes material relating to wildlife, tree farming, forests, flood prevention, trees, national parks.

417 **Frank, Bernard** (1902-1964), Papers, 16 boxes. Personal records, manuscripts, printed materials, relating largely to Frank's writings and to watershed management.

418 **Frederick, Karl Telford** (1881-1963), Papers, 5 cartons. Includes correspondence, reports, some articles and newspaper clippings, relating to the American Forestry Association, Association for the Protection of the Adirondacks, New York Conservation Council, National Wildlife Foundation, and other organizations.

419 **Gabrielson, Ira Noel** (b. 1889), Papers. Largely material relating to wildlife management and Gabrielson's other conservation interests. Includes correspondence, manuscripts, speeches, pamphlets, documents, miscellaneous materials.

420 **Graves, C. Edward,** Photographs, 660 items. Transparencies of national park scenes, used to accompany lectures for the National Parks Association.

421 **Haanstad, Albert,** Photographs, 3,500 items. 2,500 black and white and 1,000 color negatives of Rocky Mountain scenic areas, from ca. 1910.

422 **Hill, Ralph R.,** Papers, 3 boxes. Correspondence, manuscripts, publications, research materials, clippings on wildlife and forestry. Topics include ponderosa pine silviculture, national forests, land planning, multiple use, forestry handbook, Bureau of Land Management and Forest Service grazing fees, proposed reorganization of the Colorado Game, Fish and Parks Department, watershed use of the Rocky Mountain National Park and national forests in Colorado, etc.

423 **Johnson, Fred,** Papers. Includes articles, manuscripts, reports, and photographic negatives relating to Johnson's career with the U.S. Forest Service, Region 2. Included is a chronological history of the Nebraska National Forest.

424 **Kelly, George,** Papers. Letters, reports, and other materials relating to the Dinosaur National Monument-Echo Park Dam controversy. Not open for research use at present (1974).

425 **Keplinger, Peter,** Papers, 6 inches (2 boxes). Manuscripts and U.S. Forest Service publications developed by Keplinger relating to personnel management; also scrapbook of letters received by Keplinger at the time of his retirement (1946).

426 **Leopold, Aldo** (1886-1948), Papers, ca. 1943-1948. Correspondence, biographical materials, drafts of writings.

427 **Miscellaneous.** Materials of varied provenance, cataloged as single items or small collections in the manuscript card catalog, and stored partly in the manuscript boxes, and partly in vertical file drawers. Included are manuscripts, letters, clippings, typed and processed reports and studies, speeches, hearing transcripts, cartoon drawings, photographs, and tape recordings. There are memoirs of U.S. Forest Service personnel, small groups of personal papers, technical forestry studies, national forest histories, public statements of U.S. Forest Service and National Park Service officials, and other items relating to such topics as wildlife management and land use in national forests, mining laws, tree diseases, the lumber industry, wilderness areas, grazing in the national forests, human impact in national parks, National Park Service interpretative programs, forest insect damage and control, watershed management in national forest, etc. Geographical references are largely to forestry in Colorado, but there is considerable material relating to national policies. There are reports on forest fires in Idaho and Montana; a history of the Oregon and California revested lands; materials relating to formation of the Redwood National Park; the Boundary Waters Canoe Area; forest economics of Knott County, Kentucky (1931); the Dixie National Forest, Utah; and hearings on wilderness proposals for the Great Swamp National Wildlife Refuge, New Jersey (1967), and the Wichita Mountains National Wildlife Refuge, Oklahoma. Organizations and government agencies represented include the Adirondack Mountain Association (reports, 1953); Advisory Committee on Conservation minutes, 1948-56); Arizona Wool Growers Association (minutes, 1897-1923); Citizens Committee for the Outdoor Recreation Resources Review Commission Report (meetings, 1963-65); Colorado Audubon Society; Colorado Game, Fish and Parks Commission (meetings, 1938-72); Colorado State Forester (annual report, 1941-60); Conservation Council of Colorado; Conservation Foundation; Federation of Western Outdoor Clubs; Resources for the Future; Society of American Foresters, Central Rocky Mountain and Denver Sections; U.S. Bureau of Land Management; U.S. Forest Service; and the Wilderness Society. Among the individuals represented are Fred B. Agee, Robert W. Ayers, Christian O. Basler, Charles R. Batten, Paul Buck, Charles G. Carpenter, Earle H. Clapp, Arthur Carhart, Guy Cox, Ronald B. Craig, F. FRaser Darling, Edward Dodd, George A. Duthie, Noel D. Eichorn, Henry E. Haefner, Ralph S. Hosmer, A.F. Hough, Royal S. Kellogg, Elers Koch, Louis R. LaPerriere, J. Stokley Ligon, Charles Lathrop Pack, Charles E. Powell, Carl P. Russell, Charles A. Scott, Len Shoemaker, A.D. Taylor, Angus M. Woodbury.

428 **Munns, Edward N.,** Papers. Manuscripts, articles, correspondence, clippings.

429 **Murie, Olaus Johann** (1889-1963), Papers, 5 boxes. Correspondence, publications and reports, printed materials relating to wildlife and national parks.

430 **National Association of State Foresters,** Records, 1½ feet. Minutes (1935-47); executive committee minutes (1946-58); proceedings (1920-69, with gaps); constitution and bylaws; president's reports (1965); list of officers; a brief historical review, etc. Also 1 reel microfilm of congressional hearings on reforestation (1923). Prior to 1963, this organization was known as the Association of State Foresters.

431 **Nature Conservancy,** Records, 1 box. Financial reports, agenda of general and executive meetings, conference attendance lists, special reports (1962-66), organizational records, bylaws, promotional materials, membership lists.

432 **Outdoor Writers Association of America, Inc.** (Columbia, Missouri), Records, 1940-1963, 14 boxes. Correspondence (1940-1963, 8 boxes); annual meetings (1941-57, 1 box); lists of members and officers (1 box); miscellaneous (3 boxes).

433 **Sandvig, Earl D.,** Papers, ca. 1923-1971, ca. 2 boxes. Correspondence, memos., reports, speeches, and miscellaneous materials, including clippings, relating to Sandvig's career in the U.S. Forest Service; much of the material relates to grazing management on national forests in Colorado.

434 **U.S. Forest Service,** Collection, ca. 1870s-1970s, ca. 70 + boxes. A collection of miscellaneous Forest Service records and papers, of diverse provenance, pertaining to the work of the Forest Service and the history of national forests. The collection largely consists of material originating in Forest Service offices in Colorado, but contains a considerable number of items from ZNational forests, ranger districts, and experimental stations throughout Region 2 (Rocky Mountain Region), with substantial amounts from Region 1 (Northern) and 4 (Intermountain), and from the Washington, D.C., office. Most other regions are represented by a few items. Record types include scattered narrative and statistical reports, studies, memoranda, units of work and rank of forests, letters, memoirs, unit histories, addresses, manuals, clippings, photographs, plats, maps, hearings, minutes, processed and printed items, and other documents. Subjects include the World Forestry Conference (Seattle, 1960), landscape architecture, wildlife, research, grazing, watersheds, mining, land utilization, fire control, reforestation, silviculture, use permits, land classification, improvements, training, fees, legislation, timber sales, insects, administration, geographical names, forest experiment stations, finance, and the Emergency Rubber Project. Included are ledgers (1871-82- + •) of the San Isabel National Forest, Colorado, showing land status, mineral and timber claims, reservoir sites, with maps; ledgers (1900-05, 11 vols.) of the Gila Forest Reserve, New Mexico; financial records (1910-33, with gaps) of the Shasta National Forest, California; blister rust control reports of the Pacific Northwest Region (1925-65) and the Eastern Region (1917-52, 10 cartons); diaries (40 vols.) of a number of U.S. Forest Service officials, including Raphael Zon (ca. 1902-30, with gaps); mounted newspaper clippings (1890-1935, 28 boxes). Unarranged. Partly cataloged in the manuscript card catalog.

435 **U.S. National Park Service,** Oral history collection, 1 box. Transcripts of interviews conducted under the National Park Service oral history project. Interviewees are cataloged in the manuscripts card file.

436 **Wheeler, H.N.,** Photographs and tape recordings. Several hundred hand-colored glass slides and tape recordings, used in lecturing on behalf of the U.S. Forest Service.

Denver Public Library, Western History Department

Head, 1357 Broadway 80203 Telephone: (303) 573-5152 Photocopy or microfilm may be borrowed on inter-institutional loan.

437 **Agee, Fred B., and J.M. Cuenin,** Papers, May 14, 1923, 1 vol. (67 l.). Early history of Cochetopa National Forest (Colorado) and surrounding region. Includes information on forest officers and timber.

438 **Boardman, Alonzo Harris,** Letters, 1863, 10 items. Photocopies of letters from Lump Gulch and Central City area, Colorado, where Boardman operated a sawmill with Doc James H. Hayford (+). Calendar.

439 **Carpenter, Farrington R.,** Papers, 1934-1950, 3 boxes and 2 scrapbooks. U.S. Director of Grazing, 1934-1938. Proceedings, directives, correspondence, memoranda, and printed materials relating to the administration of public lands under the Taylor Grazing Act of 1934. The collection is largely material selected by Carpenter to document the relationship between the Grazing Service and the Office of the Secretary of the Interior. Includes letters from Harold Ickes. Calendared on 3x5 cards.

440 **Hale, Irving** (1861-1930), Papers, 1869-1937. Limited correspondence, assorted clippings and printed materials relating to conservation movement (ca. 1905-1910). Guide and index with papers.

441 **Lake, David** (b. 1878), Notes and reminiscences, 1905-1940, 26 pp. Ranger for the U.S. Forest Service in Montana, 1916-1940. The collection includes historical information on people and events in his district in the Snowy Mountains region.

442 **Mills, Enos Abijah** (1870-1922), Papers, 1897-1922, 2 boxes. Correspondence, speeches, biographical data, clippings, and manuscripts of Mills. Deals with conservation of resources, especially in Colorado. Mills' correspondence is mainly about his publications or speeches. Calendared.

443 **Moore, Charles Cornell,** Papers, 1930-1970, 2 boxes. Wyoming rancher, lawyer, legislator and conservationist. Correspondence, speeches, records and papers primarily dealing with western ranching interests and such subjects as public lands, Shoshone National Forest, Jackson Hole and Dinosaur National Monument, Grand Teton National Park, wilderness areas, Izaak Walton League. Inventory and index.

444 **Renze, Dolores C.,** "A brief study of the lumber industry in Colorado, 1858-1948," 36 l. Carbon copy of a University of Denver course paper.

445 **Teller, Henry Moore** (1830-1914), Correspondence, 1877-1879, 578 items. U.S. Senator from Colorado, 1876-1882, 1886-1909; U.S. Secretary of the Interior, 1882-1885. Mostly letters from constituents. There are letters concerning timber policies. Calendar gives names of correspondents but not subjects.

Federal Archives and Records Center

Chief, Archives Branch, Building 48, Denver Federal Center 80225. Records not accessioned into the archives branch remain under the jurisdiction of the agency of origin. Part of the records in the Denver Federal Archives and Records Center are described in *Research Opportunities,* Denver: General Services Administration, Federal Archives and Records Center, Archives Branch, 1973. More detailed descriptions of some record groups are contained in a draft "Preliminary Guide to the Research Records in the Denver Federal Archives and Records Center," 1969. Also available is a *List of the Denver Federal Archives and Records Center's Collection of National Archives Microfilm Publications,* 1973, mimeographed, which describes microfilm available in the center and on inter-institutional loan.

Regional Archives Branch

446 **District Courts of the United States,** Records, 2,776 feet. District court records include case files from Colorado, New Mexico, Utah, and Wyoming. Criminal cases include prosecutions under federal law for timber depredations on the public domain; bankruptcy cases may include records relating to lumber companies.

447 **Forest Service,** Records, 29 cubic feet (Record Group 95). Atlas binders (ca. 1920-53, 20 cubic feet) relate to land classification, boundaries, administrative sites, and land use/planning, mostly for national forests in Colorado, but also in Wyoming, Nebraska, Oklahoma, South Dakota, and Kansas. They include narrative descriptions of the terrain, timber, mineral resources, wildlife, past and present uses, roads and

trails, streams, and rainfall, and annotated maps and charts. Civilian Conservation Corps operations files (1933-42, 6 cubic feet) include correspondence of forest supervisors and regional foresters, reports, inspections, and work plans for camps, chiefly in Colorado, but also in South Dakota, Nebraska, and Wyoming. Oversize binders (ca. 1920-41, 4 vols.), contain annotated maps and charts, narrative reports, etc., concerning soil erosion in Colorado (1935-41, 1 vol.) land classification in proposed additions to national forests (2 vols.), and a master plan report for the proposed Custer Game Sanctuary, Harney National Forest, South Dakota (1927, 1 vol.).

448 **Indian Affairs Bureau,** Records, 919 cubic feet (Record Group 75). General records of the Mescalero Indian Agency include log scale books (1923-45, 21 feet [351 vols.]) showing amount and kinds of timber removed from logging units on the Mescalero Indian Reservation, New Mexico. These records are described in the *Preliminary Inventory of the Records of the Mescalero Indian Agency,* compiled by Robert Svenningsen, Denver: National Archives and Records Service, Federal Records Center, 1971, 14 pp.

449 **Land Management Bureau,** Records, 2,759 cubic feet (Record Group 49). Local land office abstracts (1863-1908) created in more than 70 federal land offices in what are now the states of Colorado, New Mexico, Montana, North and South Dakota, Utah, and Wyoming include timber culture and other entries under the general land laws. For the period 1908 through 1955 there are serial registers for each of the above states, except Wyoming, that document all classes of land transactions. There are also closed timber sales contracts. Draft guide, 15 pp.

Federal Records Center

450 **District Courts of the United States,** Records, 7,920 cubic feet (Record Group 21). Records of U.S. District Courts, similar to those described above, but remaining under the custody of the agency of origin.

451 **Forest Service,** Records, 10,841 cubic feet (Record Group 95). The Denver center has custody of records created by three Forest Service Regional offices (Rocky Mountain, Denver, Colorado; Southwestern, Albuquerque, New Mexico; and Intermountain, Ogden, Utah), two forest and range experiment stations (Intermountain. Odgen. Utah; and Rocky Mountain, Fort Collins, Colorado), and 37 national forests in Arizona, Colorado, Idaho, Nevada, New Mexico, Utah, and Wyoming. The bulk of these records are subject files, dating from ca. 1910, with correspondence, reports, land classification studies, photographs, maps, land acquisition files, fire reports, range inventories, and newspaper clipping files.

452 **Indian Affairs Bureau,** Records, 3,385 cubic feet (Record Group 75). Among the 714 cubic feet of Bureau of Indian Affairs records to be permanently retained are timber cutting records, scale books, and reports of logging operations.

453 **National Park Service,** Records, 1,813 cubic feet (Record Group 79). National Park Service records in the Denver Center consist of records of the Southwest Regional Office, Santa Fe, New Mexico (1926-67); the Denver Service Center, Denver, Colorado (1954-67); and national parks and monuments (1908-66) located in Arizona, Colorado, New Mexico, Utah, and Wyoming.

State Historical Society of Colorado

Curator of Documentary Resources, Colorado State Museum, 200 Fourteenth Avenue 80203 Telephone: (303) 892-2305

454 **Bell, William Abraham** (1841-1921), 850 items. Organizer, with William Jackson Palmer, and others, of the Denver & Rio Grande Railway Co. The collection includes correspondence (1871-1913), personal and business records (1871-1921), memorabilia and photographs. Includes statements of resources and liabilities and report to the stockholders of the Sloan Saw Mill & Lumber Co. (1874-75, 3 items); annual reports of the Colorado Pinery Trust, Philadelphia (1872 and 1876, 2 items); and correspondence and business records of more than 30 firms, chiefly railroads, land and mining companies. Mimeographed calendar, 84 pp.

455 **Carpenter, Farrington R.,** Papers, 4 items. U.S. Director of Grazing, ca. 1934-1938. The collection contains typescripts of taped interviews (1956, 1959) at the Old Timers' Meetings in Steamboat Springs, Colorado.

456 **Colorado State Forestry Association,** Papers, 1884-1947, 135 items. Correspondence (1912-16, 94 items), chiefly letters between William G.M. Stone and John H. Gabriel and other members, discussing meetings, elections, and policies of the association, its position on forestry matters, state and federal legislation, and taxation of forest lands; business papers (1884-1947), including minutes, membership lists, lists of officers, programs of meetings, resolutions, and newsletters; newspaper clippings (1913-1916). Calendar.

457 **Hallack, Erastus F.** (d. 1897), Papers, 5000 items. Lumber retailer, Denver. The collection contains correspondence (1882-1889) and business records, mostly bound journals and ledgers (1873-1916). The materials pertain to E.F. Hallack, E.F. Hallack Lumber and Manufacturing Co., L.H. Butcher, and Hallack and Howard Lumber Co.

458 **Holliday, J.T., Lumber Company** (Greeley, Colorado), Account book, 1872-1874, 2 vols.

459 **Jackson, William Henry** (1843-1942), Papers and photographs, 1864-1949, 2 feet (260 items). Photographs, correspondence, miscellaneous printed material and memorabilia, relating to Jackson's pioneer work in photographing natural wonders of the western United States, including the Grand Tetons and the Yellowstone. Most of the correspondence relates to trips in the Rocky Mountains during the 1930s. Calendar.

460 **McCreight, Isreal** (1865-1948), Papers, 1906-1936, 141 items. Includes letters from McCreight, historian of Dubois, Pennsylvania, to government officials including Theodore Roosevelt and Gifford Pinchot, regarding forest conservation.

461 **Nachtrieb, Charles** (1833-1881), Papers, ca. 1862-1882, 1 foot. Correspondence containing many references to the supply of timber for the Atchison, Topeka, and Santa Fe Railroad, timber cutting rights along its right of way, lumber, shingles, cord wood, etc. (all ca. 1879-81); financial records and time sheets of sawmills (ca. 1876-80); court records and contracts relating to disputed ownership of logs, timber cutting rights, supply of lumber to railroads, etc. (ca. 1879-80). Nachtrieb was a resident of Northrop, Colorado. Calendar, 25 l. Restricted.

462 **Palmer, William Jackson** (1836-1909), Papers, 2,500 items. The collection contains material on logging operations and construction activities during the railroad building era in Colorado. Included are papers pertaining to the Colorado Pinery Trust, the Denver and Rio Grande Western Railroad, and other firms.

463 **Pence, Kingsley A.,** Scrapbook, 1 vol., 1910-1920. Chairman of the Denver Mountain Parks Joint Committee composed

of representatives of the Chamber of Commerce, Denver Real Estate Exchange, and the Rocky Mountain Motor Club. Scrapbook contains correspondence, newspaper clippings, and photographs about Denver's mountain parks system.

464 **Porter, Henry Miller** (1838-1937), Papers, 1885-1935, 51,000 items. The collection includes material on investment in Washington State timber lands. Described in *An Index to the Henry Miller Porter Papers,* compiled by Paul Ton, Denver: Western Business History Research Center, State Historical Society of Colorado, 1968. 225 pp., which lists correspondents alphabetically; there is also an unpublished catalog (8 vols.) with the collection.

465 **Railroad Collection,** 100,000 items. The collection contains correspondence, business papers, and other material relating to the construction and operation of various railroads, mainly in Colorado, but also throughout the American West. Organized by name of eleven major railroads, and thereunder generally chronologically. The bulk of the collection pertains to the Denver and Rio Grande Western Railroad Company (1872-1963, 75,000 items). There are scattered materials relating to tie cutting and the production of mining and bridge timbers.

466 **Robinson, E.W.,** Papers, 104 items. Proprietor of the E.W. Robinson Lumber Co. (Denver). The collection includes correspondence and business papers (1892-1932), family history, and scrapbooks.

467 **Rogers, James Grafton** (1883-1971), Papers, 62,000 items. The collection includes much material relating to the Colorado Mountain Club and Rocky Mountain National Park; included are correspondence of the Sierra Club, American Alpine Club, U.S. Forest Service, Department of the Interior, the Mountaineers, and Charles S. Thomas, U.S. Senator. There are also related photographs and scrapbooks (ca. 1912-19). Guide, typed, 73 l.

468 **Schomburg, Thomas A.,** Oral history interview. Lumber business in Trinidad, Colorado; Eagle, Colorado (?); and Raton and Cimarron, New Mexico.

469 **Thomas, Charles Spaulding** (1849-1934), Papers, 4,200 items. U.S. Senator from Colorado, 1913-1921; governor of Colorado, 1899-1901. The collection includes correspondence relating to the Senate committees on Agriculture and Forestry and the committee on Public Lands. Inventory, 81 pp., mimeographed.

470 **Weitbrec, Robert F.,** Papers, 1848-1931, 1,600 items. Correspondence (70 pieces), notebooks (28 vols.; about 1,400 entries), miscellaneous documents (83 items), 10 maps or profiles, and 78 photographs, dealing primarily with construction and development of the Denver & Rio Grande and other railroads. The notebooks contain references to sources of supply, prices, etc., of timber for bridge construction, ties, and lumber for general construction; also references to timber lands (ca. 1876-89). Mimeographed calendar, 67 pp.

FORT COLLINS

Colorado State University, Library

Forestry and Agricultural Sciences Librarian, Special Collections 80521

471 **Colorado State Forestry Association,** Archives, ca. 8 feet (3 cartons, 12 boxes, 9 volumes). Correspondence and operational records covering a period of about 10 years, including correspondence of Gifford Pinchot. To be opened for research when processing is complete, planned for 1975.

Connecticut

MYSTIC

Marine Historical Association, Inc.

Manuscript Librarian, The G.W. Blunt White Library 06355
Telephone: (203) 536-2631.

472 **Balch, Hiram A.,** Papers, 1792-1896, 4,977 pieces. Shipowner and agent, Trescott and Lubec, Maine. The material pertains to the coastwide trade, including timber cargoes.

473 **Beckwith, J. & D.D.** (New London, Connecticut), Papers, 1845-1884, 848 pieces and 1 volume. Merchants and shipbuilders.

474 **Burgess, James,** Papers, 1855-1864, 103 items. Shipmaster and agent, of Boston. Correspondence, accounts, bills, receipts, crew lists, bills of lading, and charter parties for the barks *Ocean Plide* and *Kremlin,* of Boston, carrying lumber and other cargo.

475 **Carpenter & Edson Co.** (Canterbury, Connecticut), Papers, 1902-1908, 259 pieces. Mast hoops.

476 **Cottrell, Gallup & Co.** (Mystic, Connecticut), Papers, 1836-1873, 118 pieces. Lumberyard.

477 **Fish Family Papers,** 1812-1887, 2150 pieces and 8 volumes. Shipbuilders, Mystic, Connecticut.

478 **Gildersleeve Shipbuilding Co.** (Portland, Connecticut), Papers, 1880-1931, 59 pieces. Shipbuilding company.

479 **Hillman Shipyard** (New Bedford, Massachusetts), Papers, 1827-1862, 56 pieces. Shipbuilding.

480 **Holmes, Joseph,** Papers, 1797-1887, 4,838 pieces. Shipbuilder, owner, and merchant, Kingston, Massachusetts.

481 **Holmes, Josiah,** Papers, 1810-1880, 11,285 pieces and 15 volumes. Shipbuilder, owner and merchant, Mattapoisett, Massachusetts.

482 **Hotchkiss Collection,** 1851-1882, 13 volumes. Shipmaster, Boston, Massachusetts. Volumes 2, 3, 7, and 8 are logbooks relating to cargoes of lumber.

483 **Latham, James A.,** Papers, 1848-1897, 10 pieces and 13 volumes. Master carpenter, Noank, Connecticut.

484 **Lawrence & Co.** (New London, Connecticut), Papers, 1822-1904, 5,700 pieces and 133 vols. Ship owners and agents. Vol. 130 relates to cargoes of lumber.

485 **Lee, George William** (1776-1850), Papers, 1788-1843, 1,444 pieces (3 boxes). Shipmaster and merchant shipowner of New York, New York, and Norwich, Connecticut. Correspondence, accounts, bills, and receipts relating to vessels he commanded, including material relating to the lumber trade. Inventory.

486 **Logbooks and Journals,** 654 volumes. Bound or boxed volumes smaller than 16 inches include journal of Frank E. Young on the bark *Sarah,* Portland, Maine, during a coastwise and South American passage carrying lumber (1876-77); log of the brig *Zebra,* Boston, Massachusetts, during a coastwise, South American, and West Indies voyage carrying cargo of lumber and shingles (1845); log kept on the ship *St. John,* New York, on a coastwise, South American, trans-Pacific, and trans-Atlantic voyage, carrying redwood lumber (1875-78).

487 **Lowell, James,** Papers, 1817-1855, 145 pieces (1 box). Shipmaster and shipowner of Bath, Maine. Bills, accounts,

receipts, and letters concerning vessels commanded by Lowell in the coastwise and West Indies trade, with cargoes including lumber.

488 **Lupton, Edward,** Papers, 1862-1917, 115 pieces. Williamsburg, New York, Shipbuilder.

489 **Miscellaneous Volumes,** 418 volumes. Cashbooks, daybooks, letterbooks, etc. Included are a ledger of the schooner *William Jones,* Boston, Massachusetts, carrying lumber between coastal points (1873-1878) ; and a ledger of Nelson Ingalls, merchant shipowner of Machiasport, Maine, including accounts for vessels he owned, carrying primarily pilings on coastal voyages (1861-79).

490 **Morse, Asa P.,** Papers, 1855-1879, 9532 pieces. Cooperage, Boston, Massachusetts.

491 **Noyes, Benjamin F.,** Papers, 1860-1880, 1125 pieces and 5 volumes. Shipmaster, Mystic, Connecticut. Volumes 3 and 6 are logbooks for voyages with lumber cargoes.

492 **Smalley, Anthony,** Papers, 1870-1879, 234 items. Shipmaster and agent of Nantucket, Massachusetts. Correspondence, charter parties, and accounts mentioning cargoes including lumber. Correspondents include Samuel C. Cobb and Dudley Hall of Boston; Baring Brothers of London, England; and Joseph J. Rider.

493 **Strickland, Peter,** Papers, 1864-1920, 20 vols. Shipmaster, New London, Connecticut. Volume 1 is logbook relating to a lumber cargo.

494 **Treat, E.P., & Co.,** Papers, 1840-1894, 2017 pieces (4 boxes). Merchant shipowners of Frankfort, Maine. Includes letters, accounts, bills, receipts, charter parties, and bills of lading dealing with vessels owned or managed by the company, and relating to the lumber trade. Inventory.

495 **Vertical File Manuscripts,** 6000 pieces. Included are scattered letters to John B. Stuart, Richmond, Maine, regarding cargoes of lumber for the West Indies in the schooner *Mechanic* and the brig *Harriet* (1826-47, 10 items) ; bill of lading for a cargo including timber shipped by Ropes & Ward, Boston, Massachusetts, to Smyrna, Turkey, in the brig *Pacific* (1822) ; papers of the schooner *P.T. Barnum*, Bridgeport, Connecticut, carrying lumber (1890-1905, 5 pieces) ; letters from Joseph Warren Holmes, master of the ship *Charmer,* New York, taking on a load of lumber in San Francisco, California, and Woodyville, British Columbia, for Cape Town, South Africa (1900-01, 2 items) ; letters and bills of lading of Rufus Daniel, Bangor, Maine, for lumber shipped to Cottrell and Hoxie, Mystic, Connecticut (3 pieces, 1846-47) ; statements of earnings for the schooners *Clara A. Phinney,* Edgartown, Massachusetts, carrying lumber between coastal ports (1890-92, 22 items) and letters from Cameron, Miller & Company, Liverpool, England, to William Bennett & Brothers, Hopewell, New Brunswick, regarding the lumber market (1850, 3 items).

NEW HAVEN

Yale University, Yale University Library, Manuscripts and Archives

Associate Librarian for Archives and Manuscripts, Sterling Memorial Library 06520. The following entries were compiled from *Forest History Sources of the United States and Canada,* comp. by Clodaugh M. Neiderheiser, St. Paul: Forest History Foundation, Inc., 1956; the *National Union Catalog of Manuscript Collections,* Washington: Library of Congress, 1962-1974); "Papers Relating to Forestry," processed brochure issued by the Yale University Library, Manuscripts and Archives, 1972, 2 l.; and accessions announcements in *American Archivist* and *Forest History*.

496 **Abbot, Samuel** (1765-1856), Daybooks, 1820-1825, 1828-1846, 2 volumes. One daybook (1820-25) deals with the activities of Abbot's lumber wharf.

497 **Bradbury, Reuben,** Daybook, 1809-1815, 1 volume. Daybook of a lumber dealer of Portland and Standish, Maine. Entries deal with the sale of boxes, lumber, rum, etc. (Partly pasted over as a scrapbook.)

498 **Brown, William Robinson** (1875-1955), Papers, 1907-1957, 6 feet. Forester and businessman, of Berlin, New Hampshire. Business correspondence, articles, speeches, galley sheets of Brown's book *Our Forest Heritage: A History of Forestry and Recreation in New Hampshire* (1958), photos, forestry studies, and congressional bills. Includes papers relating to the Brown Company, Berlin, New Hampshire, manufacturer of paper products, chemicals, and other wood-using products; American Forestry Association; New Hampshire Disaster Emergency Council; New Hampshire Forestry Commission; North East Forest Research Council; Society for the Protection of New Hampshire Forests; Society of American Foresters; and the Timber Lands Mutual Fire Insurance Company. Register.

499 **Chapman, Herman Haupt** (1874-1963), Papers, 1896-1956, 20 feet. Professor of the Yale School of Forestry. Correspondence, articles, photos, and other records relating to forest mensuration, valuation, and management, forest policy, forest improvements and surveying, the American Forestry Association, Connecticut Forest and Park Association, Northeastern Agriculture Experiment Farm, Society of American Foresters, U.S. Department of Agriculture, and U.S. Forest Service. Persons represented include Samuel Trask Dana, Bernhard E. Fernow, William B. Greeley, Gifford Pinchot, Carl A. Schenck, and Henry A. Wallace. Register. The Minnesota Historical Society has 5 reels of microfilm of some of the papers of this collection.

500 **Goodwin, James Lippincott** (1881-1967), Papers, 1882-1967, 50 feet. Forester, landscape architect, timber executive of Hartford, Connecticut. Correspondence, business accounts, timber records, and maps, relating to his work for his own forest properties in Connecticut and North Carolina, as administrator of the Talcott Mountain Forest Protective Association, and as President of the Connecticut Forest and Park Association, 1958-61.

501 **Graves, Henry Solon** (1871-1951), Papers, 1895-1951, 102 boxes (ca. 25,000 items). Forester; director, Yale Forestry School, 1900-1910: chief, U.S. Forest Service, 1910-1920; dean, Yale Forestry School, 1923-1927. The papers include correspondence, diaries, notes, articles, maps, photos, and other papers relating mainly to American forestry. Subject files (44 boxes), include such topics as agricultural policy, erosion, forest policy, state parks, wildlife, etc. There is also correspondence (35 boxes), arranged by author. Diaries cover Graves' career as forest engineer in France, World War I. There are also series on war papers, and the biography and history of forestry, which contain autobiographical notes, and series on legislation, lecture material, maps, and photographs. Represented in the collection are the American Forestry Association, Society of American Foresters, and the Yale Forestry School. Unpublished register, 7 pp.

502 **Hawes, Austin Foster** (1879-1962), Papers, 1940-1947, 2 feet (ca. 1000 items). State Forester of Connecticut. Correspondence, diary, manuscripts of writings, photographs, newspaper clippings, and other papers, relating to Hawes' student

life, his career as State Forester and his controversial retirement from that position, his job as State Forester of Vermont and as professor of history at the University of Vermont. Persons represented include Henry Graves and Gifford Pinchot. Unpublished register.

503 **Heintzleman, Benjamin Frank** (1888-1965), Papers, 1940-1965, ca. 3000 items. Forester and governor of Alaska. Correspondence, diaries, speeches, statements, writings, tape recordings, printed material and maps on Alaska, and a political scrapbook, relating chiefly to Heintzleman's years as governor (1953-1957) and his retirement, which he devoted to the economic development of Alaska. Includes materials on the history, forests and forestry, and travels in Alaska. Unpublished register.

504 **Kent, William** (1864-1928), Papers, 1801-1961, ca. 75,000 items. U.S. Representative from California, 1910-1916. Correspondence, scrapbooks, clippings, memorabilia, and printed matter relating to Kent's activities as a politician and as a private citizen interested in conservation programs, particularly with respect to public control or ownership of water power sites. Unpublished guide.

505 **King & Norton Lumber Co.** (Suffield, Connecticut), Papers, 1833-1846, 1860-1863, 2 boxes and 1 folder. Letters and papers dealing with the affairs of the King & Norton Lumber Co., Seth King, Jr., and Daniel W. Norton, proprietors. Includes letters, contracts, a small account book, 3 real estate maps, and miscellaneous papers. The firm apparently supplied rudder stocks for the U.S. Navy, bridge timber for the New Haven and Hartford Railroad, shipbuilding timber, and other lumber demands.

506 **Loomis and Norton** (Windsor Locks, Connecticut), Records, 1866-1878, 27 boxes and 5 volumes. The records include bills, shipping receipts, checks, promissory notes, business letters, 2 general journals (1868-77), 1 cash journal (1868-75), and 2 order books (1869-71, 1872-75). The Franklin Paper Mills, Windsor Locks, Connecticut, which specialized in the manufacture of Manila paper, was purchased by John Hughes Norton and Byron Loomis, both of Suffield, Connecticut, on May 1, 1871, and became Loomis and Norton.

507 **Mason, David Townsend** (1883-1973), Diaries, 1906-1966.

508 **Munger, Thornton Taft** (1883-1975), Papers, 1911-46, ca. 1000 items. Forester and Chief of Forest Management Research, U.S. Forest Service. Personal and business correspondence, diaries, forestry analysis reports, articles, and commentaries relating mainly to his service, 1908-46, as Assistant Chief of Forest Management in Oregon, Director of the Pacific Northwest Forest Experiment Station, and from 1938 as Chief of Forest Management Research. Letters dealing with his association with the Society of American Foresters and his service as a member of its executive council concentrate on two significant reports, the survey of industrial forestry in the U.S. (1928-31) and the report of the Joint Committee on Forestry of the National Research Council and the Society of American Foresters, *Problems and Progress of Forestry in the United States* (1942-46).

509 **Newlands, Francis Griffith** (1848-1917), Papers, 1891-1917, ca. 45,000 items. U.S. Representative from Nevada, 1893-1903; member of the Irrigation of Arid Lands Committee; U.S. Senator, 1903-1917, serving on the Conservation of Natural Resources Committee. The papers include correspondence, speeches, reports, annotated bills, and scrapbooks, relating to irrigation, waterways, forestry, conservation, and other matters. Unpublished register. Part of this collection has been published in *The Public Papers of Francis G. Newlands,* ed. by Arthur B. Darlington, 2 vols.; New York: 1932.

510 **Parmele, Norman L.,** Account book, 1864-1890, 1 volume. The account book comprises a daybook of family expenses in the vicinity of Deep River, Connecticut (1864-67) and a ledger (1871-90) dealing with a lumber business in which timber, possibly from Parmele's own farm, was sawed at E.W. Norton's mill. It also includes charges for farm work, use of horse, oxen and wagon, and the sale of wood and some farm produce.

511 **Potter Family,** Papers, 1795-1887, 2 feet. Includes accounts relating to the sale of wood. Hamden, Connecticut. Unpublished register.

512 **Rockwell, Thomas H.** (1776-1865), Ledger, 1820-1859, 1 volume. Ledger of a cabinetmaker and farmer of Ridgefield, Connecticut. Entries indicate that he made coffins, chairs, tables, stools, shelves, cradles, bureaus, etc., and sold farm produce, meat and wool.

513 **Schrenk, Hermann von,** Papers. Forest pathologist. Lectured at Yale School of Forestry, 1902-1906. Pioneer student of wood preservation.

514 **Stillman, Levi** (1791-1871), Ledger, 1815-1834, 1 volume. Ledger of a cabinetmaker of New Haven, Connecticut. Entries deal primarily with household furniture, but also with coffins and other articles, and with sales of farm produce and wood.

515 **Winer, Herbert Isaac** (b. 1921), Papers, 1950-1962, 11 boxes. Records of southern logging and mill field data, including reports, data sheets, statistics, and working papers, collected by Dr. Winer of the Pulp & Paper Research Institute, Pointe Claire, Quebec.

Yale University, Yale University Library, the Beinecke Rare Book and Manuscript Library, Western Americana Collection

Curator, 1603A Yale Station 06520. Holdings are described in *A Catalogue of Manuscripts in the Collection of Western Americana Founded by William Robertson Coe, Yale University Library,* compiled by Mary C. Withington, New Haven: Yale University Press, 1952, and *A Catalogue of the Frederick W. and Carrie S. Beinecke Collection of Western Americana,* vol. I: *Manuscripts,* compiled by Jeanne M. Goddard and Charles Kritzler, edited by Archibald Hanna, New Haven: Yale University Press, 1965. The following descriptions were largely compiled from these two works.

516 *Albion* (Ship), Documents and Letters, 1849-1850, 25 items. Material relating to the seizure of the British ship *Albion* in Puget Sound, April 22, 1850, for cutting spars on the American side of the Strait of Juan de Fuca in violation of the revenue laws. (Withington, #3).

517 **Brewer, William Henry** (1828-1910), Papers, 1830-1927, 33 feet. Correspondence, diaries, reports, essays, lectures, addresses, articles, and other scientific papers, and personal and family papers, printed material, memorabilia, and photographs. The papers reflect Brewer's varied scientific interests including forestry and botany. Unpublished register. Letters relating to the California Geological Survey were published as *Up and Down California in 1860-1864,* edited by Francis P. Farquhar, New Haven: Yale University Press, 1930; the Rocky Mountain letters were published as *Rocky Mountain Letters, 1869,* Denver: Colorado Mountain Club, 1930. (Withington, #47).

518 **Carter, William Alexander** (1820-1881), Journal of transactions of the post trader at Fort Bridger, Wyoming,

August 18 to December 15, 1869, 285 pp. Carter established a sawmill at Fort Bridger. (Withington, #64).

519 **Ellis, William** (1821-1905), Letters, 1884-1891, 5 items. Letters to Nathan Witter Williams, from Washington Territory, describing the country around Grays Harbor and the opportunities in farming and the lumber industry. (Withington, #163).

520 **Gorrill, William H.** (1836-1874), Diary, 1869, 180 pp. Gorrill visited the big trees and Yosemite, December, 1869. Typewritten transcript with the diary. (Withington, #222).

521 **Greene, Francis Vinton** (1850-1921), Letters, 1875, 2 items. The letters describe Greene's trip with Lieutenant Doane through Yellowstone National Park in July and August, 1875. (Withington, #231).

522 **Hauser, Samuel Thomas** (1833-1914), Diary, 1870, 55 pp. Hauser's diary of the Washburn-Langford-Doane expedition to the vicinity of Yellowstone National Park, August 17-September 4, 1870. It covers 23 pages in pencil, and is followed by miscellaneous material and maps. (Withington, #249).

523 **Keller, Josiah P.** (1812-1862), Collection, 1853-1862, 106 items. Letters, documents, and two charts, relating to the establishment and early years of the Puget Mill Company at Teekalet (Port Gamble), Washington Territory, chiefly written by Keller to his partner, Charles Foster of East Machias, Maine. There are also copies of two letters by William C. Talbot. (Withington, #280).

524 **McLoughlin, John** (1784-1857), Letter to Peter Ogden, February 20, 1850. McLoughlin asks Ogden if he knows of anyone to build a sawmill at Port Discovery (Oregon Territory). (Withington, #320).

525 **Stuart, Granville** (1834-1918), Journal, 1873, 39 pp. Journal of a trip to the National Park. The Yellowstone Expedition of 1873. In pencil; includes newspaper clippings, and a sketch map of the trail at Fire Hole River. (Withington, #450).

526 **Wilson, James S.R.**, Journal of a trip to Yellowstone Park, 1875, 56 pp. The expedition was commanded by William Ludlow. (Withington, #528).

Delaware

DOVER

*Division of Archives and Cultural Affairs

Hall of Records, P.O. Box 796 19901 Telephone: (302) 678-4651. The following entries are taken from Neiderheiser's *Forest History Sources of the United States and Canada,* St. Paul: 1956; the *National Union Catalog of Manuscript Collections,* Washington: 1962-74; and from the Eleutherian Mills Historical Society's unpublished "Union Catalog of Business and Economic History Manuscripts."

527 **Account Books.** Included are accounts (1850-69, 1 vol.) of Benjamin Burton of Millsboro, Delaware, for lumber and other goods; accounts (1880-1919, 2 vols.) of F. & R.H. Davis (Sussex Co.) for a sawmill; accounts (1897-1903, with gaps, 2 vols.) of John Donaldson of New Castle County, Delaware, for timber sales; ledger (1 vol.) of John Reece, with accounts for the sale of lumber, brick, and logs.

528 **Ridgely Family,** Papers, 1742-1889, 8 feet (ca. 4,000 items). Correspondence, accounts, legal documents, miscellaneous personal papers of five generations of a Delaware family prominent in business and public life. The letters contain scattered references to phases of the lumber industry, mostly in the mid-19th century. Included are mentions of claering of land by burning, the destruction of timber and timberland, and thefts of wood and timber. Partly described in *Calendar of Redgely Family Letters,* ed. and comp. by Leon de Valinger, Jr., and Virginia E. Shaw, 3 vols.: Dover: 1948-1961.

529 **Stockton Family,** Business records, 1790-1815, ca. 450 items. Papers, chiefly 1800-1807, relating to the mercantile activities of Thomas Stockton and others. Includes records of trade in lumber, spars, shingles, and other goods, between Wilmington, Delaware, and ports in New England, eastern Canada, and the West Indies.

530 **Townsend, Samuel** (1812-1881), Papers, 1829-1922, 1861 items. Correspondence, business records, and other papers relating to the agricultural interests of various members of a Delaware family. There are scattered references to the purchase and handling of fruit containers (ca. 1870-81). Also included are papers of a business in ship timbers in the 2nd quarter of the 19th century (66 items).

GREENVILLE

Eleutherian Mills-Hagley Foundation, Inc., Eleutherian Mills Historical Library

The Director 19807 Telephone: (302) 658-2401. Holdings of the Eleutherian Mills Historical Library are described in *A Guide to Manuscripts in the Eleutherian Mills Historical Library: Accessions through the Year 1965,* by John Beverly Riggs, Greenville: 1970. The library also has an unpublished "Union Catalog of Business and Economic History Manuscripts," compiled by Hugh Gibb, which lists holdings of neighboring institutions, including the American Philosophical Society, Historical Society of Pennsylvania, Maryland Historical Society, Rutgers University Library, New Jersey Historical Society, Delaware State Archives, Hopewell National Historical Site, Glassboro State College, Chester County Historical Society, Schwenkfelder Library, Wagner Free Institute of Science, Historical Society of Delaware, Lebanon County Historical Society, Bucks County Historical Society, Salem County Historical Society, New Jersey State Library, University of Delaware Library, and Paley Library of Temple University.

531 **Barnard, William J.,** Accounts, 4 vols. Newark, Delaware. Accounts of The Barnard Co. (Westtown, Pennsylvania), and Sandusky & Co., including lumber accounts.

532 **Barnard, William J., Lumber Company** (Newark, Delaware), Records, 1905-1956, 10 boxes and 1 volume. Business records including one account book, with index.

533 **Cohen, Charles J.,** Papers, 1842, 1879, 1893-1952, ca. 200 items. Philadelphia, Pennsylvania. Wholesale stationer and manufacturer of envelopes.

534 **Curtis & Bro. Co.** (Newark, Delaware), Business records, 1823-1942, 40 volumes and about 2,000 loose items. Correspondence, journals, ledgers, cashbooks, bill books, invoice book, customer orders, cost sheets, freight records, time books, roll sizes book, machine record book, record of materials and supplies, bylaws, agreement, judgment bonds, and other records of paper manufacturing firm.

535 **Dubois, Abraham** (d. 1807), and Family, Letterbook, 1803-1826. Includes material relating to the sale of lumber for shipbuilding in the Carolinas.

536 **Du Pont, Alfred Victor,** Papers, 1833-1893, 340 items. Wilmington, Delaware. Business and personal correspondence, relating to various interests, including the Louisville Paper Mill.

537 **Du Pont, Charles Irenee,** Papers, 920 items. Wilmington, Delaware. Include personal and business papers, some concerning the Louisville Paper Mill.

538 **Du Pont, Eleutheria Bradford,** Collection, 1799-1834, 1,835 items. Contains early records of E.I. du Pont de Nemours and Co., including material relating to Erich Bollmann's charcoal process, supply of charcoal, casks, and willow.

539 **Du Pont, Irenee** (1876-1963), Papers, ca. 108,000 items. Vice President files (1912-18, 2,564 items) contain scattered items relating to the Hudson River Wood Pulp Plant (1914, 2 items); manufacture of nitrocellulose and powder from wood pulp (1914-1917, 9 items); the sale of potash (1914, 3 items); etc. Restricted.

540 **du Pont de Nemours, E.I., & Co.** (Wilmington, Delaware), Records, 1801-1902, ca. 500,000 items and 1,050 volumes. Records concerning the gunpowder business of the Du Pont family. Among the letters received are references to the manufacture, purchase, sale, and prices of wood, including willow, poplar, and maple, lumber, staves, keg heads, kegs, barrels, wood spirits, shaved shingles, posts and rails, charcoal, potash, wood tar, turpentine, Spanish oak for tanning, yokes, timber, hoop poles, hickory poles, a sawmill at Chesapeake City, Maryland, damage to trees at Brandywine Springs, Pennsylvania, caused in cutting willow (1851), machinery for peeling bark (1884-1887), woodworking machinery, etc. General Accounts, Miscellaneous Records, contain a file on raw materials (1827-1884), including willow wood (1827), muriate potash (1877), etc. Purchase and Receiving Records include willow and poplar books, concerning wood for charcoal (1812-1895, 21 vols.); among miscellaneous account books are a volume for logs (1848-1852), and one for lumber purchases (1869-1873).

541 **Du Pont Family,** The Longwood Manuscripts, Group 5: E.I. Du Pont De Nemours & Co., Papers, 1782-1902, ca. 25,575 items. Company correspondence (1802-1891) includes letters received relating to such subjects as lumber, cypress shingles, staves, charcoal, wood prices, and various types of wood, including willow and maple. Accounts and related papers (1800-1894) contain bills, receipts, and miscellaneous accounts for wood used for charcoal. Special Papers (1782-1902) include patents for mitrocellulose, notes for experiments with distilled charcoal, etc.

542 **Du Pont Family,** The Longwood Manuscripts, Group 10: Papers of Pierre S. Du Pont (1870-1954), ca. 1,000,000 items. The Personal and Business Papers include a file pertaining to the American Forestry Association (1906-1930, ca. 53 pieces). Restricted.

543 **Eyester, Weiser Company** (York, Pennsylvania), Records, 1884-1970, 162 vols. and 53,000 items. Makers of parts for paper making machinery.

544 **Gale, Luther,** Patents, 1811, 1815, 2 items. Lenox, Massachusetts. Two patents for improvements in the "Bark Mill."

545 **Haldeman Family,** Papers, 1801-1885, ca. 7,600 items. Lancaster County and Harrisburg, Pennsylvania. Includes production books of woodcutting (1802-1818).

546 **Harper's Ferry Paper Co. and Shenandoah Pulp Co.,** Production graph (1909-1913), and chart showing financial positions on a comparative basis (1911-13), both signed by H.H. Thayer, Jr.

547 **Harper's Ferry Paper Co. and Shenandoah Pulp Co.,** Miscellaneous papers, 1886-1887, 1920-1923, 20 items. Includes certificates of incorporation of the two companies (1886-1887); financial notes, reports, and charts on operation costs, inventories (1920-1923); and correspondence with Horace H. Thayer, Jr., of Philadelphia (1922-23, 7 items).

548 **Harper's Ferry Paper Co. and Shenandoah Pulp Co.,** Papers, 1909-1920, 4 items. Includes a letter of Josiah Marvel to Thomas Savery relating to a notice to bondholders; list of payments to stockholders; other items.

549 **Hearn, George W.,** Accounts, 1850-69, 1 reel microfilm. Millsboro, Delaware. Lumber mill and general store.

550 **Jackson, David W.,** Accounts, 1890-1909, 1930-1939, 2 vols. Bartville, Lancaster County, Pennsylvania; lumber dealers, saw and grist mill accounts.

551 **Kent & Weeks** (Wilmington, Delaware), Business Papers, 1886-1929, 49 items. Lumber and coal dealers, succeeded by Brosius & Smedley. The papers include articles of agreement, receipts, balance sheets, minutes of shareholders' meetings, etc.

552 **Miscellaneous Account Books.** Not fully identified items, including Philadelphia mercantile accounts for lumber, etc. (1773-1776, 1 vol.); ledgers of a sawmill in Philadelphia area (1828-32, 1 vol.), and Bucks or Lehigh Co., Pennsylvania (1832-43, 1 vol.).

553 **Moon, Charles** (d. 1888), **and Charles Henry Moon,** Papers, 1855-1925, 46 vols. Woodbourne, Bucks County, Pennsylvania. Accounts, letterbooks of farmers, surveyors, include sale of bran, coal, lumber.

554 **Phillips, John J.,** Accounts, 1850-69, 1 reel microfilm. New Castle ½ County, Delaware; miller. Includes accounts for flour, hay, sawing boards.

555 **Pusey & Jones Corporation** (Wilmington, Delaware), Records, 1851-1960, 240 feet. Articles of agreement, deeds, and property papers, minute books (1906-08, 1918), register of ships built or outfitted, list of papermaking machines built, employee contracts, drawings, blueprints, photos, and other records of a firm of shipbuilders, founders, and machinists. The bulk of the papers consists of drawings and photos, including records of specifications, work rules of ships, displacement calculations, and engine numbers. Includes correspondence, patent papers, financial data, and other records (ca. 1888-1935) relating to pulp-saving machines in the manufacture of paper.

556 **Savery, Thomas H.** (1837-1910), Collection, 1880-1939, 9 boxes. Correspondence, miscellaneous, relating to Parsons Engineering Co., Harper's Ferry Paper Co., Pusey & Jones Corporation, etc. Also original and published patents issued to him for papermaking machinery (1868-1906, 2 vols.).

557 **Sawmills.** Notebook listing names and addresses of mills in certain states, including Pennsylvania, in the United States, and in provinces of Canada, ca. 1860, 80 pp.

558 **Smith's Mills** (Milton, Delaware), Ledger, 1829-1889. Contains sawmill account book, 1829-1833.

559 **Stockly & Rowland** (Smyrna, Delaware), Accounts, 1810-1929, 78 items. General store accounts, including lumber.

560 **Tatnall, Walter G.,** Accounts, 1902-1905, 1 vol. Wilmington, Delaware. Engineer, Du Pont Keg Mill.

561 **Thompson, Andrew,** Accounts, 1vol. Lancaster County, Pennsylvania (?). Accounts of a miller, including lengths and types of wood sawed.

562 **Tweed Mill** (Hockessin, Delaware), Account books, 1881-1895, 3 vols. Accounts receivable books relating to sawmill, grain mill, and cider sold.

563 **Weissport Planing Mill and Lumber Co.** (Weissport, Pennsylvania), Account book, 1876-77.

564 **West Branch Coal, Iron Ore, and Lumber Co.** (Philadelphia, Pennsylvania), Records, 1864-93, 2 vols. Records during the presidency of C.P. Bayard (1864-93, 2 vols.), including journal (1870-80), with letters and a stock ledger (1864-91).

565 **Whitaker, Joseph** (1789-1870), Papers, 1819-1888, 60 items. Ironmaster, of Phoenixville and Mont Clare, Pennsylvania, including ledger of Davis & Whitaker, containing sawmill accounts (1846-59); daybook of Mont Clare accounts for lumber and coal (1853-60).

NEWARK

University of Delaware, the University Library

19711

566 **Albertson Family,** Papers, 1800-1873, ca. 1200 items, including 4 account books. Personal and business letters and receipts of a family of lumber and lime merchants residing in Plymouth, Montgomery County, Pennsylvania, and in Philadelphia, Pennsylvania.

567 **Bodey, William H.,** Daybook, 1876, 1 vol. Lumber merchant, Norristown, Pennsylvania.

568 **Fitzwater, George** (d. 1831), Papers, 1801-1831, ca. 500 items. Receipts and legal documents of a merchant in lumber and lime, Montgomery County, Pennsylvania.

569 **Hall, John Wood** (1817-1893), Papers, 1861-1889, ca. 500 items. Business papers of Hall's coastwise shipping firm at Frederica, Delaware, including invoices of cargoes, reports from schooner captains, etc., carrying grain and lumber.

570 **Maretta, John Clement,** Papers, 1836-1838, 1 vol. Lumber accounts and carpentry receipts of a Lancaster County, Pennsylvania, carpenter.

571 **Marie, M.R.,** Receipts, 1831-1833, 1 vol. Receipts for coal and lumber at Norristown, Pennsylvania.

572 **Maulsby, Samuel,** Papers, 1792-1838, ca. 825 items. Business records and receipts in the coal and lumber trades for the firms of Maulsby and Moore (M.R. Moore), and Thomas and Maulsby (Jonathan Thomas).

573 **Nichols, A.T. & W.,** Accounts, 1856-60, 1 vol. Lumber accounts receivable at Philadelphia, Pennsylvania.

574 **Philadelphia Customs House,** Papers, 1779-1930, 61 boxes (ca. 17,500 items). Papers of the United States Customs Service of the Port of Philadelphia, including bills of lading for the lumber and forest products trade.

575 **Potts and Yocum** (Kensington, Pennsylvania), Accounts, 1840-1844, 1 vol. Account with A. Manderson, Jr., and Co., lumber merchants.

WILMINGTON

Historical Society of Delaware

Manuscript Librarian, Old Town Hall 19801

576 **Camden Mills** (Camden and Newport, Delaware), Records, 1827-1870, 10 vols. Daybooks and journal for a sawmill in Camden and retail outlet in Newport, Delaware. Includes a sawmill book (1860-63).

577 **Hemphill Family**, Papers, 1775-1829, 12 vols. Wilmington, Delaware; accounts, correspondence, miscellaneous, records of a shipping firm plying to East and West Indies, carrying building materials, etc.

578 **Jessup & Moore** (Wilmington, Delaware), Accounts, 1880-1899, 2 vols. Paper manufacturer.

579 **McCallmont & Belvill** (New Castle, Delaware), Daybook, 1794-1797, 1 vol. Board yard.

580 **Sharpless & Brothers** (Fairville, Pennsylvania), Records, 1853-1894, 8 feet. Gristmill, general store, sawmill.

District of Columbia

WASHINGTON

Georgetown University, The University Library

University Archivist, Special Collections Division, 37th and O Streets N.W. 20007

581 **Wagner, Robert Ferdinand** (1877-1953), Papers, 1927-1949, ca. 400 file drawers. U.S. Senator from New York, 1927-49; served as chairman of the Public Lands and Surveys Committee. The papers contain his office records during his term as senator, including legislative files. Restricted. Partly unprocessed.

Library of Congress, Manuscript Division

Chief 20540. In the Library of Congress Annex. Many of the division's holdings are described in the following publications: *Handbook of Manuscripts in the Library of Congress,* Washington: Government Printing Office, 1918, 750 pp.; "List of Manuscript Collections in the Library of Congress to July, 1931," compiled by Curtis Wiswell Garrison, *Annual Report of the American Historical Association,* 1930, I, Washington: Government Printing Office, 1931, pp. 123-233; "List of Manuscript Collections Received by the Library of Congress, July 1931 to July 1938," compiled by C. Percy Powell, *Annual Report of the American Historical Association for the Year 1937,* I, Washington: Government Printing Office, 1939, pp. 113-145; "Annual Report on Acquisitions: Manuscripts," in the *Quarterly Journal of the Library of Congress,* 1943 + ; Grace Gardner Griffin, *Guide to Manuscripts Relating to American History in British Repositories, Reproduced for the Division of Manuscripts of the Library of Congress,* Washington: The Library of Congress, 1946; *A Guide to Archives and Manuscripts in the United States,* edited by Philip M. Hamer, New Haven: Yale University Press: 1961, pp. 85-121; *The National Union Catalog of Manuscript Collections,* 12 vols. + , Washington: The Library of Congress, 1962 + ; Library of Congress, Manuscript Division, *Manuscripts on Microfilm: A Checklist of the Holdings in the Manuscript Division,* compiled by Richard B. Bickel, Washington: 1975; and *Manuscript Sources in the Library of Congress for Research on the American Revolution,* Washington: Library of Congress, 1975, 372 pp. There is very likely forest history material in some of the division's holdings in addition to those listed below, but the finding aids do not reveal their subject content. Such holdings would include the papers of a number of Presidents of the United States, Secretaries of the Interior and of Agriculture, United States Senators, Members of Congress, and others.

582 **Ahern, George Patrick** (1859-1942), Papers, 1911-1932, 32 items. Typed transcripts. Army officer, conservationist, and Director, Philippine Bureau of Forestry. Primarily letters exchanged between Ngan Han, Chief Forester of China, and Ahern, concerning the Philippine Forest School, Los Banos, the establishment of a forest school in China, and education and military training in China. Includes an article, "Restoring China's Forests," by Thomas H. Simpson, *Review of Reviews,* March, 1916, concerning Ahern's travels in China, his forestry career, and the reforestation movement.

583 **American Society of Landscape Architects,** Records, 1900-1966, 32 boxes (11,000 items). Correspondence, subject files, reports from chapters and committees, letterheads of society members, and books. The bulk of the material is from the period 1925 to 1955. There are records for the Committee on National Parks and Forests, and State Parks; and records relating to the National Park Service in the files of the Committee on Professional and Governmental Relations. Draft guide, 10 pp.

584 **Anderson, Clinton Presba** (1895-1975), Papers, 1941-1960, 140,000 items (675 boxes). U.S. Representative from New Mexico, 1941-1945; U.S. Secretary of Agriculture, 1945-1948; U.S. Senator from New Mexico, 1948-1973. Congressional and Senatorial papers, arranged by Congress, include files relating to the administration of the National Park Service, the Forest Service, the Bureau of Outdoor Recreation; also the Mescalero Indian Lumber Project (89th Congress). There are files for the 87th-92nd Congresses relating to Senate committees on which Anderson served, including Agriculture and Forestry, Interior and Insular Affairs, including its Public Lands subcommittee; also the Public Land Law Review Commission, and the Outdoor Recreation Resources Review Commission. Files on specific legislation include material on the Wilderness bill (87th Congress, S. 174). Restricted: permission for use must be requested through the Chief of the Manuscript Division. Draft finding aid, 56 pp.

585 **Ballinger, Richard Achilles** (1858-1922), Papers, 1907-1911, Microfilm. Copies of originals at the University of Washington, Seattle.

586 **Beard, Daniel Carter** (1850-1941), Papers, 1798-1941, 120 feet (261 boxes; 72,000 items). Artist, author, editor, and Boy Scout official. Correspondence, diaries, speeches, articles, research notes, photographs, memorabilia, and printed matter, relating to the Boy Scouts of America and other outdoor organizations. Draft register, 62 pp.

587 **Beveridge, Albert Jeremiah** (1862-1927), Papers, ca. 1890-1927, 135 feet (458 boxes; ca. 98,000 items). U.S. Senator from Indiana, 1899-1911; served on the Committee on Conservation of Natural Resources. His office files pertain only to patronage, but in the general correspondence are substantial groups of correspondence with Theodore Roosevelt (1901-1919) and Gifford Pinchot (1902-1927). Typed list of contents.

588 **Borah, William Edgar** (1865-1940), Papers, 1890-1907, 24 feet (77 boxes; 250,000 items). U.S. Senator from Idaho, 1907-1940; served on the Irrigation and Reclamation of Arid Lands Committee, and the Public Lands and Surveys Committee. The collection includes correspondence, legal papers, drafts of addresses, and notes relating mainly to Borah's legal practice. A large series on "Land" relates to Idaho, 1913-15. There are papers relating to the Barber Lumber Company and the Potlatch Lumber Company, and, among miscellaneous papers, material pertaining to the Hetch Hetchy Dam site. Typed list of contents, 46 pp.

589 **Brant, Irving Newton** (b. 1885), Papers, 7 boxes. Includes material dealing with conservation. Unprocessed. Not open for research.

590 **Coolidge, Calvin** (1872-1933), Papers, 1921-1929, ca. 210 feet (ca. 178,600 items; also available on microfilm, 190 reels). U.S. President, 1923-1929. The collection includes executive office correspondence (283 boxes); additional correspondence (1 box); reception lists (3 vols.). Subject files in the executive office correspondence include material on American Forests Week, American Forestry Association, Appalachian Park Commission, Butler Lumber Company, conservation, forest fires, Forest Protection Week, U.S. Forest Service, International Forestry Congress (1925), International Congress on Forestry Experimental Stations (1919), forests, game protection, U.S. General Land Office, Interior Department, National Park Service, Outdoor Recreation Conference, reforestation, Yellowstone National Park. The collection is described in *Index to Calvin Coolidge Papers,* Washington, D.C.: Library of Congress, Manuscript Division, 1965, 34 pp.

591 **Cutting, Bronson Murray** (1888-1935), Papers, 1899-1950, 45 feet (ca. 33,000 items). U.S. Senator from New Mexico, 1929-1935; member of the Senate committees on Agriculture and Forestry, Irrigation and Reclamation, and Public Lands and Surveys. Correspondence, speeches, essays, articles, biographical information, clippings, scrapbooks, printed material, photographs, and other records. General correspondence (1889-1941), arranged chronologically, includes scattered files on such forest history topics as the Civilian Conservation Corps (1933), conservation (1929), mining in forest reserves (1929), national forests (1928-35), and public lands bills of John Joseph Dempsey, U.S. Representative from New Mexico, 1935-41. Typed container list, 23 pp.

592 **Dern, George Henry** (1872-1936), Papers, 1933-36, 2 feet (ca. 1,000 items). U.S. Secretary of War, 1933-1936. The collection includes files on the Civilian Conservation Corps, Boise, Idaho (1933), and on the Forest Festival, Elkins, West Virginia, October 6, 1934. Typed container list, 4 pp.

593 **Dock, Mira Lloyd,** Papers, 1814-1947, 3 feet (10 boxes; 3,000 items). Correspondence, printed material, clippings, photos, and maps, dealing mainly with forestry, gardening, park development, conservation, and nature study in Pennsylvania and elsewhere in the United States, as well as in Germany. Some papers concern Mira Dock's work as member of the State Forest Commission of Pennsylvania and as chairman of forestry in the State Federation of Pennsylvania Women. Besides Lloyd and Dock family letters, there are letters of Sir Dietrich Brandis, Joseph T. Rothrock, Gifford Pinchot, J. Horace McFarland, and others.

594 **Douglas, William Orville** (b. 1898), Papers, 1925-1951, 105 boxes (30,000 items). Personal correspondence arranged chronologically and alphabetically. Most of the collection seems to relate to legal matters. Restricted. Container list, 14 pp.

595 **Fisher, Walter Lowrie** (1862-1935), Papers, 1879-1935, 19 feet (ca. 14,000 items). U.S. Secretary of the Interior, 1911-1913. Correspondence, manuscripts of speeches and articles, memoranda, receipts, photos, memorabilia, scrapbooks, clippings, and printed matter. Among the files relating to the Interior Department is material pertaining to Richard Achilles Ballinger, national parks, Hetch Hetchy Valley, Louis R. Glavis, National Forest Reservation Commission, National Irrigation Congress, water power, conservation, American Civic Association, American Forestry Association, etc. The collection is described in: *Walter L. Fisher: A Register of His*

Papers in the Library of Congress, Washington: Library of Congress, Manuscript Division, 1960, 10 pp.

596 **Forest History Society, Inc.,** Oral history interviews, 1958-1961, 25 items. Twenty-two transcripts and some related correspondence. The transcripts, all typed carbons, are indexed. The interviews were conducted by Elwood R. Maunder, unless otherwise noted. Subjects include Royal S. Kellogg (1955, 54 pp.), Charles S. Cowan (1957, 55 pp.), Leonard Costley (1957, 20 pp., by Bruce C. Harding), George W. Dulany (1956, 37 pp.), Arthur B. Recknagel (1958, 46 pp.), Clinton H. Coulter (1958, 21 pp.), William Joseph Griffin (1958, 26 pp., by Charles D. Bonsted), Wendell R. Becton (1958, 19 pp.), Donald Mackenzie (1957, 18 pp.), James Stevens (1957, 33 pp.), J.A. Mathieu (1957, 10 pp., by Harding), Francis Frink (1958, 13 pp.), George Drake (1958, 32 pp.), Elis Olsson (1959, 28 pp.), L.T. Murray, Sr. (1957, 58 pp.), Reuben B. Robertson and E.L. Demmon (1959, 30 pp.), Emmit Aston (1958, 41 pp.), Stuart Moir (1958, 18 pp.), E.L. Demmon (1958-59, 66 pp.), Inman F. (Cap) Eldredge (1959, ca. 100 pp.), and Mrs. William B. Greeley (1960, 21 pp.).

597 **Garfield, James Rudolph** (1865-1950), Papers, 1879-1950, 98 feet (ca. 70,000 items). U.S. Secretary of the Interior, 1907-1909. Correspondence, diaries, professional and office records, speeches, legal case files, articles, books, scrapbooks, and memorabilia relating to Garfield's activities as a lawyer, politician, and civil servant. Correspondents include Gifford Pinchot, Theodore Roosevelt, Woodrow Wilson, Franklin K. Lane, and others. In Garfield's secretarial office records are files on the U.S. Forest Service (1907); North American Conservation Conference (1909); a subject file including material on the Department of the Interior (1907-1924); Bureau of Reclamation (1923-24); National Conservation Association (1913-1921); and public lands (1931). The correspondence between Pinchot and Garfield (1911-1912) remarks upon J. Horace McFarland's proposal for a National Parks Bureau, and discusses the work and finances of the National Conservation Association, and a projected book by Pinchot. Draft guide, 22 pp.

598 **Great Britain. Colonial Office 390: Board of Trade, Miscellaneous.** Vol. 6, Custom House accounts: Naval stores imported into England, 1701-1723. Photostats of originals in the Public Record Office. Described in Grace Gardner Griffin, *Guide to Manuscripts Relating to American History in British Repositories, Reproduced for the Division of Manuscripts of the Library of Congress,* Washington: The Library of Congress, 1946, p. 42.

599 **Great Britain. House of Lords, Committee Books.** Vol. 10: (March 21, 1733), Cultivation of naval stores in the American colonies. Vol. 67: (June 16, 1820), Leith merchants on duties on timber and the trade with American colonies; (June 30), Importation of timber from the colonies in North America. Photocopies. The recollection is described in Grace Gardner Griffin, *Guide to Manuscripts in British Repositories, Reproduced for the Division of Manuscripts of the Library of Congress,* Washington: The Library of Congress, 1946, pp. 176-177.

600 **Ickes, Harold L.** (1874-1952), Papers, 1906-1952, 158 feet (117,000 items). U.S. Secretary of the Interior, 1933-1946. Correspondence, legal papers, memoranda, reports, articles, press releases, speeches, clippings, scrapbooks, and memorabilia relating to Icke's work as a lawyer and political figure in Chicago, his service as Secretary of the Interior, and later career. The Chicago files include legal cases involving the Sample Lumber Company (1911-1912). Office files as Interior Secretary include conservation (1935-1940), conservation conference (1946), reforestation (1933), reorganization (1933-1945), trip to the Great Smoky Mountains (1937). Late files (1946-1952) include materials on proposed Alaska pulp project, conservation, the Emergency Conservation Committee, Isle Royale National Park, Jackson Hole National Monument, the National Park Service, and the Save-the-Redwoods League. Also included are diaries, with index. Typed register, 59 pp.

601 **Krug, Julius Albert** (1907-1970), Papers, 1936-1950, 47 feet (34,816 items). U.S. Secretary of the Interior, 1946-1949. Chiefly a subject file of papers relating to Krug's government service, 1941-1949, on the War Production Board and as Secretary of the Interior, with correspondence, memoranda, reports, speeches, articles, statements before congressional committees, records of trips, appointment schedules, newspaper clippings, and other printed matter. Interior Department subject files include minutes of weekly staff meetings, Bureau of Land Management, Office of Land Utilization, Bureau of Reclamation, Office of the Solicitor, Alaska, international conferences on conservation and natural resources, National Park Service, etc. Typed register, 11 pp.

602 **La Follette Family,** Papers, 1781-1970, 96,000 items. Papers of Robert M. La Follette, Sr. (1855-1925), U.S. Senator from Wisconsin, 1906-1925; Robert M. La Follette, Jr. (1895-1953), U.S. Senator from Wisconsin, 1925-1947; and other family members. General correspondence for each member of the family is arranged alphabetically. Subject files of Robert M. La Follette, Sr., include conservation (1910), notes on conservation and natural resources (1924), lumber trusts (undated). His speeches and writings include an editorial on the lumber monopoly (1923). Typed draft register. Restricted: permission for use of this collection must be requested through the Chief of the Manuscripts Division.

603 **McGee, William John** (1853-1912), Papers, 1882-1916, 12 feet (32 boxes; 7,000 items). Geologist, anthropologist, and hydrologist. Correspondence, letter books, speeches, articles, scientific papers, lectures, notes, geological notebooks, scrapbooks, bibliographical notes, and memorabilia, relating to the organization of the U.S. conservation movement.

604 **McNary, Charles Linza** (1874-1944), Papers, 1921-1944, 29 feet (20,000 items). U.S. Senator from Oregon, 1918-1944; member of the committees on Agriculture and Forestry, Irrigation and Reclamation, and the Special Committee on Conservation of Wildlife Resources. The collection contains correspondence, speeches, memoranda, extensive legislative files arranged by subject, notebooks, pamphlets, clippings, and other manuscript and printed materials. In the legislative files are 2 boxes on forestry, including files on the McSweeney-McNary Act, reforestation in Puerto Rico, protection of watersheds of navigable streams, appropriations for fire protection, reforestation, revision of boundaries of various national forests in Oregon, jurisdiction over Crater Lake National Park, importation of shingles, Reconstruction Finance Corporation loans to timberland owners, timber sales on Indian lands, and other forestry legislation. Legislative files relating to public lands include material on the Oregon and California Revested Railroad Lands, Crater Lake National Park boundaries, cutting timber on revested lands, and other legislation relating to forestry and timber lands in Oregon. Reciprocal trade files relate to lumber, lumber tariff, and lumber data. Typed container list, 16 pp.

605 **MacVeagh, Franklin** (1837-1934), Papers, 1860-1928, 9 feet (ca. 8,000 items). U.S. Secretary of the Treasury, 1909-1913. The subject file of correspondence and other materials

connected with Treasury Department operations and subjects of personal interest to MacVeagh contains material on forestry. Typed draft register.

606 **Merriam, John Campbell** (1869-1945), Papers, 1899-1938, 96 feet (70,000 items). Paleontologist, educator, author. The collection includes correspondence, articles, memoranda, notes, reports, speeches, drawings, maps, memorabilia, photographs, and other manuscript and printed materials. General correspondence, arranged alphabetically by topic or author, relates to such matters as, among others: Newton B. Drury, national parks, Union Lumber Co., Save-the-Redwoods League, state parks, State Parks Association, timber conservation, Robert Sterling Yard, and Yosemite National Park. Typed list of contents.

607 **Miscellaneous Manuscripts Collection.** Shoemaker, Theodore, "Some of My Experiences in the Forest Service, 1907-1938." Manuscript (98 pp., typed).

608 **Norris, George William** (1861-1944), Papers, 1861-1944, 256,000 items. U.S. Representative, 1903-1913, and U.S. Senator, 1913-1943, from Nebraska; chairman of the Senate Agriculture and Forestry Committee. The collection contains correspondence, notebooks, memoranda, speeches, memorabilia, scrapbooks, subject files, maps, charts, photographs, and miscellaneous printed materials. Congressional "case files" include materials on Gifford Pinchot (1914-1931), U.S. Forest Service, the Civilian Conservation Corps (1933-1936), public power and the Tennessee Valley Authority, the Committee on Agriculture and Forestry (75th-77th Congresses), and the U.S. National Park Service. Typed list of contents.

609 **Olmsted, Frederick Law** (1822-1903), Papers, 1822-1903, 20 feet (24,000 items). Landscape architect and author. The collection includes correspondence, letterbooks, journals, notebooks, financial records, contracts, proposals, drafts of reports, speeches, lectures, articles, essays, maps, drawings, scrapbooks, clippings, and drafts of Olmsted's writings. In part, they are photocopies of originals at the Kansas State Historical Society. Subject files include material on *Garden and Forest* journal (1886-1889); files on parks, including the Adirondacks, New York (1884-1886), and Yosemite, California (1863-1891); and the Vanderbilt Estate, North Carolina; miscellaneous files include scenic reservations (1876-1887). The collection is described in *Frederick Law Olmsted: A Register of His Papers in the Library of Congress,* Washington, D.C.: Library of Congress, Manuscript Division, 1963, 13 pp.

610 **Olmsted Associates, Inc.** (Brookline, Massachusetts), Records, 1868-1950, 255 feet (170,000 items). Olmsted Associates is the successor to Olmsted and Vaux, landscape architects, the firm founded by Frederick Law Olmsted and Calvert Vaux in 1858. This collection consists largely of the business files of the firm. The bulk of the material is a numerical "job file" (1871-1950, 523 boxes), with a microfilm copy of the related card index and Xerox copies of the job books (3 vols.), and a numerically arranged index. These files include information on landscape design, layouts, work arrangements, and financing of projects throughout the United States. In part, they relate to state, national, and other parks. There are also letterbooks, arranged chronologically (1884-1899) and general correspondence, arranged alphabetically. Included under miscellany are copies of correspondence relating to the origin of the U.S. National Park Service (1910-1916). Typed draft register, 37 pp.

611 **Pinchot, Amos Richards Eno** (1873-1944), Papers, 1863-1943, 273 boxes (180,000 items). Family correspondence (1863-1941, 6 boxes), and general correspondence (1898-1942, 68 boxes) are arranged chronologically. The subject file (1907-1942, 139 boxes) includes materials relating to the Ballinger case, reforestation (1931), etc. Draft register, 7 pp., incomplete.

612 **Pinchot, Gifford** (1865-1946), Papers, ca. 1830-1947, 1,411 feet. Chief, U.S. Division of Forestry, Bureau of Forestry, and Forest Service, 1898-1910. Correspondence, diaries, memoranda, speeches, articles, reports, financial papers, memorabilia, photographs, subject files, bulletins, pamphlets, clippings, and other printed and manuscript materials relating primarily to Pinchot's activities in the fields of conservation and forestry and to his terms as governor of Pennsylvania. There is material on forests and forestry (40 boxes), conservation (67 boxes), flood control, and public utilities (43 boxes). Included are Pinchot's diaries (1889-1946), and his forestry journals (1889-1907). Within Pinchot's correspondence is material from Philip P. Wells' files (ca. 1890-1910), material on the Biltmore Forest in North Carolina (ca. 1892-1895), the Forest Service (ca. 1898-1914), and a Confidential File Forest Service (ca. 1906-1909). There are also correspondence and other papers of Philip P. Wells (ca. 1908-1929, 42 boxes), Forest Service attorney; Eugene S. Bruce (ca. 1900-1915, 22 boxes) and Herbert A. Smith (ca. 1917-1936, 2 boxes), both Forest Service employees; and papers of M.E. Gregg (6 boxes, ca. 1931-1936), personal secretary to Pinchot. In Pinchot's Personal Subject File, among many other forest-related topics, there is material on the American Forestry Association, American Reforestation Association, Appalachian Forest Reserve, the Ballinger case, Barber Lumber Company, Biltmore Forest, Boy Scouts, Capper Bill, National Conservation Association, Hetch Hetchy, water power, Frederick Coville, forest devastation, Douglas-Fir region, Dutch Elm disease, farm woodlands, fires, transfer of the Forest Service, Pennsylvania forests, Pisgah Forest, San Dimas Experimental Forest, Yale Forestry School, Southern forest experimental stations, Robert Marshall, Public Lands Commission, Rothrock Testimonial Fund, Ferdinand A. Silcox, Weeks Law, West Coast Lumbermen's Association. Draft register, 305 pp.

613 **Pittman, Key** (1872-1940), Papers, 1898-1951, 80 feet (191 boxes; ca. 55,000 items). U.S. Senator from Nevada, 1913-1940; served on the Senate Public Lands Committee and the Special Committee on Conservation and Wildlife. Correspondence, which comprises the bulk of the collection, relates mainly to Pittman's service in the Senate, and contains materials on U.S. Department of Agriculture, Department of the Interior, Forest Service, and wildlife; the legislative file contains material on national forests in Nevada, grazing fees, Mt. McKinley National Park, timber, wildlife, and reclamation. Draft register, 34 pp.

614 **Redington, George** (1798-1850), Papers, 1811-1907, 7 feet (28 boxes, ca. 5,000 items). Correspondence, legal documents, business records, ledgers, and photos, dating largely from 1825-1850 and relating to Redington's lumber mills and business activities in the Waddington-Ogdensburg, New York, area. Includes material dealing with the Northern Railroad and family papers.

615 **Roosevelt, Theodore** (1858-1919), Papers, 1759-1920, 276,000 items (also on 485 reels microfilm). U.S. President, 1901-1909. The collection, generally arranged chronologically within series, contains letters received, letterpress copy books, letters sent, speeches and executive orders, press releases and proclamations, articles, personal diaries, desk diaries, White House reception books, record books of letters received at the White House and forwarded to other agencies, shorthand

notes, miscellany, scrapbooks, etc. The materials cover Roosevelt's entire career, and contain many references to such topics as conservation, forest reserves, national parks, etc. The collection is described in *Index to the Theodore Roosevelt Papers,* Washington, D.C.: United States Library of Congress, Manuscript Division, 1969, 3 vols. Index entries are only to writer and recipient. See also *The Letters of Theodore Roosevelt,* edited by Elting E. Morison, Cambridge: Harvard University Press, 1951-1954, 8 vols.

616 **Schurz, Carl** (1829-1906), Papers, 1842-1932, 55 feet (214 boxes; ca. 23,110 items; also on microfilm, 121 reels). U.S. Senator from Missouri, 1869-1875; Secretary of the Interior, 1877-1881. The collection contains correspondence, speeches, articles, newspaper clippings, printed matter, and scrapbooks (chiefly 1860-1906). The general correspondence, arranged chronologically, includes 40 containers covering March 12, 1877, to March 4, 1881, Schurz's term as Secretary of the Interior. There are 6 containers of letterpress copybooks for this period; there is also Department of the Interior material in the subject file. The collection is described in: *Carl Schurz: A Register of His Papers in the Library of Congress,* Washington, D.C.: Library of Congress, 1966, 17 pp. There are other finding aids available with the collection. Part of the material has been published in *Speeches, Correspondence, and Political Papers of Carl Schurz,* selected and edited by Frederick Bancroft, New York: G.P. Putnam's Sons, 1913, 6 vols.

617 **Straus, Michael W.** (1897-1970), Papers, 1829-1965, 31 boxes. U.S. Reclamation Commissioner, 1946-53. Included are correspondence and other records relating to reclamation and conservation problems; reports and lists on the Mid-Century Conference on Resources for the Future; speeches and statements of Harold L. Ickes, Julius Krug, etc. Draft register, 7 pp.

618 **Taft, William Howard** (1857-1930), Papers, 1810-1930, 700,000 items (also on 658 reels microfilm). U.S. President, 1909-1913. The collection includes family and personal correspondence; general correspondence, including 6 boxes for the presidential period; correspondence with Theodore Roosevelt (8 boxes); executive office correspondence (1909-1910, 151 boxes); executive office correspondence (1910-1913, 215 boxes); the president's personal file (17 boxes); letterbooks (1872-1921, 222 volumes); speeches, articles, and messages; professional diaries; family diaries; several series of legal papers; scrapbooks; family financial papers; and miscellaneous materials. The correspondence is generally arranged alphabetically within series, except for the executive office correspondence and the president's personal file, arranged in case or subject files. Subject files (1909-10) include Appalachian Forest Reserve, Appalachian National Forest Association, Appalachian Park Commission, Richard Achilles Ballinger, Louis R. Glavis, National Conservation Association, Gifford Pinchot, Yellowstone National Park Notes. Subject files (1910-1913) include Alamo Forest; Alaska Coal Claims; Richard A. Ballinger; Ballinger-Pinchot Investigation; Coeur d'Alene National Forest Reserve; Columbia National Forest Reserve, elimination of certain lands; Conservation of Natural Resources; Estes National Park, Colorado; U.S. Forest Service; Glacier National Park; Louis R. Glavis; Hetch Hetchy Water Supply Grant; Weldon Brinton Heyburn—Idaho National Forests; Investigation of the Interior Department; Platt National Park; Yellowstone National Park; Yellowstone National Park Bears—museum, natural history. In the personal file are files on Walter L. Fisher and Louis R. Glavis. The collection is described in *Index to the William Howard Taft Papers,* Washington, D.C.: Library of Congress, Manuscript Division, 1972, 6 vols.

619 **U.S. Forest Service,** Collection, 1902-1936, 10 boxes (100 items). The Louis A. Barrett Papers (1904-1936, 3 boxes) contain field diaries and a few miscellaneous papers of an assistant regional forester in charge of lands and recreation, California region. Richard L.P. Bigelow Papers (1902-1936, 3 boxes) include diaries and field notes relating to the western United States; one volume is concerned with the Civilian Conservation Corps (1933). Louis Dorr Papers (1908-1932, 4 boxes) include diaries and field notes relating to the western United States. E.A. Holcomb Papers (1939-1940, 1 portfolio) contain memoirs, and correspondence with Gifford Pinchot.

620 **Walsh, Thomas James** (1859-1933), Walsh-Erickson Papers, ca. 1910-1934, 208 feet (524 boxes; 262,000 items). Papers of Thomas James Walsh, U.S. Senator from Montana, 1913-33, and John Edward Erickson (1863-1946), U.S. Senator from Montana, 1933-34. Walsh was a member of the Committee on Irrigation and Reclamation of Arid Lands; Erickson served on the Committee on Public Lands and Surveys. The papers include correspondence, speeches, legislative files, extracts, pamphlets, scrapbooks, clippings, and photos. Legislative files comprise the bulk of the collection. Subject files (1913-33) include material on national forests; grazing permits; national parks, including Yellowstone; forest reserves (1928); forestry (1926-28); and lumber. The forestry materials relate largely to Montana. Unpublished guide, 29 pp.

621 **White, Wallace Humphrey** (1877-1952), Papers, 1915-1948, 35 feet (ca. 24,000 items). U.S. Representative, 1917-1931, and Senator, 1931-1949, from Maine; served on the Senate Special Committee on the Conservation of Wildlife Resources. The papers contain correspondence, scrapbooks, printed matter, manuscripts of speeches and articles, primarily relating to White's career in the House, during which he was concerned with legislation affecting conservation. Departmental files (mostly ca. 1917-27) include correspondence with the U.S. Forest Service, the Department of the Interior, etc. The collection is described in *Wallace H. White: A Register of His Papers in the Library of Congress,* Washington, D.C.: Library of Congress, Manuscript Division, 1958, 23 pp.

622 **Wilson, Woodrow** (1856-1924), Papers, ca. 300,000 items (also on 540 reels of microfilm). U.S. President, 1913-1921. The collection includes diaries, family and general correspondence, letter books, executive office file, peace conference correspondence and documents, speeches, financial materials, scrapbooks, social, family, and other miscellaneous material. The papers are generally in chronological order within series, except for the executive office file, arranged in case or subject files. Subject file titles include Horace M. Albright, American Mining Congress, Conservation, Federal Power Commission, forest fires, Hetch Hetchy Bill, Interior Department, Robert Underwood Johnson, Loyal Legion of Loggers and Lumbermen, Mount Desert Island, National Game Sanctuaries, Yellowstone National Park—Troops. The collection is described in *Index to the Woodrow Wilson Papers,* Washington, D.C.: Library of Congress, Manuscript Division, 1973, 3 vols. See also: Ray Stannard Baker, *Woodrow Wilson: Life and Letters,* Garden City: Doubleday, Page & Co., 1927-1939, 8 vols.; *The Papers of Woodrow Wilson,* edited by Arthur S. Link, Princeton, N.J.: Princeton University Press, 1966+, 14 vols.+; and *The Public Papers of Woodrow Wilson,* edited by Ray S. Baker and William E. Dodd, New York: Harper and Bros., 1925-26, 6 vols., which print some of Wilson's papers.

623 **Wright, Elizur** (1804-1885), Papers, 1817-1895, 4,000 items. Reformer, actuary, and publisher. Includes letters to

Wright's daughter, Ellen Martha Wright (ca. 60 items), mainly from the Metropolitan Park Commission relating to parklands for Boston, Massachusetts, on Middlesex Fells and to her father's interest in conservation. Other correspondents include Warren Higley of the American Forestry Congress.

Library of Congress, Music Division

Reference Department 20540

623a **Archive of Folk Song.** Included are tapes and discs (ca. 16 items, 1937-1961) of lumberjack songs, ballads, dance tunes, and monologues recorded in California, Maine, Michigan, Nova Scotia, Pennsylvania, Vermont, Wisconsin, the National Folk Festival, Chicago (1937), and other places. There are also copies (51 reels) of recordings by the Northeast Archives of Folklore and Oral History of the University of Maine. The collection is supplemented by printed volumes and collections of lumberjack songs. Some of the recorded items are available for sale on Library of Congress Folk Recording L56, Songs of the Michigan Lumberjacks.

National Archives

Central Reference Division 20408 Telephone: (202) 963-6411. Holdings are described in: National Archives and Records Service, *Guide to the National Archives of the United States,* Washington, D.C.: Government Printing Office, 1974. However, the National Archives, *Guide to the Records of the National Archives,* Washington, D.C.: Government Printing Office, 1948, often provides more thorough subject information on those records which it includes. Complete citations to the published finding aids cited in the entries below are listed in the bibliography of the present work. Those currently in print may be obtained upon request to the Publications Sales Branch, National Archives, Washington, D.C. 20408. Microcopies available for purchase are described in *Catalog of National Archives Microfilm Publications,* Washington, D.C.: National Archives Trust Fund Board, National Archives and Records Service, 1974. Federal Records Centers have microfilm collections of many series of records held in the National Archives. Researchers may find it worthwhile to consult the guides to microfilm holdings cited in the entries for the various centers.

Aeronautics Bureau (Record Group 72)

624 *Records of predecessors, 1911-1925, 559 feet.* Among the aeronautical records of the Bureau of Construction and Repair are records of the Office of Spruce Production for the Navy, Boston (1918-19, 10 feet). P.I. No. 26.

Agricultural and Industrial Chemistry Bureau (Record Group 97)

625 *Records of the Bureau of Chemistry and Soils, 1901-1942, 574 feet.* General records include correspondence (1935-39, 36 feet), relating to all phases of Bureau activity, including information concerning chemical and technological research on naval stores, pitch, and turpentine. General correspondence of the Chemical and Technological Research Branch (1927-35) includes materials on naval stores. Records of the Naval Stores Research Division, mainly records of predecessor units, include materials concerning the development of methods for improving the production of rosin, turpentine, and other chemical products made from pine wood; there are correspondence of the Industrial Farm Products Division relating to naval stores research (1903-35, 6 feet), records relating mainly to the work of G.P. Shingler in demonstrating methods for improving production, quality, and reducing costs of naval stores (1914-36, 5 feet), records relating to F.P. Veitch's work on the Committee on Naval Stores of the American Society for Testing Materials (1915-39, 4 feet), manuscripts by staff members (1928-40, 3 feet), and records relating to naval stores patent applications (1934-41, 11 inches). Records relating to Trail smelter fumes investigations include records of an investigation of smelter fumes by Canadian National Forest Council for the Department of External Affairs (1929-30, 2 inches), partly relating to fume damage to forests. P.I. No. 149.

Agriculture, Office of the Secretary (Record Group 16)

626 *Records of the Commissioner and the Secretary of Agriculture, 1879-1964, 5,006 feet.* Chiefly subject files of the Office of the Secretary (1906-64), and letters sent by the Secretary (1893-1941). Letters sent are partly available on National Archives Microfilm Publication M440, 563 rolls, of which 2 rolls contain copies of letters prepared for the signature of the Secretary by the Office of Forest Appeals (1913-21); and 77 rolls contain copies of routine public letters prepared by the Forest Service (1906-1929). Subject files contain many references to such topics as naval stores, including naval stores under the Agricultural Adjustment Act; ashes; Civilian Conservation Corps camps; cedar; charcoal; chestnut blight; forest reserves (1906-07); the U.S. Forest Service; cooperative farm forestry act; such aspects of national forests as antiquities, permits, homesites, and trespass; the Norris-Doxey Act; shelter belt; International Forest and Lumber Exposition (1930); International Wood Conference (1932); Plumas Forest Reserve Investigation (1906-07); International Congress on Forestry (1937); Jackson Hole (Wyoming); lumber and timber; National Recovery Administration Lumber Code (1933-35); National Recovery Administration Paper and Pulp Code; wood and paper pulp; reforestation; sawdust; Timber Conservation Board (1931-37); timber and stone entry; turpentine; white pine blister rust; willows; wood distillation; wood preservation; World's Forestry Congress (1925-27). Folder titles are indexed in *Subject-Numeric Headings of the Correspondence Files of the Office of the Secretary of Agriculture (Record Group 16),* Washington: 1962, 157 pp. Also see pamphlet accompanying Microcopy No. 440.

627 *Records of the Office of the Solicitor, 1891-1945, 1,700 feet.* The bulk of the Solicitor's records are case files relating to the violation or alleged violation of the various acts enforced by the department. The acts and the dates for which there are closed case files include the National Forest Acts of 1891, 1897, and 1905 (1910-1936, 58 feet); and the Weeks Act (1911-1937, 68 feet). There are also records relating to naval stores legislation (1914-1926); correspondence and reports concerning proposed forest legislation, and minutes of the National Forest Reservation Commission (1911-1937, 17 feet); and records concerning Forest Service contracts and litigation (1909-1910, 17 inches). Draft checklist (1945, 11 pp.).

628 *Records of the Office of Civilian Conservation Corps Activities, 1933-1942, 24 feet.* Correspondence, memoranda, reports, and other materials arranged under the following subject headings: claims, communications, cooperation, disbursement, information, inspection, organization, personnel, publications, safety, supervision, and supply.

629 *Records of other units, 1862-1948, 433 feet.* Records of the Chief Clerk and various successor units, including deeds, indentures, transcripts, abstracts, and related papers covering title to lands acquired by the Department of Agriculture (1912-1943, 150 feet). Most of them cover Forest Service acquisitions.

Alien Property Office (Record Group 131)

630 *Records relating to activities arising from World War I, 1917-1957, 1,578 feet.* The records of enemy-owned businesses seized by the Alien Property Custodian include part of the records of the German-American Lumber Company (1890-1921, 48 feet), consisting chiefly of correspondence, minute books, contracts, and lumber tract records. Typed checklist (1945, 18 pp.).

Army Air Forces (Record Group 18)

631 *Records of the Spruce Production Division, 1917-1946, 239 feet.* The records consist of issuances, reports, general correspondence, messages, rosters, and other documents relating to functions mainly during World War I. There are also records of the Spruce Production Section of the Bureau of Aircraft Production, districts, and units, and of the U.S. Spruce Production Corporation. Draft P.I. (1967, 11 pp.).

632 *Audiovisual Records, 1901-1964, 452,414 items.* Includes photographs of national parks, and logging and other activities of the Spruce Production Corporation (1917-1922).

Boundary and Claims Commissions and Arbitrations (Record Group 76)

633 *Records relating to international claims.* Included are records of the Trail Smelter Arbitration between the United States and Canada under the Convention of April 15, 1935, relating to fume damage to forest trees and timber in Washington state caused by a smelter at Trail, British Columbia. There is a copy of the U.S. statement, with appendixes (1936, 2½ feet); hearings (1937-40, 56 vols., with gaps); reports on scientific investigations (1926-40, 2 feet); maps and charts (ca. 1929-30, 3 inches); photographs (ca. 1930, 1 inch); printed documents (ca. 1930-41, 4 inches); solicitor's memoranda (ca. 1928, 2 inches); report of distribution of Trail Smelter award (n.d., 2 inches). Described in *Preliminary Inventory of Records Relating to International Claims,* compiled by George S. Ulibarri, Washington: National Archives and Records Service, General Services Administration, 1974, 73 pp.

Census Bureau (Record Group 29)

634 *Administration records of the Census Office, 1820-1905, 35 feet.* Records of the tenth census relating to personnel include a list of special agents and experts (1879-1881, 1 vol.) who prepared reports on such subjects as the lumber industry.

635 *Administrative records of the Bureau of the Census, 1882-1965, 573 feet.* Records of the Assistant Director for Economic Fields include, among records of the Industry Division, scrapbooks relating to the lumber and timber industry (1908-11, 1918-45, 4 vols.), with copies of schedule forms, preliminary reports, press releases, and newspaper clippings.

636 *Census schedules and supplementary records, 1,275 feet and 27,988 rolls of microfilm.* Includes census schedules for the census of industry (1840, 5 feet), with statistical tables including data for geographical areas on products of the forest.

637 *Cartographic and audiovisual records, 1850-1960, ca. 57,514 items.* The published record set of decennial census maps includes items showing the forest cover and distribution of major forest trees in the United States, Canada, and Alaska (1880); the atlas summarizing the results of the 1937 census of manufactures contains maps showing, by county, the number and location of manufacturing establishments, numbers of wage earners, and the value added by such manufactures for the forest products industries. P.I. No. 103.

Civil and Defense Mobilization Office (Record Group 304)

638 *Records of the National Security Resources Board, 1947-1953, 437 feet.* Records of the Mobilization Planning Staff, Materials Office, include the security-classified general office subject file of the Forest Products Division (1948-50, 3 feet), and the security-classified office subject file of John D. Mylrea and Mathias Niewenhous, directors of the Forest Products Division (1947-51, 2 feet). Draft P.I. (1966, 49 pp.).

Civilian Conservation Corps (Record Group 35)

639 *General records of the Civilian Conservation Corps, 1933-1943,* 432 feet and 193 rolls of microfilm. Central correspondence files of the Directors' Office relate to personnel, location of camps, cooperation with schools and colleges, cooperation with federal government agencies, annual reports, funds, supplies and equipment, and regulations. Other files pertain to related matters and to surplus property, minutes of the Advisory Council to the Director (consisting of representatives of the Departments of the Interior, Agriculture, War, and Labor), work progress reports, camp and personnel reports, accident summary reports, resolutions of the Advisory Committee on Allotments, camp directories, organization charts, finance forms, blueprints, etc. A reference data file (1933-42, 2 feet [18 vols.]) consisting chiefly of copies of correspondence, memoranda, forms, and other papers, was compiled by John D. Guthrie and C.H. Tracy, U.S. Forest Service officials, during the liquidation of the C.C.C., and relates to many aspects of the organization and work of the corps, forming a basic tool for research in these records. P.I. No. 11.

640 *Divisional records, 1933-1943, 457 feet.* Correspondence, reports, and other records of the Investigations, Safety, Planning and Public Relations, Research and Statistics, Selection, and Automotive and Priorities Divisions of the Civilian Conservation Corps relate to the recruitment of personnel and to the maintenance and operation of C.C.C. camps and equipment. P.I. No. 11.

641 *Records of the C.C.C. Liquidation Unit, 1942-1948, 34 feet.* Reports, regulations, correspondence concerning terminal operations of the C.C.C. and National Youth Administration and the disposition of their property.

642 *Audiovisual records, 1933-1943, 10,852 items.* Photographs of the office of the Director of the C.C.C. and camp facilities.

Farmers Home Administration (Record Group 96)

643 *Central Office records, 1931-1959, 920 feet.* General correspondence maintained in the Washington office of the Administrator, Farm Security Administration (1935-38, 34 feet) has considerable material relating to the conversion of submarginal land to forests and wildlife preserves. P.I. No. 118.

644 *Field Office records, 1934-1947, 1,629 feet.* The correspondence files of Region 1 (1935-37, 110 feet), with headquarters at New Haven, Connecticut, relate to all phases of the work in New England, the Middle Atlantic states, and Maryland and Delaware, including the conversion of submarginal lands to forests and wildlife preserves. Records of Region 4 Resettlement Administration, include correspondence and other records concerning the Shenandoah Homesteads (1934-37, 2 feet), an agricultural community project for the resettlement of 250 destitute and low-income families from an area later included in Shenandoah National Park. P.I. No. 118.

Federal Extension Service (Record Group 33)

645 *Correspondence, 1906-1949, 879 feet.* General corres-

pondence of the Extension Service and its predecessors (1907-43, 402 feet) concerns all phases of extension work, including forestry. There are author and subject card indexes. General correspondence (1943-46, 57 feet) is similar in content, but classified according to a subject numeric scheme. Material on forestry relates to farm forestry and farm wood lots, forest fire protection, lumbering, and marketing. There is material on naval stores. P.I. No. 83.

646 *Audiovisual records, 1906-1952, 5,565 items.* Includes motion pictures of U.S. Forest Service activities, Civilian Conservation Corps work, national parks, etc.

Federal Trade Commission (Record Group 122)

647 *Records of the Bureau of Corporations, 1903-1914, 543 feet.* A numeric file (1903-14, 484 feet), with reports and correspondence pertaining to special investigations of particular industries and corporations, includes records relating to an investigation of the lumber industry (1908-1913, 10 feet). There is a card index to the investigations (1903-14, 35 feet). P.I. No. 7.

Foreign Agricultural Service (Record Group 166)

648 *Records, 1901-1954, 1,028 feet.* Forestry and forest products reports (1901-1941, 15 feet), evidently maintained by the Forest Service prior to 1935, and mostly for the period 1920-41, pertain to all phases of forestry in foreign countries, including planting, lumbering, protection, legislation, silviculture, and forest products.

Foreign and Domestic Commerce Bureau (Record Group 151)

649 *Records of the Bureau, 1913-1958, 1,526 feet.* Included are records of annual forest industry conferences (1920-33, 1 foot), held to discuss matters of interest to the lumber industry and lumber users. Draft P.I. (1963, 11 pp.).

650 *Records of the National Committee on Wood Utilization and of the Timber Conservation Board, 1925-1933, 22 feet.* The records of the National Committee on Wood Utilization, with indexes, relate to the administration and operations of the Committee, the organization and financing of the lumber industry, and the processing and distribution of lumber products. The records of the Timber Conservation Board consist of minutes of the Board and its committees; a general file with information about the administration and work of the Board, world timber supply, laws relating to forest conservation, and conservation methods used in the lumber industry; outgoing correspondence kept by the Board's secretary; and records relating to the Board's research projects, including correspondence, tabulations, and statistical data. Typed checklist (1946, 4 pp.).

Forest Service (Record Group 95)

651 *General records, 1882-1958, 268 feet.* Records of the Office of the Chief and other general records include letters received (1888-99, 13 feet) and letters sent (1886-99, 5 feet [52 vols.]) by the Division of Forestry; general correspondence (1898-1908, 22 feet); records of the Office of the Chief (1908-47) relating to administration, supervision, planning, and general activities; letters sent by the Office of the Associate Forester (1905-08, 1 foot [7 vols.]) concerning general administrative matters; correspondence (1906-08, 10 feet) and reports (1906-08, 14 feet) of the Section of Inspection; minutes of the Service Committee (1903-35, 3 feet) relating to the formulation of administrative policies for the entire service; correspondence of the Office of Law (1905-09, 16 feet); circulars and orders from the Office of the Chief (1903-07, 3 inches [2 vols.]); field office records of forest supervisors (1898-1904, 3 feet [27 vols.]) in Montana, Colorado, and New Mexico; correspondence of the Field Office Title Attorney, Frederick W. Goshen (1913-16, 2 feet); appointment and authorization records (1901-14, 1 foot); executive orders (1908-13, 2 feet); copies of letters sent by the Commissioner of the General Land Office (1903, 1 vol.), relating to forest administration; accounts and vouchers of the General Land Office (1898-1903, 2 feet [10 vols.]) concerning expenses, wages, leave, and days employed of forest supervisors and rangers; records of the Northeastern Timber Salvage Administration (1938-41, 25 feet) regarding cooperation with other government and private agencies in the salvage, conservation, and marketing of the timber felled by the New England Hurricane of 1938; records of the Emergency Rubber Project (1942-46, 36 feet); records relating to the Ballinger-Pinchot controversy (1904-10, 5 feet), from the Office of the Forester and the Region 6 Office in Portland, Oregon; selected records relating to the administration of Gifford Pinchot (1905-10, 2 feet), used in the preparation of *Breaking New Ground;* reports and clippings relating to the first convention of the American Forestry Congress (1882, 1 vol.) in Cincinnati, Ohio; questionnaires, correspondence, and memoranda relating to the work of W.T. Cox and R.S. Kellogg for the National Conservation Commission (1908-09, 2 feet); records relating to the National Conservation Congress (1909, 1 foot); records relating to the National Conservation Exposition (1912-13, 1 foot); records relating to the National Forest Reservation Commission (1911-58, 27 feet); records maintained by O.W. Price and J.B. Adams relating to the Keep Commission on Department Methods (1904-08, 5 feet); records of Henry S. Graves (1911, 2 inches) and Earl W. Loveridge (1913-54, 10 inches); General Integrating Inspection Reports (1937-55, 4 feet [27 vols.]); district allotment and appropriation estimates (1916-21, 2 feet) by supervisors of each national forest; annual reports on federal aid for national forest roads and trails (1926-33, 1 foot) under the Federal Highway Act of 1921; land use planning reports (1934, 2 feet [26 vols.]) prepared by the Forest Service for the Lands Committee of the National Resources Board; papers relating to the history of forestry in the United States (1882-1909, 6 inches); and miscellaneous items. There are also official diaries of the work of district rangers, assistant district rangers, nursery superintendents, and guards, selected to exemplify normal routines and procedures (1906-44, 11 feet). P.I. No. 18.

652 *Records relating to administrative management and information, 1900-1951, 302 feet.* Records of the Division of Operation include general correspondence (1901-39, 92 feet); correspondence of the Office of Organization (1905-09, 10 feet); records of the Office of Accounts (1900-19, 7 feet); records relating to supervision of the district offices by the Washington Office (1908-10, 17 feet); fire reports, monthly reports, and correspondence of forest supervisors to the Office of Forest Reserves (1905-07, 3 feet); records relating to the 10th and 20th Engineer Regiments, U.S. Army (1917-18, 24 feet); reports, studies, and related records of the Division of Operation (1936-44, 23 feet) on fiscal matters and record keeping practices, communications, etc.; and miscellaneous records (6 feet). Records of the Division of Fiscal Control (1927-1951, 12 feet) include general correspondence, chiefly with other offices in the Forest Service. Records of the Division of Information and Education include general correspondence (1907-41, 51 feet); correspondence of the Office of Editor (1908-1917, 12 feet) concerning publications; news articles and press releases (1900-33, 1937-46, 25 feet); publications of

national forest regions (1910-41, 15 feet); correspondence, scripts, photographs, and memoranda relating to the production of motion picture films to promote Forest Service programs (1920-49, 1 foot); radio scripts (1932-44, 3 feet [17 vols.]); correspondence relating to the participation of the U.S. Division of Forestry in the World's Columbian Exposition (1892-93, 3 inches); and correspondence relating to forestry exhibits in the Cotton States and International Exposition (1895-96, 1 foot). P.I. No. 18.

653 *Records relating to the administration of national forest resources, 1896-1952, 591 feet.* Records of the Division of Engineering include correspondence and related records (1932-52, 95 feet); correspondence relating to the Forest Atlas (1906-08, 9 inches); correspondence of the Office of Geography relating to land alienation, classification, boundaries, and other matters (1906-09, 3 feet); records of Regional Office No. 7 relating to forest engineering work in the eastern part of the United States (1920-35, 19 feet). Records of the Division of Fire Control include general correspondence (1909-37, 51 feet) and statistical data (1935-41, 2 feet) relating to fire costs, burned areas, and the like. Records of the Division of Range Management (1905-52, 192 feet) include general correspondence relating to regulation of grazing on national forest ranges, general protection and development of range resources, cooperative projects, regional inspections, legislation, and administration. Records of the Division of Watershed Management (1939-50, 22 feet) pertain chiefly to the Emergency Rubber Project. Records of the Division of Timber Management include general correspondence relating to insect control, disease control, planting, timber sales, legislation, forest supervision, general administration, etc. (1905-52, 132 feet); correspondence of the office of Forest Extension concerning planting stations and nurseries (1899-1908, 15 feet); correspondence of the Office of Forest Management with field agents (1901-09, 4 feet); correspondence of the Office of Federal Cooperation relating to work with the Indian Office of the Department of the Interior (1908-11, 6 feet); records of the Office of State and Private Cooperation relating to the examination of state and private timber lands and the preparation of working plans for their management (1896-1908, 3 feet); letters sent and several ledgers relating to timber sales in forest reserves under the General Land Office of the Department of the Interior (1898-1905, 5 feet [42 vols]); records relating to timber sales on national forests (1908-37, 36 feet); records relating to timber sales in the Black Hills National Forest (1898-1912, 3 feet); and records relating to surveys of timber conditions in national forests (1908-30, 6 feet). Records of the Division of Wildlife Management (1914-1950, 42 feet) contain general correspondence relating to wildlife in national forests. P.I. No. 18.

654 *Records relating to land acquisition and administration, 1906-1965, 736 feet.* Records of the Division of Land Acquisition include general correspondence relating to almost all aspects of land matters in national forests (1901-40, 27 feet); records relating to the exchange (1922-46, 41 feet), purchase (1933-46, 178 feet), and donation (1924-46, 5 feet) of lands for national forests (1946-51, 33 feet); records relating to the condemnation of land for national forests under the Weeks Act of 1911 in Regions 7, 8, and 9 (1939-46, 14 feet); deeds to land acquired and related records (1913-50, 186 feet); deeds for land in Bedford, Botetourt, and Page Counties, Virginia (1772-1833, 1 inch); records relating to the Northern Pacific land grant suit affecting national forest lands in Wyoming, Montana, Idaho, and Washington (1906-40, 9 feet); correspondence of W.L. Hall (1910-17, 4 inches); and maps and statistical charts of states and counties relating to the wildland study (1937, 2 inches [1 vol.]). Records of the Division of Recreation and Lands include general correspondence (1906-51, 72 feet), relating to mining claims, special uses, contracts, settlement, trespass, claims, boundaries, and administration; and records relating to Homestead cases (1909-37, 4 feet) under the Forest Homestead Act of 1906. Records of the Division of Land Utilization contain records inherited from predecessor units, including correspondence relating to the acquisition of land (1935-43, 6 feet); records of dropped or cancelled exchange cases (1934-44, 30 feet); records relating to flood control, erosion control, and land purchase projects (1935-42, 9 feet); records relating to projects submitted to the Forest Service by the Soil Conservation Service for approval (1934-44, 25 feet); forms showing property taxes and other liens on property purchased by the government (1935-43, 2 feet); statistical tabulations showing the status of condemnation cases, title clearances, and projects (1934-43, 8 feet); and records relating to recommendations on land use projects (1937-43, 7 feet). Records of the Division of Lands (1907-48, 35 feet) include correspondence and other records inherited from predecessor units concerning land-use planning, boundary adjustments, public domain lands, natural resources, parks, cooperative projects, the Public Lands Commission, and the Committee on Conservation and Administration of the Public Domain. Records of the Division of Land Adjustments (1935-1965, 45 feet) include deeds, judgments, and condemnations pertaining to land acquired by the United States for national forest purposes under the Soil Conservation Service, the Resettlement Administration, and the Forest Service. P.I. No. 18.

655 *Records relating to state and private forestry cooperation, 1913-1951, 56 feet.* Records of the Division of White Pine Blister Rust Control contain reports of the Office of Blister Rust Control of the Bureau of Entomology and Plant Quarantine (1916-38, 2 feet), relating largely to blister rust in foreign countries, but including reports for some states and regions of the United States. Records of the State and Private Forestry Divisions include records from the central files of the Divisions of Private Forestry, State Forestry, and State Cooperation, relating to state and private cooperative activities under the Weeks Law of 1911 and the Clarke-McNary Act of 1924 (1913-44, 12 feet); and compilations of state laws on forestry (1923-49, 8 feet). Records of the Division of Cooperative Forest Management relate to forest conservation projects (1935-51, 12 feet). Records of the Division of Cooperative Forest Protection include records relating to federal legislation for cooperative forest fire protection, largely communications between the division and regional foresters, many pertaining to the Weeks Law of 1911 (1915-45, 3 feet); annual statistical fire reports relating to needed expansion of cooperative fire protection under the Clarke-McNary Act of 1924 (1925-48, 7 feet); state budget allocations for fire protection (1926-49, 5 feet); records relating to expenditures and appropriations on state and federal cooperative forest protection programs (1930-45, 6 feet); correspondence relating to the Facility Security Program of the Office of Civil Defense (1942-46, 1 foot); and articles, studies, and speeches by Austin Cary relating to forestry (1918-27, 4 inches). P.I. No. 18.

656 *Records relating to research activities, 1890-1954, 493 feet.* Records of the Forest Research Divisions include correspondence of the Office of Silvics (1907-13, 5 feet), relating to requests for information and publications and to various types of forest investigations; memoranda of the Office of Silvics (1907-12, 3 inches); correspondence, relating to cut-over lands in national forests and reforestation (1907, 3 inches), the use of basket willows in reforestation (1907-17, 3 feet), eucalyptus experiments in Florida (1910, 3 inches), and volunteer pheno-

logical observations on forest trees (1911-13, 1 foot); correspondence of the Office of Measurements (1902-10, 1 foot), and of the Office of Forest Statistics (1909, 6 inches); research compilation file of important silvical information (1897-1935, 205 feet), with index (3 feet); records of the central files of the forest research divisions (1930-49, 34 feet) relating to all research activities of the Forest Service; reports of research investigations (1912-33, 8 feet); correspondence of the Office of Products (1912-23, 6 feet); reports and correspondence assembled in preparation of the Copeland Report on public forest policy and federal and state cooperation (1923-33, 6 feet); records of the forest taxation inquiry under the Clarke-McNary Act of 1924 (1926-37, 66 feet); clippings (ca. 1890-1940, 1 foot); records relating to rain-producing experiments under the general supervision of the Forest Service (1892, 2 inches); records relating to taxation study of Wisconsin timberlands (1910, 1 foot); and miscellaneous records (1901-26, 2 feet). Records inherited from predecessor units by the Division of Forest Management Research include correspondence relating to dendrology (1910-54, 4 feet), research programs of experiment stations and of the Washington Office (1930-52, 3 feet), forest genetics, mensuration, and naval stores (1931-52, 2 feet), and plant regeneration (1931-51, 2 feet); also records relating to silviculture (1940-54, 1 foot) and to special studies and reports concerning the effects of diseases, insects, and animals on trees and plants (1930-52, 1 foot); volume tables indicating mensurate quantities, such as age, spacing, height, diameter, and number of trees per acre (ca. 1901-40, 5 feet); and a card file (1 foot) relating to topography, soils, climate, and trees. Records of the Division of Range Management Research (1909-54, 18 feet) contain general correspondence and reports inherited from predecessor units relating chiefly to plant identification, but also to research programs, artificial revegetation, and administration. Records of the Division of Watershed Management Research (1925-51, 2 feet) include reports and studies, inherited from the Branch of Research and the Division of Forest Influences, relating to research on climatic problems, soil stabilization, erosion control, and flood control. Records of the Division of Forest Insect Research, inherited from the Bureau of Entomology and Plant Quarantine of the Department of Agriculture, include annual progress reports relating to entomological studies conducted by Forest Insect Field Stations (1918-22, 1 foot); published studies (1946-53, 7 feet); survey reports and related correspondence between the experiment stations and other units of the department (1928-57, 7 feet); reports of the chief of the division and internal reports (1912-53, 16 feet); studies concerning the dendroctonus beetle (1915-52, 3 feet); and miscellaneous records (1909-53, 8 feet). Records of the Forest Products Laboratory, Madison, Wisconsin, include general records (1907-22, 6 feet) and reports and studies (1917-35, 1937-49, 55 feet) relating to forest products research. P.I. No. 18.

657 *Records relating to Civilian Conservation Corps work, 1933-1942, 371 feet.* General correspondence (1933-42, 162 feet) of the office of Fred Morrell, Assistant Chief of the Forest Service and supervisor of Civilian Conservation Corps work; records of representative C.C.C. camps under Forest Service supervision (1933-42, 207 feet); and correspondence of regional office No. 10 (1937-42, 1 foot), relating mainly to projects for the restoration of primitive totems and houses, and work at the Annette Island Air Field in Alaska. P.I. No. 18.

658 *Cartographic and audiovisual records, 1890-1962, 19,400 items.* Cartographic records (16,032 items) include topographic, planimetric, and special maps of the Division of Engineering, including published atlases relating to individual national forests (1908-25); project files containing manuscript, photoprocessed, and published maps, graphs, and charts used as illustrations (1910-59); large-scale standard published maps of national forests, game refuges, and other reservations, (1911-60); maps relating to fire control activities and road systems (1931-62); and aerial survey film negatives of major forest areas (1934-38). Map files of other Forest Service Divisions include base maps illustrating lumber production or consumption (1909-35); forest planning (1910-52); tree species, timber surveys, and timber sale areas in national forests (1913-40); grazing districts, rangelands, land use, shelterbelt project areas, and range management plans (1915-45); recreational facilities in national forests (1917-30); and county range surveys (1935-38). Audiovisual records (3,368 items) include photographs, posters, and prints of posters illustrating Forest Service activities (1898-1941); photographs of forest reserves in Arizona, California, Colorado, Idaho, Montana, Oregon, and South Dakota (1898-1900); illustrations of war-related activities of the Forest Products Laboratory (1917-18); pictures of forest cover, streams, lookouts, fires, reforestation, and trails (1933-39); and photographs illustrating the Emergency Rubber Project (1942-45). P.I. No. 167; P.I. No. 18.

Geological Survey (Record Group 57)

659 *Records of the Geological Division, 1867-1951, 269 feet.* Records of Arnold Hague include files relating to proposed legislation for forest reserves (1890-97); and to his interest in Yellowstone National Park (1869-1912).

660 *Records of the Topographic Division, 1880-1948, 75 feet.* Reports concerning U.S. Forest Service mapping projects (1834-38).

661 *Cartographic and audiovisual records, 1868-1971, 77,940 pieces.* Cartographic records include maps of national parks and forests.

Gift Collection, National Archives (Record Group 200)

662 *Personal papers and historical manuscripts, 1837-1960, 459 feet.* Private papers (1880-1909, 33 feet) of Ethan Allen Hitchcock, U.S. Secretary of the Interior, 1898-1907, contain correspondence of Theodore Roosevelt (1900-1907); correspondence relating to land frauds in Oregon (1907, 2 folders) and California (1901-04, 1 folder); letters from U.S. Senator Moses Edwin Clapp relating to White Earth timber sales, Minnesota (1905, 1 folder); and correspondence of Francis J. Heney, U.S. Senator John H. Mitchell, and others relating to land frauds (1903-1908, 3+ folders); correspondence of W.B. Parsons relating to timber frauds in Montana (1895-1903, 1 folder); also letters and memoranda regarding the San Francisco Mountains Forest Reserve, Arizona (ca. 1900, 1 envelope). (Typed box list). Personal papers (ca. 1905-1970, 7 feet) of Earle H. Clapp consist mainly of miscellaneous speeches, notes, and memoranda relating to the Forest Conservation Society of America, the Forest Research Institute, the American Forestry Association, U.S. and world forestry programs, especially the Copeland Report and the National Recovery Administration lumber code; records closed until July 11, 1987, include an autobiography, materials relating to the proposed transfer of the Forest Service to the Interior Department under Interior Secretary Harold Ickes, and a reprimand of Clapp by the President of the United States. Typed list of contents.

House of Representatives of the United States (Record Group 233)

663 *Committee records, 1789-1962, 11,415 feet.* For the Public Lands Committee there are papers accompanying bills and resolutions referred to the committee, petitions referred to the committee, and other records, 9th to 79th Congresses (1805-1947). There are similar records of the Committee on Agriculture, 16th to 80th Congresses (1819-1949). Restricted. P.I. No. 113.

Housing Expediter Office (Record Group 252)

664 *Central Office records, 1941-1953, 1,061 feet.* Included are minutes, reports, correspondence, and related records of the Western Log and Lumber Administration inherited by the Office of the Housing Expediter from the Civilian Production Administration (1942-47).

Indian Affairs Bureau (Record Group 75)

665 *General correspondence and other records of the Bureau, 1801-1939, 14,007 feet.* Letters received (1881-1907) contain letters from bureau employees relating to timber rights. Special case files (1821-1907) contain letters removed from general incoming correspondence and grouped by subject; they concern such matters as timber trespasses, logging, etc. Incoming and outgoing correspondence (1907-1939) is filed according to a decimal-subject classification system; subjects include forest reserves, forests—logging and lumber operations, forest fires, Emergency Conservation Work (Civilian Conservation Corps), forestry, nurseries. P.I. No. 163.

666 *Records of the Land Division, 1797-1967, 534 feet.* General records include material relating to timber operations; surveying and allotting records show value of timber on the White Earth Reservation in Minnesota (1907-33, ¾ inch), with individual allotments to Chippewa Indians; records relating to land sales and leases include appraisal of land and timber on surplus lands of the Yakima Reservation in Washington, compiled by the Yakima Appraising Commission (1910, 4 vols.); records relating to Choctaw and Chickasaw Segregated Coal and Asphalt Lands include appraisal of timber on Choctaw and Chippewa land in Indian Territory (1913-14, 8 inches); records relating to leases include timber contracts of the La Pointe Agency, between individual Chippewa Indians and contractors (1883-89, 5 vols.), and renewals (1888-89, 1 vol.); other records include registers of letters relating to bonds required for timber sales and approved by the Secretary of the Interior (1904-18, 2 vols.); miscellaneous records and reference materials include copies of records of the Bureau, the General Land Office, and of Klamath and Lake Counties, Oregon, relating to a dispute concerning an exchange of land between the United States and the California and Oregon Land Company (1906-27, 2 inches). P.I. No. 163.

667 *Records of the Forestry Division, 1908-44, 6 feet.* Records of predecessor units include records concerning sale of Choctaw timber lands (1910-14, 2 inches); reports of inspection and appraisal of timber lands on the Flathead Reservation (1908-09, 2 feet [24 vols.]); field notebooks relating to an examination by topographers and timber cruisers of the Red Lake Reservation in Minnesota under the direction of the U.S. Forest Service with the cooperation of the Bureau (1909-10, 1 foot [17 vols.]); schedules of timber appraisement of the Colville Reservation (1914, 1 foot); maps and plats (ca. 1914-26, 5 vols.) and bound papers (10 inches) showing location and amounts of timber on the Flathead, Klamath, Menominee, and Spokane Reservations; section plats of the Spokane Reservation showing timber and other lands (1910, 4 inches [8 vols.]). P.I. No. 163.

668 *Records of the Alaska Division, 1877-1940, 161 feet.* Includes correspondence concerning the so-called Metlakahtla controversy, which in part related to the management of a sawmill (1897-1931, 4 feet). P.I. No. 163.

669 *Records of the Finance Division, 1817-1949, 501 feet.* Records concerning tribal and individual moneys and payments to Indians include annuity rolls for timber sales (1841-1949). P.I. No. 163.

670 *Records of the Inspection Division, 1873-1948, 148 feet.* Inspection reports (1908-1940, 42 feet) include reports relating to timber and mill operations. P.I. No. 163.

671 *Records of the Statistics Division, 1885-1948, 604 feet.* Superintendents' annual narrative reports (1910-38, 45 feet) on agencies and other field jurisdictions include material on forestry. The accompanying statistical reports (1920-35, 80 feet) also include a section on forestry. P.I. No. 163.

672 *Records of the Civilian Conservation Corps—Indian Division/ 1933-1944/ 98 feet.* Include general records (1933-44, 87 feet) and special files (1933-36, 10 inches), with correspondence, memoranda, reports, photographs, and other records concerning beetle control, blister rust control, leader camps, health, and forestry. P.I. No. 163.

673 *Field Office records, 1794-1956, 5,698 feet.* This subgroup, primarily records of discontinued field offices, includes records of the Consolidated Chippewa Agency, with correspondence of Charles D. Wilkinson, forest guard at Beaulieu and White Earth (1919-29, 3 inches); of John D. Morrison, field clerk at White Earth (1930-35, 1 foot), partly regarding timber matters; and of Clyde W. Flinn, field clerk at Mahnomen on the White Earth Reservation (1923-44, 2 inches), including work of the Civilian Conservation Corps; also letters received by Frank Fisher, farmer in charge at Leech Lake (1922-30, 2 feet), including letters relating to timber. Records of the Leech Lake Agency include letters sent concerning logging on Indian lands (1900-1901, 5 inches [4 vols.]); copies of letters sent to the Commissioner of Indian Affairs, Commissioner of the General Land Office, Auditor of the Interior Department, and the Forester (1904-14, 1 foot [10 vols.]), relating to logging operations, receipt and deposit of payments for timber, and use of money in accounts of various Indians; also letters sent concerning timber and allotments (1904-14, 3 feet [15 vols.]); miscellaneous letters sent (1899-1914, 6 feet [59 vols.]), which partly concern logging; and contracts for the sale of timber on Indian Allotments (1900-17, 7 inches). Records of the Nett Lake Agency contain data books (1908-14, 2 inches [2 vols.]) with a record of labor on forest fires by the Chippewa Indians of the Bois Fort Reservation. Records of the White Earth Agency, successor to the Chippewa Agency, contain correspondence (1885-1922, 19 feet), partly relating to logging; letters sent to the U.S. District Attorney at St. Paul (1907-1908, 1 inch [1 vol.]), relating chiefly to legal cases involving transfer of land and removal of timber; letters sent concerning land (1907-14, 2 feet [17 vols.]), partly relating to timber cutting; miscellaneous letters sent (ca. 1878-1914, 9 feet [90 vols.]), including letters to lumber companies, relating to logging; logging contracts (1890-1908, 3 inches); contracts (1914, 3 inches) between Indian allottees and the Nichols-Chisholm Lumber Company for the sale of timber in allotments; related unsigned certificates of completion of timber contracts (ca. 1914-17, 1 inch); cashbooks (1875-1907) which include one volume for timber transactions (1898-99). P.I. No. 163.

674 *Cartographic records, 1800-1944, 37,779 items.* Cartographic records maintained separately from the textual records. The central map file (1800-1939) includes maps showing timber lands. The map file of the foresty branch consists of fire control maps relating to ranger and fire guard stations, lookout stations, and reserved lands (1920-44). S.L. No. 13. Also see *Reference Information Paper No. 71: Cartographic Records in the National Archives of the United States Relating to American Indians,* compiled by Laura E. Kelsay, Washington: National Archives and Records Service, 1974, pp. 16-18.

Interior, Office of the Secretary (Record Group 48)

675 *Records of the Division of Finance, 1849-1935, 158 feet.* Records include appropriations ledgers (1853-1923, ca. 45 vols.) which contain accounts representing all funds appropriated by Congress to be expended in the Department of the Interior, and contain references to various national parks, protection of forest reserves (ca. 1898-1905), restoration of lands in forest reserves (ca. 1906-23), surveys of forest reserves (ca. 1897-1905), national monuments, National Park Service (ca. 1917-23), the accounting service of the National Park Service (ca. 1917-23), resurvey of lands in suit against the Sierra Lumber Company of California (1888-1893), timber depredations (ca. 1872-1897), protection of timber (ca. 1892-1923), surveying of timber (1878-1881). S.L. No. 18.

676 *Records of the Patents and Miscellaneous Division, 1849-1943, 392 feet.* Among miscellaneous letters received (1881-1907, 50 feet) is material relating to advertisements concerning public lands and timber; records relating to forest reserves (1891-1902, 6 inches) include letters received, reports, printed copies of proclamations, maps, petitions, lists, and printed congressional reports and documents, relating to establishment of reserves, protection, sheep and other stock grazing, settlers, surveys by Gifford Pinchot and by the U.S. Geological Survey, and other subjects; miscellaneous letters sent (1849-1906, 15 feet [88 vols.]) contain material relating to national parks and reservations. Draft P.I.

677 *Records of the Lands and Railroads Division, 1849-1907, 570 feet.* There are incoming communications concerning timber trespasses (1877-1882, ca. 8 feet); miscellaneous letters received (1849-50, 5 feet), in part relating to timber depredations; letters received from the Commissioner of the General Land Office (1849-82, 20 feet), including material on timber depredations; letters received (1881-1907, 163 feet), partly relating to depredations on the public doman, especially of timber, and to forest reserves; records relating to timber trespasses and fraudulent land entries (1877-82, 6 feet), chiefly letters received and reports, which in part concern illegal removal of timber from public and Indian lands; records relating to timber trespasses in Oregon (1890-1899, 2 inches), chiefly reports of agents relating to investigations; register of actions concerning timber trespass cases (1900-01, ¼ inch); letters, telegrams, and letters received from Inspector Caleb F. Davis (1889-93, 1 inch), in part relating to timber depredations; letters and reports received from Special Land Inspector Eugene F. Weigel (1891-92, 1 inch) including inspections of national parks and timber reservations; letters sent (1849-1904, 53 feet [307 vols.], available as microfilm publication M620, 310 rolls), containing material on establishment of forest reserves, and depredations on the public domain, especially of timber; press copies of letters sent (1854-1907, 64 feet [606 vols.]), partly relating to forest reserves, sales of timber, timber depredations; press copies of confidential letters sent to inspectors and others (1886-93, 1 vol. and 1 inch of unbound material) containing material on national parks; records concerning contest for islands in the Maumee River (1900-02, 4 inches), relating to lands claimed as forest reserve lieu selections. Typed checklist, 9 pp.; various registers and indexes; pamphlet accompanying Microcopy No. 620.

678 *Records of the Indian Division, 1849-1907, 308 feet.* Among the special files of letters received (1898-1907, 31 feet) are records concerning the timber and stone act in Indian Territory. Draft P.I.

679 *General classified files, 1907-1953, 1,982 feet.* There are correspondence, reports, and related materials concerning forests and conservation policies and programs. Subject and name index.

680 *Special files of department officials, 1918-1961, 165 feet.* Includes records of Secretaries of the Interior Hubert Work (1923-28), Harold Ickes (1933-42), Oscar L. Chapman (1933-53), and Douglas McKay (1952-56), under secretaries, assistant secretaries, and other officials. Records of Assistant Secretary C. Girard Davidson concerning the National Park Service (1946-50, 2 feet) relate to appropriations, concessions, and specific park areas. Draft P.I.

681 *Records of the Office of the Solicitor, 1906-1959, 163 feet.* Records of the Associate Solicitor (1928-48) include material relating to Hetch Hetchy Dam. Records of the Assistant Solicitor for Land Matters (1933-40, 2 feet), for the periods of service of Rufus J. Poole and David J. Speck, relate to such matters as the administration of revested land grants in Oregon, forestry, and grazing. Draft P.I.

682 *Records of the Division of Land Utilization, 1935-1956, 183 feet.* The classified files (1937-56) include correspondence and other records relating to conservation, the Oregon and California revested lands, etc. Draft P.I.

683 *Cartographic Records, 1849-1923, 616 items.* General cartographic records include maps relating to states and territories (1875-1914, 21 items), among which are General Land Office maps, dated 1897-1905, annotated to show national parks and forest reservations in Montana, Oregon, and Washington, and in parts of Idaho, South Dakota, and Wyoming. P.I. No. 81.

684 *Audiovisual records, 1862-1964, 3,000 items.* Motion pictures (1929-1962, 179 reels) include films concerning state parks, Civilian Conservation Corps work in national and state parks, forest fires and firefighting, etc. Sound recordings (1936-52, 504 items) relate to the national park system, the work of the C.C.C., etc.

Joint Committees of Congress (Record Group 128)

685 *Records, 1799-1969, 589 feet.* Records of the Joint Committee to Investigate the Interior Department and the Bureau of Forestry, 61st Congress (1909-11), include typewritten transcript of the testimony before the committee, and the signed report of the committee. The testimony was printed as S. Doc. 719, 61st Congress, 3rd session. The report was also printed, but does not carry a congressional report number.

Justice Department (Record Group 60)

686 *Letters received, 1849-1904, 2,690 feet.* "Source-chronological files" of letters (1870-84), filed by source from which received, and "year files" of letters (1884-1904) filed by number in subject or case files, relate to a variety of subjects, such as protection of the public domain, especially timber and mineral resources.

Labor Department (Record Group 174)

687 *Records, 1907-1968, 1,426 feet.* The central file of the Office of the Secretary of Labor (1913-33, 112 feet) includes material on the participation of the Labor Department in Emergency Conservation Work and Civilian Conservation Corps programs (1933-39), and material on the Loyal Legion of Loggers and Lumbermen, Portland, Oregon (1922). (Card subject index.) Records of the President's Mediation Commission (1917-18) contain two typed reports, "Labor in the Inland Empire Lumber Territory," by Willard E. Hotchkiss (ca. 1918, 24 pp.), and a report by the Lumbermen's Protective League on industrial conditions in the Pacific Northwest lumber industry (ca. 1918, 71 pp.) (Index.) Records of the United States Commission on Industrial Relations (1912-15, 7 items) include unpublished economic and social studies of the lumber industry, with emphasis on working and living conditions, prepared in 1914 by H.E. Hoagland, Frederick C. Mills, David J. Saposs, Peter Speek, and others. Draft P.I. (1964, 9 pp.); Draft checklist (1945, 31 pp.).

Labor Standards Bureau (Record Group 100)

688 *Records, 1934-1957, 877 feet and six rolls of microfilm.* Case files for registrations of labor organizations (1946-57) contain forms submitted in accordance with the terms of the Labor Management Relations (Taft-Hartley) Act of 1947, providing detailed information concerning the officials, membership, constitutions, bylaws, and policies of the unions involved, including those in the lumber, wood products, furniture, and pulp and paper products industries. Microfilm copy (6 rolls) of a card index for the years 1946-1957. Draft P.I. (1965, 7 pp.).

Labor Standards Bureau (Record Group 100)

689 *Records, 1885-1945, 286 feet.* Collective bargaining agreements filed with the bureau by employers, trade associations, and trade unions (1912-45) include agreements for such industries as lumber and wood products, furniture, and pulp and paper products. Draft P.I. (1964, 6 pp.).

Land Management Bureau (Record Group 49)

690 *General records, 1796-1938, 6,429 feet.* General correspondence (1796-1908) includes telegrams, reports, and other communications sent, pertaining to all functions of the General Land Office, including forest reserves, timber and stone claims, and timber trespasses. Letters received (1803-1908), with registers and indexes, include separate entries for timber-cutting permits, timber trespasses, timber sales, etc. There are also records concerning federal reimbursements for tax revenues lost by counties when the Oregon and California railroad land grant was revested in the United States (1916-1931); and records relating to the Kaweah Cooperative Colony in California (1934-35). See the draft inventory, especially for Divisions "P" and "FS".

691 *Records relating to public land surveys, 1796-1962, 728 feet.* Records of the General Land Office Division of Surveys (Division "E") include records relating to surveys of national parks; and homestead entry and forest exchange surveys in national forests (1910-53).

692 *Records relating to management of the public domain, 1818-1946, 40 feet.* Included are the journal and correspondence of James L. Cathcart and James Hutton, agents appointed under an act of March 1, 1817, to report on the reservation of public lands to supply timber for naval purposes between the Mermentau River, Louisiana, and the Mobile River, Alabama (1818-19, 1 vol.; available as National Archives Microfilm Publication M8, 1 reel); a reports register of special (timber) agents (1882-1903); correspondence, accounts, contracts, and other records of the General Land Office in Washington, D.C. (1897-1938, 16 feet) and of the Office of Superintendent of Logging at Cass Lake, Minnesota (1903-38, 13 feet) relating to logging on the Chippewa ceded lands in Minnesota. Shelflist; P.I. No. 22.

693 *Records relating to administration, 1813-1950, 125 feet.* Records of the Administrative "A" Division consist of the records of Division "A", containing letters relating to forestry personnel (1900-03), and records of Division "M", including accounts relating to timber depredations. Draft P.I.; shelflist.

694 *Records relating to public land disposals, 1796-1951, 32,950 feet.* Control records include Executive Orders and proclamations relating to national parks and forests, wildlife refuges, etc. (1906-49). There are records of the Forestry Division (Division "R"), relating to use of military bounty land warrants and script on the public domain (1788-1908). P.I. No. 22; draft P.I. for Division "R".

695 *Cartographic and audiovisual records, 1785-1966, 114,451 items.* Cartographic records include township plats showing naval timber reserve requirements in Alabama, Florida, Louisiana, and Mississippi; and maps and plats of forest reserves and national parks surveyed by the U.S. Surveyor. S.L. No. 19.

Mines Bureau (Record Group 70)

696 *Division and field records, 1895-1954, 382 feet and 230 rolls of microfilm.* Among the many records that are not clearly identified by administrative origin are some relating to the production of mine lumber.

697 *Cartographic and audiovisual records, 1908-1955, 576 items.* There are motion pictures about the natural resources and scenery of Arizona and Texas, and national parks, including Yellowstone, Yosemite, Grand Canyon, Rocky Mountain, and Shenandoah (1925-55).

National Park Service (Record Group 79)

698 *Records of the Office of the Secretary of the Interior relating to national parks and monuments, 1872-1916,* 38 feet. Records of the Office of the Secretary of the Interior relating to the administration of parks and monuments before the establishment of the National Park Service, including letters received (1872-1907, 37 feet) concerning establishment of parks and related legislation, appropriations, rules and regulations, surveys and boundaries, park facilities, concessions, fire control, protection, intruders, claims, investigations, adminstration, etc., arranged alphabetically by name of park. P.I. No. 166.

699 *Records of the National Park Service, 1901-1964.* General records contain central files (1907-39, 163 feet) and central classified files (1907-49, 1,571 feet), both including correspondence of the Office of the Secretary of the Interior until 1917 and thereafter of the National Park Service, memoranda, narrative and statistical reports, contracts, bonds, permits, maps, and many other types of records relating to national parks and monuments; also among the general records are processed issuances (1940-47, 9 feet), organization charts, records concerning the National Capital Park and Planning Commssion, printed materials, and clippings. Financial records relate to expenditure of appropriations, and other disbursements and allotments (ca. 1915-32, ca. 2 feet). Among records of key officials are office files of Horace

M. Albright (1927-33, 3 feet), Arno B. Cammerer (1922-40, 7 feet), Newton B. Drury (1940-51, 11 feet), and Roger W. Toll (1928-36, 5 feet). Records of the Office of the Chief Counsel include a legislative file (1932-50, 49 feet). Records of the Branch of Engineering include general records (1917-26, 8 feet), records relating to the Hetch Hetchy Project (1901-1934, 1 foot), and to the water supply of San Francisco (1902-12, 6 feet); contracts, proposals, and specifications for construction work (1920-26, 1 foot); road survey reports (1925-39, 2 feet); construction reports (1934-42, 5 inches); and monthly narrative reports of Bureau of Public Roads engineers assigned to projects in Park Service areas (1936-37, 3 inches). There are classified files of the Field Headquarters in San Francisco (1925-36, 23 feet); records of the Branch of Plans and Design (1914-41, 41 feet); forest fire reports of the Branch of Forestry (1928-49, 3 feet); and records of the Wildlife Division (9 feet), including records of the Washington Office (1934-36), of David J. Madson (1930-39), and of the Supervisor of Fish Resources (1935-39). Records of the Branch of Recreation, Land Planning, and State Cooperation include material relating to state parks (1933-47, ca. 300 feet), including parks in the Tennessee Valley Authority area, Emergency Conservation Work of the Civilian Conservation Corps, Civil Works Administration, and Work Projects Administration programs in national, state, and local park areas; also records concerning recreational demonstration areas (ca. 1934-47, ca. 114 feet), the Recreation-Area Study (ca. 1936-47, ca. 20 feet), and records of the Development Division (ca. 83 feet), concerning Civilian Conservation Corps and W.P.A. projects, including Civilian Public Service Camps. P.I. No. 166.

700 *Records of regional offices, 1865-1954, 341 feet.* Records of regional offices include records of Region I, Richmond, Virginia (ca. 1935-52, ca. 338 feet) including central classified files, records relating to the state park program, records concerning Work Projects Administration, Civilian Conservation Corps, and Public Works Administration projects, records of the regional wildlife technician, records of the regional supervisor of Fish Resources (1935-39). Records of the Branch of recreational demonstration areas. There are also records of the National Capital Region (2½ feet). P.I. No. 166.

701 *Audiovisual records, 1871-1965, 17,330 items.* Photographs and sound recordings include much material collected for its historical interest, but also pictures of engineering activities of the service in various parks (40 items), and pictures of Yellowstone, Glacier, Great Smoky Mountains, and other parks, and the "Mather Collection" of photographs of park personnel. There are motion pictures (1930-37, 15 reels) of Shenandoah and Great Smoky Mountains National Parks, parks in Georgia and Washington, and Camp Roosevelt. P.I. No. 166.

National Production Authority (Record Group 277)

702 *Records, 1950-1953, 564 feet.* Records of the Chemical, Rubber, and Forest Products Bureau consist of policy files and regulations (1942-53) of the Containers and Packaging Division, and files of the Lumber and Wood Products Division (1941-43), including correspondence relating to N.P.A. dealings with foreign governments and statistical records.

National Recovery Administration (Record Group 9)

703 *General records of the National Recovery Administration, 1933-37, 4,241 feet and 186 rolls of microfilm.* Records of the Coordinator for Industrial Cooperation include a subject file with reports on production, wages, and employment in the forest products industries (1935-57). Records maintained by the General Files Unit are arranged in two consolidated files. Consolidated files on industries governed by approved codes (1933-36, 2,200 feet) contain records dealing with the administration of N.R.A. code authorities, trade associations, trade practices, labor problems, price lists, and other materials, including records on the lumber and timber industry, the paper and pulp industry, and other wood products and related industries, including the match industry, furniture manufacturing, pulp and paper machinery, retail lumber, sawmill machinery, wood flooring, wood flour, wood turning and shaping industries, woodworking machinery, etc.; related consolidated graphic materials (1933-36, 5 feet) include charts showing organization of the Lumber Code Authority, costs of forest products materials with value added by manufacture, organization of the construction industry and its code authority, and prices, wages, exports, consumption, employment, price trends, and other aspects of the lumber and timber industry; there are also charts for the paper and pulp industry. Consolidated reference materials include detailed studies of various aspects of the U.S. economy made by the Research and Planning Division (1933-35, 11 feet), containing documents relating to lumber code prices, lumber in construction work, economic trends of the lumber industry, railroad ties, research in forest economics, various statistical reports, pulp and paper industry, and paper and pulp industry prices. Code administration studies prepared by the Research and Planning Division on select industries while the codes were in operation (1935, 2 feet) include documents on the lumber and forest products industries and the paper and pulp industry. Evidence studies prepared by the Division of Review (1935-36, 10 inches) largely contain statistical information on how industries gather raw material, manufacture, and market it; they include reports on the furniture manufacturing industry, lumber and timber products industry, and retail lumber industry. Work materials prepared by the Division of Review (1935-36, 7 feet), and consolidated typescript drafts (1935-36, 22 feet), include studies on the foreign trade of forest products and on the economic problems of the lumber and timber products industry. Price studies prepared by the Consumers' Division include a report on the price of paper (1936). Miscellaneous reports and documents (1933-37, 38 feet) include reports and studies concerning the Southern Pine Division, homework in furniture manufacturing, lumber and timber products information, the lumber and timber products code, interviews with secretaries of lumber associations, sales of southern pine and southern hardwoods, the paper and pulp code, and the retail lumber code. P.I. No. 44; S.L. No. 12.

704 *Division records, 1929-37, 1,068 feet.* Records of the Legal Division include copies of weekly reports by Bernice Lotwin, legal advisor on Lumber and Timber Products Codes, and her correspondence with other members of the legal division (1933-35, 3 inches). Records of the Compliance Division, Coordinating Branch, include miscellaneous summaries (1934-35, 3 inches), with special statistical reports on complaint adjustments in the lumber and timber industry. Miscellaneous office files (1932-35, 15 feet) in the records of the Research and Planning Division include a report on lumber exports from Europe, and a proposed study of the paper and pulp industry; records of the Statistics Section, Research and Planning Division, contain charts of general economic data and statistical information (1923-35) on the effects of N.R.A. codes on various industries, including lumber and millwork, lumber and timber products, Douglas-fir, common boards, and shingles; records of the Import Section relating to complaints (1933-35, 6 feet) include formal complaints filed under

Section 3 (e) of the National Industrial Recovery Act concerning red cedar shingles and paper products, and correspondence relating to cordwood, lumber, paper, potash, pulpwood, red cedar shingles, turpentine, wood pulp, and wooden containers. Records of the Division of Review contain code histories for industries under approved codes (1935-36, 31 feet), including the match, furniture manufacturing, lumber and timber products, paper and pulp, paper and pulp machinery, retail lumber, lumber products, building materials, building specialties trade, sawmill machinery, wood floor contracting, wood heel, wood preserving, wood turning and shaping, and woodworking machinery industries; records of the Industries Studies Section, Division of Review, contain the office files of Peter A. Stone (1934-35, 2 feet), with memoranda, reports, outlines, and trade association publications used in compiling studies of the lumber and construction industries, including a report on public forest acquisition in the southern states, prepared by the Conservation Division, Southern Pine Association; general records of the Foreign Trade Studies Section, Division of Review, include some statistical material on the lumber and timber industry. Records of Industry Divisions of the N.R.A. include those of the Chemical Division, Paper Section, relating to mill products agreements (1933-34, 2 inches), with correspondence and memoranda of W.W. Packard, deputy administrator, that pertain to the formation of an association in the kraftboard and paperboard industry, and records relating to the formulation of a code for the pulpwood industry (1934-35, 10 inches). P.I. No. 44.

705 *Records of boards and committees, 1933-1937, 171 feet.* Records of the Labor Advisory Council include office files of Henry H. Collins (1933-35, 2 feet) with a report of the Negro Advisory League concerning the lumber and timber products code (1933, 10 pp.). P.I. No. 44.

706 *Records of Code Authorities, 1927-1935, 149 feet.* Records of the Code Authority for the Builders' Supply Industry include questionnaires (1933-34, 4 inches), with information on sales in 1933, labor and maintenance costs, delivery and overhead expenses, etc. Records of the Virginia Lumber and Building Supply Dealers Association (1933-35, 3 feet) include budgetary materials, lists of Virginia companies subject to the code, and correspondence with various firms and with the Washington headquarters of the Retail Lumber and Building Materials Code Authority regarding suggested amendments to the code and unpaid assessments due under the code. Unarranged. P.I. No. 44.

National Resources Planning Board (Record Group 187)

707 *Records of the National Resources Planning Board Central Office, 1933-1943, 1,200 feet.* Central office classified files include correspondence with state planning boards and the U.S. Forest Service, and other records relating to the conservation and development of forest resources in Alabama, Alaska, Pacific Northwest, and the South. Card indexes. P.I. No. 50.

708 *Records of the National Resources Planning Board Regional Offices, 1934-43, 410 feet.* Most regional offices of the National Resources Planning Board had some concern with forest resources. The records of Region 3 (Southeastern Region) include administrative and technical records concerning the Southern Forest Resources Survey (1937-40, 3 feet), conducted by a special committee representing Alabama, Florida, Georgia, North Carolina, Tennessee, Mississippi, Arkansas, Louisiana, South Carolina, and Texas. The survey was concerned with the pulp and paper industries in the South, but also with all forest industries in that region. P.I. No. 64.

National War Labor Board (World War I Record Group 2)

709 *Case files, 1918-1919, 75 feet.* The files include disputed labor cases concerning the building materials industry, lumber mill workers, lumber products, cooperage, food containers, paper and pulp mill machinery, paper and pulp products. P.I. No. 5.

National War Labor Board (World War II) (Record Group 202)

710 *Records of the National War Labor Board, 1941-1947, 4,220 feet.* Headquarters case records include voluntary wage and salary adjustment case files (1942-45, 46 feet) relating to various industries including furniture, lumber and wood products, and paper and paper products (Card index). Among the records of Board Members Representing the Public are correspondence relating to the West Coast Lumber Commission, memoranda, and reports relating to the work of the commission; also press releases, copies of opinions and reports of the commission, and a report prepared by the Wage Stabilization Division on the wage stabilization activities of the commission (1943-44, 3 inches). Records of cases heard by the commission are in the dispute case files of commissions and panels (1942-46, 249 feet). Records of the West Coast Lumber Commission (1942-46, 12½ feet), which heard and determined dispute cases that threatened to interrupt the production of lumber and lumber products in Oregon, Washington, Idaho, Montana, and California, include policy documentation files, with minutes of the commission's enforcement division and wage rate committee, copies of general orders and correspondence relating to their interpretation, application, and revision, case analysis memoranda for cases on appeal to the national board, and copies of the commission's calendar; general records, with correspondence, memoranda, reports, etc., relating to the establishment, jurisdiction, policies, and activities of the commission, to particular issues and cases, and to production, the manpower situation, wages, and labor disturbances in the lumber industry; case materials, with copies of directive orders, decisions, opinions, and referee reports in cases heard by the commission; transcripts of hearings in dispute cases and occasionally of conferences; card indexes to dispute case files in the case records of commissions and panels; card index to voluntary wage and salary adjustment case files in the case records of commissions and panels, and to the regional case files of the National Wage Stabilization Board; and a job classification card index, giving wage rates established by the commission. P.I. No. 78 and S.L. No. 10.

711 *Records of the National Wage Stabilization Board, 1943-1947, 444 feet.* Among regional case records are West Coast Lumber Commission voluntary wage and salary adjustment case files (1945-46, 2 feet), relating to all applications handled by the West Coast Lumber Commission, which had jurisdiction over all wage adjustments affecting employees in the lumber and lumber products industries in Washington, Oregon, California, Idaho, and Montana. Rulings in these cases were issued by Region XII National Wage Stabilization Board over the title of the commission. There is a card index to these case files in the records of the National War Labor Board, records of the West Coast Lumber Commission. Records of Region XII National Wage Stabilization Board include records relating to the West Coast Lumber Commission (1943-45, 2 feet), with correspondence, memoranda, telegrams,

reports, resolutions, papers containing wage data, notes and working papers, job classification studies, copies of lumber general orders, forms, copies of case materials, press releases, and reference materials, relating mainly to the jurisdiction of the commission, wage scales in the lumber industry, the commission's wage policy, proceedings in particular cases, and the interpretation of lumber general orders. General correspondence of the legal division (1943-47, 2 feet) includes similar materials relating to all phases of the enforcement work of the West Coast Lumber Commission. P.I. No. 78.

Naval Records Collection of the Office of Naval Records and Library (Record Group 45)

712 *Records of the Office of the Secretary of the Navy, 1776-1913, 1,127 feet.* Incoming correspondence includes letters from Navy agents and naval storekeepers (1843-1865, 26 vols.; Microfilm Publication M528, 12 rolls). Miscellaneous records include reports from Charles Haire and Thomas F. Cornell, agents to the Secretary of the Navy, on live-oak lands adjoining Escambia Bay and Escambia River, Florida (1828, 1 vol.). Draft P.I.; pamphlet accompanying Microcopy No. 528.

713 *Records of the Board of Navy Commissioners, 1794-1842, 109 feet.* Includes records relating to shipbuilding; inventories of stores in navy yards (1814-1816); summary of information respecting the live-oak timber of the Carolinas and Georgia (1815-1817, 1 vol.); journals of timber expeditions to Georgia (1817-1818, 1 vol.), and the Gulf Coast (1818-1819, 1 vol.). Draft P.I.

714 *Records of offices and bureaus of the Navy Department, 1811-1914, 48 feet.* Records of the Bureau of Yards and Docks relating to naval timber reservations include letters from timber agents (1828-1836, 1839-1859, 16 vols.) relating to Georgia, Florida, Alabama, Mississippi, and Louisiana; register of letters from timber agents (1845-1860, 2 vols.); letters relating to live-oak plantations (1829-1861, 1 vol.), chiefly from the Commissioner of the General Land Office; maps of timber reservations (1831-1857, 1 vol.); letters to timber agents (1840-1861, 3 vols.); letters of expenses of timber agents (1845-1860, 1 vol.); letters from the Secretary of the Navy relative to timber agencies (1845-1855, 1 vol.); approved bills submitted by timber agents (1845-1861, 2 vols.); reports on experiments on preserving timber against marine worms (1850-1855, 1 vol.). Draft P.I.

Plant Industry, Soils, and Agricultural Engineering Bureau (Record Group 54)

715 *Office and divisional records of the Bureau of Plant Industry, 1881-1953, 2,773 feet.* Records of the Division of Forest Pathology (1901-1938, 132½ feet) contain material on investigations of timber and forest tree diseases, including problems of decay in lumber, wood products, and structural timber; letters sent (1901-1908), relating to all types of investigations conducted by the division and predecessor units; general correspondence (1907-1938), relating to all activities of the division, including research of such diseases as white-pine blister rust, chestnut blight, heart rot of Douglas-fir, and blue stain deterioration of timber; also correspondence with field personnel (1907-38), miscellaneous administrative records (1907-38), office files of George C. Hedgecock (1901-07), letters received by Hermann von Schrenk (1902), office file of Perley Spaulding (1913-1924), correspondence of a field station at Missoula, Montana (1911-1920), project file (1912-1937), field notes (1913-1924), reference file (1915-1938). P.I. No. 66.

716 *Other records, 1899-1936, 415 feet.* Records of the Soil Survey Division contain reports on field investigations, including forest land classification and soil investigations in Maryland, New Jersey, and New York (1914-1919).

Presidential Committees, Commissions, and Boards (Record Group 220)

717 *Records of the National Conference on Outdoor Recreation, 1924-1929, 21 feet.* General files (1924-29, 15 feet) of the executive secretary, Arthur Ringland, include proceedings and other records of conferences reflecting the cooperation of the Forest Service, National Park Service, state park agencies, other government agencies, and private organizations; also records relating to particular national parks and national forests. Records of the Joint Committee on Recreational Survey of Federal Lands of the American Forestry Association and the National Parks Association (1924-26, 8 inches); minutes, correspondence, and other records on state parks, national forests, and wilderness areas. Draft P.I. (1962, 28 pp.).

718 *Cartographic records, 1918-1931, 35 items.* Maps prepared by the Committee on the Conservation and Administration of the Public Domain, relating to national forests, public lands, and other matters in the western states.

Price Administration Office (Record Group 188)

719 *Records of the Enforcement Department, 1941-1947, 417 feet.* The centralized files of the Enforcement Department include weekly reports of the Lumber Units or sections for Regions 4, 5, and 8 (1946). Procedural records of the Office of the Deputy Administrator for Enforcement include correspondence and reports relating to the progress of enforcement in the building materials and construction industries. Records of the operating divisions include case files of the Nashville district office relating to permanent injunction and treble-damage cases resulting from price violations in lumber and consumer goods (1944-46, 2 feet), and records relating to inspections of lumber shipments and mills (1943-45, 11 feet), handled chiefly by the Southern Pine Inspection Bureau. In the records of the Industrial Manufacturing Enforcement Branch relating to regulations (1942-44, 3 inches) are reports, issuances, memoranda, and other types of correspondence, and work papers showing the development and administration of the enforcement provisions of price and rationing regulations pertaining to paper and paper products. In the records of the Industrial Materials Enforcement Branch relating to regulations (1942-46, 1 foot) is similar material relating to lumber, building materials, etc. Records of Field Enforcement Divisions, Region VIII, relating to significant enforcement problems (1943-45, 2 inches) show the difficulties encountered in maintaining effective enforcement control in the fishing and lumbering industries of the Pacific Northwest. P.I. No. 120.

720 *Records of the Accounting Department, 1940-1947, 1,844 feet.* The records of the Industrial Accounting Division include, among records relating to the chemical industry, survey reports on the naval stores industry; also materials relating to the work of the lumber accounting branch and the paper accounting branch. In the records of the Lumber Accounting Branch (1941-46, ca. 27 feet) are financial studies and records of surveys taken to determine cost data for the western box shook industry, the manufacture of crates and boxes, the agricultural container industry, the wirebound box industry, barrels and keg staves, the cooperage industry, battery separators, the red cedar shingle industry, the ash and hickory handle industries, northern hardwood flooring, the maple, birch, and beech flooring industries, southern oak

flooring, the logging industry (West Coast, Pacific Northwest, Great Lakes, and New England regions), the railroad cross tie industry, the rotary-cut box industry, the commercial plywood and veneer industry, and for various species of timber, including northern and northeastern softwood and hardwood, southern hardwood, southern pine, western pine, western red cedar, south central hardwood, north central hardwood, Appalachian hardwood, Douglas-fir, Sitka spruce, redwood, cypress, walnut, mahogany, and miscellaneous lumber industries; also price adjustment case files, and miscellaneous records relating to the lumber industry. In the records of the Paper Accounting Branch (1942-46, ca. 21 feet) are case files relating to requests for price determinations, applications for price adjustments, and financial studies of the producers and distributors of such commodities as various paper products, different types of paper, and pulpwood; yearly profit and loss data of firms, 1936-45; monthly data about processors of sulphate, sulphite, and groundwood pulp in the Northeastern, Southern, Lake Central, and West Coast regions (1944-46); miscellaneous records relating to the paper industry; and work papers. Records of the Field Accounting Divisions include survey files relating to the lumber industry (1942-46), containing case files relating to community or statewide surveys of shook and box manufacturers, lumber retailers, wholesalers for stock millwork and softwood plywood, lumber producers and distributors, and other wood products manufacturers or distributors. Also case files (1942-46), containing records of company investigations, industry surveys, ration currency audits, and administrative matters, relating to similar industries. Arranged by region, and thereunder by district office. P.I. No. 32.

721 *Records of the Price Department, 1940-1947, 4,652 feet.* Among records of the Price Legal Division (1941-46, 2 feet) are briefs, correspondence, memoranda, and studies relating to O.P.A. control over freight rates for intercoastal lumber shippers. Records of the Building and Construction Price Division, Building Materials Price Branch, include records of the Millwork Section, with several series relating to price control for doors, molding, plywood, stock and special millwork (1941-46, ca. 12½ feet); in the records of the Distribution and Construction Price Branch are records of the Lumber Distribution Section (1942-46, ca. 26 feet), including memoranda concerning administrative and procedural matters and the administration of regulations affecting the price of fir boards and salvage lumber, correspondence with regional offices and with private industry, and records relating to proposed price regulations for lumber dealers, distribution-yard sales of softwood and hardwood lumber, used lumber, millwork, molding, and plywood, and applications for permission to operate lumber yards. Records of the Fuel Price Division include records of the Solid Fuels Price Branch, containing material relating to the formulation and operation of regulations pertaining to price policies for firewood. Records of the Industrials Price Division include records of the Lumber Price Branch, with memoranda of the Price Executive reflecting Peter A. Stone's administration (1941-46, 19 feet), correspondence with regional and district offices (1942-46, 1 foot), and records of the chief counsel (1941-46, 4 feet). There are also records of sections of the Lumber Price Branch. In the records of the Hardwood, Veneer, Plywood, and Miscellaneous Wood Products Section (1942-46, 21 feet) is correspondence of the Section Head, Harry B. Krausz, price action file on veneers, records relating to commodity programs, concerned mainly with development and administration of maximum price regulations for plywood, imported woods, turned and shaped wood products, hardwood small dimension products, and miscellaneous wood products. In the records of the Eastern Softwoods Section (1941-46, 12 feet) are records relating to similar commodity programs for the northeastern and northern softwood lumber, tidewater red cypress, and southern pine lumber. In the records of the Hardwood Section (1941-46, ca. 18 feet) are files of the section head and commodity program records for southern hardwood lumber, Appalachian hardwood lumber, south central and north central hardwood lumber, northern hardwood lumber, northeastern hardwood lumber, flooring lumber, and aromatic red cedar lumber. In the records of the Primary Forest Products Section (1941-46, ca. 16½ feet) are correspondence with regional offices, price action files relating to maximum price regulations for primary wood products (mine, railroad, and other types of lumber) in several western states; records relating to commodity programs for west coast logs (Douglas-fir, western red cedar, western hemlock, western white fir, noble fir, and Sitka spruce) produced west of the Cascades crest in Oregon, Washington, California, and Canada; records relating to commodity programs for railroad ties, mine materials, posts, poles, and piling; stumpage and logs; western timber; wood preservation by pressure methods; and miscellaneous primary forest products. In the records of the wooden containers section (1941-46, ca. 24 feet) are applications for price adjustments, letter orders in response to applications for price adjustments, reports on the wood container industry, and records relating to commodity programs for used egg cases, box-grade veneer, industrial wooden containers, the wooden cooperage industry, wirebound boxes, agricultural wood containers, used agricultural wood containers, and miscellaneous container products. In the records of the Western Softwoods and Services Section (1941-46, 28 feet) are records relating to commodity programs for Douglas-fir lumber, western pine lumber, aircraft lumber, red cedar shingles, redwood lumber, Sitka spruce, western red cedar lumber, and records relating to the development and operation of regulations for the custom milling and kiln drying of softwood and hardwood lumber. The records of the Industrials Price Division also include records of the Machinery Price Branch, Processing Equipment Section, with material relating to commodity programs for woodworking machinery (1942-46, 5 feet). Also in the records of the Industrials Price Division are records of the Paper and Paper Products Price Branch, with memoranda and letters of the price executive regarding price policies on various commodities (1942-46, 2 feet), general correspondence (1941-42, 7 feet), correspondence of the legal staff interpreting regulations concerning price stabilization (1942-47, ca. ½ foot). Records of the Paper and Paper Products Price Branch also include records of the printing and fine papers section (1941-46, 34½ feet), concerning newsprint, tissue paper, writing and book paper, etc.; records of the paper products section (1941-46, ca. 56 feet) concerning paperboard, paper boxes, converted paper products, industrial papers, etc.; and records of the raw materials section (1942-46, 33 feet), with records relating to the formulation and administration of price regulations for pulpwood and pulp manufacturing costs, economic data on the pulpwood industry, and pulpwood subsidy claims. Records of the Field Price Divisions include records of the Building Materials and Construction Price Division of each region; the central files of region II (Mid-Atlantic states) include records relating to commodity programs for firewood. P.I. No. 95.

722 *Records of the Rationing Department, 1940-1947, 801 feet.* In the general records of the Automotive Supply Rationing Division, Tire Rationing Branch (1941-45, 44 feet), are quota and statistics files on the U.S. Forest Service and the

Logging Survey and lumber and logging problems. The records of the Fuel Rationing Division, Northwest Solid Fuels Branch (1942-45, ca. 15½ feet), relate to supply and rationing of coal and firewood in Oregon, Washington, and the Idaho panhandle, including records concerning demand, distribution, and consumption of firewood, registration of firewood dealers, lists of dealers, statistics on firewood supplies, demand, production, consumption, and sales, dealers' monthly reports, and tabulations of sales; also a manuscript narrative history of the Branch (1946). P.I. No. 102.

Prison Industries Reorganization Administration (Record Group 209)

723 *Records, 1935-1940, 54 feet.* Records (1935-37, 1 foot) relating to industrial work for prisoners and maintained by Fred Holloday, investigator, contain correspondence, reports, and pamphlets relating to forestry. Records of the Library Division (1935-40, 2 feet) include files, mostly reference materials, concerning the Civilian Conservation Corps. Draft P.I. (1966, 10 pp.).

Public Land Law Review Commission (Record Group 409)

724 *General records, 1965-1970, 81 feet.* Textual records include materials relating to nearly all aspects of the commission's work, including programs affecting national parks, mining laws, outdoor recreation, wilderness, timber, and national forest revenues. Described in *Records of the Public Land Law Review Commission: Inventory of Record Group 409* (1973).

Public Roads Bureau (Record Group 30)

725 *Cartographic and audiovisual records, 1900-1970, ca. 97,850 items.* Among the cartographic records are maps of the 11 Western states (1921-1946, 34 items), including maps showing highways administered by the U.S. Forest Service; one of the maps is annotated to show timber access roads. Photographic records illustrating the activities of the Bureau of Public Roads (1900-1953, 89 feet) include illustrations of national forest and national park roads. P.I. No. 134.

Railroad Administration (Record Group 14)

726 *Records of the Division of Purchases, Forest Products Section, 1918-1920, 1 foot.* Included are records relating to the production and procurement of railroad crossties (1 inch). Draft P.I. (31 pp.).

Reclamation Bureau (Record Group 115)

727 *General records, 1891-1960, 2,431 feet.* General records of the Bureau contain files of the Drafting Division (1891-1945, 72 feet), mostly correspondence, concerning the withdrawal and restoration of public lands, including records relating to proposed forest reserves. Miscellaneous photographs include pictures of, or relating to, national parks and Civilian Conservation Corps activities sponsored by the Bureau. Records relating to Civilian Conservation Corps activities pertain to work on reclamation under the supervision of the Bureau (1935-43, ca. 42 feet). P.I. No. 109.

Reconstruction Finance Corporation (Record Group 234)

728 *Records of the Office of the Controller-Treasurer of the RFC, 1932-1957, 1,465 feet.* Included are reports and surveys of the Industrial Analysis Branch (1948-53, 7 inches), with reports on the wood pulp industry, showing annual production rates, trend surveys in commercial fields such as lumber, and special surveys of commercial failures in the building materials and furniture industries. P.I. No. 173.

729 *Records of the Defense Supplies Corporation, 1940-1949, 198 feet.* Included are exhibits to the minutes (1941-45, 4 vols.), containing copies of and amendments to agreements and contracts between the Defense Supplies Corporation and private manufacturers and distributors undertaking to produce and supply the government with pulpwood and other strategic and critical materials. P.I. No. 173.

Senate of the United States (Record Group 46)

730 *Committee records and reports, 1789-1968, 3,134 feet.* There are bills and related papers, executive communications, correspondence, reports, petitions, and memorials relating to forestry and forest lands among the records of the Committee on Forest Reservations and Protection of Game, 57th-62nd Congresses (1901-1913); the Committee on Agriculture and Forestry, 57th-79th Congresses (1901-1947); and the Committee on Public Lands and Surveys, 14th-79th Congresses (1815-1947). Restricted. P.I. No. 23.

Shipping Board, United States (Record Group 32)

731 *Records of the Shipping Board and the Shipping Board Bureau, 1917-36, 4,301 feet.* Records pertaining to legal matters, including claims, contain copies of contracts for the construction of wooden vessels (1919-21, 2 feet), assembled by the legal department as background for suits arising from cancellations of contracts or from too high costs. P.I. No. 97.

732 *Records of the Construction Organization, Fleet Corporation, 1917-1928, 2,047 feet.* Records of the Construction Organization, Ship Construction Division, include records of the Technical Section (1917-19), which served both the wood and steel ship divisions; and records of the Wood Ship Section (1917-19, 40 feet), relating to all aspects of the construction of wooden ships (indexed). Records of the Fir Production Board include general correspondence of the Portland office (1917-19, 4 feet), relating to lumber prices, board policies, relations with other government agencies, expediting of payments and collections, and general policy as to lumber orders; correspondence of the secretary's office at Portland (1917-19, 2 feet), regarding lumber orders for West Coast shipyards and for airplanes for the Army Signal Corps, interoffice communications with the Seattle and Washington, D.C., offices, and correspondence of Jay S. Hamilton, Secretary of the Board at Portland; reports of field supervisors to the Portland office (1918, 1 foot), relating to supply and quantity of logs at each sawmill in the Portland area for fir lumber for aircraft; ledgers of orders and shipments (1917-19, 2 feet) kept by the auditor at Portland; circulars (1918-19, 8 inches), kept at the Seattle office; records of J.H. Bloedl, chairman of the Fir Production Board in Seattle (1918, 10 inches); records of C.W. Stimson, Secretary of the Fir Production Board in Seattle (1918, 3 feet); general correspondence of the Seattle office relating to the lumber embargo and releases regarding Douglas-fir lumber for shipbuilding (1917-18, 2 feet); reports of field supervisors to the Seattle office (1918, 2 feet); correspondence of H.E. Post, office manager and auditor of the Seattle office (1918-19, 8 inches); records of Lynde Palmer, representative of the Fir Emergency Committee (or Bureau), and Washington, D.C., representative of the Fir Production Board, transmitting fir lumber orders to the board from the War Industries Board and the Emergency Fleet Corporation (1917-19, 3 feet). P.I. No. 97.

Ships Bureau (Record Group 19)

733 *Records of the Bureau of Construction and Repair, 1794-1943, 11,915 feet.* Correspondence of the Office of the Chief of the Bureau includes letters sent (1862-68, 2 vols.) relating to advertisements of proposals for bids to supply timber and other materials, and construct or repair vessels. There are also miscellaneous correspondence and reports (1887-1911, 5 feet) including material on ship construction, balance sheets of naval stores and materials on vessels; contracts for supplies (1868-69, 1 vol.), including naval stores, and a record of contracts (1869-74, 2 vols.), including contracts for lumber; and returns of stores at naval yards and stations (1842-86, 299 vols.). Fiscal records include records of receipts and expenditures of funds and stores at naval yards (1856-67, 1 vol.). P.I. 133.

Smaller War Plants Corporation (Record Group 240)

734 *Records of the Central Office, 1941-1947, 306 feet.* Among the records of the Board of Directors are records of S. Abbot Smith, SWPC representative on the Log and Lumber Policy Committee of the War Production Board relating to lumber (1943-1944, 10 inches). Records of the Chairman of the Board of Directors include general records of Laurence F. Arnold (1945-1946, 1 foot), consisting of correspondence, memoranda, and other materials relating to the lumber industry. Among the records of the Chairman and General Manager are histories of the SWPC (1945, 21 vols.), with chapters on Small Lumber Concerns, the Smaller War Plants Corporation, and Lumber Production for the War Effort (34 pp.); records, relating to lumber (1942-1946, 10 inches), of Nathaniel Dyke, Technical Consultant in the Lumber and Lumber Products Division of the WPB until 1943, and thereafter Special Lumber Consultant to the Chairman of the SWPC; and records of M. Rea Paul relating to the SWPC lumber project (1943-1945, 5 inches). In the records of the Office of Reports is the reports file of Director John M. Blair relating to loan applications, partly concerning wood products (1945-1946, 5 inches). Records of the Operations Bureau include records of the Lumber Section, consisting of inspection reports of lumber mills, progress reports of field activities and other general records (1945-1946, 10 inches); and miscellaneous records relating to lumber activities (1943-1945, 2 inches). P.I. No. 160.

Soil Conservation Service (Record Group 114)

735 *Central Office records, 1900-1952, 1,974 feet and 290 rolls of microfilm.* General records (1,047 feet) contain classified files relating to all phases of the work of the Soil Erosion Service of the Department of the Interior and the Soil Conservation Service of the Department of Agriculture; among records relating to specific projects are materials on woodland management. Land utilization records include project files and job plans (1934-39, 480 feet), which contain statements of the purposes of projects, preliminary and final plans, and other papers relating to some 270 projects for converting lands generally unsuited for farming into forests, game preserves, and other uses.

736 *Field records, ca. 1933-1942, 1,021 feet.* The records of the Southern Great Plains Region (Region 6), Amarillo, Texas (1935-42, 276 feet) include correspondence pertaining to forestry. Selected records of project offices (1933-41) include the records of the Prairie States Forestry Project (1934-41, 20 feet), originally the Shelterbelt Project, which was established under the supervision of the U.S. Forest Service to plant protective belts of forest in the plains region and thereby to ameliorate drought conditions, and was transferred to the Soil Conservation Service in 1942.

737 *Records of the Civilian Conservation Corps Camps, 1933-42, 79 feet.* The records of area offices consist largely of correspondence with and reports of local units subject to area-office supervision, including Civilian Conservation Corps camps and farm forestry projects.

738 *Cartographic and audiovisual records, 1915-1969, 216,992 items.* Aerial photographs and photo indexes of areas in the United States (1933-39), and manuscript and published maps of the United States and its regions and states compiled by the SCS and its predecessors (1915-69), showing various natural resources, including forests.

Solicitor of the Treasury (Record Group 206)

739 *Case files and suit papers, 1791-1929, 110 feet.* Among the case files and suit papers are closed case files (1812-1915, 12 feet) relating to public timber trespasses. There are registers and indexes. P.I. No. 171.

Territories Office (Record Group 126)

740 *General records of the office and its predecessors, 1907-1951, 710 feet.* In the general records are central classified files (1907-51, 608 feet) with correspondence, memoranda, reports, circulars, clippings, press releases, printed and processed matter, and other records relating to territorial administration. These records were created and maintained as part of the files of the Office of the Secretary of the Interior from 1907-1934, and were thereafter continued by the Division of Territories and Island Possessions. Subjects relating to Alaska include national forests, Alaska-Asiatic Lumber Mills and Alaska Wrangell Mills, forest fires, general forestry, paper pulp, national parks and monuments; relating to Arizona are materials on the Grand Canyon Forest Reserve toll road (1907) and national forests; relating to Hawaii are annual reports of the Civilian Conservation Corps, and annual reports of Hawaii National Park; relating to New Mexico is material on the investigation of land transactions of the American Lumber Company (1907-1913), sale of timber on territorial lands (1908), and proceeds from national forests (1909). There is material on the lumber industry in the Philippine Islands (1941-45); on Luquillo National Forest, general forestry, and the Tropical Forest Experiment Station in Puerto Rico; and on industrial forestry and naval stores in the Virgin Islands. There is also a separate classified file (1916-51, 76 feet) relating to the Alaska Railroad, containing similar materials pertaining to Mount McKinley National Park and to lumber used in the operation and maintenance of the railroad. P.I. No. 154.

War Labor Policies Board (Record Group 1)

741 *Records, 1918-1919, 14 feet.* Correspondence of the chairman and of the executive secretary includes membership badges and buttons, and other issuances of the Loyal Legion of Loggers and Lumbermen; papers relating to the President's Mediation Commission (1917-18) include correspondence of Felix Frankfurter, memoranda, and reports on labor conditions in the Pacific Northwest lumber industry. P.I. No. 4.

War Manpower Commission (Record Group 211)

742 *Records of the Reports and Analysis Service, 1939-1946, 105 feet.* Records of the Historical Analysis Section contain draft chapters of the War Manpower Commission History and related documents (1944-46, 1 foot), with a chapter on the lumbering industry in the West. *Inventory of the Records of*

the War Manpower Commission, N.A.R.S. Inventory Series No. 6 (1973).

743 *Records of the Legal Service, 1942-1945, 36 feet.* Office files of general counsels Bernard C. Gavit and Charlees M. May (1942-44, 6 feet) in part relate to proposed programs for the lumber industry. *Inventory,* No. 6 (1973).

744 *Records of the Bureau of Placement, 1939-1946, 68 feet.* General records (1943-46) include War Production Board reports on the production performance of prisoners of war employed in pulpwood mills in New England and New York. There are also records of the War Industries Division of the Bureau, which was involved in planning programs for the recruitment and placement of labor for logging, lumbering, and pulpwood production. *Inventory,* No. 6 (1973).

745 *Records of the Bureau of Manpower Utilization, 1936-1945, 21 feet.* Records of the Division of Occupational Analysis and Manning Tables include issuances and publications (1936-45, 5 feet), with job descriptions for certain occupations in the lumber and lumber products industries, prepared by the United States Employment Service, Department of Labor. *Inventory,* No. 6 (1973).

War Mobilization and Reconversion Office (Record Group 250)

746 *Records of the Office of War Mobilization and Reconversion, 1942-1947, 174 feet.* Records of the Office of the Deputy Director for Reconversion contain reports and correspondence on construction problems (1944-1946, 2 feet) maintained by the Construction Division, relating to building materials and lumber production, and include minutes of the Interagency Committee on Lumber (1945), questionnaire returns on conditions in the Southern pine industry (1946), and Federal Works Agency reports on construction and lumber exports. Memoranda and other documents on construction (1946-1947, 9 inches), a general reference file maintained by Frederick Remington, Program Analyst in the Office of Economic Advisor and later chief of the Production Division, relates to building materials, Army lumber requirements, woodworking machinery, and diversion of lumber from normal channels to provide more materials for housing. P.I. No. 25.

747 *Records of the Office of Economic Stabilization, 1942-1947, 61 feet.* Correspondence on price stabilization (1945-1946, 5 feet) includes material relating to determinations regarding proposed increases in prices of lumber. Correspondence on price control and subsidies (1946-1947, 1 foot) includes correspondence with other government agencies regarding price ceilings on lumber and other commodities (indexed). Correspondence of John R. Steelman and John C. Collett (1946, 4 inches) relates to the government's economic stabilization policy, and includes references to lumber workers' wages. Records relating to construction matters (1945-1946, 5 inches), a subject file of Philip H. Coombs, Construction Consultant, consisting of correspondence with lumber associations and other non-governmental individuals and organizations, include Civilian Production Administration orders on priorities for lumber, memoranda on the problems of producing lumber and lumber products during the reconversion period, on the shortage of pattern-grade lumber, on the possible increase in United States supplies of lumber from abroad, on Douglas-fir, northeastern softwoods, northern hardwoods, northern softwoods, southern pine, softwood moldings, fir stock millwork, hardwood stock stair parts, and other lumber and lumber products. Records relating to housing and building materials (1946, 5 inches), a subject file of the Construction Consultant on problems relating to housing during the reconversion period, includes monthly reports on factors affecting lumber production, surveys of building materials, etc. Among Office of Economic Stabilization directives (1944-1946, 3 inches) are copies of directives, published in the *Federal Register,* which pertain to subsidies and prices for wood pulp, gum rosin, southern pine lumber, and other commodities (indexed). P.I. No. 25.

War Production Board (Record Group 179)

748 *Records of the War Production Board, 1940-47, 1,640 feet.* In the general records is a policy documentation file (1939-47, 1,220 feet), with correspondence received and sent, reports, memoranda, and other documents selected from the records of the various organizational units of the War Production Board, the Civilian Production Administration, and their predecessor agencies, and brought together in one file to document the growth and development of the policies, administration, and functions of the agencies responsible for the control of scarce and critical materials and commodities during World War II. This file includes materials of the Forest Products Bureau, War Production Board; Lumber and Building Materials Branch, Office of Price Administration; Lumber and Lumber Products Division, War Production Board; and Lumber and Timber Products Group, NDAC. It contains, among files relating to the construction industry, material on hardwood plywood and veneer, lumber and lumber products, lumber machinery and equipment, plywood, and walnut lumber; among files relating to commodities is material on pulp and paper, groundwood, rosin, and turpentine; and among the files on the national war economy are materials on forest protection. There is a card index (300 feet) to the policy documentation file. P.I. No. 15.

Work Projects Administration (Record Group 69)

749 *Central Files of C.W.A., F.E.R.A., and W.P.A., 1933-1944, 1,576 feet.* Topics include forestation, tree preservation, forest fire hazard reduction, etc. Typed checklist (1946, 30 pp.).

Yards and Docks Bureau (Record Group 71)

750 *General correspondence, 1842-1942, 2,437 feet.* Records relating to the procurement of timber for naval use are scattered throughout. P.I. No. 10.

Smithsonian Institution, Archives

Archivist 20560. Holdings of the Smithsonian Institution Archives are described in *Preliminary Guide to the Smithsonian Archives,* Washington, D.C.: Smithsonian Institution Press, 1971.

751 **Walcott, Charles D.** (1850-1927), Papers, 1870-1939, ca. 18 feet. This collection contains diaries (1870-1927); professional and family correspondence, including official correspondence from Walcott's work with the United States Geological Survey, mostly outgoing letterpress, but also incoming (1880-1898); draft reports on certain forest reserves (ca. 1898-99); also an unpublished biography by Adele Jenny. Unarranged.

Florida

GAINESVILLE

The University of Florida, P.K. Yonge Library of Florida History

Librarian 32611

Manuscripts

752 **Cary, Austin** (1865-1936), Papers, 1905-1936, 17 boxes. Assistant Professor of Forestry, Harvard College, 1905-09; Superintendent of State Parks, New York, 1909-10; logging engineer, U.S. Forest Service, 1910-35. A portion of this collection (ca. 433 items, 1918-1936) has been described in the Historical Records Survey, *A List of the Materials in the Austin Cary Memorial Forestry Collection in the University of Florida* (Tallahassee, 1941), and consists of manuscripts relating to such topics as forest botany, silviculture, protection, utilization and lumbering, technology, management, and economics; there are printed articles, letters, reports, minutes, field notebooks (ca. 1915-35, 103 vols.), and photographs.

753 **Miscellaneous Manuscripts** Collection. Included are papers (1833-1916) of Robert T. Boyd, lumber merchant of Palatka, Florida; letters (1866-76) of Ambrose B. Hart discussing lumbering on the St. Johns River and in the Palatka region; a letter (January 26, 1846) of Daniel G. McLean, Euchee Anna, Florida, to Perkins and Allen, New York, concerning the price of red cedar; and an agreement between Annie S. Owens, Harry L. Reed, and Joseph Owens for operation of a sawmill in Bartram, Alachua County, Florida (August 3, 1888).

754 **Morgan, Arthur Ernest** (b. 1878), Papers, 1909-1954, 2 boxes. Civil Engineer, Ohio and Tennessee. Chairman of the Tennessee Valley Authority; associated with Everglades drainage.

755 **Southern Cypress Manufacturers Association,** Records ca. 1915-1960s. Unsorted.

756 **Will, Thomas E.,** Papers, 1906-1936, 40 boxes. Author, pioneer settler, and developer of the Florida Everglades.

757 **Yonge, Philip Keyes** (1850-1934), 1906-1933, 23 boxes. The collection apparently includes some records of the Southern States Lumber Company (Pensacola), which was concerned with pine lumbering, paper making, etc.

Theses

758 **Owen, Frank Royal.** "Cypress Lumbering on the St. Johns River from 1884 to 1944," Gainesville: University of Florida, course paper, n.d., 28 pp.

759 **Tyson, Willie Kate.** "History of the Utilization of Longleaf Pine (*Pinus Palustris* Mill.) in Florida from 1513 until the Twentieth Century," Gainesville: University of Florida, M.A. thesis, 1956.

760 **White, Roy Ring.** "Austin Cary and Forestry in the South," Gainesville: University of Florida, M.A. thesis, 1960.

761 **Witmer, Richard Everett.** "Multiple Use of the Ocala National Forest: A Study in the Distribution and Utilization of Resources," Gainesville: University of Florida, M.S. thesis, 1964.

PENSACOLA

Pensacola Historical Museum, Lelia Abercrombie Library

Curator, 405 South Adams Street, Seville Square 32501

762 **Brackenridge, Henry Marie,** Letters, 1827-1832. Letters to Brackenridge's wife Caroline, containing description of forests of northwest Florida. Restricted.

763 **Miscellaneous.** Photographs of lumber mills in northeast Florida and southern Alabama (1880s+); file of newspaper clippings relating to local lumber activities (1937-1970).

764 **Wilson, A.H., & Co.**, Letters, 37 items. Manufacturers of yellow pine and cypress lumber.

University of West Florida, John C. Pace Library, Special Collections

Head, Special Collections 32504 Telephone: (904) 476-9500

765 **Alger-Sullivan Lumber Company** (Century, Escambia County, Florida), 612 pieces. Letterbooks, ledgers, cash books, sales journals, mill cut books, maps, and financial statements of a firm established by Russell A. Alger (1836-1907) and Martin H. Sullivan, Pensacola, Florida, in 1900.

766 **Blount Family,** Papers, 1726-1965, ca. 6,780 pieces. Family papers, correspondence, maps, and other documents relating primarily to W.A. Blount and his son F.M. Blount of Pensacola. Maps show live oak lands along the Gulf Coast. Inventory.

767 **Cary Family,** Papers, 1900-1961, 739 pieces and 82 volumes. Includes papers and records of the Southport Lumber Co. (1909-20).

768 **Jones, J. McHenry.** Draft of a brief relating to the Live Oak Reservation on Santa Rosa Island. Includes miscellaneous documents, legal cases and abstracts.

769 **Murphy** Papers, 1802-1960, 1,375 pieces. Includes materials relating to lands and timber in Santa Rosa County, Florida, chiefly the lumber business of Gideon Murphy (ca. 1860s to 1880s).

770 **Pace Family,** Papers, 1890-1920, 263 pieces. Family papers, and business records, including a variety of documents (ca. 1890-1920) relating to various companies owned by James G. Pace of Pensacola, Florida, involving timber products and naval stores. There are contracts and leases for timber cutting and for turpentine lands.

771 **Rolfs, Gerhard,** Papers, 1898-1954, 449 pieces. Materials relating to the German-American Lumber Company.

772 **Rosaco Brothers** (Pensacola, Florida), Papers, 1895-1958, 62 ledgers and 1 foot correspondence. Business records and related papers of a ship brokerage firm, chiefly financial records, including accounts with lumber companies; also ship charters (1923-39, 1 vol. [ca. 100 forms]) for timber vessels; also papers (1944-61) relating to the Aiken Towing Co. and the Santa Rosa Lumber Co.

773 **Runyan, William Burgoyne,** Papers, 1899-1938, 375 pieces. Includes file on the Runyan-Burgoyne Lumber Co., Manistee, Alabama, with a corporation agreement (1910), inventories, and list of equipment.

774 **Taylor, Francis William,** Papers, 1892-1944, 1,972 pieces. Bound volume, J.F. Taylor; payroll ledger, mill accounts, general ledger of the Bohemia Shingle Mill Co. (1892-93, 1895).

775 **Weddell, Justin R.,** Papers, 1950-1969, 304 pieces. Lumberman, Pensacola, Florida. The papers are largely printed brochures. Inventory.

776 **Yonge Family,** Papers, 1781-1934, 5,821 items. Contains papers of C.C. Yonge and his son Philip Keyes Yonge (1850-1934), lumbermen of Pensacola, Florida, with material on the lumber industry along the Gulf Coast, including timber for naval use. Inventory.

TALLAHASSEE

Division of Archives, History, and Records Management

Chief, The Capitol 32304. Some records are described in

Preliminary Catalog No. 2: Accessions of the Florida State Archives, Tallahassee : 1974.

777 The papers of Governors Reubin Askew (1971-1974), Millard Fillmore Caldwell (1945-49), and Doyle Elam Carlton (1929-33), contain material relating to political matters regarding the Department of Natural Resources, educational programs on natural resources, etc.; none of these collections is well organized, and they are likely to deal with a multitude of things regarding forestry. Secretary of State miscellaneous papers (largely 1960s) contain similar material.

Florida State Library

Reference Librarian, Supreme Court Building 32304

778 **Seton, George Sibbald,** Letterbook, 1850-1868, 1 vol. The letterbook contains copies and memoranda of letters sent relating mainly to the boundaries of a Spanish grant on the Nassau River, Nassau County, Florida; logging operations (ca. 1851-1856) on the tract; efforts to patent a "Round Timber Road and Carriage," and negotiations for the sale of the tract (1866-1868).

Florida State University, Robert Manning Strozier Library, Special Collections

Head, Special Collections 32306

779 **German-American Lumber Company,** Papers, ca. 1899-1918, 437 items. Included are 106 letter files and 331 ledger books.

780 **Hankins, William Foster** (1930-1969), Papers, 1923-1965, 1564 items. Research material accumulated for a book on the lumber and timber industry of Florida. Included are a biographical sketch of Hankins, material relating to the economic development of Florida; notes; looseleaf book listing the names, addresses, and pertinent information about sawmills in Florida; photos (61 items), some aerial-topographic, of timberlands and sawmills in south Florida; maps of 23 Florida counties; Edison flat discs relating to the Florida timber and lumber industry; clippings from newspapers and the *Southern Lumber Journal* concerning forests, sawmills, and equipment for sawmills in Florida; and a description and history of river boats in Florida with the names of their captains. In part, photocopies. Restricted. Finding aid, typed, 6 pp.

781 **Leonard, Wade Hampton,** Papers, ca. 1914-1938, 252 items. 89 letter files and 163 account books, daybooks, timebooks, and store inventory. Provides information about Florida's naval stores industry.

782 **Rosasco, William S.** (1895-1962), Papers, 1880, 1916-1963, 4656 items. Businessman, of Pensacola and Milton, Florida. Correspondence, biographical sketch, scrapbooks, land contracts, logging contracts, legal and other records pertaining to the affairs of the Santa Rosa Lumber Company, land management in timber and naval stores, reforestation, operation of sawmills and planing mill, naval stores, pulpwood, the export of lumber and naval stores, and papers relating to other business enterprises. Other persons represented in the collection and related in business affairs include Rosasco's father, William S., Sr., his son, William S., III, and his uncle, Albert Thomas Rosasco. Finding aid, 9 pp. typed.

Georgia

ATLANTA

Department of Archives and History

Director, 330 Capitol Avenue, S.E. 30334 Telephone: (404) 656-2358. Information: 656-2381

783 **Georgia Forestry Commission,** Records, 1922-1972, 1 box. Biennial and annual reports of the Georgia Forestry Commission and its predecessor agencies (1922, 1926-30, 1933-38, 1945-72, 31 vols.); also bulletins and leaflets published by the State Forestry Department.

Emory University, The Robert W. Woodruff Library for Advanced Studies

Reference Archivist, Special Collections 30322

784 **Herty, Charles Holmes** (1867-1938), Papers, 1884-1938, ca. 60 cubic feet (184 boxes). Industrial chemist, researcher in field of forestry and forest products, and professor of chemistry. The major interest of Herty's career was in the development of forest production and forest products, particularly the development of a timber industry and related industries in Georgia. The papers cover every aspect of his career. Record types include professional and personal correspondence (1884-1938); financial papers; manuscripts and notes concerning Herty's research and writings; statistics and data from the results of that research, concerning the pine and pulp industry, farm chemurgy, chemical warfare, and the dye industry; records of his work with professional organizations; clippings and miscellaneous printed matter relating to Herty's many scientific and professional interests; photographs; and memorabilia. In process (1974).

785 **Woodward, Emily Barnelia** (1885-1970), Papers, 1918-1966, 6 feet (2504 items). Member of the board of the Georgia Forestry Association, chairman of the Herty Memorial Commission to honor the work of Dr. Charles Herty in developing the pine pulp industry in Georgia. Three folders (ca. 204 items) relate to forestry and the Georgia Forestry Association; these are mainly press releases from the association (1937-43), with some clippings and typed addresses. Three additional folders (230 items) relate to Dr. Charles Holmes Herty, industrial chemist and developer of Georgia's pine industry; this material includes correspondence (1933-47), primarily concerning the Herty memorial, and clippings. Inventory.

Georgia State University, Southern Labor Archives

Archivist, Southern Labor Archives 30303. Telephone: (404) 658-2477

785a The Southern Labor Archives collects records of labor organizations and papers of labor leaders and union members, particularly in the South. Holdings include records (1943-1959) of the Woodworkers, District 4, with extensive office files on organizing activities, locals, contracts, and other activity in the southeastern United States; and records (1951-1968, 875 leaves) of the Woodworkers Local 5-51, Louisville, Kentucky, relating to local union activity.

EAST POINT

*Federal Archives and Records Center

Chief, Archives Branch, 1557 St. Joseph Ave. 10344. In addition to original materials, each Regional Archives Branch holds microfilm copies of selected significant records at the National Archives. These films may be borrowed on interinstitutional loan. The holdings of the Federal Archives and Records Center, East Point, are described in *Research Opportunities* (East Point: Archives Branch, Federal Records Center, National Archives and Records Service, General Services Administration, 1972, 5 pp.).

Regional Archives Branch

786 **District Courts of the United States** (Record Group 21), 26,536 cubic feet. Includes records for District Courts of the United States in Alabama, Florida, Georgia, Mississippi, North Carolina, South Carolina, and Tennessee. Among the records for the Eastern District of North Carolina (1789-1942) are 4 volumes of copies of letters written by the North Carolina Lumber Company, Tillery, North Carolina, and an original journal of the receivers of that concern.

787 **Tennessee Valley Authority** (Record Group 142). Project history files (1933-35) from the Forestry, Fisheries and Wildlife Development Division of the TVA containing information about erosion control, tree planting, and other land improvement efforts in the states served by the TVA.

Federal Records Center

788 **Forest Service** (Record Group 95), 8137 cubic feet. Primarily land, administrative, and fiscal files, and also other records, of U.S. Forest Service offices in Atlanta and Gainesville, Georgia; Asheville, North Carolina; Cleveland, Tennessee; Columbia, South Carolina; Jackson, Mississippi; and Montgomery, Alabama.

789 **National Park Service** (Record Group 79), 171 cubic feet.

Hawaii

HONOLULU

Public Archives

State Archivist, Iolani Palace Grounds 96813

790 **Hawaii, Board of Commissioners of Agriculture and Forestry, Division of Forestry,** Records, 4½ feet. Records of the Planters' Association Special Committee on Forestry (1 inch) include correspondence (1902-04), miscellaneous documents, and clippings. Records (1903-15, 9 inches) of Ralph S. Hosmer (1874-1963), superintendent of the division, 1904-14, and chairman, Territorial Conservation Commission, 1908-1914, include correspondence and reports from other officials, the U.S. Forest Service, etc., largely arranged by correspondent, but with some subject files relating to lumbering, forest reserves, tree planting, etc. Records (1906-41, 3½ feet) of Charles S. Judd, successor to Hosmer, contain correspondence and reports of the board, of assistant foresters, rangers, and other officials, including files relating to the Clarke-McNary Act, Emergency Conservation Work (Civilian Conservation Corps), eradication of goats, fencing, forest reserves, Hawaii National Park, legislation, permits, tree planting, and other matters.

Idaho

BOISE

*Idaho State Historical Society

610 North Julia Davis Drive 83706. The following list of holdings pertaining to forest history was compiled from *Guide to Archives and Manuscripts in the United States,* ed. Philip M. Hamer, New Haven: 1961; the *National Union Catalog of Manuscript Collections,* Washington: Library of Congress, 1962; and earlier reports in the Forest History Society's office files. The Idaho State Historical Society also holds the papers of a number of past governors of the state, and of other Idaho political figures, and these records may also contain material of interest to forest history.

791 **Boise Cascade Corporation,** Records. Included are records of several associated firms. There are a journal, cashbook, ledgers, etc., of the Barber Lumber Company (ca. 1904-14, 16 vols.); ledgers, vouchers, property tax notices, sales registers, specifications of mills at Emmett and Barber, and sawing reports of the Boise Payette Lumber Co. (ca. 1914-53, ca. 73 vols., 8½ boxes, and 21 feet); monthly financial statements and other material of the Intermountain Railway Company and the Boise Payette Lumber Company; ledgers, journal register, check register of the Payette Improvement and Boom Company (1903-18); land records and tax records of the Payette Lumber & Mfg. Co. (4 vols.); and other records.

792 **Borah, William Edgar** (1865-1940), Papers, 1890-1907, 24 feet. U.S. Senator from Idaho, 1907-40. The collection includes correspondence, legal papers, drafts of addresses, and notes relating mainly to Borah's law practice. Includes papers relating to the Barber Lumber Company and the Potlatch Lumber Company, the Pacific and Idaho Northern Railway, and other matters.

793 **Cobb, Calvin,** Papers. Papers of Calvin Cobb and his daughter Margaret Cobb Ailshie, Idaho newspaper publishers, relating substantially to forest history.

COEUR D'ALENE

Museum of North Idaho

P.O. Box 812 83814

794 **Oral History Collection**. Interviews with foresters and others connected with the lumber industry in north Idaho, conducted by the North Idaho College in conjunction with the Museum, relate to such topics as logging, camps, log construction, Indian lands, sawmills, 1910 fire, logging railroads, Priest River log drives, labor conditions, strikes, beginnings of the Potlatch lumber industry. (17 tapes; others will be added to the collection.)

MOSCOW

University of Idaho, Library

Head, Department of Special Collections 83843

795 **Clearwater and Potlatch Timber Protective Association,** Reports, 1966+.

796 **Clearwater Timber Protective Association,** Annual reports, 1915-1965.

797 **Craig Mountain Lumber Co. and Craig Mountain Railroad** (Winchester, Idaho), Papers, 145 boxes.

798 **Jewett, George Frederick** (1896-1956), Papers, 1853-1963, 61 boxes. Lumber company executive. Correspondence, personal and family papers, account books, ledgers, financial records, reports, minutes, tax records, legal papers, personal bills of Frank P. Davies, personal business papers, and other papers of Jewett and of the Edward Rutledge Timber Company, of which he was manager and vice president, and Potlatch Forests, Inc., of which he was vice president. Other companies represented include the Bonners Ferry Lumber Company, Coeur d'Alene & St. Joe Transportation Company, Coeur d'Alene Timber Protective Association, Lake Creek Navigation Company, Perry Lyon Navigation Company, Red Collar Line, St. Joe Boom Company, Weyerhaeuser Sales Company, and the White Star Navigation Company. The papers relate to the lumber industry and trade, particularly in Idaho, forestry, fire fighting, prevention of forest fires, conservation, the Depression, labor difficulties, the Industrial

Workers of the World, the Loyal Legion of Loggers and Lumbermen, and Jewett's connections with or interest in the National Lumber Manufacturers Association, its Committee on Forest Conservation, and the North Idaho Conservation Association. Inventory published by the library in 1969. Restrictions. Additions to the collection are anticipated.

799 **Perrin, Robert Anthony, Jr.** "Two Decades of Turbulence: A Study of the Great Lumber Strikes in Northern Idaho (1916-1936)," Moscow: University of Idaho, M.A. thesis, 1961.

800 **Potlatch Timber Protective Association,** Reports, 1918, 1921, 1923-24, 1927-32, 1947-65.

801 **Priest Lake Timber Protective Association,** Biennial reports, 1922-23, 1927-30.

802 **Richards, Laura Barbara.** "George Frederick Jewett: Lumberman and Conservationist," Moscow: University of Idaho, M.A. thesis, 1969.

Illinois

CARBONDALE

Southern Illinois University, Morris Library

Curator of Manuscripts, Special Collections 62901. Theses in the collection are available on inter-institutional loan.

803 **Runyon, Kenneth L.** "The Pulp and Paper Industry in the United States: Data Generation, and Measurement of Locational Change and Trends, with Emphasis on the Southern Region, 1900-1965," Carbondale, Illinois: Southern Illinois University, M.S. thesis (Forestry), 1967, 242 pp.

804 **Johnson, Larry Karl.** "Policies relating to the Development and Management of State Park Systems in the United States," Carbondale: Southern Illinois University, M.S. thesis (Forestry), 1967, 135 pp.

CHICAGO

Chicago Historical Society

Manuscript Department, North Avenue and Clark Street 60614. There is a processed guide to the collections.

805 **Blackwell, Samuel B.,** Daybooks, 1859-1863?, 2 vols. Records relating to the operation of a Chicago planing mill.

806 **Douglas, Paul Howard** (b. 1892), Papers, ca. 700 feet. U.S. Senator from Illinois, 1949-1967. Senatorial papers, chiefly correspondence with constituents, the general public, and governmental agencies, including scrapbooks and the related research files of a congressional office, along with recordings, tapes, films, photographs, etc. There are small scattered segments of files concerning the conservation of natural resources. Restricted: anyone wishing to use these papers must present Mr. Douglas' written permission.

807 **Eastman and Kellogg** (Chicago), Account book, 1858, 1 vol. Lumber business.

808 **Illinois** Collection. A collection of miscellaneous items containing material descriptive of the state of Illinois. Included are letters written to Francis Craig describing lumber rafting on the Mississippi River to Galena, by H. More (May 7, 1845, 3 pp.) and George E. More (Novemeber 2, 1845, 2 pp.).

809 **Mears, Charles** (1814-1895), Papers, 1835-1950, ca. 1,300 items, including 25 vols. Correspondence, diaries, financial records, bills and receipts, and other records and printed matter relating to the lumber business established by Mears, including lumberyards and business offices in Chicago after 1850; lumber camps and villages in Michigan, especially in the White Lake area after 1838; and at Lincoln, Hamlin, and Ludington, Michigan, in the 1870s; and Mears' extensive lumber trade on the Great Lakes, comprising fifteen mills and five harbors, in the 1880s. The correspondence between Mears and his managers describes in detail the business transactions of his various enterprises, including the yards in Chicago, and the settlements, including general stores, boardinghouses, etc., in Michigan. There is also a great deal of material which provides information on costs, prices, and wages during the period 1849 through 1895. Finding aid.

810 **Medill, Joseph** (1823-1899). Letter (February 5, 1872) to Schuyler Colfax protesting blockage by lumber interests of a bill authorizing refund of duties on imported building materials used to rebuild the area in Chicago burned in the fire. 8 pp. Newspaper clipping attached.

811 **Menard and Vallé,** Time books, 1796-1799, 4 vols. Two vols. list wood sawed or cypress logs carried by workmen.

812 **Pulsifer, S., and Company** (Plainfield, Conn.), Account book, 1811-1816, 1 vol. (271 pp.). Account book of a general store, with clippings covering the first 44 pages. Includes agreements with raft hands for lumbering operations (1813-14), on p. 268.

813 **Simon, Seymour F.** (b. 1915), Papers, 1927-1967, 4 feet (9 boxes). President of Cook County Board of Commissioners, 1962-1966. The papers include three folders relating to parks and forest reserves (1961-66), including many requests to use their facilities. Inventory.

814 **Throop, G.A.** Letter (August 11, 1847) to D.D.W.C. Throop, discussing the lumber business, among other matters. 4 pp.

815 **Whitney, Daniel,** Lumber invoice, 1836, 2 items. Daniel Whitney's bill and L.J. Daniel's draft paid for lumber in Cassville, Illinois (October 7, 1836).

816 **Wisconsin** Collection. Miscellaneous items containing material descriptive of the state of Wisconsin. Included are a plat book of Knapp, Stout and Company, showing Wisconsin counties with outlines of the firm's holdings (n.d., 1 vol.); and a typescript copy of a letter of John L. Bracklin to H.E. Knapp (September 28, 1898, 9 pp.), describing a forest fire on lands owned by Knapp, Stout and Company.

Federal Archives and Records Center

Chief, Archives Branch, 7358 South Pulaski Road 60652. Records not accessioned into the Regional Archives Branch remain under the jurisdiction of the originating agency.

Federal Records Center

817 **U.S. Forest Service,** Records, 1913-1970, 2,545 cubic feet. Includes records from national forests and field offices in Illinois, Wisconsin, Minnesota, Indiana, Michigan, and Ohio. Includes administrative files from the Carbondale, Illinois, Research Center; subject files; statistical and accounting records; correspondence; reports on tree planting, nurseries, and timber stand improvement; land purchase files; engineering records; permits and leases; inspection records; timber management records; scale tickets; aerial photographs, etc.

Newberry Library

Special Collections, 60 West Walton Street 60610 Telephone: (312) 943-9090

818 **Chicago, Burlington and Quincy Railroad Company,** Records, ca. 1,000,000 letters, ca. 1,500 bundles, and ca. 2,000 vols. Scattered items relate to forest history. Records of the Chicago, Burlington and Quincy Rail Road Company, a predecessor of the present firm, include, in a bundle of miscellaneous papers (1864-78), a letter (1865) from the Office of Locks and Canals, Lowell, Massachusetts, on experience in preserving lumber. Records of the Chicago, Burlington and Quincy Railroad Company, the present firm, contain letters of officials, including in-letters (1882-1903, 65 boxes) from minor officials and outside men and firms, concerning freight charges, lumber companies, miscellaneous matters, etc.; subject files arranged by decades, including a lease book (1873-78, 1 vol.) with leases of property for lumber yards and other purposes (index); miscellaneous papers (1852-79, 1 bundle) containing a statement of rates received on lumber (1879); miscellaneous papers (1876-81, 1 bundle) including letters concerning the substitution of iron for wood in cars and a copy of a resolution of the Lumber Association of Chicago for cars for lumber shipments; legal papers relating to interstate commerce (1885-89, 1 bundle), with a copy of a letter from Senator S.M. Cullom to Howell Lumber Co., Omaha (February 11, 1887); miscellaneous papers (1878-80, 1 bundle) including a memo about lumber rates. Records of the Burlington and Missouri River Railroad Company (Iowa), letters of officials, contain a letter (February 14, 1862) to David Rorer from E.L. Baker about cottonwood trees. Records of the Burlington and Missouri River Railroad Company in Nebraska relating to New Lines and Other Roads contain material on 'Foreign' Relations (1874-76, 1 bundle) including telegrams (10 items) between Robert Harris and C.E. Perkins about CB&Q rate agreements on lumber for Mennonites, and on trees, and other commodities; miscellaneous papers (1872-74, 1 bundle) including statements of the amount expended planting trees west of Lincoln (1872-73) and a report on red oak lands (July 20, 1874); miscellaneous statements, including statements (1892) on carload shipments of lumber. Records of the Land Department, Private Land Papers, include in-letters (1867-71, 1 bundle) partly relating to planting trees. Records of the Chicago, Burlington & Northern Railroad Company include miscellaneous statements (1880-88, 1 bundle), including lumber tonnage, Southwestern Lumber Association (1884-85), and a Wisconsin lumber statement (November 13, 1885). The collection is described in Elisabeth Coleman Jackson and Carolyn Curtis, *Guide to the Burlington Archives in the Newberry Library, 1851-1901,* Chicago: Newberry Library, 1949.

819 **Illinois Central Railroad Company,** Records, 400,000 letters, 126 bundles or boxes, and 2,000 vols. Items pertaining to forest history are scattered throughout the collection. Legal Reports include contracts (1875-82, 1 vol.) relating to leases of land from the railroad for lumber yards and other purposes (index). Accounting records contain construction contracts (1851-56, 1 bundle), with agreements for ties; reports on inspection of bridges (1896-97, 1 bundle), including reports on materials used; lumber accounts (1896-1909, 1 vol.); and miscellaneous construction accounts (1893-1902, 1 vol.), with accounts for renewals of ties. Records of the New Orleans, St. Louis & Chicago Railroad Company include out-letters (1875, 1 vol.) of W.M. Francis, with many forms of vouchers and payments, and a few letters to local officials and citizens, mostly about wood products. Records of the Yazoo & Mississippi Valley Railroad Company contain timber contracts (1899-1901, 1 vol.). Land Department records include prospectuses, maps, advertisements, etc., of lands, including woodlands, in Illinois offered for sale by the Illinois Central Railroad Company (1854-57, 2 vols.). The collection is described in the *Guide to the Illinois Central Archives in the Newberry Library, 1851-1906,* compiled by Carolyn Curtis Mohr, Chicago: Newberry Library, 1951.

The University of Chicago, The Joseph Regenstein Library, University Archives

Assistant Curator for Manuscripts and Archives, Department East 57th Street 60637

820 **Coulter, John Merle** (1851-1928), Papers. Includes two papers on the exploration of Yellowstone National Park by the Hayden expedition of 1872, and letters (1872-73), from Coulter to Caroline E. Coulter and Georgie Gaylord (later Mrs. Coulter) which add a further dimension to the formal expositions of the exploration.

821 **Pierce, Bessie Louise** (b. 1888), Papers. Contain extensive research notes on the lumber industry in Chicago and its environs, arranged systematically with reference to various aspects of that subject as treated in her *A History of Chicago,* New York: Knopf, 1937-57.

University of Illinois at Chicago Circle, The Library

Curator, Rare Books, Box 8198 60680 Telephone: (312)996-2742, Holdings are described in *Guide to the Manuscript Collections in the Department of Special Collections, the Library,* ca. 1974 +, processed.

822 **Butler, Walker** (1889-1969), Papers, 1921-1969, 19 feet. Attorney of Chicago, Illinois; member of State Senate, 1942-1953. The papers contain some material relating to Wolf Lake State Park. Unpublished guide.

823 **Meine, Franklin J.** (1898-1968), Collection, 1824-1968, 12 feet. Meine was editor of the *American People's Encyclopedia,* and a collector of American regional and folk humor. The collection contains some material pertaining to Paul Bunyan. Unpublished guide.

824 **Save Our Resources and Environment** Collection, 1969-1973, ½ foot. Unprocessed.

SPRINGFIELD

Illinois State Historical Library

Curator of Manuscripts, Old State Capitol 62706 Telephone: (217)782-4836

825 **Castor, William H.,** Correspondence, 1855, 19 items. Lumberman, Chicago. Letters to and from S.W. Hall, business partner and operator of the mill at Oconto, Wisconsin, where lumbering was done; letters from John Volk, Chicago shipper and creditor to Castor.

826 **Fields, Samuel H.** (b. 1832), Correspondence, 1864-1874, 24 items. Incoming business letters of a grain and lumber merchant of Atlanta, Illinois.

827 **Hopkins-McVay Family,** Papers. Includes papers (1935-55, ca. 500 items) of Joel W. Hopkins, Granville, Illinois, relating to lumber business in Coos County, Oregon.

Illinois State Archives

Director 62756

828 **U.S. Surveyor,** Field notes, 1804-1868, 495 vols. Original field notes for land office surveys of townships and ranges for the entire state of Illinois. Includes notations of the predominant species of trees, and other significant land features.

829 **U.S. Surveyor,** Township plats, 52 vols. Drawn from the

information in the field notes, showing forest areas, the prairies, and other natural features. Indexed.

URBANA

University of Illinois, Illinois Historical Survey

Director, Illinois Historical Survey, 1A Library 61801

830 **Delano, Columbus,** Letterpress copybooks, 1869-1873, 5 vols. U.S. Secretary of the Interior, 1870-1875. Vol. 4 contains scattered letters relating to trees and timber, including a letter to a Mr. Welch (January 25, 1873), mentioning a timber contract; to Dr. John Warder (January 27, 1873), applauding the recipient's connections with a group planting trees on the plains; to Frederick Watts (May 10, 1873), mentioning the latter's reception of Scotch firs for distribution in the West for trial. Vol. 5 contains a letter to Rev. George Whipple (October 8, 1873) relating in part to a timber-cutting contract on Indian lands by A.H. Wilder.

831 **Williams, Amos** (1797-1857), Williams-Woodbury papers, 1820-1900, 16 vols. and 190 folders. Includes business accounts, receipts, and plans and specifications for the machinery of a sawmill on the North Fork of the Vermilion River near Danville, Vermilion County, Illinois (ca. 1840s-1850s, ca. 50 items). Unpublished inventory.

University of Illinois at Urbana-Champaign, University Archives

University Archivist, Room 19, University Library 61801

Archives

Agriculture

832 *Extension Service, Records.* Included are 4-H Camping, Conservation, and Forestry Materials (1943-74, ca. ⅓ foot), containing printed brochures, instructional material, issuances, etc., for conservation camps (1966+) and Illinois Boys' Farm Forestry camps (1963+).

833 *Experiment Station, Records.* Agricultural research project file (1925-1960, 6 feet), includes records of research projects in forestry. Botanist's correspondence (1901-10, ca. ½ foot) consists of letterpress copybooks of Thomas J. Burrill, including material on forestry and tree planting, especially catalpa and black locust, and fence posts.

834 *Forestry Department, Records.* Processed publications (1941-1965, ⅓ foot) including film lists, lists of forestry publications, marketing directories for wood, lists of sawmills, research reports, and recommendations for growing windbreaks and seasoning wood. Departmental publications include printed and processed materials (1911, 1961, 1964, 1/10 foot) including a list of wood-using industries of Illinois, forests of southern Illinois, identification of forest trees, a list of sawmills and grading procedures. Forestry research reports (1948+, ⅓ foot) are duplicated monthly accounts of technical experiments and methods of forest protection. Illinois Technical Forestry Association publications (1947-59, 1/10 foot) are duplicated publications including forestry plans, objectives and practices, proceedings and papers of annual meetings, and issuances of the Illinois Christmas Tree Growers Association. Course materials (1965 +), include materials from regular and short courses. Illini Foresters and Xi Sigma Pi material (1966 +, 1/10 feet) includes processed issuances.

Commerce and Business Administration

835 *Economics Department, Records.* Papers (1921-66, 15 feet) of Horace M. Gray (b. 1898), professor of economics, contain material on natural resources policy and economic concentration, water resources and conservation, and related topics.

Fine and Applied Arts

836 *City Planning and Landscape Architecture Department, Records.* Papers (1914-63, 13 feet) of Karl B. Lohmann (1887-1963), professor of landscape architecture and city and regional planning, include material on forests, land use, parks, etc.

Liberal Arts and Sciences

837 *Botany Department, Records.* Papers (1880-1944, 1 foot) of William Trelease (1857-1945), professor of botany, include correspondence with botanists, curators, and collectors in North and Central America, Europe and Jamaica about research on oaks and other plants. Papers (1854-1931, ca. ½ foot) of Thomas G. Burrill (1839-1916), professor of botany, include a letterpress copybook (1904-10) containing Burrill's private correspondence relating to drainage and timber on his Arkansas lands.

838 *Zoology Department, Records.* Papers (1901-1965, ca. 3½ feet) of Victor E. Shelford, professor of zoology, include correspondence, reports, and other papers relating to the organization, development, membership, and functions of the Ecological Society of America and its committees (1937-45), preservation of natural areas as sanctuaries for the ecological study of biotic and animal communities, the political involvement of ecologists in preserving natural areas, etc.

Natural History Survey

839 *Applied Botany and Plant Pathology, Records.* Forester's correspondence (1919-29, 4 feet) includes correspondence of Robert B. Miller, State Forester, and Clarence J. Telford, Survey Forester and Extension Forester, with U.S. Forest Service officials, school administrators, state foresters of Ohio, Indiana, Pennsylvania, and with James W. Toumey, Henry C. Cowles, George D. Fuller, Percy Risdale, Robert Ridgeway, and the public concerning surveys and forestry circulars, Arbor Day and memorial tree planting, advice on planting and growing trees, strip mine tree planting, timber prices, nut trees, forest fire protection, state park sites, forest preserves, training for forestry work, and administrative matters. Forester's workpapers (1920-25, ca. ½ foot) include inventory sheets, lists, maps, tables, worksheets, clippings, press releases, questionnaires, and correspondence relating to surveys of timber resources in Illinois counties, the New York State College of Forestry, and related subjects.

840 *Wildlife Research, Records.* Papers (1932-70, 7.3 feet) of Ralph E. Yeatter (1896-1971), wildlife specialist, include correspondence and other papers relating to conservation, forestry, rural land use, ecology, and related topics.

President

841 *Selim H. Peabody, Papers.* Land correspondence (1883-89, ⅓ foot) relates to university lands. Peabody was a regent of the university.

Manuscripts

842 **Long-Bell Lumber Company** (Kansas City, Missouri), Records, 1928-1940, 5 cubic feet. Documents, correspondence, and other papers pertaining to the reorganization of this firm.

843 **Pickering Lumber Company** (Kansas City, Missouri), Records, 1934-1940, ½ cubic foot. A file of papers pertaining to the reorganization of this Standard, California, company.

Indiana

BLOOMINGTON

Indiana University, The Lilly Library

Curator of Manuscripts 47401 Telephone: (812) 337-2452

844 **Barlow, Ames, and Company,** Memorandum book, July, 1839-March, 1844, 1 vol. (282 pages). This memorandum book of a lumber company in Princeton, Indiana, includes accounts, records of orders, time sheets of employees, tables of measures, etc.

845 **Hoffman, J.R., & Company,** Papers, 1868-87, 3 vols. and 1 folder of illustrations. The papers include an order book (1885-87), a letterpress book (1885-86), and an account book (1868-69) of a Fort Wayne, Indiana, firm that manufactured the Hoffman Patent Band Saw Mill, and also manufactured hardwood lumber.

846 **Howard Ship Yards & Dock Company,** Papers, 1834-1842, 250,000 items. These papers of a builder of steamboats, barges, and towboats at Jefferson, Indiana, include journals, daybooks, ledgers, time and payroll books, lettercopy books, and incoming letters, as well as blueprints, drawings, and some photographs.

847 **Indiana Forestry Association,** Records, 1910-1917. Correspondence, papers, and printed materials relating to the association, its organization, and activities, from the files of Charles Warren Fairbanks (1852-1918), the first president.

848 **Ralston, Samuel Moffett,** Papers, 1868-1928, 20,000 pieces. Governor of Indiana, 1913-17; U.S. Senator from Indiana, 1923-25. There are letters (1913-15, 44 items) relating to expenses, appointments, and problems of the Indiana Department of Conservation, Division of Fish and Game; letters (1915-16, 7 items) relating to the purchase of land, containing timber, for the Turkey Run State Park.

849 **Scott, Emmet Hoyt,** Papers, 1851-1924, 2488 items. Railroad man, lumberman, manufacturer, and mayor of La Porte, Indiana. Includes correspondence, letterpress books (1866-1919, 1921-1922, 63 vols.), account books, miscellaneous papers, and an autobiography. Scott was occupied with railroad construction in New York, 1863-1867, and in Indiana, 1867-1869. He was a partner in Niles and Scott, manufacturer of wheels for agricultural machinery, 1876-1902, began the development of Munising, Michigan, and built the railroad from Munising to Little Lake in 1894. From 1869-1887, Scott was engaged in the lumber industry in East Saginaw, Michigan.

GREENCASTLE

DePauw University, Roy O. West Library, The Archives of DePauw University

Archivist 46135

850 **Hargrave, Frank Flavius** (1878-1962), Papers, 1830-1960, ca. 4 feet (916 items). Professor of economics at Purdue University. The collection includes a scrapbook (1 vol.) containing many original documents which in part pertain to the New Albany and Salem Railroad (later the Louisville, New Albany, and Chicago Railroad); there is an expense account of John C. Sullivan, timber agent (1864-65, p. 213), and bills and receipts for the purchase of crossties and planks, stating quantity, price, type of wood, dimensions, etc. (1855, 1860-65, 32 items, pp. 165-73, 175-79, 181). Indexed.

INDIANAPOLIS

Indiana Historical Society Library

Manuscripts Librarian, State Library and Historical Building, 140 N. Senate Avenue 46204 Telephone: (317) 633-4976

851 **Account Books—Rush County, Indiana.** General store, Melrose, Union Township, Ledger and daybook (1853-61, 2 vols.). Daybook contains records of a sawmill, 1871, and perhaps earlier. Also Felix Frazee's accounts (1869-70) of work at the mill. John Abernathy may have been the proprietor.

852 **Aiken, Oliver Perry,** Papers. Two letters (October 14, 1858 and August 22, 1870) relate to lumber bills; one letter (February 19, 1872) quotes prices on yellow pine lumber. Vanderburgh County, Indiana.

853 **Byerly, Alexander,** Papers, 1883-1903, 10 items. Crawford County, Indiana. Byerly conducted a lumber business. The papers contain a manuscript, 6 printed forms filled in, 2 receipt books, 1 printed item. Includes some work records.

854 **Elliott, John Bennett,** Account book, 1 vol. microfilm. New Harmony. Operation of a sawmill (1868, 1871).

855 **Emmons, Francis Whitefield,** Account book, January 1, 1835-December, 1840, 1 vol. Noblesville, Indiana. The volume gives details about measuring timber (p. 88); and mentions cabinetmakers from whom he bought furniture (pp. 92, 121).

855a **Fairbanks, Charles W.,** Papers, 1876-1917, 20 feet. Included are 23 letterbooks, 16 letter boxes, 19 scrapbooks, and photograph albums. There are letters (1910-1912) and a certificate of incorporation relating to the formation of the Indiana Forestry Association, forestry legislation in the state, and the preservation of forest lands belonging to John Lusk in Parke County for park purposes. Among the correspondents are Addison C. Harris and Stanley Coulter.

856 **Kapp, George,** Papers. The papers contain a letter from Joseph F. Ancoin, Assumption Parish, Louisiana, to George Kapp, Dubois County, Indiana (January 27, 1860), about a flatboat load of hoops and poles.

857 **Kennedy, George,** Papers. The papers contain a letter from Kennedy, Fairfield Township, Franklin County, to Leonard Pickel, Lancaster County, Pennsylvania (September 13, 1837), discussing business prospects, and the prices of different varieties of barrels.

858 **Netz, John,** Account book, 1867-77, 1 vol. Microfilm. Ashland, Henry County, Indiana. Lumber, sawmill. Indexed.

859 **Perine, Peter R.,** Papers. Typed transcripts. The papers consist of brief autobiographical notes on Perine, and a letter from him (April 2, 1854) to his wife, telling of his job in Indianapolis making pork barrels, of an offer of a job on a circle saw in a planing factory, and discussing wages.

860 **Perry, Alexander,** Notebook, microfilm, Vanderburgh County, Indiana. Work records, getting out timber and hewing (1873-1894, pp. 1-25, 40).

861 **Purssell, John T.,** Manuscript, 1838-1841. Purssell was a carpenter of Cass County, Indiana, working on a house, making furniture, etc. The manuscript mentions Taber's, Morgan's, and Martin's sawmills of Logansport (1839).

862 **Schramm, August,** Papers, originals and microfilm. Included are a list of timber (February 2, 1874, 2 pp.) and lumber bills (ca. 1874). There may be other references in diaries (1878-1880) of a carpenter building stables and corn cribs; and carpenter's accounts (1867-1876).

863 **Vinnedge, Moore P.,** Papers. Bills, receipts, etc. (1861-76, 21 items), some in connection with Vinnedge's business as lumber dealer, Kokomo. Also notebook with accounts (ca. 1867).

Indiana State Library, Indiana Division

Manuscript Librarian, 140 North Senate Avenue 46204
Telephone: (317) 633-5440

864 **Banks, Burr,** Ledger, microfilm. Ledger of a flour and sawmill, Washington County, Indiana. The original is in the Washington County Historical Society, Salem, Indiana.

865 **Branham, David C.,** Account book, 1840. Contains prices of building materials in Madison, Indiana.

866 **Brimfield Sawmill & Lumber Co.,** Account book, 1871-1884, microfilm. Noble County, Indiana. The original is owned by H.C. Schlictenmyer, Cromwell, Indiana.

867 **Butler, Amos,** Account book. Brookville, Indiana. Relates to the sale of lumber.

868 **Chamberlin and Chase** (Terre Haute, Indiana), Account books, 1811-1846. The books contain accounts for lumber, labor, and commodities.

869 **Coats, Nellie M.,** Correspondence, 1961, 6 items. Correspondence of Orin Nowlin, Jack McCormick, and Fred Meyer on the Pine Hills Natural Area addition to Shades State Park.

870 **Cobb, Dyar,** Reminiscences. Reminiscences of a sawmill operator near Aurora, Indiana, in the 1830s.

871 **County File.** Fayette County Account books include a volume (1837-1839) recording prices of building materials in Attica, Fountain County, Indiana.

872 **Cox, Elisha,** Receipts, 1820-1847. Receipts giving prices of lumber and wagon-beds, in Dearborn County, Indiana.

873 **Crosier, Adam,** Collection, 1840-1902. Includes information on lumber prices and on shipping lumber by raft down the Ohio River (1870).

874 **Davis, Oliver P.,** Letters, 1846-1892. Information on price of cord wood, and on the price of poplar trees.

875 **Deam, Charles Clemons** (1865-1953), Papers, 1889-1952, 8500 items. Forester. The materials relate to scientific and political questions including forests and forestry in Indiana, and botany in Indiana.

876 **Dole, William Palmer,** Papers, 1836-1861. Xerox copies. Mention is made of prices of hoop poles and of lumber.

877 **Dowling, Thomas,** Collection. Includes a bill for lumber (1864).

878 **Embree, Lucius C.,** Collection. Includes a letter (1892) relating to prices for trees to be used for lumber.

879 **Humphreys, Thomas,** Account books, 1852-1900, microfilm, 1 reel (6 vols.). Floyd County, Indiana, shipbuilder. The records show the types of lumber used and prices of construction materials. The originals are in the New Albany Public Library, Indiana.

880 **Jackson, Edward,** Letters. Governor of Indiana. Includes letters (1925-51, 8 items) relating to his contribution to the state park system. Authors include Richard Lieber, Everett R. Gardner, Kenneth Kunkel.

881 **Lieber, Richard,** Correspondence and speeches, 1905-1944, 18 boxes. Commissioner of State Parks, Indiana Division of Conservation. The papers include material on Indiana Dunes State Park; National Park Service Advisory Board (1936-44); McCormick's Creek, Spring Mill, Turkey Run State Parks, Indiana; forests in national parks. Inventory.

882 **Lieber, William,** Speech, April, 1971. Richard Lieber and Turkey Run State Park acquisition.

883 **McCormick, Jack,** Papers, 1919-1969, ca. 1500 items. Includes material on Shades State Park and Pine Hills natural reserve. Maps, photographs. Index.

884 **McCutchan, William,** Papers. The papers contain a letter from Samuel McCutchan (July 19, 1832) stating that land is valuable in Floyd County because of the timber.

885 **Manter, F.,** Account book, 1850-1862. Sawmill and lumber company at Ladoga, Indiana.

886 **Markle, Augustus R.,** Account book, 1853. Records for carpentry work done in Terre Haute, Indiana.

887 **Nail, William,** Daybook, 1854-1874. Coffin and cabinet-maker. Costs of lumber, coffins, cabinets are mentioned.

888 **Norris, William F.,** Letters. Wood prices are mentioned (1865-66).

889 **Reynolds, William,** Account book, 1866-1869. Account book showing the sale of lumber, Brookville, Indiana.

890 **Schweinitz (Von), Ludwig Davis.** Travel account describing large trees and forests in southern Indiana (1831).

891 **State Park,** Collection. Includes material on the history of pioneer village, Spring Mill State Park; story of Turkey Run State Park; historical data on Turkey Run (1915-16).

892 **Thompson, Lewis G.,** Papers. The papers mention a sawmill built by a Judge Manning (1842).

893 **Tipton, John,** Letters. Timber thieves are mentioned (1833, 1838, 1842).

894 **Unthank, Joseph,** Records, 1837-1850. Spiceland, Indiana. The records show the amounts received for cedar, lumber, and logs.

895 **Willis, Raymond E.** (1875-1956), Papers, 1936-1950, 3 boxes. U.S. Senator from Indiana, 1941-1947; member of committees on Agriculture and Forestry, and Public Lands and Surveys. 17 letters refer to the U.S. Forest Service or to conservation.

896 **Young, W.B.,** Daily journal, 1854-1867. Fulton and Rochester, Indiana. Mentions sawmills and hauling wood.

WEST LAFAYETTE

Purdue University, Forestry-Horticulture Library

47907

897 **Society of American Foresters, Central States Section,** Records, ca. 4½ feet (2 vertical file drawers). Records of the establishment of the Ohio Valley Section, predecessor of the Central States Section (1922); secretary-treasurers' reports, committee reports, newsletters, etc. (1922-42); minutes of Central States Section meetings and news notes (1943-71); correspondence from section officers (1943-71). Records of the Indiana Chapter, now in possession of Roy C. Brundage, West Lafayette, Indiana, are expected to be deposited later.

Purdue University, Main Library

47907

898 **Indiana Hardwood Lumbermen's Association,** Records, 1899-1955. Record book of the secretary-treasurer, containing reports, clippings, etc. (1899-1934); an envelope

containing a membership list (1923), a booklet on the 50th Anniversary Meeting (1949), and clippings; a copy of the *Northeastern Logger* (January, 1961) containing historical articles.

Iowa

AMES

Iowa State University of Science and Technology, Library

Manuscript Curator, Department of Special Collections 50010

899 **Gwynne, Charles S.** (1885-1972), Papers. Professor of earth science, Iowa State University, 1927-58. From 1907-23 Gwynne was employed in industry and with the U.S. Forest Service. The collection contains some material on parks.

900 **Pammel, Louis Hermann,** Papers, 1889-1931. The collection includes considerable material on the early history of forestry in Iowa and the central states. There are correspondence files on national conservation (1902-27, 6 folders) including letters concerning the Central States Forestry Conference (1921) and material on the National Conference on State Parks (1925-27); files on state parks and state conservation (1894-1919, 14 folders), including material on state forestry and park legislation; articles, addresses, and notes on state parks; clippings; files on Iowa State Parks (1919-27, 26 folders); Iowa Park and Forestry Association and Iowa Forestry and Conservation Association (1901, 1914, 1 folder); Iowa Conservation Association (1917-26, 4 folders); State Board of Conservation (1918-27, 9 folders); Iowa and U.S. conservation legislation (1919-25, 1 folder). Inventory.

DES MOINES

Iowa State Department of History and Archives

Curator, Historical Building, East 12th Street and Grand Avenue 50319

901 **Hough, Emerson** (1857-1923), Papers, 1862-1923, ca. 10 items and 131 vols. Business and personal correspondence, manuscripts of Hough's writings, and other papers are in the collection, including material reflecting his interest in the conservation movement and his editorship of *Forest and Stream*. Finding aid.

902 **Lacey, John Fletcher** (1841-1913), Papers, 1853-1912, ca. 60,000 items. U.S. Representative from Iowa, 1889-91, 1893-1907; member of the Public Lands Committee. Lacey was a leader of the conservation movement in the Congress and sponsor of important legislation including the Antiquities Act and national wildlife acts. The papers include business, political, and personal correspondence.

IOWA CITY

State Historical Society of Iowa, Library

Manuscript Librarian, P.O. Box 871 52240 402 Iowa Avenue. Telephone: (319) 338-5471. Holdings are described in *Guide to Manuscripts,* comp. by Katherine Harris, Iowa City: State Historical Society of Iowa, 1973, ca. 332 pp., processed.

903 **Beall, Walter H.,** Papers, 1913-1943, 30 folders. West Union, Iowa. Miscellaneous papers and letters showing broad interest in civic improvements and political affairs. Letters and papers relating to the Northeastern Iowa National Park Association, his work to establish a national park for northeastern Iowa and to Effigy Mounds National Monument and Pikes Peak State Park.

904 **Dolliver, Jonathan Prentiss** (1858-1910), Correspondence, 1860-1920, 12 vols., 87 boxes, 2 vertical file drawers, and 2 cases. U.S. Representative from Iowa, 1889-1900; U.S. Senator, 1900-10. Sponsor of the Antiquities Act, 1906. The collection includes letters (1902-10, 16 items) from Gifford Pinchot. Portions of the collection are indexed.

905 **Farley-Loetscher Manufacturing Co.** (Dubuque, Iowa), Records, 1893-1960, 193 vols. and unbound material. Originally a planing mill; manufacturer of sashes, doors, cabinets, and other mill work. Records include vouchers, trial balance books, cash books, cash receipts, check records, inventory journals and records, general journals, ledgers, purchase ledgers, orders, private ledger, secretary's book, index, contracts, patent certificates, trademarks, copyrights, employees' earning records, accounts receivable, bonds, photographs.

906 **Muscatine Sash and Door Co.** (Muscatine, Iowa), Records, 1884-1927, 15 vols. Account book (1918-27), 6 ledgers (1891-1924), descriptions of products, sales books, pictures of products, and journals of transactions (1902-05).

907 **Musser Lumber Company** (Muscatine and Iowa City, Iowa), Records, ca. 1880s-1920s. Primarily ledger books. Unprocessed.

908 **North-Western Cabinet Company** (Burlington, Iowa), Papers, 1899-1962, 75 items. Account books (22 vols.), cashbooks (2 vols.), journals (7 vols.), check registers (10 items), payroll books (3 items), ledgers (4 items), orders, sales books, with pictures of furniture (2 boxes).

The University of Iowa, Iowa University Libraries

Head, Special Collections 52242. Holdings are described in the *Alphabetical Index to Manuscript Collections,* 3rd. ed., 1969, processed, 19 1.

909 **Biermann, Frederick Elliot** (1884-1968), Papers, 1930-1967, 6 feet. U.S. Representative from Iowa, 1933-39. Served on the Agriculture Committee. The papers include correspondence, speeches, lists and clippings, and include material reflecting congressional concern with forestry and conservation.

910 **Carhart, Arthur Hawthorne** (b. 1892), Papers, 1916-1959, 155 items. Conservationist, landscape architect, and author. Correspondence and drafts of books.

911 **Curtis Companies, Inc.,** Records, 1869-1948, ca. 60 feet. Correspondence, ledgers, journals, cashbooks, stockbooks, and minute books of Curtis Brothers and Company and of its affiliates, concerning the lumber and woodworking industries. Correspondents include George Martin Curtis.

912 **Darling, Jay Norwood** (1876-1962), Papers, 1909-1954, 8 feet and 100 map case drawers. Cartoonist and conservationist. Correspondence with leading conservationists and others, a large file of original cartoons, speeches, scrapbooks, and other papers. Includes material relating to the National Wildlife Federation which Darling helped found. Chronological list of cartoons.

913 **Hoeven, Charles Bernard** (b. 1895), Papers, ca. 1925-1965, 20 feet. U.S. Representative from Iowa, 1943-1965. Served on the Agriculture Committee. The papers include correspondence, office notes and related materials, including material reflecting congressional concern with forestry and conservation.

914 **Meredith, Edwin Thomas** (1876-1928), Papers, 1894-1928, 38 feet (ca. 30,000 items). U.S. Secretary of Agriculture,

1920-21. The collection contains correspondence, speeches, articles, scrapbooks, pamphlets, clippings, photos, and other materials dealing with politics, publishing, farm relief, and land development. Includes items relating to the U.S. Forest Service; forestry; Iowa State College forestry course; Henry S. Graves; William B. Greeley. Unpublished index.

915 **Poyneer, Frederick Julian** (1885-1973), Papers, 1939-1972, 1 foot. Member of the Iowa State Conservation Commission, 1939-1951. Collection includes correspondence, reports, newspaper clippings, photographs, and memorabilia. Correspondents include Jay N. Darling and Ewald G. Trost. Unpublished inventory.

916 **Roach and Musser Company** (Muscatine, Iowa), Records, 1902-1963, 9 feet. Correspondence, financial reports, employee insurance reports, and stock proxies for the sale of Roach and Musser, a lumber manufacturing company, to the Empire Millwork Company.

917 **Trost, Ewald George** (1898-1972), Papers, 1945-1968, ca. 111 items. Member of the Iowa State Conservation Commission. The collection includes correspondence, speeches, special reports, subject files, and two large scrapbooks. Correspondents include Jay N. Darling. Unpublished inventory.

918 **Wallace, Henry Agard** (1888-1965), Papers, ca. 1920-1965, 70 feet (also on microfilm, 67 reels) Iowa editor; politician; U.S. Secretary of Agriculture, 1933-1940; Vice President, 1941-45; Secretary of Commerce, 1945-46. The papers include correspondence, speeches, appointment books, clippings, magazines, and photographs concerning his career and interests. Holdings for the years Wallace was Secretary of Agriculture are relatively thin, but there is material concerning the U.S. Forest Service. The microfilm of the Wallace Papers at the University of Iowa is available through interinstitutional loan, or for purchase, and is described in *Guide to a Microfilm Edition of the Henry A. Wallace Papers at the University of Iowa,* University of Iowa Libraries, Iowa City, 1974, 19 pp. The University of Iowa Libraries has published a joint index to the microfilm editions of Wallace Papers at the University of Iowa, the Library of Congress, and the Franklin D. Roosevelt Library, 2 vols., 1975.

919 **Young, W.J., & Company,** Records, 1858-1920, ca. 200 feet. In part, photocopies of originals in the Cornell University Collection of Regional History and University Archives. Correspondence, letterbooks, estimates, order books, sales records, cashbooks, ledgers, daybooks, journals, payroll books, account books, invoices, freight bills, loading slips, time tickets, cancelled checks, trial balances, balance sheets, statistics on purchases and output, contracts, towboat logs, a pine land register, a manifest, and other records of the home office in Clinton and branch offices in Belle Plaine, Council Bluffs, Tama, and Wheatland, Iowa, and in Omaha. Also includes records relating to the Ann River Logging Co., the Mississippi River Logging Co., and the Clinton Bridge Co.; correspondence and personal papers of William John Young; material on the mining interests of Young's sons; and some Young family correspondence. Correspondents include Douglass Boardman and Frederick Weyerhaeuser.

WEST BRANCH

Herbert Hoover Presidential Library

Director 52358 Telephone: (319) 643-5301. The collections held by the Hoover Library are described in *Historical Materials in the Herbert Hoover Presidential Library,* West Branch: 1973, 37 pp.

920 **Albright, Horace M.,** Oral History Transcript, 99 pp. Albright was Assistant Director, National Park Service, 1917-1919; Superintendent, Yellowstone National Park, 1919-1929; Director, National Park Service, 1929-1933. The transcript deals with conservation here and there, and with the U.S. Forest Service (pp. 90-91).

921 **Hoover, Herbert Clark** (1874-1964), Commerce Papers, 1921-28, 275 feet. U.S. Secretary of Commerce, 1921-28. The bulk of the unpublished material relating to forests is correspondence, with some printed material as a result of committee work, press releases which may or may not have been used, etc. The Commerce Official File (1920-28) contains folders on Agriculture-Forest Service (17 pp.), conferences-lumber (128 pp.), conferences-wood (forest) utilization (125 pp.), conservation (1 p.), forests (27 pp.), lumber (132 pp.), lumber service committee (8 pp.), lumber quarrel (15 pp.), National Association of Wood Turners (10 pp.), National Hardwood Lumber Association (13 pp.), National Lumber Manufacturers Association (20 pp.), National Retail Lumber Dealers Association (6 pp.), National Wholesale Lumber Dealers Association (7 pp.). The Commerce Personal File (1920-28) includes folders on Agriculture-Forest Service Policy (9 pp.), conservation (8 pp.), paper and pulpwood (1 p.). Container list available.

922 **Hoover, Herbert Clark** (1874-1964), Presidential Papers, 1929-1933, 564 feet. U.S. President, 1929-33. The papers contain materials similar to those in the Hoover Commerce Papers, described above. In the Presidential Subject File (1929-32) are files on conservation (28 pp.), forests-fires and reforestation (150 pp.), National Committee on Wood Utilization (33 pp.), National Timber Conservation Board (75 pp.), pulpwood (20 pp.), tariff-commodities-lumber (1,100 pp.), tariff-commodities-wood pulp and pulpwood (600 pp.). In the Presidential Secretary's File (1929-32) there are approximately 60-75 cross-references to correspondence from individuals in regard to such topics as conservation, forest, lumber, and wood. Container list.

923 **Hoover, Herbert Clark** (1874-1964), Post-Presidential Papers, 1933-1964, 380 feet. This collection contains materials similar to those in the Hoover Commerce Papers and Hoover Presidential Papers, described above. The Post-Presidential Subject File (1933-64) includes files on conservation (20 pp.), Butano Forest Associates (2 pp.), Committee on National Resources and Conservation (44 pp.), Resources for the Future, Inc. (20 pp.), Save-the-Redwoods League (10 pp.), Lou Henry Hoover-Memorial Forests (7 pp.). Container list. There are approximately 95 boxes of additional material (1933-64) closed during processing and review. This material is organized in a post-presidential subject file, and is being incorporated into existing series and files as it is processed.

924 **U.S. Committee on the Conservation and Administration of the Public Domain,** Records, 1930-31, 7 feet. Records of an agency established to study the disposal of the unreserved public domain, including material furnished by many federal and state agencies, individuals, and private organizations concerned with public lands, among them U.S. Forest Service reports (1922-31, 4 folders [400 pp.]). Container list; also a draft P.I. (1965, 9 pp.) in the National Archives.

925 **Wilbur, Ray Lyman** (1875-1949), Papers, 1916-1948, 8 feet. U.S. Secretary of the Interior, 1929-1933. Correspondence, memoranda, reports, clippings, printed materials, and maps, chiefly relating to Wilbur's service as Secretary. Subject files include national parks. There are also files relating to the U.S. Forest Service (1919-32, ca. 50 pp.). Container list. This collection was formerly a portion of the Wilbur papers in the

Hoover Institution on War, Revolution, and Peace, Stanford University.

Kansas

ABILENE

Dwight D. Eisenhower Library

Director 67410 Telephone: (913) 263-4751. Holdings are described in *Historical Materials in the Dwight D. Eisenhower Library,* Abilene: Dwight D. Eisenhower Library, National Archives and Records Service, General Services Administration, 1974, 57 pp.

926 **Adams, Sherman** (b. 1899), Papers, 1952-1959, 19 feet. Assistant to the President, 1953-1958. There is correspondence, arranged alphabetically (4 boxes); special correspondence relating to maple sugar, maple syrup, and maple products (3 folders); and speeches before the American Forestry Congress (October 30, 1953), Soil Conservation Society of America (November 5, 1953), the American Paper and Pulp Association (February 18, 1954), and the National Watershed Congress (December 5, 1955). Typed preliminary inventory, 19 pp.

927 **Anderson, Jack Z.,** Papers, 1952-1968, ca. 500 pp. Special Assistant to the Secretary of Agriculture, 1955-56; Administrative Assistant to the President, 1956-61. The daily diary contains material relating to Hells Canyon legislation.

928 **Anderson, Jack Z.,** Records, 1956-1961, 1 foot. There are memoranda to Sherman Adams relating to legislation, including Hells Canyon, Public Law 480; also the Barrett bill, S. 863, on watershed projects (1957); Hells Canyon legislation (1957); working papers with the Department of the Interior, including material on mining, flood control dams, including Isabella, Glen Canyon, Public Law 480, Hells Canyon, etc. Shelf list.

929 **Bennett, Elmer F.,** Papers, 1953-1961, 6 feet. Legislative counsel, Department of the Interior, 1953-56; Assistant to the Secretary of the Interior, 1956-57; General Counsel, Department of the Interior, 1957-58; Under Secretary of the Interior, 1958-61. The papers include briefing books on current issues; clippings and drafts relating to Columbia River Basin projects; a correspondence log (1956-61); copies of reports and correspondence, clippings, and press releases on the "giveaway issue" in the Al Sarena case; miscellaneous materials on Glen Canyon, National Park Service legislation, the Barrett bill (S. 863), Bonneville Power Administration projects, and similar legislation and proposed legislation in the Eisenhower administration. The major subject of the papers is the Columbia River Basin Project, and the U.S.-Canadian treaty over this project. Shelf list.

930 **Benson, Ezra Taft** (b. 1899), Papers, 1952-1961, ca. 400 items and 38 reels microfilm. U.S. Secretary of Agriculture, 1953-61. Printed and Xerox copies of Benson's speeches and statements as secretary, including testimony before congressional committees, press releases, and other papers. The microfilm includes speeches, some correspondence, clippings, fact papers, pamphlets, etc., index of appointments and interviews, along with trip itineraries. The originals are held by the church archives of the Church of Jesus Christ of Latter-day Saints, Salt Lake City, Utah.

931 **Bragdon, John S.,** Records, 1949-61, 37 feet. Special Assistant to the President to coordinate public works planning. The records mostly relate to the President's Advisory Committee on Water Resources Policy, the Columbia River basin, and other interstate compacts, and related legislation, etc. There are files on the Alaska Rampart Dam; the Amistad (Diablo) Dam (1959-60); Hells Canyon feasibility; 3 typed reports (carbons) relating to the present status and problems of the national park system (1955); a copy of the National Park Service's "Mission 66" program (120 pp.), with manuscript notes (1956). Shelf list.

Eisenhower, Dwight David (1890-1969), Records as President: White House Central Files, 1953-1961, 3,241 feet.

932 *Official File.* Material on the U.S. Forest Service (1953-57) includes correspondence and processed material relating to audits; the service's 50th anniversary; hydrologic research in New England, largely concerning Sherman Adams. Department of the Interior files (1952-61) relate to staffing. Files on the conservation of natural resources (1952-60) relate to the attitude of the administration to conservation in general, and include public reaction, with materials of the National Wildlife Federation, Bernard De Voto, Horace Albright, Friends of the Wilderness, Sherman Adams, Howard Zahniser, etc. Files on forestry contain correspondence relating to administration programs to aid state and private forestry; timber production, including the proposed Trinity Sustained Yield Unit; and National Forest Products Week. Files on national forests include material related to the reservation of national forest lands under the Bankhead-Jones Farm Tenant Act, recreation on New Hampshire national forests, National Lumber Manufacturers Association interest in salvage of insect damaged timber in the Pacific Northwest (ca. 1953), the inclusion of Tennessee Valley Authority lands in Nantahala National Forest (1954), Superior National Forest land purchases (ca. 1955), prison camps in Cleveland National Forest (ca. 1956), Operation Outdoors, boundary changes and development of national forests, executive orders or proposed executive orders, and matters requiring the attention of Sherman Adams. There are also files relating to appointments to the Outdoor Recreation Resources Review Commission and its Advisory Council (1958-59). Shelf list.

933 *General File.* The general file is largely correspondence from members of the public. Files relating to the conservation of natural resources contain materials relating to general policies of the administration, with correspondence of Sherman Adams, printed materials, etc. Topics include tree planting, watershed reforestation, etc.; specific national forest areas; appointments to the President's Quetico-Superior Committee; the Quetico-Superior Wilderness Area; the National Wilderness Preservation System legislation (1958-59); public domain involved in the proposed North Cascades National Park; national park campground conditions, proposed new parks, park administration, etc.; and similar materials on national monuments. Shelf list.

934 **Else, John Hubert** (1907-1968), Papers, 1925-1968, 4 boxes. National affairs counsel (lobbyist) for the National Retail Lumber Dealers Association, 1951-65. The correspondence, all in box 1, contains scattered references to work for the National Retail Lumber Dealers Association, chiefly thank-yous for good work (1957), and material relating to Else's resignation (1965). There are some letters from political leaders. Shelf list.

935 **Hamlin, John H.,** Records, 1956-1959. Hamlin was executive assistant to the president, serving as liaison between the executive branch and the Republican National Committee. He compiled analyses of the activities of various agencies, designed to inform the White House staff. Included are copies of a typed report, "Federal Conservation Activities" (n.d., 8 pp.), and

correspondence, memoranda, and reports of each bureau of the Department of the Interior (ca. 1958). Inventory.

936 **Harlow, Bryce, N.,** Records, 1953-1961, 20 feet. Special Assistant to the President, 1953; Administrative Assistant, 1953-58; Deputy Assistant for Congressional Affairs, 1958-61. Relating to congressional liaison are files of the Department of Agriculture, containing reports by Ezra Taft Benson and other material relating to legislation on the administration of national forests and other lands by the department (1953); files on the Department of the Interior include a memorandum by Horace Albright (November 27, 1953, ca. 10 pp.) proposing remarks for President Eisenhower at the Mid-Century Resources Conference on Conservation; also memoranda on Hells Canyon and power policy (ca. 1954, 3 items). File on Reorganization Plan No. 1 (1959), includes a mimeographed memorandum relating to proposed transfer of Department of Interior authority over national forests to the Department of Agriculture. There is also a file relating to legislation affecting timber sales policy (1956), information compiled on the antitrust difficulties of the Crown-Zellerbach Corporation when James D. Zellerbach was considered for a political appointment, and draft speeches of President Eisenhower to the Natural Resources Conference (December 2, 1953), on watershed legislation (August, 1954), and Montana forestry, Missoula (September 22, 1954). Shelf list.

937 **Kestnbaum, Meyer** (1896-1960), Records, 1955-1960, 11 1/3 feet. Special Assistant to the President and Chairman, Commission on Intergovernmental Relations. Major subjects of these records include conservation, forestry, etc., as dealt with in the recommendations of the 2nd Hoover Commission. Records relating to the implementation of these recommendations include reports concerning natural resources and conservation in the Department of Agriculture (1955), and correspondence on proposed legislation for grants-in-aid for reforestation (1955-56). General subject files relating to the Department of Agriculture include correspondence, memoranda, etc., concerning the administration of forest lands by a proposed Department of Natural Resources (1956-58), and a copy of the Hoover Commission recommendations; also reports, correspondence, etc., relating to conservation of land, forest, and water resources (1958). Typed list of contents.

938 **Lambie, James M., Jr.,** Records, 1952-1961, 25 feet. Special Assistant to the President in charge of the White House advertising liaison office, 1952-61, coordinating government claims for the use of the Advertising Council's facilities for public service campaigns. The records include files on forest fire prevention (1954-60) under the Department of Agriculture, including correspondence, reports, clippings, photographs, etc., relating to Smokey the Bear dolls, radio allocation plan schedule, Fire Prevention Week, Southern Forest Fire Prevention Conference (1956), and other regional fire prevention activities. Container list.

939 **Merriam, Robert E.,** Records, 1955-1961, 7 feet. Deputy Assistant to the President for Interdepartmental Affairs, 1958-1961. The records include some correspondence, reports, etc., relating to the President's Quetico-Superior Committee (ca. 1953-60). Container list.

940 **Morgan, Gerald D.,** Records, 1953-1961, 21 cubic feet. Special Assistant in the White House, 1953; Administrative Assistant to the President, 1953-55; Special Counsel to the President, 1955-58; Deputy Assistant to the President, 1958-61. Files on the administration's legislative program for the Department of Agriculture contain occasional references to the national forests. There is also a file on Reorganization Plan No. 2 for the Department of Agriculture (1953).

941 **Pyle, Howard,** Records, 1955-1959, 17 1/3 cubic feet (52 boxes). Deputy Assistant to the President for Intergovernmental Relations, 1955-59. The speech and engagement file contains correspondence, telegrams, etc., relating to Pyle's speeches, and a few copies of speeches, including those delivered to the American Paper and Pulp Company, New York City (February 3, 1956) and the Southern Forest Fire Prevention Conference (April 13, 1956). Shelf list.

942 **Seaton, Frederick Andrew** (b. 1909), Papers, 1946-1972, 100 cubic feet. U.S. Secretary of the Interior, 1956-61.

943 **Tudor, Ralph A.,** Papers, 1953-1954, ca. 1000 pp. Under Secretary of the Interior, 1953-54. The collection contains a copy of "Notes Recorded While Under Secretary," a newsletter written by Tudor for members of the staff of Tudor Engineering Corp., San Francisco, commenting on his activities in Washington; also a file (ca. 500 pp.) of carbons of letters written by Tudor as Under Secretary, or dictated by him for the signature of Secretary McKay, which concern a wide variety of federal conservation problems handled by the Interior Department, including the quality of the National Park Service, the proposed Echo Park Dam, proposed reorganization of the National Park Service, proposed Hells Canyon Dam, etc.

944 **U.S. Commission on Intergovernmental Relations,** Records, 1953-1955, 44½ feet (Record Group 220). The records consist of working papers of the commission and its various special study subcommittees. Among committee and meeting records relating to federal aid for conservation and natural resources are files of Samuel Trask Dana, and reports, memos, etc., regarding payments in lieu of taxes for the revested Oregon and California railroad grant lands in Lane County, Oregon. Records on natural resources (1953-55) include files relating to grants to states for natural resource management, including state and private forestry. Records concerning shared revenues contain files relating to grazing fees on national forests and grazing districts. Shelf list.

LAWRENCE

University of Kansas, Kenneth Spencer Research Library, Kansas Collection

Curator 66044 Telephone: (913) 864-4274

945 **Hixon Lumber Company** (Atchison, Kansas), Records, 1872-1928, 99 vols. Accounting records, including ledgers, journals, cashbooks, and accounting books.

TOPEKA

Kansas State Historical Society

State Archivist, 10th and Jackson Streets 66612 Telephone: (913) 296-3251

946 **Hodges Brothers Lumber Co.** (Olathe, Kansas), Records, 1888-1922, 11 boxes and 13 ledgers. Ledgers and daybooks.

947 **Kelsey, S.T.,** Papers, 1865-1866, 5 items. The papers consist of five lectures, in manuscript form, on trees and tree-growing. Some have been published in the *Western Home Journal* of Ottawa, Kansas.

Kentucky

BOWLING GREEN

Western Kentucky University, Kentucky Library and Museum

Manuscript Librarian 42101

948 **Green Family,** Papers, 1814-1971, 18 feet (7719 items). Correspondence, account books, and business papers of the Green family, of Falls of Rough, Kentucky. There are letters relating to the lumbering business (1885-97, ca. 120 items), and sawmill accounts (1896-98, 1903-06, 2 vols.). Unpublished guide, and card catalog.

FRANKFORT

Department of Library and Archives, Division of Archives and Records

Director, 851 East Main Street 40601 Telephone: (502)564-3616

949 **Kentucky. Department for Natural Resources and Environmental Protection, Forestry Division,** Records. Inactive administrative folders (1960-67), including correspondence and various reports on projects; fire control folders (1936-64); Cooperative Forest Management Program folders (1950-1968); map files (1932-65); director's files (1945-1965); assistant director's files (1940-1962); and reference and information materials.

LOUISVILLE

The Filson Club, Inc.

Office of the Secretary, 118 West Breckinridge Street 40203

950 **Charleson, Max,** Manuscript, 1931, 1 item (527 pages). "The Shakers of Kentucky, the Story of a Strange Sect." Indexed. Accompanied by a collection of photographs and photostats (73 items). Included are photographs of buildings and industries of the Shaker colony at Pleasant Hill, Mercer County, Kentucky; photostats of its manuscript records; photographs of buildings of the Shaker colony at South Union, Kentucky.

951 **Clark, Jonathan,** Papers, 1786, 1805-1810, 3 vols. The papers consist of an account book (1805-10, 10 pp.), recording charges for work on wagons, wheels, and plows; a notebook (June, 1786, 64 pp.), containing notes on lands and improvements of settlers in the Northern Neck of Virginia on the Narrow Passage survey, Denton and Palmer's tract, South River survey, and the Powell's Fort tract; and a notebook (July-August, 1786, 50 pp.), containing notes on lands and improvements of settlers in the Northern Neck of Virginia on the North Mountain tract, Opequan Creek, the Great Pond Branch, Swan Pond, Potomac River, Lick Branch, and the Shenandoah River. The notebooks give the names of settlers, number of acres cleared, descriptions of houses, barns, and other buildings, of bark, round logs, hewn logs, or scalped logs, with roofs of board shingles, or straw, etc. Sawmills are noted.

952 **Green, Willis,** Papers, 1818-1862, 3 boxes. Lumberman and U.S. Representative from Kentucky, 1839-45.

953 **Harding, John,** Papers, 1859-1914, 5½ boxes. The papers include a contract for building a two-story addition to a house, and statements of account and receipts for expenses of building (1896-1898).

954 **Journeymen Oak Coopers of the City of Louisville and Vicinity,** Papers, 30 l. Bylaws, names of officers, names of members, proceedings of meetings (March 16-October 13, 1839, and January 25, 1841), German translation of the constitution and bylaws.

955 **Joyes, Thomas,** Papers. Joyes cleared the woods where New Albany and Jeffersonville, Indiana, now stand, and supplied fuel for the early Ohio River steamboats.

956 **Needham, Charles K.,** "The du Pont Paper Mill and Artesian Well." A paper read at a meeting of the Filson Club, March 3, 1924.

957 **Thurman, B.H.,** Records, 1855-1871, 3 vols. Records of a lumber dealer of Louisville, Kentucky, consisting of two ledgers (1855-63; 1866-71), and a journal (1855-67).

958 **United Order of Believers (Shakers),** Records, 1815-1917. Journals (1815-1917) and account books (1874-1891) which contain accounts for Shaker industries and the construction of their buildings at Pleasant Hill.

Louisiana

BATON ROUGE

Louisiana State University and Agricultural and Mechanical College, Department of Archives

Archivist 70803

959 **Anonymous,** Lumber Company Account Book, 1841-1842, 1 vol. Daybook of an unidentified lumber dealer, possibly Andrew Brown, in Natchez, Mississippi, of accounts of sales of lumber and shingles, giving kind of lumber, sizes, prices, and quantity sold.

960 **Anonymous,** Lumberyard Daybook, April-August 1904, 1 vol. Daybook of a lumberyard probably in East Feliciana Parish. Entries give date, name of person placing order, itemized list of purchases, such as flooring, siding, pickets, lumber dressed or rough, price, and amount. Part of the East Feliciana Parish Archives Collection.

961 **Affleck, Thomas** (1812-1868), Papers, 1807-1938, 991 items. Includes a record book (1859-76), with entries of costs for the establishment of a sawmill and woodworking mill in Washington County, Texas, with entries of operating costs, and sales of lumber and related products.

962 **Batchelor, Albert A.,** Papers, 1852-1930, 18,322 items. This collection of papers of a planter and legislator from Pointe Coupée Parish, Louisiana, includes correspondence, bills, receipts and invoices with lumber merchants in New Orleans pertaining to the purchase of lumber for Negro cabins and cribs (1871); and contracts and other papers pertaining to the sale and cutting of timber (1906, 1908).

963 **Bruce, Seddon and Wilkins,** Plantation Records, 1741-1865, 605 items, and 6 vols. Papers from a sugar plantation near Convent, St. James Parish, Louisiana, include a cashbook and daybook recording business operations of a sawmill, including wages paid to slaves for cutting wood, orders, footage and other items.

964 **Butler, Judge Thomas, and Family,** Papers, 1793-1950, 9490 items, 165 vols. Personal and business papers of 3 generations of the Butler family of West Feliciana and Terrebonne Parishes, Louisiana, including letters and related papers (1913-1919), concerning the sale of cypress timber in Terrebonne Parish to lumber companies.

965 **Chaffe,** Record Books, 1887-1895, 3 vols. These business records of a cotton and sugar firm in New Orleans, which also manufactured cypress shingles and lumber in St. John the Baptist Parish, Louisiana, include a volume of daily shingle and lumber form reports (1893), giving amount cut, record of shipments, record of expenses, and related information.

966 **Corbin, Robert A.** (d. 1906), Papers, 1835-1917, 1212 items. Dealer in timber and farm lands in Tangipahoa and Livingston Parishes, Louisiana. Unpublished inventory.

967 **Crosset Paper Mill,** Records, ca. 1900-1961. The collection includes material relating to the Roaring River Logging Company, Wauna and Knappa, Oregon (ca. 1942-1948) ; and the Fordyce Lumber Co. (ca. 1926-61) . Names of persons include E.G. Gates, C.W. Gates; places include Crosset, Arkansas. The collection relates to forest management and to labor problems.

968 **Fordyce Lumber Company** (Fordyce, Arkansas) , Records, 1926-1963, 64 vols. Company records include inter-office correspondence; production, promotional and sales records; wage schedules; and photographs.

969 **Franklin Parish,** Lumber Photographs, 1923, 10 items. Photographs of logging operations near Gilbert, Franklin Parish, Louisiana.

970 **Garland, Henry L.,** Collection, 1881-1882, 1905-1909, n.d., 1 item and 11 vols. Account books of Garland and Garland, general merchants of Mamou, Evangeline Parish, and Garland, St. Landry Parish, Louisiana, include records of lumber purchases and ledger entries concerning sawmill expenses of the Garland Lumber Company (1904-1909) .

971 **Hawkins, J.E.,** Papers, 1880-1900, 4464 items, 184 vols. Professional and personal papers of a physician and farm owner of Evangeline Parish, Louisiana, which include correspondence and related papers concerning intermittent purchases of lumber (1880-1908) ; rent of land to lumber companies to "bank" timber (1903-1905) ; and litigation over timberlands (1906-1907) .

972 **Hursey, Asa H., and Family,** Papers, 1824-1903, n.d., 589 items and 1 vol. Papers of a sawmill operator and postmaster of Pearlington, Mississippi, include materials pertaining to the lumber industry in Canada, Mississippi, and Louisiana.

973 **Jackson Lumber Company** (Lockhart, Alabama), Records, 1902-1953, 20 vols. Annual reports, forestry data, and photographs.

974 **Johns-Manville Products Corporation,** Timberland records, 1939-1972 and n.d., 305 items, 2 vols., and 25 feet. Correspondence, reports and other materials pertaining to timberland management, fire control, timber sales, wood procurement, and other aspects of the management of the Johns-Manville timberlands in southwest Mississippi. Included is an unfinished unpublished history of these lands.

975 **Kellogg, Walter, Lumber Company, Inc.,** Records, 1918-1961, 7 feet. Plant, mills, yard and offices located in Alexandria and Monroe, Louisiana. Records include appraisals, minute books and records, accountant reports, monthly statements, and income tax returns.

976 **Kent, Amos, and Family,** Papers, 1770-1906, 128 items. Business and personal papers of a family of New England settlers in Louisiana, including items pertaining to the purchase of timberlands for Kent's mill and brickyard in Tangipahoa Parish (1867-1906) .

977 **Kilbourne, James Gilliam, Family,** Papers, 1869-1939, 108 items and 2 vols. Papers of planter family of East Feliciana Parish, Louisiana, include a record book (1870) pertaining to the sawing and hauling of lumber.

978 **Killian and Harry,** Account book, 1839-1852, 1 vol. Ledger of dealers in lumber, St. Helena Parish, Louisiana, listing kinds of lumber, sizes, prices, and quantities sold.

979 **Koch, Christian D., and Family,** Papers, 1845-1900, 3324 items, 11 vols. Papers of a family of Danish settlers in Hancock County, Mississippi, which include records of local transportation of lumber by schooner, and, after 1882, the operation of a steam planing mill, and a sash, door, and moulding factory by two members of the family in Bozeman, Montana.

980 **LeBlanc, Sam A.,** Judicial File, 1921-1955, ca. 1,875 items. Papers of a Louisiana judge including correspondence and other papers mentioning the cypress industry (1951, 1953).

981 **Liddell, Moses and St. John R., and Family,** Papers, 1838-1870, 6261 items, 49 vols. Plantation records, personal correspondence, business and legal papers, account books, notebooks, and plantation diaries of the Liddell family, cotton and sugar planters and owners of a sawmill in Morehouse Parish, Louisiana. A lumber record book (1856-1959) shows sales of pine and cypress lumber; plantation diaries (1839-1844) , for the Catahoula Parish plantation have entries concerning felling timber and rolling, hauling, and splitting logs.

982 **Lords Commissioners for Trade and Plantation,** Letterbook, 1700-1721, 1 vol. Copies of letters, memorials, and reports to the British colonial commissioners, including items pertaining to the log wood industry and trade in Middle America and the West Indies.

983 **Lyons Lumber Company** (Garyville, St. John Parish, Louisiana) , Records, 1905-1925, ca. 100 feet. Records include correspondence, financial records, logging records and other materials.

984 **McCollam, Andrew, and Ellen E. McCollam,** Diaries, 1839-1867, 3 items, 2 vols. Records of a surveyor and planter of Ascension and Assumption Parishes, Louisiana, and of his wife. Ellen McCollam's diary contains entries pertaining to the operation of a plantation sugar mill, sawmill, and brick kiln.

985 **McCutchon, Samuel,** Papers, 1832-1874, 82 items and 8 vols. Records of a sugar planter of St. Charles Parish, Louisiana, and the manager of a sugar plantation near Belize, British Honduras. Diary entries (1838-1842) for the Louisiana plantation include records of work assignments of slaves cutting timber, particularly cordwood. Entries (1868-1871) for the Honduras plantation include reports on the operation and management of the sawmill.

986 **Moreland, William F.,** Records, 1834-1867, 3 vols. The records of a Georgia planter comprising 2 account books and a diary of a journey from Alabama via New Orleans to East Texas in 1850, with frequent mention of the nature of the soil and timber.

987 **Newsom, M.S., and Family,** Papers, 1870-1910, ca. 3000 items, 163 vols. Business and some personal papers of M.S. Newsom and his sons, Maston S. and Samuel M. Newsom, owners of the Newsom Brick and Lumber Company, Tangipahoa Parish, Louisiana, which include journals (1891-1905) and letterbooks (1886-1907) of the lumber company.

988 **Pharr, John N., and Family,** papers, 1848-1941, 15,000 items, 350 vols. Business papers and account books of a St. Mary Parish sugar planter, including materials pertaining to timber enterprises, and the establishment of a sawmill in 1874.

989 **Shamrock Lumber Company** (Natchitoches Parish, Louisiana) , Account book, 1905-1908, 1 vol. A record of the assets and liabilities of the Brown Lumber Company purchased as stock, and entries of debit and credit of the Shamrock Company for lumber accounts, freight, logging, payrolls, and similar expenses.

990 **Southern Pine Association,** Records, ca. 1903-1948, ca. 300 feet. Correspondence, press releases, minutes, form letters, photographs, and related material concerning admin-

istrative work and services of the association (now known as the Southern Forest Products Association) to its subscribers, including the provision of statistical information on production, shipments, stocks on hand; the establishment of uniform grades, and maintenance of an inspection service to grade lumber; scientific and engineering research; trade promotion and advertising; conservation, reforestation; traffic; labor information; mechanical efficiency; legislation and governmental relations; and similar matters. The collection includes some records of predecessor organizations such as the Yellow Pine Manufacturers Association (ca. 1903-1915) and a manuscript copy of the memoirs of James Boyd.

991 **Stewart, Robert H.,** Records, 1834-1904, 140 items, 51 vols. Business records of a Natchez, Mississippi, mortician and cabinetmaker and dealer, including ledgers (1834-52) containing accounts for making and repairing furniture; a ledger (1834-1857) containing accounts with employees, principally cabinetmakers; daybooks (1859-1861) of a furniture store in New Orleans; and related account books.

992 **Taliaferro, James G., and Family,** Papers, 1787-1934, 895 items, 4 vols. Personal and business papers principally of a Louisiana judge, but including also business papers pertaining to operation of a sawmill in Virginia (1804-1807) and in Catahoula Parish, Louisiana (1807-27, 1856-60).

993 **Taylor, Calvin, and Family,** Collection, 1813-1913, 1,975 pieces and 17 vols. Business records and personal papers of Calvin Taylor and family of Satartia, Yazoo County, Mississippi. The business records are concerned with the lumber industry in southeastern Mississippi and trading connected therewith; land speculation; and salt-making. Among the lumbering records are material on the prospects for establishing a sawmill near Bayou Macon, Louisiana (1827); a partnership with Gordon Davis in a sawmill (1846); letters describing prospects for a sawmill in Texas (late 1840s); formation of the firm of Taylor and Fowler (1850-53); Taylor's lumber interests after the Civil War, and a patent for an improvement in trucks for hauling logs (1869). The 13 volumes of diaries also contain account books (1831-1906).

994 **Tucker, J.M.B.,** Papers, 1835-1925, 592 items, 11 vols. Personal papers of a Natchitoches Parish, Louisiana, judge, including items pertaining to timber and timberlands in northwest Louisiana and Texas (1890-99).

995 **Verret, Theodule,** Papers, 1850-1870, 627 items, 7 vols. Business papers relating principally to a sawmill and lumber business in St. Tammany Parish, Louisiana. Partly in French.

Louisiana State University and Agricultural and Mechanical College, Library, Louisiana and Rare Book Rooms

Louisiana and Rare Book Rooms 70803. Doctoral dissertations and theses in all subjects done at Louisiana State University are listed in *LSU Forestry Notes,* Louisiana State University & A & M College, School of Forestry and Wildlife Management, Baton Rouge, Louisiana, Numbers 17, 22, 36, 45, 54, 62, 64, 73, 84, 96, 103.

Theses

996 **Cole, Fred C.** "The Early Life of Thomas Affleck, 1813-1842," Baton Rouge: Louisiana State University, M.A. Thesis, 1936.

997 **Fickle, James.** "History of the Southern Pine Association, 1914-1920," Baton Rouge: Louisiana State University, M.A. thesis (History), 1963.

998 **Lowry, Stanley Todd.** "Henry Hardtner, Pioneer in Southern Forestry: An Analysis of the Economic Basis of His Reforestation Program," Baton Rouge: Louisiana State University, M.A. thesis (Economics), 1957?

999 **McBride, James Patrick.** "The Evolution of the Louisiana Hardwood Lumber Industry: The Effect of Changes in Resource Factors on Size and Number of Mills," Baton Rouge: Louisiana State University, M.A. thesis (Economics), 1972.

1000 **Wimbush, Solomon Mitchell.** "Louisiana: The Civilian Conservation Corps Black Camps in District 'E', 1933-1942," Baton Rouge: Louisiana State University, M.A. thesis (History), 1972.

LAFAYETTE

*University of Southwestern Louisiana, Southwestern Archives and Manuscripts Collection

The following entry was taken from the *National Union Catalog of Manuscript Collections* (item #MS 67-2034).

1001 **Live Oak Society,** Papers, 1900-1950, 96 items. Correspondence, speeches, and other papers, relating to an organization at Alexandria, Louisiana, for the study and preservation of Louisiana's live oak trees, collected by Horace A. White, of Alexandria, Louisiana.

NATCHITOCHES

Northwestern State University of Louisiana, Eugene P. Watson Memorial Library, Division of Special Collections

Director 71457

1002 **Dormon, Caroline,** Collection, 1890s-1972, 61½ feet. Forestry instructor, Louisiana State University. The collection contains 5,000 manuscripts, including diaries, unpublished writings, letters, etc.; also 50 maps; 170 original paintings, sketches, and drawings; 300 colored prints; 500 slides. Associated with the collection are 50 miscellaneous artifacts, and an additional 46 linear feet of books, magazines, and other printed materials.

NEW ORLEANS

Tulane University Library, Howard-Tilton Memorial Library, Special Collections Division

Director 70118. In addition to the collections noted below, the division holds other forest history sources, not cataloged as such, in the form of bills of lading of lumber shipments on the Mississippi River, scattered throughout the steamboat collection, and references to timber cutting by Negro slaves scattered throughout the plantation records.

1003 **Diary,** 1868, 1 vol. (84 pp.). The diary contains the daily jottings of a lumberman who hauled lumber near Pearlington and Logtown, Mississippi.

1004 **Otis, Frank G.,** Papers, 1888-1961, ca. 6,500 items. Personal and business papers of Otis and of the Otis Manufacturing Company, a lumber mill in New Orleans specializing in tropical hardwoods, but also dealing in domestic lumber.

RUSTON

Louisiana Tech University, Prescott Memorial Library, Louisiana Tech Archives

Archivist 71270

1005 **Louisiana Forestry Commission,** Records and research reports, ca. 1913-1940, 5 file boxes. Largely research reports on forestry in Louisiana, together with occasional correspondence, clippings, pamphlets, statistical records, and other forestry information materials.

Maine

AUGUSTA

Maine State Archives

State House 04330

1006 Included in the records of the Maine Land Office and the Maine Forestry Department, the agencies charged with inspection and care of public lands, preservation of the masts, timbers and trees thereon, and inquiry into trespasses, are deeds and related materials (1794-1949); field notes (1803-1890), containing information about boundaries, forest growth, etc.; maps; and miscellaneous other records. The Black Papers, containing records of Colonel John Black, agent in Maine for the Bingham Estate, are available on microfilm.

Maine State Library

Librarian 04330 Telephone: (207) 289-3561

1007 **Avery, Myron H.** (b. 1899), Manuscript, 96 pp. A bibliography of Mt. Katahdin, compiled by Edward E.S.C. Smith of Biddeford and Myron H. Avery of North Lubec; second revision by Myron Avery (1926). Annotated. Refers to Appalachian Mountain Club, wildlife, forest fires, game preserve, lumbering, proposed national park, state park, trails, and other aspects of Mt. Katahdin and vicinity. Additional manuscripts are included pertaining to the further revision of the bibliography. Typed copy; originals in possession of Avery.

1008 **Bangor, Maine,** Tax Records, 1859, 1 vol. Valuation of property and tax lists of Bangor, including stocks of lumber, saw and shingle mill.

1009 **Carmel, Maine,** Manuscript report, July 19, 1836, 1 vol. Contains an account of the lumbering operations in Carmel, Maine. Records made to the topography of the township, location, vegetation, population, size in acreage, streams, ponds, roads, history including sketches of a few land deeds, description and valuation of lots, record of land sales and proposed canal and railroad. Tax lists (1812-1836), accounts of sales, and extracts from business letters are included in the notebook.

1010 **Draper, Mrs. Edward** (Collector), Photographs, 1900-1926? Slides and photographs of Maine scenes, including lumbering. There are slides (10 cases), photograph albums (5 vols.), glass and film negatives, some loose photographs, and a Delineascope.

1011 **Pike, Charles F.** (1797-1886), Papers, 1828-1864, 257 pieces. Kingfield, Maine. Miscellaneous papers of a merchant and lumberman, including ledgers, notes on sales, receipts, materials concerning court cases, etc.

1012 **Storer, John,** Journal, September 29-December 6, 1740. Copy of report of repairing the fort at Richmond, Maine, on the Kennebec River, including cutting and hewing timber, and construction of the fort.

1013 **Union, Maine,** Tax List, December 22, 1848. Valuation made on May 1, 1849, by the assessors of Union. Includes appraisal of sawmills, stave mills, shingle mills, bark mills, lumber, etc.

1014 **Vassalborough, Maine,** Assessors records, 1939, 1 vol. True copy attested by Isaac Fairfield, Oliver Prescott, and Oliver A. Webber, assessors of the town of Vassalborough. Includes valuation of sawmills, shingle machines, etc.

1015 **Westbrook, Maine,** Tax List, 1829, 1 vol. True copy of tax list of Westbrook, attested by local assessors. Among the business enterprises mentioned are lumbering, ship yards, sawmills, box factories.

Maine State Museum

Director, State House 04330 Telephone: (207) 289-2301

1016 **Emerson and Stevens Axe Factory** (Oakland, Maine), Papers, 1870-1960s, 30 + feet. Letters, receipts, ledgers, account books, advertising, photographs, etc. The collection is accompanied by a short motion picture of axe making at the factory, and such artifacts as axe heads in various stages of completion, a working model of a triphammer, some axe samples, and a small number of axe-making tools.

1017 **Lombard Company** (Waterville, Maine), Papers, ca. 2 feet. Manufacturer of the Lombard Log Hauler. The papers include numerous patents, miscellaneous correspondence, advertising materials, abstracts of court cases, records of sales, photographs, and similar materials. There is also a motion picture of the Lombard operating in the woods.

1018 **Maine. State Land Agent,** Letters, ca. 1830s-1840s, ca. 100 items. Letters received from various individuals, mostly concerned with the northern Maine region.

1019 **Stetson Company** (Bangor, Maine), Papers, ca. 1890-1920, 20 + feet. Business letters, account books, receipts, freight bills, stumpage permits, field notes, etc., along with a collection of books, pamphlets, and periodicals, mostly early 20th-century, relating to lumbering, that were used by the company. A quarter to a third of the materials in the collection are devoted to lumbering.

BANGOR

Bangor Historical Society

Secretary, G.A.R. Memorial, 159 Union Street 04401

1020 The holdings of the society were hastily boxed during removal from the Bangor Public Library, and have not yet been reorganized (1975). From time to time various materials pertaining to forest history have been reported in *A Reference List of Manuscripts Relating to the History of Maine,* edited by Elizabeth Ring, Orono: 1938; *Lumbering in the Maine Woods: A Bibliographical Guide,* compiled by David C. Smith, Portland: 1971; and in accessions announcements in *Forest History.* Among these items are account books of Bangor business firms, including an account book of lumber carried by ships owned by James Treat of Winterport, J.A. Buck of Orland, and Joseph Buck of Bucksport (1868); records of Hinckley and Egery, a foundary which manufactured sawmills and traded in lumber (1828-95); papers of Ira Pitman, including agreements with James F. Ranson to cut timber in Penobscot County (1857); and hearings and testimony of the St. John River Commission, containing information on boundary disputes and wood practices since 1870 (1909-16, 6 vols.).

BRUNSWICK

Pejepscot Historical Society

Brunswick Public Library 04011

1021 The holdings, uncataloged, have not yet been arranged after their removal to the Brunswick Public Library (1975). Among items pertaining to forest history are a record book (1835-38, 1 vol.) of Tibbets & Howland (Brunswick, Maine), containing accounts of log rafting and sales of lumber and shingles; miscellaneous papers relating to the lumber operations of the Wilson family, Topsham, Maine (1807-08); and two or three other account books, origins unknown, which refer to timber surveys and sales (ca. 1796, 1799, and 1822).

CALAIS

Calais Free Library

Librarian 04619 Telephone: (207) 454-3223

1022 **Hayden, Richard V.,** Diaries, 1824-1867, 60 vols. Both originals and Xerox copies of notebooks kept by a surveyor in Robbinston, Maine, and surrounding country, containing entries on lumbering, shipping, surveying, and personal matters.

1023 **Pike, James Sheppard** (1811-1882), Collection. Included are a deed from Benjamin Young and Munroe Hill of Calais to J.S. Pike for 1/8 of a privilege of cutting timber on the Indian township (1858); extracts (1877) from Joseph Porter papers and account books (1788-1810) relating to lumbering and mills on the St. Croix River; and papers relating to the Calais Boom Corporation (1856-60, 9 items). There are also bills and notes (1789-1817, 9 items) of Joseph Porter relating to the lumber business in Calais, and account books (1797-1835, 2 vols.) of William Pike (d. 1818) of Pownalborough, Wiscasset, and Calais, Maine, including accounts for lumber.

ELLSWORTH

The Colonel Black House

Chairman of the Black House Committee 04605

1024 **Black House Papers,** 1797-1880, ca. 600 vols. and 25 cubic feet. Financial journals, ledgers, and cashbooks of the Black family enterprises in lumbering, shipping, mills, dry goods, and land agency; also memorandum books used in the mills, on trips, and during logging drives. Letterbooks (11 vols.) include details of excursions into unsettled areas on the Kennebec tract and the Penobscot tract. Small diaries kept by subordinate personnel relate to collection of payments and to investigation of timber thefts. There is related family business correspondence, and many unrelated records of individuals who lived in Ellsworth. In 1974-76 the collection was in the process of being sorted and microfilmed. The originals were to be returned to the Black House, but not opened for research, while the Maine State Archives, Augusta, will have a microfilm copy available for use.

ORONO

University of Maine at Orono, Raymond H. Fogler Library, Special Collections Department

Head, Special Collections 04473 Telephone: (207) 866-7328.

Holdings are listed in a brochure, *The Special Collections,* Orono: The University of Maine, the Raymond H. Fogler Library, 1974, 16 pp.

Collections

1025 **Baker, Gregory,** Collection, 1900-1940, 2 boxes. Photographs and photographic negatives of lumbering operations in Maine in the early 20th century.

1026 **Baldwin, Henry Ives,** Collection, 1900-1950, 2 feet. Correspondence, project reports, radio scripts, relating to conservation, forestry, wood utilization, and his career as a research forester in New Hampshire.

1027 **Black & Co.** (Ellsworth, Maine), Records, 1839-1890, 9 vols. and 1 box. Records of a lumber company, including account books (1839-60, 9 vols.) and receipts and correspondence (1839-90. 1 box).

1028 **Cary, Austin,** Papers, 1865-1936, 1 folder (10 items). Included are statistics on production of lumber, lath, and shingles by states and species (1904-06); census of forest trees in Maine (1880); and timber sale forms (ca. 1900).

1029 **Cary Family Collection,** 1856-1919, 1 box. Contains diaries of various members of the Cary family, including Seth Cary (1856-1919, 32 vols.), Topsham, Maine.

1030 **Chamberlain Family,** Papers, 1800-1914, 1 box. Included is material relating to lumbering operations in Maine during the 1800s, with hand-drawn maps by Joshua Chamberlain (ca. 1890-1910), relating to lumbering areas and towns in Maine.

1031 **Chandler Family,** Papers, 1775-1930, 7 boxes. Correspondence and other records. Chandler, of New Gloucester, Maine, had substantial lumber interests in Franklin County, 1870-1890.

1032 **Cronk, Corydon P.,** Papers, 1900-1967, 17 feet. Forester and lumberman. Included are correspondence, notes, clippings, publications, U.S. Office of Price Administration papers, etc., containing information about lumbering, wood, logging operations, and forestry.

1033 **East Middlesex Canal Township Grant,** Maps, 1850-1950, 40 items. Logging and lumbering maps, most relating to Great Northern Paper Co., from the firm of I.K. and I.G. Stetson (ca. 1940).

1034 **Eckstorm, Fannie Hardy,** Collection, 1730-1947, ca. 12 boxes. Research notes, correspondence, drafts of articles, photocopies of articles, and related materials pertaining to Eckstorm's writings. On loan from the Bangor, Maine, Public Library.

1035 **Fellows, Oscar,** Papers, 1841-1925, 1 box. Contains material relating to the St. Johns River Commission in the early 1900s.

1036 **Fife, Hilda M.,** Papers, 1940-1970, 5 boxes. A collection of student papers by a former professor of English at the University of Maine, Orono. Many items relate to the Maine woods, woodsmen, and woodsmen's songs.

1037 **Gilbert, Fred Alliston,** Papers, 1900-1973, 25 boxes. The papers of a family from Bangor, Maine. Contains records relating to the Great Northern Paper Co., paper making, and lumbering. Uncataloged.

1038 **Guild, Daniel,** Papers, 1800-1900, 2 boxes. Papers, and some account books, of the 1800s and early 1900s, many relating to ships and shipping. Included is a diary and information on trips into forests around Bridgetown, Nova Scotia.

1039 **Hamlin Family,** Papers, 1823-1920, 4 file cabinets. The collection consists almost entirely of the letters of Hannibal Hamlin (1809-91). U.S. Senator from Maine, 1849-61, 1869-81; Vice President of the U.S., 1861-65. Included is much information about the state's forests.

1040 **Holden, Margo,** Papers, 1965 +, ca. 30 items. Manuscripts of Holden's writings, including *Down the Allagash,* short stories, and articles on the Maine woods.

1041 **Ingersoll, George W.** (1803-1860), Papers, 1835-1856, 100 items. Lawyer, businessman, and Collector of the Port of Bangor, Maine. The collection includes letters, bills, and receipts, mostly pertaining to logging in Maine.

1042 **International Brotherhood of Paper Makers,** Records, 1927-1967, 1 box. Lewiston, Maine.

1043 **Ireland, Charles R.,** Collection, 1806-1894, 1 box. Included is a cashbook of Ferguson and Conner, containing many entries relating to timber and lumber (July, 1866-August, 1868).

1044 **Ledgers and Account Books,** 1800-1930. Includes ledger and diary (1856-71) of Samuel G. Bachelder, relating to forestry and timber cutting; account book (1866-68, 1 vol.) of Blunt and Hinman (Bangor, Maine), relating to lumbering in Maine; account book (1881) of Hiram Davis of Winn, containing estimates of timber in Aroostook County, Maine; ledger (1869-91) of A.L. Harmon (?), recording surveys of timber lots; sawmill record book (1835-44) of Kilby and Foster; ledger of newspaper clippings (1873-1910) on lumbering, wildlands, tax assessments, with permits (1874-97) for cutting, belonging to Prentiss and Carlisle; ledger of the Seboois Log Driving Co., account book of White and Wadleigh (Old Town, Maine), containing accounts (1865-66) for lumbering operations on Indian Purchase No. 3.

1045 **Lumbering Operations in the State of Maine,** Photographs, 88 items. Photographs of lumbermen and lumber operations in Maine during the latter part of the 19th and early part of the 20th centuries.

1046 **McIntire, Clifford Guy** (b. 1908), Papers, 1951-1965, 42 feet. U.S. Representative from Maine, 1951-65; served on the Agriculture Committee. The collection includes records from his service in the House. There is also much information concerning Maine's forests.

1047 **Mattawamkeag Log Drive Co.,** Charter and bylaws, 1885.

1048 **Melcher, William D.,** Article. "The Bulldog Sluice," the story of lumber cutting and driving lumber to the river by means of a sluice (6 pp.).

1049 **Peirce Family,** Papers, 1800-1930, 9 feet. The business papers of Hayford and Waldo T. Peirce, relating to lumbering, real estate, and land titles. Included are a copy of woods scale of logs cut on lands belonging to W.B. Hayford (1872-73); W.B. Hayford journal (1871-87); William B. Hayford, stumpage account book (1870-1922); Hayford and Stetson, journal (1877-1882); Seboois Log Driving Co., records of incorporation and directors' records (1880); timberland ledger and account book (1869-1912, 1 vol.); woods scale record book (1927-29).

1050 **Penobscot Boom Corporation,** Charter, 1832.

1051 **Penobscot East Branch Log Driving Company,** Charter and bylaws, 1897.

1052 **Penobscot Log Driving Company,** Charter and bylaws, 1846.

1053 **Penobscot Lumbering Association,** Records, 1850s, 28 items. Records kept for the operation of the Penobscot Boom.

1054 **Penobscot Lumbering Association,** Ledgers, 1849-1913, 200 vols. Included are scaler reports, daily journals, bills paid, wangan expenses, and other items relating to the day to day activity of the Penobscot Boom. Associated materials include minutes of a Masonic lodge and an insurance agent's letterbook (1897-1906).

1055 **Pike, James Sheppard** (1811-1882), Papers, 1850-1880, 634 items. Correspondence, including letters formerly in the Calais Free Library, discussing lumbering on the St. Croix River, Maine (1861-63, 3 items).

1056 **Soule, Harris W.,** Papers, undated, 1 folder. Research materials gathered for his book, *Northwoods Tales and Unusual Recipes.*

1057 **Sprinchorn, Carl,** Papers, 1887-1971, 10 feet. Artist who painted the Maine woods.

1058 **Stetson, Clarence,** Papers, 1916-1920s, 1 box. Business correspondence, receipts, and personal correspondence relating to the Stetson estate, timberlands, and property at Hancock, Maine.

1059 **Wakefield, O.H.,** Papers, 1869-1938, 2 boxes. Papers and ledgers relating to a sawmill on the Passadumkeag River at Oxford, Maine, and other material relating to lumbering, covering from the mid-1800s to the mid-1900s.

Theses

1060 **Bearce, Arthur B.** "Regional Elements in the Novels and Poems of Holman F. Day." M.A. thesis, 1965.

1061 **Carter, Harland H.** "A History of the Cumberland and Oxford Canal." M.A. thesis, 1950.

1062 **DiMeglio, John E.** "Civil War Bangor." M.A. thesis, 1967. Includes information on lumbering.

1063 **Eastman, Joel W.** "A History of the Katahdin Iron Works," M.A. thesis, 1962.

1064 **Goode, Robert Donald.** "The Economic Growth of the Pulp and Paper Industry in Maine," M.A. thesis, 1934.

1065 **Herrick, Rebecca B.** "A Century of Shipbuilding in Blue Hill, Maine, 1792-1892." M.A. thesis, 1965.

1066 **McGuire, Harvey P.** "The Civilian Conservation Corps in Maine, 1933-1942." M.A. thesis, 1966.

1067 **Riley, George A.** "A History of Tanning in the State of Maine." M.A. thesis, 1935.

1068 **Roundy, Charles Gould.** "Changing Attitudes Toward the Maine Wilderness." M.A. thesis, 1970.

1069 **Sewall, Joseph Herbert.** "A Comparative Study of the Development of Forest Policy in Maine and New Brunswick." M.S. thesis, 1957.

1070 **Sherman, Ivan C.** "The Life and Work of Holman Francis Day." M.A. thesis, 1932.

1071 **Sherman, Rexford B.** "The Bangor and Aroostook Railroad and the Development of the Port of Searsport." M.A. thesis, 1966.

1072 **Stanley, Robert D.** "The Rise of the Penobscot Lumber Industry to 1860." M.A. thesis, 1958.

1073 **Stevens, Thalia O.** "The Saint John River Commission, 1909-1916." M.A. thesis, 1970.

1074 **Wescott, Richard R.** "Economic, Social, and Governmental Aspects of the Development of Maine's Vacation Industry, 1850-1920." M.A. thesis, 1959.

1075 **Wilde, Margaret F.** "History of the Public Land Policy of Maine, 1620-1820." M.A. thesis, 1932.

University of Maine at Orono, Northeast Archives of Folklore and Oral History

Director, Northeast Archives of Folklore and Oral History, Department of Anthropology, Stevens Hall South 04473
Telephone: (207) 581-7466. The collections are cataloged in

The Northeast Archives of Folklore and Oral History: A Brief Description and A Catalog of Its Holdings, by Florence Ireland, Orono: The Northeast Folklore Society under the Auspices of the Department of Anthropology, University of Maine, 1973 (*Northeast Folklore,* XIII, 1972).

1076 **Northeast Archives of Folklore and Oral History,** 1962-1972, 744 collections. Included are student papers, taped interviews, transcripts; also photographs of lumbering operations and handwritten memoirs collected by students. The Lumbermen's Life Project contains some 40 collections and parts of many others, over 95 hours of taped interviews, and an index of lumbering terms (in progress, 1975). The subjects of the interviews were men in their eighties and nineties, usually workmen in the northeastern woods between 1880 and 1920. There are indexes for personal names and place names, and a lumbering index, complete for 545 collections, containing all references to lumber operations, including data on equipment, food, accidents, songs, hours, salary, and every aspect of lumbering mentioned. Lumbering topics include ballads, songs, poems, tales and stories, the "Lumberman's Alphabet," Miramichi fire, river drives, Paul Bunyan, labor unions, camp social life, the Great Depression, camp food, wood carving, medicine, life as a forest service warden, biographical sketches, the Great Northern Paper Company, the Industrial Workers of the World, International Brotherhood of Pulp, Sulphite, and Paper Mill Workers, and International Brotherhood of Paper Makers. Places mentioned include Maine, New Hampshire, and New Brunswick; among the specific localities are Hainesville, Moosehead Lake, Great Pond, Ellsworth, Machias, and Topsfield, Maine; also the Penobscot, Union, and Machias Rivers, Maine.

1077 **Edward D. Ives Collection,** 1956-1970, 70 reels of tape plus transcripts. Transcripts of interviews conducted by Ives, Director of the Northeast Archives of Folklore and Oral History, including folksongs and information on the people who wrote them and sing them, collected in connection with books by Ives; also tapes dealing with all aspects of woods work. Most tapes have transcriptions. Geographical areas include Maine, Nova Scotia, and New Brunswick; among specific places mentioned are the St. Croix River, Union River, Rumford, and Rangeley, Maine. Topics include ballads (sung by Charles Sibley), musical instruments, entertainment in lumber camps, relations between the French and English in New Brunswick, river driving, booms, camp living arrangements, lice, food, yarding, log marks, Larry Gorman, prices of wood, lumbering terms, sawmills around Veazie, Bangor, and Brewer, Maine, Joe Scott, and Charles DeWitt. Copies of some of the tapes in this collection are on deposit at the Archives of Traditional Music, Maxwell Hall, Indiana University, Bloomington, Indiana, and at the Archive of Folksong, Music Division, Library of Congress, Washington, D.C.

PORTLAND

Maine Historical Society

Curator of Manuscripts, 485 Congress Street 04111

1078 **Adde & Company** (Portland, Maine), Records, ca. 1899-1969, ca. 33 feet. Records of a firm of machinists and manufacturers of lumber meters and picket fence pointers, including letterbooks, journals, cashbooks, salesbooks, purchase registers, bills receivable, time and cost distribution, trial balances, invoices, and correspondence files. Unprocessed.

1079 **Babb, Henry,** Letters patent, January 13, 1824. Letters patent on parchment containing a description by Henry Babb of an invention for construction of booms for enclosing logs in a raft. Granted and signed by President James Madison, and Secretary of State John Quincy Adams, Attorney General William Wirt, Jonathan Morgan, and John Cloutman.

1080 **Barron, John,** Papers, 1830-1859, 1 folder. Correspondence and papers of John and William Barron of Topsham, Maine, regarding their lumber and shipping business.

1081 **Bean, Wayne R.,** Manuscript, 1972, 37 l. "Calais [Maine]: Its Lumber Hinterland, 1840-1923." Typescript, including maps and illustrations.

1082 **Boston and Eastern Land and Mill Company,** Records, 1835-1850, ca. 800 items. Correspondence, business and legal papers, maps, and other papers concerning the company's lumbering operations in and around Machias, Maine. Includes material on the purchase of standing timber, logging operations, the problems of maintaining the camps for lumberjacks, milling the lumber, and the shipping and sale of the finished product. The firm was originally known as the Machias West River Land and Mill Company.

1083 **Bradley, Rev. Caleb** (1771-1861), Miscellaneous papers, 1797-1828. Portland, Maine. Includes accounts, bills, and receipts for labor, lumber, and other goods.

1084 **Bragdon, Solomon** (1710-1778), Account books, 1745-1762, 2 vols. Scarborough, Maine. The first book (1745-1759) includes accounts for timber hauled to ships.

1085 **Cary, Shepard** (1805-1866), Papers, 1832-1895, 1 foot. Papers and letters on politics and the lumber business.

1086 **Conlogue, Eugene James,** Manuscript, 24 l. "Riparian History of the Aroostook River." Carbon typescript.

1087 **Dodge Family,** Papers, 1809-1909, 12 boxes. Miscellaneous papers of a family of Blue Hill and Ellsworth, Maine, and Worcester, Massachusetts, including material relating to lumbering. Inventory.

1088 **Emery, Moses, et al.,** Articles of agreement, March 3, 1773. Agreement between Moses Little, Samuel Gerrish, Edmund Bagley, Robert Sargent, and Jonathan Bagley, all of Essex County, Massachusetts, a committee chosen by the Proprietors of Bakerstown (Poland), Maine, and Moses Emery, by which Emery agrees to build a sawmill in Bakerstown.

1089 **Fitch, Luther** (1783-1870), Miscellaneous papers, ca. 40 items. Portland, Maine. The papers include bills, receipts, invitations, etc., among them a list of sale of logs, sold by John Merritt (?) for William Fitch, Westbrook, Maine (December 28, 1824).

1090 **Fitch, William,** Papers, ca. 1829-30, 6 items. Miscellaneous papers pertaining to the lumbering activities of William Fitch and Son, and Benjamin Chadbourne, including survey lists and accounts.

1091 **Foss, James,** Account book, 1826-1857, 1 vol. Surveys of timber, logs hauled, boards sawed, Limington, Maine. Also included are War of 1812 land warrants.

1092 **Freeman, Enoch** (1706-1788), Journal, 1742-1753. Falmouth, Maine. Includes accounts of the rent of a mill at Presumpscot and invoices of cargoes of boards and planks shipped on board several sloops.

1093 **Goodwin, James,** Deed, April 11, 1807. Sale of 1/24 part of a double sawmill located at Salmon Falls in Berwick, Maine, to Elisha Goodwin.

1094 **Greenleaf-Merrill Family,** Papers, 1819-1917, 3 boxes, 8 vols. and 1 package. Included are business papers of the

Adams H. Merrill Family of Williamsburg, Maine, and Bangor, Maine, relating to lumbering and other activities. Inventory.

1095 **Hobbs, Elizabeth K.,** Collection, 1636-1809, 327 items. Documents and papers relating to the history of the Piscataqua River settlements in Maine and New Hampshire, including lumber accounts and other papers. Inventory.

1096 **Howard, Samuel and William,** Account book, 1773-1793. Hallowell, Maine. Includes accounts for lumber.

1097 **Hubbard, D.,** Agreement, February 15, 1806. Permission to Elisha Goodwin, Berwick, Maine, to cut timber on Hubbard's land, Lebanon, Maine.

1098 **Ilsley, Daniel** (1740-1813), Account book, 1790-1807. Portland, Maine. Includes accounts for shipbuilding, lumber and launching ships.

1099 **Johnson Family,** Papers, 1755-1855, ca. 100 items. Stroudwater (Westbrook), Maine. Includes business agreements, some concerning lumber, of John Johnson (1755-1838).

1100 **King, Richard** (1718-1775), Daybook, 1772-1774. Scarborough, Maine. Includes accounts for labor, lumber, a new vessel being built, etc.

1101 **King, William** (1768-1852), Papers, 1760-1834, 24 boxes, 4 packages. Bath, Maine. Included is a bundle of papers (1801-1802), mostly relating to shipments of lumber to Liverpool. There is also a letter from E. Wheelock Snow (March 27, 1820), about timber and agriculture in Atkinson, Maine, and an account book containing many charges against General Henry Knox for building materials.

1102 **Knight, Joyce Wilson,** Collection, 1803-1897, 42 vols. Daybooks and ledgers of several general merchandisers of Benton, Maine, and other places. Included is information on the cost of timber rafting and the price of lumber and timber products. Inventory.

1103 **Knight, Samuel M.** (d. 1871), Papers, 1831-1884, 236 pieces. Memoranda, contracts, agreements, bills, receipts, and insurance policies of a Falmouth, Maine, shipbuilder, including items pertaining to labor costs, procurement of ship timbers, etc.

1104 **Knox, Henry** (1750-1806), Papers, 1715-1806, 12 boxes, 8 vols. Papers (1715-1788, 1 box), are largely those of Samuel Waldo, Isaac Winslow, and Thomas Flucker relative to land transactions in the Waldo Patent. Knox's papers relate primarily to his business enterprises, including lumbering and shipping in and around Thomastown, Maine. The eight volumes comprise one entitled "Lands" (1788), a "Farm Journal" (1799-1806), and ledgers and personal account books (1791-1804).

1105 **Larrabee, Benjamin** (1700-1784), Account book, 1751-1761. Falmouth, Maine. Includes accounts for wood. Stephen Longfellow and Benjamin Preble are mentioned.

1106 **Lewis, John** (1717-1803), Papers, 1730-1843, 2 boxes. North Yarmouth, Maine. Includes a memorandum of deed of land and sawmill on Royal's River.

1107 **Littlefield, Edmund** (d. 1661), Contract, February 27, 1655. Giving Littlefield the right to erect sawmills in Wells, Maine, and to fell trees on unfenced lots; verso contains a notice (July 5, 1656) by Littlefield which specifies the prices to be charged for boards and planks of various sizes.

1108 **Longfellow, Stephen** (1723-1790), Account book, 1749-1782. Portland, Maine. Includes accounts for building a house, lumber, and other materials.

1109 **Longfellow, Stephen, Jr.** (1776-1849), Scrapbook, 1736-1850. Scrapbook folio of deeds, indentures, leases, writs, obligations, and other papers. Includes deed of Jonathan Winslow to Daniel Conant for a double sawmill at Saccarappa in Falmouth, Maine, on the Presumpscot River (1801).

1110 **Lord, Major,** Papers. Portland, Maine. Account books of a lumber dealer, including bills and receipts.

1111 **Meader, Job W.,** Miscellaneous papers, 1836-1882, 119 pieces. Included are deeds, memoranda, and leases for a sawmill in Westbrook.

1112 **Meserve, Clement** (1746-1805), Account book, 1769-1794. Scarborough, Maine. Contains accounts for lumber, labor, etc.

1113 **Moulton, Jonathan** (1767-1830), Ledgers, 1793-1815, 2 vols. North Yarmouth, Maine. Include accounts for wood.

1114 **Mussey, John** (1751-1823), Papers, 1780-1800. Mostly bills, letters, and miscellaneous business papers, including memorandum of an agreement (October 5, 1800) of Timothy Hamblin of Gorham and John Mussey, to furnish Mussey with lumber; agreement between Mussey and William Cobb of Portland for building a vessel; dimensions of timber for John Mussey by Robert Huston.

1115 **Patterson, William Davis** (b. 1858), Papers. Wiscasset, Maine. Includes letters from Maine business firms and individuals to Capt. Joseph Cargill, Boston, concerning shipments of lumber and other materials (1870-1871).

1116 **Pepperell, Sir William** (1696-1759), Papers, 1716-1755. Kittery, Maine. Includes a bill of lumber to Stephen Seavey, Louisburg, Cape Breton, Nova Scotia (September 13, 1745).

1117 **Porter, Aaron,** Papers, 1629-1847, 2 vols. miscellaneous papers. The papers contain information regarding land grants and early settlements at Biddeford and Saco, Maine. Many of them (1798-1838) relate to the purchase of a mill and mill privilege on the Saco River, and a related suit against Daniel Cole.

1118 **Pratt, Levi H.,** Papers, 1840-1856. The papers are concerned with ownership of lands and logging operations in the Chase Stream Tract in Somerset County, Maine, adjoining Indian Pond. Levi Pratt of North Yarmouth was the principal owner.

1119 **Rideout, Joshua M.,** Deeds, 1828-1847, 1 package. Included is a power of attorney from Henry Pennell, John Merrill, and Eliphalet Clark to Rideout (September 10, 1839), to sell timber on Moose River, north of Bingham Purchase, Somerset County.

1120 **Sewall, Charles** (1794-1884), Account book, 1834-1842. Bath, Maine. Includes accounts for lumber, shipbuilding materials, etc.; mentions the North Stream Mill.

1121 **Smith, St. John** (1799-1878), Papers, 1835-1880, 15 vols. Letterbook of St. John Smith, of Portland, relating to land and lumber investments in Centre, Clearfield, Elk, and Jefferson counties in Pennsylvania, and other enterprises.

1122 **Smith, Stephen** (d. 1806), Account books, 1762-1780. Ledger A (1771 +) includes accounts for lumber.

1123 **Southgate, Robert,** Miscellaneous papers, 1754-1799. The papers contain two letters (1772-1773) from Robert Pagan to Richard King concerning a cargo of boards and shingles.

1124 **Spencer, Thomas, and Humphrey Chadbourn,** Deed, April 8, 1651. Deed for Tom Tinker's swamp and timber, Kittery, Maine.

1125 **Stanley, Robert Dana** (b. 1936). "The Rise of the Penobscot Lumber Industry to 1860," Orono: University of Maine, M.A. thesis (History), 1963, 81 pp.

1126 **Stevens, Nathaniel** (1786-1865), Papers, 1827-1857, 96 pieces. Correspondence of Captain Nathaniel Stevens of North Andover, Massachusetts, from Isaac J. Stevens, Moses B. Stevens, William Henry, and others concerning lumber and timber operations in Washington County.

1127 **Tebbetts, Reliance** (1786-1856), Account book, 1813-1835. Topsham, Maine. Included are accounts for hauling wood.

1128 **Thacher, George,** Deeds, 1813-1853. Deed from Mayo to Willis for land in Monroe, Maine, with sawmill (October 21, 1835).

1129 **Thompson, D.C.,** Agreement, May 29, 1779. Obligation of Thompson to Nathaniel Parsons to pay for lumber in York, Maine.

1130 **Usher, Ellis Baker** (1785-1855), Papers, 1790-1877. One carton of papers relate to lumbering operations in Maine and New Hampshire (1827-1859); one box contains miscellaneous papers, mostly bills and receipts, and a few letters, relating to lumbering (1790-1877), and a small daybook. Usher was a resident of Hollis, Maine.

1131 **Waldo, Samuel** (1696-1759), Papers, 1631-1824. Includes information concerning a sawmill on Stroudwater River in Maine, and six acres of land.

SACO

*York Institute

375 Main Street 04072. The following entries were taken from *A Reference List of Manuscripts Relating to the History of Maine,* University of Maine Studies, 2d Series, No. 45, ed. by Elizabeth Ring, Orono, Maine: 1938, 3 vols.(The *Maine Bulletin,* XLI-XLIII, August, 1938, August, 1939, February, 1941).

1132 **Jordan, Tristram,** Record and account books, 1759-1799, 6 vols. Saco, Maine. The two account books (1759-1792) are for lumber and for labor by men hired out by Jordan.

1133 **King, Richard** (1719/20-1775), Account books, 1748-1787, 3 vols. Account books kept by Richard King, Saco merchant, largely for lumber and labor supplied by King.

1134 **Milliken, Mulbery,** Account book, 1769-1775, 1 vol. Account book of a Pepperellborough (Saco) lumber firm, chiefly a record of lumber sold.

1135 **Moulton, Edward L.,** Papers, 1834-1858, 12 bundles. Letters, invoices, and receipts relating to the business affairs of the Austin Stream Company, a lumber firm of which Moulton was treasurer and a large shareholder. The correspondence is mainly from employees of the company. The collection includes material concerning disputes between the company and other firms in the state.

WATERVILLE

Waterville Historical Society

President, Redington Museum, 65 Silver Street 04901

1136 **Colby Collection,** 1839-1904. Miscellaneous papers concerning Colby College (1895-1904) including sale reports of timberlands; less complete than, and in most cases duplicates of, the collection held at the Colby College Library.

1137 **Flood, George,** Diary, 1860-1894. Contains accounts of purchases of fuel wood for the Androscoggin and Kennebec Railroad, and of rail transportation of sawed lumber.

1138 **Lombard Log Hauler.** A large collection of photographs and documents concerning the Lombard Log Hauler.

1139 **Stackpole Collection,** 1763-1918. Collection of several hundred pieces, consisting of diaries, letters, receipts, notes, etc., of the Stackpole family of Winslow, later of Waterville, Maine. Includes bills for lumber; there are many references to lumber operations, especially to rafting down the Kennebec River.

Maryland

BALTIMORE

Maryland Historical Society

Curator of manuscripts, 201 West Monument Street 21201
Telephone: (301) 685-3750

1140 **Blackford, John,** Diary, 1829-1831, 1 vol. Diary of a Hagerstown, Maryland, resident, discussing lumbering.

1141 **Clarke, Ambrose** (d. 1810), Papers, 1793-1829, ca. 3 feet. Merchant and shipowner of Baltimore. Personal and business correspondence, letterbook, business and cargo auction accounts and receipts relating to trade in lumber and other goods shipped from North Carolina to Amsterdam, Bremen, Havana, New Orleans, New York, the West Indies, and other places.

1142 **Cohn and Bock Company** (Princess Anne, Maryland), Records, 1882-1934, 28 vols. and 3 boxes. Correspondence, ledgers, journals, daybooks, cashbooks, merchandise book, cancelled checks, and receipts of a company in Talbot County, Maryland, concerning its various enterprises, including lumbering and manufacturing of boxes.

1143 **Douglass, Richard H.** (1780-1829), Letterbook, 1816-1818, 1 vol. (ca. 1600 items). Merchant, of Baltimore. Letters of the Baltimore trading firm of Richard H. and William Douglass to correspondents in American, West Indian, and European ports relating to commercial activities, including trade in lumber. Some of the letters are in French.

1144 **Kirk, William** (fl. 1758-1813), Business records, 1758-1813, 1 box. Merchant of Baltimore, Maryland. Correspondence, daybooks, and other accounts. Included is an account book of the Baltimore Lumber Company.

1145 **Leakin-Sioussant Family,** Papers, ca. 1650-1960, ca. 20 feet. In part, photocopies of originals in the possession of the American Philosophical Society and the Library of Congress. Subjects include the Maryland Forestry Association. A few of the papers are indexed.

1146 **Leib, J.C., and Company** (Baltimore, Maryland), Records, 1895-1910, 250 items. Produce and commission merchants. Bulk of collection consists of receipts for goods purchased, including lumber.

1147 **Lumber Exchange of Baltimore City,** Minutes, 1875-1960, 12 vols., and 1 box. Organization founded in 1875 to correct abuses and differences between members of the lumber trade, and to advance the trade. Collection includes minutes of general Exchange mettings plus some minutes of the Managing

Committee. Minutes include reports concerning the following: development of lumber inspection regulations and licensing of inspectors, relations with other lumber societies in the city and nationally, statistics of lumber stock on hand, etc.

1148 **Marine, Matthew** (1777-1854), Account books, 1822-1870, 15 vols. Includes Cratcher's Ferry account book (1822-1846), recording shipments and sales of lumber and other commodities.

1149 **Maryland State Planning Commission,** Reports, 1934-1936, 16 items. Typed reports on various aspects of Maryland, including conservation and land use.

1150 **Pringle, Mark U.** (d. 1826), Correspondence, 1796-1798, 1811-1818, 2 vols. Businessman, of Baltimore. The correspondence relates to the lumber trade and other business matters. In part, transcripts.

1151 **Ridgely,** Papers, 1732-1900, ca. 9 feet and 106 vols. Correspondence and other records, including financial records, relating largely to Captain Charles Ridgely (b. 1705) and dealing, in part, with lumber. In part, photocopies.

1152 **Scharf, Thomas G., & Son** (Baltimore), Account book, 1861-1865, 1 vol. Daybook of a lumber merchant, showing transactions in shingles, posts, rails, scantling, and other items.

1153 **Shriver Family,** Collection, 1712-1944, ca. 9 feet and 40 vols. Correspondence and business papers of a family of Union Mills, Carroll County, Maryland, which operated a sawmill and other business concerns. In part, photocopies and transcripts of originals in the possession of the Shriver family. Indexed in part.

1154 **Spear Papers,** 1858-1866, 17 items. Correspondence of Robert Spear, New Haven, Connecticut, mainly relating to lumber trade, Nanjemoy and Havre de Grace, Maryland; and supplies for U.S. navy yards.

1155 **Vanneman, J.P.,** Receipt book, 1850-1855, 1 vol. Receipts, mainly for purchases of lumber made at Port Deposit, Maryland.

1156 **Whitaker Iron Company,** Records, 1839-1938, 6 boxes. Company owned by George P. Whitaker. Included are agreements for the cutting of wood for charcoal and for telegraph poles.

SUITLAND

Washington National Records Center

Director, General Archives Division, Washington, D.C. 20409 Telephone: (301) 763-7000. Records shown in archival custody are listed in the National Archives and Records Service, *Guide to the National Archives of the United States,* Washington: Government Printing Office, 1974.

Archives Branch

Court of Claims (Record Group 123)

1157 *Case files for General Jurisdiction Cases, 1855-1939, 2,060 feet.* The records include a few suits arising out of claims against the United States for timber and lumber alleged to have been taken from the claimants' land by the United States military forces. P.I. No. 58.

District Courts of the United States (Record Group 21)

1158 *New York, Southern District, 1789-1911, 3,710 feet.* Records of the Vice Admiralty Court of the Province of New York contain case papers (1757-1775, 2 feet) including proceedings initiated by the Surveyor of His Majesty's Woods in American against individuals for cutting down white pine trees reserved for masts of vessels of the Royal Navy (1758-1771). Records of the United States Circuit Court for the Southern District of New York include cases after 1859 under federal laws providing for punishment of persons committing depredations on timberlands of the United States. P.I. No. 116.

1159 *Pennsylvania, Eastern District, 1789-1915, 986 feet.* Among criminal records for the United States Circuit Court for the Eastern District of Pennsylvania are case files (1791-1883, 6 feet) and other records concerning almost every crime and misdemeanor punishable under the laws of the United States, including depredations on timberlands of the United States after 1859. P.I. No. 124.

Federal Trade Commission (Record Group 122)

1160 *Records of the Federal Trade Commission, 1914-1959, 8,086 feet.* Records of the Docket Section (1916-52) include material pertaining to investigations conducted at the request of the U.S. Attorney General on the lumber industry. P.I. No. 7.

Interstate Commerce Commission (Record Group 134)

1161 *Case files, 1887-1934, 5,456 feet.* Included are cases involving rail rates applying to forest and wood products. Some files contain background data on economic conditions in the forest and wood products industries. Draft P.I. (1964, 5 pp.).

Land Management Bureau (Record Group 49)

1162 *Records relating to public land disposals.* Nonmilitary land entry papers (1788-1951), containing records accumulated by the General Land Office prior to issuing patent for entries under the timber culture act, the timber and stone act, and other general land laws. P.I. No. 22.

National Labor Relations Board (Record Group 25)

1163 *Headquarters records of the National Labor Board, 1933-1935, 56 feet.* Unfair labor practices and collective bargaining representation case files under the National Industrial Recovery Act of 1933, including cases concerning the lumber industry (1934-35). Indexed. Restricted.

1164 *Records of the Second National Labor Relations Board, 1935-1959, 5,728 feet.* Unfair labor practices and collective bargaining representation case files under the National Labor Relations (Wagner) Act of 1935, and the Labor Management Relations (Taft-Hartley) Act of 1947. Includes NLRB cases concerning the lumber industry (1935-37). Indexed. Restricted.

Naval Districts and Shore Establishments (Record Group 181)

1165 *Records, 1783-1948, 10,875 feet and 3 rolls of microfilm.* Records of the Washington, D.C., Navy Yard include a journal of expenditures for lumber (1867-70, 1 vol.). Probably many other references to lumber purchases are scattered throughout the records of navy yards. Unpublished P.I. (1966, 81 pp.).

Public Roads Bureau (Record Group 30)

1166 *Records, 1892-1967, 2,959 feet.* General correspondence (1893-1912, 40 feet) includes material concerning cooperation with government agencies in promoting roads, including files relating to forest reservations. Office files of Captain P. St. J. Wilson, Chief Engineer (1916-1934, 3 feet) include policy statements concerning federal aid for forest roads. P.I. No. 134.

Federal Records Center

Forest Service (Record Group 95)

1167 *Forest Service records, 5,024 cubic feet.* Scheduled for permanent retention are correspondence, land records, financial, and market research records, records of training programs, diaries, subject files, and miscellaneous records, from the Washington, D.C., office and national forests and field offices in Virginia and West Virginia.

Outdoor Recreation Bureau (Record Group 368)

1168 *Outdoor Recreation Resources Review Commission, Records, 1958-1962, 41 cubic feet.* Printed documents, minutes, pamphlets, reports, publications, and correspondence files of the O.R.R.R.C. General correspondence includes files on the U.S. Forest Service, wilderness areas, wilderness study, correspondence of various O.R.R.R.C. officials, and O.R.R.R.C. inventory data relating to the Forest Service, National Park Service, Indian lands, etc.; files relating to contracts let by the O.R.R.R.C., mostly printed material; and material related to O.R.R.R.C. administration. Shelf list (32 pp.).

Massachusetts

BARRE

Barre Historical Society, Inc.

Curator, Common Street 91005

1169 **Broad, Willard** (1796-1868), Papers, 1820-1836, 2 feet. Businessman, of Barre, Massachusetts. Correspondence, bills of lading, waybills, invoices, and other papers relating to Broad's mercantile business in Miramichi, New Brunswick, which engaged in import-export, the sale of lumber and the outfitting of ships.

1170 **Eaton, Marshall Durock** (1807-1885), Papers, 1830s-1870, 2 feet. Lumber dealer and public official, of Barre, Massachusetts. Legal and family records, deed, receipts, and miscellaneous town papers. Records relate to the handling of estates, and the general operations of the lumber business—buying and selling logs, boards, etc., and payments for cutting and sawing.

BOSTON

Boston Public Library, Department of Rare Books and Manuscripts

Curator of Manuscripts 02117

1171 Among the holdings are a few letters and manuscripts, including a letter of Samuel Eliot of Boston to George Thatcher [Thacher], (January 17, 1785, 3 l.) concerning timber trespassing on Deerwander, the estate of Joseph Green near Saco Falls, Maine; two letters by François André Michaux to Thomas Pennant Barton (November 21, 1835, 2 pp., and November 26/27, 1835, 2 pp.), recommending works on forestry and sending chestnuts for planting in America; also a manuscript by Michaux, "Account of study of American forests," with recommendations on oaks to be introduced in France (September 6, 1831, 3 pp.).

Harvard University Graduate School of Business Administration, Baker Library

Curator of Manuscripts and Archives, Soldiers Field 02163. Collections are described in *List of Business Manuscripts in Baker Library,* compiled by Robert W. Lovett and Eleanor C. Bishop, 3rd edition; Boston: Baker Library, 1969, 334 pp.

1172 **Anderson, Hugh,** Ledger, 1819-1849, 1 vol. Accounts of sales of elemental wood products. Warren, Maine.

1173 **Andrews, Nathaniel, Jr.,** Account book, 1864-1865, 1 vol. Pertains to forest products.

1174 **Bellamy Grist Mill** (Dover, New Hampshire), Records, 1868-1886, 1 foot (15 vols.). Ledgers and daybooks of a grist and sawmill. William Hale, Jr., was connected with this business. One ledger is labelled G.F. Rollins Co.

1175 **Burleigh and Frost Families,** 1727-1884, Papers, 8 feet (84 vols., 1 case, 1 box). Durham, New Hampshire. Included in the collection are wood and lumber accounts (1806-1833).

1176 **Capen, John,** Papers, 1849-1867, 4 vols. Daybooks and ledger for a sawmill in Goshen, Vermont.

1177 **Chapin Family,** Papers, 1782-1866, 5 vols. Springfield, Massachusetts. Includes a ledger (1796-1827) of William Chapin 2d, including accounts for sawmills. There are also accounts for lumbering in the daybooks (1806-1852, 1866) of William Chapin 3d.

1178 **Clindennin, John,** Account book, 1822-1836, 1 vol. Salem, New Hampshire. Pertains to wood products.

1179 **Crafts, Samuel C.,** Ledger, 1820-1832. Craftsbury, Vermont. The accounts are, in the main, those of a sawmill and gristmill, operated in partnership with Newell Conant.

1180 **Crosby, William Chase** (1806-1880), **and Associates** (Bangor, Maine), Papers, 1818-1898, 1 foot (4 vols., 4 boxes). Letters, deeds, maps, and other papers relating to timber holdings in Maine.

1181 **Cushing, Nathaniel,** Account books, 4 vols. Included are accounts (1827-59) of Elijah Cushing of Hanover and Pembroke, Massachusetts, relating to a sawmill and wooden box manufactory.

1182 **Cushing, Pyam,** Ledger, 1837-1845, 1 vol. Medford, Massachusetts. The accounts show primarily cost of timber of various kinds used in shipbuilding.

1183 **Dudley Manufacturing Company** (Dudley, Massachusetts), Records, 1827-1845, 28 vols., 1 box. Also included are accounts of a sawmill.

1184 **East Boston Timber Company** (Boston, Massachusetts), Records, 1834-1840, 1 box. Unbound papers consisting of letters sent, estimates and contracts.

1185 **Fellows, Moses A.,** Daybook, 1883, 1 vol. Relates to wood products. Ipswich, Massachusetts.

1186 **Forest Iron Works** (Marquette, Michigan), Records, 1853-1865, 4 vols. Includes a ledger (1863) and a daybook (1863-1865) of a sawmill at Marquette.

1187 **Foster, Hopestill,** Ledger, 1759-1772, 1 vol. Relates to wood products. Boston, Massachusetts.

1188 **Gardner Family,** Records, 1780-1934, 4 feet (33 vols., 1 box). Includes material relating to lumbering. Boston, Massachusetts.

1189 **Gates, John, & Sons Company** (Worchester, Massachusetts), Records, 1835-1907, 12 feet (64 vols., 2 boxes, 2 cases). Records of a lumber dealer, including journals, ledgers, and incoming and outgoing letters. The firm was at various times known as John Gates & Company and John Gates & Sons Company; it seems to have had connections with Earle & Turner and P.J. Turner & Company.

1190 **Hall, Dudley P.** (1820-1885), Business records, 1841-1893, 38 vols. and 9 cartons. Correspondence and account books relating to Hall's interests in Vermont and Michigan lumber. Includes references to Hall & Chase, Island Pond Lumber Company, and Lyndon Mill Company, of Vermont; Erastus Corning, of New York; and E. & T. Fairbanks, of St. Johnsbury, Vermont. Inventory.

1191 **Henniker Crutch Co.** (Werner, New Hampshire), 1936-1961, 1 case. Includes material on the Kearsarge Wood Products Co. Weekly reports, monthly reports, balance sheets, correspondence.

1192 **Hinckley, Daniel B.,** Papers, 1871-1892, 5 boxes, 1 foot. Bangor, Maine. Letters, bills, and permits. Contains material pertaining to lumbering in Maine.

1193 **Holt, Deming & Johnson** (Bath, New Hampshire), Records, 1844-1860, 7 vols. (10 notebooks in 1 vol.). Account books, including daybooks, ledgers, cashbooks, and notebooks, pertaining to wood products.

1194 **Johnson, William C.,** Papers, 1897-1908, 8 feet (18 vols., 1 box, 5 cases). Letterbooks and incoming letters dealing principally with insurance on lumber. Johnson was president of the Lumber Mutual Fire Insurance Company, Boston. He was also engaged in the wholesale lumber business at Fitchburg, Massachusetts, and was attorney for various lumber companies. He owned the W.C. Johnson Lumber Company at Leominster, Massachusetts.

1195 **Lowell Machine Shop** (Lowell, Massachusetts), Records, 1845-1912, 45 feet (447 vols., 9 boxes, 5 cases). Correspondence, directors' minutes, dividend books, general accounting records and other papers relating to the manufacture of paper mill machinery and other equipment. Inventory.

1196 **Marrett, Edward,** Records, 1750-1780, 4 vols. Records of a tailor, Cambridge, Massachusetts. Includes an account book (1768-1773), relating to lumber dealings.

1197 **Miller, Nathaniel J.,** Papers, 1854-1859, 1 folder. Pertains to lumbering. Portland, Maine.

1198 **Morse, Asa P., & Company** (Boston, Massachusetts), Papers, 1846-1880, 2 boxes. Miscellaneous business papers of a dealer in shooks, including foreign letters (1860-77), and domestic letters (1860, 1864, 1870-73).

1199 **Newhall, Asa Tarbell,** Daybook, 1810-1822, 1 vol. Relates to wood products. Lynn, Massachusetts.

1200 **Patten, Amos, & Associates** (Bangor, Maine), Records, 1828-1849, 1 foot (5 boxes). Deeds, memoranda, and personal papers. Includes material relating to lumbering.

1201 **Peirce, Job & Ebenezer** (Freetown, Massachusetts), Papers, 1798-1852, 1 foot (14 vols.). Daybooks (10 vols.) and ledgers (4 vols.). The Peirces were part owners of a sawmill, among their other business concerns.

1202 **Proprietors of the Locks and Canals on the Merrimack River** (Lowell, Massachusetts), Records, 1792-1947, 13 feet (355 vols., 2 cases). Includes information on the kyanizing of lumber. Inventory and indexes.

1203 **Sag. & Bay Salt Company** (Michigan), Daybook, 1870-1900, 1 vol. Includes material relating to lumbering.

1204 **Shawmut Fibre Company** (Boston, Massachusetts, and Somerset Mills, Maine), Records, 1885-1889, 1 vol. Letters to Alexander H. Rice, mainly from Theodore P. Burgess, G.A. Phillips, officials of the company, and various firms concerning the paper mill business.

1205 **Smith, Joseph,** Records, 1806-1810, 3 vols. Daybook and lumber ledger and journal. Dover, New Hampshire.

1206 **Stetson Family Interests** (Bangor, Maine), Records, 1822-1880, 18 feet (178 vols. and 1 box). Included are records of Edward Stetson; Isaiah Stetson; Simeon Stetson; George Stetson; Brown and Stetson; Emery & Stetson; Emery, Stetson & Co.; Stetson and Co.; Bruce Mills (1849-80); Kenduskeag Log Driving Co. (1852-61, 1 folder); Bangor & Brewer Railroad (1879-80). Inventory. Records of the Stetson Family interests dating after 1880 have been deposited at the Maine State Museum, Augusta.

1207 **Stinson, Stephen S.,** Daybook, 1818-1857, 1 vol. Topsham, Maine. Entries for transport of logs, sawing, surveying of timber, shoemaking.

1208 **Turners Falls Lumber Company** (Turners Falls, Massachusetts), Records, 1872-1908, 5 feet (18 vols., 2 cases). Incomplete series of account books, letters and other records of a firm bringing spruce logs down the Connecticut River. There is much unbound material on lumber holdings in New Hampshire and Vermont, and on logging and driving operations. Some material relates to the Turners Falls Company and to the New England Fibre Company, a subsidiary.

1209 **Unidentified** (Paradise, Nova Scotia), Journal, 1863-1864, 1 vol. Includes month-end accounts for labor (lumbering) and board.

1210 **Union Mills** (Webster, Massachusetts), Papers, 1827-1907, 19 boxes, 176 vols. The collection, which pertains primarily to textile manufacturing, contains the accounts of a sawmill.

1211 **Usher, Ellis Baker** (1785-1855), Papers, 1800-1868, 5 feet (8 vols., 22 boxes). Lumberman of Hollis, Maine. Records of lumbering transactions in the Saco River area; a ledger and unbound papers, including letters, deeds, bills, and receipts.

1212 **Wilder & Company** (Boston, Massachusetts, and Olcott, Vermont), Records, 1892-1894, 1 box. Letters, mostly between Herbert A. Wilder and Charles T. Wilder. The firm manufactured newsprint. It was sold to the International Paper Company after 1897.

Massachusetts Historical Society

Director, 1154 Boylston Street 02215. Holdings are listed in the Massachusetts Historical Society, *Catalog of Manuscripts of the Massachusetts Historical Society,* Boston: G.K. Hall & Co., 1969, 7 vols., which reproduces the society's card catalog, primarily a list of individual items by personal and corporate names.

1213 **Belknap Family Papers.** Included is a letter (November 1, 1791) of Manasseh Cutler to Jeremy Belknap discussing trees of New England; also an item relating to the British government's reasons for the preservation of pines.

1214 **Bowdoin-Temple Family,** Papers. Letter (September 3, 1767) of Governor Wentworth to John Temple discusses the need for close supervision of exports of naval stores.

1215 **Coffin Family,** Papers, 1769-1818. Letters (March 4, April 11, April 16, 1783; March 30, 1786) of Thomas Aston Coffin, New York, discuss trees. A letter (December 3, 1792) of Thomas Aston Coffin in Quebec concerns curled maple wood.

1216 **Davis, Caleb,** Papers, 1684-1892, 18 vols. Boston merchant engaged in trade with North Carolina, the West Indies, and Spain. Lumber trade is mentioned in scattered letters (ca. 1787-93, ca. 6 items) from Machias and Portland, Maine; Trinidad, British West Indies; Georgetown, Massachusetts; Bristol, Maine?; New York; Charleston, South Carolina?. Various accounts and receipts (1777, ca. 7 + items) concern lumber and ship timber in Boston. There is mention of the market for naval stores in a letter of John Burguine (1787) from Wilmington, North Carolina?. Letters (1792, 2 items) from Cadiz, Spain, by Capt. Noah Stoddard of the ship *Fabius,* concern timber.

1217 **Eustis, Benjamin,** Account book, 1749-1757, 1 vol. Account book of a carpenter.

1218 **Miscellaneous.** Petition (1708) for appointment of additional surveyors of naval stores in Boston; Obadiah Gill and Joshua Gee, concerning price of timber (1699); bill of lading relating to naval stores (1715). Photostats of documents of the General Court of Massachusetts on preservation of wood and timber (1670); petition of Wharton and Saffrin concerning turpentine (1671); vote of Massachusetts deputies concerning timber (1670; original in the Massachusetts Archives); shipment of naval stores by Massachusetts (1694). Cataloged individually in the *Catalog of Manuscripts.*

1219 **Sedgwick Family,** Papers (54 feet). Includes correspondence, diaries, accounts, and documents of Henry Dwight Sedgwick (1785-1831), containing letter of J.M. to Lord Halifax (1753) concerning potash, and Benjamin Frobisher to John Morke about manufacture of potash (1753).

1220 **Shattuck Family,** Papers, Letter (1836) of Benjamin Shattuck, Calais, Maine, concerning lumber; letter (1824) of Roswell Shurtleff? concerning mountain ash trees; letters (1825, 2 items) of George C. Shattuck, Jr., Northampton, Massachusetts, concerning wood.

1221 **Wendell Family,** Papers, Letter (1771) of Joseph Locke to Oliver Wendell about making of potash.

1222 **Wentworth Family,** Papers, 1656-1941, ca. 100 items. Includes accounts and receipts for a sawmill in Berwick, Maine. Unpublished guide.

CAMBRIDGE

The Farlow Herbarium of Harvard University, Farlow Reference Library of Cryptogamic Botany

Librarian, 20 Divinity Ave. 02138

1223 **Faull, Joseph H.,** Papers. Botanist and forest pathologist, Canada, and Massachusetts. Included are letters from Faull to William Gibson Farlow (1901-19, 9 items); letters from Faull to Roland Thaxter (1901-31, 36 items); accession books of Faull's herbarium (2 vols.), currently on loan to the U.S. Department of Agriculture, Plant Science Research Division, Beltsville, Maryland.

MEDFORD

*Medford Historical Society

10 Governors Avenue 02155. The following item was listed in the Historical Records Survey, *Preliminary Edition of Guide to Depositories of Manuscript Collections in Massachusetts,* Boston: 1939.

1224 **Sprague and James,** Account books, 1816-1830. Accounts for lumber fittings and supplies for various ships.

NORTHBOROUGH

Northborough Historical Society, Inc.

Curator, Main Street 01532

1225 **Northborough, Massachusetts, Forest Warden,** Ledger, 1908-1914, 1 vol. The volume contains information on local pines, oaks, etc.

1226 **Proctor-Felt Lumber Company** (Northborough, Massachusetts), Records, 1901-1947. Account books and other records. This collection is cited in the *National Union Catalog of Manuscript Collections 1968,* Washington: Library of Congress, 1969, item #MS 68-1367. It has not been confirmed by the repository.

PETERSHAM

Harvard University, Harvard Forest

Librarian 01366 Telephone: (617) 724-3285

1227 **Cary, Austin,** Papers, 1897-1935. Forester on U.S. Forest Service staff. The papers include notebooks (70 vols.), a series of reports, and some correspondence, and relate chiefly to the Maine woodlands.

Petersham Historical Society

Librarian 01366

1228 **Wheeler, Joel** (1742-1816), Papers, 1770-1850, ca. 200 items. Sawmill owner and farmer of Petersham, Massachusetts. Sawmill accounts, bills, insurance policies, estate inventories for Joel and Jacob Wheeler.

SALEM

Essex Institute, James Duncan Phillips Library

Librarian 01970

1229 **Abbot, Benjamin,** Records, 1825-1872, 27 vols. The records consist of daybooks (4 vols.) and account books (23 vols.) of a Boston cooper.

1230 **Adams, Paul,** Papers, 1761-1856, 1 box. Blacksmith and miller of Newburyport, Massachusetts. Included are papers concerning the purchase by Adams of a sawmill and, in 1809, purchase of a patent right in a bark mill.

1231 **Andrews, John H.** (b. 1775), Papers, ca. 1719-1874, 2 vols. and 3 boxes. Merchant, Salem, Massachusetts. Some of the correspondence concerns ironworks in Bath, Maine, and Franconia, New Hampshire, and the acquisition of woodland there, in connection with the New Hampshire Iron Company (Danvers, Massachusetts), of which Andrews became a director in 1809. The bulk of the material concerns his shipping operations.

1232 **Bailey, George** (b. 1800), Papers, 1818-1846, 2 boxes. Sea captain of Salem, Massachusetts. Accounts and shipping papers of Bailey's voyages on the bark *Rosabella* and the brigs *Charlestown, Deposit, Fair American, New Union, Virginia,* and *Wizard,* carrying lumber and other goods to the West Indies and South America.

1233 **Brown, George F., and Samuel Brown,** Papers, 1840-1890, 9 vols. Account books (1840-1890, 8 vols.), of George F. and Samuel Brown, lumber merchants of Salem, Massachusetts, together with a diary and personal accounts.

1234 **Curwen Family,** Papers, 1652-1889, 4 vols. Salem, Massachusetts. The last three volumes contain records (1659-

1692) of the sawmills at Wells, Maine, originally called the Cape Porpoise Mills, and later the Mousam Mills, including a copy of the land grant made by the town of Wells for the erection of the mills (1669), articles of agreement between the owners and managers, accounts, contracts, letters, a ledger (1679), and legal records of the disputed ownership of the mill between Jonathan Curwen and Mary Sayward (1682). Joseph Storer of Wells was for a time owner and thereafter manager of the mills.

1235 **Cushing, Isaac,** Papers, 1826-1859, 2 boxes. Merchant and lumber dealer of Salem, Massachusetts. Correspondence, bills, and accounts. Most of the papers relate to voyages and maintenance of the schooners *Invincible* and *Helena,* and brigs *America* and *Sam. Small,* of which vessels Cushing was part owner, engaged in a coastal freight trade in lumber and other goods.

1236 **Derby Family,** Papers. Included are references to purchases of lumber from the Mousam sawmills at Wells, Maine.

1237 **Jonesboro, Maine,** Papers, 1784-1829, 1 vol. Included are statements concerning the sawmill and the quantity of lumber in the township.

1238 **Miscellaneous Account Books.** Included are a large number of account books and daybooks relating to wood products industries, mostly in Massachusetts. Daybooks of box makers include those of Tappan Pearson, Byfield and Newbury, Massachusetts (1825-52), and Smith and Smart, Haverhill (1863-64). Daybooks and account books of cabinetmakers include those of Joseph Brown, Newbury (1725-1783); Richard Gerrish, Salisbury (1778-98); Joshua B. Grant, Salem (1860-79); Jonathan Kettell, Newburyport (1781-94); Joshua and Abraham Lunt, Newbury (1736-72); Daniel Ross, Ipswich (1781-1804); Enos Runnels, Boxford (1793-1840); Joseph and George Short, Newburyport (1804-19); Joseph Symonds, Salem; and Miles Ward, Salem (1713-1789). Daybooks and account books of carpenters include those of Joshua Buffum, Salem (1672-1709); Edward Dole, Ipswich (1837-58); Richard Gerrish, Salisbury (1778-98); John and Stephen Jaques, Newbury (1712-94); Joseph Kimball, Rowley and Georgetown (1823-60); Joshua and Abraham Lunt, Newbury (1736-72); Isaac Porter, Wenham (1773-1824); Addison Richardson, Jr., Salem (1803-39); Enos Runnels, Boxford (1793-1840); Joseph Symonds, Salem; Amos Trask, Danvers (1765-82); Miles Ward, Salem (1713-89); Frank P. Todd, Rowley (1855-92); O.H. Rundlett, Rowley (1853-94). Account books of carriage makers include volumes of James Bott, Salem (1803-30), and Robert Boyes, Georgetown (1829-54). Account books of coopers include those of Joshua Hills, Jr., Newburyport (1818-49); Arthur Jeffry, Salem (1723-68); Joshua Phippen, Salem (1783-1801); Benjamin Ropes, Salem (1751-84); Stephen Runnels, Boxford (1757-70); John Sawyer, Salisbury (1772-1832); Joseph Stacey, Marblehead (1756-71); Robert Stone, Salem (1723-60). Account books and journals of lumbermen include those of Isaac Cushing, Salem (1841-44); William S. Felton, Salem (1843-68); John Flint, Salem (1679-82); Nathaniel Peaslee, Haverhill (1725-26); George Ash, Franklin, Maine (1827-53); Eben Putnam, Salem (1797-1803); Samuel Brown and Sons, Salem (1842-46). Account books and daybooks relating to sawmills include those of Eben Pearson, Newbury and Byfield (1818-58); and Elisha Story, Ipswich (1771-1850). Account books relating to shipbuilders include those of Abner Dole, Newbury (1730-60); James Beckett, Salem (1731-1803); Richard Hackett, Salisbury (1737-68); James Horton & Co., Newburyport (1781-1825); David Lowell, Amesbury (1781-1835); John Lee, Danvers (1773-83); Elijah and Ezra Morrill, Salisbury (1738-82); Peter Papillon, Boston (1713-25); James Topsham, Marblehead (1782-1835). Account books of ship carpenters include those of Benjamin Webb, Jr., Salem (1837-57); and John Woodwell, Newburyport (1802-22). An account book of a wood carver is that of Joseph True, Salem (1809-67).

1239 **Pingree, David,** Papers, 175 boxes. Salem, Massachusetts. Included are papers of the Penobscot boom, Maine lands, timber, townships, accounts of stumpage, and eastern lands, records of the Pingree lumber interests in Maine, principally in Penobscot and Somerset Counties (1832-1862, 18 boxes). Most of the material consists of letters to the Maine managers of the business, chiefly S.K. Howard and John Winn. There are also accounts, vouchers, stock reports, deeds, bonds, and reports of surveys.

Peabody Museum

Curator of Maritime History, 161 Essex Street 01970

1240 **Fernald & Petigrew** (Portsmouth, New Hampshire), Records, ca. 1840-ca. 1860, 33 boxes. Shipbuilders.

1241 **Fox, Josiah,** Papers, 1794 folders, 4 vols. and 960 letters. U.S. naval constructor of Massachusetts. The collection relates to shipbuilding.

SUDBURY

*Goodnow Library

21 Concord Road 01776. The following material was reported in the *Preliminary Edition of Guide to Manuscript Collections in Massachusetts,* Boston: The Historical Records Survey, 1939.

1242 **Thompson, James, and Captain Israel Haynes,** Account books, 1769-1839, 2 vols. Accounts for general store supplies, farm produce, lumber, and labor.

WALTHAM

Federal Archives and Records Center

Chief, Archives Branch, 380 Trapelo Road 02154. Records not accessioned into the Archives Branch remain under the jurisdiction of the originating agency.

Archives Branch

1243 **District Courts of the United States** (Record Group 21), 9,610 feet. Holdings of District Court records relate to the states of Connecticut (1789-1948, 1,262 feet); Maine (1789-1950, 1,010 feet); Massachusetts (1789-1948, 5,807 feet); New Hampshire (1789-1952, 404 feet); Rhode Island (1790-1945, 821 feet); and Vermont (1792-1945, 306 feet). Included are records of cases involving the acquisition of land for national forests, cases involving the violation of environmental laws, and bankruptcies of lumber companies.

Federal Records Center

1244 **Forest Service** (Record Group 95), 143 cubic feet. Among the records scheduled for permanent retention are records of the Green Mountain National Forest, Vermont, including special use permits (1935-63, 1 carton), donations (2 cartons), right-of-way acquisitions (1 carton), condemnation cases (2 cartons), donations and purchases with abstracts (27 cartons); correspondence, grazing statistical and allotment analysis, Appalachian Trail correspondence, etc. (1957-71, 1 carton). For the Durham, New Hampshire, Forestry Sciences Laboratory and the Laconia Research Unit there are general administrative records (1950, 6 cartons). For the

Northeastern Forest Experiment Station at the University of Maine, Orono, there are research studies (1945-66, 7 cartons). For the Northern Forest Insect and Disease Control zone, and Field Office, Amherst, Massachusetts, there are management records (1962, 1 carton) and administrative files and reports (1962-64, 2 cartons).

John F. Kennedy Library

Director, 380 Trapelo Road 02154 Telephone: (617) 223-7250

1245 **Agriculture Department,** Records, 1961-1963, 5 feet of microfilm. Copies of official records of the department for the period of the Kennedy administration, containing a small amount of material on the U.S. Forest Service.

1246 **Carver, John A., Jr.,** Personal papers, 1961-1963, ca. 1 foot. Assistant Secretary of the Interior for Public Lands Management, 1961-1963. Copies of addresses and public remarks.

1247 **Democratic National Committee,** Records, 1952-1963. Included are records of the 1956 election campaign of the Natural Resources Division and the Advisory Committee on Natural Resources (2 feet).

1248 **Freeman, Orville Lothrop** (b. 1918), Papers, 5 cubic feet. Secretary of Agriculture, 1961-1969. Memoranda, White House reports, appointment files, transcripts of press conferences and speeches, and a historical file. Microfilm copy of diary; original at Minnesota Historical Society. Restricted.

Kennedy, John Fitzgerald (1917-1963), Presidential Papers, 1961-1963

1249 *President's Office Files,* 80 feet. Originally a set of working files kept in the office of the President's personal secretary, this collection includes material on the Department of Agriculture (5 folders) and Interior (8 folders), conservation (1 folder), the National Wildlife Federation (1 folder), and also includes material concerning Kennedy's conservation trips throughout the United States.

1250 *White House Central Subject File.* Included is correspondence to and from the President, and, to a somewhat greater extent, members of the White House staff. While a considerable amount of the material deals with routine inquiries from the general public, also included are memoranda, reports and correspondence relating to various federal government programs, legislation, and other activities of special interest to the White House. Index. Major subject categories include federal government agencies, natural resources, parks and monuments, and recreation. Included is information concerning the U.S. Forest Service, National Forest Products Week, the National Forest Reservation Commission, the Senate Committee on Agriculture and Forestry, and forests.

1251 *White House Staff Files.* The personal papers of Lee White (10 feet) contain material relating to legislation, including conservation and the development of natural resources.

1252 **Oral History Transcripts.** Interviews with the following subjects include some reference to topics such as conservation and the U.S. Forest Service: William Blatt, Administrator, Area Redevelopment Administration (206 pp.); D. Otis Beasley, Administrative Assistant Secretary, Department of the Interior (24 pp.); Kenneth Birkhead, Assistant to the Secretary of Agriculture for Congressional Liaison (91 pp.); John Blatnik, U.S. Representative from Minnesota (34 pp.); Edmund Brown, Governor of California (19 pp.); Edward Cliff, Chief, U.S. Forest Service (17 pp.); William Douglas, Associate Justice of the Supreme Court (39 pp.); Michael Kirwan, U.S. Representative from Ohio (16 pp.); Gaylord Nelson, Governor of Wisconsin and U.S. Senator (15 pp., restricted); Maurine Neuberger, Senator from Oregon (33 pp.); Elmer Staats, Deputy Director, Bureau of the Budget (36 pp.); Donald A. Williams (23 pp.); Wayne Aspinall (6 pp.). There are also interviews, soon to be opened for research, with Orville Freeman, Secretary of Agriculture; Stewart Udall, Secretary of the Interior; Edward Crafts, U.S. Forest Service official and director, Bureau of Outdoor Recreation; John Carver, Assistant Secretary of the Interior for Public Land Management; James K. Carr, Under Secretary of the Interior; Kenneth Holum, Assistant Secretary of the Interior for Water and Power; Philleo Nash, Commissioner of Indian Affairs; Orren Beaty, Assistant to the Secretary of the Interior; Walter Pozen, Assistant to the Secretary of the Interior; and Lee White, Special Counsel to the President.

1253 **Smith, Frank E.,** Personal Papers, 1960-1961, 2 feet, Director of the Kennedy-Johnson Natural Resources Advisory Committee. Correspondence, reports, recommendations, and other papers of the committee.

WORCESTER

American Antiquarian Society

Curator of Manuscripts, 185 Salisbury Street 01609

1254 **Account Books,** Collections, ca. 500 vols. Account book (1740-78, 1 vol.) of Benjamin and Richard Cowan, North Providence, Rhode Island, contains records of sales of lumber. Account books and ledger (1852-79, 3 vols.) of Henry Gleason, Dana?, Massachusetts, include prices of lumber, sawing boards, drawing wood, cutting shingles. Record book (1835, 1 vol.) of the Kennebeck Lumber Co., Maine, contains records of the trustees, Rejoice Newton and William Lincoln, of Worcester. Account books (1848-61, 2 vols.) of David H. Sumner, Hartland, Vermont, include prices for cutting lumber, shingles, clapboards, and information on the navigation of the Connecticut River. Account books (1846-50, 2 vols.) of Oliver C. Bullard and Leonard Streeter, Sutton, Massachusetts, contain lumber and sawmill accounts. Sawmill accounts are contained in the account book of John Couse?, Worcester, Massachusetts (1789, 1 vol.); Josiah Dean, Dudley, Massachusetts (1804-26, 1 vol.); Enoch Lincoln, Worcester, Massachusetts (1820-21, 1 vol.); Levi Lincoln, Worcester (1800-15, 2 vols.); Lemuel? Rice, Worcester (1824-28, 2 vols.); Samuel Smith, Mendon, Massachusetts (1785-1828, 1 vol.); and in unidentified account books from Worcester (1822-27, 4 vols.) and Douglas, Massachusetts (1853-54, 1 vol.).

*Worcester Historical Society

39 Salisbury Street 01608. The following collections were reported in the Historical Records Survey, *Guide to Manuscript Collections in the Worcester Historical Society,* Boston: The Historical Records Survey, 1941.

1255 **Beaman, Ezra,** Papers, 1728-1866, ca. 3000 items. The records of Ezra Beaman, Sr., include daybooks and journals of a store, tavern, and mill, and miscellaneous accounts (1754-98, 13 vols.); lumber, store, and tavern accounts (1784-1808, 5 vols.); books for hardwood ashes shipped to a Boston soap factory (1784-1814); and a saw book (1804). The records of Ezra Beaman, Jr., include daybooks and journals (1816-1853, 11 vols.); and saw books (1839-1843, 4 vols.). The collection is described in more detail in the Historical Records Survey *Guide,* pp. 3-8.

1256 **Kendall Collection,** 1743-1847, 648 items. The papers include journals (1806-1814, 2 vols.) and a saw book (1828). The family were mill owners of Sterling, Massachusetts.

1257 **City of Worcester,** Manuscript Collection, 1707-1934, ca. 3000 items. This collection, which comprises correspondence, many types of business records, diaries, genealogical materials, municipal and political papers, includes the records of Tolman and Hunstable, carriage makers. These records consist of correspondence, contracts, deeds, price lists and inventories (1825-1880), a ledger (1833-1837), and an inventory book (1842-1845). The collection also contains the following lumber records: Goulding, Gregory, Thompson & Company, ledger (1854-1855); Ellery B. Crane, letter copybooks of lumber accounts (1886-1896, 2 vols.), a journal (1893-1901), cashbooks (1897-1901, 2 vols.), ledgers (1899-1901, 3 vols.), and daybook (1899-1901).

1258 **County of Worcester,** Manuscript Collection, 1667-1899, ca. 1200 items. Papers from the towns of Holden, Templeton, and Uxbridge contain pertinent records. For Holden there is a sawmill account book (1792-1829) of Ezra Hastings; from Templeton, a carpenter's account book (1827-1873) and diaries (1829-1887, 4 vols.) of Silas Norcross; from Uxbridge, an account book and ledger for sawing and labor (1789-1819) and a record of lumber sold (1806-1811) of William Aldrich.

Michigan

ALBERTA

Michigan Technological University, Ford Forestry Center

Manager of Operations 49946 Telephone: (906)524-7335.

1259 Holdings include research records (since 1955); correspondence and other records pertaining to the establishment of Alberta town and sawmill as the first of several projected Ford Motor Company subsistence communities; sustained yield management of the Ford northern hardwood forest of the Upper Peninsula; Alberta and Trout Creek mill scale studies; photographs (1930s); anecdotal material from former Ford Motor Company employees who worked for the Ford Forestry Center; data on the original survey of the western half of the Upper Peninsula; data on the original forest of the Keweenaw Peninsula and Baraga and Marquette Counties.

ANN ARBOR

University of Michigan, Michigan Historical Collections

Director, 1150 Beal Avenue, Bently Historical Library 48105 Telephone: (313) 764-3482. Holdings are described in Robert M. Warner and Ida C. Brown, *Guide to Manuscripts in the Michigan Historical Collections of the University of Michigan,* Ann Arbor: 1963; and J. Fraser Cocks, III, *A Bibliography of Manuscript Resources Relating to Natural Resources and Conservation in the Michigan Historical Collections of the University of Michigan,* 1970, processed.

1260 **Barbeau, Peter,** Papers, 1789-1909, microfilm, 8 rolls. Correspondence and business papers of Barbeau dealing with lumbering, and other businesses in the northern peninsula, particularly in Sault Ste. Marie.

1261 **Bassett, Ray E.,** Scrapbook, 1914-1919, 1 vol. City Forester of Ann Arbor. Scrapbook of clippings and other printed material about Bassett's work as city forester and his other interests.

1262 **Baxter, Dow Vawter** (b. 1898), Papers, 1941-1944, 6 items. Forest pathologist, University of Michigan. Manuscript articles and book reviews by Baxter.

1263 **Bellaire, John Ira** (b. 1871), Papers, n.d., 2 items. Manuscripts, one entitled "Hiawathaland, The Upper Peninsula of Michigan"; the other concerning logging in Michigan.

1264 **Bohr, Joseph, Lumber and Grain Mill** (Westphalia, Michigan), Business records, 1850-1866, 2 vols. Daybook and ledger.

1265 **Brown, John Wesley** (b. 1875), Reminiscences, n.d., 3 pp. Reminiscences of a dentist, recorded by Michael Church, in which Dr. Brown describes his life in Bad Axe and the great forest fire of 1881.

1266 **Bury Family,** Papers, 1831-1931, 21 items. The papers include a photostat of a letter (1842), and a telegram to Richard A. Bury, a lumber dealer of Adrian, Michigan, from Edward D. Gregory.

1267 **Carpenter, Daniel,** Papers, 1857-1884, 105 items. Businessman of Warfield, Massachusetts. Correspondence and business papers originating from agents in Michigan, particularly Livonia, Detroit, and Ann Arbor, dealing with lumbering and other affairs.

1268 **Carton, John J.,** Papers, 1897-1920, 17 feet. Contains information on the conservation of natural resources.

1269 **Case, Charles M.** (ca. 1859-1940), Manuscript articles, 1925, 3 items. One article on Dr. George W. Earle; two historical accounts of Hermansville. Includes information on lumbering.

1270 **Case, William L.** (1856-1933), Papers, 1860-1937, 85 items. Case was a partner in the Case Brothers Lumber Company, Benzonia, Michigan. The collection includes personal and business correspondence, business and legal papers, and account books. There are articles of association of the company (1889); account books of the local sawmill (1866-67, 1881-82, 2 vols.); and miscellaneous papers, including land grants (1860), contracts, bills, and business reports.

1271 **Chapman, J. P.,** Business records, ca. 1872-1885, 5 feet. Logger of Bay City, Michigan. The records pertain to the lumber business, including cashbooks, accounts of men in camp, store account books, time books, and teams and teamster accounts.

1272 **Chesbrough, A., Lumber Company** (Thompson, Michigan), Ledger, 1891-1893, 1 vol.

1273 **Cole, Martha Knapp** (1831-1901), Papers, 1837-1910, ca. 150 items and 30 vols. Included are receipts for lumber (1886-1900, 13 items).

1274 **Cook, Marshall, L., and William Randolph Cook,** 1880-1945, 9 feet and 51 vols. Included are records of the Cook Brothers wood-lot business (1 vol.).

1275 **Copley, Alexander** (1790-1842), Papers, 1814-1881, 260 items and 7 vols. Included is an account book of a lumber business (1839-1840, 1 vol.)

1276 **Coulter, Clinton H.,** transcript of interview, 1958, 1 item. University of Michigan forestry student. The transcript mentions Clarence C. Little and Samuel T. Dana; concerns Florida forestry and pine plantations.

1277 **Craig, Robert** (1882-1962), Collection, 2 vols. Scrapbook (1910-1923, 2 vols.), compiled by Robert Craig and Shirley W. Allen, containing correspondence, clippings, typescripts of speeches relating to the career of Filibert Roth as chairman of the University of Michigan Department of Forestry.

1278 **Cramton, Louis Convers** (1875-1966), Papers, 1896-1966, 6 feet. U.S. Representative from Michigan, 1913-31; member, Public Lands Committee. Correspondence, speeches and clippings pertaining to Cramton's service in Congress, particularly his relations with the Interior Department and the National Park Service. The collection includes correspondence of Horace Marden Albright (1927-66), Stephen Tyng Mather (1925-29), Gifford Pinchot (1928, 1940, 3 items), Conrad Louis Wirth (1951-62), and other conservationists. Inventory.

1279 **Crapo, Henry Howland** (1804-1869), Papers, 1820-1907, ca. 9 feet and 78 vols. Land speculator, lumberman, state legislator, and governor of Michigan. Correspondence, diaries, notes, business and legal papers, speeches, memorandum books, and miscellaneous papers of Crapo and his son, William Wallace Crapo (1830-1926), advisor to his father and U.S. Representative from Massachusetts, 1875-83. Much of the correspondence and other papers relate to Crapo's Michigan career as a land speculator, lumberman, and Michigan politician between 1855-69. There is material relating to the Lapeer, Michigan, pine lands, land dealers, viewers, difficulties of transporting and marketing logs, Crapo's sawmill at Flint, Michigan, and prices of lumber and labor. There is correspondence relating to a sash and door factory, a planing mill, and correspondence with Crapo's son-in-law, Humphrey Henry Howland Crapo Smith, manager of the Detroit mills. Crapo's diaries and cash account books (1854-69, 11 vols.), contain entries under barter, land, loggers, lumber, mill men, mills, prices, railroads, shipping, ships, staves, and taxes. There are also statements of prominent citizens and lumbermen on lumbering in Michigan.

1280 **Dana, Samuel T.** (b. 1883), Papers, 1893-1970, 10 boxes. Forester; professor of forestry and dean of the School of Natural Resources at the University of Michigan; fellow and president of the Society of American Foresters. The papers contain correspondence, reports, newspaper clippings, and other materials relating to his activities in forestry and conservation. There is correspondence of many foresters and conservationists, including Herman Haupt Chapman, Henry Edward Clepper, the Conservation Foundation, Edward Clayton Crafts, James B. Craig, Emanuel Fritz, William B. Greeley, Ralph Sheldon Hosmer, Myron Krueger, Richard E. McArdle, Laurence S. Rockefeller, Raphael Zon, and others. Inventory.

1281 **Davis, Charles Albert** (1861-1916), Papers, 1900-1903, 38 items. Instructor in forestry, University of Michigan. The papers include lecture notes, manuscripts, and letters (31 items) to Davis discussing geology and problems of forestry.

1282 **Deadman, Richard Hector** (b. 1872), Reminiscences, 1956, 2 pp. State representative, lumberman of Alpena, Michigan. Reminiscences, describing early lumbering experiences in Michigan.

1283 **Demmon, Elwood Leonard,** Interviews, 1958 and 1959. University of Michigan forestry student. Refers to Filibert Roth and scientific forestry in the United States.

1284 **Den Bleyker Family,** Papers, 1828-1936, 7 feet and 14 vols. Kalamazoo and Ottawa Counties, Michigan. Correspondence (1851-1856) concerns, in part, a sawmill. Partly in Dutch language.

1285 **Downey, David** (b.1859), Notebooks, ca. 1900-1932, 18 items. Timber-cruiser notebooks kept by Downey, of Hermansville, Michigan, while working for the Wisconsin Land and Lumber Company.

1286 **Dustin, Fred** (1886-1957), Papers, 1901-1932, 12 items and 1 vol. Businessman and archeologist. The collection includes notes on old boom houses on the Tittabawassee River and Saginaw areas (1 p.).

1287 **Earle, G. Harold,** Notes, 1886, 2 pp. Lumberman. Copy of notes on lumber land in Wisconsin.

1288 **Evans, Oscar** (b. 1878), Letter, 1958, 1 item. Forester. Letter to F. Clever Bald giving brief autobiographical data.

1289 **Fairbanks, Erastus,** Letter, 1864, 1 item. Letter to R.M. Richardson describing pine lands owned by Saint Marys Falls Ship Canal Company.

1290 **Fargo and Fargo General Store and Flour Mill** (Manchester, Michigan), Records, 1833-1841, 4 vols. Includes ledger of James H. Fargo's sawmill (1835-39, 1 vol.).

1291 **Farrar Family,** Papers, 1817-1886, 110 items and 4 vols. Includes the records of Charles S. Farrar and Company, containing the cashbook of a wood turning shop (1855-1857), the daybook of a cooper's shop (1876-1877), and accounts of Cyrus S. Farrar with the Armada Mill Company for cooperage products (1874-1876).

1292 **Ferry Family,** Papers, 1822-1911, 169 items and 3 vols. William Montague Ferry (1796-1867) was a missionary on Mackinac Island, lumberman, and founder of Grand Haven. The papers are chiefly correspondence, including some letters to Thomas White Ferry (1827-1898), lumberman in the Grand River Valley, Michigan, and U.S. Representative, 1865-71; U.S. Senator, 1871-83.

1293 **Field Family and Buck Family,** Papers, 1835-1923, 2 feet. Included are papers of Myron Buck, chiefly concerning lumbering. There are notebooks containing lumbering and sawmill accounts (12 vols.).

1294 **Fitzmaurice, John,** Reminiscences, 1889. Life in lumber camps.

1295 **Fletcher, Addison, Lumber Company** (Alpena, Michigan), Records, 1868-1925, 13 vols. Ledgers, journals, cashbook, and checkbook.

1296 **Fletcher, Frank Ward** (1853-1922), Papers, 1871-1922, ca. 1 foot and 4 vols. Lumberman of Alpena, Michigan. Correspondence of Fletcher, mainly with his father, George N. Fletcher, relating to the lumbering and paper manufacturing business at Alpena and Detroit (1880-98, 423 items). There is also correspondence (1887-98, 32 items) from Allen Fletcher concerning the lumber business, a letterpress book (1887-99), and a letterbook (1871) of G.N. Fletcher. The collection includes photographs (ca. 100 items) of lumbermen, logging activities, and railroads operating on Fletcher's premises.

1297 **Fletcher, Pack and Company** (Alpena, Michigan), Records, 1879-1898, 25 vols. and 200 items. Correspondence on taxes and land records, ledgers, and daybooks, covering their lumber business.

1298 **Forbes, Darwin C.** (b. 1871), Papers, 1866-1938, 335 items and 1 vol. Engineer and pioneer surveyor of northern Michigan. The records include the articles of co-partnership in the lumbering business of Forbes and Frank Hopkins (1893), and other business papers.

1299 **Fordney, Joseph Warren** (1853-1932), Papers, 1919-1927, 11 items and 1 vol. Lumberman, U.S. Representative, 1899-1923. The papers include notes on lumbering (1 vol).

1300 **Garfield, Charles William** (1848-1934), Papers, 1899-1929, 50 items and 1 vol. President of the Michigan Forestry Commission, member of the State Board of Agriculture. Correspondence of Garfield dealing with the Michigan

Forestry Association and conservation and forestation in Michigan; drafts of speeches or articles on forestry and conservation, probably written by Garfield (8 items); letterbook (1900-1903, ca. 630 letters), of the Michigan Forestry Commission; also information on origin of University of Michigan Forestry School. Correspondents include Julius C. Barrows, Arthur Hill, Burke A. Hinsdale, Alfred C. Lane, William B. Mershon, Gifford Pinchot, Hazen S. Pingree, Filibert Roth, Volney M. Spalding.

1301 **Gilchrist Family,** Papers, 1871-1945, 6 feet. The collection contains material on lumbering.

1302 **Gillette, Genevieve** (b. 1898), Papers, 1920-1972, 10 feet. Landscape architect, speaker and lobbyist on behalf of conservation and beautification, president of the Michigan Parks Association, and member of the Citizens' Advisory Committee on Recreation and Natural Beauty during the Lyndon B. Johnson administration. The papers contain correspondence, reports, and printed materials concerning her interest in state and national conservation legislation. Correspondents include Orville L. Freeman, Phillip A. Hart, Jens Jensen, and Laurence Rockefeller. Contents list.

1303 **Hakala, D. Robert,** Manuscript article, 1953, 54 pp. Ranger, Isle Royale National Park. The article is entitled: "Isle Royale—Primeval Prince, a History."

1304 **Hannah and Lay Company** (Traverse City, Michigan), Records, 1846-1931, 10 feet and ca. 400 vols. Correspondence, vouchers, daybooks, invoices, cashbooks, journals, ledgers, and other records of a lumber company and general merchandise business.

1305 **Hascall, Charles C.,** Papers, 1833-1860, 3 vols. The collection includes steam mill accounts (1854-1855), logging accounts (1857), and correspondence. Michigan.

1306 **Hastings, Walter E.** (1888-1965), Papers, 1921-1950, 100 items and 3 vols. Michigan Department of Conservation officer. Correspondence, scrapbooks, and other papers relating to Hastings' work as photographer, naturalist, and conservation officer. Correspondents include Fred W. Green, John A. Hannah, Clarence C. Little, Frank Murphy, and Alexander G. Ruthven.

1307 **Helme, James W.** (1860-1938), Papers, 1844-1938, ca. 550 items and 8 vols. Includes papers of his father, James W. Helme (1817-1883), containing a letter from Charles Helme about cutting cypress timber in the swamps, Yazoo, Mississippi.

1308 **Hetherington, Mary Elizabeth.** "A Study of the Development of Journalism During the Lumbering Days of the Saginaws, 1853-1882," Evanston, Illinois: Northwestern University, M.A. Thesis, 1933, 90 pp.

1309 **Hill, George D.** (1820-1881), Papers, 1843-1876, 2 feet and 1 vol. Ann Arbor businessman. The collection contains material on lumbering.

1310 **Hotchkiss, Everett S.** (b. 1857), Reminiscences, n.d., 88 pp. Lumber supplier, Bay City, Michigan. The typescript describes his boyhood and youth in Bay City in the 1860s and 1870s.

1311 **Hotchkiss, George W.** (1831-1926), Papers, 1857-1927, ca. 50 items and 1 vol. Lumberman, of Bay City and Saginaw, Michigan, and editor of the *Lumberman's Gazette*. The collection includes an autobiography and miscellaneous clippings and other papers relating to Hotchkiss' career.

1312 **Hubbell, Julian Bertine** (1847-1929), Papers, 1881-1883, 1948, 33 items. In part photostat and typewritten copies. Some of the letters from Hubbell to Clara Barton (25 items) discuss the Michigan forest fire of 1881.

1313 **Isle Royale,** Papers, 1921-1955, 2⅓ feet. Correspondence, newspaper clippings, and miscellaneous articles relating to the *Detroit News* campaign to make Isle Royale a national park and to secure land for it; manuscripts and notes of an article by Martha M. Bigelow (1955); correspondence of Arno B. Cammerer, William P.F. Ferguson, Harold L. Ickes, Chase S. Osborn, Alexander G. Ruthven, Albert Stoll, Arthur H. Vandenburg, Lee A. White.

1314 **Johnson, F.O.,** Manuscript, 1949, 1 page. Biographical sketch of Louis Sands (1825-1905), Manistee, Michigan, lumberman.

1315 **Leech, Carl Addison,** Papers, 1928-1940, ca. 475 items and 26 vols. Largely notes on various Michigan topics, including lumbering; and a manuscript on lumbering by Bert Harcourt.

1316 **Loud, Henry Nelson** (1850-1938), Letterpress book, 1901-1904, 1 vol. Lumberman of Au Sable, Michigan. The letters relate to conservation and other matters.

1317 **Loud, Marian V.,** Manuscript, n.d., 446 pp. "No Winter Came," a biography of Henry Nelson Loud, Au Sable, Michigan, lumberman.

1318 **Lovejoy, Parrish Storrs,** Papers, 1918-1941. Contains information on University of Michigan School of Forestry.

1319 **Lyon, Lucius** (1800-1851), Papers, 1826-1851, 72 items. Territorial delegate, U.S. Representative, 1833-35, 1843-45, and Senator, 1837-39, from Michigan. The papers contain a letter from Lyon to Jonathan White concerning the purchase of timber land.

1320 **Mackinac Island State Park,** Annual Report, 1899, 4 pp. Typescript report of superintendent S.B. Poole.

1321 **Maltby Lumber Company** (Bay City, Michigan), Records, 1899-1901, 1 vol.

1322 **Manwaring, Joshua** (1824-1903), Papers, 1867-1905, 50 items and 3 vols. Michigan state legislator, farmer, and lumberman. Includes a diary of a trip to the Pikes Peak region of Colorado (1859, transcript made in 1869), legal documents, business papers, scrapbooks of newspaper clippings (2 vols.).

1323 **Matthews, Donald Maxwell,** Photographs, 1924-1926. University of Michigan forestry professor. Part of the collection relates to lumbering in Louisiana.

1324 **Mead, James, Lumber Company** (Saginaw, Michigan), Records, 1872-1874, 2 vols. Account book (1872-1874) of Mead Lumber Company; ledger (1873) of Mead, Lee, & Company.

1325 **Mears, Charles** (1814-1895), Papers, 1822-1876, 1945, 40 items and 20 vols. Sawmill operator and Michigan legislator. Correspondence; diary; report; legal papers; business records (20 vols.) consisting of daybooks, inventories, time books, ledgers from various camps and mills owned by Mears, including an account book (1837-1843) showing his expenditures in his first lumbering venture in the White Lake area in Michigan; biographical sketch (1945).

1326 **Merrill Lumber Company** (Lexington, Michigan), Records, 1854-1922, 5 feet and 198 vols. Correspondence, daybooks, ledgers, journals, cashbooks, invoices, and miscellaneous records of Daniel C. Merrill.

1327 **Mershon, William Butts** (1856-1943), Papers, 1887-1925, 78 feet and 195 vols. Saginaw, Michigan, lumberman, businessman, and Michigan State tax commissioner. Correspondence, including letterpress books (63 vols.), dealing with Michigan wildlife conservation, Michigan politics, business investments, lumbering and mining interests, and personal affairs; diaries (4 vols.) and a book of notes, concerning hunting and fishing trips; and business records, including cashbooks, time books, ledgers, journals, and other business papers concerning Mershon's personal accounts, investments, and the lumbering business. Correspondents include Waldo A. Avery, William F. Baker, Aaron T. Bliss, Woodbridge N. Ferris, Joseph W. Fordney, Charles W. Garfield, Alexander J. Groesbeck, Emerson Hough, George M. Humphrey, Watts S. Humphrey, Walter J. Hunsacker, George B. Morley, William R. Oates, Chase S. Osborn, Filibert Roth, Robert H. Shields, and Albert E. Sleeper.

1328 **Meyer, Charles J.,** Papers, 1885-1886, 20 items. Founder of the Wisconsin Land and Lumber Company.

1329 **Michigan. Public Domain Commission,** Records, 1911-1913, 1 vol. Minutes, reports, extracts of the Commission and financial records.

1330 **Michigan-California Lumber Company,** Records, 1872-1933, 28 feet and 27 vols. The company was owned by the Blodgett family. The records consist of letterpress books (1899-1913, 27 vols.) and other records relating to the operation of the company.

1331 **Michigan Forestry Association,** Records, 1905-1942, 250 items and 1 vol. Correspondence, financial records, constitution, membership lists, and reports. Correspondents include Robert Craig, Jr.

1332 **Michigan State Federation of Women's Clubs,** Papers, 1898-1951, 2 feet. Includes material on conservation.

1333 **Michigan, University. Forester's Association,** Records, 1925-1932, ca. 100 items.

1334 **Midwest Foresters' Conclave,** Papers, 1954-1960, ca. 100 items. Meeting of 6 midwestern forestry schools.

1335 **Mitchell and McClure Lumber Company** (Saginaw, Michigan), Records, 1889-1927, 33 feet and 71 vols. Correspondence, deeds, work reports, timber cutting reports, contracts, patents, tax statements and other papers of the company.

1336 **Munster, Norman,** Papers, 1 vol. Forestry class notes compiled while Munster was a student at the University of Michigan during the 1920s. Dendrology class notes include sketches of cones, leaves, and seeds. Other notes show the student's view of basic courses in surveying, pathology, mensuration, and entomology.

1337 **Mutton, Charles A.** (d. 1956), Diaries, 1899-1956, 38 vols. Lumber dealer, Millington. The diaries contain brief entries of everyday events, including notations on the weather.

1338 **Nevue, Wilfred** (b. 1886), Papers, 9 items and 1 vol. Lumberman. Manuscript history entitled "Logging in the Huron Mountains," a detailed description and history of the lumbering industry in the Huron Mountain region of the northern peninsula, with particular attention to the Oconto Company. There are also miscellaneous manuscripts dealing with lumbering and biographical information on Nevue.

1339 **Osborn, Chase Salmon** (1860-1949), Papers, 1889-1949, 179 feet and 267 vols. Author, newspaper editor, prospector, fish and game warden, governor of Michigan, 1911-1912. The collection contains much material on conservation activity in Michigan (1894-1913); also correspondence with the manager of Osborn's plantation and forest reserve in Georgia, discussing the conditions of the trees and the measures taken to insure their health.

1340 **Otis, Charles Herbert** (b. 1886), Papers, 1913-1958, 50 items. Author of *Michigan Trees;* Curator, the University of Michigan Arboretum, 1910-1913; Professor, Bowling Green University. The papers include correspondence about *Michigan Trees,* other publications of Otis, and University of Michigan Arboretum and School of Natural Resources.

1341 **Overpack, Roy M.,** Letter, 1953, 3 items. Manistee, Michigan, businessman. Letter and 2 enclosures discussing the Michigan logging wheels manufactured by S.C. Overpack of Manistee.

1342 **Peters, Kenneth E.** "Forest Conservation in Michigan, 1866-1903: A Social and Legislative History." University of Michigan, Seminar Paper, 1967. 155 pp. + 12 pp. bibliography.

1343 **Pond, Cornelius V.R.** (b. 1836), Report, 1885, 1 item, photostat. Report of Pond, Michigan Commissioner of Labor, to Governor Russell A. Alger on the strike in the Saginaw sawmills.

1344 **Proctor, Joseph F.** (1835-1925), Papers, 1868-1925, 44 vols. Lumber sawyer and scaler, farmer of Hersey, Osceola County, Michigan. The papers include diaries, with discussion of activities in the lumber camps and mills, and account books (1872-1915, 2 vols.).

1345 **Raymond, Uri,** Papers, 1848-1910, 31 items and 6 vols. Records of a general store in Port Sanilac, Michigan, including letterpress books (1859-69, 1894-99), relating to wholesale orders for tanner's bark and shingle bolts.

1346 **Reichert, Rudolph Edward** (b. 1889), Papers, 1907-1962, 18 feet. Contains information on the Friends of the Land movement.

1347 **Rice, Justin,** Correspondence, 1824-1888, 20 items. Family correspondence of Rice, a pioneer in the lumber business.

1348 **Roth, Filibert** (1858-1925), Papers, 1893-1951, 1 foot and 5 vols. Professor of Forestry, University of Michigan. Correspondence (1893-1951), including letters from Roth to President Harry B. Hutchins concerning the School of Forestry (10 items); letters written by Roth's friends and former students on his retirement (1 vol.); drafts of speeches on forestry and conservation; notes and manuscripts by Roth and Robert Craig on various research projects and courses in forestry; a memorial upon Roth's death; a diary and daybook (1910); biographical sketches.

1349 **Ruggles, Rademaker Lumber Company** (Manistee, Michigan), Records, 1881-1907, 49 feet and 239 vols. Letterbooks, invoices, ledgers, daybooks, account books, land records, maps, and miscellaneous records.

1350 **Russell, Curran Northrum** (1873-1961), Papers, 1957-1960, 6 items. Director, Manistee Museum. Included is a typescript of tape recordings on lumbering and Great Lakes shipping around Manistee.

1351 **Salling-Hansen Lumber Company** (Grayling, Michigan), Records, ca. 1887-1927, 250 vols. Business records, consisting chiefly of a variety of account books covering the operations of the firm.

1352 **Sands, Louis** (1825-1905), Records, 1863-1933, 262 vols. Lumberman and banker of Manistee, Michigan. Included are records of Louis Sands Salt and Lumber Company (1863-1933, 116 vols.), among others.

1353 **Satterwaite, Joseph C.**, Papers, 1835-1885, ca. 100 items and 44 vols. Business papers of the proprietor of a flour and lumber mill at Tecumseh, Michigan. Includes daybooks (1850-54, 1856-65, 8 vols.), cashbooks (1848-62, 6 vols.), ledgers (1848-60, 4 vols.), special accounts (1860-65), accounts for rebuilding the sawmill, and sawmill boarders' records (1852-57).

1354 **Sawyer, Carl J.**, Manuscript, 1949, 43 pp. Article entitled "History of Lumbering in Delta County."

1355 **Shay, Ephraim** (1839-1916), Papers, 1875-1914, ca. 185 items. Waring, Michigan. The collection includes notes on the Shay locomotive, and other business papers.

1356 **Sligh Family**, Papers, 1825-1960, 21 feet and 125 vols. The collection contains a series of articles on reforestation by Arthur W. Stacy of the Grand Rapids, Michigan, *Herald.*

1357 **Stearns Salt & Lumber Company** (Ludington, Michigan), Records, 1885-1925, 250 vols. Ledger, account book, and general business records.

1358 **Stroebel, Ralph W.**, Letter, 1962, 4 items. Letter and enclosures relative to the Tittabawassee Boom Company.

1359 **Sturgeon River Lumber Company** (Hancock, Michigan), Records, 1872-1902, 13 items and 1 vol. Articles of association, proxies, minutes of directors, list of stockholders.

1360 **Thunder Bay Boom Company** (Alpena, Michigan), Records, 1868-1912, 8 feet. Cashbook, daybooks, time books, journals, ledgers, constitution and bylaws, and a miscellaneous book.

1361 **Tibbets, J.F., Lumber and Hardware Store** (Ravenna, Michigan), Ledger, 1880-1893, 1 vol.

1362 **Tittabawassee Boom Company** (Saginaw, Michigan), Record Book, 1864-1881, 1 vol. Contains articles of association and minutes of meetings.

1363 **Titus, Harold** (1888-1967), Papers, 1910-1967, 7 feet. Prominent member of the Izaak Walton League; conservation editor of *Field and Stream.*

1364 **Tower Family**, Papers, 1841-1937. Ionia, Michigan. The collection contains information on lumbering.

1365 **Weaver, Richard Lee** (1911-1964), Papers, 1937-1965, 31 feet. Associate professor of conservation and conservation education, University of Michigan.

1366 **Wheeler, Frederick M.** (b. 1862), Papers, 1859-1941, 1 foot and 21 vols. President of the Michigan Forestry Association. Correspondence, business papers, articles, and miscellaneous notebooks (21 vols.). The early papers pertain to the Wheeler family and much of the later papers deals with family affairs, conservation, and the Michigan Forestry Association.

1367 **White, Alma**. "Governor Warner and Conservation." University of Michigan: Seminar paper, n.d., 21 pp.

1368 **White, Stewart Edward** (1873-1946), Papers, 1901, 1912, 1941, 2 items and 2 vols. The papers relate to writings by White, and include a letter concerning the historical accuracy of the main episode in *The Riverman,* and manuscripts of *The Blazed Trail* (1902) and *The Riverman* (1908).

1369 **Whittemore, Gideon Olin** (1800-1863), Papers, 1817-1938, 3 feet. Business and family correspondence, relating to lumbering and many other matters.

1370 **Whittemore and Whittemore General Store and Lumber Company** (Tawas City, Michigan), Records, 1854-1858, 4 vols. Daybooks and ledger of general store and lumbering accounts of George O. and James O. Whittemore.

1371 **Williams, Gerhard Mennen** (b. 1911), Papers, 1948-1973, 974 feet and 27 vols. Governor of Michigan, 1949-1960. Correspondence, speeches, reports, memoranda, and notes dealing with state and national politics, and the operations of the state government of Michigan; and scrapbooks of newspaper clippings (27 vols.), and other material. Index to correspondents.

1372 **Wilson, Bethany (Mrs. Hugh E. Wilson)**, Manuscripts, 2 items. One item, entitled "It's an ill wind," concerns the Michigan forest fire of 1881.

1373 **Wisconsin Land and Lumber Company**, Records, 1871-1914, 16 feet and 492 vols. Correspondence, ledgers, daybooks, journals, and other business records of the H.A. Jewell Company, the William Mueller Company, the Fond du Lac division, the Hermansville, Michigan, division, and the C.J.L. Meyer division of the Wisconsin Land and Lumber Company.

1374 **Wise, Margaret E. Turner (Mrs. Charles H. Wise)** (b. 1865), Reminiscences, 1956, 1 item. Reminiscences of Mrs. Wise of early lumbering days near Alpena where she was a cook in the lumber camps and where her husband also worked.

1375 **Young, Leigh Jarvis** (1883-1960), Papers, 1908-1960, 250 items and 6 vols. Professor of forestry at the University of Michigan, Director of the Michigan Department of Conservation, and mayor of Ann Arbor. Included are student notebooks on university courses in botany, economics, geology, psychology, forest engineering, entomology, and dendrology. Correspondents include Gifford Pinchot.

University of Michigan, The Engineering-Transportation Library

Librarian, 312 Undergraduate Library Building 48104

1376 **Brown, George M.** (b. 1843), Papers, 1895-1901, 3 vols. and 395 pieces. The papers contain a letter to Brown from H.A. Gould, St. Simons Mills, Georgia (1877), about details of transporting logs, shipping and exporting lumber, wages, and types of laborers; and two letters (1891) from S.S. Wilhelm, East Saginaw, a dealer in pine lands.

1377 **Shay, Ephraim** (1839-1916), Papers, 1881-1900, 7 pieces. Mimeographed copies of letters from various lumber companies to the Lima Locomotive and Machine Works recommending the Shay locomotive.

University of Michigan, William L. Clements Library

Curator of Manuscripts 48104. The manuscript holdings of the Clements Library are listed in *Guide to Manuscript Collections in the William L. Clements Library,* 2nd ed., compiled by William S. Ewing, Ann Arbor: University of Michigan Press. 1953, and supplement, 1959. The 1st ed. of this work, compiled by Howard H. Peckham, Ann Arbor: 1942, has more complete lists of correspondence than the 1953 edition, but gives little information on subjects.

1378 **Lyon, Lucius** (1800-1851), Papers, 1812-52, 37 vols. U.S. Senator and Representative from Michigan, and surveyor general for Ohio, Indiana, and Michigan. The correspondence refers to the extensive tracts of land Lyon owned in Michigan and Wisconsin, and to lumbering on those lands.

1379 **Shelburne, Earl of [Sir William Petty]** (1737-1805), Papers, 1663-1797, 179 vols. (ca. 11,000 pieces). Four items concern forest history: letter from Walter Patterson to the Earl of Shelburne relative to the preservation of His Majesty's timber in America (ca. 1766-68) [49: 533-541]; Sir William Moore to Great Britain Board of Trade (January 10, 1767), abstract of a dispatch relative to the cutting of pine trees [51: 693]; William Tryon, abstract of a dispatch (December 3, 1766) relative to the timber in North Carolina [52: 102]; William Popple to Great Britain Solicitor General (April 24, 1710), relative to trees fit for masts in Massachusetts Bay [61: 219-220].

BAY CITY

Bay County Historical Society, Museum of the Great Lakes

Director, 1700 Center Street 48706

1380 A collection of material (ca. 1819-1934) pertaining mostly to the lumbering era of Bay County, Michigan, which was highlighted during the 1880s. Included are photographs of local lumber mills, local forests, and of workers in the mills and in the woods; also some letters and notes from lumbermen and lumber companies.

DEARBORN

Greenfield Village and Henry Ford Museum, Ford Archives

Archivist, Ford Archives, Edison Institute 48121

1381 Holdings include lumbering records (1881-1946, 24 feet) of Ford subsidiaries and companies controlled by Henry Ford. These records are scattered through a number of accessions, and relate largely to operations in Michigan in the 1920s and 1930s. Included are land acquisition records, minute books of lumber companies, correspondence, reports, and photographs, including a very extensive collection of copies of 19th-century lumbering photographs, representing all operations involved in the industry.

DETROIT

Detroit Public Library, Burton Historical Collection

Chief, Burton Historical Collection, 5201 Woodward Avenue 48201 Telephone: (313) 321-1000

1382 **Askins Family,** Papers. Papers of Charles Askins contain a lumber record book of George Jacobs, Sandwich, Ontario (July 24-November 22, 1933).

1383 **Edgar, James, and Company,** Photostat negative, 1 item. The photostat pictures a scale showing the number of 12, 14, and 16 foot logs. The original is owned by Albert Smith, Newaygo, Michigan.

1384 **Harrow Family,** Papers, 1779-1891, 22 vols. Papers of a St. Clair, Michigan, family. Material relating to lumbering is included in the journal and letterbook of Alexander Harrow (1755-1811).

1385 **Heald, Joseph** (b. 1823), Papers, 1830-1906, ca. 600 items. Correspondence and business and family papers, including papers relating to Heald's lumber interests in Maine and Michigan.

1386 **Holmes, Arthur Logan** (1862-1916), Papers, 1878-1920, 4 vols. and 9 wallets. Businessman of Detroit. Correspondence and other papers chiefly relating to Holmes' activities as secretary of the Michigan Retail Lumber Dealers Association, including correspondence (1878-1917); cashbooks (1900-12, 2 vols.); a record book (1900-04); and a registration book (1907).

1387 **Hubbard, Bela** (1814-1896), Papers, 1835-1909, 32 vols. and 3 wallets. Businessman, of Detroit. Maps of Michigan counties and townships (30 vols.); indentures, land contract accounts (1906-09) of the Bela Hubbard estate; records of tax sales, taxes, and lumber lands; and other papers relating chiefly to Hubbard's real estate activities.

1388 **Ingersoll, Mary A.,** Papers, 9 letters. The papers contain a letter from George Ingersoll to Martha Ingersoll regarding lumber interests at Cherry Creek, Michigan (July 9, 1850).

1389 **Jenks, William Lee,** Collection, 1824-1917, 22 boxes. Included are ledgers, daybooks, and account books of the Black River Steam Mill Company (1832-48); Zephaniah W. Bunce, sawmill (1822-67); Captain John Clark, sawmill (1832-71); Andrew and Joseph B. Comstock; James Haynes; and Ralph Wadhams, Henry Howard, and W.S. DeZeug (1831); also Port Huron Executive Committee for Relief, report on the work of relieving the sufferers of the Huron Peninsula by the great fires of 1881, by M.H. Allardt, Secretary. On loan to the Burton Historical Collection from the St. Clair County Library, Port Huron, Michigan.

1390 **Kent, George** (1796-1849), Papers, 1803-1849, 2 wallets. Lawyer, of Concord, New Hampshire. Chiefly legal papers (1815-42), together with a few family and friendly letters. Some of the letters deal with the lumber trade.

1391 **Marquis, James Winchester** (1812-1906), Papers, 1814-1924, 70 items and 4 vols. Includes the record book (1814-19) of sawmill activities at Prince Edward Island, Canada, and records of shipbuilding activities at Nova Scotia and Prince Edward Island (1838) and Detroit, Algonac, and St. Clair, Michigan (1849-68).

1392 **Mears, Charles,** Papers, 1854-1878, 16 pieces. Record of Mears' lumber interests in Michigan, including deeds, land warrants, printed table for measuring logs, form for hiring workmen.

1393 **Michigan Lumber Industry,** Photographs, 5 boxes. Included is an indexed collection (1870-1901, 11 vols.) of Carl Addison Leech.

1394 **Michigan Pine Land Association,** Records, 1863-1886, 1 box. Correspondence, survey reports, and other papers relating to the business activities of an association founded to manage and sell Michigan timber lands in behalf of persons who had acquired these lands through exchange of shares they held in the St. Marys Falls Ship Canal Company. Included in the collection are the papers of Cyrus Woodman, of Boston, who served as agent for the association.

1395 **Palmer Family,** Papers, 35,000 items. Thomas W. Palmer was a Michigan lumberman and U.S. Senator, 1883-1889. Papers deal with his personal life to 1871.

1396 **Palms, Francis** (1809-1886), Papers, 1852-1909, ca. 50 items, 30 vols., 2 boxes, and 1 folder. Lumberman of Michigan. Correspondence, deeds, ledgers, cashbooks, and other records from the estates of Palms and Herbert Book. Included are plat books of property in the Upper and Lower Peninsulas of Michigan, Canada, and Wisconsin, and other records of the Peninsular Land Company.

1397 **Slocum, Elliott C.,** Papers, 21 items. The papers contain a letter from F.M. Hentig to Slocum concerning a lumber mill at Slocum's Grove, Michigan.

1398 **Spooner, H.L.,** Papers, 2 items. The papers contain an agreement of Hyde and Rose to deliver lumber at Newell's Bay, Muskegon Lake (July 23, 1842).

1399 **Taylor, Knowles,** Papers, 1836-1871. These papers deal with speculation in lumber lands in Michigan, Illinois, and Wisconsin. Included is a letter to Taylor from Oshea Wilder concerning speculation in pine lands in Allegan County, Michigan.

1400 **Thompson, Charles Donald** (1873-1956), Papers, 1851-1945, 10 boxes. Lawyer of Bad Axe, Huron County, Michigan. Correspondence and other papers of Thompson, and his father, Charles E. Thompson (1845-1907). Box 5 contains papers dealing with the forest fire of 1881 and assistance given by the American Red Cross and Michigan Fire Relief Commission. Container list.

1401 **Thompson, Charles E.,** Papers. The papers include record books and correspondence of John Lund Woods, relating to his lumber interest (1852-1929).

1402 **Trowbridge Family,** Papers, 1781-1922, 1 vol. and 1 box. Papers of members of the Trowbridge family, of Southfield and Detroit, Michigan. Included are the Detroit Agency journal (1865-69) of the New York and Michigan Lumber Company, a firm operated by John Hubbard and Levi Trowbridge; 2 wallets and several record books concerning the lumber career of Francis C. Trowbridge (1844-1925), and material relating to plank roads in New York, and pine land speculation in Wisconsin.

1403 **Truesdail, Wesley** (1812-1886), Papers, 1867-1879, 2 vols. and 1 box. Banker and timber businessman. Ledger and journal of operations of Truesdail's farm in St. Clair County, Michigan.

1404 **Webb, Jefferson Bonson** (1882-1935), Papers, 1908-1912, 4 boxes. Lumber dealer, of Grand Rapids and Detroit, Michigan. Correspondence and other papers relating to Webb's activities as Michigan agent of the Morgan Sash and Door Company in Grand Rapids; also material relating to the Concatenated Order of Hoo Hoo.

1405 **Wexford County,** Papers, 1 package. The papers contain records of the Stronach Lumber Company (1881-90), including deeds and miscellaneous material for Lake and Wexford Counties, Michigan.

1406 **Wilson, Etta Smith Wolfe** (1857-1936), Papers, 1837-1935, 1 wallet. Journalist and naturalist. The collection contains correspondence with state officials concerning reforestation projects as well as material relating to Mrs. Wilson's activities as a journalist for Grand Rapids and Detroit newspapers.

1407 **Wing, Nelson H.** (1807-1877), Papers, ca. 1833-1890, 24 vols. Real estate investor, of Detroit, and afterwards of Greenville, New York. Correspondents include Hiram Bean, of Pentwater, Michigan, relating to logging operations and land prices and sales in the 1860s and early 1870s.

1408 **Wolfe, C.J.,** "Hannah, Lay, & Company: A Study in Michigan Lumber Industry," Detroit: Wayne State University, M.A. thesis, 1938.

1409 **Woodbridge, William,** Papers, 1780-1860, 145 wallets, 32 vols., 7 boxes. The papers contain letters to William Woodbridge from Thomas Scott concerning the cost of lumber for building a house at Hamtramck, Michigan (1833).

Wayne State University, Archives of Labor History and Urban Affairs

Director, 144 General Library 48202 Telephone: (313) 577-4024. Holdings are described in *A Guide to the Archives of Labor History and Urban Affairs,* compiled and edited by Warner W. Pflug, Detroit: Wayne State University Press, 1974.

1410 **Industrial Workers of the World,** Papers, 1905-1972, 80 feet, including 45 reels of microfilm. Includes records relating to activities among the lumberjacks of the Pacific Northwest. There are proceedings, trial records and evidence, newspapers, pamphlets, and correspondence. Subjects include the Centralia, Washington, incident; the Everett, Washington, incident; labor conditions; trials of various members.

EAST LANSING

Michigan State University, University Archives—Historical Collections

Director 48824. Holdings are listed in *A Selective Guide to the Michigan State University Manuscripts Collection,* East Lansing (?): The Museum, Michigan State University, 1961, mimeographed, 10 l., and William H. Combs and Anthony Zito, *A Guide to the Historical Collections of Michigan State University,* East Lansing: n.p., 1969.

1411 **Allcott Family,** Correspondence, 1835-1850, 7 folders. Correspondence between Sidney S. Allcott of Calhoun County, Michigan, and his brother William, of New York City, relating to their financial operations, land speculation in Eaton and Calhoun Counties, Michigan, and Sidney Allcott's saw and flour mills at Marshall, Michigan.

1412 **Gerrish, Jacob,** Papers, 1836-1867, 2 folders and 6 vols. Diaries (1836-1850, 5 vols.) and letters (23 items) concerning Berrien County life, with comments on lumbering; scattered entries regard lumbering and timber supplies in the area around Three Oaks, in southwestern Michigan.

1413 **Hackley & Hume Papers,** 120 cubic feet. Journals, ledgers, cashbooks, correspondence, of major and minor companies relating to the investments of Charles Hackley and Thomas Hume, chiefly concerning the lumber industry. There are also photographs, land maps, blueprints, a log-marked tree section, personal correspondence, and notes dealing with family history.

1414 **Halladay, Herman H.,** Papers, 1922-1931, 1 folder. Letters and newspaper clippings pertaining to the development of Walter J. Hays State Park in Michigan, giving details on the early efforts made by Halladay and others to acquire land for the project.

1415 **Harvey, Mrs. Carrie,** Reminiscences, n.d., 1 folder. Reminiscences of pioneering in Michigan recording the history of Dennis Usewick and his family who settled in Wexford County in 1867. Included are references to tapping maple trees, clearing the land of trees, and rivermen and logging operations.

1416 **Herrick, E.B.,** Papers, 1887-1921, 1 folder and 1 vol. This collection contains a volume kept by L.S. Warren of Howell, Michigan, of records of a loggers' lodging house. The folder contains loose pages from this volume as well as miscellaneous receipts belonging to J.R. Herrick and W. Denker, both of Clare, Michigan.

1417 **Huron Log Booming Company,** Articles of Association, 1864, 1 item. The company was to operate on the Cass, Shiawassee, and Saginaw Rivers in Michigan.

1418 **Littlefield, Josiah,** Papers, 1850-1895, 1 folder and 4 vols. Included are Xerox copies of newspaper clippings and a speech dealing with the career of Littlefield, a lumberman, in the Farwell, Michigan area. Four ledgers record lumber transactions and the cost of labor and supplies.

1419 **McKibbin, May,** Papers, 1826-1832, 2 boxes. Included are letters (1910-15) from Clifford McKibbin to his mother, written while serving with the U.S. Forest Service in Arizona and New Mexico.

1420 **Moore, Horace D.** (b. 1821), Papers, 1851-1925, 53 vols. and 1 folder. Included are account books and miscellaneous papers concerning Moore's lumbering activities, chiefly in Saugatuck, Michigan (ca. 1857-75).

1421 **Overlease, William,** Collection. One manuscript on the history of logging in Benzie County, Michigan; and an edited work based on an oral history of the lumber industry in the county.

1422 **Parsons, Ivan,** Papers, 1 folder. Includes an account by Thomas Wright of Fentonville, an agent of the Fire Relief Commission of Michigan, which describes the area in the aftermath of a forest fire, primarily in Caro, Bad Axe, Verona, Ubly, and Cass City, Michigan (n.d.).

1423 **Pere Marquette Lumber Company,** Records, 1860-1912, 9 boxes, 17 vols. Correspondence, deeds, ledgers, cashbooks, tax receipts, and miscellaneous legal papers, relating to a company with property holdings in Lake, Mason, Mecosta, Muskegon, Newaygo, Oceana, and Osceola Counties, Michigan, largely obtained by buying land grants from War of 1812 veterans. Includes a journal (1886-92) and cashbook (1882-90) of the company store, miscellaneous papers of the Flint and Pere Marquette Railroad, and a cashbook (1882-90) of the Ludington Water Supply Company. Persons represented include Delos Filer, Luther Foster, John Loomis, and James Ludington, founders of the company. Inventory, 2 pp.

1424 **Pyatt, Frank** (b. ca. 1876), Papers, n.d., 1964, 8 folders. Manuscript copies of two poems and an unpublished novel, "Woodsman's Trail," relating to Michigan lumbering, both authored by Pyatt; also notes (1964) describing Pyatt's career as a lumberman at the turn of the century and after.

1425 **Saunders, Harvey C.,** Papers, 1878-1934, 1 folder. The collection consists of Xerox copies of notes by Saunders about his life and work as a river-driving foreman with lumber companies in Schoolcraft County, Michigan, in the early 20th century; describes the work of these foremen and life in the logging camps; details of river improvements, the building of dams, the formation and operation of companies and the growth of settlements; includes material on the Chicago Company, F.N. Cookson, Consolidated River Improvement Co., and John Moran.

1426 **Slafter Family,** Papers, 1855-1891, 1 box. Personal and business correspondence and other papers of the family of William Slafter. Included are an agreement (1865) relating to logging in Tuscola County, Michigan, and to log prices; a letter (1873) relating to logging around Farwell, Michigan; and a letter (1873) relating to wages of loggers.

1427 **Snow, Orin D.,** papers, 1847-1852, 1 folder. Included is a brief description of the building of a sawmill by a resident of Niles, Michigan.

1428 **Stelzer, J.,** Papers, 1871-1891, 2 folders. Included are receipts giving the cost of lumber products for building, and diaries (4 vols.) containing occasional references to lumbering, forestry, and the cost of lumber products as they relate to agriculture and farm operations. Stelzer was a resident of Oak Grove, Michigan.

1429 **Tittabawassee Boom Company,** Records, 3 items. Listings of various lumbering companies and their log marks taken from the books of the Tittabawassee Boom Company, located in the Saginaw Valley of Michigan.

1430 **Tubbs, Harold,** Collection, 1852-1912, 1 folder. Included is one typed interview describing the Grayling, Michigan, vicinity during the lumbering era.

1431 **Walker, Jay,** Interviews, 1940, 1 folder. Included are typescripts of interviews by Walker with old-timer residents of Oceana County, Michigan, relating to logging in the area, 1863-1926. Subjects include Mr. and Mrs. W.S. Beattys who worked in Newaygo County, giving details of equipment and life in lumber camps; Minne Beadle, cook in logging camps near Ludington, Michigan; Irving C. Harwood, lumberjack, Oceana County, 1879-80; Fred Skinner, describing river drives; Michael Freshett, sawmill operator; and other short interviews.

1432 **Warren, Mrs. R.V.,** Papers, 1844-1889, 4 items and 1 vol. Included is an account book (1884-89) of a wholesale business in lumber and coal in central Michigan.

1433 **Whitman, Merton J.,** Collection, 1900-1949, 5 folders and 1 vol. Letters (1946-49) by Whitman giving his recollections of lumbering in Ogemaw and Arenac Counties, Michigan, in the 1880s and 1890s; a history of the town of Delano, Whitney Township, Arenac County, Michigan; and a volume of accounts and minutes (1900-21) of Delano Camp No. 8732, Modern Woodmen of America; and a folder of its miscellaneous papers (1904-19).

GRAND RAPIDS

Grand Rapids Public Museum

Director, 54 Jefferson Ave., S.E. 49502
Telephone: (616)456-5494

1434 Holdings (ca. 50,000 items) include papers on lumbering and photographs of early Michigan lumbering operations.

HOLLAND

*Netherlands Museum

49423. The following collections were described in the *Guide to the Dutch-American Historical Collections of Western Michigan,* edited by Herbert Brinks, Grand Rapids and Holland: Dutch-American Historical Commission, 1967.

1435 **Boone and Company** (Holland, Michigan), Records, 1874-1881, 2 vols. Daybook and ledger of the Boone Lumber Company.

HOUGHTON

Michigan Technological University, Institute of Wood Research

Director 49931 Telephone: (906)487-4264

1436 **Lake States Logging Congress,** Records, 1946-1974. Correspondence and other organizational materials concerning the founding and on-going activities, 1st to 18th Congresses.

Michigan Technological University, Library

Director 49931 Telephone: (906)487-2507

1437 **Cleveland-Cliffs Iron Company, Michigan District** (Ishpeming, Michigan), Forest survey notes, 1919-1943, 84 vols. This collection describes land owned by the company on Michigan's Upper Peninsula. Along with the survey notes are copies of notes prepared by Henry S. Graves in 1896 as part of his working plan for the management of the company's forest lands, and a copy of a report on forest management of the area in the 1930s, written by J.W. Toumey. The land area described in these records extends from the edge of the Ottawa National Forest to Sault Ste. Marie.

1438 **Stockley, William W.,** Notebooks, 1894-1921, 25 vols. Numbered vol. 28-62. Survey notes, reports, estimates, etc., of various plats, properties in Hancock and Houghton vicinity, Houghton County, Michigan.

1439 **U.S. General Land Office,** Field Notes, 1846, 29 vols. Copies from the collection of W.W. Stockley, Hancock, Michigan, relating to townships 51, 52, 54-59 of Houghton and Keweenaw Counties, Michigan.

KALAMAZOO

Western Michigan University, University Archives and Regional History Collections

Director 49001 Waldo Library

1440 **Adams, Charles Christopher** (1873-1955), Papers, ca. 1910-1943, ca. 40 feet. Ecologist and student of animal ecology and biogeography, particularly of the boreal spruce-fir forest of North America. The collection was formerly located at the Charles C. Adams Center for Ecological Studies, Department of Biology, Western Michigan College. Unprocessed.

1441 **Hodgman, Francis,** Papers, 1833-1907. Included is a transcript (ca. 1860-70, 1 vol. [541 pp.]) of U.S. government surveyors' field notes for surveys conducted ca. 1826-31 in Kalamazoo County, Michigan, by John Mullett, Robert Clark, Jr., and George W. Harrison.

LANSING

Michigan History Division, State Archives

State Archivist, 3405 North Logan Street 48918

1442 **Michigan. Department of Natural Resources,** Records, 1899-1966. Prior to 1966, this department was known as the Department of Conservation. Minutes (1899-1909, 2 inches) of the State Forestry Commission are carbon copies, primarily policy statements and reports of activities; also included is a brief history of the creation of the commission, and a list of its expenditures in 1900. Records of the Forestry Division include County Forest Development Plats (1914-46, 1¼ feet), containing township maps of selected Michigan counties, chiefly prepared in 1918 by Marcus Schaaf, State Forester, with later additions and corrections, showing cultural and physical features, land owned by the state, and sometimes nature of vegetation cover; maps and correspondence concerning proposed postwar work projects in state forests (1942-44, 10 inches); monthly summaries of receipts by state forests (1948-62, 5 inches); summary of forest fires containing statistics of occurrence in districts and counties (1928-47, 1 folder); forest fire reports (1965-66, 3 feet, 2 inches); and Clarke-McNary Budget Reports (1960-65, 2 inches), with financial plans for state and federal cooperative forestry programs. Records of the Forest Fire Division include location guide to towers for fire control (ca. 1936-39, 1 vol.); and records of towers and stations (n.d., 2 folders). Records of the Field Administration Division include form reports of forest fires (1949-50, 1952, 6 feet).

MARQUETTE

Marquette County Historical Society, The J.M. Longyear Research Library

Executive Secretary, 213 North Front Street 49855

1443 **Barbeau, Peter Boisdoré,** Diaries, 1866-67, 2 vols. Diaries kept at Sault Ste. Marie, Marquette, and Negaunee, Michigan.

1444 **Brotherton, Ray A.,** Manuscript, 2 pp., illustrated. "Oxen, Big Wheels, and Roistering Lumberjacks Made Early Logging History in the Upper Peninsula of Michigan and Northern Wisconsin."

1445 **Harvey, Charles Thompson,** Papers, 1870-71, 2 items. An agreement relative to furnishing facilities for manufacturing lumber to Sidney Adams at Harvey, Michigan (1870); and a letter (1871), to Adams, proposing a new sawmill at Marquette.

1446 **Hebard, Charles,** Collection. Several cartons of papers of a man active in logging on the Keweenaw Peninsula.

1447 **King, Dorothy D.,** "Seney Stockades," 1951, 3 pp.

1448 **Longyear, John Munro,** Papers. Papers of a businessman interested in timber lands.

1449 **Paul, Helen Longyear,** Manuscript, 1950. "Carp River Sawmill," compiled from correspondence (1857) of J.H. Anthony and Peter Boisdoré Barbeau.

1450 **Pendill, James P.,** Inventory of mill property at Pendill's Creek, May 1, 1854, 23 pp.

1451 **Read, F.W.,** Collection. Several cartons of materials relating to the sawmill and logging industry in the Marquette vicinity.

1452 **Taylor, Charles S.,** Manuscript, 1949, 4 pp. "Pendill's Sawmill on the Tahquamenon."

MOUNT PLEASANT

Central Michigan University, Clarke Historical Library

48859

1453 **American Fur Company,** Records, 1810-1848, 924 items and 7 vols. Correspondence, account books, bills of lading, receipts, and other papers relating to the conduct of business between company managers and agents at Detroit, L'Anse, Sault Ste. Marie, Mackinac Island, Michigan, La Pointe, Wisconsin, Montreal and New York City, and other places, and their customers regarding the company's many business interests, including lumbering. Forms parts of the library's Sault Ste. Marie Collection. Summary.

1454 **Bailey, J.B.,** Letters, 1837, 1840, 2 items. Kalamazoo, Michigan. Included is a letter to Miss Mitchell, New York, with estimates of cutting and delivering lumber to Chicago (1840).

1455 **Barbeau, Peter B.** (1800-1882), Papers, 1834-1872, 2980 items and 6 vols. Sault Ste. Marie, Michigan. Correspondence, account books, bills, receipts, articles of agreement, land transfer papers, and other papers relating to Barbeau's business transactions in lumber and other concerns. Forms part of the library's Sault Ste. Marie Collection. Library also holds a positive microfilm copy of most of the collection.

1456 **Beck, Earl Clifton,** Papers, ca. 1930, 225 items. Collection of typewritten shanty-boy songs, ballads, etc., collected by Beck during the 1930s. Name of collector and place where collected mentioned.

1457 **Bliss, Charles S.,** Business Records, 1879-1935, 40 feet and 129 vols. Lumber merchant in Saginaw, Michigan. Ledgers and other account books, correspondence, financial records, bills, receipts, and other papers relating to the lumber trade and industry in Michigan. Records concern transactions in lumber camps in Gladwin, Isabella and Midland counties. There is also family correspondence. Unpublished guide.

1458 **Boughey, Herbert F.,** Papers, 1911-1934, 118 feet and 16 vols. Lumber merchant, real estate broker in Traverse City,

Michigan, treasurer of the Carp Lake Lumber Company, Bingham, Michigan. Personal, family and business correspondence, ledgers, account books, letterpress copies, bills, receipts, stock market reports (1917-18), relating chiefly to his lumber and real estate business.

1459 **Boyce, Jonathan** (1856-1902), Business Records, 1878-1955, ca 14,000 items and 54 vols. Manufacturer and wholesale and retail dealer in lumber and salt in Muskegon, later Bay City and Essexville, Michigan. Included are legal documents and business correspondence relating mainly to the Boyce Lumber Company. Shelf list.

1460 **Buckley & Douglas Lumber Company** (Manistee, Michigan), Papers, 1902, 2 items. Two illustrated bank notes for $26.67 and $42.34 payable to the First National Bank of Manistee.

1461 **Canier, A.,** Letters, 1857-1859, 1 vol. Copies of brief business letters from the owner of a Detroit lumber firm to his business associates concerning shipping, sale and purchase of all types of timber.

1462 **Case, Charles** (1853-1935), Recollections, 1942, 1 item (33 pp.). Typewritten recollections of his life and lumbering experiences in Benzonia, Michigan.

1463 **Central Michigan University.** Student term papers. Relating to forest history are some 23 items (1957-58 and n.d., 5 to 37 pp. each) dealing with such topics as lumbering in Michigan, local histories, evolution of log marks, logging railroads, logging equipment, company histories, Indian timber lands, chiefly in Michigan.

1464 **Cleveland-Cliffs Iron Company** (Negaunee, Michigan), Business Records, 1871-1895, 23 vols. Bound volumes with invoices, receipts, shipping notes of various company purchases and sales, arranged chronologically with indexes in front of each volume. 1886-90 volume includes material relating to forestry and lumbering.

1465 **Collen, James A., and George W.,** Bill of sale, 1946, 2 pp. Bill of sale of sunken logs, known as deadheads, now lying in the Manistee River, or any of its tributaries, sold by James A. & George W. Collen of Grayling to Eric R. Steenburg of Roscommon County, Michigan. Text followed by symbols of log marks. Photostat of the original in the County Court House, Grayling, Michigan.

1466 **Cook & Jones Company** (Morley, Michigan), 1867-1883, 4 vols. Included are journals, part of a cashbook, and a weekly timebook of a company which manufactured sawed shingles.

1467 **Dinius, Sylvester,** Papers, 1886-1910, 322 items and 24 vols. Dealer in logs and lumber at Ashton, Michigan, Leroy, Michigan, and Fort Wayne, Indiana. Business and family correspondence, invoices, legal papers, and pension papers related to Dinius' Civil War service. Shelf list.

1468 **Dodge, Roy L.,** Paper, 1968, 1 item (7 pp.). "Anatomy of a shingle mill, 1887-1892." Typewritten story of the lumberman, Philip Cory, in Dodge, Hamilton township, Clare County, Michigan.

1469 **Flint & Pere Marquette Railroad,** Business Records. Principally the records of early railroads absorbed by the Flint & Pere Marquette Railroad during the 1860s and 70s. Unprocessed.

1470 **French Land & Lumber Company[?]** (Rose City, Michigan), Ledger, 1893-1896, 1 vol. Expenses of lumbering operations and lumber camps.

1471 **Gould, Amos** (1808-1882), Papers, 1828-1931, ca. 65,000 items and 153 vols. Owosso, Michigan. Correspondence and other papers relating to banking, investments in land, lumber, and railroads; and legal, political, and family affairs. Unpublished guide.

1472 **Hoffmann, Jacob R.,** Papers, 1864-1888, 43 items. Fort Wayne, Indiana. Patents issued to Hoffman for his invention of an "Implement in Saw-Mills." Includes drafts and illustrations of sawmill machinery.

1473 **Hubbard, Gordon S.,** Deed, 1882, 1 item. Warranty deed on land sale from Hubbard's annex to the national park, Island of Michilimackinac. The reverse of the deed has illustrated plat map of Hubbard's annex.

1474 **Leech, Carl Addison,** Article, n.d., 1 item (8 pp.). "Deward: a lumberman's ghost town." Typewritten article of a lumbering town in the western part of Crawford County, Michigan, and of its "timber king," David Ward.

1475 **Lumbering Diary,** 1893. Diary kept by an unknown lumber dealer of Saginaw, Michigan. Contains day by day accounts of business plus inventories and bills payable. Most of the business conducted with Ayres Lumber and Salt Company and E. Hibbard & Sons.

1476 **McCallum, George P.,** Papers, 1931-32, 12 items. Group of mimeographed papers relating to forest taxation in Michigan, reforestation, forest crop laws, and D.M. Matthews' "Suggestions for the Selective Cutting of a Hardwood and Hemlock Forest in the Lake States," assembled by McCallum as chairman of the Michigan Land Utilization Conference, in response to Henry Schmitz of the University of Minnesota.

1477 **Main, Sydney S.,** Papers, 1889-1893, 20 items. County surveyor, Brutus, Emmet County, Michigan. Letters relating to land and timber, telegrams, and checks sent to him.

1478 **Malloch, Douglas** (1877-1938), Scrapbooks, 4 reels microfilm (23 vols.). Journalist, poet, writer, editor and publisher of the *American Lumberman* in Chicago. Correspondence, speeches, short stories and poems devoted to the forest, newspaper clippings telling of his success, his speaking engagements throughout the United States, etc. Originals are in the Muskegon County Historical Museum, Muskegon, Michigan.

1479 **Montague, Irene,** Articles, 1969, 1 item (21 pp.). "Michigan's final great push in pine logging." Unsigned, typewritten account of lumbering in the northeastern part of the Lower Peninsula.

1480 **Olds, Millard D., Lumber Company** (Cheboygan, Michigan), Records, 1885-1955, 119 feet and 309 vols. Correspondence, work reports, contracts, tax statements, ledgers, journals, cashbooks, log books, bills, receipts, records of the steamer *Schoolcraft,* and other papers of the company dealing in lumber and coal. Includes the records of the Cheboygan & Presque Isle Railroad which the company owned and operated. Shelf list.

1481 **Onekama Lumber Company** (Onekama, Michigan), Records, 1865-1898, 36 vols. A.W. Farr, president, George A. Barslow, secretary. General store, lumber manufacturers and bark dealers. Includes ledgers, journals, and cashbooks (1869-72, 1874-96, 33 vols.) and letterpress books (1865-66, 1875-83, 1894-98, 3 vols.), concerning purchases and sales of dry goods, groceries, hardware, bark and lumber.

1482 **Quinnin, Louis C.,** Papers, 1880-1902, 23 items. Owner of the Michigan Lumber Company of Saginaw, Michigan. Included are various legal documents, such as mortgages, leases, checks, notes, shares of stock, club membership certificates, and a few business letters.

1483 **Rouse, William B.,** Business Correspondence, 1885-1889, 66 items. Correspondence of a dealer in salt and a lumber manufacturer, Bay City, Michigan, concerning the sale and delivery of goods.

1484 **Ruggles, Charles F.,** Business Records, 1869-1936, ca. 397 items and 30 vols. Dealer in timber lands, owner of the Calaveras Timber Company. Daybooks, ledgers, journals, collection registers, contracts, correspondence, etc., of a real estate dealer in Manistee, Michigan.

1485 **Salling, Hanson & Company,** Scale Book, 1883-1884, 1 vol. Record of logs put in by the company and sold to C.B. Lewis & Company, giving size and amount of lumber. No place mentioned.

1486 **Smith, Elias W.** (1831-1882), Papers, 1853-1897, 109 items and 1 vol. St. Louis, Gratiot County, Michigan. Business letters, account book (1862-73), land contracts, and papers relating to a timber sale lawsuit and wage payments to teachers of St. Louis, Michigan.

1487 **Stahl, Ferdinand,** Letters, 1884-1885, 7 items. Included are letters (in German language) from Fort River lumber camp no. 10, Delta County, Michigan, and from Escanaba, Michigan, to "a friend," requesting him to send a German newspaper and his passport.

1488 **Stevenson, A., & Son** (Adrian, Michigan), Correspondence, 1881, 128 items. Included are business letters from many firms in and outside Michigan to a coal and lumber dealer concerning the purchase and sale of lumber, coal, and shingles; also telegrams and receipts.

1489 **Stroebel, Ralph W.,** card file, n.d., 700 cards. Typewritten card file recording lumbering and related industries in the Saginaw, Michigan, area.

1490 **Tittabawassee Boom & Raft Company** (Saginaw, Michigan), Records, 1873-1905, 6 items. Photostatic copies of portions of stockholder meetings, directors' meetings, and log scaling report of the C.K. Eddy & Son Company; a Xerox copy of a business letter to Day Dean, concerning titles to certain lands and a boom privilege clause; also a price list of the company for 1905. Originals of the photostatic copies are in the Michigan Historical Collections, University of Michigan, Ann Arbor.

1491 **Vincent, Edward,** Daybook, 1865, 1 vol. Concerns lumbering in the Port Huron, Michigan, area.

1492 **Ward, Gerrit Smut** (1843-1916), Papers, 1860-1864, 1879-1927, 1960, 512 items and 1 vol. Lumberman, Alma, Michigan. Family and business correspondence, including material pertaining to land transactions. Also certificates, mortgages, and newspaper clippings.

1493 **Watson, Amasa Brown** (1826-1888), Papers, 1855-1923, 1744 items. Included is business correspondence from Watson's agents and others relating to lumbering around Muskegon and Newaygo, Michigan; also deeds, mortgages, and receipts. Shelf list.

1494 **Whittemore, James Olin,** Diary, 1866, 1 item (33 pp.). Tawas City, Iosco County, Michigan. Excerpts of his personal diary, 1866, reproduced from an unsigned typewritten copy. Concerns lumbering and other business and social life.

MUSKEGON

Hackley Public Library

Reference Librarian, 316 West Webster Ave. 49440

1495 **Bowles, Captain H.G.,** Log order books, 1884-1887, 3 vols. Records of movements of log rafts for various Muskegon lumber companies, kept by a tug boss.

1496 **Getty, H.H.,** Log marks, 1885, 1 vol. Log marks of most of the Muskegon lumber companies.

1497 **Harkins,** Log tally books, 1870-83, 9 vols. Log tally books kept for various lumber companies of Muskegon and vicinity by Harkins, a scaler.

SAGINAW

Hoyt Public Library

Reference Chief, Public Libraries of Saginaw, 505 Janes Avenue 48605 Telephone: (517)754-6541.

1498 **Barnard, Newell** (1825-1883), Business records, 1882-1902, ca. 2 feet. Records of the N. & A. Barnard Company, a lumbering and salt business operated by Newell Barnard and his son, Arthur Barnard, at Saginaw, Michigan. Includes daybooks containing bylaws of the company, ledgers listing accounts of log purchases, index of an account book, check stub book, and agreements.

1499 **Hardin, Daniel** (1820-1901), Papers, 1848-1901, ca. 2 feet. Businessman of Saginaw, Michigan. Correspondence, tax receipts, and other papers relating to Hardin's interests in lumbering and other businesses in Saginaw and Leonardsville, New York.

1500 **Tittabawassee Boom Company,** Records, 1863-1893, 9 vols. and 1 reel of microfilm (negative). Record book containing articles of agreement, minutes of directors and stockholders' meetings (1864-80), and notes on the lumbering industry in Saginaw Valley, Michigan, by Rockwell M. Kempton; log books listing owners' marks and names (1889-93); and storage book of the Melbourne Boom listing marks of logs stored for the Tittabawassee Boom Company. The original of the microfilmed record book is in the Michigan Historical Collections, University of Michigan.

TRAVERSE CITY

Con Foster Museum

49684

1501 **Hannah, Lay & Company** (Traverse City, Michigan), Records, ca. 1850-1909. Records of a sawmill and lumberyard engaged in pine lumbering in the Traverse City area, ca. 1851-90. There are journals or cashbooks (1850-54, 1878-82, 1897-1909, 3 vols.), daybooks (1882-97, 2 vols.), a ledger, photographs, and miscellaneous items.

Minnesota

DULUTH

St. Louis County Historical Society

Director, 2228 East Superior Street 55812

1502 **Manuscripts.** Included are an assortment of student papers and other manuscripts relating to local forests and forest resources. Among the holdings are Anonymous, "From the Forest to the Writing Desk," (n.d.), concerning paper making; Betty Bachmann, "Minnesota Forestry Service Personalities of the Past" (1968); John A. Bardon, "First Saw Mills at the Head of Lake Superior" (1920); George G. Barnum, "Shipment of Lumber to Duluth in 1869" (1932); Fred W. Bessette, "Logging and Lumber Industry through

118 Years" (1940); Garfield Blackwood, "History of Lumbering at Fleetwood, Minnesota" (1929); William Culkin, "A Forest Tragedy" (1936); W.E. Culkin, "Removal of the Arrowhead Forests" (1937); Duane Eichholz, "Virginia and Rainy Lake Logging Company" (1954), with map and pictures; Benjamin Finch, "Early Logging in the Northwest" (1940), a Work Projects Administration project; John Fritzen, "Lumbering on the North Shore and Early Saw Mills" (1945); R.L. Griffin, "Forests and Logging in the Vicinity of Hibbing, Minnesota" (1930); John Stone Pardee, "In the Treasure Country" (n.d.), and "Lumbering: The Vanishing Industry: Lumberjack Days" (n.d.); W.E. Scott, "The Forest History of Lake County" (1929); S. George Stevens, "Twin Ports Municipal Forest" (n.d.); William Trygg, "Ecological and Physiological Traits of Quetico Superior" (1963); Fred D. Vibert, "Conservation of Forest Lands and Reforestation" (1924) and "History of Early Lumbering in Minnesota" (n.d.); J.C. Wadd, "Lumbering in the Buhl, Kinney & Great Scott Township District" (n.d.), which mentions the Civilian Conservation Corps; Otto Wieland, "North Shore and Superior National Forest" (ca. 1930) and "The Lumber Industry at the Head of the Lakes" (1936); E.J. Wilkinson, "Birds of National Forest" (1936), "Present Status of Wildlife in the Superior National Forest" (1936) and "Superior National Forest in Cook County" (n.d.); O.A. Wiseman, "The Lumbering Industry at Tower" (1926); and Julius Wolff, "Quetico Superior Country" (1964) and "Forest Fires of the Sawbill Country" (1958).

1503 **Merritt, Leonidas** (1844-1926), Papers, 1870-ca.1926, 12 cartons. Explorer, pioneer, timber cruiser, and discoverer of iron ore on the Mesabi Range. Correspondence, personal writings, family documents, business papers, pamphlets, photographs, and memorabilia. Restricted.

1504 **Richardson, H.W.,** Papers. Includes a report on a forest fire (1918).

1505 **Swan Lake Lumber Company,** Account book, 1903-1904, 1 vol.

University of Minnesota at Duluth, Library

Librarian 55812

1506 **Nute, Grace Lee,** Papers. There is material relating to the lumbering history of Minnesota, including photographs. Uncataloged.

FARIBAULT

Rice County Historical Society, Buckham Memorial Library

Curator 55021

1507 **Brown-Martin Lumber Company** (Northfield, Minnesota), Records, 1884-1900. Lumber journals, cashbooks, order books, etc.

LAKE CITY

*Wabasha County Historical Society

204 Soak Street 55041

1508 According to *A Guide to Archives and Manuscripts in the United States,* edited by Philip M. Hamer, New Haven: Yale University Press, 1961, p. 313, the Wabasha County Historical Society held an unknown quantity of business records, including those of a lumber mill and lumber yards (1857-93).

MINNEAPOLIS

University of Minnesota, University Libraries, The University Archives

Archivist 55455

Archives and Manuscripts

1509 **Blegen, Theodore Christian** (1891-1969), Papers, 1918-1969, 33 feet. Correspondence and papers relating to the Advisory Board of the National Park Service (1945-64, 16 folders), largely mimeographed announcements, minutes, etc., with a small amount of correspondence, including some with Bernard De Voto; also correspondence and papers relating to the Forest History Society (1951-67, 17 folders).

1510 **Kaufert, Frank** (b. 1905), Papers, 1964-1971, 1 foot. Correspondence and papers relating to the Boundary Waters Canoe Area (1966-72), Superior National Forest Advisory Committee (1966-71), and Voyageurs National Park (1964-70).

1511 **Roberts, Thomas Sadler** (1858-1946), Papers, 1872-1946, 31 feet. Director of the University of Minnesota Museum of Natural History. Correspondence concerning natural history, professional and personal affairs; personal diaries and notebooks (1874-1930); speeches and radio talks, and manuscripts of writings. Includes material relating to the conservation of wildlife, Itasca State Park, and other matters. Correspondents include Carlos Avery, James Ford Bell, Walter John Breckenridge, Major Alan Brooks, Mabel Densmore, William Finley, Herbert Gleason, C. Judson Herrick, Clarence Herrick, Francis Lee Jacques, William Kilgore, Olga Lakela, Henry Francis Nachtrieb, Robert Nord, and Olin Sewall Pettingill.

1512 **University of Minnesota. College of Agriculture, Forestry and Home Economics,** Papers, 1911-1960. Included is correspondence concerning the curriculum and general matters relating to the College of Forestry and forest-related organizations in Minnesota.

1513 **University of Minnesota. College of Forestry,** Papers, 1898-1971, 74 feet (24 feet unprocessed). These papers cover forestry education at the university from its beginnings in the Department of Horticulture. There is information relating to all phases of the history of the college, including its growth, development, and experimental work, including the Itasca Forestry and Biological Station, the Cloquet Forestry Center, and the John H. Allison Forest; reforestation and conservation problems on both the state and national levels; the work of the Civilian Conservation Corps in blister rust control and other activities; the commercial aspects of forestry, such as research in wood preservation, wood uses, wood strength; forest legislation; the work of national and international organizations in which Henry Schmitz (director, 1925-47) and Frank Kaufert (director, 1947-74) were active. Some of Kaufert's correspondence relates to the Forest History Society.

1514 **University of Minnesota. Department of Horticulture,** Papers, 1888-1948. Included are papers of Samuel B. Green (1888-1910, ca. 25 boxes and 25 vols.), first dean of the University of Minnesota Department of Forestry, and some papers of his successor E.G. Cheyney.

1515 **University of Minnesota. Lake Itasca Forestry and Biological Station,** Papers, 1927-1964, 10 feet. Most of this collection relates to the Biological Station and the summer Biological Institute.

Tape Recordings

1516 **Allison, John Howard** (b. 1883), Tape recording,

1968 (T.R. 388, Reel 24). Interview on his life and experiences in the University of Minnesota School of Forestry, and in the development of the John H. Allison Forest, Lake Vadnais, and in the field of forestry. April 19, 1968. (Audio-dub from KTVA *Emeritus* Series).

1517 **King Timber,** Tape recording, 1951 (T.R. 320, Reel 6). KUOM, Minnesota Mid-Century Series.

1518 **Lake Itasca Forestry and Biological Station,** Tape recording, 1959 (T.R. 110, 8 reels). 50th anniversary of forestry training at the station. Speakers include: John H. Allison, Henry Clepper, Austin A. Dowell, Frank H. Kaufert, Thorwald Schantz-Hansen.

1519 **Schantz-Hansen, Thorwald** (1891-1971), Tape recording, 1968 (T.R. 388, Reel 31). Interviewed on his experiences in the University of Minnesota School of Forestry, including Cloquet Forest Experiment Station. June 14, 1968. (Audio-dub from KTCA-TV *Emeritus* Series).

1520 **The World—Its Natural and Human Resources.** Tape recording (T.R. 250). Five-member panel including Henry Schmitz (forestry) and Arthur Wilcox (horticulture). (*World We Want,* radio series, 1943-45).

1521 **The World's Timber Resources.** Tape recording (T.R. 304). Panel includes: J.H. Allison, Russell Cunningham, Frank Kaufert, Henry Schmitz. (*World We Want,* radio series, 1943-45).

University of Minnesota, University Libraries, Department of Special Collections

Curator 55455

1522 **Paul Bunyan Collection,** 3 feet. Included are manuscripts of James Stevens, many in longhand, correspondence with Alfred Knopf, H.L. Mencken, Stuart Chase, Louis Adamic, and William Green of the American Federation of Labor, and many clippings and photographs; books, pamphlets, and photographs from Professor W.W. Charters of Ohio State; the William B. Laughead papers consisting of original drawings, correspondence with Max Gartenberg, Louis A. Maier, Franklin J. Meine, and Archie Walker, sketches and final drawings, a file of Red River Lumber Company advertisements from *American Lumberman* and *Mississippi Valley Lumberman* (1915-49), and stories both in pamphlet and manuscript form; and material from Mrs. Dorothy Moulding Brown of Madison, Wisconsin.

NORTHFIELD

The Norwegian-American Historical Association

Secretary, Norwegian-American Historical Association, St. Olaf College 55057

1523 **Østerud, Ole Olson** (1820-1909), Papers, 1833-1909, 170 items. Correspondence (1854-94), includes references to Wisconsin pineries.

SAINT PAUL

James Jerome Hill Reference Library

Executive Director, Fourth Street at Market Street 55102 Telephone: (612) 227-9531

1524 **Hill, James Jerome** (1838-1916), Papers. Included are materials relating to the lumbering interests of the Great Northern Railroad. Closed until December 31, 1981.

Minnesota Historical Society

Curator of Manuscripts, 690 Cedar Street 55101 Telephone: (612) 296-2747. Some of the holdings are described in *Guide to the Personal Papers in the Minnesota Historical Society,* compiled by Grace Lee Nute and Gertrude W. Ackermann, St. Paul: Minnesota Historical Society, 1935; *Manuscripts Collections of the Minnesota Historical Society: Guide No. 2,* compiled by Lucile M. Kane and Kathryn A. Johnson, 1955; and *Guide to the Public Affairs Collection of the Minnesota Historical Society,* compiled by Lucile M. Kane, 1968.

1525 **Adams, Elmer Ellsworth (1861-1950), and Family,** Papers, 1860-1951, 61 boxes (including 98 vols.) and 5 vols., 3 oversize folders, and 2 items in reserve. Businessman and state legislator of northwestern Minnesota. The papers consist of correspondence, diaries, scrapbooks, financial records, clippings, and printed materials. There is a review by Grover M. Conzet on a forest policy for Minnesota; letters with information on the sale of state timberlands of the Leech Lake Reservation; lumber legislation (ca. 1904-06); and material on flood control. Restricted in part (post-1919 only).

1526 **American Immigration Company** (Chippewa Falls, Wisconsin), Records, 1903-1940, 88 boxes (including 7 vols.) and 44 vols. Correspondence, minutes, tax registers, abstracts, sales contracts, and financial records of an organization, located at Chippewa Falls, Wisconsin, formed by lumber companies in Wisconsin and Minnesota to sell cutover lands. There is much information about land utilization, taxation, land companies, land prices, forest products industries, and the movement of population. Correspondents include Fred S. Bell, the British Land Investment Company, James L. Gates, the Homeseekers Land Company, the Laird, Norton Co., the Polish-American Colonization Co., the Sawyer County Taxpayers' Association, the Soo Line Catholic Colonization Co., Frederick E. Weyerhaeuser, the Wisconsin Advancement Association, and the Wisconsin Colonization Company. Restricted.

1527 **Anderson, Albert N.,** Papers, 1936-1955, 2 folders (including 41 items and 39 l.). Anderson was concerned in flood control projects on the Red Lake and Clearwater rivers, Clearwater County, Minnesota. The collection contains correspondence and clippings, and drafts of short historical articles in part relating to these projects, and including information on logging in Sinclair Township, Beltrami County, and on Robert Neving, an early logger in the area.

1528 **Andrews, Christopher Columbus** (1829-1922), Papers, 1843-1930, 8½ feet (13 boxes, including 105 vols., and 21 additional vols.). Correspondence, diaries, newspaper clippings, and other papers of an army officer, diplomat, and chief fire warden of Minnesota, 1895-1905, state forestry commissioner, 1905-11, and secretary of the forestry board, 1911-22. The papers include information on the development of forestry in Minnesota.

1529 **Ann River Land Company,** Papers, 1875-1933, 1 box, 3 vols. Land book, journal, and ledger (1901-31), correspondence, deeds, agreements, and miscellaneous records of a Stillwater, Minnesota, firm include transactions between Isaac Staples and Frederick Weyerhaeuser and associates.

1530 **Atwood Lumber Company** (Stillwater, Minnesota), Papers, 1891-1914, 2½ inches (1 box). Financial statements, inventories, and letters. There are also papers of the Barronett Lumber Company of Barronett, Wisconsin (1891-94), and the Atwood Lumber and Manufacturing Company, Park Falls, Wisconsin (1909-14). The records contain information on land holdings, lumber production, wages and salaries, lumber and log prices, insurance, sawmills, store buildings, dams, river improvements, logging, timber land and log purchases, and the purchase of a mill site in Park Falls, Wisconsin. The company formed part of the Weyerhaeuser Company.

1531 **Auerbach, Maurice** (1835-1915), Records, 1858-1918, 12 boxes. Business and financial papers relating in part to the C.N. Nelson Lumber Company at Cloquet, Minnesota.

1532 **Ayer, Lyman Warren** (b. 1834), Papers, 1862-1903, 9 items, and 1 reel microfilm. The collection includes a microfilm copy of a journal (1902-03, 175 pp.) entitled "Report on Spruce and Other Timber Tributary to International Falls, Minnesota," kept by Ayer while surveying timber for the Backus-Brooks Comapny in Minnesota and Ontario, and containing data on timber, sawmills, logging and railroad camps, dams, sluices, and logging roads, etc. The original is in the possession of the Minnesota and Ontario Paper Company, International Falls, Minnesota. Typed description, 3 pp.

1533 **Backus-Brooks Company** (Minneapolis, Minnesota), Records, 1892-1911, 1 box and 5 vols. Minutes of meetings, ledgers, cashbooks, and a journal of a Minnesota lumber firm known, until 1899, as the E.W. Backus Lumber Company.

1534 **Bailey, Everett Hoskins** (1850-1938), Papers, 1839-1954, ca. 5 boxes and 8 vols. Banker. Included is material relating to Bailey's work in creating a national park in the Isle Royale, Michigan, area. Unpublished description.

1535 **Bardwell-Robinson Co.** (Minneapolis), Minute book, 1892-1932, 1 vol. Minute book, with copies of articles of incorporation and bylaws, of a firm manufacturing lumber, sashes, doors, blinds, moldings, and other lumber products.

1536 **Bartholomew, Paige** (1860-1944), Correspondence, 1884-1889, 44 items. Xerox copies of letters to Bartholomew from family and friends in New York and Minnesota, dealing, in part, with lumbering. Originals are in the Collection of Regional History, Cornell University.

1537 **Beef Slough Manufacturing, Booming, Log Driving, and Transportation Co.** (Alma, Wisconsin), Papers, 1867-1905, 1 box (including 2 vols.) and 1 vol. Articles of association, minutes, and stock certificates of a firm which formed part of the Weyerhaeuser Company. There is information on meetings of the directors, boomage, agreements concerning the use of the company's booms, purchase and sale of land, financial accounts, cooperation with the Mississippi River Logging Company and sale of property to that firm, improvement of the Chippewa River and Beef Slough, and Wisconsin state legislation relating to the company. Restricted.

1538 **Bell, James Hughes** (1852-1892), Papers, 1845-1900, 6 items, 2 reels microfilm. The papers include an autobiographical sketch; information on Bell's work as a lumberman in Wisconsin; and his life in Winona, where he worked in railroad shops. A brief diary (1889-92) covers the period of Bell's residence in Winona.

1539 **Bonners Ferry Lumber Company, Ltd.** (Bonners Ferry, Idaho), Papers, 1904-1931, 7 inches (2 boxes). Financial statements, inventories of logs, lumber, lands, stock, and milling and logging supplies, and a small amount of correspondence concerning the financial statements. The company formed part of the Weyerhaeuser Company.

1540 **Bonness, Frederick W.**, Records, 1893-1909, 42 vols. and 10 boxes. Correspondence, accounts, bills, receipts, time checks, financial statements, timber and land contracts, cancelled checks, and 30 vols. of ledgers, cashbooks, and journals of lumbering and logging concerns that operated in Aitkin, Crow Wing, and Cass counties, Minnesota. Companies represented include Aitkin Investment Co., F.W. Bonness and Co., Pokegama Lumber Company, Sandy River Lumber Company, and West and Bonness.

1541 **Bovey-DeLaittre Lumber Company,** Records, 1870-1911, 2 boxes and 30 vols. Labor, land, and logging contracts, ledgers, journals, daybooks, and some correspondence.

1542 **Boy River Logging Co.** (Minneapolis, Minnesota), Papers, 1891-1915, 2 vols. A minute book (1891, 1893-1915), including articles of incorporation, and a volume of stock certificates (1891-93), of a company organized to improve the Boy River, Cass County, Minnesota; drive logs and construct booms. This firm became part of the Weyerhaeuser Company. Restricted.

1543 **Boyd, Robert Knowles,** Papers, 1925-1930, 1 box, including 6 vols. Includes a manuscript, "A Mechanical Wonder: The Sheer Boom," 1926, 4 l., with charts, describing a boom patented by Levi Pond, ca. 1860, and first used on the Chippewa River at Eau Claire, Wisconsin.

1544 **Boyle, Dennis,** Papers, 1865-1869, 2 items. Includes Xerox copy of a diary (1865-69, 104 pp.) which chiefly concerns work as a cook on logging rafts on the Mississippi River between Minnesota and Iowa and Illinois, and also in lumber camps; includes the wages Boyle received. Original is in the possession of Ed and Mabel Boyle, Stillwater, Minnesota. There is also an autobiographical sketch of Dennis Boyle (2 pp.).

1545 **Brandborg, Charles W., and Family,** Papers, 1887-1970, 1 box. Includes papers of Guy M. Brandborg (b. 1893), employee of the U.S. Forest Service in Montana, 1915-55, among them copies of a statement to the Public Land Law Review Commission (ca. 1967); lecture on land management, University of Montana (1970); letter to Senator Lee Metcalf proposing a joint forestry committee (1970); and related items. Typed list of contents, 3 l.

1546 **Bronson and Folsom Towing Company** (Stillwater, Minnesota), Papers, 1866-1929, 17 boxes and 88 vols. Correspondence of a log rafting firm (1905-15); records of steamboat equipment and of trips on the St. Croix and Mississippi Rivers; and rafting and financial records, including ledgers, cashbooks, and time books.

1547 **Brown-Burt Logging Co.** (Minneapolis, Minnesota), Papers, 1907-1912, 1 box, including 1 vol. Certificate of incorporation (1907); schedule of logs (1908-10) scaled to various firms by the Mississippi and Rum River Boom Company and the surveyor general, 2nd district, Minnesota; record of sunken logs (1907-12, 1 vol.). The firm was incorporated to improve streams, drive logs, and deal in and manufacture timber in its various forms, and formed part of the Weyerhaeuser Company. Restricted.

1548 **Butler, Nathan** (1831-1927), Papers, 1850-1923, 5 boxes, including 99 vols. Civil engineer. The collection relates largely to surveying in Minnesota for railroads and for the U.S. government. It includes an article on lumbering in Minnesota.

1549 **Campbell, J.C., Company** (Duluth, Minnesota), Papers, 1887-1960, 1 foot (3 boxes, including 15 vols.). Correspondence, contracts, orders, price lists, wage work records for lumber camps, shipping records, tax certificates, deeds, and maps of a lumber company in northern Minnesota, mostly relating to contracts for railroad ties, posts and pulpwood, and to timberland holdings (primarily 1927-49). There is also information on the formation of a timberman's union in northern Minnesota (1957-58).

1550 **Chapman, Herman Haupt,** Manuscript autobiography. There is a microfilm copy at the Forest History Society, Santa Cruz, California.

1551 **Chapman, Herman Haupt** (1874-1963), Papers, 1900-1957, 1 box, 2 reels microfilm. Articles on forest legislation and on methods used to preserve Minnesota forests, and a copy of a scrapbook with clippings and letters relating to the origin of the Chippewa National Forest. Originals are in the possession of the School of Forestry, Yale University, and of H.H. Chapman, New Haven, Connecticut.

1552 **Chase, Ray Park** (1880-1948), Papers, ca. 1897-1944, 44 boxes. Lawyer, politician, and state official, of St. Paul and Anoka, Minnesota. The papers include material relating to Chase's duties as U.S. Representative from Minnesota, 1933-35 (he served on the House Agriculture Committee), and his work as a researcher, concerning conservation of natural resources, hydroelectric power developments, and other matters. Unpublished inventory.

1553 **Cheny, Mary Moulton, and Family,** Papers, 1841-1929, 3 boxes, including 40 vols. and 7 oversize items. Includes diaries (1841-80, 39 vols.) of Bradbury C. Morrison (b. ca. 1807) which contain entries describing work done in lumber mills in New Hampshire and, after 1856, in Minnesota.

1554 **Chippewa Logging Co.** (Nelson, Wisconsin), Papers, 1874-1922, 1 box, 2 vols. Minutes of meetings (1881-1908); testimony on the flooding of dams, land maps, tax statements, timber estimates, agreements, patent claims, tax receipts, indentures, annual statements (1891-1909), and articles of incorporation of a company organized to purchase timberlands and conduct a general logging business, and which formed part of the Weyerhaeuser Company. There is information on costs of log driving, prices of timberland, quantity of timber, land taxes, land acquisition, and logging. Restricted.

1555 **Chippewa Lumber and Boom Company** (Chippewa Falls, Wisconsin), Papers, 1870-1929, 3 boxes (including 5 vols.) and 7 additional vols. Accident reports (ca. 1895-1909), annual statements (1894-1929), minutes (1879-1929), field notes on Wisconsin land, printed reports (1910) of board of audit for Medford Township, Taylor County, Wisconsin, land conveyances, abstracts, contracts, profit and loss statements, and other papers of a company which conducted a general lumbering business, and which formed part of the Weyerhaeuser Company. Most of the papers deal with log driving, booms, and land matters. Restricted.

1556 **Chippewa River Improvement and Log Driving Co.** (Nelson, Wisconsin), Papers, 1869-1930, 2 boxes (including 3 vols.) and 3 additional vols. Income tax reports (1909-30), minute book (1877-1929), articles of association (1876), journal (1894-1901), notices, land plats, correspondence, a list of dams, articles of incorporation of the Union Lumber Company (1869), indentures, resolutions, lists of tax sales, agreements, abstracts, stock book (1877-1929), and miscellaneous papers of a company which operated on the Chippewa River and which formed a part of the Weyerhaeuser Company. There is information on logging expenses, dam construction and maintenance, labor costs, flowage rights, taxation, location of dams, booms, and log driving. Restricted.

1557 **Christensen, Otto Augustus, and Family,** Papers, 1854-1964, 3 boxes (including 4 vols.), and 1 reel microfilm. Three diaries (1872-75), contain information about Christensen's experiences working at Pine City and Stillwater, Minnesota, in lumber camps, and on the St. Croix Boom.

1558 **Clarke, Hopewell** (1854-1931), Notebook, 1863-1883, 1 vol. Notes of a survey of Cass County, Minnesota, including information on the location of lumber camps, the names of loggers cutting in the area, and the types of trees on the land.

1559 **Clement, Paul,** Papers, ca. 1930-1972, 48 feet (39 cartons). Printed material, correspondence and miscellaneous papers related to the conservation, political, and business concerns of Minneapolis insurance executive Paul Clement. He was at various times president of the Minneapolis chapter, the Minnesota Division, and the national Izaak Walton League of America, president of the IWLA Endowment Fund, chairman of the U.S. President's Committee on the Quetico-Superior Area, treasurer of the Quetico-Superior Foundation, and member of the Minnesota Wildlife Federation, Keep Minnesota Green, The Nature Conservancy, Minneapolis Bird Club, Save the St. Croix Committee, Ducks Unlimited, Natural History Society, and the natural resources committees of the Minneapolis Junior Chamber of Commerce, Minneapolis Chamber of Commerce, U.S. Chamber of Commerce, and the National Association of Manufacturers.

1560 **Cloquet Lumber Company** (Cloquet, Minnesota), Papers, 1891-1944, 8 inches (2 boxes). The collection includes annual reports, annual statements, trial balances, and production and sales reports. There is information on a number of other firms, including the St. Louis River Logging Company, the Cloquet Tie and Post Company, the Duluth Logging and Contracting Company, the Duluth and Northeastern Railway Company, the Coast Lumber Company, the St. Louis River Dam and Improvement Company, the Johnson-Wentworth Company, logging, the purchase and sale of logs, the Cloquet Lumber Company's general store, maintenance of sawmills, timberlands, production, and marketing. The Cloquet Lumber Company formed a part of the Weyerhaeuser Company.

1561 **Coast Lumber Company,** Records, 1898-1903, 1 vol. Articles of incorporation, subscription agreement, and minutes of meetings of an Eau Claire, Wisconsin, lumber company.

1562 **Cole, Emerson** (1838-1907), Papers, 1869-1940, 1 vol. and 2 boxes. Businessman. Personal cards, memoranda, bills, receipts, reports, legal documents, clippings, and other papers relating to Cole's business activities including his membership in the firm of Cole and Hammond, a Minneapolis lumber company.

1563 **Cole and Hammond** (Minneapolis, Minnesota), Records, 1877-1897, 4 boxes, 22 vols. Business records of a Minnesota lumber company, including correspondence, bills, accounts, journals, ledgers, inventories, and records of real estate owned in Minneapolis (1881-92). The firm was known as Cole and Weeks after 1882, and after 1889 it became a real estate and loan company. Members of the firm were Emerson Cole, Charles Hammond, and Charles W. Weeks.

1564 **Cray, Willard Rush** (1853-1938), Papers, 1853-1904, 2 boxes (including 2 vols.), 1 additional vol. Legal briefs, summaries of testimony and court proceedings, notes, and documents regarding the Washburn-Crosby Company, the C.C. Washburn Flouring Mills Company, the Merritt brothers, Laura A. Day's investments in iron and timberlands, and other business interests.

1565 **Cross, Judson N., and Family,** Papers, 1850s-ca. 1926, 2 feet. Miscellaneous papers, including material relating to Cross's service on the Minnesota State Forestry Board (ca. 1890s, 2 folders). Uncataloged and unarranged.

1566 **Cross Lake Logging Company** (Minneapolis, Minnesota), Papers, 1890-1915, 2 vols. Minute book, including articles of incorporation, and stock certificate book of a firm empowered to cut, haul, and drive logs, and which formed part of the Weyerhaeuser Company. Restricted.

1567 **Crow Wing River Log Driving and Improvement Company** (Little Falls, Minnesota), Papers, 1899-1917, ¾ inch (2 folders, including 1 vol.). Articles of incorporation, by-laws, and minutes of an organization incorporated to improve for log driving parts of the Crow Wing, Shell, Fishhook, Potato, and Straight Rivers, Blue Berry and Hay Creeks, Eagle Island and Two Inlet Lakes. The cryptic minutes (1899, 1908-17) have information on the election of officers and directors, and the dams on the Crow Wing River and Potato Lake. The firm formed part of the Weyerhaeuser Company.

1568 **Crowe, Isaac,** Papers, 1843-1875, 12 boxes. Correspondence, legal documents, and business papers of a real estate dealer who settled in St. Anthony, Minnesota, in 1857 and acted as agent for Abraham Becker, a banker and broker of South Worcester, New York. The papers relate to land transactions in New York to Crowe's land holdings in Minnesota, and to land which he located for others. There are some lumbermen's accounts, and correspondence regarding lumbering at Little Falls, Minnesota, and a sawmill at St. Cloud, among other matters. Correspondents include Benjamin F. Daggett, George E.H. Day's banking house, Daniel S.B. Johnston, Orlando C. Merriman, and others.

1569 **Dally, Charles,** Account book, 1835-49, 1 vol. Accounts kept by a cabinetmaker in Ohio and Illinois.

1570 **Davidson, William Fuson, and Family,** Papers, 1817-1919, 156 boxes (including 226 vols.), and 34 vols. and 2 oversize folders. Business and personal papers of a St. Paul, Minnesota, family. Among the correspondence relating to the real estate business of Watson Pogue Davidson (1907-19, 40 boxes) and his associates are files concerning the Oregon and Western Colonization Company, purchaser of timber lands in Oregon. There are cruise reports, correspondence discussing timber prices, and documents and letters concerning fire prevention (ca. 1912-18). Typed inventory.

1571 **Davis, Cushman Kellogg** (1838-1900), Papers, 1863-1903, 1 vol. and 13 boxes. Lawyer, legislator, Governor of Minnesota, and U.S. Senator (1887-1900); member of Senate Select Committee on Forest Reservations. The collection contains correspondence and scrapbooks relating to politics, forestry, and other matters. Inventory, 9 pp.

1572 **De Graw, John,** Papers, 1867-1921, 22 vols. and 2 boxes. Correspondence, daybooks, journals, and other business records (1878-1910), of the St. Paul lumber firm of John De Graw and Sons, including a record of insurance policies on mills and lumber.

1573 **DeLaittre, John** (1832-1912), Reminiscences, 1910, 1 vol. An account by a Minneapolitan, giving data on lumbering in Maine and in St. Anthony, Minnesota; and other subjects.

1574 **Delong, George Palmer** (b. 1853), Papers, 1899-1910, 1 box (2 vols.). Included are a letterpress book (1899-1901) of the Hudson Saw Mill Company (Hudson, Wisconsin) relating to orders received from dealers in Minnesota, Wisconsin, and Iowa; and a letterpress book (1907-10) containing material on DeLong's Lumber interests in Idaho and around Nickerson, Duquette, and Kerrick, Minnesota.

1575 **Dimock, Gould and Company** (Moline, Illinois), Papers, 1927-1950, 2½ inches (1 box). Annual financial statements of a retail lumber yard. The company formed part of the Weyerhaeuser Company.

1576 **Dinius, Sylvester,** Papers, 1880-1910, 2 boxes. Legal papers, correspondence, invoices, orders, broadsides, advertisements, patent records, and miscellaneous personal records of a dealer in logs and lumber who operated in Fort Wayne, Indiana, LeRoy, Michigan, and Ashton, Michigan. Files concerning Andrew Ely Hoffman's invention of a feed mechanism for sawmill carriages are also included in the papers.

1577 **Dispatch-Pioneer Press** (St. Paul, Minnesota), Papers, 1862-1948, 4 boxes. Miscellaneous correspondence includes a letter from Governor Theodore Christenson on the sale of timber lands (1930), and a memorandum on the Quetico-Superior National Forest (1931).

1578 **Dole, Alexander M.** (1841-1888), Business records, 1855-1887, 100 items and 2 vols. Minneapolis businessman, engaged in real estate, lumber, and oil. Correspondence, business cards, advertising circulars, articles of agreements, and miscellaneous papers relating to Dole's interests in Minnesota and Canada. Included are materials on Dole's sale of fence and telegraph poles, mill supplies, construction of log booms below the Falls of St. Anthony, legislation regarding the St. Louis River Dalles Improvement Company in Minnesota, and other matters. Inventory.

1579 **Donnelly, Ignatius** (1831-1901), Papers, 1850-1909, 38 vols. and 95 boxes (also on microfilm, 167 reels). Minnesota politician. Lieutenant governor, 1859-63; U.S. Representative, 1863-69; member of state senate, 1874-78. Letters (1883) from James Fergus of Montana contain Fergus' claim to have originated the idea of a national park in the West. There is correspondence (1874) relating to an investigation of illegal timber sales on state lands. Library pamphlets (vols. 72 and 77) contain material gathered to support Donnelly's sponsorship of a bill for tree-planting on the western plains. There is material relating to Donnelly's encouragement of tree planting through contests in the *Anti-Monopolist,* and correspondence with Joseph O. Barrett and Leonard B. Hodges, author of *Forest Tree Planters Manual,* concerning tree planting on railroad rights-of-way, and the Minnesota State Forestry Association (1878). Correspondence (1891) includes material on forestry bills in the Minnesota legislature. Letters of George W. Day (1893) relate to investigations of the sale of state pinelands. The collection is described in *Guide to a Microfilm Edition of the Ignatius Donnelly papers,* by Helen McCann White, St. Paul: Minnesota Historical Society, 1968, 34 pp.

1580 **Dorer, Richard J.,** Papers, 1944-1967, ½ foot. Includes correspondence (ca. 1962-67) relating to the Minnesota Memorial Hardwood Forest; also correspondence, printed items, relating to memorial forests (ca. 1953-54). Uncataloged and unarranged. Not open for research.

1581 **Drew Timber Company** (Muscatine, Iowa), Papers, 1906-1940, 1¼ inches (2 folders). Lists of lands, financial statements, correspondence concerning the financial statements, and reports to stockholders of a company which owned timberland in Oregon and formed part of the Weyerhaeuser Company. There is information on forest fires, county and state fire-fighting organizations, estimated stumpage on company lands, land purchases, taxation, land exchanges, valuation studies designed to precede a merger of timber interests in Oregon, and logging contracts.

1582 **Dyer, Harry G.,** Papers, 1844-1915, 15 items. Bills of lading for goods shipped by steamboat on the Mississippi and St. Croix Rivers; a letter listing the first boats of the season arriving at St. Paul, 1844-61; and a list of raft boats on the Upper Mississippi, 1864-1915.

1583 **Ellison, Smith,** Papers, 1848-1902, 1 box (including 9 vols.), and 27 additional vols. The papers consist of woods, time, day, log, and land books, a journal, a ledger, and diaries relating to Ellison's various partnerships in logging operations in the St. Croix Valley.

1584 **Evesmith, Hansen,** Papers, 1906-40, 22 items. The papers contain information on Frank E. Higgins, a missionary to lumberjacks; Andreas M. Miller, lumbering and mining promoter; and Minnesota forest fires of 1918.

1585 **Flint, Francis S., and Schuyler Flint,** Papers, 1855-1881, 63 items. Included are a few letters from Francis Flint recounting his logging ventures in northern Minnesota in 1876.

1586 **Folsom, William Henry Carman** (1817-1900), Papers, 1836-1944, 8 boxes (including 22 vols.). Folsom was a lumberman in the St. Croix Valley of Wisconsin and Minnesota, owning, at various times, mills at Stillwater and Taylor's Falls, Minnesota, and St. Croix Falls, Wisconsin; an incorporator of the St. Croix Boom Company, 1851; a founder of the Interstate Park in the Dalles of the St. Croix; and author of a history of St. Croix lumbering.

1587 **Forest History Foundation,** Interviews, 1953-1957, 1 box (16 items). Copies of transcripts of interviews of pioneer lumbermen of the Upper Midwest by members of the Forest History Society Staff. Originals are in the possession of the Forest History Society, Santa Cruz, California. Subjects include Paul Caplazi (1953, 4 pp.); Leonard Costley (1957, 20 pp.); Walter Ernest Dexter (1953, 10 pp.); George W. Dulany (1956, 37 pp.); Herman Heitman (1953, 11 pp.); Julius Joel (1953, 6 pp.); James Arthur Mathieu (1957, 10 pp.); Hope Garlick Mineau and Maud Mullan Carlgren (1955, 27 pp.); Wirt Mineau (1955, 17 pp.); George Neils (1953, 6 pp.); Walter Neils (1953, 9 pp.); L.J. Olson (1953, 4 pp.); Maggie Orr O'Neill (1955, 18 pp.); Hugo Schlenk (1955, 5 pp.); Orrin W. Sinclair (1954, 10 pp.); and James Stevens (1957, 33 pp.). Interviewers included John Larson, Bruce Harding, Elwood R. Maunder, Helen M. White.

1588 **Foster, Ellery,** Papers, 1940s-1960s, ¾ foot (1 bacase). Articles, reports, reprints, pamphlets, etc., regarding ecology, conservation, land use, forest planning, etc. Unarranged and uncataloged.

1589 **Foster and Roos,** Papers, 1856-76, 1 vol. An account book (1856-57) and a tax record (1868-76) of a lumbering firm that operated in Chisago County. After the firm was dissolved in 1856, its business was carried on by Bagley and Roos. Members of the firm were Thomas Foster, David T. Bagley, and Oscar Roos.

1590 **Fraser, Donald Mackay** (b. 1924), papers, 1944-68, 84 feet. Lawyer, state senator, and U.S. Representative from Minnesota, 1963+. Correspondence, reports, memoranda, financial records, campaign material, newspaper clippings, and other material, relating to Fraser's political career. The bulk of the collection consists of congressional legislative and campaign files. Includes material on conservation of natural resources, state and national parks. Unpublished container list. Restrictions.

1591 **Friend, Andrew,** Diary, 1857-80, 69 l. A typewritten account of employment in sawmills and gristmills in Minnesota.

1592 **Friends of the Wilderness,** Papers, 1917-1974, ca. 8 feet (19 boxes). Correspondence, printed materials, clippings, and miscellaneous items related to the activities of William H. Magie as executive secretary (mostly 1949-68). Subjects include opposition to use of aircraft in Superior National Forest roadless areas, federal acquisition of land in the area, opposition to mining, logging, and other commercial activities, the Wilderness Act, designation of the Boundary Waters Canoe Area as a wilderness, and creation of Voyageurs National Park. Correspondents include conservationists, congressmen, officials of the U.S. Forest Service, Minnesota Conservation Department, Izaak Walton League, Minnesota Emergency Conservation Committee, Save the St. Croix Committee, Voyageurs National Park Association, and the Wilderness Society.

1593 **Gillespie and Harper Company** (Stillwater, Minnesota), Records, 1874-1890, 500 items and 2 vols. Correspondence, time book, journal, agreements, deeds, receipts, invoices, and other legal, financial, and miscellaneous papers, giving information on contracts between Gillespie and Harper and various lumber firms in Minnesota, Wisconsin, and Iowa for the cutting and delivering of logs and lumber, and tracing the record of property in Stillwater, Minnesota, bought by the company. Descriptive inventory.

1594 **Gillmor, Frank H.** (b. 1875), Papers, 1910-1928, 1948, 1 box (including 3 vols.). Business records kept by the superintendent of logging for the Virginia and Rainy Lake Lumber Company (1910-28); correspondence (1948), with the Forest Products History Foundation, St. Paul, Minnesota, about forest products industries; and business records relating to food, equipment, labor, timberlands, and logging in Minnesota.

1595 **Glaser, Emma,** Manuscript, 1943-1950, 1 box (ca. 274 l.). "Logs, Lumber and Settlement in the St. Croix Valley."

1596 **Gleason, Sarah** (d. 1930), Papers, 1878-1930, 79 items. Includes two letters relating to the Hinckley forest fire of 1894.

1597 **Glenn, Andrew W., and Family,** Papers, 1842-1919. 1 box (including 1 vol.). Included are letters (1898, 1901) from Horace H. Glenn relating to his activities with a surveying crew in South Dakota and with a logging crew near Two Harbors and Marcy, Minnesota, and describing the life and character of lumberjacks. Inventory.

1598 **Godfrey, Ard** (1813-1894), Papers, 1839-1945, 3 boxes. Settler and sawmill operator, of Minneapolis, Minnesota. Correspondence, diaries, reminiscences, deeds, mortgages, powers of attorney, wills, newspaper clippings, and other papers, relating to Godfrey, his family, and business ventures. Includes Godfrey's business agreements with Franklin Steele, material on logging ventures in the Dakota Territory, and other matters. Unpublished inventory.

1599 **Graber, Albert** (1867-1959), Papers, 1857-1955, 200 items and 7 vols. A group of letters, written by Graber's son Horace while he worked on various jobs in the Pacific Northwest (1924-26), describe working conditions in logging and railroad construction camps. Inventory.

1600 **Great Northern Railway Company,** Records, 1854-1970, 3,155 feet. Correspondence, minutes, stock records, maps, photographs, and accounts of Great Northern and 250 affiliated companies. There is much information about the timber industry in Minnesota and in the American and Canadian Pacific Northwest, including correspondence to and from logging companies and land and townsite companies. Extensive files relate to the railroad's concerns with the companies of Frederick Weyerhaeuser. Many of these records contain information on Glacier National Park, its development, operation, and future; including corporate records of the Glacier Park Company, Glacier Park Hotel Company, and other firms active in the park. There are ledgers (1883-95, 4 vols.) of the Mille Lacs Lumber Company, relating to a logging firm active in the Rum River area of central Minnesota; records (1892-

1928, 5 vols.) of the Duluth, Mississippi River and Northern Railroad Company, including minutes and stock certificates of a railroad in north central Minnesota, primarily concerned with logging; records (1899-1912, 2 boxes and 5 vols.) of the Swan River Logging Company, including journals, ledgers, balance sheets, vouchers, and miscellaneous financial information of a logging firm active in northern Minnesota; and records (1900-45, 81 vols. and 9 boxes) of the Somers Lumber Company (known as the John O'Brien Lumber Company until 1907), containing minutes, stock certificates, correspondence, and financial records of a Montana lumbering firm. Additions to the collection are anticipated. Restricted.

1601 **Grittner, Karl F.** (b. 1922), Papers, 1965-1967, 3 boxes. Reports and study papers relating to the work of the Minnesota legislature, 1967, including the work of the Committee on the Public Domain. There is material on wildlife, report of the Minnesota Water Pollution Control Commission, reports of the Minnesota Department of Conservation on the Memorial Hardwood Forest, and on organizational changes and internal improvements, and U.S. Federal Water Pollution Control Administration recommendations for the Upper Mississippi (all ca. 1966-67, and n.d.).

1602 **Hagen, Harold C.** (1901-1957), Papers, 1923-1957, 18 feet (43 boxes including 1 vol.). U.S. congressman from Minnesota, 1943-55. The papers contain information on his work as congressman, particularly on flood control, conservation, and other subjects. Inventory, 18 pp.

1603 **Hale, William Dinsmore** (1836-1915), Papers, 1819-1913, 99 vols. and 19 boxes. Army officer, postmaster, and businessman. Correspondence and other papers relating to the milling, lumbering, banking, and railway interests of Hale and of William D. Washburn, U.S. Representative and Senator from Minnesota. There are letters by Washburn (ca. 200 items); letterbooks of the Washburn Mill Company (1886-93); and a log book of Alger Smith and Company (1911).

1604 **Haycraft, Isaac G.,** Reminiscences, n.d., 17 items. Accounts (1865-1907) of a boyhood spent along the Watonwan River, of farm life in Minnesota, and of an expedition undertaken in 1907 to settle on timber claims in St. Louis County.

1605 **Hayes, Moses P.,** Papers, 1859-88, 8 boxes, and 43 vols. Correspondence, order books, account books, receipt and cashbooks for C.R. Bushnell & Company, a concern which built and installed steam engines and sawmill machinery in various Minnesota towns.

1606 **Heerman, Edward Edson,** Papers, 1855-1929, 2 boxes (including 14 vols.), 6 additional vols. Correspondence, newspaper clippings, steamboat lists, historical skteches, cabin registers, freight books, portage books, cashbooks, and diaries of a steamboat captain and owner. The papers have information on steamboating on the Chippewa, Mississippi, and Missouri Rivers, and on Devils Lake in North Dakota. There is also information on the Minnesota Boom Company, the Pioneer Rivermen's Association, and other subjects.

1607 **Hersey, Staples and Company** (Stillwater, Minnesota), Records, 1794-1934, 11 boxes (including 3 vols.), and 2 vols. and 141 oversize items. Correspondence, maps, deeds, land patents, invoices and other papers of a Minnesota lumber company and affiliated firms. Includes information on land acquisition, logging, lumber manufacturing, marketing of lumber products in Minnesota, Iowa, Michigan, Pennsylvania, and Canada. Also includes records of Hersey, Staples and Hall; Hersey, Staples and Bean; Hersey, Staples and Doe; Hersey, Bean and Brown; North Western Saw Mills; Hersey and Smith (Keithsburg, Illinois); Eastern Trust; Hersey, Bronson, Doe, and Folsom; Hersey and Bean; and other firms. Inventory.

1608 **Hobe, Engebret H.** (1860-1940), Papers, 1874-1940, 104 boxes (including 113 vols.), and 101 additional vols. Personal correspondence, financial records, and legal papers of Hobe, who settled in Minnesota in 1883. Included in the collection is information about logging, lumbering, and timberlands, chiefly in correspondence of the following companies: E.H. Hobe, Western Lands, with headquarters at St. Paul, Minnesota, and branch headquarters and a sawmill at Knox Mills, Wisconsin; Hobe Peters Land Company, which dealt in timberlands, farmlands, and city real estate; E.H. Hobe Lumber Company, dealers in hardwood, pine, and hemlock at Knox Mills and Brantwood, Wisconsin; E.H. Hobe Land and Lumber Company; and Cooke and Hobe Company, a land company with headquarters at Superior, Wisconsin.

1609 **Houlton, William Henry** (1840-1915), Papers, 1792-1914, 60 vols. and 60 boxes. Businessman, county official, and Minnesota State senator. Correspondence, business records, and other papers are in the collection. Those after 1860 relate chiefly to Houlton's activities as a lumber and flour manufacturer, at Elk River, Minnesota. There are ledgers, journals, daybooks, sales books, correspondence, and other records, 1880-1904, of lumbering and logging in Sherburne County, Minnesota. The collection also contains information about the Minneapolis Retail Lumberman's Association.

1610 **Humbird Lumber Company, Ltd.** (Sandpoint, Idaho), Papers, 1903-1944, 1¼ feet (3 boxes). Profit and loss statements, annual statements, annual reports, and a few letters concerning the reports. The company had offices at Sandpoint, and operations at Kootenai, Idaho, and Newport, Washington. It formed part of the Weyerhaeuser Company.

1611 **Humphrey, Hubert Horatio** (b. 1911), Papers, 1943-1967, 245¼ feet (589 boxes). U.S. Senator from Minnesota, 1949-64, 1970+, serving on the Agriculture and Forestry Committee; U.S. Vice President, 1965-69. The portions of the collection relating to Humphrey's career as mayor of Minneapolis and his terms as U.S. Senator, 1949-64, have been processed and partially inventoried. The Vice Presidential records are not open to research. General correspondence includes folders on the U.S. Forest Service (1951). Legislative correspondence contains material on the ban of aircraft from the Boundary Waters Canoe Area (1953); appropriations for national parks (1953); appropriations for the Forest Service (1953-54); conservation in national parks (1952-54); public power controversies (1952-57); Minnesota wilderness areas (1952-53); aircraft ban in wilderness area (1954); Dinosaur National Monument (1953-55); forest products (1955); timber marketing bill (1955); U.S. Forest Service timber sales (1955); watershed conservation (1955-56); wildlife refuge bills (1955); grazing on public lands (1954-55); Lookout Mountain National Park bill (1955); wilderness areas (1956); appropriations for Operation Outdoors (1957); appropriations for Forest Service research (1957); forest resources (1957). Research files include material on the wilderness bill (1955-58); and there are other files on the conservation of natural resources.

1612 **Immigration Land Company,** Records, 1898-1951, ca. 6 feet (13 boxes). Correspondence and other records of the Immigration Land Co., and the Pine Tree Manufacturing Co. The collection contains information on the purchase and sale of timber land, logs, and stumpage; logging arrangements with contractors; the sale or disposition of logging camps and railroad cars; fire hazards; the sale of property of the Northland

Pine Co. (1902); abolishment of hospital tickets; sale by the U.S. government of timber on the Red Lake Indian Reservation; protests of farmers because of overflowed lands; reports on water levels; land donations to towns and villages; mineral lands; leases and options; taxation; relations with the U.S. Forest Service; game preservation; and sheep grazing during World War I. Descriptive inventory.

1613 **International Timber Company,** Papers, 1908-22, 4 vols. Minute book, including articles of incorporation and bylaws, cashbook (1908-13), ledger (1908-13), and cash receipt book (1908-22) of a company with headquarters in Phoenix, Arizona, but with operating headquarters at a branch office in Minneapolis. Information about stock distribution, and the sale of lands and a mill in Hubbard and Morrison Counties, Minnesota, to the Hennepin Lumber Company is contained in the papers.

1614 **Irvine, Thomas** (1841-1930), Papers, 1899-1940, ca. 4 feet (4 boxes and 24 vols.). Lumberman, of St. Paul, Minnesota. Correspondence, journals, ledgers, minute books, and financial records, of the lumber and investment companies in St. Paul, and vicinity, with which Irvine and his son, Horace Irvine, were associated, including the Irvine Family Investment Company and its timber lands in Oregon and Washington, as well as the Hills Investment Company, Riverside Builders Supply Company, St. Paul Brick Company, St. Paul Building Material Exchange, and Thomas Irvine Lumber Company. Unpublished inventory.

1615 **Izaak Walton League, Duluth Chapter,** Papers, 1963-1970, ¾ foot (2 boxes). Correspondence, clippings, reports, booklets, financial statements, fact sheets, and promotional literature relating to membership and activities, wildlife and resource management, water resources and water pollution, Superior National Forest, Boundary Waters Canoe Area, Voyageurs National Park, conservation education, and the Quetico-Superior Council. Included are minutes (1964, 1968-70).

1616 **Izaak Walton League, Minnesota Branch,** Papers, ca. 1928-1954, 6¾ feet (6 bacases). Chapter files (ca. 1940-54); correspondence (1941-46); correspondence by subject (1941-54); publications; corporate records, including certificate of incorporation, bylaws, minutes; financial records (1945-47?); telegrams (1947-53); nursery file; news releases and newsletters; miscellaneous papers. Uncataloged. Closed to research.

1617 **Johnson, Abraham,** Papers, 1862-1901, 1 box (including 8 vols.), 27 additional vols. Accounts of a lumber dealer at Marine, Minnesota, showing the amount of lumber sold, articles charged against lumberjacks, and wages paid (1862-1901); accounts of a livery stable at Marine (1879-84), and of the Marine, Minnesota, and Osceola, Wisconsin, stage (1879-86); and identification cards (1897-99) issued by employment agencies in St. Paul and Minneapolis.

1618 **Johnson, Henry A.** (b. 1896), Papers, 1931, 57 items. Member of the Minnesota Legislature from Minneapolis. The papers include material relating to forestry legislation. Inventory description.

1619 **Johnson, Lawrence Henry** (1862-1947), Correspondence, 1906-1941, 1 box. Letters written to Johnson while he was representative in the Minnesota legislature from Hennepin County and speaker of the House, dealing with appropriations for Itasca State Park, and other matters.

1620 **Johnson-Wentworth Company** (Cloquet, Minnesota), Papers, 1903-1931, 7 inches (2 boxes). Annual statements, annual reports, letters relating to statements and to timberland holdings, lumber inventories, balance sheets, and lumber production and shipment reports. The company formed part of the Weyerhaeuser Company.

1621 **Judd, Walter Henry** (b. 1898), Papers, 1918, 1942-62, 212 boxes (including 1 vol.). U.S. Representative from Minnesota, 1943-63. Legislative files contain material on conservation of wildlife; flood control; national parks and conservation policies; National Resources Planning Board; House committee on public lands; Senate committee on Agriculture and Forestry. Inventory. Restrictions.

1622 **Keep Minnesota Green,** Papers, ca. 1953-1967, 1 foot. Minutes, reports, financial records, publications, correspondence, membership records, promotional literature, records of tree farm committees, and other papers of an organization which supported forest conservation and reforestation in Minnesota. Folder list. Uncataloged.

1623 **Kegley, Charles V.,** Reminiscences, 1934, 5 l. Typewritten. Reminiscences containing information on early settlement at Le Sueur Center, 1877, where Kegley operated a sawmill and built many of the homes.

1624 **Kennerson, Fred,** Reminiscences, 1934, 10 l. Typewritten. An account of life in the lumber camps of Aitkin, Pine, and Itasca Counties, Minnesota.

1625 **King, Stafford** (1893-1970), Papers, 1898-1969, 29 feet (39 boxes [including 20 vols.], and 65 additional vols.). King served in the Minnesota Forest Service after 1919 as a patrolman, and held various other state offices. There are files on forestry (1925-28 and undated, 5 folders), and a memo book (1919) relating to the forest service. Container list. Closed to research until 1987.

1626 **Kinney, Jay P** (b. 1875), Papers, 1910-1941, ca. 3 feet. Correspondence and reports kept by Kinney while directing the forest management activities of the Indian Service, 1910-33. The collection is divided into two sections containing information on the Consolidated Chippewa Agency, Cass Lake, Minnesota, and on the Red Lake Indian Reservation, Minnesota. The correspondence is between Kinney and the Commissioner of Indian Affairs and by Kinney and the superintendents of the Indian agencies, lumbermen, and other individuals. Included are information on allotments of lands to Indians, sale of reservation timber to lumber companies, trespass on timberlands, logging on the Grand Portage Reservation, road building, administration of the agencies, types of timber on the reservations, operations of the Red Lake Agency sawmill, and attitudes of the Indians toward the sawmill operations. Descriptive inventory.

1627 **Koll, Mathias N.** (1874-1934), Papers, 1890-1936, 48 boxes. The papers include correspondence, accounts, clippings, and miscellaneous papers kept by an officer of various organizations among which are the Northern Minnesota Development Association, the Northwestern Minnesota Historical Association, the Minnesota Scenic Highway Association, and other organizations. There is information about Koll's real estate business, the Chippewa National Forest, and agitation for a dam at the outlet of Cass Lake.

1628 **Laird, Norton Company** (Winona, Minnesota), Papers, 1855-1905, 74 boxes and 322 vols. Letterpress books, incoming correspondence, daybooks, journals, and ledgers of a lumber company. There is extensive information on land acquisition; taxation, management, and sale of cutover lands; logging; transportation of logs and lumber, the manufacture of lumber; the marketing of bricks, lime, cement, paper, glass, lumber, and other building materials at retail yards in Minnesota and North and South Dakota; lumber trade associations; legislation relating to logs, grading, and inspection of lumber, fire prevention, and timber trespass; labor in the woods and mills; scaling; relations with railroads; insurance; cooperation with the Wisconsin state and county boards of immigration; pineland investigations; and the improvement of streams. Among the firms with which the Laird, Norton Company had extensive dealings were the American Colonization Company, the American Immigration Company, the Beef Slough Boom Company, the Bronson and Folsom Towing Company, the Chippewa Logging Company, the Chippewa Lumber and Boom Company, the Gull River Lumber Company, the Musser-Sauntry Land, Logging and Manufacturing Company, the North Wisconsin Lumber Company, the Pine Tree Lumber Company, the St. Croix Boom Corporation, Weyerhaeuser and Denkmann, and the Wisconsin Central Railway Company. Restricted.

1629 **Lammers, Albert John** (1859-1920), **and George August Lammers,** Papers, 1894-1933, 2 boxes. Correspondence, legal papers, minutes of meetings, log scales, financial data, and other papers relating to the Lammers brothers' business enterprises in mining, logging and lumbering, coal and coke, real estate, and railroad construction. Among the companies represented are the Grand Forks Lumber Co., East Grand Forks, Minnesota; Adams River Lumber Co., Ltd., Chase, British Columbia; Florida Lands, De Soto County, Florida; and Port Mann Syndicate, Vancouver, British Columbia. Inventory.

1630 **Langen, Odin** (b. 1913), Papers, 1958-1970, 66 feet (53 boxes). U.S. Representative from Minnesota, 1959-70; served on Interior and Insular Affairs Committee. His papers include committee files (1959-70), correspondence, bills and subject files on the work of the committee, the U.S. Department of the Interior (1961-70), the U.S. Forest Service (1963-70), the Soil Conservation Service (1959-64), Voyageurs National Park (1965-70), Boundary Waters Canoe Area, reclamation projects, wilderness and other legislation, and conservation issues and legislation in general. Restrictions.

1631 **Langford, Nathaniel P.** (1832-1911), **and Family,** Papers, 1819-1942, 27 vols. and 3 boxes. One of the organizers of the expedition to Yellowstone, 1870; superintendent of Yellowstone National Park, 1872-77. Relating to the park are only printed bills and excerpts from the *Congressional Record,* and a scrapbook with information on the park. Inventory, typed, 6 pp.

1632 **Lind, John** (1854-1930), Papers, 1870-1912, 1917-1933, 4½ feet (12 boxes, including 9 vols.). U.S. Representative from Minnesota, 1887-93, 1903-05; served on Public Lands Committee. The papers for the period 1919-33 include comments on logging in northern Minnesota and on the Red Lake Indian Reservation, forest reserves (1926-28), management of the Superior National Forest (1926). Typed description.

1633 **Little Falls Dam,** Account book, 1892-1899, 1 vol. An account book recording purchases of foodstuffs, household equipment, tools, and other items. The dam was apparently owned by the Weyerhaeuser Company. Restricted.

1634 **Lowell, Mrs. R.M.,** Papers, 1850-55, 3 items. Letters describing the hauling of logs in Anoka County, Minnesota, and the death of two men in a lumber camp.

1635 **Lyon, Daniel Brayton,** Papers, 1854-97, 1 box, including 1 vol. The papers include a bill of sale (1882) for 1,000,000 feet of lumber from Wakefield and Troy, Merrillan, Wisconsin, to Lyon, and to C.E. Keller and Company, Minneapolis.

1636 **McCarthy, Eugene Joseph** (b. 1916), Papers, 1947-70, 390 feet. U.S. Congressman from Minnesota, 1949-58; U.S. Senator, 1959-71; served on the Agriculture committee in the House and the Agriculture and Forestry Committee in the Senate. His papers contain correspondence, reports, government documents, campaign materials, newspaper clippings, memoranda, and other materials, dealing with McCarthy's congressional career, 1949-71. Included are materials on congressional committees and their activities, the executive departments and agencies of the federal government, political campaigns, legislation, public and constituent relations, speeches and speaking engagements, and various subjects of government policy, among them forest lands, grazing in national forests, national parks, and the National Park Service. Closed to research.

1637 **McClure, Samuel,** Letter, September 2, 1882, 16 pp. Photostat. Letter from McClure to his parents in Ireland describing his life as a lumberman on the St. Croix River.

1638 **McDonald, John F., Lumber Company** (Minneapolis, Minnesota), Papers, 1857-1945, 2 boxes (including 25 vols.). Correspondence, financial statements, inventories, newspaper clippings, auditors' reports, and miscellaneous papers relating to the John F. McDonald Lumber Company, the Thompson McDonald Lumber Company, and the McDonald Lumber Company, and the "McDonald System" for establishing lumber stores dealing in plans, financing, and lumber products for the home in cities of more than 100,000 population. A daybook (1867-91) of John S. McDonald of Fond du Lac, Wisconsin, is included in the collection.

1639 **McDonald, John Stewart** (1831-1916), Papers, 1908-16, 1 box (including 1 vol.). The papers include correspondence about lands in Michigan and Wisconsin; and an undated plat book of Michigan indicating lands owned by McDonald, as well as timber and mineral lands which he wished to acquire.

1640 **MacGregor, Clark** (b. 1922), Papers, 1958-1971, 168 feet (134 boxes). U.S. Representative from Minnesota, 1961-70. His congressional office files include correspondence with constituents and federal officials, news releases and clippings, and background materials on current issues and legislation relating to programs and policies of the U.S. Department of the Interior and the U.S.Forest Service; recreation, mining and logging use of the Boundary Waters Canoe Area; and particularly to the establishment of Voyageurs National Park.

1641 **McKellar, Peter, and Family,** Papers, 1839-97, 72 items. The papers include correspondence between McKellar, an Iowa farmer, and his son, Archibald, who worked for A.C. Sevey, a blacksmith at Taylor's Falls, and for the St. Croix Boom Company.

1642 **McKusick and Company** (Stillwater, Minnesota), Account books, 1848-1906, 3 vols. Account books of John McKusick's general store and lumber business, which included a sawmill at Stillwater, 1844-72.

1643 **Marin, William A.,** "The Prairie Schooner Passes," n.d., 1 vol. A reminiscent account of the experiences of the author's father as a logging contractor and hotelkeeper in Michigan, his trip from Michigan to Crookston, Minnesota, in 1879, and his experiences as a homesteader in the Red River Valley.

1644 **Merritt, Andrus R. and Jessie L.,** Reminiscences, 1934, 1 vol. The account tells of the Merritt family's travels from New York to Ohio and then to Minnesota, where they settled about 1856 at Oneota, now part of Duluth. Included is information on frontier life in northern Minnesota, the discovery of the Mesabi Range, and northern pinelands.

1645 **Minnesota Boom Company,** Papers, 1889-1900, 2 folders, including 1 vol. Articles of incorporation, minutes of meetings of a Winona, Minnesota, company organized to construct booms and dams on the Mississippi River.

1646 **Minnesota State Forestry Association,** Papers, 1876-1903, 1948, 1 box (including 1 vol.). Minutes of meetings, newspaper clippings, and correspondence relating to the preservation of Minnesota forests; and a certificate of membership book (1876-87).

1647 **Mississippi and Rum River Boom Co.** (Minneapolis, Minnesota), Papers, 1868-1920, 1 box (7 vols.). Log books and stock certificates of a firm organized to manage the movement of logs from the Upper Mississippi River to the Falls of St. Anthony. The firm formed part of the Weyerhaeuser Company. Restricted.

1648 **Mississippi Boom Corp.** (St. Paul, Minnesota), Papers, 1876-1916, 2 vols. Minute book, including articles of incorporation, and a volume of stock certificates of a firm empowered to construct and operate booms between the Falls of St. Anthony and the mouth of the St. Croix River. The firm formed part of the Weyerhaeuser Company. Restricted.

1649 **Mississippi River Logging Co.** (Davenport, Iowa), Papers, 1870-1931, 6 boxes, 9 vols. Articles of incorporation, minutes, reports on operations, resolutions, tax records, agreements, trial balances, correspondence, annual statements, miscellaneous papers, and other records of a firm organized to drive, buy, and sell logs, and buy land. The firm formed part of the Weyerhaeuser Company. Restricted.

1650 **Mississippi River Lumber Co.** (Clinton, Iowa), Papers, 1892-1946, 1 box (including 2 vols.) and 19 additional vols. Minute book (1893-1946), log books, stock certificate book, cashbook, journals, ledgers, land book, plat books, lien book, balance sheets, annual statements, and correspondence of a firm which formed part of the Weyerhaeuser Company. Restricted.

1651 **Morrison, Dorilus** (1814-1897), Papers, 1706-1913, ca. 2 feet (4 boxes [including 12 vols.]). Businessman, mayor of Minneapolis, and Minnesota state senator. Correspondence, diaries, deeds, leases, financial statements, genealogical data, newspaper clippings, notebooks, account books, daybook, and record books. The papers of Dorilus Morrison, starting in 1844, and those of his son, Clinton, starting in the 1860s, deal with many phases of the Minnesota career of the family: land transactions, the Minneapolis Mill Co., water power at the Falls of St. Anthony, pine land acquisitions, Morrison's sawmill, and other matters. Inventory.

1652 **Mullery-McDonald Lumber Company** (Duluth, Minnesota), Records, 1904-1934, ca. 3 feet (21 vols.). Minute books, ledgers, journals, lumber contracts and shipments, and check journals. Inventory.

1653 **Mulvey and Carmichael Company** (Stillwater, Minnesota), Records, 1883-1912, 3 boxes and 25 vols. Ledgers and journals of the Mulvey and Carmichael Company; records of Mulvey and Son (1904-10); also ledger, journal, records of incorporation, and stock sales of Clearwater Logging Company (1897-1912, 5 vols.), of Stillwater.

1654 **Murdock, Hollis R.** (1832-1891), Papers, 1837-1891, 1 box. Lawyer of Stillwater, Minnesota. The collection includes papers on lumbering on the St. Croix River, 1875-90.

1655 **Murphy, Raymond T.,** Reminiscences, 1969, 19 l. Recollections of life in northern Minnesota lumber camps ca. 1900. Copy of original in private possession.

1656 **Musser, Sauntry and Company** (Muscatine, Iowa), Papers, 1886-1897, 2½ inches (1 vol.). Contains miscellaneous data on logging accounts with other lumber firms in Minnesota, log marks, etc.

1657 **Musser-Sauntry Land, Logging and Manufacturing Company,** Papers, 1857-1934, 2 boxes (including 2 vols.), and 7 vols. Correspondence (1886-1902), land and logging records, agreements and contracts, journals (1885-89, 1905-10), a ledger (1903-10), and other papers of a Muscatine, Iowa, company which operated in Wisconsin and Minnesota. Restricted.

1658 **Nature Conservancy, Minnesota Chapter,** Records, 1958-1971, 1 reel of microfilm. Minutes and treasurer's reports relating to acquisition and management of tracts of land in Minnesota, taxation of Nature Conservancy land, finances, relationships with other organizations interested in natural areas; the chapter's organization, membership, and relations with the national Nature Conservancy, and discussion of action needed to preserve natural areas. Microfilmed from originals loaned by Edmund C. Bray, Minneapolis, chairman of the chapter.

1659 **Nebagamon Lumber Co.** (Chippewa Falls, Wisconsin), Papers, 1882-1934, 1 box and 4 vols. Land deeds (most of them for the American Colonization Company), articles of incorporation, statements, minutes, journals, and stock certificates of a firm formed to buy and sell timberland, build and acquire sawmills, logs, transport logs, and manufacture lumber. The firm formed part of the Weyerhaeuser Company. Restricted.

1660 **Nickerson, John Q.A.** (1825-1917), Papers, 1832-1919, 3 boxes (including 81 vols.), 15 additional vols. Account books (1851-1915) showing personal living expenses and supplies for logging camps, and promissory notes for logs cut along the Platte River in Minnesota are included in these papers of an early Elk River resident who operated a hotel and a farm and engaged in the lumber business.

1661 **Nolan, William Ignatius** (1874-1943), Papers, 1900-1943, ca. 3 feet (17 boxes [including 4 vols.] and 4 vols.). State legislator, Lieutenant Governor, and U.S. Representative from Minnesota, 1929-33; served on the House Public Lands Committee. The papers include correspondence, speeches, reports, scrapbooks of clippings referring to Nolan's political career, miscellaneous political pamphlets, and other material relating to his activities as Speaker of the Minnesota House of Representatives, and his career as congressman. Includes records of investigation and hearings of the Minnesota Reforestation Commission (1927), of which Nolan was chairman. Unpublished inventory.

1662 **North Wisconsin Lumber Company,** Papers, 1874-1933, 4 boxes, 1 vol., and 1 oversize item. Correspondence (1886-1911), land and tax records, and a map of Hayward, Wisconsin, showing the location of company mills. Restricted.

1663 **Northern Boom Co.** (Minneapolis, Minnesota), Papers, 1890-1920, 4 vols. Articles of incorporation, log book, cashbook, journal, and ledger of a company formed to improve the Mississippi River from Brainerd to Pokegama Falls, to drive logs and construct and maintain booms. The firms formed part of the Weyerhaeuser Company. Restricted.

1664 **Northern Lumber Company** (Cloquet, Minnesota), Papers, 1897-1944, 2 boxes. Annual statements, annual reports, inventories, and correspondence. The letters contain data explaining the annual statements and have information on the log pool of the Northern Lumber Company, Cloquet

Lumber Company, and the Johnson-Wentworth Company. There is an inventory of logging and railroad equipment and on other lumber firms. The company formed part of the Weyerhaeuser Company.

1665 **Northern Mississippi Railway Co.** (Minneapolis, Minnesota), Papers, 1890-1915, 2 vols. Minute books, including articles of incorporation, and stock certificate book of a firm incorporated to build a railway in Cass County, Minnesota. The firm formed part of the Weyerhaeuser Company. Restricted.

1666 **Northern Pacific Railway Company,** Records, ca. 10,000 linear feet. Include data on natural resources; management of the railroad's land grant, including timber and mineral contracts; and Yellowstone National Park. Largely uncataloged, but information in the various records series is available upon request. Restricted in part.

1667 **Northern Pine Manufacturers Association** (Minneapolis, Minnesota), Records, 1890-1960, 5 boxes, 8 vols. Correspondence, minutes of meetings, financial papers, membership data, printed matter, and other records. The papers contain information on the government's prosecution of the timber trust (1907), wages paid in pine and saw mills, various types of government regulation of the lumber industry, the National Recovery Administration, rulings on the type of lumber used in Federal Housing Administration projects and in buildings in army camps, food rationing, price control measures in effect from 1941-53, conferences held by the association, the grading and inspection of lumber, and the American Lumber Standards Committee. Includes minutes of meetings of the Railroad Committee and of the Board of Directors; and reports of the proceedings of the Bureau of Grades of the association and its predecessors, the Wisconsin Valley Lumbermen's Association and the Mississippi Valley Lumbermen's Association.

1668 **Northern White Pine Manufacturers Association,** Papers, 1897-1960, 1 foot. Correspondence and miscellaneous papers (1897-1960), memo invoices (1952-60), auditor's reports (1933-60), weekly and monthly production and shipping records (1939-58). Folder list. Uncataloged.

1669 **Northland Pine Co.** (St. Paul, Minnesota), Papers, 1899-1931, 2 boxes (including 2 vols.) and 10 additional vols. Articles of incorporation and minutes (1899-1925); accounts, journals, and ledgers (10 vols.); annual statements and miscellaneous financial reports (1907-23) of a company formed to deal in lands, stumpage, logs, and other property, and to manufacture and deal in lumber, improve streams, build and operate logging and other railways, and drive logs, originally in northern Minnesota. The company was consolidated in 1925 with the Mississippi Lumber Company and formed part of the Weyerhaeuser Company. Restricted.

1670 **Northwestern Lumber Company** (Eau Claire, Wisconsin), Correspondence, 1870-1899, 160 items. Correspondence of the Northwestern Lumber Company with its subsidairy, the Montreal River Lumber Co. and with other companies. Includes information on the routine business of a lumber firm and on the visits of Sumner T. McKnight, president of the company, to California and England.

1671 **Nowell, James Albert** (1866-1937), Papers, 1910-20, 1 box (including 1 vol) and 2 additional vols. Correspondence and miscellaneous records pertaining to Minnesota timberland locations, and land records are included in these papers.

1672 **Oberholtzer, Ernest C.** (b. 1884), Interviews, 1960, 1963, 1964, Tapes, 12 reels (27 hours and 25 minutes). The interviews contain the reminiscences of Oberholtzer, a Ranier, Minnesota, conservationist. He discusses the nature of the Quetico-Superior wilderness area on the Minnesota-Ontario border; the efforts, beginning in the 1920s, to secure federal and state legislation for the protection of the area; and the work of the Quetico-Superior Council. Interviewers include Russell M. Fridley, Evan A. Hart, F. Peavey Heffelfinger, Lucile M. Kane, and George Monahan. Closed to research.

1673 **Oldenburg, Henry** (1858-1926?), Papers, 1888-1934, 1 box. Newspaper clippings, legal documents, and other papers of a Carlton, Minnesota, lawyer who was interested in forestry; and tributes to Oldenburg after his death. Oldenburg was a member of the State Forestry Board, 1911-17, and legal representative for various Minnesota lumber companies in the Duluth area. Included is information on the utilization of waste materials by the lumber industry, and on Samuel S. Johnson, a Michigan and Minnesota lumberman.

1674 **Olson, Floyd Björnstjerne** (1891-1936), Papers, 1923-1936, 4 boxes. Press releases, speeches, and correspondence, chiefly for 1930-36 when Olson was governor of Minnesota. There is information on the conservation of natural resources.

1675 **Olson, Sigurd F.,** Papers, 1920s-1974, 72½ feet. Literary manuscripts (1920s-1972) and conservation correspondence and miscellaneous materials (mostly 1940s-1974) of Olson, an author, conservationist, and wilderness ecologist. Conservation materials include data on his activities supporting the preservation of the Quetico-Superior area and on his work with the Wilderness Society, the National Parks Association, and the Advisory Board on National Parks, Historic Sites, Buildings and Monuments of the U.S. National Park Service.

1676 **O'Neal Timber Co., Ltd.** (St. Paul, Minnesota), Stock Certificate Book, 1905-1907, 1 vol. The company was incorporated to deal in lands, timber, and timber products, and conduct a general logging, lumbering and manufacturing business, and formed as part of the Weyerhaeuser Company. Restricted.

1677 **Pattee, Edward Sidney** (1861-1930), Papers, 1839-1899, ca. 1 foot (1 box, including 2 vols.). Accountant, auditor, and bookkeeper for various business firms in Minneapolis, Minnesota. Correspondence, diaries, memoranda, and other papers of Pattee, his wife, Dora (Jewett) Pattee (1853-1939), her father, Samuel Albert Jewett (1816-1893), and grandfather, Samuel Jewett (1772-1865), of Pittston, Maine. Subjects represented include the lumber trade of Maine with England. Unpublished inventory.

1678 **Pearson, Charles Gottfried** (b. 1875), Papers, 1911-1931, 1 box. Biographical sketch and account books of a man who was born in Sweden and who settled at Dassel, Minnesota, in 1892. The account books list Pearson's expenses while working in sawmills in Minneapolis and on a farm near Dassel.

1679 **Pederson, Thomas,** Reminiscences, 1934, 89 l. An account of pioneer life among Norwegians in La Crosse County, Wisconsin, in the 1860s; of Pederson's experiences as a lumberjack in northern Wisconsin, and at Randall, Minnesota, where he witnessed a forest fire in 1894, and other experiences. Published in *Wisconsin Magazine of History,* 21 (December, 1937-December, 1938).

1680 **Peterson, Alf Arvid,** Reminiscences, 1947, 47 l. Reminiscences of the life of the Peterson family in Sweden and of their experiences after immigrating to Red Wing, Minnesota, in 1868, including some information on log rafting.

1681 **Peyton, Hamilton Murray** (b. 1835), Papers, 1857-1935, 2 boxes (including 52 vols.). Peyton's diaries (1857-

1924, 52 vols.) include brief entries describing his work with the May, Ingalls, Hollenstein and Parameter Sawmill (1864); and the establishment of the Peyton, Kimball, and Barber Lumber Company (1883), manufacturers of lumber, lath, and shingles. The folder of miscellaneous papers includes invoices and bills of the Peyton lumber firms (1867-1927), business correspondence (1866-1905), and clippings.

1682 **Pillsbury, John S.,** Papers, 1883-1934, 10 boxes. A book of deeds, listing lands belonging to Pillsbury in the following Minnesota counties: Beltrami, Itasca, Cass, St. Louis, Crow Wing, Aitkin, Hubbard, Carlton, and Becker; also scattered items relating to timberlands and logging.

1683 **Pioneer Rivermen's Association,** Papers, 1915-35, 2 vols. Minutes of meetings (1915-22), letters, notices, and other papers of an organization composed of Minnesotans and others who formerly had been employed in steamboating or rafting on the Mississippi River, its tributaries, and the Red River. Included are biographical sketches of members, and a large number of letters written by Fred A. Bill.

1684 **Polk, Asa D.** (1850-1940), **and Family,** Papers, 1853-1930, 2 boxes. Lawyer of Brainerd, Minnesota. Correspondence, clippings, and other papers. Includes patents for lands at St. Cloud, Minnesota, and in Kentucky, granted to the Day Lumber Co., Minneapolis, a firm with which Polk was associated.

1685 **Potlatch Timber Co., Inc.** (St. Paul, Minnesota), Papers, 1900-1934, 1 box (including 2 vols.) and 10 additional vols. Indentures, agreements, correspondence, minutes, journals, ledgers, and miscellaneous papers of a company formed to conduct a general lumbering business in Minnesota and later in Wisconsin. There is information on land acquisition, the estate of Frank C. Walker, taxation, conveyance of land to the American Immigration Company, and operating expenses. The firm formed a part of the Weyerhaeuser Company. Restricted.

1686 **Powell, Oliver Stanley, and Family,** Papers, 1832-1954, ca. 3 feet (6 boxes, including 25 vols.). Papers relating to family matters and to the business ventures of Oliver Stanley Powell (1831-88) in River Falls, Wisconsin. They contain information on water power on the Kinnickinnic River and on the building of sawmills. Inventory.

1687 **Powell, Ransom J.,** Papers, 1906-1922, 7 boxes. The correspondence of a Minneapolis attorney who acted for persons claiming title to White Earth Indian Reservation lands including data on litigation against the Nichols-Chisholm Lumber Company of Minnesota.

1688 **Preus, Jacob Aall Ottseon** (1883-1961), **and Family,** Papers, 1853-1946, 38 boxes and 1 vol. Correspondence, transcripts of speeches, printed materials and other papers relating to the career of Preus, governor of Minnesota, 1921-25. Arranged in chronological subgroups. Included is correspondence of S.G. Iverson, state auditor, relating to the operation and maintenance of Minneopa Falls State Park, and of J.B. Hodge, superintendent of the park (1905-10); material collected by Iverson on the Minneopa Falls State Park, the establishment of the Ramsey State Park, and the sale of timber on the Chippewa Indian Reservation (1912); Iverson material on Minneopa Falls State Park (1914); and material relating to timber cutting on Indian reservations (1915). Typed description.

1689 **Preus, Jacob A.O.** (1883-1961), Interviews, 1960, 1 box (114 l.; also 4 reels tape [2 hours]). Governor of Minnesota, 1921-25. Subjects include Richard A. Ballinger, Theodore Roosevelt, other political figures, and the creation of the Superior National Forest. These are sample tapes saved from a more extensive series, all of which are available in transcript. Interviewers include June D. Holmquist and Lucile M. Kane. Inventory, 3 pp.

1690 **Quetico-Superior Council,** Records, 1906-1964, ca. 55 feet (108 boxes, 23 vols.). Correspondence, publications, memoranda, reports, speeches, newspaper clippings, articles, budgets, and other materials, documenting the history of the council which was organized in 1928 to work for the preservation of wilderness values in the area of northeastern Minnesota and Ontario comprising the Rainy Lake Watershed. The bulk of the papers date from 1927-50. Organizations represented include the American Forestry Association, the American Legion, the Izaak Walton League of America, and the Minnesota Wildlife Federation. Includes personal papers of Ernest C. Oberholtzer, conservationist, outdoorsman, and wilderness philosopher, of Ranier, Minnesota. Correspondents include Frances E. Andrews, John Bakeless, Lawrence Burpee, Arthur Hawkes, Frank Brookes Hubachek, Harold L. Ickes, O.L. Kaupanger, Charles Scott Kelly, Robert Marshall, Barrington Moore, Ernest C. Oberholtzer, Sigurd F. Olson, Henrik Shipstead, Sewell Tappan Tyng, Chester S. Wilson, and Frederick S. Winston. Unpublished inventory. Restrictions.

1691 **Quetico-Superior Council,** Additional papers, 1920-1962, 18 feet. Dockets and classified files selected from the Chicago law offices of Hubachek & Kelly. Many seem to be the correspondence and background files of Charles S. Kelly. They include correspondence, clippings and articles, speeches, pamphlets, brochures and other literature, publicity materials, legal briefs and other documents. Major correspondents include Ernest C. Oberholtzer, Fred Winston, Sewell Tyng, congressmen and legislators, government officials, and members of the President's Quetico-Superior Committee. Includes information concerning the Izaak Walton League, forest and land policies in northern Minnesota, Shipstead-Nolan Act, legal cases and suits, flood control on the Canadian border lands, Minnesota and Ontario Paper Company, Edward Wellington Backus, the International Joint Commission-United States and Canada, recreation policies, wilderness management, wildlife, flora, ecology in Quetico-Superior Area, Pigeon River Lumber Company, federal land acquisition in the Quetico-Superior Area, Canadian Quetico-Superior Committee, control of aircraft use of the Quetico-Superior Area. Uncataloged. Closed until cataloged.

1692 **Quie, Albert H.** (b. 1923), Papers, 1943-1971, 125 feet (134 boxes). Member of Minnesota legislature, 1955-58; U.S. Representative, 1958 + . His committee service included Agriculture. The papers consist of legislative and general files, with extensive data on agriculture, water pollution, and other subjects. Included are files (1943-57) of August H. Andersen, Quie's predecessor as congressman from the 1st district, who also served on the Agriculture committee. The papers consist of legislative, general, departmental and subject files. Uncataloged. Not open for research.

1693 **Railway Fire Prevention Papers,** 1914-1940, 8 boxes. Correspondence and papers from the office files of the Commissioner of Forestry and Fire Prevention of the Minnesota Department of Conservation with railroad companies of the Middle West regarding the outbreak of fires, conditions of engines, appointments of fire wardens, and general reports and instructions by Commissioner G.M. Conzet and Assistant Commissioner William M. Byrne.

1694 **Rainy Lake Investment Co.** (Wilmington, Delaware), Minute book, 1927-1937, 1 vol. Minutes, including articles of incorporation, of a firm incorporated to deal in land and other property. The firm formed part of the Weyerhaeuser Company. The book includes information on the Virginia and Rainy Lake Lumber Company. Restricted.

1695 **Red Stone Paper Mill,** Papers, 1815-18, 1 vol. Time accounts of employees and a record of the output of a Red Stone, Pennsylvania, mill.

1696 **Robertson, Russel K.,** Correspondence, 1852-54, 9 items. Photostats of correspondence of Robertson with relatives in Minnesota, Missouri, and Ohio, containing information on land claims, life in New Cambridge (Cambridge?), and Minnesota City, soils, forest cover, wildlife, and plans of several families to move from Massachusetts to Minnesota. Originals in the Regional History Collection, Cornell University, Ithaca, New York.

1697 **Rock Island Lumber and Coal Company** (Davenport, Iowa), Papers, 1903-1935, 7 inches (2 boxes). Statement regarding the company's incorporation (1903); annual statements; annual reports to the stockholders. There is a great deal of information on lumber marketing. The firm had yards in Wichita, Kansas, and formed part of the Weyerhaeuser Company.

1698 **Rock Island Lumber and Manufacturing Company** (Rock Island, Illinois), Papers, 1892-1935, 9 inches (2 boxes). Balance sheets, annual statements, annual reports, trial balances, letters relating to financial statements, and minutes. There is information on the company's yards at Rock Island and East Moline, Illinois, and Princeton and West Branch, Iowa; the sale of cutover lands in Sawyer and Ashland counties, Wisconsin, to the Potlatch Timber Company; and the sale of property in Kansas and Oklahoma to the Rock Island Lumber and Coal Company in 1904. The Rock Island Lumber and Manufacturing Company formed part of the Weyerhaeuser Company.

1699 **Rock Island Sash and Door Works** (Rock Island, Illinois), Papers, 1902-1946, 1 box. The papers include a catalog, lists of logs sawed, correspondence relating to annual statements, annual reports, and financial statements of a firm which formed part of the Weyerhaeuser Company. There is information about the prices of logs and manufactured wood products, labor costs, marketing, insurance, production, freight rates, etc.

1700 **Rotary Mill Company** (St. Paul, Minnesota), Accounts, 1854, 5 items. Miscellaneous accounts of the company for building materials used in the Mendota area, particularly for the Mendota House and the buildings of the fur traders.

1701 **Rutledge, Fred S.,** Papers, 1888-1961, 2 boxes (including 1 vol.). The papers contain information on logging, sawmills, and tie cutting in Minnesota; and on agricultural machinery, farming, and railroads.

1702 **Rutledge Lumber and Manufacturing Co.** (Kettle River, Minnesota), Papers, 1891-1906, 2 vols. Minute book, including articles of incorporation, and stock certificate book of a firm that was organized for the general lumbering business, and formed part of the Weyerhaeuser Company. Restricted.

1703 **St. Croix River Association,** Records, 1917-1966, 1 box. Correspondence, reports, press releases, and clippings relating to conservation problems in the St. Croix River area. Included are correspondence of John Warner Griggs Dunn, vice-president, and officials of the Wisconsin and Minnesota departments of conservation, the U.S. Department of the Interior, the U.S. Army Corps of Engineers, and various conservation and political figures. Issues discussed include fish and game matters, water power development, control of blister rust, effect of dam construction on water levels, and two Minnesota Senate bills, on wild rivers (S. 1446) and the St. Croix Scenic Waterway (S. 897), etc. The organization was originally known as the St. Croix River Improvement Association. Inventory.

1704 **St. Louis Country Club and Farm Bureau Association,** Papers, 1915-1962, 8 boxes (including 5 vols.) and 1 additional vol. Financial records, reports, clippings, correspondence, constitution, bylaws, minutes, resolutions, and other papers of an organization founded to ease the transition of St. Louis County, Minnesota, through periods of concentration on the mining and lumber industries into an era of more varied economic interests. Includes information on such topics as land classification, land clearing, forest fire control, enlargement of the Superior National Forest, forest conservation, wilderness preservation, encouragement of woodworking industries, reforestation, aircraft use in the roadless area. Typed description, 6 pp.

1705 **St. Louis River Mercantile Company** (Cloquet, Minnesota), Papers, 1906-1915, 2½ inches (1 box). Financial statements of a firm organized to carry on a wholesale business in lumbermen's supplies, which may also have engaged in logging transactions with the Northern Lumber Company, the Cloquet Lumber Company, and the Johnson-Wentworth Company. The firm formed part of the Weyerhaeuser Company.

1706 **St. Louis Sash and Door Works** (St. Louis, Missouri), Papers, 1931-1945, 1½ inches (2 folders). Financial statements and annual reports to stockholders. There is information on labor policies, marketing problems, government regulations, and relationships to other Weyerhaeuser companies.

1707 **Sauntry and Tozer** (Stillwater, Minnesota), Papers 1881-90, 4 vols. Ledgers, a journal, and a cashbook of a logging firm. Restricted.

1708 **Sauntry, William, and Company** (Stillwater?, Minnesota), Papers, 1893-1913, 2 inches (2 vols.). Ledger and journal of an obscure firm which formed part of the Weyerhaeuser Company.

1709 **Secombe, David,** Papers, 1848-79, 2 boxes. The collection contains legal papers used by a Minneapolis attorney in a variety of lawsuits including cases involving lumber businesses.

1710 **Senkler, George Easton, and Family,** Papers, 1844-1955, ca. 1 box and 7 vols. Includes letters (1921-30), of Albert Easton Senkler (1907-47) describing his work in lumber camps in Montana and Washington. Inventory.

1711 **Shaw, Daniel, Lumber Company** (Eau Claire, Wisconsin), Papers, 1861-1911, 1 box (including 4 vols.), 19 additional vols. Letterpress books (1877-1911), contracts, and miscellaneous papers of an Eau Claire, Wisconsin, lumber company, with information on logging, lumber manufacturing, transportation, and the marketing of logs and lumber. Included is information on business trips made by Daniel and Eugene Shaw.

1712 **Shaw, Percy M., Jr., and Company** (Duluth, Minnesota), Correspondence, 1895-1925, 1 box, photostats. Letters relating to the business of a Duluth, Minnesota, lumber broker, giving the names of lumber vessels leaving the port of Duluth, the amount of lumber shipped by Shaw, dates of shipment, and the names of other Minnesota lumber companies with which he did business.

1713 **Shedd, Charles, and Family,** Papers, 1857-1933, 1 roll microfilm. Includes letters (1859-70) by Cornelius W. Shedd, describing his work in sawmills in Alabama and in Columbus and Meridian, Mississippi. Originals were in the possession of Mrs. George E. Cook, Minneapolis, Minnesota.

1714 **Sheehan, Timothy J.** (1835-1913), Papers, 1857-1913, 7 boxes (including 8 vols.). Accounts and correspondence in the collection (1885-89) give information on the cutting of timber on the White Earth Indian Reservation where Sheehan was agent, and on a smallpox epidemic in the lumber camps, 1885.

1715 **Shell Lake Lumber Company** (Shell Lake, Wisconsin), Papers, 1891-1905, 1¼ inches (2 folders). Inventory of lumber, shingles, and lath; balance sheets; comparative statements of cut and sales; annual statements; trial balances; and letters concerning the annual statements. The firm formed part of the Weyerhaeuser Company. Also included are a few letters, minutes, and a financial statement of the Lumbermen's Bank (Shell Lake).

1716 **Sibley, Henry Hastings** (1811-1891), Papers, 1815-1930, 90 vols. and 18 boxes (also 32 rolls microfilm). Fur trader and Governor of Minnesota. Transcripts and originals of correspondence and other personal and business records relate to various interests, including lumbering and early sawmills. There is a lumber account book (1852-53), kept at Mendota, Minnesota Territory. Guide to the microfilm edition.

1717 **Sibley State Park Improvement Association** (New London, Minnesota), Records, 1919-1961, 1 box. Correspondence, newspaper clippings, and other papers relating to the illegal trespassing on park property, the agitation to put the park under state control, the establishment of a Civilian Conservation Corps camp in the park in 1938, the raising of funds to secure new campsites and to improve old ones, and other matters concerning Sibley State Park. O.A. Nelson was one of the leading figures in the association. Inventory.

1718 **Simonds, Chauncey** (ca. 1823-1905), Reminiscences, 1889, 8 l. Brief reminiscences of Simonds, who settled at Marine in 1843. Included is information about the Stillwater Lumber Company, Stillwater, Minnesota; and early settlers, lumber camps, and sawmills in the St. Croix Valley.

1719 **Smart, John F.,** "Daggett Brook Sketches," 1886-1916, 61 l. Reminiscences of life in Crow Wing County, Minnesota, which contain some information on lumbering.

1720 **Soil Conservation Society of America, Minnesota Chapter** (St. Paul, Minnesota), Papers, 1955-1973, ¾ foot (2 boxes [including 1 vol.]). Included are some articles and other materials on watershed and woodland management, especially for water conservation, and on land use planning.

1721 **Soule, Benjamin,** Papers, 1847-71, 13 items. An agreement (1851) for cutting timber in Maine; deeds for land in Benton County and in the townsite of Princeton, Mille Lacs County, Minnesota (1863, 1868); and miscellaneous letters.

1722 **Sound Timber Company** (Davenport, Iowa, and Seattle, Washington), Papers, 1900-1955, 5 inches (1 box). Statements, annual reports, and a few letters and minutes of a firm that did a logging and lumbering business in Washington and Oregon, and formed part of the Weyerhaeuser Company. There is information on timberland holdings, land and stumpage prices, logging contracts, logging operations, trespass, right-of-way agreements, lawsuits, leases, the company's financial status, the work of the Washington Forest Fire Association, and taxation.

1723 **Standard Lumber Company** (Dubuque, Iowa), Papers, 1872-1948, 16 pieces. Clippings, annual statements, agreements, letters, and miscellaneous papers collected by Charles Ingram which relate chiefly to the Standard Lumber Company, and Ingram, Kennedy and Day, both of Dubuque, Iowa.

1724 **Steele, Franklin** (1813-1880), Papers, 1839-1888, 62 vols. and 13 boxes. Correspondence, deeds, bills, legal papers relating to Steele's interests, including local politics in Minnesota, financial affairs of the St. Anthony Falls Water Power Co., the building and financing of railroads in Minnesota and various parts of the U.S.; land sales; taxes on property; records of logs scaled for Steele; correspondence about the sale of lumber.

1725 **Stuntz, Albert C.** (1825-1914), Diary, 1858-1882, 1 reel microfilm. A diary kept by a surveyor in the Lake Superior area of northwestern Wisconsin, in the St. Croix River Valley, and in Wisconsin lumber camps. Information is included about land surveys and acquisition in Wisconsin and Minnesota, and about lumbering practices. The original is in the possession of the State Historical Society of Wisconsin, Madison.

1726 **Superior National Forest,** Records, 1903-1969, 17 reels microfilm. Copies of originals loaned by the U.S. Forest Service, Superior National Forest, Duluth, Minnesota, selected for historical interest. Topics include land classification reports, legislation concerning creation of and proposals for enlargement of the Superior National Forest, and consolidation of lands (1903-65); proposed hydroelectric projects (1910-68); proposed paper mill (1939-40); the Boundary Waters Canoe Area; honors paid to Arthur H. Carhart; international joint commissions on floods; mineral policy; roads; Cloquet and Northern Lumber Company logging railroad; Superior National Forest Recreation Association; Quetico-Superior Council; Northern Great Lakes Area Council; George Washington Memorial Forest Plantations; Land and People Conference, Duluth (1963); proposed Voyageurs National Park; Superior National Forest Advisory Committee; natural areas; Herman H. Chapman; motorized transportation in wilderness areas, including aircraft; wilderness legislation and regulations; clippings. Unpublished guide.

1727 **Superior Timber Co.** (Chippewa Falls, Wisconsin), Papers, 1885-1947, 7 boxes (including 1 vol.) and 8 additional vols. Deeds (1886-1934), correspondence (1885-1944), annual statements (1901-43), abstracts of title (1908-35), annual reports of corporations issued by the Wisconsin Secretary of State (1910-44), tax returns (1911-44), trial balances (1912-43), minutes, stock certificates and statements of assets and liabilities (1918-28) of a firm organized for mining, manufacturing, mercantile operations, operating sawmills, manufacturing and selling lumber, and which, after 1907, dealt mainly in the sale of unimproved cutover lands, with operations centering in Douglas County, Wisconsin. This firm formed part of the Weyerhaeuser Company. Typed container list, 10 pp. Restricted.

1728 **Tawney, James Albertus** (1855-1919), Papers, 1876-1919, 4 vols. and 15 boxes. U.S. Representative from Minnesota, 1893-1911. Tawney served on the House Committee on Irrigation of Arid Lands. The papers include correspondence, diaries, campaign literature, drafts and printed copies of speeches, scrapbooks, and newspaper clippings. It includes letters from manufacturers regarding tariffs on lumber, and two letters from Theodore Roosevelt on the tariff and conservation (1904).

1729 **Tozer, David, Company** (Stillwater, Minnesota), Records, 1857-1940, 12 boxes (including 28 vols.) and 26 vols.

The records relate to the numerous partnerships Tozer had with other lumbermen, the David Tozer Company, and the St. Croix Timber Company. Many records relate to the acquisition and management of land in Skagit and Snohomish counties in Washington; Polk, Burnett, Ashland, Barron, and Douglas counties in Wisconsin; and Washington, Pine, Kanabec, Mille Lacs, and Aitkin counties in Minnesota. There are many agreements, mostly logging contracts, with numerous individuals and firms; information about the Stillwater Lumber Company; and other records. There are descriptions of Tozer lands by Arthur J. Mulvey (4 vols.), other land descriptions (5 vols.), notes on land examinations (6 vols.), timber estimates (2 vols.), pine land book (1 vol.), tamarack stumpage (1 vol.), log scale record (1 vol.), log accounts (1880-81, 1 vol.), records of land and timber estimates (1900), stock certificates of the Lower Tamarack Improvement Company (1898, 1901, 1905), articles of incorporation and minutes, land book, cashbook, journal, and ledgers of the St. Croix Timber Company (1907-18, 7 vols.); stock certificates, articles of incorporation and minutes, pine land book, journals, cashbooks, and ledgers of David Tozer and the David Tozer Company (1887-1926, 24 vols.). Inventory, 6 pp.

1730 **Union Lumber Company** (Chippewa Falls, Wisconsin), Papers, 1869-78, 1 vol. Articles of incorporation and minutes of a company organized to do a general lumber business, buy lands, build dams, and carry on a mercantile business. Restricted.

1731 **Walker, Judd, and Veazie** (Marine, Minnesota), Account books, 1849-1948, 125 vols. Cashbooks, daybooks, ledgers, and journals of a Marine, Minnesota, lumber company that conducted logging operations in the St. Croix Valley, manufactured lumber, and marketed logs and lumber. The company also conducted a general store and gristmill, and had a part interest in a steamboat. The collection includes the camp book of Bergman and Company (1881-84), and the mill accounts of Burris and Champeau (1894-96).

1732 **Walker, Thomas Barlow, and Family,** Papers, ca. 1860-ca. 1951, ca. 270 feet. Business and personal papers relating to the lumber, real estate, and other business interests of the Walker family, primarily Thomas B. Walker, Archie D. Walker, and Harriet G. Walker. Many of the papers are concerned with the operations of the Red River Lumber Co. of Akeley, Minnesota, and Westwood, California; and with Walker real estate interests in Minnesota. The collection includes correspondence, miscellaneous papers, reports, legal papers, clippings, land and cruise records, lumber camp records, financial records, scrapbooks, maps and drawings, letterpress books, minute books, stock record books, journals, ledgers, cashbooks, and other financial volumes. Unprocessed. Closed until cataloged, then open under restrictions.

1733 **Warren, George Henry** (1829-1891), Papers, 1856-1929, 1 box. Lumberman's diaries (1868, 1888, 1891); log marks used by Warren; and a record of the efforts of his daughter Annie Warren to hold the homestead claim filed by Warren when the land was claimed by the Northern Pacific Railway Co.

1734 **Washburn, Algernon S.** (1814-79), **and Family,** Papers. 1833-82, 1 box. Letters to Washburn in Hallowell, Maine, some from his brother William D. Washburn, Minneapolis, Minnesota, about his lumbering and flour milling ventures there, and his work as surveyor general of Minnesota, 1861-65.

1735 **Washburn, Cadwallader Colden** (1818-82), Correspondence, 1869, 19 items. Letters from William D. Washburn, chiefly concerning lumbering in Minnesota, the acquisition of pinelands and possible dam sites with half-breed scrip, and trespass on pinelands in the St. Croix Valley. Photostats, made from originals in the possession of the State Historical Society of Wisconsin, Madison.

1736 **Wefald, Magnus,** Papers, 1946-1961, 42 boxes. Includes information on Itasca State Park, Garrison Dam, and forestry.

1737 **Wells, Cyrus W.,** Papers, 1865-1917, 6 vols. and 1 box. Real estate agent of Minneapolis. Business records, including accounts (1884-85) of the lumber firm of Ellis and Leikem, of which Wells was an assignee.

1738 **Weyerhaeuser, Frederick, and Company,** Papers, 1892-1940, 14 boxes (including 10 vols.). Correspondence (1902-26); agreements, contracts, and other legal papers of F. Weyerhaeuser and Company, St. Paul, Minnesota, and various associated companies. Information is included on the logging situation in Clark County, Washington, 1903, and on timber purchase, and plans for logging railroads in the Pacific Northwest. From 1892 to about 1910 the records deal largely with land ownership, booming, scaling, rafting, and taxation matters. After 1910, most of the papers deal with affairs of the Superior Timber Company, the Homeseekers' Land Company, taxes, cutover lands, etc.

1739 **Weyerhaeuser and Denkmann Company** (Rock Island, Illinois), Records, 1891-1934, 2 boxes. Financial statements of a company organized by Frederick Weyerhaeuser and Frederick C.A. Denkmann, which manufactured lumber, lath, and shingles. Inventory.

1740 **Weyerhaeuser and Rutledge Co.** (Chippewa Falls, Wisconsin), Papers, 1882-1938, 1 box (6 vols.). Schedules of property, statements, lists of bills receivable, trial balances, purchases, income tax returns, minutes, tax records, journals, ledger, and stock certificates of Weyerhaeuser and Rutledge Company and Weyerhaeuser and Rutledge firms carrying on a general lumbering business. Restricted.

1741 **Weyerhaeuser Family,** Papers, 1860-1961, 104 feet (252 boxes). Correspondence (mostly business) and lumber company financial records, chiefly 1900-57, of Frederick Weyerhaeuser (1834-1914), Frederick Edward Weyerhaeuser (1872-1945), Frederick King Weyerhaeuser (b. 1895), John Philip Weyerhaeuser (1858-1935); and Edwin Weyerhaeuser Davis. Among the firms represented are the Humbird Lumber Company, General Timber Service, Northwest Paper Company, Potlatch Forests, Potlatch Lumber Company, Potlatch Yards, Rilco Laminated Products, Rock Island Lumber Company, Thompson Yards, Timber Securities Company, Weyerhaeuser Sales Company, and Weyerhaeuser Timber Company. Inventory and container list; also a detailed folder list (526 pp.), prepared by the Forest History Society, covers the first 250 boxes. Restricted until 1990.

1742 **Whipple, Henry B.,** Papers, 1833-1934, 45 boxes (including 77 vols.). Papers of the first Protestant Episcopal bishop of Minnesota, who was active in Indian affairs, containing some information about timber frauds.

1743 **White, Henry Gilbert** (1908-1967), Papers, ca. 1925-1967, 31¼ feet (33 boxes and 1 oversize bacase). White served as surveyor, fire lookout, and forest economist with the U.S. Forest Service, 1927-50, and as a forest and resources economist with the Allied military government of Japan (SCAP), and in various other governmental posts. The papers cover all phases of his career. Included are U.S. government documents and other printed and research materials on forestry (1930s-60s); notes and miscellany (ca. 1930s? and 40s) related to studies at

Yale School of Forestry; miscellaneous materials related to Lake States Experiment Station and forestry studies (ca. 1930s); White's thesis (1948), on "Forest Regulation"; miscellaneous subject files (1930s-50s), including forestry in Japan. Closed to research until cataloged. Box list.

1744 **White Yard** (Davenport, Iowa), Papers, 1918-1929, 3 inches (1 box). Annual reports and statements of a retail lumber yard operated by the Weyerhaeuser and Denkmann Company.

1745 **Wier, Roy William** (1888-1963), Papers, ca. 1920-1963, ca. 3 feet (6 boxes). State legislator and U.S. Representative from Minnesota, 1949-61. Correspondence, newspaper clippings, voting records, and other papers, including material on conservation and other issues. Unpublished inventory.

1746 **Wilkin, Alexander and Westcott,** Papers, 1770-1894, 3 boxes. The papers include miscellaneous correspondence about the Mississippi Boom Company (1852).

1747 **Winton, David J.** (b. 1897), Papers, 1929-1962, 30½ feet (73 boxes). Lumberman of Minneapolis. He served as chief of the pulp and paper division of the War Production Board, head of the WPB lumber mission to England, and director of Region 12 of WPB during World War II. His papers have data on these activities as well as on his work with or interest in such groups as the National Planning Association, the National Policy Committee, the Atlantic Union Committee, and the United States National Commission for UNESCO (United Nations Economic and Social Council). Draft inventory, 40 pp. Restricted.

1748 **Winton Lumber Co.** (Minneapolis, Minnesota), Records, 1894-1963, 90 feet. Correspondence, memos, telegrams, financial records and reports, operating records, timber and logging statistics, contracts, tax records, stockholders' information, and related records of a company owned by the Charles J. Winton family, and of various affiliated or subsidiary companies. They document the Winton land and timber holdings, lumbering and milling operations, lumber and lumber products companies, primarily in California, Oregon, Canada, and Wisconsin. The major firms represented are the Winton Timber Co. (1901-32), Winton-Oregon Timber Co. (1919-52), Eagle Lake Spruce Mills (1917-40), Winton Lumber Co. (1907-63), and The Pas Lumber Co. (1919-62). The majority of the papers date from the 1920s-30s. Restricted.

1749 **Yetka, Frank,** Papers, ca. 1931-1933, 1 foot. Papers relating to fire claim suits against the federal government resulting from the Cloquet fire of 1918, consisting primarily of correspondence of Yetka with congressmen and individual fire claim applicants. Uncataloged. Closed to research until cataloged.

1750 **Zon, Raphael** (1874-1956), Papers, 1887-1957, ca. 6 feet (15 boxes). Forester and director of the Lake States Forest Experiment Station, St. Paul, Minnesota, 1923-44. Correspondence, articles, memoranda, minutes and reports of organizations, and certificates and appointments, relating to Zon's career in the Forest Service, other employees in the service, forestry legislation, conservation in Minnesota, relations of the Service with the University of Minnesota, and federal regulation of private forests. Includes material relating to the Zon family. Correspondents include Louis Adamic, Herman H. Chapman, B.E. Fernow, Walter Lippman, George Marshall, Robert Marshall, Arthur Newton Pack, Charles Lathrop Pack, Gifford Pinchot, Carl A. Schenck, F.A. Silcox, Oswald Garrison Villard, and Henry A. Wallace. Unpublished inventory. Some items restricted until 1980.

Minnesota State Archives

117 University Avenue 55101 Telephone: (612) 296-2747

1751 **Minnesota. Auditor, Land Department,** Records. Includes State Timber Correspondence (1885-1933, 31 cartons); State Stumpage Land Correspondence (1891-1933, 21 vols.); State Timber Permits (1891-1933, 111 vols.); State Logging Permits and Bonds (1873-78, 1891-1933, 10 vols.); State Timber Trespass Reports (1891-1933, 23 vols.); State Timber Cases Settled (1891-1933, 7 vols.); Forestry Record (1886-1915, 4 vols.), containing record of bounties paid for tree planting; Stumpage Record (1888-1914, 3 vols.); Timber Appraisal Reports (1913-31, 15 vols.); Timber Sales Book (1927-32, 1 vol.); Stumpage Journal (1873-75, 1 vol.); Stumpage Ledger (1873-75, 1 vol.); State Timber Remittance Statements (1891-1933, 2 vols.); State Timber Appraisers Weekly Reports (1920-26, 2 vols.); State Timber Permits [Cruiser Field Books] (1891-1933, 1 vol.); Timber Maps (n.d., 1 inch); Attorney General Opinions (1912-32, 1 vol.); Miscellaneous Maps and Notes (1891-1933, 2 vols.); Miscellaneous Timber Records (1918, 1927, 1 carton); and 2 boxes of Stumpage Reports (1865-89), Trespass Reports (1882-90), and Miscellaneous Papers (1874-80).

1752 **Minnesota. Conservation Department, Commissioner's Office,** Records. Contains Commissioner's Correspondence (1937-68, 99 cartons), including a considerable amount on forestry matters; monthly reports (1935-41, 4 cartons); quarterly reports (1941-42, 1 carton); miscellaneous reports (1946-53, 2 cartons).

1753 **Minnesota. Conservation Department, Forestry Division,** Records. Stumpage Record (1934-57, 13 vols.); Timber Trespass Record (1943-57, 1 vol.); Land Classification Record (1927-28, 2 vols.); Forest Management Studies (1942-46, 3 cartons); Forestry Division Director's File (1953-61, 4 cartons); National Forests Correspondence (1925-56, 1 carton); Itasca Park Forest Experimental Station Correspondence (1933-59, 1 carton); subject correspondence (1927-62, 15 cartons); Association of State Foresters Correspondence (1954-60, 1 carton); lands correspondence (1941-60, 2 cartons); circular letters (1925-59, 2 cartons); Civilian Conservation Corps correspondence (1940-42, 1 carton); correspondence with area supervisors (1938-40, 1 carton); timber correspondence (1940-41, 1 carton); Timber Products War Project (1945, 1 carton); correspondence and reports, project files and maps from the Faribault, Brainerd, and Hibbing Area Offices (1932-42, 1957-64, 22 cartons).

1754 **Minnesota. Forestry Board,** Records. Minutes (1899-1923, 3 vols.); Letters Received (1895-1925, 4 boxes); Letters Sent (1899-1913, 7 vols.); Correspondence (1911-25, 15 cartons); National Forests Correspondence (1909-25, 1 carton); Fire Warden Reports (1895-96, 1910, 21 vols.); Itasca State Park records (1875-1921, 13 vols.), including diaries, reports, surveys, and field notes; Forestry Records (1890-1915, 13 vols.), including diaries, reports, surveys, and field notes; Miscellaneous Papers and Correspondence (1887-1925, 3 cartons); Published Reports (1895-1924, 15 vols.); and Reports of the State Forestry Board (1910-25) and Reports of the Forest Fire Relief Commission (1908-21), filed in Official Reports and Communications, Records of the Office of the Governor.

Minnesota. Surveyor General of Logs and Lumber, Records, 1854-1927, 36 boxes, 18 cases, and 529 vols.

1755 *District 1 (Stillwater), Records, 1854-1922, 163 vols.* Journals (1863-1917, 52 vols.); ledgers (1865-1918, 75 vols.); records of log marks and log mark transfers (1854-1922, 8 vols.) ; bills of sale, agreements, mortgages (1865-68, 1871-1910, 13 vols.) ; journal and ledger of the Lumber Board of Exchange of Stillwater (1885, 2 vols.) ; Danbury Mill log records (1918-19, 2 vols.). There are also records of the St. Croix Lumbermen's Board of Trade (1873-1912, 6 vols.), including minutes (1871-1910), journals (1880, 1885, 2 vols.), ledgers (1878-80, 2 vols.), treasurer's book (1889-1912).

1756 *District 2 (Minneapolis), Records, 1854-1927, 30 boxes, 15 cases, and 273 vols.* St. Anthony Boom letterpress scale books for the Mississippi and Rum River Boom Company (1876-1905, 17 vols.) ; log ledgers (1857-80, 22 vols.) ; journals (1854-58, 1866-68, 1875, 6 vols.) ; records of contract (1856-98, 8 vols.) ; letterpress books (1888-1914, 8 vols.) ; incoming correspondence (1871-78, 8 vols.; 1879-84, 1887-1904, 23 boxes) ; orders, liens, writs, correspondence (1905-18, 7 boxes) ; scale bills (1854-80, 22 vols.) ; log mark record books (1858-1901, 8 vols.) ; miscellaneous Minneapolis lumber companies' tally and scale books (1876-1927, 17 vols.) ; scale orders, woods scale records, rafts and strays, timber appraisals, logging and cutting permits, etc. (16 vols.) ; St. Paul Boom brail books (1896-1911, 1913-14, 47 vols.) ; pond scale and gathering books (1867-1909, 94 vols.) ; tally books (15 boxes.).

1757 *District 4 (St. Cloud), Records, 1862-68, 2 vols.* Log Marks Book A (1862-68) and Bills of Sale and Agreements (1862-68).

1758 *District 5 (Duluth), Records, 1883-1919, 79 vols.* Letterpress log scale books (1884-1909, 25 vols.) ; log ledgers (1890-1917, 6 vols.) ; records of log marks (1875, 1883-1919, 8 vols.) ; books of log liens (1891-1913, 6 vols.) ; logging permits and bonds (1890-1917, 7 vols.) ; scale reports of the Rainy Lake River Boom Corporation (1901-11, 23 vols.), including records of timber ownership of the Rat Portage Lumber Company, the Rainy River Lumber Company, the Namakan Lumber Company, and Shevlin Mathieu.

1759 *District 7 (Crookston), Records, 1885-1913, 2 boxes and 3 vols.* Correspondence, records of log marks (1885-1910), log liens, logging permits, bills of scale, and a record book (1885-1913).

1760 **U.S. General Land Office, Surveyor General** (St. Paul), Records, 1854-1908, 40 cubic feet. Field notes (1854-1908, incomplete, 170 vols. and 9 boxes) ; letters sent (1854-1908, 28 vols.) ; letters received (1857-1908, 51 vols.). There are many letters dealing with the location of pinelands and with timber trespass.

TWO HARBORS

Lake County Historical Society, Railroad and Historical Museum

Museum Administrator 55616

1761 Manuscripts include William E. Scott, "The Forest History of Lake County" (1929), in North Shore Lore, I, 7; Otto Wieland, "Lumber industry at the Head of the Lakes" (July 13, 1936), in General, I, 171; John F. Coggswell, "River Full of Logs," General, I, 211 (appeared in the *Saturday Evening Post,* July 20, 1940) ; John Fritzen, "History of North Shore Lumbering" (October, 1968) ; and Report of the Wilderness Reseach Foundation on the operations of the Quetico-Superior Wilderness Research Center for 1949-69.

Mississippi

JACKSON

Mississippi Department of Archives and History, Archives and Library Division

Director, P.O. Box 571 39205. Records of state and local agencies held by the Department are described in *Guide to Official Records in the Mississippi Department of Archives and History,* compiled by Thomas W. Henderson and Ronald E. Tomlin, Jackson, 1974.

1762 **Finkbine Lumber Company** (Wiggins, Mississippi), Collection, 1905-1906, 34 items. Picture postcards and photographs taken in 1905 and a letterhead of the Finkbine Lumber Company. The pictures show the dry sheds, commissary, roundhouse and machine shops, planing mill, company-owned dwellings, and other scenes of sawmill operations; also oxen snaking the timber out of the woods for loading, and unloading at the mill.

1763 **Logging in the Mississippi Delta,** Photographs, ca. 1910, ca. 20 items. Photographs by Polland of Memphis, given to the archives by Mrs. C. Bryant Young, 1972.

1764 **Reynolds, L.P.,** Papers, 1847-1913, 1100 items. Included are a few letters, receipts, and orders for Reynolds' lumber business (1875-99). Jacinto, Alcorn County, Mississippi.

1765 **Weston, H., Lumber Company** (Logtown, Mississippi), Lumber logs, 1910-1911, 2 vols. Record books showing number of logs, number of feet, and price paid to individuals selling timber to H. Weston Lumber Company.

1766 **Weston, Henry, and Family,** Papers, 1846-1965, 65 items. Henry and Horatio Weston founded the H. Weston Lumber Company of Logtown, Mississippi, near the Gulf Coast on the Pearl River. The collection includes, along with family correspondence and genealogical material, letters concerning the lumber industry in the states of Washington and Mississippi.

STATE COLLEGE

Mississippi State University, Mitchell Memorial Library, Special Collections

Reference Librarian, P.O. Box 5408 39762

Telephone: (601) 325-4225

1767 **Abbott, Liberty C.** (1837-1894), **Family,** Papers, 1882-1930, 2750 items. Includes letters to and from Frank Gearhart, manager of a timber firm, and record books (1902-07), of Thompson & McClure, timber operators, Leflore County, Mississippi.

Mississippi State University, Mitchell Memorial Library, University Archives

Archivist, P.O. Box 5408 39762

Telephone: (601) 325-4225

1768 **College of Forest Resources,** Records, 1907-1967, 275 items. Miscellaneous printed material and reports concerning history of forestry in Mississippi (ca. 1941-42), and correspondence concerning the University of Mississippi Lumber Archives (1956-64) ; correspondence of Mississippi State University officials planning for the School of Forestry (1945-67, 57 items) ; booklets, papers, and clippings of S.W. Greene, Director, McNeill Experiment Station, McNeill, Mississippi, concerning annual grass and forest fires (1932-44, 19 items) ; S.W. Greene correspondence relating to the establishment, maintenance, and operations of the McNeill Experiment Station (1907-47, 41 items) ; copy of a synopsis of the work at McNeill by J.R. Ricks to S.W. Greene (1934, 2 items) ; miscellaneous correspondence of Greene and reports relating to forestry at McNeill (1935, 25 items) ; correspondence relating to work at McNeill and S.W. Greene's resignation (1936, 50 items) ; correspondence and other materials relating to the establishment of the Forest Products Utilization Laboratory at Mississippi State University (1961-67, 52 items).

Mississippi State University, Mitchell Memorial Library, The John C. Stennis Collection

Curator, P.O. Box 5408 39762 Telephone: (601) 325-4225

1769 U.S. Senator from Mississippi, 1947 + ; served on the National Forest Reservation Commission. Forest history materials include a 2-hour taped interview of Senator Stennis by T.R. Clapp on the subjects of forestry, the Senator's interest in forestry and forest legislation, background of the Stennis-McIntire Forestry Legislation, etc.; a souvenir compilation of materials relating to the Stennis-McIntire bill; a sizable collection of government documents and secondary sources relating to forestry, assembled by Senator Stennis; considerable units of correspondence relating to state and national forestry in the Senator's offical files. Closed to research.

UNIVERSITY

University of Mississippi, The Library

University Archivist and Special Collections Librarian 38677

1770 **Brown, Andrew, Lumber Company** (Natchez, Mississippi), Papers, 1830-1880, 23 boxes.

1771 **Dantzler, L.N., Lumber Company,** Papers, 1906-1936. A south Mississippi company. Correspondence (ca. 37 items); cashbooks, journals, time books, shipping and sales record books (ca. 200 items); miscellaneous documents ranging from land deeds to purchase slips (ca. 100 cubic feet). Also included is a manuscript history of the company written by Walter Barber.

1772 **Learned, R.F., and Sons** (Natchez, Mississippi), Business Records, 1870-1908, 92 boxes.

1773 **Pettibone, Charles J.,** Papers, 1935-1958, 1 box. Correspondence relating to the Edward Hines Lumber Company operations in Mississippi, including descriptions of various aspects of the business written by Pettibone.

1774 **Weston, H., Lumber Company,** Papers, 1929-1943, 63 boxes. Business and personal correspondence, cashbooks, journals, ledgers, and other business records, and memoirs. There are also a few letters written by Henry Weston, founder of the company, relating to the antebellum era and his brief manuscript history of the company from 1845-1900.

Missouri

COLUMBIA

University of Missouri Library, Western Historical Manuscript Collection and State Historical Society Manuscripts, Joint Collection

Director. Some of the holdings are described in the *Guide to the Western Historical Manuscripts Collection,* Columbia: University of Missouri, 1952 (*University of Missouri Bulletin,* 53, No. 33, 1952); and *Guide to the Western Historical Manuscripts Collection,* by John A. Galloway, Columbia: University of Missouri, 1956 (*University of Missouri Bulletin,* 58, No. 13, 1956).

Western Historical Manuscripts Collection

1775 **Bell, Charles Jasper** (b. 1885), Papers, 1934-48, 9022 folders. U.S. Representative from Missouri, 1935-49; served on the Public Lands committee in the 80th Congress. The papers include correspondence and other materials covering his congressional service. The more important part relates to his committee work. Public correspondence includes a file on opinions on transferring the Forest Service from the Department of Agriculture; subject correspondence contains files on Department of the Interior vacant public lands, Jackson Hole National Monument controversy, conservation reports of the National Wildlife Federation, including summaries of proposed legislation (1947-48), pertaining to forests, reclamation, parks, public lands, etc.; legislation files include letters concerning Forest Research Center, Columbia, Missouri, reforestation of strip mining areas, purchase of Ha-ha-tonka as a state park, etc.

1776 **Bennett, Marion Tinsley** (b. 1914), Papers, 1941-1948, 1942 folders, 70 vols. U.S. Representative from Missouri, 1943-49. Correspondence, speeches, bulletins, and other papers, including material on efforts to locate a forestry research center near Springfiled, Missouri; there are other files of correspondence from constituents concerning forestry research.

1777 **Bradshaw, J.D.,** Papers, 1904-1946, 11 folders, 46 vols. Inventory records and ledgers of Bradshaw and Proctor Lumber Company, Barnett, Missouri, plus some records of the Leeton Lumber and Hardware Company, Leeton, Missouri. The latter include ledgers (1931, 1935, 1936); purchase journals (1937, 1944); and freight bills (1939, 1943, 1945).

1778 **Caledonia Account Books,** 1857, 1871-1873, 1882-1883, 3 vols. Two volumes contain the accounts of a store which dealt in lumber and merchandise in or near Caledonia, Washington County, Missouri.

1779 **Cannon, Clarence** (1879-1964), Papers, 1896-1964, 3,300 folders, 124 vols. U.S. Representative from Missouri, 1923-64. Served on the Appropriations committee, of which he became chairman. The papers include correspondence, speeches, records, campaign materials, printed works, and papers related to conservation and flood control, and many other matters. There are correspondence, project plans, and related photographs for a park on the Cuivre River, including National Park Service correspondence; correspondence relating to the establishment of an Erosion Control Nursery in Elsberry, Missouri; clippings and correspondence concerning the Missouri State Park Board; and materials concerning the Missouri Valley Authority.

1780 **Carnahan, Albert Sidney Johnson** (1897-1968), Papers, 1944-1960, 5,081 folders. U.S. Representative from Missouri, 1945-47. Subject files include material on the Interior Department, parks, etc.; among legislative files for the 81st Congress is material on the Forest Service.

1781 **Drew-Stiff Lumber Company** (Willow Springs, Missouri), Order list, 1890, 1 item, photostat. List of stock and prices of a Howell County company.

1782 **Dulany, George W.,** Papers, 1900-1949, 1 folder. Includes a letter from G.W. Dulany to Dean W. Francis English at Colubmia, Missouri, discussing the history of the Dulany Family and the lumber industry at Hannibal, Missouri.

1783 **Dunlap, Frederick** (1881-1968), Papers, 1891-1937, 425 folders, 4 vols. Professor of forestry at the University of Missouri, secretary of the Missouri Forestry Association, and editor of *Forest Quarterly*. Family and professional correspondence, diaries, lecture notes for forestry courses, notes concerning experiments and forestry projects, forestry legislation, and surveys of timber lands, and pamphlets and bulletins on forestry, lumbering, forest and wood preservation, conservation, and wildlife. Card Catalog.

1784 **Forest History Foundation,** Papers, 1956-1958, 1 folder. Transcript of an oral history interview with George W.

Dulaney, Jr. (1956), concerning lumbering in early Missouri; also newsletters.

1785 **Gauldin, Martin A.,** Diary, 1845, 5 l., typed copy. Diary kept on a journey from Marshall, Saline County, Missouri, to Austin, Texas (November 27-December 25, 1845), describing the timber and water resources and fertility of the country.

1786 **Hennings, Thomas Carey, Jr.** (1903-1960), Papers, 1934-1960, 474 boxes, 45 vols. U.S. Representative from Missouri, 1935-40; U.S. Senator, 1951-60. In his legislative correspondence are files for forest conservation (1953-56); and forests (1957-60). Among service case files are correspondence referring to forest land policies, exchange or purchase of forest lands, and grazing permits and fees, consisting of letters by constituents, Hennings, and Forest Service officials (1951-60); miscellaneous files include photographs of Missouri forests (1937-42), and of Missouri national forests (1955). Restricted. Inventoried on cards.

1787 **Hyde, Arthur Mastick** (1877-1947), Papers, 1886-1949, 1828 folders. Governor of Missouri, 1921-25; and U.S. Secretary of Agriculture, 1929-33. Files on conservation and reforestation (ca. 1919-25) include chiefly correspondence on formation of a Missouri state Department of Conservation and reports on conservation conferences; there is correspondence concerning selection of state fish and game commissioner; correspondence (ca. 1919-25) concerned with the establishment of state parks in Missouri; correspondence and other material on irrigation projects (ca. 1922-36); assorted material relating to forestry (ca. 1927-40), including minutes of arboretum meeting (1927), bill to establish a national arboretum (1940), material relating to forest management, the Clarke-McNary Act, comments by Hyde on Michigan forestry problems, etc.; speech on national forest policy by Hyde at dedication of part of Huron National Forest (1929); clippings concerning the land utilization conference (1931), which related to the reconversion of submarginal land to forest or pasture. Shelf list.

1788 **Kelleter, Paul Delmar** (1881-1950), Speeches, 1935-1940, 6 folders. Speeches and miscellaneous reports by Kelleter, Supervisor of the Clark National Forest, Missouri, chiefly concerning the U.S. Forest Service in Missouri, the Clark National Forest, and the rehabilitation of the Ozark region.

1789 **Kem, James P.** (1890-1965), Papers, 1946-1952, 12,780 folders. U.S. Senator from Missouri, 1947-53; member of the Committee on Agriculture and Forestry. Most of Kem's committee papers seem to concern farm products, but there are also reports from the Missouri Forest Survey (1948), and general correspondence of the committee (ca. 1949-52).

1790 **Louisiana Central Lumber Company,** Records. Unprocessed.

1791 **Lumber Prices and Shipping Lists,** 1869-1894, 5 items. Price lists for various kinds and grades of lumber as quoted by G.C. Hixon and Company, Hannibal, Missouri; Judson Lyon, St. Joseph, Missouri; and I.M. Howell and Company, Atchison, Kansas. Also includes information concerned with shipping and a schedule of logging expenses for various logging centers.

1792 **Missouri Lumber and Mining Company,** Papers, 1853-1945, 7,096 folders, 308 vols., 1/3 roll microfilm. Papers of a land and lumber company located at Grandin, Carter County, which operated in Shannon, Ripley, Butler, Carter, Reynolds, and Wayne Counties. Includes letters, hospital record books, deed index books, coupon receipt books, iron sales and sample analysis books, index to statement books, insurance books, land sales books, lumber and freight record books, personal injury reports, pine land record books, railroad claim books, record of trespasses on land, stock claim books, and trial balance books. There are also papers concerned with the articles of association, minute books of directors' meetings, estimates of business, and manufacturing statements. Other items include blueprints of the company land holdings, vouchers and invoices, journal entries, lumber inspection reports, and payroll receipt books. Persons mentioned include John Barber White.

1793 **Missouri National Forests' Association,** Books, 1933-35, 2 vols. Compiled by Robert Good, president of the association. Included are clippings (1933-35, 1 vol.), and minutes and correspondence (1933-34, 1 vol.).

1794 **Missouri Rural Area Development Committee,** Records, 1957-1967, 143 folders. Correspondence, reports on projects in various counties, U.S. Department of Agriculture leaflets, and other papers, relating to committee meetings, regional conferences, program objectives, and responsibilities and functions of various committees, of an organization established under the direction of the Department of Agriculture to coordinate efforts to revitalize agricultural and natural resources in Missouri's rural areas. Includes material relating to wood-using industries, construction of recreation areas, and conservation programs. Card catalog.

1795 **Morris Family,** Papers, 1840-1946, 36 folders, 22 vols. Papers of the Morris family of Rockbridge, Ozark County, Missouri. Byron Vaughn Morris moved from Benton County, Iowa, to southern Missouri, where he operated a lumber mill and other businesses. The papers include miscellaneous business and personal papers, and Mrs. Eliza Morris' diary (19 vols.).

1796 **Morton, Stratford Lee,** Papers, 1943-1945, 379 folders. Correspondence and other papers relating to the St. Louis and St. Louis County Committee for the Revision of the Constitution, containing material on state and local parks.

1797 **Ozark Land and Lumber Company** (Winona, Missouri), Account Books and Papers, 1887-1933, 688 folders and 99 vols. Records and correspondence of a Shannon County, Missouri, company including cashbooks, general ledgers, general journals, and miscellaneous items including a minute book of stockholders' meetings, order books, tax record books, store sales books, bank material, bills-of-lading, freight bills, and abstracts, lumber records, maps of land to be sold, sorrespondence of stockholders, records of expense accounts of the mill, planer, and railroad, and an account of the company's consolidation with the Hershey Land and Lumber Company of Muscatine, Iowa.

1798 **Park, Guy B.,** Papers, 1932-36, 2,255 folders. Governor of Missouri. Files on the state Fish and Game Department (1933-36), relate primarily to wildlife conservation and administration, but include protests against the use of license funds for state park work; correspondence relating to state parks (1925-36) contains material concerning staffing, purchase of land in the state for a national park, and reports on state park management. Restricted.

1799 **Slayton, Edward M.,** Papers, 1943-44, 118 folders. Bulletins, papers, and correspondence of the Committee on Agriculture and Conservation of the Missouri Constitutional Convention; includes files relating to the Missouri Conservation Commission, including papers on forestry work.

1800 **Slusher, H.E.,** Papers, 1944-1958, 1,676 folders. President, Missouri Farm Bureau Federation. Included are pamphlets on forests (1957), correspondence, reports, publications, etc., of the U.S. Forest Service (1949, 1954-57). Card fiel inventory.

1801 **Sparks, John Nathan** (b. 1876), Papers, 1901-1947, 337 folders. Included is some business correspondence dealing with the Missouri Lumber and Mining Company (1914-43).

1802 **Swiggy, A.W., and A.C. Jones** (Green City, Missouri), Mercantile Records, 1895-1903, 8 vols. Records of a Sullivan County firm which sold lumber, shingles, sashes, doors, blinds, lime, cement, and paints. The books contain records of sales, bank deposits, accounts receivable, accounts payable, and cash on hand.

1803 **University of Missouri. Forestry School**, Papers, 1936-1968, 293 folders. The papers are primarily correspondence during the Westveld administration; however, budget reports, committee reports, inventories, are included, as well as papers from the Duncan administration. The major sections of the collection are extension (1965-68), Forestry School administration (1947-68), personnel (1950-68), research (1943-68), teaching (1936-68), university administration (1949-62).

1804 **Westveld, Ruthford Henry** (b. 1900), Papers, 1956-1965, 209 folders. Professor of forestry at the University of Missouri. Correspondence and other papers relating to the background, passage, and enforcement of the McIntire-Stennis act for cooperative forestry research, forestry research and facilities at land grant colleges and other institutions, and the U.S. Forest Service. Correspondents include state forestry officials, U.S. senators and representatives, and the U.S. Department of Agriculture. Card catalog.

1805 **Whitlow, W.B.** (1893-1942), Papers, 2864 folders, 15 vols. Letters and papers pertaining to liquidating the assets and liabilities of the Peoples Bank of St Charles (1933-39) include numerous letters regarding the sale of land in Wayne County to the U.S. Department of Agriculture for the Forest Service.

State Historical Society Manuscripts

1806 **Lackland, James C.**, Papers, 1769-1897, ca. 1000 items. The papers contain documents concerning a steam sawmill in St. Charles, Missouri, owned jointly by James Lackland and Hugh H. and Benjamin R. Wardlaw; bills due for work at the mill (1837-41); and books of mill accounts (1846-56).

1807 **MacKenzie, Kenneth** (1797-1861), Papers, 1796-1918, ca. 1500 items. The collection includes the papers containing specifications for machinery necessary to operate a sawmill in St. Louis (1849, 4 letters).

1808 **Martin, William**, Correspondence, 1839-1847, 53 items. Alton, Illinois, attorney. Legal correspondence including one letter concerning illegal cutting of timber on a client's property.

1809 **Missouri, University. Forestry Department**, Papers, 1904-1921, 57 folders. Correspondence, leases, maps, and miscellaneous material regarding the sale and leasing of land owned by the Department of Forestry. At that time, the Department of Forestry was part of the College of Agriculture, University of Missouri. The land was apparently granted to the College of Agriculture under the Morrill Act of 1862.

INDEPENDENCE

Harry S Truman Library

Director 64050. Holdings of the Truman Library are listed in *Historical Materials in the Harry S Truman Library*, Independence: Harry S Truman Library, National Archives and Records Service, General Services Administration, 1973.

1810 **Anderson, Clinton Presba** (b. 1895), Papers, 1945-1948, 16 boxes. U.S. Secretary of Agriculture, 1945-48. This collection contains official and personal correspondence, memoranda, reports, and other papers related to his service as secretary. There is an alphabetical correspondence file (4 boxes); official correspondence (8 boxes), including a file on the Oregon and California revested lands; and a personal file (ca. 2 boxes) containing some letters from Horace Albright. Shelf list, 4 l.

1811 **Chapman, Oscar Littleton** (b. 1896), Papers, 1931-1953, 50 feet. U.S. Assistant Secretary of the Interior, 1933-46; Under Secretary, 1946-49; Secretary, 1949-53. The papers contain correspondence, memoranda, speeches, reports, statements, and other papers relating to Chapman's work in the Interior Department and his political activities. Correspondence (1933-49) includes letters of Horace Albright, Charles F. Brannan, Arthur H. Carhart, Gifford Pinchot, etc.; correspondence (1949-53) includes letters of Harold Ickes; general correspondence (1949-53) contains files on Agriculture-Interior relations, luncheon and dinner for A.E. Demaray (November 18, 1951), Federal Power Commission hearings on the Kings River project (1950), Jackson Hole, etc.; public addresses and statements include some on natural resources issues; also there is a reading file of correspondence (1946-52). Shelf list, 20 l. Restricted: this collection may be consulted only with permission from Mr. Chapman.

1812 **Doty, Dale E.**, Papers, 1939-54, 8 feet. Assistant Secretary of the Interior, 1950-52; Commissioner, Federal Power Commission, 1952-54. Subject files relating to the Department of the Interior include folders on conservation; Conservation Conference (1947); Dinosaur National Monument-Echo Park; National Park Service Correspondence (3 folders), including files on Newton Drury and John Elliott; and the World Conservation Conference. There is also a reading file (1950-51, 4 boxes) relating to the Department of the Interior. Federal Power Commission files include a folder on Hells Canyon. Shelf list, 4 pp.

1813 **Gardner, Warner Winslow** (b. 1909), Papers, 1937-1947, 2 feet. Solicitor of the Department of the Interior, and Assistant Secretary, 1946-47. Conservation files relate to Alaska, the Columbia River dam, and the Jackson Hole area. Restricted.

1814 **Jackson, Charles W.**, Papers, 1946-52, 10 feet. Special Assistant in the White House Office. Included are several files relating to the U.S. Forest Service, and to forest fire prevention (chiefly 1949-52).

1815 **Murphy, Charles S.**, Papers, 1947-67, 18 feet. Special Counsel to the President, 1950-53; Under Secretary of Agriculture, 1960-65. Included are files relating to several aspects of natural resource conservation, including the proposed Columbia Valley Authority, the Missouri River Basin, Central Arizona Project, Olympic National Park, etc.

1816 **President's Materials Policy Commission**, Records, 1951-1952, 55 feet (Record Group 220). Drafts of the report to the president and other records of a commission appointed to study the problem of assuring an adequate supply of production materials for the nation's long-term needs. Included are questionnaires, correspondence, drafts of narrative and statistical reports and other working papers relating to forest resources and products, forest taxation, lumber operations, the domestic and world timber markets, etc. There are correspondence and proposals by the U.S. Forest Products Laboratory and the U.S. Forest Service. P.I., mimeographed, 1962, 16 pp.

1817 **President's Water Resources Policy Commission**, Records, 1950-1951, 25 feet. Correspondence, administrative files, conference proceedings, reports, memoranda, legal files, and consultants' studies. Subject files include material on forests. Files relating to a survey of federal agencies include correspondence and topical reports of the U.S. Forest Service.

1818 **Truman, Harry S** (1884-1973), Papers as President of the United States from the Central Files of the White House, 1948-1953, 2,688 feet. Correspondence, memoranda, speeches, notes, press releases, scrapbooks, clippings, and other papers. Included in the Official File are materials relating to the U.S. Forest Service, Superior National Forest, National Park Service, forests, forest conservation, game conservation, reclamation, lumber, the Civilian Conservation Corps, the White House Conference on Resources for the Future, flood control, Tongass National Forest, forest highways, aircraft in Superior National Forest, Shasta National Forest, national monuments, forestry in the Bureau of Mines, trees, Arbor Day, disasters, lumber, forest fires. Restrictions.

1819 **Truman, Harry S,** Post-Presidential Papers, 1953-1973, 29 feet. Included are files on forestry and A.F. Weston. Closed.

1820 **Wolfsohn, Joel D.** (1900-1961), Papers, 1926-1961, 29 feet. Correspondence, memoranda, speeches, reports, clippings, and other papers relating to Wolfsohn's service as Executive Secretary of the National Power Policy Committee, 1934-47, and in the Department of the Interior, 1941-53. Included are files relating to the National Power Policy Committee and General Land Office (1934-39), including reports on the Oregon and California revested lands (1 folder). Subject files pertaining to the General Land Office and Bureau of Land Management (1939-47) contain bills relating to forestry administration (1 folder), and material relating to national forest areas proposed for inclusion in national parks and monuments (1 folder). Files as Assistant to the Secretary of the Interior and Assistant Secretary of the Interior include material on the Oregon and California lands (1 folder), the National Commission for Conservation (1 folder), and other records relating to Departmental policies. Shelf list, 16 pp. Restricted.

KANSAS CITY

Federal Archives and Records Center

Chief, Archives Branch, 2306 East Bannister Road 64131 Telephone: (816) 926-7271. Some of the holdings are described in General Services Administration, National Archives and Records Service, Region 6, *Guide to the Records in the Federal Records Center,* compiled by Harry L. Weingart, 1966, processed, 46 pp. For records held in the Federal Records Center, address inquiries to Chief, Reference Service Branch.

Regional Archives Branch

Indian Affairs Bureau (Record Group 75)

1821 *Field Office Records, 1860-1866, 620 cubic feet.* These records were created mainly by area offices, agencies, and nonreservation schools within Minnesota and the Dakotas. The kinds of records maintained by the area offices, agencies, and schools varied little between jurisdictions, although there are great differences in the quantities that survived. For the most part they consist chiefly of correspondence and financial documents. Records for the periods indicated for the following area offices include materials documenting forests, forestry, and conservation: Aberdeen (1931-62), and Minneapolis (1887-1961). Area office records include administration decimal files, correspondence with agencies and schools, area office monthly narrative reports, agency annual reports, tribal enterprise correspondence, annual forest and grazing reports, general forestry correspondence, sawmill correspondence, blister rust reports, timber contracts, general forestry files, and paper company contracts. Among the Minneapolis Area Office records are records of the Menominee Tribe in Wisconsin; these include Menominee Indian Mills transfer documents; records of Court of Claims Case 44303 containing exhibits, study reports, surveys, and plans of Menominee Reservation and timber operations; and Menominee Garment Factory Enterprises records. Agency records of the periods indicated include forest, forestry, and conservation documentation: Cheyenne River (1862-1967); Consolidated Chippewa (1888-1963); Crow Creek (1875-1963); Fort Berthold (1864-1955); Fort Totten (1875-1949); Lower Brule (1875-1963); Minnesota (1896-1963); Pierre (1918-1963); Pine Ridge (1867-1961); Ponca (1860-1966); Red Lake (1894-1963); Rosebud (1860-1966); Sisseton (1875-1965); Standing Rock (1862-1967); Turtle Mountain (1884-1955); and Yankton (1860-1966). Records of Indian agencies include alphabetical subject files, annual forest and grazing reports, annual reports, correspondence, daily journals, decimal code files, fish and wildlife records, forestry and grazing correspondence, forestry and nursery records, forestry contracts, forestry officer's diaries, forestry records, Indian mills finance records and log scale books, indexes to land sales, and allotments and indexes, land entry records, land sales ledgers, land transaction records, permits, range unit records, records of timber cuts, statistical reports, timber cutting permits, timber sales records, timber surveys, township and county plats and maps, and tract books and journals. Included in these records are files from the Civilian Conservation Corps—Indian Division which contain administrative files, camp maps and plats, correspondence, contract reports, decimal files, land acquisition, photographs, project files, and weekly reports. Nonreservation schools having forestry-related records for the time periods indicated are Flandreau School (1898-1951) and Pierre School (1890-1951). These records consist of C.C.C.—Indian Division correspondence and subject files, decimal correspondence files, and general correspondence.

U.S. Attorneys and Marshals (Record Group 118)

1822 *Records of the White Earth Investigation, 1874-1921, 90 cubic feet.* These records were created by the Special Assistant Attorney General assigned to investigate allegations of fraud perpetrated against Chippewa Indians of the White Earth Reservation in Minnesota, and consist primarily of investigations into the unlawful sale of allotted lands by mixed-blood Indians to timber interests in Minnesota. Included are timber trespass dockets (1878-80); letters received by the U.S. Attorney (1874-78, 1889-1910); copies of letters sent by the U.S. Attorney (1884-85, 1887-1915); Attorneys General correspondence (1911-14); White Earth civil dockets (1910-17); White Earth equity case files (1910-20); allotment settlement folders (1918-22); reports of settlement (1916-20); Special Assistant Attorney General's subject file (1913-21); and Special Assistant Attorney General's correspondence (1914-15).

Federal Records Center

1823 **Forest Service** (Record Group 95), 1839-1963, 666 cubic feet. Administrative correspondence; accounting records; historical data; case files for grazing permits, land appraisal and acquisition, timber sales; research records, such as publications, manuscripts, and station papers; and other studies, reports, surveys, etc. Topics include growth and harvesting of timber, protection from fire, insects, and disease, rangeland management, economical utilization of forest products, research in forest economics and taxation, watershed management, and forestry survey of the United States. Included are records of national forests in South Dakota, Minnesota, Missouri, and Nebraska; also of the Lake States Forest Experiment Station.

1824 **National Park Service** (Record Group 79), 1834-1962, 192 cubic feet. Correspondence and subject files created by the Midwest Region and the parks and monuments under its jurisdiction; subjects include development and maintenance, forestry, history and archaeology, information, lands and planning, laws and legal matters, and natural sciences, etc.; there are also fiscal and accounting records, civil defense records, Operation Alert files, and correspondence relating to concession.

ST. LOUIS

National Personnel Records Center (Civilian Personnel Records)

Chief, Civilian Management and Technical Staff,
111Winnebago Street 63118

1825 Holdings consist of personnel folders or their equivalent of separated civilian federal employees deposited in the records center approximately thirty days after separation from service. Holdings for U.S. Forest Service personnel include records dating back to before 1900. Personnel folders are not always established for seasonal employees, pick-up firefighters, and those employed under informal appointments. Availability of access to these records is governed by U.S. Civil Service Commission regulations, which specify that information from official personnel folders may be made available to a person engaged in research for historical or educational purposes when the person has a researcher's pass to the administrative records of the agency from which the employee was separated. The folders of some of the more prominent men in the Forest Service such as Bernhard E. Fernow and Gifford Pinchot were retained in the National Archives.

Missouri Botanical Garden

Librarian, 2315 Tower Grove Avenue 63110
Telephone: (314) 865-0440

1826 **Engelmann, George** (1809-1884), Papers. Botanist. The papers include notes and sketches (1835-84, 58 vols.), on various genera or families of plants; of the notebooks, 12 are on conifers, 3 on oaks, and 3 on yucca, containing drawings, sketches, and notes, descriptions and correspondence. There are also several hundred letters to Engelmann from Asa Gray (1841-83), and several thousand letters from other botanists (1841-84), making frequent references to trees.

1827 **Schrenk, Hermann Von,** Letters, 158 items. Letters to Schrenk by various foresters, naturalists, and others.

Montana

BOZEMAN

Montana State University, Library

Special Collections Librarian 59715 Telephone: (406) 587-3121. Holdings are described in *Guide to Manuscripts in Montana Repositories,* compiled and edited by Brian Cockhill and Dale L. Johnson, Missoula: University of Montana Library, 1973.

1828 **Belgrade Company,** Records, 1906-1947, 11 feet. Gallatin Valley, Montana, general merchandise firm and lumber yard. Letter press books and sales slips. Lumber records are scattered throughout.

1829 **Cowan and Son** (Havre, Montana), Records, 1888-1966, 27 feet. Correspondence and business records of David Cowan, William T. Cowan, and William E. Cowan who maintained various businesses, including a lumber yard. There are fragments of lumber records pertaining to the Ft. Assiniboine Indian Reservation (1897-1901); letters of Henry Good of Kalispell relating to lumber (1935, 11 items); and other lumber records throughout.

1830 **D'Ewart, Wesley A.** (b. 1889), Papers, 1955-1968, 4 1/5 feet. U.S. Representative from Montana, 1945-54; served on the Public Lands, Irrigation and Reclamation, and Interior and Insular Affairs committees. These materials are filed by date and by committee. There is little unpublished material, mostly relating to land use, environment and water resources. Also a transcript of his reminiscences for the Columbia Oral History project.

1831 **Flanders, George W.,** Interview, 1966, 1 item. Typescript of reminiscences concerning Flanders' Mill in Hyalite Canyon, Montana, and George Flanders, Sr. Notes on a ninety-minute interview.

1832 **Hodson's Mill,** Records, 1881-1893, 1 foot, 10 vols. Gallatin Valley, Montana, lumber mill owned by Enoch Hodson. Financial records.

1833 **Langohr, Don, Sr.,** Interview, 1970, 1 reel (90 minutes). Bozeman, Montana, resident. Reminiscence about his father, Michael Langohr, as a U.S. Forest Service ranger.

1834 **Langohr, Michael,** Diaries, 1899-1914, 3 vols. U.S. Forest Service ranger, and supervisor of the Gallatin National Forest. Typescript copies of diaries relating to work in the Gallatin National Forest, the Crazy Mountains (1911), and the Upper Yellowstone Valley (1913-14).

1835 **National Forest Preservation Group,** Correspondence, 1970. Correspondence from the Forest Board of Appeals, with information about Big Sky of Montana and the land exchanges 2 and 3.

1836 **O'Neil, Charles I.** (b. 1869), Manuscript, 1 item. Experiences as a lumberman near Kalispell, Montana, 1894-1912. Thermofax duplicate of a rough draft (50 pp.).

1837 **Whipps, Samuel William Carvoso,** Manuscript, 1933, 1 item. Reminiscences relating to the history of Kalispell and the origin of Glacier National Park.

1838 **Wilson, Milburn Lincoln** (1885-1969), Papers, 1933-1960, 6 feet. Montana State University professor, Under Secretary, U.S. Department of Agriculture, 1933-1940, and national Director of Extension, 1940-1953. General correspondence, reports and speeches and writings.

1839 **Wylie, William Wallace,** Unpublished manuscript, 1926, 1 item. Yellowstone National Park concessionaire. History of the park.

1840 **Yellowstone National Park,** Collection, 1866-1961, 84 items, 2 feet. Diaries, reminiscences, and correspondence of explorers, administrators, and travelers. Xerox copies of originals in Yellowstone National Park Museum, Mammoth, Wyoming.

1841 **Yellowstone National Park,** Records, 1884-1924, microfilm (41 reels). Administrative records arranged chronologically. Includes information about wildlife and ecology.

HELENA

Montana Historical Society

Archivist, 225 North Roberts 59601 Telephone: (406) 449-2694

1842 **Ahern, George P.,** Scrapbook, 1882-1901, microfilm, partial reel. A conservationist, soldier stationed at Ft. Missoula and Ft. Custer, professor of military science at the College of Montana (Deer Lodge), 1892. The scrapbook contains correspondence, orders and clippings reflecting Ahern's activities at the college; conservation efforts, mapping the Rockies, military activities and his work while Director of the Philippine Forestry Bureau. A segment of the W. Bertsche Collection.

1843 **Berge, Ole,** Interview, 1970, 4 reels (70 minutes). Retired logger, sawyer, and lumber worker from the Hamilton area of the Bitterroot Valley, Montana. Reminiscences (1919-40) of logging practices, camp and labor conditions, the influence of the Industrial Workers of the World, and mill development.

1844 **Fullerton, Neil,** Collection, 1950s-1958, 6 feet. Retired forester and amateur local historian, Thompson Falls, Montana. The collection consists of historical research files on the history of northwestern Montana, including material on the early U.S. Forest Service. The files contain correspondence, notes, photographs, clippings, transcribed interviews, maps, and miscellany. Unprocessed.

1845 **Gebauer and Horton** (Helena, Montana), Records, 1880-1882, ½ foot. Financial records of a carpentry and construction firm.

1846 **Gillette, Warren Caleb** (1832-1912), Diary, 1870, 1 vol. Diary of experiences on the Washburn-Doane expedition to the Yellowstone Park area.

1847 **Hannon, Champ** (b. 1900), Interview, 1 reel (50 minutes). Retired U.S. Forest Service employee. Reminiscences pertaining to the arrival of his father, Thomas Benton Hannon, in the Bitterroot Valley in the 1880s and early day forestry practices in the area. General comment on the development of mills, logging industry, and conservation and forestry practices.

1848 **Hauser, Samuel Thomas** (1833-1914), Papers, 1832-1941, 29 1/10 feet. Governor of Montana Territory, 1885-87, and Helena businessman. The correspondence deals with Hauser's business activities, including railroad building, the development of electric power in Montana territory, and with politics. It reflects the utilization of timberland in Montana and northern Idaho in conjunction with early mining and smelting operations (1866-95), and Hauser's involvement in promotion and construction of the Northern Pacific and Montana branch lines, including material documenting the use of forest resources on the public domain and granted land. Unpublished guide.

1849 **Hedges Family,** Papers, 1828-1945, ca. 5 feet (1263 items). Correspondence, diaries, financial records, legal documents, and manuscripts of writings, chiefly of Cornelius Hedges (1831-1907), concerning his participation in the Washburn-Doane Expedition (1870) to the Yellowstone Park area, and other matters. Unpublished guide.

1850 **Hoglen, T.M.,** Diary, 1882, 1 vol. Journal of a trip to Yellowstone National Park from Billings, 1882.

1851 **Holter, Anton M.,** Records, ca. 1885-1930, 36 feet. Helena, Montana, pioneer lumberman and hardward wholesaler and retailer. Records of several of Holter's companies and business endeavors, including Holter Lumber Co. The records are largely financial, with some organizational materials and correspondence.

1852 **Holter, Norman B.** (1868-1957), Interviews, 1953, 1955, 2 cassettes (120 minutes). Helena, Montana, pioneer, and son of pioneer Montana businessman and capitalist, A.M. Holter. The taped interviews include discussions of A.M. Holter's career.

1853 **Howard, L.J.,** Papers, 1939-1940, 1/10 foot. U.S. Forest Service employee in the Helena National Forest and Lewis and Clark National Forest, Montana. The material includes one letter to Gifford Pinchot, and two from Pinchot to Howard relating to a proposed history of the Lewis and Clark National Forest, by Howard.

1854 **Langford, Nathaniel P.,** Papers, 1863-1933, 1/10 foot. Pioneer Montana miner, banker, and the first superintendent of Yellowstone National Park. Miscellaneous correspondence; diary (1868); notes and writings.

1855 **Langhor, Michael,** Diary, 1899-1906, 244 pp. Typescript copy. An early forest ranger and supervisor of the Gallatin National Forest. The handwritten original is held by the Gallatin National Forest office.

1856 **Largey, Patrick A.** (1836-1898), Records, 1867-1935, 215 feet. Butte, Montana. The collection consists of financial and operational records of firms owned by P.A. Largey and his son, including Largey Lumber Co., P.A. Largey and Co., Missoula Lumber Co., Beaverhead Lumber Co., Coeur d'Alene Lumber Co., Passmore Paper Co., and others.

1857 **McAdow and Dexter** (Andersonville, Montana), Records, 1881-1883, 1/10 foot. General correspondence, financial records, legal documents, and miscellaneous items, of a sawmill.

1858 **McDonald, Charles Haskin** (b. 1900), Interview, 1970, 1 reel (50 minutes). Retired U.S. Forest Ranger. Reminiscences concerning his career in Utah, Idaho, and Montana, from 1919 to the early 1960s. Biographical information, forestry practices, conservation, wilderness areas, the Montana Study, Lee Bass, and logging in the Stevensville, Montana, area.

1859 **MacKay, Edward** (b. 1892), Interview, 1970, 1 reel (50 minutes). Retired U.S. Forest ranger. Reminiscence concerning his career as a ranger in the Bitterroot Valley near Hamilton, Montana. General comment on railroad and mill development, logging practices, conservation and timber management, 1903-70.

1860 **May, George, and Co.** (Bear Gulch, Montana), Records, 1888-1890, 1 vol. Payroll ledger of a sawmill.

1861 **Metcalf, Lee** (b. 1911), Papers, 1953-1969, 297 feet. U.S. Representative from Montana, 1952-60, and Senator, 1960+. His committee assignments included both Senate and House committees on Interior and Insular Affairs. The collection includes congressional files including biographical material, incoming and outgoing correspondence, subject files, printed materials, campaign materials, speeches and writings, and clippings, created during 1953-69. These files contain material relating to the use and conservation of natural resources both in Montana and nationally. Guide available in the library. Usable only with permission of Senator Metcalf.

1862 **Miles, Daniel N.,** Reminiscence, 1962, 1 item. An employee and partner in the Wylie Camping Company, a concessionnaire in Yellowstone National Park. Miles' reminiscence discusses the companies operating in the park, methods of operations, and the relationships of the companies' operators, ca. 1903-20. Restricted.

1863 **Montana. Attorney General,** Public records, 1889-1969, 238 feet. As chief legal officer of the state, the Attorney General has been concerned with all complaints, investigations and litigations involving state-owned forest land. This service is documented in this record series, particularly after 1948. Included are docket registers (1893-1962); typescript opinions (1905-69); administrative correspondence files, docket and case files; subject files; complaints and requests for opinions; legal briefs and investigation reports. usable with permissions from the Montana Attorney General. Unprocessed.

1864 **Montana. Legislative Assembly,** Records, 1957+, Microfilm. Standing committee minute volumes, House and Senate, for each session, 1957 to the present. Included are Natural Resources and Fish and Game Committee minutes. This is a continuing program and each sesstions' minutes are filmed immediately upon completion of the business of that session.

1865 **Montana. Register of State Lands,** Records, 1891-1931, 34 feet. This office is responsible for administration of all state-owned lands, including state forest lands (in conjunction with the State Forestry Office) and any lease, sale or other use of these lands is reflected in this series. Included are general correspondence; applications to lease, buy or sell state lands; financial records; appraisal and description records; and glass plate negatives of survey maps.

1866 **Murphy, John T.** (1842-1914), Records, 1865-1934, 25 feet. Helena, Montana, banker, stockman and mining investor. Correspondence and business records, both personal and of several Murphy businesses. Included are materials reflecting Murphy's investment in mining and its attendant effects on surrounding timber and other resources. Murphy was also an investor in lumbering operations in Montana and northern Idaho, 1875-1905.

1867 **Power, Thomas Charles** (1839-1923), Collection, 1868-1950, 550 feet. Montana merchandise, transportation, mining, ranching, banking and real estate magnate, and U.S. Senator from Montana, 1890-95. Power's investments in transportation, marketing, banking, retailing, etc., were a major factor in development of the Northwest, including the lumber and timber operations of the area. Power's investments in lumbering enterprises included the Oregon and Montana Timber Company, Skykomish Lumber Co., Columbia Valley Lumber Co., etc. Unprocessed. Preliminary inventory.

1868 **Slack, George W.** (b. 1874), Papers, 1917-1919, 1/10 foot. Kalispell, Montana, area lumberman and a captain of engineers during World War I, operating military sawmills in France. The materials are relfective of Slack's military service and include biographical material, general correspondence, reports and orders, legal documents, and miscellany.

1869 **Slayton, Daniel Webster** (1862-1927), Papers, 1884-1918, 1/5 foot. Lavina, Montana, sheep rancher, logger. General correspondence, biographical material and miscellaneous items. Indexed to logging and lumber industry, Montana. Unpublished guide.

1870 **Yellowstone National Park,** Records, 1882-1934, 9 reels microfilm. Administrative records, largely of the period when the U.S. Army administered and patrolled the park, including general correspondence, financial record items, reports, and miscellany. The many problems of early administration are illustrated in the materials. The great bulk of the material falls in the period, 1882-1916. Loaned to Montana State University by Yellowstone National Park Museum.

1871 **Yellowstone Park Co.,** Records, 1886-1935, 3 vols. Cashbook for Child and Anceney (1910-15); laundry account ledger for Yellowstone Park Assoc. (1886-1925); journal for Yellowstone Park Camps Co. (1921-35).

MISSOULA

University of Montana

Archivist 59801 Telephone: (406) 243-0211

1872 **Anaconda Forest Products Co.** (Bonner, Montana), Financial records, 1896-1955, 300+ vols. Journals, ledgers, and other books of account; also the financial records of the mills at Bonner, Hamilton, Darby, and St. Regis, Montana; and Hope, Idaho. There are also financial records of the company stores at Hamilton and Darby.

1873 **Brown, Frank D.** (1845-1931), Papers, 1886-1935, 4 feet. Montana pioneer and Philisburg, Montana, miner. General correspondence (1892-1930); financial records (1887-1905); legal documents (1897-1923); reports; and miscellaneous items. Included is material on the logging and lumber industry, Montana.

1874 **Glacier National Park,** Records, 1907-1949, microfilm (2 reels). Correspondence, reports, and legislation dealing with the establishment and formation of the park. Includes extensive wildlife studies and surveys, and a short paper, "A History of Glacier National Park, 1670-1903," by Dr. Ralph L. Beals (1935).

1875 **Hash, Charles J.,** Diaries and correspondence, 1917-1919, 14 items. U.S. Forest Service employee.

1876 **Hayes, Ed. S.,** Diary, 1927, microfilm (1 reel). Missoula, Montana, logger.

1877 **Kirkwood, J.E.,** Reminiscence, 1 item. Typescript of "Notes on Observations in National Forests and Elsewhere during the summer of 1910."

1878 **McLeod, Charles Herbert** (1859-1946), Papers, 1880-1953, 122 feet. McLeod was manager of the Missoula Mercantile Company, Missoula, Montana, a corporation affiliated with Andrew B. Hammond's lumber interests on the Pacific Coast. There are materials pertaining to the Hammond Lumber Co. and the Largey Lumber Co.

1879 **McLeod, Walter H.** (1887-1963), Papers, 1937-1963, 53 feet. Included are materials pertaining to the Missoula Mercantile Co. and its affiliations in the West Coast lumber industry.

1880 **Montana Conservation Council,** Records, 1952-1964, 6 feet. General correspondence, minutes of meetings, scrapbooks, publications and miscellaneous items. Not processed.

1881 **Northern Montana Forestry Association,** Records, 1911-1963, 80 feet and 6 reels of microfilm. Records of a forest fire protection association in Flathead and Lincoln Counties, Montana, also called the Northern Montana Forest Protective Association. Included are correspondence, fire reports, financial records, and miscellaneous items.

1882 **O'Neil, Charles L.** (b. 1869), Papers, 1896-1905, ½ foot. Kalispell, Montana, lumberman. Letterpress book (1898-1904, 1 vol.) and financial records, including an estimate book (1896-97, 1 vol.), journal (1902-05, 1 vol.), and ledger (1901, 1 vol.).

1883 **Photograph Collection.** The archives department has about 500 pictures of the lumber industry in Montana, Idaho, and Washington. No index.

1884 **Rankin, John,** Records, 1888-1904, 2/5 foot. Missoula, Montana, lumberman and hotelkeeper. Financial records (1888-1904), and hotel register (1891-1904).

1885 **Ross, Kenneth Forbes,** Reminiscences. Manager of Anaconda Company Lumber Mill at Bonner Mountain, Montana.

1886 **Yellowstone Park Co.,** Records, 1892-1963, microfilm (34 reels). Concessionaires in Yellowstone National Park. Financial records. Includes records of Harry W. Child, Yellowstone Park Hotel Company, and other concessions under the control of the Yellowstone Park Co.

WEST GLACIER

Glacier National Park

Chief Park Naturalist 59936

1887 An extensive collection of miscellaneous items dealing with the various phases of park history, development, and use, dating from the 1880s to the present time. Uncataloged.

Nebraska

LINCOLN

Nebraska State Historical Society

Director, 1500 R Street 68508 Telephone: (402) 432-2793. Holdings are described in *A Guide to the Manuscript Division of the State Archives, Nebraska State Historical Society,* Bulletin No. 5, July, 1973, Lincoln: 1974.

Government Records

1888 **Nebraska. Board of Education Lands and Funds,** Records, 1871 + .

1889 **Nebraska. Centennial Commission,** Records. The commission promoted tree planting.

1890 **Nebraska. Office of the Governor,** Records. Included are State Parks Board, minutes (1921-29) and reports (1926-28); and Committee on Reforestation reports (1913-20).

1891 **Nebraska State Horticultural Society,** Records. Probably includes material relating to tree planting.

Manuscript Collections

1892 **Aldrich, Benton.** Letters and essays concerning trees on the Aldrich farm.

1893 **Allen, William Vincent** (1847-1924), Papers, 1889-1934, ca. 2 feet. U.S. Senator from Nebraska, 1893-1901; chairman of the Committee on Forest Reservations and Protection of Game and member of the committees on Agriculture and Forestry and Public Lands. The collection includes some correspondence relating to Allen's interest in forest conservation. Calendar of correspondents.

1894 **Beck, John V.** (1890-1969), Notebooks, etc., 1949-67. Conservationist.

1895 **Bruner Family,** Papers, 1869-1956, 4 cubic feet (ca. 1000 items). Included is correspondence between Lawrence and Hudson Bruner and the Forestry Division of the U.S. Department of Agriculture concerning tree planting on the Bruner brothers' Holt County plantation, said to be the first instance of tree planting in the Nebraska sand hills (1891-94, 2 folders); letters of Uriah Bruner, special U.S. land agent, discussing the conservation of forests and the Timber Culture Act (1883-93, 4 items).

1896 **Butler, Hugh Alfred** (1878-1954), Papers, 1941-1954, 195 cubic feet (ca. 150,000 items). U.S. Senator from Nebraska, 1941-54; member of committees on Agriculture and Forestry, Public Lands, and Interior and Insular Affairs. 13 boxes relate to service on these committees. There are several files concerning the supply of newsprint, wood pulp, and paper pulp. Interior and Insular Affairs files relate mainly to the subcommittee on territories, but do contain notes on the House hearing on the Upper Colorado River Storage Project (January 18-28, 1954). There are also legislative files, including folders on plywood and on the U.S. Forest Service nursery program (ca. 1949-50). Unpublished inventory.

1897 **Forests. Nebraska.** Miscellaneous letters pertaining to forests, shelterbelts, and tree planting in Nebraska.

1898 **Furnas, Robert Wilkinson** (1824-1905), Papers, 1844-1905, ca. 8000 items (also on microfilm, 13 rolls). Governor of Nebraska, 1872-74, and secretary of the State Board of Agriculture, 1870s-90s. The majority of the papers relate to Furnas' promotion of agriculture in Nebraska, including some mention of tree planting. There are a few letters (1885-86, 8 items) of Nathaniel H. Egleston to Furnas, relating to administrative problems Egleston faced as chief of the Division of Forestry of the U.S. Department of Agriculture, and to Senator George F. Edmunds' bill for a forest reserve in Montana. The collection is described in *Guide to the Microfilm Edition of the Robert W. Furnas Papers, 1844-1905,* edited by Douglas A. Bakken, Lincoln: Nebraska State Historical Society, 1966.

1899 **Harrison, Robert D.** (b. 1897), Papers, 1952-1958, 12 feet. U.S. Representative from Nebraska, 1951-1959; served on the Agriculture Committee. Correspondence, speeches, and other papers relating to patronage, and campaign material, but there is some material relevant to forest history.

1900 **Jenkins, Myron Bonnie, Sr.** (1884-1965), ca. 23 cubic feet (13,800 items). Includes correspondence and other records relating to Jenkins' participation in Work Projects Administration tree planting and the shelterbelt program (largely 1936-41). Inventory, 10 l.

1901 **Kuska, Val** (1886-1964), Papers, 597 feet. Colonizing agent and agricultural development agent for the Chicago, Burlington and Quincy Railroad; active in irrigation, public power and conservation endeavors.

1902 **Miscellaneous.** Manuscripts and other material stored in vertical files include "Forests and Forestry," a map of the shelterbelt area and pamphlets on forestry and the encouragement of tree planting; published matter, pamphlets, and reports on the Nebraska National Forest; Zaccheus Stratton papers (1885-97, 10 items) relating to tree planting, and Hitchcock and Lancaster Counties patent of Thomas Stratton for a tree planting machine (1887).

1903 **Morton, Julius Sterling** (1832-1902), Papers, 1844-1904, 125 feet (also on 78 rolls microfilm). U.S. Secretary of Agriculture, 1893-97. The papers contain correspondence, a diary (51 vols.), financial records, and scrapbooks. Most of the material pertaining to the Department of Agriculture relates to patronage and general administration. There are a few letters (ca. 1885-86) from Nathaniel H. Egleston of the Division of Forestry, seeking Morton's support in influencing congressional legislation for the regulation of the lumber industry and the practice of forest conservation. Other material (ca. 1883-87) relates to Morton's lumber business. There is also material pertaining to Arbor Day and tree planting. The collection is described in *Guide to the Microfilm Edition of* the J. Sterling Morton Papers, 1849-1902, edited by Douglas A. Bakken, Lincoln: Nebraska State Historical Society, 1967.

1904 **Nebraska. Executive Papers.** The governors' papers contain material on Arbor Day (1874-1912), such as requests for information, and some proclamations. Some of this material has been published in *Messages and Proclamations,* 3 vols.

1905 **Past, John Comly,** Papers, 1836-1894, ca. 60 items. Includes business papers, mostly receipts and vouchers (1871-80, 1 folder) of the Past and Marsh Steam Saw Mill and Lumber Yard, Beatrice, Nebraska.

University of Nebraska, University Archives

University Archivist, The University Libraries 68508
Telephone: (402) 472-2526

1906 **Bessey, Charles Edwin** (1845-1915), Papers, 1870-1915, 45 feet. Professor of Botany, University of Nebraska. Correspondence, notes, speeches, and articles, relating to botany and other scientific interests, including conservation, forestry, and agricultural education. Correspondents include Bernhard E. Fernow, and Louis H. Pammel. Index of correspondents.

1907 **College of Agriculture and Agricultural Experiment Station,** Records, 1883-1974, ca. 118 cubic feet. College deans' files, farmers' institutes letterbooks, agricultural experiment station records and publications, and other papers, including some materials relating to forestry.

1908 **College of Agriculture, Department of Forestry,** Records, 1907-1915, 1¼ feet. Forestry Club Program (1907) and Annual (1909-15, 6 vols.); Department of Forestry Chairman's Correspondence (1912-15, ca. 1 foot), consisting of correspondence of Olenus Lee Spensler (1912) and Walter Jean Morrill (1914-15) with other University officials concerning department programs, academic policies, and curriculum, and with the U.S. Forest Service regarding employment of students, acquisition of wood samples, and with community organizations. Included are letters regarding the Forestry and Park Association (1912).

Nevada

RENO

Nevada Historical Society

Curator of Manuscripts, P.O. Box 1129 89504 1650 North Virginia Street Telephone: (702) 784-6397. The collections, including several items relating to the lumber industry but not listed here, are described in *A Guide to the Manuscript Collections at the Nevada Historical Society,* by L. James Higgins, 1975, 298 pp., processed.

1909 **Carson & Tahoe Lumber and Fluming Company,** Record book, 1890-1896, 1 vol. (96 pp.). Time book for the company's subsidiary Lake Valley Railroad, and the company's Lake Tahoe steamers.

1910 **Clover Valley Lumber Company,** Right-of-way indentures, 1927, 1938-1939, 4 items. Plumas County, California.

1911 **Haynie, J.W., and Company** (Carson City, Nevada), Records, 1872-1880, 1 vol. Accounts of a lumber and wood company.

1912 **Kappler, Charles J.,** Correspondence, 1 box. Secretary of Senator William Morris Stewart. Kappler conducted some work for Stewart with D.L. Bliss at Lake Tahoe, regarding a proposed Lake Tahoe National Park.

1913 **Pease, Robert G.,** Manuscript, 1973. "The History of the Crystal Peak and Verdi Lumber Industries."

1914 **Raymond, H.C.,** Letters, 1875-1877, 1 folder. Letters concerning the lumber business at Boca, California.

1915 **Reno Mill and Lumber Company** (Reno, Nevada), Records, 1901-1918, 1 box and 11 vols.

1916 **Smith, Ida Sauer,** Tape recording, 1957, 2 reels. Interview by Clara S. Beatty, partly relating to lumbering in Washoe County, Nevada.

1917 **Snell Lumber Company** (Reno, Nevada), Letter, August 17, 1917, 1 item.

1918 **Stewart, William Morris** (1827-1909), Papers, 1886-1904, 20 boxes. U.S. Senator from Nevada, 1864-1875; 1887-1905. There is some data in the collection relating to Stewart's efforts to have Lake Tahoe established as a national park.

1919 **Thompson, A.B.,** Correspondence, 1876-1878, 1 folder. Correspondence concerning lumbering, lawsuits, and water rights.

1920 **Townley, John Mark,** Collection. Contains a study by Townley for the U.S. Forest Service on Historical and Architectural Research, South Lake Tahoe Estates (1974), with an extensive bibliography, and notes on the study.

1921 **Whitney, George A.,** Letters, 1862-1887, 2 folders. Includes items on lumbering in Esmeralda County, Nevada.

University of Nevada, The University Library

Special Collections Librarian 89507

1922 **Carson & Tahoe Lumber & Fluming Company,** Records, 1874-1936, 67 vols., 5 boxes, and 8 cartons. Correspondence (1900-36); cashbooks, journals, ledgers, and daybooks (1874-1930); lumber shipment books (1888-90); time book (1877); and a large group of miscellaneous checks and vouchers covering the entire period.

1923 **El Dorado Wood & Flume Company,** Records, 1875-1935, 5 vols., 1 box and 1 carton. Letterpress copy books of letters sent; legal and accounting records.

1924 **Huff, John.** Contract to chop wood, Austin, Nevada, October 10, 1865, 2 pp. This contract is for John Huff and Wilson Gale to chop wood at Harker's Ranch, Reed Canyon, for C.B. Raymond, agent for the Eureka Steam Quartz Mill at Canyon City.

1925 **Knapp & Laws** (Hawthorne, Nevada), Correspondence, 1884-1887, 97 items. Knapp & Laws had extensive interests in mining, railroad construction, lumbering, etc. The majority of the letters were written by Sewell Alvin Knapp, Jr., to his father and to his brother, A.E. (Eugene), with a few letters to his brothers Ed and Will, and chiefly discussing mining and lumbering interests.

1926 **Markleeville Flume Company** (Markleeville, California), Records, 1873-1876, 1 box. Organized to supply wood for Nevada mines. Officers included H.M. Yerington, George R. Foard, and Christopher Spratt. The collection includes organization papers, agreements, notes, etc. (1873); and miscellaneous bills, receipts, and records (1874-76).

1927 **Pacific Wood, Lumber and Flume Company,** Records, 1874-1880, 9 vols. Correspondence (1875-89); wood and lumber ledgers and account books (1874-80); a time book (1875-76); and a volume listing real estate holdings of the company (1876).

1928 **Southern Development Company of Nevada** (Carson City, Nevada), Correspondence and Records, 1884-1919, 5 vols., 1 box and 1 carton. The Southern Development Company was a consolidation of the Walker Lake Wood and Lumber Company and other firms, and was one of the companies associated with H.M. Yerington. The collection consists largely of general business correspondence, with a few financial and legal records.

New Hampshire

CANDIA

*Smyth Public Library

Box 306, R.F.D. 1, Manchester 03104

1929 According to the Historical Records Survey, *Guide to Depositories of Manuscript Collections in the United States: New Hampshire (Preliminary Edition)*, Manchester: 1940, p. 4, a collection relating to a sawmill and a hat factory was deposited in Smyth Public Library in 1937. Included were daybooks (1833-83, 10 vols.), ledgers (1837-57, 3 vols.), and a fee book (1830-57, 1 vol.).

CONCORD

New Hampshire Historical Society

Manuscript Librarian, 30 Park Street 03301 Telephone: (603) 225-3381

1930 **Chandler, William Eaton** (1838-1917), Papers, 1829-1917, 22 feet (ca. 25,000 items in 58 containers). Journals and diaries, letterbooks, correspondence, miscellaneous material. There are scattered materials relating to state and national parks and forests.

1931 **Foster, John Harold** (1880-1956), Papers, 1000 items. New Hampshire State Forester.

1932 **Gallinger, Jacob H.** (1837-1918), Papers, 1862-1919, 4 boxes (1000 items), 12 vols., 1 brief case. U.S. Representative, 1885-89; and Senator, 1891-1918, from New Hampshire; member and chairman of the National Forest Reservation Commission. Included are scattered materials on state and national parks and forests. Inventory.

1933 **Hill, Ivory B.** (1833-1906), Family papers, 1833-1890, ca. 175 items. Shoemaker and lumber businessman, of New Hampshire, chiefly at Sandwich, near Tamworth. Included are a diary (1857), records of Hill's lumber business; accounts (1880-85) of Hill & Wardell (Ivory B. Hill and Phineas Wardell), manufacturers and dealers in lumber; etc. Inventory.

1934 **Moses, George Higgins** (1869-1944), Correspondence, 1888-1944, 4 boxes (ca. 1,800 items). U.S. Senator, 1918-33. The collection contains scattered materials relating to state and national parks and forests. Inventory.

1935 **Nims, Brigham** (b. 1811), Papers, 1763-1912, 2 feet (ca. 400 items). Diaries (1840-88, 1834-59) of Nims and his brother, Kendall Nims, give details of the operations of a sawmill. Roxbury, New Hampshire.

1936 **Partridge, William, and Benjamin Jackson.** Report to the Lords Commissioners for Trade and Plantations on the state of His Majesty's provinces of Massachusetts Bay and New Hampshire (1699).

1937 **Wentworth, John** (1737-1820), Letters, 1767-1778, 3 vols. (ca. 650 pp.). Royal Governor of New Hampshire, 1767-75; later governor of Nova Scotia. Included are letters relating to lumbering. Photocopies of originals in the Public Archives of Nova Scotia, Halifax.

HANOVER

Dartmouth College Library

Archivist 03755 Archives Department, Baker Memorial Library

1938 **Adams, Sherman,** Papers. Files relating to forestry include American Pulpwood Association, Committee on Imports, with letters and other papers; correspondence relating to the Marcalus Manufacturing Company, Lincoln, New Hampshire, revealing the fortunes of the company, and some of Adams' efforts on its behalf (1949-50); correspondence relating to the American Pulpwood Association (1950-51, 1 folder); Northeast Forest Fire Protection Compact (1951-52, 1 folder); "Forest Policy for New Hampshire, Report of Governor's Forest Policy Committee" (December, 1952; 1 folder). Restricted: permission for access must be obtained from Adams, Lincoln, New Hampshire.

1939 **Bass, Robert Perkins** (1873-1960), Papers, 1900-1950, 108 feet. Correspondence, speeches, personal, business and legal papers, inventories, clippings, scrapbooks, and other papers, reflecting Bass' activities in conservation, government, and politics. Includes material relating to his term in the legislature, as forestry commissioner, state senator, and Governor of New Hampshire (1911-13), including his reforms affecting conservation, and other matters. Inventory. Restrictions.

1940 **Dartmouth College Forester,** Papers. This collection is in the process of being accessioned into the Archives (1975).

1941 **Drew, Irving Webster,** Papers, 1898-1903, 1 box, 4 vols. The papers which are concerned with the case of Dartmouth College vs. International Paper Company include the brief of the defendant's argument, memoranda for the brief's preparation, and correspondence (1899-1903). Two of the volumes contain abstracts of titles to wild lands of the Connecticut River Lumber Company (1898).

1942 **MacKaye, Benton** (1879-1975), Papers. Employee of U.S. Forest Service (1905-18) and U.S. Department of Labor (1918-19). Correspondence, clippings, pamphlets, reprints and articles concerning MacKaye's activities in conservation politics, forestry and related topics. Includes material related to his proposal for the Appalachian Trail (1921) as well as his interest in conservation, forestry geotechnics, highways, New England, wilderness, water policy and his seat on the regional planning staff of Tennessee Valley Authority. Many articles written by MacKaye and others as well as complete manuscript of his book *Geotechnics in North America.*

1943 **Morrill, Folsom, and Joseph Thompson,** Accounts, 1849-1869, 1 envelope. The accounts are of lumber operations with shipbuilders of Portsmouth and Boston, including Donald McKay.

1944 **Weeks, John Wingate** (1860-1926), Papers, ca. 1877-1928, 20 boxes, 2 portfolios, and 2 scrapbooks. U.S. Representative from Massachusetts, 1905-13; U.S. Senator, 1913-19; Secretary of War, 1921-25; served on the Senate Conservation of Natural Resources Committee and Forest Reserves and

Protection of Game Committee; and the House Agriculture Committee. Papers contain general correspondence, arranged alphabetically by years (1877-1926, 5 boxes); correspondence arranged by subject (4 boxes), including a file, chiefly carbons of outgoing letters, but also incoming letters, on the Weeks Act of 1911 (1907-13, ca. 150 items), much of it relating to the protection of the White Mountain area. There are also miscellaneous speeches, clippings, scrapbooks, printed matter, etc. Inventory, 14 pp.

1945 **Winnipiseogee Paper Company,** Typescript, 1892-1894, 2 vols. The typescript contains testimony, arguments, findings, etc., in the case of Winnipiseogee Paper Company vs. New Hampshire Land Company.

New Jersey

BAYONNE

Federal Archives and Records Center

Chief, Archives Branch, Building 22—Military Ocean Terminal, Bayonne 07002 Telephone: (201) 858-7164. Microfilm holdings are described in *National Archives Microfilm Resources in the Archives Branch,* New York: General Services Administration, National Archives and Records Service, 1973, processed, 15 pp.

Federal Records Center

1946 **Forest Service** (Record Group 95), 125 cubic feet. General files from Forest Service offices in Albany, New York; Saratoga Springs, New York; and Rio Piedras, Puerto Rico; largely concerning administrative matters. Permission of the Forest Service is required for access.

CAMDEN

Camden County Historical Society, Library

Librarian, Park Boulevard and Euclid Avenue 08103 Telephone: (609) 964-3333

1947 **Mickle, Capt. John W.,** Ledgers, 1832-1838, 2 vols. Camden, New Jersey, lumber merchant.

MORRISTOWN

The Joint Free Public Library of Morristown and Morris Township

Director, 1 Miller Road 07960 Telephone (201) 538-6161

1948 **Young, Stephen,** Diaries, 1793-1867, 5 vols. Describe the daily life of a local carpenter and sawmill owner.

Morristown National Historical Park

Park Historian, P.O. Box 1135R 07960. The holdings are described in *A Guide to the Manuscript Collection, Morristown National Historical Park,* compiled by Bruce W. Stewart and Hans Mayer, Morristown: n.d., processed.

1949 **Cobb Collection,** ca. 25 boxes. The collection relates to the economic and legal history of post-Revolutionary and early 19th century Morris County, New Jersey, particularly to the iron industry in western New Jersey, including limited material relating to land and forest use, as reflected by the Morris County iron industry, land surveys, and other business records.

NEW BRUNSWICK

*Rutgers University, University Library

Holdings are described in *A Guide to the Manuscript Collections of the Rutgers University Library,* compiled by Herbert F. Smith, New Brunswick: 1964. The following list was drawn from that source, and from the *National Union Catalog of Manuscript Collections.*

1950 **Buttler Company** (New Brunswick, New Jersey), Records, 1852-1917, 19 vols. and 1 folder. Accounts, estimates, orders, of sash and blind factory. Includes a ledger and daybook (1852-55); estimates and orders, with specifications, sketches, and prices for millwork (1858-65, 1880-81, 1904-11, 1917); stock accounts and time books. The name of the company was Brokaw & Buttler, 1852-59; Buttler Johnson, until the middle 1880s; A.J. Buttler & Co.

1951 **Card, Theodore,** 1886-1910, 1 item. Accounts of an uncertain character, including references to posts and rails.

1952 **Connet, S.E., and Son** (Waterville [Brookside], New Jersey), Records, 1853-1924, 20 vols. Sawmill and lumberyard accounts; blotters (1853-1906, incomplete); ledger (1890-1924), daybook (1894-1915); labor cost ledgers (1856-94). The business was continued by Connet's son, Earl F. Connet, after 1877.

1953 **Cooper, Leonard,** Papers, 1823-1845, 1 vol. Montville, Maine. Sawmill ledger (1823-32); miscellaneous accounts (1844-45).

1954 **Crane, John,** Account book 1822-32, 1 vol. Rathway, New Jersey. Accounts for saw and grist mill and general store belonging to the Crane family.

1955 **Dumont, Peter,** Journal, 1768-1788, 1 vol. South Branch, New Jersey. Accounts for general store, sawmill, and gristmill.

1956 **Ferris, Apollo,** Accounts, 1848-76, 1 vol. Gristmill and sawmill daybook, Pompton, New Jersey (1848-49), of the partnership of Ferris and Pewtner. Pompton is now part of the borough of Riverdale.

1957 **Grubbs, Peter,** Accounts, 1745-1753, 1 item. Merchant shipping accounts; journal (1749-50/51), dealing with the operations of the shallop *Cornwall,* carrying shingles, lumber, grain, etc., to and from points in south Jersey and elsewhere.

1958 **Herbert Family,** Papers, ca. 1800-ca. 1900, 31 vols. and 6 boxes. Papers of the Herbert family of New Brunswick, New Jersey, including the account books and other papers relating to wood, land, and other businesses of Obadiah Herbert (1775-1856).

1959 **Iszard, Joseph** (1798-1865), Papers, 1812-1864, 6 vols. and 2 folders. Glassboro, New Jersey, and vicinity, and Burlington County, New Jersey. Daybooks (1833-40, 1858-64) include entries for sawmill products; sawing accounts (1835-58).

1960 **Marcy, Matthew,** Accounts, 1805-1875, 7 vols. Cape May County and Buckshutem, New Jersey. Accounts of general store; dealings in sawmill and gristmill products, cordwood (1826-75).

1961 **Mitchell, George,** Letters, 1785-1790, 6 items. Daggsbury, (Dagsboro?), Delaware. Letters received from Lambert Cadwalader concern manufacture, disposal, shipping, and storing of shingles during Mitchell's term as a member of Congress.

1962 **Smock, John Conover,** Papers, ca. 1899-1902, 6 items. Monographs on relation of forestry to public health, forest reservations in the pine district of New Jersey, geology, etc.

1963 **Wholesaler of Naval Stores,** Ledger, 1829-1834, 1 vol. New York.

NEWARK

The New Jersey Historical Society

Keeper of Manuscripts, 230 Broadway 07104 Telephone: (201)483-3939

1964 **Connet, Earl Fairchild,** Record books, 1862-1891, 12 vols. Records of the business transactions of the Byram sawmill at Mendham, Morris County, New Jersey, of which E.F. Connet was a co-owner.

1965 **Connet, Stephen E.,** Record books, 1839-1889, 6 vols. A record of goods sold and payments received by S.E. Connet at the Byram sawmill of Mendham, Morris County, New Jersey.

1966 **Northern Liberties Company,** Record book, 1829-1830, 1 vol. Principally a record of the Northern Liberties Company's lumber products sold and prices received. About 1/3 of the volume has been used as a scrapbook.

SALEM

*Salem County Historical Society

79-83 Market Street 08079. The following entries were reported in the Eleutherian Mills Historical Society's unpublished "Union Catalog of Business and Economic History Manuscripts."

1967 **Ballinger, John G.,** Accounts, 1840-1849, 1 vol. Allowaytown, Salem County, New Jersey.

1968 **Salem County, New Jersey, Lumber Yard,** Accounts, 1798, 1 vol. Accounts of an unidentified lumber yard.

TRENTON

State Library

185 West State Street, P.O. Box 1898 08625

1969 **Batsto Works Estate,** Accounts, 1830-1859, 2 vols. Washington Township, Burlington County, New Jersey. Lumbering.

1970 **Miscellaneous Manuscripts.** Included are a letter to Samuel Smith, Burlington, New Jersey, from William Foster protesting the waste of timber on Indian land going for rum and strong drink (n.d.); letter (1749) from Mr. Hackett stating that he is prevented from cutting timber for a slitting mill until the right is deeded in England; account of timber cut in certain sawmills, and a description of the floating of lumber down the Delaware River (n.d.); petition (April 10, 1746) to the General Assembly from the inhabitants of New Jersey, especially those of Essex County, requesting the repeal of the act for preserving the timber in the Province of New Jersey; petition (1801) to the New Jersey Legislature from the inhabitants of Lamberton on the Delaware discussing regulation of the floating of logs, timber, and lumber down the Delaware River.

1971 **New Jersey. Department of Defense Records.** Revolutionary War documents relating to requests for lumber to erect shelters, etc.

New Mexico

ALBUQUERQUE

University of New Mexico, General Library, Special Collections Department

Director, Special Collections Department 87131 Telephone: (505) 277-4241. Some of the holdings are described in Albert James Diaz, *Manuscripts and Records in the University of New Mexico Library*, Albuquerque: University of New Mexico Library, 1957.

1972 **Blazer's Mill** (Mescalero, New Mexico), Records, 1871-1896, 10 vols. Time books, blotters, salesbooks, gristmill accounts, journals, and other records of a sawmill operated by Joseph H. Blazer (1828-98) and A.N. Blazer.

1973 **Breece Lumber and Supply Company** (Albuquerque, New Mexico), Records, ca. 1890-1925, 1 box (41 packets). Contracts, state patents, deeds, abstracts, certificates of land titles, contracts with the Santa Fe Railroad, minute books, and other documents.

1974 **Calvin, Ross** (1889-1970), Papers, 1799-1969, 7 feet (4600 items). Clergyman and author. The papers include a chronological record (8 vols.) of observations, field trips, travels, climate, botany and wildlife in Illinois, New England, Pennsylvania, and the Southwest, especially New Mexico; manuscripts; files including field notes and newspaper clippings related to such subjects as Indians, old-timers, sheep, and forests. Unpublished index.

1975 **Catron, Thomas Benton** (1840-1921), Papers, ca. 1878-1915, 259 boxes, 105 vols. and 7 file drawers. Delegate and U.S. Senator from New Mexico, 1912-17. Papers include Southwestern Lumber and Railway Company minute books (1875-1901, 3 vols.).

1976 **Fall, Albert Bacon** (1861-1944), Papers, 1916-1927, 10 file boxes, 6 scrapbooks, and 42 reels of microfilm. U.S. Senator from New Mexico, 1912-21; Secretary of the Interior, 1921-23. The original material consists of correspondence, Senate office files, Department of the Interior papers, and other materials relating to oil policy and scandals, Fall's Senate activities, his Tres Rios Ranch, and his activities as Secretary of the Interior. These papers (but not the microfilm) are described in *List of the Albert B. Fall Papers,* by Barbara Anthes (Albuquerque: University of New Mexico, 1957), which is out of print (1973), but available in photocopy. The originals of the microfilmed material are in the Huntington Library, San Marino, California.

1977 **U.S. Soil Conservation Service,** Reports relating to Arizona and New Mexico, ca. 1934-1946, 23 feet (ca. 1000 items). Soil and conservation surveys, range and erosion condition reports, range management plans, ground water reports, land management reports, maps, charts, and other material relating to areas in the Southwest, particularly Arizona and New Mexico.

New York

ALBANY

The New York State Library

Librarian, Manuscripts and History Library 12224

1978 **Gardiner, James Terry** (1842-1912), Papers, 1791-1886, 338 items. Correspondence, business papers relating mainly to land in Petersburg and Troy, New York, and LaGrange, Illinois, and papers dealing with state surveys and state parks. There is additional Gardiner material (ca. 850 items) in the William Croswell Doane Papers, mostly published writings and clippings, but including some photographs of Yosemite Valley, California. Unpublished indices and finding aids.

1979 **Hough, Franklin Benjamin** (1822-1885), Collection (ca. 1800-1883), ca. 72 vols., 18 packages, and 40 boxes. Hough was employed by the U.S. Department of Agriculture to compile the *Report Upon Forestry,* 1876-1881; Chief of the Division of Forestry, 1881-1883. The collection contains originals of many of Hough's writings on forestry and other subjects; diary (1863-83, with gaps, 17 vols.); typed transcripts of diary (1 folder); letters to Hough (1846-53, 5 vols.); letters by Hough (1874-83, 1 vol.); miscellaneous papers relating to forestry, etc. (ca. 150-200 items), including a folder of material relating to a speech on forestry by Mark Dunnell, U.S. Representative from Minnesota, before the House of Representatives, March 9, 1882. There is also material relating to Hough's interest in the University of the State of New York, and in local history.

1980 **Law, Samuel A.,** Papers, 1795-1887, 10,000 items. Records of a land agent, farmer, and businessman in Delaware County, New York. Included are statements of accounts and receipts for payments made to Law by persons to whom he sold the lands in Franklin Patent; maps of lands in Franklin Patent showing surveys, lot divisions, roads, etc. (35 items); correspondence and daily journals, containing data on Law's lumbering, farming, and other activities.

1981 **New York. Conservation Commission,** Records, 1886-1921, 1721 vols. Assessment rolls of towns in which the state owns land, including data on wild or forest lands.

1982 **New York. Governor,** Records, 1919-1920, 1923-1928. Correspondence, reports, and speeches relating to Governor Alfred E. Smith's activities, including the field of conservation. Unpublished finding aid.

1983 **New York State Association for the Protection of Game,** Papers, 1867-1933. Minutes, clippings, financial records, invitations, pamphlets, reports and correspondence. The Association campaigned for forest preservation. Unpublished finding aid.

1984 **Redington, George** (1810-1888), Papers, 1810-1885. Manager of the Ogden Plantation, St. Lawrence County, New York, an attorney and assemblyman, involved in logging and the lumber trade. Papers include correspondence, accounts, legal papers, receipts. Unpublished finding aid.

1985 **Van Bergen, Martin Gerritsen,** Papers, 1734-1802, 1 box (225 items). Correspondence, accounts, and receipts of an Albany, New York, merchant and ship owner, who shipped pipe, barrel staves, mast timbers, and potash from Albany to New York City.

1986 **Weeks, Walter N., and Company** (Whitehall, New York), 1912-24, 23 boxes. Business records of Burleigh and Weeks and W.N. Weeks, firms buying pulpwood in Canada and selling it to pulp and paper companies in New York. The records contain correspondence, bills of lading and accounts. Unpublished finding aid.

BLUE MOUNTAIN LAKE

The Adirondack Museum

Director 12812

Manuscripts

1987 **Adirondack Land and Lumber Company,** Certificate, July, 1901.

1988 **Association for the Protection of the Adirondacks,** Records, 1902-1964, 9 reels of microfilm. Minutes of the meetings and documents.

1989 **Borton, David.** "Adirondack Lumbering: The Jessup Operation." 13 pp., maps. Also includes references to logging in southeastern Alaska.

1990 **Brown, Ralph Adams.** "The Lumber Industry in the State of New York, 1790-1830," New York: Columbia University, M.A. Thesis, 1933, 135 pp. (on microfilm).

1991 **Childwood Park Hotel Company** (Massawepie, New York), Papers, 1892-1909, 97 pieces. Miscellaneous records of a tourist hotel in the Adirondacks. Some early letter copybooks indicate the Childwood Company also engaged in a logging business.

1992 **Elliott, Louise.** "The Lumber Industry in New York State and Its Relationship to the Development of the Adirondack Forest Region," Genesco, New York: Genesco State College, student paper, 1968, 23 pp.

1993 **Emporium Forestry Company** (Cranberry Lake, New York), Records, 1912-1971, 266 boxes + maps. Records chiefly relating to the New York state operations of a lumber company which had holdings in Pennsylvania, New York, Vermont, and Tennessee. Financial, legal, and business records relating to lumbering, timberlands, and wood products. Register, 39 pp., typed.

1994 **Kellog, William N.,** Diaries, 1911-1922, 9 pieces. Relate to lumbering.

1995 **Meigs, Ferris J.** "The Santa Clara Lumber Company, 1888-1941," 2 vols. (vol. I, 1888-1912; vol. II, 1912-38) (on microfilm, 1 reel). Includes photographs.

1996 **Reed, Frank,** Narration, September 11, 1963, 13 pp. Narration for the logging films in the museum.

1997 **Shepard, Edward M.,** Letters, 1904, 2 pieces. Correspondence between Shepard, former Forestry Commissioner of New York, and Professor N.L. Britton concerning a Cornell University Forestry School project and forestry in New York state.

1998 **Snyder** Papers. Papers of Charles Snyder (1863-1941) and David Snyder (1890-1964), Herkimer, New York, attorneys. The papers contain case files in the Adirondack area, mostly concerning lumbering rights, infringements on state property, land titles, etc. Typed container list.

1999 **Wolf, Robert E.** "A Partial Survey of Forest Legislation in New York State," Syracuse: New York State College of Forestry, thesis, 1948, 353 pp. (on microfilm).

BRONX

New York Botanical Garden, Library

Reference Librarian, 200 Street & Webster Avenue 10458.

Holdings are described in New York Botanical Garden Library, *Catalog of the Manuscripts and Archival Collections, and*

Index to the Correspondence of John Torrey, compiled by Sara Lenley, et al., Boston: G.K. Hall & Co., 1973.

2000 Holdings (ca. 220,000 items) include letters, research notes, and manuscripts of most of the scientists associated with the garden since its foundation. There are also papers of other persons prominent in the fields of botany, horticulture, and natural history. The primary emphasis of the collection relates to major 20th-century developments in these fields. Among the items relevant to forest history are William Richard Barbour (b. 1890) letters (1933-34, 5 pp.) to N.L. Britton, discussing Luquillo Forest in Puerto Rico; Nathaniel Lord Britton (1859-1934) addresses, agreement, records, lists (1895-1933, 22 l.) on the Scientific Alliance, the forests of Puerto Rico, and on nomenclature; Ephraim Porter Felt (1868-1943) letters (1919-40, 59 items) to Fred J. Seaver, Bernard Ogilvie Dodge, and others, on tree diseases and insect pests; Clayton Dissinger Mell (1875-1945) notebook (1904, 208 l.) containing information on tree seeds and tree planting for forestry and lumbering purposes; John Kunkel Small (1869-1938) manuscript (157 l.) on the pine and yew families, and junipers; Norman Taylor (1883-1967) papers (1640-1933, 513 items) containing notes on forests, bogs, grasslands, and marshes of Long Island, New York, and correspondence concerning state parks in the area; John Torrey (1796-1873) letters (ca. 200 items, on permanent loan from Columbia University) from most outstanding 19th-century botanists, including George Engelmann, Asa Gray, and others; and R.L. Wegel manuscripts (1937, 148 l.) on tree-ring dating of chestnut trees; Harry Nichols Whitford (1872-1941) notebooks (1900-19, 31 vols.) on his forestry work in Chicago, the Philippines, Canada, Central America, and South America.

CALEDONIA

*Caledonia Library Association

14423. The following collection was listed in the Historical Records Survey, *Guide to Depositories of Manuscript Collections in New York State (Exclusive of New York City),* Albany: 1941, p. 76.

2001 **Garbutt, P.,** Business records, 1843-1859, 7 vols. Mumford, New York. Includes daybooks, ledgers, cashbooks, containing sawmill accounts, expenditures for lumber and farm products, etc.

COOPERSTOWN

New York State Historical Association

Curator of New York History 13326
Telephone: (607)547-2533

2002 **Bascom and Gaylord,** Collection, 1838-1880, 72 vols. and 1 box of miscellaneous materials. Ledgers, daybooks, blotters, journals, shipping and invoice books, orders, notes, drafts, bills, books of accounts, and miscellaneous records of mercantile, shipping and lumber concerns of Whitehall, New York. Names of firms prominent in this collection include Bascom, Eddy & Gaylord; Bascom and Gaylord; Bascom, Vaughn & Co.; Eddy and Chubb; Eddy, Bascom & Co.; Eddy & Co.; Eddy, Rice and Co.

2003 **Cabinet Makers,** Account books. The supply and cost of wood is occasionally mentioned in the following account books of cabinet makers, cataloged separately in the repository: Miles Benjamin, ledger of a cabinet maker of Cooperstown, Otsego County, New York (1821-28, 1 vol.); Ezekial Bennet, daybook of a carpenter and cabinet maker, Weston, Connecticut, and West Laurens, Otsego County, New York (1794-98, 1802-12, 1 vol.); William Bentley, ledger of a cabinet maker of Westfort, Otsego County, New York (1813-44, 1 vol.); John Brewer, daybooks of a carpenter and wagon maker of Cooperstown (1833-49, 2 vols.); Joseph Brownell, ledger of a carpenter, wagon maker, woodcutter, etc., Otsego County (1821-50, 1 vol.); Mortimer Mapes, ledger of a carpenter of Florida, New York (1860-68, 1 vol.); Robert C. Scadin, ledger and daybook of a cabinet maker of Cooperstown (1829-31, 2 vols.); Chauncey Strong, ledgers and daybooks of a cabinet maker of Laurens, Otsego County (1815-69, 4 vols.).

2004 **Collins-Merriam,** Collection, 1786-1922, 9 boxes. The collection contains some material concerning the sale of timberlands and logs in Lewis County, New York (1904-20, ca. 25 items).

2005 **Fairman, Sarah Amelia,** Diary, 1819-21, 1 vol. Includes a description of maple sugar making at Butternuts (Morris), Otsego County, New York (April 16, 1820, 2 pp.).

2006 **Hubbard, Martin D.,** Records, 1854-1885, 2 vols. Daybooks of a lumber dealer at White Creek and Cambridge, Washington County, New York.

2007 **Low, John Hatch,** Collection, 1797-1870, 43 vols. Miscellaneous records (1833-70, 4 vols.) include lumber receipts.

2008 **Odell, Jacob,** Invoice and account books, 1867-1871, 2 vols. Lumber records, Tarrytown, Westchester County, New York.

2009 **Ross, A.J.B.,** Collection, 1811-1900, 3 vols. and 5,000 pieces. The collection includes an account book (1813-30) of Chauncey D. Tuttle, of Rutland, Vermont, concerning the lumber business.

2010 **Sawmills,** Account books. The following sawmill accounts are cataloged separately in the repository: Joseph Blanchard, ledger with accounts for sawmill, Maryland, Otsego County, New York (1866-68, 1 vol.); Barrilla Bradley, ledger for sawmill and lumber accounts, Middlefield, Otsego County, New York (1853-65, 1 vol.); Henry Bulkley, ledger for sawmill, n.p. (1848-52, 1 vol.); Ephraim Carr, ledger with accounts of sawmill, Hartwick, Otsego County (1821-48, 1 vol.); William Eddy, ledger for sawing, Hartwick (1829-52, 1 vol.); William Griffin, ledgers of a sawmill, Little Lakes, Herkimer County, New York (1833-60, 4 vols.); Parley Johnson, ledger for sawmill, lumber yard, etc., Oaksville, Otsego County, New York (1844-79, 1 vol.); Levi Kelley, daybook of a sawmill, Copperstown (1816-27, 1 vol.); John J. Rose, ledger of a farmer and sawyer, Maryland, Otsego County, New York, with entries for sawing timber, hemlock siding, etc. (1862-66, 1 vol.); Rufus Stone, daybooks including records of a sawmill, Dansville, New York (1842-67, 4 vols.); Elijah Wilbur, ledger of a sawmill, South Worcester, Otsego County (1851-56, 1 vol.); John Williams, ledger of sawmills, Lebanon, Connecticut, and Pierstown, Otsego County, New York (1784-1804, 1 vol.).

2011 **Weed, Joseph,** Papers, 1812-1860. New York; state legislator. Material in the collection concerns the lumber industry in New York state.

HUNTINGTON

Huntington Historical Society

Librarian, P.O. Box 506, High Street and New York Avenue 11743

2012 **Sammis, Daniel,** Account books, 1817-1847, 3 vols. Included are accounts (1825-47) of a wind-driven sawmill operated by Sammis in Huntington.

HYDE PARK

Franklin D. Roosevelt Library

Director 12538 Telephone: (914) 229-8114. Holdings are described in *Historical Materials in the Franklin D. Roosevelt Library,* Hyde Park: n.d., 16 pp. If no volume figure is given, the quantity of material in the collection is less than one linear foot.

2013 **Brown, Nelson C.,** Papers, 1930-1951. Professor of forest utilization at Syracuse University, 1921-51. The papers and correspondence relate to Brown's association with President Roosevelt in forestry practices on the Hyde Park estate.

2014 **Carmody, John M.** (b. 1881), Papers, 1900-1958, 146 feet. Papers relating to President Truman's Water Resources Policy Commission (1950-51), include material relating to the management of forested lands.

2015 **Cooke, Morris Llewellyn** (1872-1960), Papers, 1910-1959, 124 feet. Included are some materials documenting his interest in electrification, power, labor, and scientific management. General correspondence contains a small file on forestry; files on Truman's Water Resources Policy Commission include materials relating to the management of forest lands.

2016 **Delano, Frederic A.** (1863-1953), Papers, 1909-1962, 8 feet. Included is material on Delano's activities as Chairman of the National Capital Park and Planning Commission (1924-42) and as Vice Chairman of the National Resources Committee, and relating to his lifelong interest in regional planning and conservation.

2017 **Lowdermilk, Walter Clay,** Oral History Interview, 704 pp. Soil, Forest, and Water Conservation and Reclamation in China, Israel, Africa, and the United States.

2017 **Marshall, Robert,** Papers, 1933. Director of Forestry, Bureau of Indian Affairs, 1933-36. The papers consist of a copy of the Copeland Report, "A National Plan for American Forestry," submitted to president-elect Roosevelt by Gifford Pinchot, and correspondence relating to the authorship of the plan.

2019 **Mauhs, Sharon J.,** Papers, 1948-1964, 25 feet. New York State Conservation Commissioner, 1956-58.

2020 **Morganthau, Henry M.** (b. 1891), Papers, 1866-1953, 414 feet. Included are files relating to Morganthau's career as New York State Conservation Commissioner.

2021 **Recknagel, A.B.,** Oral History Interview, 46 pp. Concerns Franklin D. Roosevelt and conservation.

2022 **Roosevelt, Franklin Delano** (1882-1945), Papers as Governor of New York, 1929-1932, 72 feet. For the relevance of this collection to forest history, see the entry for Franklin D. Roosevelt, Papers as President of the United States, immediately below.

2023 **Roosevelt, Franklin Delano** (1882-1945), Papers as President of the United States, 1933-1945, ca. 5,000+ feet. The papers are arranged in an alphabetical file (3,135 feet), official file (1,174 feet), personal file (608 feet), secretary's file (139 feet), press conferences (15 feet), and map room file (81 feet). A calendar of all documents in Roosevelt's papers relating to conservation of natural and renewable resources, including soil and water, forests and other soil cover, wildlife, and scenic and wilderness areas, has been prepared on 3x5 cards, and is available on microfilm (1 reel). 1,172 of these documents, about one-third of the total number, have been published in *Franklin D. Roosevelt and Conservation, 1911-1945,* compiled and edited by Edgar B. Nixon, 2 vols., Hyde Park: Franklin D. Roosevelt Library, 1957. Other available compilations of Roosevelt's papers include *F.D.R.: His Personal Letters,* edited by Elliott Roosevelt, 4 vols., New York: Duell, Sloan, and Pearce, 1947-50; and the *Public Papers and Addresses of Franklin D. Roosevelt,* compiled by S.I. Rosenman, 13 vols., New York: Random House, 1938-50.

2024 **Tugwell, Rexford G.,** Papers, 1927-1965, 38 feet. Assistant Secretary of Agriculture, 1933-34; Under Secretary, 1934-37. Diaries, correspondence, speeches, and writings. The diaries contain references to the National Recovery Administration timber code, the proposed U.S. Department of Conservation, the Pinchot-Ballinger affair, the dispute over whether the Forest Service should be in Interior or Agriculture, and other matters relating to the U.S. Forest Service. Examples among Tugwell's speeches and writings include a radio speech, "Jobs in the Woods," and his address to the 50th Anniversary of the founding of New York's Forest Preserve. Correspondence includes a file on the Forest Service, containing letters from or to Franklin D. Roosevelt, Louis Brownlow, Ellery A. Foster, Harold Ickes, Paul Appleby, and Ward Shepard, with an "Analysis of the Private Forestry and Public Acquisition Program Presented by the Forest Service to the Joint Congressional Committee on Forestry," by Shepard, and on an "Outline of a Proposed Organic Act of Congress to Prevent Forest Degeneration and to Preserve and Rebuild Forest Resources," by Shepard. Access to material dated 1937 and later is restricted to persons having written authorization of Tugwell, Box 4068, Santa Barbara, California. Finding aid.

2025 **Welch, Fay,** Papers, 1940. Professor of Forestry, New York State College of Forestry at Syracuse. Correspondence from Welch to leading conservationists in opposition to Senate Bill 3840 (76th Congress, 3rd session), which would have established an Adirondack National Recreation Area in New York State.

ITHACA

Cornell University Library, Department of Manuscripts and University Archives

Archivist, 14850 Telephone: (607) 256-3530. Holdings are described in the *Collection of Regional History and the University Archives: Report of the Curator and Archivist, 1950-1954,* Ithaca: 1955; *1954-1958,* Ithaca: 1959; *1958-1962,* Ithaca: 1963; *1962-1966,* Ithaca, 1974. Similar reports, 1945-1950, are out of print, but available on microfilm.

Collection of Regional History

2026 **Allen, H.S.,** Account books, 1836-1890, 50 vols. Chiefly the records of the H.S. Allen Lumbering Company of Chippewa Falls, Wisconsin. Included are daybooks, journals, and ledgers.

2027 **American Nature Study Society,** Records, 1908-1966, 4⅓ feet. Records (1937-49, 3 feet [Acc. #2195]) include correspondence of Secretary-Treasurers Nellie Matlock (1937-40) and Richard L. Weaver (1943-49) and the Conservation Committee (1947-49); correspondence (1947-48) with the U.S. Department of the Interior, senators and congressmen, Izaak Walton League, Ecologists Union, National Council of State Garden Clubs, and other groups concerning legislation on national land use policy; letters dealing with annual meetings and with the printing and distribution of *Canadian Nature,* the official publication of the society; and minutes, reports, membership lists, and other mimeographed and printed items.

Also correspondence, announcements, and membership and mailing lists of the American Association for the Advancement of Science (1942-49) and the Natural Resources Council of America (1946-49) and resolutions and press releases of the National Parks Association (1941-42). Among the correspondents are Otis W. Caldwell, C. Girard Davidson, George Free, Ruth Miriam Gilmore, E. Laurence Palmer, Samuel H. Ordway, Jr., L.B. Sharp, Dwight E. Sollberger, Edwin Way Teale, William Gould Vinal, Richard W. Westwood, and Farida Wiley. Additional records (1908-66, 1⅓ feet [Acc. #2505]) contain correspondence, minutes, officers' reports, programs, accounts, ballots, constitutions, newsletters, and other papers (chiefly 1925-66), including correspondence of Nellie F. Matlock and other secretary-treasurers of the society and Ephraim Laurence Palmer, professor of rural education at Cornell University, officer and active member, with Bertha Chapman Cady, William L. Finley, George R. Green, Edith Patch, Ellen Eddy Shaw, and other members, educators, and interested persons, pertaining to the business of the society and to education in science and nature study.

2028 **Arnold Lumber Company** (Poughkeepsie, New York), Records, 1821-1938, ca. 88 feet. Journals, trial balances, and other ledgers, inventories, cashbooks, daybooks, and other account books (516 vols.); and catalogs of furniture, lumber, and building supplies; also accounts pertaining to other Poughkeepsie companies (3 vols.).

2029 **Balcom Family,** Papers, 1838-1917, 5 boxes. Includes sawmill accounts (1870-72, 1 vol.).

2030 **Barhight, Ezra** (1875-1968), Tape recordings, 1956-1957, 8 reels (9,600 feet). Seventy-five English, Irish, and American folk songs and popular tunes sung by Ezra Barhight, a resident and long-time lumberman in the Potter County region of Pennsylvania. Recorded by Ellen Stekert. Restricted.

2031 **Cady, Elias W.** (b. 1792), Papers, 1801-1857, 54 items. New York State assemblyman, 1850, 1857, and farmer. Letters to Cady deal with lumbering, and other issues raised during his terms, as well as personal and family matters.

2032 **Cobb, Cobb & Simpson** (Ithaca, New York), Records, 1872-1941, 1 foot (3 boxes). Law firm. The papers include correspondence concerning the Adah P. Horton and Randolph Horton family investments in Canadian timber lands (1913-28, 118 pieces).

2033 **Curtiss, Charles** (Collector), McGraw papers, 1870-1901, 1 foot. Papers concerning the business of the John McGraw Company, which in 1878 became Thomas H. McGraw & Company, Rough & Dressed Lumber, Shingles, Lath, and Salt, Portsmouth, Michigan; include tax assessments, mortgages, deeds, abstracts of titles, lumber price lists, timberland sales and lists, statements of assets and liabilities, timber inspection and cutting reports, agreements between McGraw & Company and the Flint & Pere Marquette Railroad as to freight rates, rebates, and the sale of right-of-way land (1870-96); two letter books of Charles Curtiss, partner of Thomas McGraw, which relate to lumber orders, shipping routes, and rates of the Rochester, New York, branch of the company; letters in these books referring to Ira Bennet, who headed the Tonawanda, New York, branch of the company after the death of John McGraw and the dissolution of his firm (1875-85). [Acc. #1229].

2034 **Dickinson, Cyril M.** (b. 1896), Correspondence, 1960-1961, 24 pieces. Copy of letter from C.M. Dickinson, Denver, Colorado, to John G. Morrison, Bemidji, Minnesota; reminiscences concerning timber stands and logging operations, transportation, and living conditions in northern Minnesota, in the late 1890s and early 1900s, after the Chippewa Reservation was opened to settlement; tributes to Jay P Kinney, Addison C. Goddard, and Morrison for their work on Indian timber lands; also Kinney's comments on Dickinson's letter to Morrison, and Dickinson's replies.

2035 **Diven, Alexander S.,** Collection, 1849-1940 [Acc. #598]. The collection includes two boxes of papers pertaining to Kentucky lands (1872-1900, ca. 707 pieces). They relate to the acquisition of Rowan County, Kentucky, lands by Diven and the management of the land by Samuel McKee. There are letters discussing the loss of acreage through lawsuits and conflicting claims; activities of timber thieves, squatters, and feuding mountaineers; and timber and ore resources. Letters from George T. Carter of Pittsburgh, Pennsylvania, discuss lumbering possibilities in Kentucky lands.

2036 **Dorsey, Bert Joseph** (b. 1883), Papers, 1877-1964, 1 foot. Chairman, Cattaraugus County Board of Supervisors, 1939-51, and farmer. Includes correspondence, clippings, and other records concerning forest conservation and other topics; also family papers.

2037 **Edwards, Hamilton,** and **Franklin Boyd Edwards,** Account books, 1840-1900, 64 vols. Included are small notebooks, with lumbering accounts (1859-1900, 48 items) from Lisle, Broome Co., New York. These include log books (n.d., 8 items), log books (1870-86, 10 items), and general account books (30 items), containing some logging records.

2038 **Edwards, Hiram K.,** Accounts, 1860-1895, 27 items. Bills paid by a lumberman of Homer, New York.

2039 **Elsbree and Jackson Families,** Papers, 1825-1875, 2 vols. and 36 items. Papers of these related families include accounts (1852-60) of Samuel Jackson (b. 1811) for logging and lumbering expenses, and other expenses, Dutchess County.

2040 **Empire State Forest Products Association,** Records, 1912-1961, ca. 5 feet. Correspondence, minutes, and reports of annual meetings and committee meetings, circular letters, articles, speeches, financial reports, membership lists, surveys, inventory of land reports, tax levy lists, scrapbook of newspaper clippings (1917-18) kept by A.B. Recknagel, photos, pamphlets, and other printed matter, and papers relating to the part the Association played in World War I, sales of land, conservation, forest fire and forest pest damage control, veteran's vocational training projects, congressional and state legislative bills, the operation of the forest practice act, and salvage projects. Includes papers by or relating to Lyman A. Beeman, Allan W. Bratton, C.O. Brown, Nelson C. Brown, A.W. Budd, William D. Comings, Harry R. Gurnow, John R. Curry, Harold D. Ellis, Clarence L. Fisher, G.M. Francis, Maurice K. Goddard, Stanley W. Hamilton, Howard A. Hanlon, Arthur S. Hopkins, S.J. Hyde, George P. Ringsley, Lawrence J. Kugelman, E.W. Littlefield, A. Augustus Low, Thomas F. Luther, Sharon J. Mauhs, Wheeler Milmoe, Charles Norris, George Ostrander, Gerard A. Pesez, A.B. Recknagel, A.M. Ross, Hardy L. Shirley, Fred C. Simmons, George Sission, Thomas H. Sterling, W. Clyde Sykes, Robert Whyland, C.J. Yops, America Forest Products Industries, Inc., Keep New York Green Committee, New York State Conservation Department, New York Tree Farm Committee, Northeastern Interstate Forest Fire Protection Commission, and Pennsylvania Department of Forests and Waters.

2041 **Gauntlett, John C.,** Papers, 1843-1928. The papers include balance sheets, general financial statement, ledger pages of the Beaver River Lumber Company, Croghan, New

York (1890-1905, 2 folders); also two contracts with Adirondack Timber & Mineral Co. for timber rights.

2042 **Griffin, William Joseph** (1863-1959), Oral history interview, by Charles D. Bonsted, 26 pp. transcript, carbon copy; original held by the Forest History Society. Reminiscences of life as a lumber camp worker, sawmill operator, farmer, and maple syrup cooperative president.

2043 **Harris, William B.,** Journal, 1837-1839, 75 pp. Typescript, carbon copy. Day-to-day account of developing a substantial farm and sawmill. Onondaga County, New York.

2044 **Hill, William Henry** (1876-1972), Papers, 1878-1964, 5 feet. New York State Senator, 1914-18; U.S. Representative from New York, 1919-21; Central New York State Parks Commissioner, 1925-33, and thereafter chairman. Included are a small amount of correspondence, minutes, reports, other papers concerning meetings of or issues facing the Central New York State Parks Commission. Unpublished guide.

2045 **Hinde and Dauch Paper Co.** (Sandusky, Ohio), Records, 1897-1953.

2046 **Horton, Emmet** (1850-1932), Papers, 1870-1932, 2 feet. Consists of correspondence, financial, and legal documents, dealing largely with the patenting, sale, and manufacture of Horton's inventions. Included among papers dealing with Horton's many other interests are maps concerning his land in Michigan and the sale of white pine thereon (1889-96).

2047 **Howe, P.C.,** Account book, 1854-1855. Records of milling at Wheeler, Steuben County, New York, and also some labor accounts.

2048 **Hulce, Martial R.** (Collector), Papers, 1786-1893, 22 inches. Concerned in part with lumbering along the Delaware River.

2049 **Kelly, Crosby,** Accounts, 1920, 1924, 3 vols. Daybook (1 vol.) and ledgers (2 vols.) of a Fleischmanns, Delaware County, New York, lumber dealer and builder.

2050 **Kimball and Nutter** (Woodsville, New Hampshire), Records, 1903-1919, 7 vols. Lumbering accounts; includes records (1 vol.) of a livery stable or horse dealer who apparently supplied horses for lumbering purposes.

2051 **Kinney, Jay P** (b. 1875), Papers, 1855-1957, 28 feet + 11 boxes. Papers (largely 1910-57, 4 feet [Acc. #1791]) include letters received and mimeographed, photostated, or typescript carbon copies of correspondence, reports, memoranda, departmental directives, and other items, dealing largely with policy and administration in the Forestry Branch of the Indian Service where Kinney served from 1910-33. Also included are correspondence pertaining to Kinney's conservation work in the same department, 1933-45, and his services as an attorney for the Lands Division of the Department of Justice, 1945-54; and letters and memoranda concerning the Society of American Foresters, 1919-44, with particular reference to questions of membership and election of officers and to the work of its committees on national forest policy, forest history, and international relations; editorial policies of the *Journal of Forestry*; awards made by the American Forest Fire Medal Board, of which Kinney was chairman, 1938-47; lumber production on Indian forests, stumpage and railroad tie prices, sustained-yield forest management, soil conservation, Indian welfare, and other matters; also photostats of official letters, treaties, contracts, Indian Council minutes, maps, and other items pertaining to the sale of Indian lands and establishment of reservations in the Pacific Northwest and to timber cutting operations and stumpage prices on Indian forests in Wisconsin; typed, mimeographed, or printed copies of speeches and articles, mostly by Kinney, on forest and range policies and practices, Indian land tenure, the Emergency Conservation Work program on Indian reservations (1933), relations between the Forest Service of the Department of Agriculture and the Forestry Branch of the Interior Department, the nature and extent of government regulation in the field of forest conservation, and other problems; typescripts of his books, *The Development of Forest Law in America* and *Indian Forest and Range—A History of the Administration and Conservation of the Redman's Heritage*; miscellaneous pamphlets; maps of Indian reservations; photographs of soil erosion in the West. Correspondents include Charles Henry Burke, Herman Haupt Chapman, John Collier, Samuel Trask Dana, Bernhard Eduard Fernow, John Clayton Gifford, C.M. Granger, Henry Solon Graves, John Dennett Guthrie, William P. Harley, C.F. Hauke, William Heritage, Fred E. Hornaday, Ralph Sheldon Hosmer, Paul Delmar Kelleter, Edgar Briant Meritt, Lee Muck, Barrington Moore, Jay B. Nash, Frank Pierce, Arthur B. Recknagel, Paul G. Redington, Franklin Reed, Charles James Rhoads, Joseph Henry Scattergood, Cato Sells, Hardy L. Shirley, Robert Y. Stuart, Robert Grosvenor Valentine, Amelia White, William Zeh. Unpublished guide available. Additional papers (29 boxes [Acc. #2948]) include some manuscripts but much processed and printed material relating chiefly to the Bureau of Indian Affairs. There is some correspondence of Kinney concerning Indian forests. Unarranged.

2052 **Knowlton Brothers** (Watertown, New York), Records, 1813-1964, microfilm, 6 rolls. Paper manufacturers. Included are indentures, correspondence (1854-95), deeds, notes, bills, biographical notes, historical data, mill plans and specifications (1912), constitution and bylaws, clippings, notebooks of George Willard Knowlton (1845-81) and diaries (1860-86, with gaps), Knowlton & Rice letter record (1831-35), account books (1833-49, 3 vols.), photographs, and other papers.

2053 **Lauman, George N.** (1874-1944), Collection, 1889-1951, 4 feet. Rural sociologist. The collection chiefly consists of diverse types of printed materials relating to various banking and investment interests. Included are Hulce family lumbering papers (1832-93) containing detailed records of timber and lumber contracts, sawmill accounts, and the rafting of lumber to the Kensington and Philadelphia, Pennsylvania, markets. Guide.

2054 **Lord, Bert** (1869-1939), Papers, 1902-1939, 15½ feet. Afton, Chenago County, New York. State legislator and U.S. Representative. Political, business, and personal papers, consisting mainly of correspondence, but including also scrapbooks, clippings, pamphlets, and other printed or processed material. Correspondence for the time Lord was a member of the New York State Assembly (1915-22, 1924-29) and the state senate (1929-35), includes material on forestry and fish and game laws. His business papers (1902-32) include letters concerning his timber land and lumber mill interests, especially the sale of railroad ties and mine props to the Hudson Coal Company and the Delaware & Hudson Railroad Company. Correspondents include, among others, the Otsego Forest Products Cooperative Association, Rock Royal Cooperative, South Coast & Northern Lumber Company. Unpublished guide.

2055 **Lyon Company** (Aurora, New York), Records, 1888-1942, 135 vols. and 2 boxes (8 inches). Accounts and letterbooks (1897-1906) of S.G. Lyon, cattle, grain, and lumber dealers.

2056 **McCray, Elizabeth** (Collector), Papers, 1810-1944, 4 inches (ca. 90 items and 1 vol.). Correspondence, deeds, and other legal documents, accounts, and miscellaneous papers (chiefly 1810-77) of the Chatfield, Foster, and McCray families, mainly concerned with the sale of land and lumber in Broome, Delaware, Otsego, and Steuben Counties, New York; and Luzerne County, Pennsylvania. Ca. 40 items are concerned with lumber.

2057 **McGraw, John** (1815-1877), Papers, 1846-1882, 223 items. [Acc. #132, #1183]. Correspondence relating to the land business of W.J. Young and Co., Ithaca, New York; to McGraw's lumber interests in Canada, Minnesota, and Wisconsin; and to other matters. Correspondents include Douglass Boardman, S.J. Peck, H.W. Sage, Daniel Shaw, William Stoddard, William E. Taylor.

2058 **McGraw Family,** Papers, 1854-1956, 114 items, 3 vols. [Acc. #2355, #2644]. Includes business correspondence (1854-58) of John Ellis Southworth, including four letters from B.F. Johnson reporting on logging and sawmilling near Caneadea, Allegany County; business correspondence of John McGraw, including a letter from Henry W. Sage (1861) concerning his lumbering operations at Lake Simcoe, Ontario; photographs of an unidentified lumber mill.

2059 **McMaster & Parkhurst** (Bath, New York), Records, 1860-1897, 2⅔ feet (8 boxes). Papers relating to the law practice of Guy Humphreys McMaster and his partner, John F. Parkhurst. The collection contains much material on the contemporary history of Steuben and neighboring New York counties, and includes papers on lumbering interests in the vicinity.

2060 **Mid-Hudson Forest Products Cooperative** (Kingston, New York), Records, 6⅔ feet (5 boxes).

2061 **Munger, John,** Account book, 1806-1812, 1 vol. Lumbering accounts from Warsaw, Wyoming County, New York.

2062 **National Forestry Program Committee,** Records, 1920-1928, 1945-1947, 1⅓ feet. This committee, on which were represented the Western Forestry and Conservation Association, the Society for the Protection of New Hampshire Forests, the American Newspaper Publishers Association, the National Lumber Manufacturers Association, the American Paper and Pulp Association, the American Forestry Association, the Association of Wood Using Industries, the Central States Forestry League, and the American Tree Association, was established in 1920 for the purpose of securing national forest conservation legislation. Its papers consist largely of correspondence between Royal S. Kellogg, chairman, and members, professional foresters, lumber dealers, manufacturers of wood and paper products and naval stores, regional and state conservation associations, farm organizations, newspaper publishers, congressmen, and others concerning efforts made to promote the Snell Bill (1920-22), and later the Clarke-McNary Act (1924), the McNary-McSweeney Forest Research Law (1928), and the McNary-Woodruff Act (1928); these letters contain comment on the provisions of the bills, plans for congressional hearings, arrangements for annual Forest Protection Weeks, the timber situation in the United States and Alaska, the role and extent of federal and state regulation of forest practice, particularly as viewed by Gifford Pinchot and opposing schools of thought, and other matters. Related to the correspondence are numerous circular letters, memoranda, resolutions, press releases, and other publicity materials, and miscellaneous mimeographed and printed items. Also included is the correspondence of Ralph S. Hosmer and other foresters concerning an article by Hosmer on the work of the committee, together with his rough notes and drafts (1945-47). Correspondents include Edgar Allen, E.T. Allen, Shirley Allen, Philip W. Ayres, Elbert H. Baker, Hugh Potter Baker, John W. Blodgett, Warren B. Bullock, Ovid M. Butler, Earle Hart Clapp, Wilson M. Compton, Samuel Trask Dana, John Foley, David L. Goodwillie, William B. Greeley, William L. Hall, Gilbert N. Haugen, Ralph Sheldon Hosmer, Charles L. McNary, Barrington Moore, George N. Ostrander, Charles Lathrop Pack, Joseph Hyde Pratt, Arthur B. Recknagel, Paul G. Redington, Arthur C. Ringland, Percival Sheldon Ridsdale, Sherman Rogers, Edward A. Sherman, George W. Sisson, Jr., Bertrand H. Snell, Robert Y. Stuart, James William Toumey, Arthur T. Upson, Raphael Zon. Unpublished guide.

2063 **Northeastern Wood Utilization Council,** Records, 1942-1959, 1⅓ feet. Letters and other papers from the files of Edgar Heermance, secretary (1942-53), Henry W. Hicock, secretary (1954-58), and A.B. Recknagel, executive secretary (1952-53). Papers pertaining to the dissolution of the council (1958-59); pamphlets and miscellaneous material (1942-59), relating to council activities and the forest products industry; incomplete runs of *Wood Notes* and the *Council Bulletin.* Unpublished guide.

2064 **Ostertag, Harold Charles** (b. 1896), Papers, 1937-1964, 200 feet. New York State Assemblyman, 1932-50; U.S. Representative, 1950-64. Early papers consist of correspondence and printed matter concerning his term in the assembly; forestry is among the subjects discussed. Papers of Ostertag's term in the House of Representatives include correspondence concerning conservation. Unpublished guide available. Restricted.

2065 **Ostrander, John, Family,** Papers, 1827-1865, 2 inches. Includes ten letters (1831-47) from John Ostrander, Kennedyville (now Kanona), to his wife, Eveline, giving an account of his travels in lower New York State, where he was selling lumber and cattle. 3 items mention lumber.

2066 **Otsego Forest Products Cooperative,** Records, 1937-1963, 30 feet.

2067 **Pack, Dan Sinclair** (1869-1959), Papers, 1910-1947, 3 items. Forester. Pack's memoirs (1946-47, 91 pp., typed, with holograph original) recount his experiences in the U.S. Forest Service, 1901-02, in Wyoming, Utah, and Idaho, where his work involved regulation of stock grazing, fire fighting, surveying and reporting on "June 11 claims" in Weiser (now Payette) National Forest, Idaho. There are also Xerox copies of letters (1910, 1912, 2 items) from Gifford Pinchot concerning the Forest Service under Henry S. Graves, and asking for Pack's cooperation in gathering information for a projected book by Pinchot.

2068 **Pearson, C.H., Hardwood Company** (New York), Records, 1929-1931, 155 items. Correspondence and considerable printed material relating to attempts of the company to amend tariff legislation which placed heavy duties on imported woods.

2069 **Pulsifer, Nathan T.,** Papers, 1886-1902, 6 vols. and 2 boxes. Manchester, Connecticut; business executive. Correspondence and papers relating to Pulsifer's interests in the American Paper and Pulp Association, Houghton Mifflin and Company, Lawson Valentine Company, Macmillan Company, Manchester Light and Power Company, Oakland Paper Company, the Outlook Company, and the Rural Publishing Company. Includes account books of the Oakland Paper Company.

2070 **Raymond Family** (Penfield, Monroe County, New York), Farm papers, 1855-1947, 16 vols., 53 pieces. Includes sawmill accounts (1894-1901, 1 vol.).

2071 **Recknagel, Arthur Bernard** (1883-1962), Papers, 1948-1961, 1 vol. and 4 inches. Monthly reports (1948-61) by Recknagel, technical director of forestry at the St. Regis Paper Company, and related materials concerning the management of timberlands in the Adirondacks, New England, Florida and Georgia, and the Pacific Northwest; conservation, taxation, research and experimentation in wood utilization, tree diseases, and cutting and planting practices, federal and state governmental policies, and other matters of interest to the forest products industry. Pamphlets, printed or mimeographed reports and speeches, clippings, and other items pertaining to forest management in general and to the Empire State Forest Products Association, the Forest Products Research Society, the Northeastern Loggers Association, the Northeast Forest Tree Improvement Association, the Society of American Foresters, the Western Forestry and Conservation Association, the 1949 United Nations Conference on Conservation, and others; black and white snapshots of forest lands and pulpwood operations in New Hampshire (8 items); snapshots of the St. Regis River area, New York (10 items). Restricted.

2072 **Recknagel, Arthur Bernard** (1883-1962), Oral History Interview by Elwood R. Maunder, October, 1958, 1 vol. (46 pp.). Original held by Forest History Society.

2073 **Rogers, Publius Virgilius** (1824-1895), Papers, 1850-1918. Papers relating to the Saginaw and Cincinnati Lumber Transportation Company, a subsidiary of the Flint and Pere Marquette Railroad. Accounts and circulars illustrating lumbering operations and sales of the lands of the Flint and Pere Marquette (1862-78).

2074 **Rutherford, Thomas R.,** Papers, 1819-1897, 4 inches (1 box). The papers include deeds, bills, etc., relating to timber investments in North Carolina (ca. 1 folder).

2075 **Sage, Henry Williams** (1814-1897), Papers, ca. 1830-1930, 75½ feet. The bulk of this collection consists of letterpress books (1857, 1867, 1871-1910) and files of incoming correspondence (ca. 1880-1930), dealing with the business interests of H.W. Sage and W.G. Grant (1854-61), lumber manufacturers at Belle Ewart, Canada; Sage-McGraw and Company (1864-69) and H.W. Sage and Company (1869-93), lumber manufacturers at Wenona, Michigan; and Henry W. Sage and John McGraw (1864-78), H.W. Sage and Company (1869-93), and the Sage Land and Improvement Company (1893-ca. 1930), investors in timber lands. There are also tax receipts and miscellaneous materials (1837-1930, 37 boxes), and scattered ledgers, daybooks, journals, land and plat books of W.H. Sage and Company (40 vols.), as well as minutes of the Loggers Boom Company (1883-1905, 1 vol.). The almost daily letters (1871-93) of H.W. Sage to his son, Dean, deal with the lumber and salt mill at Wenona; the Albany, New York City, and Toledo lumber yards of H.W. Sage and Company; the AuGres, Rifle River, Tittabawassee, and Loggers Boom Companies; the political and economic life of Wenona (West Bay City after 1877), Michigan. Letters to and from George S. Frost (1881-86), L.V. Ripley (1880-ca. 1905), and Frank Pierson (ca. 1890-1905), are relevant for Michigan timberland investments; to W.S. Patrick (1867, 1871-73), to and from D.P. Simons (1875-1909), and L.V. Ripley (1880-1905), for Wisconsin timberlands; to and from H.A. Emery (1888-93) and Martin Van Heuvel (1893-97), for timberlands in Alabama and Mississippi; from D.P. Simons (1902-06), for Oregon timberlands; from D.P. Simons (1904-09), D.P. Simons, Jr. (1909-12), and E.J. James (1912-16), for California timberlands. There is also material concerning H.W. Sage and Company's railroad, industrial, and mortgage investments, and his concern for Cornell University whose interests he served as trustee. [Acc. #1155]. Additional letters (1882-98, 8 inches) relate to Sage's activities as president of the Board of Trustees of Cornell University and to his business interests including letters concerning the Wisconsin pinelands of Cornell University, and Sage's lumber interests in Michigan, Oregon, Canada, and elsewhere. [Acc. #1388]. Also a bound, typescript copy of a diary of Henry W. Sage (1836-37); letterbook (1894), containing congratulatory messages to Sage on his 80th birthday, letters of condolence after his death, and newspaper clippings; material relating to Cornell University; letters to members of the Sage family, including several items concerning the death, 1924, of Henry Sage's son, William Henry Sage, including a printed resolution expressing the sympathy of the president and Fellows of Yale University, to which William had given the School of Forestry building; a few Sage Land and Lumber Company, Inc., and Sage Land and Improvement Company papers, including copies of reports (1924-46) and minutes of a directors' meeting (1948). [Acc. #1557].

2076 **Seymour, Horatio** (1810-1886), Papers, 1741-1920, 11 vols. and 8 boxes. Includes lumbering accounts (1 vol.) and papers dealing with logging, sawing, tanning bark in New York, Michigan, and Wisconsin (12 folders).

2077 **Skilton Family,** Papers, 1830-1940, 200 feet. The papers (1880-1938, 157 feet) of Frank A. Skilton (1860-1931) comprise the major portion of the collection. They mostly relate to genealogy, but contain accounts (1⅓ feet) of his partnership in Benson and Skilton, a lumber, coal, and wood business in Auburn, New York. Included is correspondence by or about H.H. Benson.

2078 **Spalding, Lyman A.,** Papers, 1815-1850, 320 pieces. The papers include correspondence with his brother, E.H. Spalding, concerning timber land in Michigan Territory, 1817-35.

2079 **Taber, John** (b. 1880), Papers, 1931-1959, 217 feet. U.S. Representative from New York State, 1923-63; ranking minority member of the Select Committee on Government Organization and chairman of the appropriations committee. The papers include material on forestry (1 folder); public lands (1 folder); the Department of the Interior (22 folders). Included is correspondence from conservationists, members of the public, government officials, and others concerning such matters as national park policies, federal forestry programs, etc.

2080 **Todd Family,** Business Records, 1850-1921, 267 receipt pads and 2 boxes. Included are financial records relating to the milling, lumbering, and other businesses of Zerah A. Todd and son (1876-93, 2 vols.).

2081 **Tolbert Family,** Papers. The papers include a letter (1856) from A.M. Palmer of Wisconsin, concerning a mill at Chicago, lumbering costs, and routes from Savona, New York, to Green Bay, Wisconsin.

2082 **Vanderlyn, Henry,** Letterbook, 1810-1824, 1 vol. Included are a few letters relating to lumbering in Chenango County, New York.

2083 **Virginia Blue Ridge Railway** (Piney River, Virginia), Records, 1913-1961, 79 feet. Correspondence and administrative, legal, and fiscal files of the railroad and of associated firms, the Bee Tree Lumber Company, the Hope Falls Timber Company, the Leftwich Timber Company, and the Tye River Timber Company. Includes correspondence of John W. Dwight, John W. Powell, Howard Cobb, and Fordyce Cobb, officers of the railroad company. Powell's papers include much material dealing with the operations of the Tidewater Lumber and Veneer Company in Honduras and British Honduras and with the land promotion ventures of the Georgia Company in that state and the Great Western Land Company in Montana.

2084 **West Virginia Pulp and Paper Company** (New York City), Records, 1809-1970, ca. 371 feet. Records (1884-1953, 275 feet) contain account books (144 vols.), including those of predecessor firms; business correspondence; legal papers, including deeds, contracts, leases, mortgages, and records of suits; government wartime directives, newspaper clippings, and other printed matter, maps, statistics, and miscellaneous papers pertaining to all operations of this company. Intracompany correspondence and other papers relate to activities of various departments of the home and branch offices regarding executive and personnel administration, accounting and finance, insurance and taxation; stock transfers, issues and redemptions; advertising and public relations, industrial and labor relations, operation of mills, research and purchasing of equipment and property, wartime activities, and such matters as conservation, postwar planning, housing, drought relief, and stream pollution. Restricted. [Acc. #1781]. Additional records (1925-54, 5½ feet) include business correspondence, office records, deeds, and other legal papers, temporary stock certificates, annual reports, newspaper clippings, and other materials pertaining to stock issues and redemptions, finance, property purchases and sales, employee relations, and taxation. Additional records (1887-1970, ca. 90 feet) have not yet been described. There are also photostats of maps (1809 +).

2085 **Williams, Josiah B.,** Papers. Papers (1809-83, 35 boxes, 11 vols.) contain material relating to the wholesale lumber and general produce business, including correspondence and other documents of the partnership, beginning 1834, between Henry W. Sage, New York, C.P. Williams, Albany, and J.B. Williams, Ithaca; of the partnership between J.E. Shaw, Ithaca, and H.W. Woodworth, Chicago, and of the mill at Vienna, Canada, and other material. Ithaca and Oswego Railroad Company papers include an inventory (1834) of timber and other materials, and correspondence concerning these (1833-35). [Acc. #422]. Additional papers (2 boxes) contain material on Williams' Chicago lumber business (1858-63, 1 folder). [Acc. #1148].

Cornell University Archives

2086 **Bailey, Liberty Hyde** (1858-1954), Papers, ca. 1858-1954, ca. 25 feet. Correspondence (1888-1953, 4½ feet, including 3 vols. photo scrapbooks, 680 frames microfilm, and 30 photostats) relates to the New York State College of Forestry, the Farm Bureau movement in New York state, the examination, classification, and nomenclature of plant specimens, especially palms, the preparation of the *Standard Cyclopedia of Horticulture* and the *American Cyclopedia of Horticulture,* agricultural developments in New York state, reminiscences of rural life and Cornell University. Includes correspondence of Carol Aronovici, Peter Bisset, Nathaniel L. Britton, Maurice C. Burritt, F.V. Coville, David Fairchild, Knowles A. Ryerson, C.L. Shear, Homer C. Skeels, S.C. Stuntz. [Acc. #21/2/47, 21/2/118]. Additional papers (1895-1914, 11 feet) deal with the establishment of and appropriations for the New York State College of Forestry at Syracuse University. Unpublished guides. [Acc. #21/2/541, 21/2/576]. Additional papers (ca. 1858-1954, 9 feet) are not yet arranged. [Acc. #21/2/1400].

2087 **Cornell, Ezra** (1807-1874), Papers, 1806-1927, 9½ feet and 5 reels of microfilm. Additional papers of Ezra Cornell and his family, which pertain largely to his varied interests in agriculture, business, and education, include a letter (1872) from W.S. Cornell describing in detail the agricultural, forest, and mineral resources of the Asheville, North Carolina, area; also pamphlets concerning the lumber and salt resources of the Saginaw Valley of Michigan. Unpublished guide available.

2088 **Cornell-Los Banos Contracts,** Records, 1951-1961, ca. 10½ feet. Reports (1951-60, 1¼ feet) on a project sponsored by the International Cooperation Administration in which the New York State College of Agriculture engaged in assistance programs in agricultural research and education at the Colleges of Agriculture and Forestry, University of the Philippines. [Acc. #21/2/537]. Additional records (1951-61, 9⅓ feet) concern the participation of the New York State College of Agriculture in assistance programs in agricultural research and education at the Colleges of Agriculture (1952-60) and Forestry (1957-60), University of the Philippines, under the sponsorship of the International Cooperation Administration. [Acc. #21/2/914].

2089 **Cornell University. Forestry Department,** Papers, 1912-1915, 16 inches. Include routine college and interdepartmental administrative correspondence; correspondence pertaining to commercial forestry, government regulation of forest products industries, conservation, and professional forestry associations; also, correspondence pertaining to the formal exercises at the opening of the Forestry Building at the New York State College of Agriculture (1914). Correspondents include Alfred W. Abrams, Robert Judson Aley, Philip W. Ayres, Liberty Hyde Bailey, Hugh Potter Baker, John Bentley, Jr., Robert C. Bryant, Herman H. Chapman, Earle Hart Clapp, John Henry Comstock, Samuel Trask Dana, Henry Sturgis Drinker, Beverly Thomas Galloway, Royal Gilkey, Jay P Kinney, Lewis Knudson, Clyde Leavitt, Ora Maynard Leland, Albert Russell Mann, Alica G. McClosky, Franklin Frederick Moon, Veranus Alva Moore, Walter Mulford, Clyde Hadley Myers, Charles Lathrop Pack, George W. Parker, Gustav Adolph Pearson, Clifford Robert Pettis, Gifford Pinchot, James Edward Rice, P.S. Ridsdale, Howard Wait Riley, Samuel Newton Spring, William Alonzo Stocking, Charles Tuck, Edward M. Tuttle, Ottomar H. Van Norden, Martha Van Rensselaer, Edward Albert White, Karl McKay Wiegand, Charles Scoon Wilson, Paul Work.

2090 **Cornell University. Treasurer's Office,** Western Land investment records, 1859-1926, 39 vols. and 22 feet (66 boxes). Includes correspondence, field books, examiners' reports (1894-99) for lands in Kansas, Nebraska, and Missouri; and Wisconsin lumbering records (1882-86). The correspondence between William A. Woodward, an eastern speculator in western forest and farm land, and Henry C. Putnam, a Wisconsin land agent, throws light on public land policies and the history of land speculation. Numerous letters from the resident agents in Minnesota and Kansas to officials in Ithaca describe the problems of administering a large absentee-owned estate. Includes information concerning the settler-speculator conflict, timber stealing and efforts to minimize it, disputes with taxing authorities over assessments, the fight with the Wisconsin Central Railroad over extending exemption of railroad land grants from taxation, and the growth of the Knapp, Stout and Weyerhaeuser combinations on the Chippewa and Red Cedar Rivers. Other persons represented are

John C. Gauntlett and John McGraw. [Acc. #7/4/12]. Also included are records (1858-1904, 6 inches [490 pieces]) of the H.C. Putnam lumber business, relating to lumbering in Wisconsin and containing letters from Henry W. Sage, Ezra Cornell, C.S. Sargent, W.A. Woodward, J.W. Dwight, and other officials of Cornell University to Putnam, land agent in Wisconsin. [Acc. #7/4/48 (formerly #8)].

2091 **Fernow, Bernhard Eduard** (1851-1923), Papers, 1885-1930, 4⅓ feet. Papers (1885-1922, 1 foot) contain letters concerning the development of professional forestry and forest legislation in the United States; clippings relating to forestry, the College of Forestry at Cornell University, artificial rainmaking, and the Fernow biography; speeches and articles relating to forestry and the Division of Forestry, U.S. Department of Agriculture, which Fernow headed (1886-98). Correspondents include the Adirondack League Club, Herman Haupt Chapman, Hiram Martin Chittenden, Richard T. Ely, Abram S. Hewitt, Josiah Kingsley Ohl, and Robert Shumann. [Acc. #20/1/91]. Additional papers (1895-1930, 3 feet) are largely correspondence pertaining to Fernow's position at Cornell as dean of the New York State College of Forestry (1898-1902), professional correspondence with firms in the United States and Germany and with the New York State Forest, Fish and Game Commission (1898-1901), and family correspondence. Unpublished guide, 1 p. [Acc. #20/1/561]. Additional papers (1900-30, 4 inches [1 box]) include bills, letters, notes, concerning the Axton, New York, forest plantation. [Acc. #20/1/1179].

2092 **Hosmer, Ralph Sheldon** (1874-1963), Papers, 1803-1962, ca. 115½ feet. Superintendent of Forestry, Territory of Hawaii, 1904-14; Professor of Forestry, Cornell University, 1914-42. Papers (1898-1903, 4 inches [1 box]) collected by Hosmer include correspondence between Bernhard Eduard Fernow and prospective students and other interested persons about the course of study offered by the New York State College of Forestry, admission requirements, and the future of professional forestry in the United States. [Acc. #20/1/409]. Hosmer's papers (ca. 1891-ca. 1941, 40 feet) include correspondence, memoranda, printed matter, and other papers concerning the Department of Forestry at Cornell; correspondence and other papers (1930s) pertaining to forestry education in New York State; Hosmer's student notes taken at Harvard (1891-94), notes on Central Park (1895); fragmentary diary (1897-98, 2 vols.); manuscript of radio talks on the Agricultural College Hour; papers and articles by Hosmer and others; and much correspondence and other papers relating to the Society of American Foresters, American Forestry Association, Empire State Forest Products Association, U.S. Forest Service, Institute of Forest Genetics, State College Council, New York State Forestry and Park Association, and New York State Forestry Association. [Acc. #20/1/560]. Additional papers (1803-1962, ca. 75 feet) consist of professional papers including correspondence, reports, and memoranda concerning the College of Forestry at Cornell (1898-1903); the Department of Forestry and its curriculum and administration; the Cornell Plantations; Hosmer's forestry classes; the Society of American Foresters; its New York Section; the *Journal of Forestry*; Hawaiian forestry; the National Forest Program Committee; international forestry congresses; the proposed U.S. Department of Conservation (1934); Hosmer's writings; also Hosmer's personal papers, which contain material relating to his education at Yale School of Forestry, his post as Tompkins County Fuelwood Coordinator (1943-44). [Acc. #20/1/683]. There is also a tape recording (1955, 300 feet) of a tribute by Charles L. Tebbe, U.S. Forest Service, enumerating Hosmer's many contributions to professional forestry and presenting him with a certificate of charter membership in the U.S. Forest Service; also, Professor Hosmer's response. [Acc. #21/18/tr. 21].

2093 **New York State College of Agriculture, Directors of Extension,** Papers, 1910-1935, 21 feet. Correspondence files of Directors Charles Henry Tuck (1910-35); Maurice C. Burritt (1919-24); Carl E. Ladd (1924-30); and Lloyd R. Simons (1932-35), concerning the National Farm Forestry Extension Committee, the Clark-McNary Act, and many other subjects.

2094 **New York State College of Forestry,** Records, 1895-1937, ca. 2 feet. Records (1898-1904, 1 foot) include financial records kept by Director Bernhard E. Fernow and his assistants (3 vols.), pertaining to the management of the Cornell Forestry Reserve and the Cornell Campus branch of the college; pamphlets and printed reports prepared by Fernow and the college staff (1898-1904), pertaining to the operation of the college, its experimental forest in the Adirondacks, and the *Swenson Suit* which resulted in the abandonment of the experimental forest program. Unpublished guide. [Acc. #20/1/401]. Additional records (1900-02, 8 inches [1 vol.]) contain the correspondence of Fernow and concern job recommendations, articles, and a book Fernow was having published, the acquisition of materials for N.Y.S. College of Forestry, and advice on forestry technique. Includes correspondence of Fernow with various parties, chiefly newsmen, on the management of Cornell property in the Adirondacks (1900-02), and with President Jonathan LeMoyne Snyder of the Michigan State Agricultural College (later Michigan State University) about the recruitment of a properly trained forester to supervise the creation of a forestry program at the college. A guest book (1927-40) contains snapshots of the Cornell Forestry Camp in Newcomb, Essex County, New York, and the signatures and comments of visitors. [Acc. #20/1/462, 20/1/493]. There is additional unprocessed manuscript material relating to the Axton Branch. Also slides and photographs (1895-1937, ca. 3000 items) illustrating forestry practices in Germany and Switzerland and their application to New York state forests, and illustrations of forestry practice at Cornell University and other lands in New York state. [Acc. #21/18/121, 21/18/ph.].

2095 **Wakeley, Philip Carman** (b. 1902), Oral History Interview, 131 pp. [O.H. 143]. Forester. The interview pertains to Wakeley's education in forestry at Cornell and his work at the Southern Forest Experiment Station.

2096 **Wright, Albert Hazen** (1879-1970), Papers, 1905-1965, 3 items. [O.H. 22]. Professor of Zoology, Cornell University. The collection includes an interview conducted by Gould P. Colman (1962-63, 92 pp., typescript), including comments on field work in forestry. Restricted.

Cornell University, Labor Management Documentation Center, New York School of Industrial and Labor Relations

Ives Hall 14850 Telephone: (607) 256-5435

2097 **International Brotherhood of Pulp, Sulphite and Paper Mill Workers,** Records, 1906-1957, 284 reels microfilm.

DeWitt Historical Society of Tompkins County, Inc.

The Clinton House, 116 North Cayuga Street 14850 Telephone: (607) 273-8284

2098 **Hyatt, James C.,** Account book, 1833-1852, 1 vol. Ithaca carpenter and builder. This volume is in storage and unavailable for research (1974).

2099 **Schuyler, T.B.,** Letters from C.P. Williams & Co., 1843-1856, 258 items. Concerning the lumber business.

KEENE VALLEY

Keene Valley Public Library

Librarian 12943

2100 Holdings include the John T. Loomis Collection (2 vols.) of manuscripts and correspondence of W.H.H. Murray and F.H. Comstock dealing with the history of the Adirondacks; also a letter (1903) of Max Smith to his sister, Gertrude Worman-Smith, concerning forest fires; and a manuscript (n.d.) by Adelaide Vagnarelli, "History of Lumbering in the Ausable Valley [New York]."

NEW YORK

The American Academy of Arts and Letters

Librarian, 633 West 155th Street, New York 10032 Telephone: (212) 286-1480

2101 **Burroughs, John** (1837-1921), Papers, 1904-1921, ca. 200 items. Naturalist and author. The correspondence chiefly relates to the National Institute of Arts and Letters, to the American Academy of Arts and Letters, and to Burroughs' writings. There is some material on conservation, including a letter (October 20, 1913) to Robert Underwood Johnson concerning Hetch Hetchy Valley. Restricted.

2102 **Johnson, Robert Underwood** (1853-1937), Papers, 1898-1937, 2 feet. Correspondence and other papers chiefly relating to the National Institute of Arts and Letters and the American Academy of Arts and Letters, Johnson's writings, and his activities as an editor and a conservationist. Included are materials concerning the Hetch Hetchy Valley, California, and a commemorative tribute to John Muir (8 pp.). Restrictions.

2103 **Muir, John** (1838-1914), Correspondence, 1889-1914, ca. 65 items. Correspondence chiefly relating to the National Institute of Arts and Letters, and the American Academy of Arts and Letters. There are letters to R.U. Johnson (14 items). Restricted.

Columbia University, Library, Special Collections

801 Butler Library 10027

2104 **Bradley, Joel,** Accounts, 1765-1801. Included are extensive but routine entries on the sale and cost of lumber, New Haven, Connecticut.

2105 **Johnson, Robert Underwood** (1853-1937), Papers, 1848-1937, ca. 1,600 items. Correspondence and literary manuscripts. There are writings of Johnson, John Muir, and others; subjects include the business affairs of *Century Magazine,* and Johnson's interests in conservation and literature.

2106 **Portsmouth, New Hampshire,** Carpenter's daybook, 1818-1833. Included are routine entries on lumber costs.

2107 **Wallace, Henry Agard** (b. 1888), Papers, 12 boxes. U.S. Vice President, 1941-45; and Secretary of Agriculture, 1933-40. Correspondence and papers.

Columbia University, Oral History Research Office

Director, Oral History Research Office, Box 20, Butler Library 10027. Holdings are described in Elizabeth B. Mason and Louis M. Starr, eds. *The Oral History Collection of Columbia University,* New York: Oral History Research Office, 1973.

2108 **Albright, Horace Marden** (b. 1890), Oral history interview, 1960, 851 pp. Comments on his career with the National Park Service, particularly Yellowstone National Park, park policies and problems, the Civilian Conservation Corps, and impressions of various public figures. Permission required to cite or quote. Available on microfiche.

2109 **D'Ewart, Wesley Abner** (b. 1889), Oral history interview, 1967, 136 pp. Experiences as Assistant Secretary of the Interior, Eisenhower administration; water policy and development, national parks, public lands, Indian affairs. Permission required for access.

2110 **Forest History Society,** Oral history interviews, 1957, 237 pp. Interviews on forestry and logging, containing material on conservation, woods safety, fire fighting and the development of protective associations, old Minnesota logging camps, logging methods and machinery, and the development of the Paul Bunyan legends. Impressions of H.L. Mencken are included, as are impressions of George S. Long and other lumbermen. Participants include: Charles S. Cowan (54 pp.); George W. Dulany (37 pp.); Inman F. Eldredge (12 pp.); Royal S. Kellogg (55 pp.); Donald MacKenzie (17 pp.); Maggie Orr O'Neill (19 pp.); P.J Rutledge (10 pp.); James Stevens (33 pp.).

2111 **Henrikson, Carl H., Jr.** (b. 1899), Oral history interview, 1955, 145 pp. Lumbering in Minnesota. Permission required to cite or quote. Available on microfiche.

2112 **Jackson Hole Preserve,** Oral history interviews, 1966, 1,080 pp. This project relates the history of Jackson Hole Preserve, describing the Rockefeller family's interest in preserving and protecting the area and problems encountered in acquiring the land which was eventually added to the National Park System. Included are memoirs of people who knew Jackson Hole as their home and who have experienced the transformation of the area since it became part of the National Park System in the 1940s. Participants include Horace M. Albright (238 pp.); Mrs. Struthers Burt (91 pp.); Kenneth Chorley (160 pp.); Harold Fabian (106 pp.); Clifford P. Hansen (53 pp.); Harry E. Klissold (50 pp.); W.C. Lawrence (51 pp.); Leslie A. Miller (122 pp.); Homer C. Richards (53 pp.); Laurence Rockefeller (42 pp.); Conrad L. Wirth (76 pp.); Mike Yokel (38 pp.). Restricted.

2113 **Regional Oral History Office, University of California (Berkeley),** Oral history interviews. Copies of interviews with Horace Marden Albright (1961, 49 pp.), Newton Bishop Drury (1972, 700 pp.), and Walter Clay Lowdermilk (1968, 684 pp.).

2114 **Weyerhaeuser Timber Company,** Oral history interviews, 1956, 2,981 pp. Materials on the development of the lumber industry and the lumber regions based upon the recollections of executives and employees of the Weyerhaeuser Timber Company and others. Descriptions of lumbering practices include accounts of life in the Minnesota and Wisconsin woods; labor problems; immigrants; religious practices and conflicts (including some account of the Ku Klux Klan in Washington); camp sports; camp safety practices; fire fighting in camp, mill and forest; Civilian Conservation Corps; reforestation, homesteading and land claims in Idaho about 1900; timber speculation; cooperation in the development of white and ponderosa pine stands in Idaho, Oregon, and Washington; and methods of forest transportation. Corporate developments are described in accounts of early days of the Weyerhaeuser Timber Company and the Weyerhaeuser Sales Company, the Potlatch Lumber Company, and other related or competing firms, market changes and sales problems, advertising and public relations, exploitation of the eastern market, development of intercoastal shipping and of Baltimore and other terminals for eastern distribution, effects of the change from rail to truck lots in local sales. There are impressions of members of the Weyerhaeuser and Denkmann families, George S. Long, William Deary, and others prominent in lumbering. Participants include (Vol. I): A.E. Aitchison (85 pp.); John Aram (98 pp.); David H. Barlett (59 pp.); Jack Bishop (32 pp.); Ralph Boyd (26 pp.); Hugh B. Campbell (32 pp.); Norton Clapp (32 pp.); R.V. Clute (65 pp.); T.S. Durment (45 pp.); O.D. Fisher (73 pp.); A.N. Frederickson (71 pp.); John H. Haubert (126 pp.); E.F. Heacox, C.S. Martin, and C.D. Weyerhaeuser (98 pp.); F.W. Hewitt (66 pp.); Robert W. Hunt (85 pp.); C.H. Ingram (12 pp.); R.E. Irwin (40 pp.); S.P. Johns, Jr. (46 pp.); Don Lawrence (66 pp.); George S. Long, Jr. (46 pp.); R.R. Macartney (44 pp.); Charles J. McGough (66 pp.); William L. Maxwell (112 pp.); Howard Morgan (54 pp.); C.R. Musser (27 pp.); Leonard H. Nygaard (49 pp.); Harold H. Ogle (47 pp.); Arthur Priaulx and James Stevens (75 pp.); Al Raught (54 pp.); Otto C. Schoenwerk (40 pp.); A.O. Sheldon (41 pp.); H.C. Shellworth (77 pp.); Frank Tarr (17 pp.); G. Harris Thomas (63 pp.); David S. Troy (36 pp.); Roy Voshmik (16 pp.); John A. Wahl (18 pp.); Frederick K. Weyerhaeuser (167 pp.); J. Philip Weyerhaeuser (41 pp.); Maxwell W. Williamson (38 pp.); (Vol. II): Earl R. Bullock (32 pp.); Albert B. Curtis (103 pp.); Wells Gilbert (26 pp.); Roy Huffman (68 pp.); W.K. McNair (33 pp.); Leslie Mallory (13 pp.); S.G. and C.D. Moon (32 pp.); Jack Morgan (43 pp.); J.J. O'Connell (77 pp.); R.E. Saberson (81 pp.); Hugo Schlenck (113 pp.); Gaylord M. Upington and Lafayette Stephens (75 pp.). Restricted.

New-York Historical Society

Manuscript Department. 170 Central Park West 10024. Holdings are described in *A Guide to the Manuscript Collections of the New-York Historical Society,* by Arthur J. Breton, Westport, Connecticut: Greenwood Press, 1972, 2 vols.

2115 **Anonymous.** Memorandum book kept by an unidentified person who traveled with Richard Hirst, millwright, visiting sawmills in Baltimore; Columbia and Marietta, Pennsylvania; Newark, New Jersey; New Hampshire; New York City; and Philadelphia, in 1853. There are comments on the examination of new machinery and equipment at work. The author's initials may be J.W. (32 pp.).

2116 **Bancker, Christopher** (1695-1763). New York City merchant. Journal (1718-55) records sales of a variety of merchandise, including naval stores. Ledger (1725-46) records sales of pitch, turpentine, etc.

2117 **Blackwell, Robert N., & Co.** (Astoria, New York). Correspondence (1833-44, 132 items) is mostly letters from Samuel Blackwell, many pertaining to the production and distribution of turpentine and other naval stores in the New York City area. There are also balance sheets (1837-46, 37 items).

2118 **Bliven, Charles D.** New York City lumber dealer. Daybrook (1834-42, 567 pp.) records sales of various forms of lumber to carpenters, builders, cabinet makers, etc., including Samuel Thompson & Son, J. & J.W. Meeks, Thomas Addis Emmet, John Marseilles, William Argall, Isaac Walton, Sutton & Carpenter, Eli Sanford, Winant & Degraw, Augustus Campbell, etc. Account book (1841-46, 317 pp.) contains entries for bills receivable, bills payable, an inventory, cash and merchandise accounts, etc.

2119 **Blount, Thomas, and John Gray Blount.** Merchants, Washington, North Carolina. Record book (1796-97, 200 pp.) contains copies of their letters to Peter Schermerhorn, David Allison, William C. Lake, Budden & Smithey, and other merchants and masters of trading vessels, and pertain to such matters as the lumber trade, land transactions, etc.

2120 **Curtiss, Miles.** Account book (1800-64 [largely 1800-43], ca. 400 pp.), kept by Curtiss in Burlington, Vermont, and by his descendants in Ellisburg, Jefferson County, New York, including many entries relating to sawmill work.

2121 **De Peyster Family,** Correspondence and papers, 1699-1881, 5 feet. Letters (1699-1700, 17 items) from Richard Coote, First Earl of Bellomont (1636-1701), to Abraham de Peyster concerning masts and timbers for the Royal Navy, among other matters. Letters (13 items) from Andrews and Campbell, merchants and shippers from Campobello, New Brunswick, to Frederick de Peyster, pertain to trade with New York, particularly in plaster and lumber. Daybook (1723-28, 1730-33) kept by a member of the de Peyster family in New York City, relates to sales of a wide variety of products, including tar, staves, and pitch, and contains accounts pertaining to trading voyages to the West Indies, North Carolina, Boston, England, Madeira, etc.

2122 **Despard and Ford Families.** Pembroke, Plymouth County, Massachusetts. Ledgers (1702-1809, 142 pp.) include accounts for sales of rails and shingles, and for sawing planks and timber.

2123 **Dodge, Richard** (1762-1832). Correspondence, legal, and business papers (1806-30, ca. 310 items) include reference to trade in lumber, Johnston, Montgomery County, New York.

2124 **Duane and Featherstonhaugh Families,** Correspondence and papers, 1666-1900, ca. 8 feet. The major part of the collection is the correspondence and papers of James Duane (1733-97), New York attorney and land developer. Included are contracts with millers, millwrights, and carpenters (1770), and material relating to conditions of labor in the sawmill at Duanesburgh, New York (1793).

2125 **Duer, William** (1747-1799). Correspondence and papers (mostly 1772-97, ca. 2,950 items), including letters, bills, receipts, deeds, leases, bonds, invoices, cargo manifests, inventories, memoranda, vouchers, contracts, etc., pertaining to Duer's business interests, among them sawmills, sales of masts and spars, etc. Letterbook (1774-79, 38 pp.) contains copies of correspondence, contracts, articles of agreement, and other papers relating to ventures to supply the Royal Navy and later the governments of France and Spain with timber for masts, spars, etc. Silas Deane, James Wilson, and Mark Bird were associated with Duer in this enterprise.

2126 *Experiment* (Sloop). Papers (1785-87) relating to voyage from New York to Canton, China, and return. Included are cargo lists, instructions, lists of subscribers, auction returns, ship's outfitting lists, etc. The cargo included naval stores.

2127 **Green, Edward.** Diary (1844-45, 1 vol. [200 pp.]) notes the daily activities of a lumber yard proprietor, New York City, particularly the collection of debts.

2128 **Green, Garret, and George Green.** Lumber merchants, New York City. Daybook (1833-35, 90 pp.) records sales of lumber, including names of perchasers, charges, and cartage. Several additional pages are obscured by newspaper clippings.

2129 **Hasbrouck, Abraham Joseph** (1773-1845). Journals (1797-1842 [chiefly 1797-1832], 25 vols.) record accounts of his business as merchant and freighter, Kingston Landing (Rondout), Utter County, New York, including accounts for transactions for lumber in various forms, such as joists, planks, ship knees, staves, nut wood, oak, cherry, etc. Ledgers (1799-1846, 3 vols.) relate to entries in the preceding journals. There are also other record books (1820-45, 12 vols.).

2130 **Lawrence, Leonard.** Flushing, Queen's County, New York. Daybook (1773-1821, ca. 250 pp.) relating to sale or purchase of various goods and services, including lumber, rafting timber, etc.

2131 **Livingston Family.** Correspondence and papers (1685-ca. 1885, ca. 65 feet and 100+ vols.) include an account book of a sawmill owned by the family, New York State.

2132 **Lloyd, James** (ca. 1650-1693), Papers, 1676-1731, 1 vol. Boston and Lloyd's Neck, Long Island. Includes a volume of accounts for wood and other goods and services (1676-84).

2133 **Mott, Garrit Striker, and M. Hopper Mott.** Letterbook (1845-50, 1 vol. [156 pp.]) contains copies of letters written by the Motts, relating in part to their lumber business, Motts Point, New York. Several letters are to Samuel Coles Mott and Garrit G. Mott in San Francisco, concerning shipping of lumber to San Francisco, a partnership with Agnew & Co., etc.

2134 **New York City,** Register of Tradesmen Appointed (1789-96, 1 vol. [147 pp.]). Includes wood inspectors, appointed by New York City mayor, Richard Varick.

2135 **New York State Lumbermen,** Petition (1827). Petition of 146 lumbermen from northern and western New York to the state legislature, February 26, 1827, complaining about the Canal Commission's doubling of the duty on timber rafts.

2136 **Pendleton, Nathaniel** (1756-1821). Correspondence and papers (1781-1821, ca. 150 items) include an account of the seacoast of Georgia, discussing shipbuilding, the timber trade, etc. (February 28, 1800, 17 pp.).

2137 **Poughkeepsie, New York.** Daybook (1798-99, 208 pp.) of an unidentified merchant of a firm which apparently kept a sawmill, a plastering mill, and a general store. There are accounts for boards, planks, etc.

2138 **Pryor, John Edward,** Account books, 1759-1769, 5 vols. Accounts for carpentry work, New York City and Perth Amboy, New Jersey. Includes prices for boards.

2139 **Ruthven, James.** Daybook (1792-1804), cash received book (1843-49), and ledger (1828-43, 1872-88) contain accounts for wood-turning jobs, manufacture of boxes, etc.

2140 **Sanders Family.** Papers (ca. 1700-1850, ca. 7,000 items) pertain to business and personal affairs of the family and include a large number of bills for such things as lumber, carpentry, sawmills, other goods and services. Daybook (1733-49) of Robert Sanders, Albany, includes transactions in white pine boards. Account book (1738-59) of Barent Sanders, Schenectady, contains miscellaneous records, including accounts with Major Jacob Glen for providing timber for batteaus (1746), and for firewood delivered to royal garrisons (1756-58).

2141 **Schuyler, John Bradstreet** (1763-1795). Saratoga, New York. Correspondence, bills, orders, etc. (1778-95 [largely 1790-95], ca. 60 items) include material relating to Schuyler's sawmills and lumber business.

2142 **Stevens, Ebenezer** (1751-1823). Correspondence, bills of lading, bills, accounts, military papers, and miscellaneous papers (1776-1822, ca. 525 items) include materials pertaining to his trade in lumber and staves, New York City.

2143 **Strickland, William,** Diary (1794-95, 18 vols. + index vol. [419 pp.]) relates to tour of New England and includes discussion of sawmills, waste of timber, trees, plants, wildlife, and other topics. Correspondence (1793-1827, ca. 60 items) includes references to botany and seed collecting. Correspondence record book (1797-1801) includes a letter to John Bartram, expressing dissatisfaction with Bartram's seeds and service.

2144 **Sweet, Nehemiah** (b. 1785), Lumber merchant, Fort Edward, Washington County, and Poughkeepsie, New York. Correspondence (1814-29, 87 items) primarily with his brother, Thomas Sweet, while Nehemiah was purchasing lumber in Fort Edward for sale in Poughkeepsie, and pertaining to orders for lumber; demand for lumber; availability of boards, planks, shingles, clapboards, posts, masts, etc.; quality of lumber available; prices; rafting; etc. Daybook and ledger (1818-26) relates to sales of lumber at Poughkeepsie. Account book (1833-34) contains miscellaneous accounts. Bills, receipts, vouchers, accounts, and miscellaneous papers (1810-52, ca. 400 items) include many relating to the sale of lumber and expenses such as freight and towing. Lumber account books (1811-26, 15 items) record purchases of board, plank, shingle, beams, poles, etc., and expenses of rafting, labor, traveling, drawing, piloting, etc. Receipt books (1843-55, 2 vols.) relate to the estate of Thomas Sweet.

2145 **Taylor, John** (d. 1777). Daybook (1762-63, 19 pp.) records sales of the cargo, including lumber, of the *Charming Sally* in the West Indies. Charles Nicoll may have been involved in this enterprise.

2146 **Union, New Jersey,** Roll book, 1818, 35 pp. Lists numbers of male inhabitants and their property, including sawmills.

2147 **Van Cortlandt, ?.** Daybook (1716-41, 67 pp.), kept in New York City until 1720, and thereafter in New Jersey, recording sales of loads of wood and other commodities.

2148 **Van Cortlandt, John.** Invoice book (1766-69) records shipments of tar, shingles, etc., to Philadelphia, Virginia, Bristol, the West Indies, and elsewhere, from New York.

2149 **Van Cortlandt, Stephanus** (1643-1700). New York. Ledger (1695-1701, 412 pp.) records sales of boards and other goods.

2150 **Wendell, Abraham Evertse.** Albany, New York, saw and grist mill proprietor. Ledgers and daybooks (1740-90, 5 vols.) record business and personal transactions, especially sales of boards.

2151 **Wendell, Evert** (d. 1750). Merchant, of Albany, New York. Ledger (1708-58) includes accounts for his sawmill. Daybook (1717-49) includes charges for lumber, etc.

2152 **Wendell, Philip.** Account book (1789-98, 1 vol [ca. 60 pp.]) records sales of lumber, Albany, New York.

2153 **Wilson, William.** Daybook and ledger (1813-21, 1 vol. [222 pp.]) kept at Clermont, Columbia County, New York, records accounts for sales of lumber, etc.

2154 **Witter, Thomas** (d. 1786). Merchant of New York City. Daybooks (1753-60, 2 vols. [240 pp.]) record transactions in joint account with Daniel Stiles. Many entries are for sales of naval stores.

New York Public Library, Manuscripts and Archives Division

5th Avenue & 42nd Street 10018. Major holdings are described in the New York Public Library, Research Libraries, *Dictionary Catalog of the Manuscript Division,* 2 vols., Boston: G.K. Hall & Co., 1967.

2155 **Briggs, Alanson T.,** Bills and accounts, 1858-1867, 2 boxes. New York City. Dealer in barrels and casks.

2156 **Century Company,** Records, ca. 1870-1917, 207 boxes. Letters to Clarence C. Buel, Richard Watson Gilder, Josiah G. Holland, and Robert Underwood Johnson, as editors of *The Century Magazine,* with manuscripts of proposed contributions, arranged alphabetically by author. Includes correspondence of the magazine's predecessor, *Scribner's Monthly.* There are letters to *The Century* relating to Yosemite National Park.

2157 **Church, Maria Trumbull Silliman** (1810-1880), Correspondence, 1829-1858, 2 boxes. Family correspondence, including letters from Mrs. Church's husband, John Barker Church, and other relatives, concerning the operation of his lumber mills, among other matters.

2158 **Cross, Stephen,** Account book, 1792-1834, 1 vol. Records of lumber trade and agriculture in Monson, Massachusetts. There is also a copy of the inventory of Cross' estate.

2159 **Dering, Henry Packer,** Papers, 1780-1855. Papers of a shipper and ship captain of Sag Harbor, Long Island, which include an account of taking coopers and a cargo of staves to the West Indies in the late 18th century, in an attempt to teach the natives to make casks for shipping molasses back to the United States.

2160 **Emmet, Thomas Addis,** Collection, 94 vols. and 2,500 items (10,800 pieces). Items gathered for their autograph value. Included are scattered materials pertaining to forest history, such as an agreement between William Hill and Isaac Hayne, Camden, South Carolina, for operating a sawmill and other businesses; and a letter (March 6, 1787) of Peleg Wadsworth, Portland, Maine, to Richard Devens, discussing acceptance of lumber and oak pipe staves by the General Assembly for payment of taxes. The collection is partly calendared in the catalog published by the library in 1900.

2161 **Ervin, Thomas, and Edwin J. Ervin,** Account book, 1833-1873, 1 vol. (ca. 60 pp.). Labor costs of hewing timber, logs, rails, etc. Genesee County, New York, and DeKalb County, Illinois.

2162 **Gansevoort-Lansing,** Papers. Land and miscellaneous papers include an agreement about building a sawmill (1732-33), Saratoga County, New York.

2163 **Gardiner, John Lyon,** Memorandum book, 1790-1809, 106 pp. Gardiner's Island, New York. Records sale of timber, prices received, etc.

2164 **Hardy, Joseph,** Account book, 1819-1851, 1 vol. Merrimack County, New Hampshire. Contains credit entries for labor, wood, etc.

2165 **Lichtenberger Family,** Papers, 1837-1894, 14 vols., 75 items. Correspondence, accounts, and other papers of George and Richard Lichtenberger, New Harmony, Indiana. Includes information on costs of wood, other commodities, and labor.

2166 **Little Family,** Correspondence, 1808-1860, 7 boxes. Letters received by Josiah Little (1747-1830) of Newburyport, Massachusetts, and his son, Josiah Little (1791-1860), chiefly from Edward Little (1773-1849) of Danville, Maine, containing discussion of the lumber industry in Maine.

2167 **Log Books,** Collection. Includes a log (1859-60, 130 pp.) kept by James T. Seaver, a seaman of Chelsea, Massachusetts, on board the ship *May-Flower,* from Boston to San Francisco, Port Gamble on Puget Sound, and return, with cargo of lumber from Puget Sound to Valparaiso, Chile.

2168 **Miscellaneous.** Letters (1794) by Tench Coxe, Assistant Secretary of the Treasury, concern lumber for naval ships. There is a manuscript by Thomas Stevens accompanying his pamphlet, *The Rise and Fall of Pot-Ash in America* (London, 1758). Items relating to New Hampshire include a petition of merchants of Portsmouth to the State Senate and House of Representatives for the protection of the potash industry (1785). Minutes and proceedings (1791-97) of the Commissioners for Alms House and Bridewell, New York, concern, among other matters, the distribution of firewood.

2169 **Moffatt, Thomas,** Letter book, 1714-1716, 1 vol. (65 pp.). Merchant, of Boston, Massachusetts. Correspondence with Isaac Sperrin, John Angier, of Bristol, England, and others in England and America, relating to shipbuilding, exportation of turpentine, tar, and other goods.

2170 **Phelps, Dodge and Company** (New York), Papers, 1818-1869, 38 boxes and 37 vols. Included is correspondence relating to the purchase of pine timber lands in Pennsylvania, the erection of a mill, and the manufacture of lumber.

2171 **Porter, Augustus,** Daybook, 1818-1822, 1 vol. Merchant of Niagara, New York. Transactions at his general store. Prices of lumber and other commodities.

2172 **Sackett, Russell,** Account book, 1808-34, 1 vol. (43 l.). Cost of labor, chopping wood, etc., on his farm in Connecticut.

2173 **Stow, William Sears** (1797-1880), **and De Lancey Stow** (1841-1925), Papers, 92 boxes and 16 vols. Attorneys of Clyde, New York. The papers include correspondence (ca. 1822-24) concerning shipment of timber and towing rafts from southern New York state down the Susquehanna River to Baltimore, Maryland, and include data on prices.

2174 **Vaux, Calvert,** Papers, 1865-1921, 1 box. Included are letters (1865-87) from Frederick Law Olmsted relating to parks in New York, Yosemite, etc.

2175 **Wakeman, Burr,** Account book, 1823-1848, 1 vol. New York merchant. Included are transactions in ashes (1823-38) and other commodities.

2176 **Wallace, William A.,** Papers, 1809-1815, 56 items. New York City merchant. Financial statements, bills of lading, correspondence with merchants in Newry (?), Dublin, Nancy, Madeira, and Trinidad, concerning trade in potash, etc.

PORTVILLE

Portville Free Library

Librarian 14770

2177 **Dusenberry, J.E., & Co.** (Porterville, New York), Account books, 1849-1885, 38 vols. Records of transactions of a general store and lumber company, showing customers, amounts sold, and prices.

ROCHESTER

The University of Rochester Library, Department of Rare Books, Manuscripts, and Archives

Librarian 14627

2178 **Hollister, E.H., Lumber Company** (Rochester, New York), Records, 1828-1951, 3 boxes (48 folders), and 78 vols. Correspondence (1839-51, 20 folders), legal documents (1828-1917, 12 folders) arranged by type; abstracts of title, bonds, contracts and agreements, court papers, deeds and indentures, leases, mortgage discharges, etc.; business and financial papers (1842-1951, 16 folders), arranged by type; annual reports (1889-1951); bank drafts; insurance policies; inventories; orders, invoices, bills of sale; papers pertaining to stock meetings, officers, organization; price lists; promissory notes; receipts and bills; statements of accounts; miscellaneous. Bound volumes include miscellaneous cashbooks and accounts, daybooks, journals, ledgers, letterpress copybooks, etc., of Hollister and related firms.

SARANAC LAKE

Saranac Lake Free Library, Adirondack History Room

100 Main Street 12983

2179 **Benedict, Darwin.** "The New York Forest Preserve: Formative Years, 1872-1895," Albany: State University of New York?, M.A. Thesis, 1953, 212 1.

2180 **Donaldson, Alfred L.,** Adirondack collection, 1899-1920, 903 items. Materials gathered by Donaldson in preparing his *A History of the Adirondacks,* 2 vols., New York: Century, 1921, including historical, bibliographical, and pictorial letters, questionnaires, records of interviews, notes, references to books, articles, newspapers, letters, and other documents relating to the Adirondack region; data on the minutes of meetings of the Association for the Protection of the Adirondacks; maps; and manuscripts and other materials from various persons and publishers consulted by Donaldson. There are typewritten indexes of manuscripts, questionnaires, notes, and letters in the collection, compiled by George Marshall.

STATEN ISLAND

Staten Island Institute of Arts and Sciences

Librarian, Sailors' Snug Harbor 10301. Researchers should make appointments in advance.

2181 Manuscripts of William T. Davis, James P. Chapin, Arthur Hollick, and Nathaniel Lord Britton, and others, relating to the natural history of Staten Island, and also to various park and preservation activities (1879-1945). Uncataloged.

SYRACUSE

State University of New York, College of Environmental Science and Forestry, Syracuse Campus, F. Franklin Moon Library

Archivist, Archives, Room 15 13210

2182 **Empire State Forest Products Association,** Records, 1952-1962, 3¾ feet. Includes correspondence (1952-55, 1 foot); meetings (1955-57, 1½ feet); and meetings (1958-62, 1¼ feet). These seem to be the files of Nelson Brown, Executive Secretary of the Association. Restricted: permission to consult the records must be obtained from the Association, 89 Ludlow Street, Saratoga Springs, New York, 12866.

2183 **State University of New York, College of Environmental Science and Forestry,** Records. Included are Society of American foresters, New York Section, miscellaneous papers (4 vertical file drawers); instructional materials; New York Forest Owners Association, papers, including membership files (3 drawers); papers of H.G. Williams relating to the Natural Resources Council (1 drawer); papers of Terence J. Hoverter (1 drawer); materials concerning the history of the College of Forestry (4 drawers); materials relating to the New York State Conservation Department; photographs (2 drawers); papers, notes, articles, addresses, and photographs of Hardy L. Shirley (4 drawers); miscellaneous documents of the S.U.N.Y. Faculty Senate (1 drawer); and other materials. There are also theses and reports, printed bulletins, technical publications, etc.

Syracuse University, The George Arents Research Library 13210

Manuscripts

2184 **Baldwin, John,** Papers, 1806-1849, 78 items. Papers of Baldwin who apparently had real estate and timber industry interests in the Sharon, Vermont, area.

2185 **Coleman Family,** Papers, 1794-1877, 322 items. Includes letters (1838-54) from various members of the Coleman family who had farming and timber interests in Hanover, New Hampshire, and Lunenburg, Vermont, to Israel Dewey and other members of the Dewey family.

2186 **Delaware, Lackawanna and Western Railroad Company,** Records, 1828-1960, 2,000 feet. Records of the board of managers, the president, the vice president, and of the Coal, Land and Tax, Law, Operating and Treasury Departments. Includes records of subsidiary, predecessor, and leased companies, including timber companies. Published register (1964).

2187 **Roosevelt, Nicholas** (b. 1893), Papers, 1846-1962, 32 feet. Journalist, diplomat, and conservationist. The collection includes correspondence, manuscripts of his writings and related material, photos, scrapbooks, etc. Unpublished guide.

2188 **Sumner, David H.,** Papers, 1814-1845, 175 items. Correspondence and financial papers (1815-30) relating to Sumner's mercantile and timber concerns in Hartland, Vermont.

Archives

2189 **College of Environmental Sciences,** Records, 5½ boxes. Board of trustees records, "probably minutes" (1950-65, 5 boxes). Faculty meeting minutes (1927-28, 1929-30, ½ box).

2190 **Office of the Chancellor,** Records. Papers of James Roscoe Day, Chancellor, 1894-1922, contain letters, memoranda, documents, and notes relating to the origins, establishment, and early history of the New York State College of Forestry; there are copies of letters sent by or received from Louis Marshall, president of the Board of Trustees of the College (1911-29, 75+ items) concerning the founding of the college and the acquisition of financial support; letters from or to Hugh P. Baker, first permanent dean of the college (40+ items); circular letters from Baker to the faculty; also copies of letters of Louis Marshall relating to Syracuse University, chiefly to the College of Forestry (45 items, originals in American Jewish Archives, Cincinnati, Ohio). Papers of Charles W. Flint, Chancellor, 1922-37, and William P. Graham, Chancellor, 1937-42, contain more than 3 boxes relating to the College of Forestry, including correspondence between Flint and Graham and deans Franklin F. Moon, Hugh P. Baker, and Samuel F. Spring; annual reports of the dean; minutes of meetings of the trustees; and notes, reports, lists, and many other records; also a large amount of correspondence between the Chancellor and the President of the University and New York State Commissioner of Education, relating to the College. This collection is described in: James Owens, "College Records in SU Archives," *Forestry Library Review,* I (March, 1968), pp. 7-9 (processed publication of the Moon Memorial Library).

North Carolina

ASHEVILLE

Pack Memorial Public Library

North Carolina Collection, Soundley Reference Library 28801

2191 **Biltmore Forest School,** Collection, 1897-1957, 29 items. Books, articles, and other printed materials dealing with Carl Alwin Schenck and the Biltmore Forest School, including several addresses and papers by Schenck (ca. 1900-12); a manuscript biography of Schenck (4 pp.); clippings and 15 photographs concerning Schenck's visit to the United States in 1951; three letters by Schenck (1897-98) to C.S. Sargent, D. Mauds, and Gifford Pinchot; a photograph album of an unidentified student at the Biltmore Forest School; and the diary of J.E. Benedict (1906-07, 27 pp.).

2192 **Kephart, Horace,** Papers. There is material concerning the naming of places in the Great Smoky Mountains National Park (10 items); also material concerning Kephart's burial site in the park, and the formation of a park historical museum.

CHAPEL HILL

University of North Carolina, Library, Southern Historical Collection

Director 27514. The holdings are described in *The Southern Historical Collection: A Guide to Manuscripts,* by Susan Sokol Blosser and Clyde Norman Wilson, Jr., Chapel Hill: The Southern Historical Collection, University of North Carolina Library, 1970.

2193 **Ashe, William Willard** (1872-1932), Papers, 1859-1932, 2½ feet. Botanist and forester with the North Carolina Geological Survey, 1892-1905; the U.S. Forest Service, Washington, D.C., 1905-32; Acting Chief and Secretary and Editor of the National Forest Reservation Commission, 1918-24. The collection contains personal, business, and professional correspondence (chiefly 1898-1932) concerning forestry, forest products, and soil conservation and flood control. Includes clippings, articles, photos, and 3 vols. of notes. Unpublished description.

2194 **Baldwin, John K.** (b. 1845), Diary, 1867-1878, 3 vols. Baldwin, of Bladen County, North Carolina, worked for the Bladen Land Company and Cape Fear Company doing sawing, carpentering, and other labor.

2195 **Bertie County, North Carolina,** Miscellaneous Books, 1833-1849, 1881-1893, 5 vols. Vol. 4 (1847-49) contains accounts of Gray & Pierce for shingles from Pierce's swamp, record of lands, dates of working and rate of wage, amount of work done.

2196 **Bonner, Herbert Covington** (1891-1965), Papers, 1940-1965, 55 feet including 17 vols. U.S. Representative from North Carolina, 1940-65. Three folders include correspondence with lumber interests protesting targets for bombing ranges. Unpublished description.

2197 **Borough House,** Books, 1815-1910, Microfilm (20 vols.). Records of several generations of the Anderson family, near Stateburg, South Carolina, including lumber account books.

2198 **Brown, Hamilton** (1786-1870), Papers, 1752-1907, 2 feet (including 23 vols.). Business and personal papers of three generations of a prominent North Carolina-Tennessee family. Vol. 7 contains records of a lumber business (1853-56) which probably operated in Wilkes County, North Carolina.

2199 **Bryan Family,** Papers, 1704-1954, 20 feet (including 17 vols.). Papers and record books of the Bryan family of Carteret and Craven Counties, North Carolina, including those of James Augustus Washington Bryan (1839-1923), lumber mill operator.

2200 **Carmack, Edward Ward** (1858-1908), Papers, 1850-1942, 3 feet (including 15 vols.). U.S. Representative, 1897-1901, and Senator, 1901-1907, from Tennessee. The papers include a letter book of a lumber business at Burnside, Kentucky (1894-95).

2201 **Clarke, William John** (1819-1886), Papers, 1838-1881, 813 items. Raleigh and New Bern, North Carolina. Chiefly business papers, including bills, receipts, and reports on Clarke's lumber mill and other business interests. Unpublished description.

2202 **Comer, John Fletcher** (1811-1858), Book, 1844-1847, 1 vol. Records kept by a Barbour County, Alabama, planter, including the records of his lumber and grist milling business.

2203 **Cooley, Harold Dunbar** (b. 1897), Papers, 1934-1966, 164 feet. U.S. Representative from North Carolina, 1934-66, and chairman of the House Committee on Agriculture, 1949-66. Unprocessed (1970).

2204 **Corban, Burwell J.** (b. 1853), Books, 1850-1890, 29 vols. Records of Corban and his father, of Corbandale and Palmyra, Tennessee. Included are scattered accounts for lumber milling and other work.

2205 **Coulter, John Ellis** (1861-1947), Papers, 1849-1960, 16 feet (including 49 vols.). General and family correspondence, business papers and account books of Coulter, who engaged in lumber business, farming, and other activities in Burke and Catawba Counties, North Carolina. There are also papers concerning tanbark.

2206 **Criglar, William Louis,** Papers, 1848-1885, 60 items. Lumber mill owner of Escambia County, Alabama, and Escambia and Santa Rosa Counties, Florida. Business correspondence with firms at Mobile and New Orleans relating to supplies, equipment, slaves, and financing.

2207 **Ervin, William Ethelbert** (1809-1860), Book, 1846-1856, 1 vol. Included are saw and grist mill accounts (1851-54), Lowndes County, Mississippi.

2208 **Fisher Family,** Papers, 1758-1896, 4050 items (including 51 vols.). Included is a small amount of material relating to the lumber industry. There is an account book of Thomas and Beard, Davie County, North Carolina, relating to sawing, nails, and planks (1837-38, 14 pp.); a daybook containing entries for lumber (1838-39); also scattered ledger accounts mentioning planks.

2209 **Fredericks Hall,** Books, 1727-1862, 24 vols. Included are lumber account of a Louisa County, Virginia, plantation, and some accounts for plank roads in Hanover and Louisa Counties.

2210 **Garland, William Harris** (b. ca. 1811), Papers, 1819-1873, 500 items. Personal correspondence (from 1836) of an itinerant mechanic for railroad shops, sawmills, and steamboats in various towns of North and South Carolina, Georgia, and Florida, largely concerned with the uncertainties of livelihood without capital.

2211 **Gray Family Papers,** 1722-1879, 850 items (including 4 vols.). Included are business and personal correspondence, bills, and accounts (1846-79), of George Gray, merchant and planter of Windsor, North Carolina, concerning shingle shipments, among other matters.

2212 **Harbison, Thomas Grant** (b. ca. 1860), Papers, 1905-1926, 392 items. Arboriculturist. Letters (largely 1912-19), received by Harbison from Charles Sprague Sargent (1841-1927), botanical and arboricultural projects while Harbison was southern field botanist for Harvard University's Arnold

Arboretum, working from Highlands, Macon County, North Carolina.

2213 **Harnett County,** Papers, 1763-1886, 640 items (including 4 vols.). Included are letters and papers (largely 1845-75), relating to the lumber and naval stores industries.

2214 **Hogg, Thomas Devereux** (1823-1904), Papers, 1829-1910, 2⅔ feet (including 23 vols.). The papers include accounts of a sawmill, Raleigh, North Carolina.

2215 **Jones and Patterson Family,** Papers, 1777-1907, 7 feet (including 93 vols.). Account books (from 1796) of Gen. Edmund Jones (1771-1844), Wilkes and Caldwell counties, North Carolina, include data on lumbering. There are also papers (1861-90) concerned with the lumber business of Patterson and Fries, Winston-Salem, North Carolina.

2216 **Judge, John,** 1830-1885, Business papers, 1852-1881, 3 feet (including 69 vols.). Army officer and businessman. Business correspondence, account books, bills, receipts, and other papers (chiefly 1860-73) relating to Judge's activity as owner of a paper mill in Wilmington, North Carolina, and pre-Civil War lumber business in Wisconsin. He also had interests in the naval stores industry and other businesses. Includes account books, bank books, sales and daybooks, deeds, articles of agreement, specifications for factories and machinery, and receipts. Correspondents include Henry Howe and P.H. Winston.

2217 **Kramer Brothers Company** (Elizabeth City, North Carolina), Papers, 1917-1962, 3 vols. and 3 folders. Minutes of directors and stockholders meetings, and organizational documents of a lumber company which owned and operated sawing and finishing facilities, timberlands, and related property.

2218 **Liddell, Walter James Forbes** (d. 1889), Papers, 1831-1914, 286 items (including 4 vols.). Business correspondence, accounts, receipts, promissory notes, deeds, blueprints, and printed patents (chiefly 1880s and 1890s), relating to the firm of Liddell and his son, Walter S. Liddell, manufacturing machinery, including circular saw mills, Charlotte, North Carolina.

2219 **Macay-McNeeley,** Papers, 1791-1872. Vols. 5-9 contain records (1834-72) of a sawmill and lumberyard in Rowan County, North Carolina.

2220 **McClennan, Stonebraker, and McCartney Family,** Papers, 1822-1866, Microfilm (352 frames [including 2 vols.]). Business and family papers of Jacob Stonebraker, Lincoln County, Tennessee, saw and grist mill owner, including account books (1833-66) of his mills.

2221 **McDowell, Thomas David Smith** (1823-1898), Papers, 1783-1925, 3000 items (including 5 vols.). Business and political correspondence of a Bladen County, North Carolina, planter and political leader, including letters concerning the lumber and naval stores industries.

2222 **Marrow, Goodrich Wilson,** Papers, 1884-1927, ca. 2 feet (550 items [including 13 vols.]). Business papers of a farmer and businessman of Townsville, Vance County, North Carolina (mainly 1904-19) relating to lumber, among many other enterprises.

2223 **Mercer Family,** Papers, 1854, 26 items (including 23 manuscripts and microfilm of 3 transcripts, 73 pp.). Papers of a Brunswick County, North Carolina, family, including correspondence of John Mercer (1812-63) relating chiefly to his turpentine business.

2224 **Moses, Edward Pearson** (1857-1948), Papers, 1901-1948, 570 items (including 1 vol.). Correspondence (chiefly after 1924) of Moses, concerning his interest in the establishment of Great Smoky Mountains National Park, and other affairs.

2225 **Nevitt, John,** Diary, 1826-1832, 1 vol. Journal kept by Nevitt at his plantation near Natchez, Mississippi, including entries relating to lumber.

2226 **Nicol, Walter** (1817-1861), Diary, 1845-1849, 1 vol. Diary of an employee of a New Orleans lumber exporter. Subjects include Nicol's work and a schooner trip to visit lumber plants at Pensacola.

2227 **North Carolina Geological Survey,** Papers, 1885-1914, 11½ feet. Inquiries, reports, and correspondence of Joseph Austin Holmes, state geologist, 1891-1906; and of Joseph Hyde Pratt, state mineralogist, 1897-1906, and state geologist, 1906-23. The papers concern the state's mineral resources, forest preservation, drainage problems, fisheries, and good roads. Also included are the unofficial papers of Holmes and Pratt in connection with their work as University of North Carolina professors, industrial consultants, participants in organizations, conferences, and movements in the field of conservation and development of natural resources.

2228 **Patterson, James N.,** Papers, 1793-1864, 2 items (including 1 vol.). Miscellaneous Orange County, North Carolina, records including a ledger for lumber and provisions (1839-63).

2229 **Pensacola & Georgia Railroad Account Book,** 1858-1865, 1 vol. Accounts of lumber purchased by the railroad, probably kept by an overseer of maintenance.

2230 **Phillips and Myers Family,** Papers, 1804-1928, ca. 1400 items (including 13 vols.). Includes correspondence of William Hallett Phillips (1853-97) regarding the development of Yellowstone National Park. Unpublished description.

2231 **Powell, William Stevens** (b. 1919), Collection, 1793-1924, ca. 250 items. Papers (1853-1901) of Tobias Goodman (1814-80) of Iredell County, North Carolina, including account books for sawing and lumber sales (1853-56). Unpublished description.

2232 **Pratt, Joseph Hyde** (1870-1942), Papers, 1889-1942, 7½ feet (including 13 vols.). Pratt was a geologist for the North Carolina Geological Survey, 1897-1906; professor at the University of North Carolina, 1898-1925; North Carolina State Geologist, 1906-23; and consultant for the U.S. Geological Survey in the 1930s; and organizer and promoter of movements for the conservation and development of forests and other resources, and of the North Carolina good roads movement. Correspondence relates to these activities and includes letters received, carbon copies of letters sent, circulars and mimeographed releases issued by Pratt, and scientific writings.

2233 **Ramsey, Darley Hiden** (1891-1966), Papers, 1877-1966, 6600 items (including 7 vols.). Correspondence, speeches, and writings of Ramsey, Asheville newspaper editor and politician. Correspondents include conservationists. Topics, among others, are economic development, and history of the North Carolina mountain region. The bulk of the papers are from 1940-65.

2234 **Ransom, Matt Whitaker** (1826-1904), Papers, 1845-1914, 20 feet. U.S. Senator from North Carolina, 1872-95; his committee service included Agriculture and Forestry. Correspondents include family members, constituents, and political colleagues. Unpublished description.

2235 **Ruffin, Roulhac, and Hamilton Family Papers,** 1784-1935, 7000 items (including 16 vols.). Includes a ledger of Daniel Heyward Hamilton, Jr. (b. 1838) for his naval stores business in Madison County, Florida (1865-66).

2236 **Ruffin, Thomas** (1787-1870), Papers, 1753-1896, 20,000 items (including 50 vols.). The papers include account books for sawmilling, Orange and Alamance Counties, North Carolina.

2237 **Saunders, Joseph Hubbard** (1839-1885), Papers, 1777-1921, 800 items. Included is information on the lumber business in North Carolina after the Civil War. Unpublished description.

2238 **Seawell, Aaron Ashley Flowers,** Papers, 1810-1893, 885 items (including 8 vols.). Included are records and business correspondence (1870s-80s) relating to sales of lumber, timber, naval stores, delivery of turpentine, deeds, notes, and bills, among other business matters. Moore County, North Carolina.

2239 **Small, John Humphrey** (1858-1947), Papers, 1874-1947, 18 feet (including 3 vols.). U.S. Representative, 1899-1921. Includes papers concerning Small's political career; also directors' and stockholders' minutes and financial reports of a Chowan County, North Carolina, company engaged in lumbering and the manufacture of shingles and staves (1888-1910, 1917-25).

2240 **Smith, Peter Evans** (1829-1905), Papers, 1738-1944, 1000 items (including 22 vols.). Includes papers relating to planting and lumbering before, during, and after the Civil War. Halifax County, North Carolina.

2241 **Sparrow, Thomas** (1819-1884), Papers, 1835-1871, 4 vols. and 150 items. Included is correspondence (1837) relating to a legal case involving the question of nuisance in connection with the building of a turpentine distillery. Sparrow was a Washington, North Carolina, lawyer.

2242 **Taylor, Alfred** (b. 1823), Diary, 1851-1909, microfilm (3 vols.). A daily record of Taylor's sawmills, and other concerns, near Greenville, South Carolina.

2243 **Thomas, John Warwick** (1800-1871), Papers, 1848-1893, 20 items. Correspondence concerning a sawmill and other business matters, Thomasville, North Carolina.

2244 **Thompson, Lewis** (1808-1867), Papers, 1717-1894, 2800 items (including 8 vols.). Business papers of a planter of Bertie County, North Carolina, and Plaquemines Parish, Louisiana, including lumbering accounts (1861-62).

2245 **Tift, Nelson,** Diary, 1835-1851, Microfilm (1 vol. [175 pp.]). Microfilm of typescript. Some entries concern Tift's lumbering business, Albany, Georgia.

2246 **Tucker, Robert Pinckney** (1865-1920), Papers, 1893-1932, 7½ feet (including 31 vols.). Tucker, of Charleston, South Carolina, was organizer of a dozen corporations which bought and sold large tracts of land for lumber, minerals, hunting preserves for wealthy northerners, and development, in South Carolina and at least 8 other states. The bulk of the collection consists of business letters received and simultaneous letter copybooks (chiefly 1902-10). There are also abstracts of land titles (12 vols.), financial statements (3 vols.), accounts, and land plats and maps (ca. 1 foot). Included is material on the Carolina Cypress Co., Carolina Land Improvement Co., Dry Fork Coal and Timber Corporation, Midland Timber Co., Oakland Club, Oconee Timber Co., Pine Timber Corporation, Sea Coast Development Corporation, Sea Coast Timber Co., Southern Land and Timber Co., Southern Woodland Co., and Three Runs Lumber Co. Tucker's business associates were R.L. Montague and H.F. Welch. Unpublished finding aid.

2247 **Vass, William Worrell** (1821-1896), Business records, 1834-1896, 9 feet (ca. 19,750 items, including 24 vols.). Includes letter copy and sales invoice books of the Carolina Paper Company (1895-96). Unpublished description.

2248 **Vein Mountain Mining Company,** Records, 1903-1911, 3 vols. Letter copybooks of a firm in McDowell and Rutherford Counties, North Carolina, containing letters from the local managers to the owner, Thomas Hamlin Hubbard (1838-1915), New York; correspondence with other firms and individuals; and reports and financial statements. The firm engaged in lumbering as well as gold mining.

2249 **Williams, Sarah Francis (Hicks)** (b. 1827), Letters, 1838-1868, 110 items. Letters from the wife of Benjamin F. Williams, owner of pine lands in Greene County, North Carolina, and Charlton and Ware Counties, Georgia, to her parents in New York state. There are a number of comments on turpentine production, and descriptions of the life of Negro turpentine workers. A portion of the letters was published as "Plantation Experiences of A New York Woman," edited by James C. Bonner, *North Carolina Historical Review,* XXXIII, July and October, 1956, pp. 384-412, 529-546.

2250 **Wilson, Robert** (b. 1789), Papers, 1772-1888, 13 feet (including 85 vols.). Business records, chiefly from Pittsylvania County, Virginia, and Guilford County, North Carolina, containing material relating to sawmill operations. Included is a journal of accounts of Sandy River Saw Mills (1841-61). Unpublished descriptions.

2251 **Withers, Robert W.** (d. 1854), Papers, 1794-1890, 62 items, in part microfilm (including 8 vols.). Records of a planter and businessman of Greene and Hale Counties, Alabama (chiefly 1820-53), including accounts and memoranda pertaining to Withers' lumber and flour mills and other concerns (originals, 2 vols.).

CULLOWHEE

Western Carolina University, Hunter Library, Archives and Special Collections

University Archivist 28723

2252 **Hubbard, Joel T.** (1854-1933), Papers, 1845-1932, 1735 items. In part, photocopies of originals owned by Richard W. Iobest. Hubbard was a merchant, Wilkesboro, North Carolina. Chiefly letters (1880-1931) to Hubbard, together with information concerning timber operations in the northwestern area of North Carolina, especially Wilkes County (late 1800s-early 1900s).

2253 **Kephart, Horace** (1862-1931), Collection, 1902-1938, 25 items. Correspondence, books, notebooks, and journals, including copies of Kephart's own works on the southern Appalachians, describing the life of the mountain people before it was changed by the activities of logging companies; includes a small amount of material relating to the origins of the Great Smoky Mountains National Park.

2254 **Weaver, Zebulon Vance** (1872-1948), Papers, 1850-1951, 27,401 items. Lawyer, judge, state legislator, and U.S. Representative of North Carolina, 1917-29, 1931-47. Includes correspondence (1906-46) relating to congressional affairs of North Carolina and legislation for the Great Smoky Mountains National Park and the Blue Ridge Parkway; transcripts of the

Hiwassee Lumber Company case (1912-32); letters from Josephus Daniels offering support for the proposed Blue Ridge Parkway.

2255 **Western North Carolina Associated Communities,** Collection, 1907-1969, 1,977 items. Material collected by Dr. Maurice B. Morrill relating to the work of this organization in improving the economic, educational, and social life of southwestern North Carolina. There is correspondence (chiefly 1946-67) relating to the Great Smoky Mountains National Park, and reports relating to the Blue Ridge Parkway, environmental education workshops, and the Cherokee Historical Association.

DURHAM

Duke University, Perkins Library, Manuscript Division

Curator 27706. Some of the holdings are described in *Guide to the Manuscript Collections in the Duke University Library,* prepared by Nannie M. Tilley and Noma Lee Goodwin, Durham: Duke University Press, 1947.

2256 **Anonymous,** Ledger, 1767-1773, 1 vol. Records of an unidentified merchant of New Bern, Craven County, North Carolina, relating to trade in such products as pitch, tar, turpentine, and staves.

2257 **Anonymous,** Logbook, 1767-1768, 1 vol. The major portion of the logbook is for voyages of the brigantine *Jonnah* from the Piscataqua River between Maine and New Hampshire to Grenada and other localities in the West Indies with cargoes of lumber, and to Edenton and several places in Bertie County, North Carolina, to procure a cargo of tar, pitch, and staves for London.

2258 **Armstrong, William G.,** Papers, 1848-1882, 292 items. Letters to a Columbia, Tyrrell County, North Carolina, merchant whose business consisted largely of the sale of shingles and lumber.

2259 **Baird, Robert** (d. ca. 1866), Papers, 1832-1873, 1425 items. Correspondence and other papers, largely of a business nature (1856-71), regarding the ironworks at Manchester, Virginia, operated by Baird and Peter Small, and later by James D. Craig. Products included circular saws.

2260 **Barton, Samuel R.,** Papers, 1841-1924, 69 items. Included are papers of Stephen Barton who purchased timber and manufactured plow handles and other lumber products at Bartonsville, North Carolina, before the Civil War. There are materials relating to Barton's contracts with other companies, and material relating to a suit at law, Southampton, Virginia (1858-66).

2261 **Basden, William,** Papers, 1764 (1787-1829) 1859, 6 items. Papers concerning transfer of land by the Basden family of Onslow County, North Carolina, and the renting of turpentine forest land.

2262 **Bellune, J.T.,** Diary, 1861-1862, 1 vol. Diary of a Hamburg, South Carolina, farmer, giving amounts of wood sold.

2263 **Boyd, Joseph Fulton,** Papers, 1861-1869, 12,356 items, and 16 vols. The papers of a Union army officer include information on shipbuilding, and on ships and shipyards at Wilmington, North Carolina (1865).

2264 **Branch, John P.,** Account book, 1848, 1 vol. Wood and drayage accounts, Petersburg, Virginia.

2265 **Bryan Family,** Papers, 1717-1956, 2942 items and 39 vols. Included are references to the lumber industry.

2266 **Buck, Daniel,** Papers, 1849-1900, 385 items. Correspondence, deeds, bills, receipts, and promissory notes of a Spring Garden and Philadelphia, Pennsylvania, cabinet maker and lumber dealer.

2267 **Buford, John,** Papers, 1804-1898, 604 items. Most papers in this collection relate to Virginia railroads and Virginia railroad construction, but there is also material on the North Carolina naval stores industry. Card index.

2268 **Charleston Cotton Exchange** (Charleston, South Carolina), Papers, 1880-1952, 718 items and 11 vols. Among the volumes are price quotations from several markets on rosin, turpentine, and other naval stores (1881-86). The markets are Wilmington, North Carolina; New York; London; Liverpool (to 1883); and Savannah, Georgia (after 1883).

2269 **Davidson, George F.,** Papers, 1748-1887, 1660 items and 14 vols. Personal correspondence, diaries, legal papers, and account books. Includes letters concerning the land and timber around Covington, Newton County, Georgia (1824).

2270 **Dickinson, Samuel, and David Black,** Papers, 1801, 18 items. The papers mention marketing of shingles, the cost of boat construction at Norfolk, Virginia, and lumber for the West Indian market.

2271 **Dismal Swamp Land Company** (Suffolk, Virginia), Papers, 1763 (1830-1871) 1879, 4328 items and 8 vols. Business papers, with a list of original partners including George Washington; and the subdivision of shares during the prosperour years of the company when sale of cypress shingles and staves yielded large profits. Records include accounts of low profits in the 1780s and in the panic of 1837, labor problems, trespassers, transportation problems. Includes monthly accounts of work and production and information on stockholders, including David and Richard K. Meade, the college of William and Mary, Williamsburg, and members of many leading families of Virginia. Among the volumes are check stubs (1840-63); accounts, letter books, and shingle records (1795-1843); and bankbooks (1837-53). The correspondence consists largely of letters to and from the presidents and executive agents; and a few letters from Thomas Walker, an original stockholder.

2272 **Edwards, R.P.,** Account book, 1852-1875, 1 vol. Sawmill accounts, Catawba, North Carolina.

2273 **Evans, Joseph R.,** Papers, 1822-1835, 11 items. Letters from Evans, a Philadelphia, Pennsylvania, merchant and shipper, concerning the shipment of turpentine and lumber from New York and Boston.

2274 **Fairey, Franklin William,** Papers, 1837-1880, 22 items and 6 vols. Largely 1862-65. Includes account books and ledgers of a sawmill business, Branchville, Orangeburg County, South Carolina.

2275 **Fowler, Joseph S.,** Papers, 1779-1870, 2407 items and 4 vols. Correspondence and business papers of Fowler and members of his family of Fairfield, Connecticut, and Craven County, North Carolina, including information on the shipment of lumber, naval stores, and other goods to Bermuda prior to 1820 and to New York after 1820.

2276 **Fox, Himer,** Account book, 1844-1861, 1 vol. Accounts of a sawmill owner, Randolph County, North Carolina.

2277 **Fox, John,** Papers, 1784-1892, 2139 items. The family papers concern lands and lumbering in several Southern states, and the difficulties of operating a sawmill in Alabama.

2278 **Gage, George,** Papers, 1864-1903, 4 vols. Includes a letter book (1884-90) containing letters written by Gage concerning the operation of a sawmill at Beaufort, South Carolina.

2279 **Gairdner, James,** Papers, 1771-1816, 84 items. Letters and accounts (largely 1803-09), of a merchant of Charleston, South Carolina, dealing in lumber, rum, and wine, with firms in England, France, and the British West Indies.

2280 **Garst, Henry and John,** Papers, 1830-1867, 155 items and 4 vols. Business and personal correspondence of Henry and John Garst, Roanoke County, Virginia, operators of two sawmills and a flour mill. One volume contains records of Henry Garst's sawmill.

2281 **Gennett Lumber Company** (Asheville, North Carolina), Records, 1832-1954, ca. 16,000 items, 20 vols. and 1 reel of microfilm. Correspondence, contracts, legal documents, and other records (mostly for 1920-45), including a microfilm copy of the autobiography of Andrew Gennett IV, chiefly concerned with personal experiences and family history. Subjects include the lumber business after 1890 and forest conservation. Card index.

2282 **Gorham, John C.,** Papers, 1815-1853, 1 vol. Sporadic entries concerning farm products and timber shipped on the *Tuscarora,* a local trading schooner, Palmyra, North Carolina.

2283 **Gould, John Mead,** Papers, 1841-1943, 3,300 items and 287 vols. Includes references to the lumber industry.

2284 **Green, Duff** (d. ca. 1854), Papers and records, 1817-1894, 1,795 items and 160 vols. Papers (largely 1822-75) of a merchant and manufacturer, Falmouth, Virginia, including records of wood hauled.

2285 **Green, Timothy, and Timothy R.,** Papers, 1789-1840, 67 items. A letter (October 16, 1800) from Edenton, North Carolina, concerns timber for ships and terms of construction.

2286 **Grist, James Redding, and Richard Grist** (d. 1874), Papers, 1791-1920, 3263 items and 5 vols. Relates to the lumber and naval stores trade, Washington, North Carolina. Business letters, accounts, prices current, and manifests (1827-34) relate to Richard Grist and the exportation of naval stores, barrel staves, and other goods to an agent in the West Indies in return for sugar, rum, etc. Correspondence (after 1834) concerns the turpentine and lumber business of James Redding Grist near Wilmington, North Carolina. Post-bellum letters concern his efforts to revive his trade in naval stores. Card index.

2287 **Guerrant, John W.,** Papers, 1803-1868, 10 items and 2 vols. Business letters and accounts of a general merchant who operated a flour mill in Pittsylvania County, Virginia.

2288 **Hall, Thomas William,** Papers, 1809-1894, 125 items. The papers include mortgages and deeds of Joshua and Thomas Gilpin on the Delaware Brandywine Paper Mills.

2289 **Harper, Francis** (New Bern, North Carolina), Papers, 1846-1854, 8 items. Papers of the estate of Francis Harper, referring to the renting of turpentine forest lands and the hiring of Negroes.

2290 **Harrington, John McLean,** Papers, 1760-1901, 967 items and 4 vols. Includes an undated, detailed discussion of turpentine manufacturing processes, Harnett County, North Carolina.

2291 **Harris, Josiah,** Papers, 1872, 5 items. Letters and accounts to Harris, a Thomasville, North Carolina, dealer in barrel staves, from the firm of Peters [Petus?] and Reed who handled his staves.

2292 **Hogg and Campbell** (Wilmington, North Carolina), Daybook, 1772-1773, 250 pp. Includes information on naval stores.

2293 **Jones, M.,** Account book, 1856-1859, 1 vol. Accounts of a turpentine and rosin business, Wilmington, North Carolina.

2294 **Jones, Oliver H.,** Papers, 1832-1861, 253 items. Mercantile records of Oliver H. Jones of New York, relating to trade with Wilmington, North Carolina. The chief exports from North Carolina were naval stores and turpentine. The accounts provide information on prices and supply and demand fluctuations.

2295 **Jordan, Daniel W.** (ca. 1790-1883), Papers, 1827-1913, 4250 items and 8 vols. Chiefly 1843-75. Includes letters from Benjamin Blossom & Son and De Rossit & Brown, both of New York City, concerning Jordan's turpentine business, Camden, South Carolina. Card index.

2296 **Keyser, William Judah** (1821-1877), Papers, 1809-1940, 2115 items and 24 vols. Florida lumber exporter. Correspondence, account books, daybook, and other business papers, maps and other papers of the Keyser family. Subjects include business affairs during Reconstruction, and the lumber business in Florida. Card index.

2297 **Law, William** (1792-1868), Papers, 1761-1890, 1,843 items and 20 vols. Largely 1802-70. The bulk of the papers is concerned with Law's activities as a merchant in partnership with Daniel DuBose, including records of turpentine and other goods sold, Darlington, South Carolina. Included also is an account book of lumber sold by Law and Cyrus Bacot.

2298 **Lee, Ranson,** Papers, 1841-1908, 214 items. There is some limited information on the price of turpentine and the wages of turpentine workers in Wake County, North Carolina.

2299 **Little River Lumber Company** (Montgomery County, North Carolina), Ledger, 1926-1929, 1 vol.

2300 **Livingston, Charles,** Papers, 1812-1829, 9 items and 1 vol. Letterbook and papers of the shipping firm of Livingston and Co., Greenock, Scotland, partly relating to cargoes of turpentine and naval stores, and some staves and tar, between North Carolina and Great Britain and Ireland. Some mention is made of the West Indies trade.

2301 **McCall, Duncan G.,** Journal and diary, 1835-1854, 2 vols. Plantation journal and diary of a cotton planter near Rodney, Jefferson County, Mississippi. Included is information about the operation of a sawmill by Rodney, sale of lumber to a local college or school, and the sale of wood to captains of steamboats on the Mississippi River.

2302 **MacRae, Hugh,** Papers, 1817-1943, 4,222 items and 37 vols. Records of a general commission, especially in naval stores, lumber, etc., established in Wilmington, North Carolina, in 1849 by Colonel John and Donald MacRae.

2303 **McDowell, Thomas David** (1823-1898), Papers, 1798-1891, 276 items. Includes records of land and timber transactions, Bladen County, North Carolina.

2304 **McKay, John,** Papers, 1800-1890, 63 items. Accounts and correspondence relating to the business activities of John McKay at Shoe Heel (Maxton), North Carolina, including the firms of McKay and McLean and McKay and Gilchrist. The principal business interests were naval stores and general merchandise.

2305 **McKoy, William Berry,** Papers, 1853-1916, 17 vols. and 87 items. Account books (1853-69, 2 vols.) relate to work for the Illinois Central Railroad, particularly in Iroquois County, Illinois, and include contracts for the cutting of timber.

2306 **Mason, Horatio,** Papers, 1808-1848, 1 vol. Ledger, including entries for two lumber mills at Toddy Pond, a lake in the northwest corner of Swanville Township, Waldo County, Maine (1809-1830s).

2307 **Massachusetts State Papers.** Includes photostatic copies of papers of the Massachusetts Bay Colony (1694-1779, 25 items), containing provincial laws from the English Parliament, court orders, and petitions. They concern, among other matters, the cutting of timber and distribution of ships' stores, quoting prices or evaluations.

2308 **Memorandum Book,** 1859-1862, 1 vol. Accounts of a wood dealer, possibly one A. Coleman.

2309 **Omohundro, Malvern Hill** (b. 1866), Papers, 1886-1926, 14,142 items and 1 vol. Operator of a sawmill and other businesses, Radford, Roanoke, and Richmond, Virginia. Card index.

2310 **Orr, James Lawrence** (1822-1873), Papers, 1852-1868, 6 items. Governor of South Carolina. Includes a letter (1853) to President Franklin Pierce, recommending W.H. Hickman as a timber agent for the Eastern District of Florida.

2311 **Patterson, Samuel Finley,** Papers, 1792-1939, 2141 items and 26 vols. The collection (largely 1810-1925) includes information on Rufus T. Patterson's paper factory, Winston-Salem, North Carolina.

2312 **Paw Paw Lumber Company** (Paw Paw, Kentucky), Ledger, 1903-1907, 1 vol. Accounts for the company's lumber and mercantile operations, Pike County, Kentucky.

2313 **Petty, William C., and Company** (Archdale, North Carolina), Daybook, 1889-1890, 1 vol. A lumber, building, and contracting firm, Randolph County, North Carolina.

2314 **Powell, William C.** (1846-1923), Papers, 1883-1932, 648 items and 4 vols. Papers relating to the naval stores industry in Georgia and Florida. Includes records from W.C. Powell Company, Jacksonville, Florida, which was involved in land speculation and naval stores (1919-32); Johnston, McNeill & Company, Okeechobee, Florida, naval stores (1920-23); and 2 companies speculating in Florida land: Myakka Company of Charleston, South Carolina (1910-20) and Security Investment Company of Brunswick, Georgia (1910-23). There are also scattered business papers as early as the 1880s and papers on the settlement of Powell's estate in the 1920s. Card index.

2315 **Pratt, Caleb,** Daybook, 1785-1789, 1 vol. Relates to the voyages of the schooner *Neptune* between Boston; Wilmington, North Carolina; Charleston, South Carolina; etc., with cargoes of turpentine, shingles, etc.

2316 **Pugsley, Chester Dewitt** (b. 1887), Papers, 1873-1938, 19,216 items. Includes a number of letters (ca. 1932-33) from members of the American Scenic and Historic Preservation Society, relating to the award of the Pugsley Medal for Park Service.

2317 **Renwick, William W.,** Papers, 1792-1948, 2398 items. Includes a daybook (1857-59, 152 pp.) of Renwick and Rice, Newberry County, South Carolina, containing sawmill accounts.

2318 **Respass, Isaiah,** Papers, 1787-1887, ca. 1000 items. Lumber and shingle manufacturer, Washington, North Carolina. The collection, largely 1840-1870, includes business correspondence, accounts, shipping bills and other material relating to the coastal trade. Card index.

2319 **Revolutionary War,** Collection, 1778-1810, 127 items and 1 vol. The postwar papers include material relating to timber for shipbuilding.

2320 **Richardson, William A.B.,** Papers, 1825-1869, 16 items. One letter (July, 1860) concerns the sale of turpentine at Wilmington, North Carolina.

2321 **Riddick, Richard H.** (d. 1870), Papers, 1840-1879, 408 items and 1 vol. Lumber dealer, Pantego, North Carolina. Business correspondence, accounts, legal papers, and other materials, relating to Riddick's lumber business, his position as agent for the Albemarle Swamp Company (ca. 1871), a runaway slave, railroad construction, and landholding. Card index.

2322 **Russell, John Francis** (b. ca. 1855), Family correspondence, 1808-1946, 1078 items. The letters before 1870 are from members of the family engaged in lumbering in southern Michigan. Card index.

2323 **Schrack, C., and Company** (Philadelphia, Pennsylvania), Papers, 1862-1877, 64 items. Business correspondence of a firm dealing in naval stores. Commodity prices are quoted frequently in letters from the Civil War period.

2324 **Shepherd, James Edward** (1847-1910), Papers, 1892-1907, 156 items. Legal correspondence (largely 1901-06) of a lawyer of Raleigh and Washington, North Carolina; Beaufort County superior court judge; and Chief Justice of the North Carolina supreme court. Much of the collection pertains to legal matters involving cases bearing on a lumber firm of eastern North Carolina.

2325 **Shuford, George Adams** (1895-1962), Papers, 1952-1959, ca. 38,000 items. U.S. Representative from North Carolina, 1953-59; served on the House Interior and Insular Affairs Committee. The papers include reports, speeches, and memoranda, relating to wildlife, reclamation, public lands, water resources, public power, and other legislative matters. Card index.

2326 **Simmons, Dennis, Lumber Company** (Williamston, North Carolina), Records, 1878-1927, 27,656 items and 82 vols. Correspondence, accounts, account books, ledgers, deeds, pay lists, bills, receipts, and other records, relating to all aspects of the lumbering business, from the purchase of timberlands to the shipment of finished board. Card index.

2327 **Slade, William** (Williamston, North Carolina), Papers and account books, 1751-1929, 2,750 items and 31 vols. Among the volumes are account books of Jeremiah Slade, evidently in the lumber business. Included among William Slade's accounts are records of lumber sawed.

2328 **Slattery, Harry** (1887-1949), Papers, 1901-1953, 51,508 items and 243 vols. Correspondence, writings, speeches, official papers, and printed matter, relating to Slattery's various positions during his years of public service, and reflecting his lifelong interest in conservation. The bulk of the collection relates to Slattery's positions as personal assistant to Harold L. Ickes, 1933-38, as Under Secretary of the Interior, 1938-39, and as administrator of the Rural Electrification Administration, 1939-44. Other papers concern his service as secretary to Gifford Pinchot, 1909-12, as Secretary of the National Conservation Association, 1912-23, as special assistant to Interior Secretary Franklin K. Lane, 1917-18, as a Washington lawyer, 1923-33, and as counsel to the National Boulder Dam Association, 1925-29. Includes a typescript of Slattery's autobiography, "Behind the Scenes in Washington." Card index. Restricted.

2329 **Smith, Washington M.,** Papers, 1831-1916, 8,578 items. Smith, of Selma, Alabama, exported naval stores through brokers in Mobile.

2330 **Smith, William Alexander** (1843-1934), Papers, 1765-1949, 11,573 items and 101 vols. There are about a dozen references to the lumber industry in North Carolina (1916-25). Among the volumes are the Steam Mill Account Books (1851-61) of Smith & Ingram who operated a sawmill in Anson County. Letters (1850-51) concern the acquisition of steam machinery to run the sawmill. Card index.

2331 **Smith, William R.,** Memorandum book, 1852-1855, 1 vol. Apparently the accounts of a dealer in turpentine and general merchant, Scotland Neck, North Carolina.

2332 **State Bank of Columbia** (Columbia, Virginia), Records, 1852-1933, 40,777 items and 100 vols. Records include information on lumbering and other small business activities in Fluvanna County.

2333 **Strange, Philip A.,** Account books, 1856-1883, 4 vols. Accounts of a lumber mill, Fluvanna County, Virginia, including a record of supplies advanced to hands.

2334 **Telfair, Edward,** Papers, 1762-1831, 906 items and 5 vols. The papers relate to trade in lumber and other goods by the principal commercial houses in Savannah, Georgia, in which Telfair was a partner. He was also engaged in shipbuilding and accumulated large landholdings.

2335 **Tillinghast Family,** Papers, 1765-1971, 4,860 items and 48 vols. Includes a task book (1849-51) for turpentine operations, Fayetteville, North Carolina.

2336 **Turlington, A.J.,** Papers, 1851-1877, 15 items. Personal and business correspondence between A.J. Turlington, Cumberland County, North Carolina, and his brother, W.H. Turlington, a Wilmington merchant, concerning the timber and turpentine business and other matters.

2337 **Union Manufacturing Company,** Papers, 1848-1868, 1 item and 2 vols. Includes reference to lumber industry.

2338 **Wall, Robert D.,** Papers, 1850-1857, 26 items. Includes comments on the prices of naval stores, Beaufort County, North Carolina.

2339 **Walter, William, and Company** (Boston, Massachusetts), Papers, 1806-1814, 55 items. These letters from Virginia and North Carolina merchants to a Boston, Massachusetts, mercantile house relate to the coastwise trade immediately preceding the War of 1812, and concern shipments of naval stores, lumber, barrels and staves, etc.

2340 **Whitehead, Swepson,** Papers, 1817-1833, 3 items. Business correspondence of Whitehead, apparently a lumber dealer, Portsmouth, Virginia, referring to the lumber business, land speculation, and a lawsuit to recover slaves.

2341 **Wise, William B.,** Papers, 1846-1892, 60 items. Correspondence and accounts of a Murfreesboro, Hartford County, North Carolina, merchant, relating to his naval stores business.

2342 **Wooten & Taylor Company** (Catharine Lake, North Carolina), Papers, 1846-1884, 111 items. Business correspondence of a firm of distillers and dealers in turpentine and other naval stores. Includes many letters from mercantile houses in New York and Wilmington and Beaufort, North Carolina, discussing prices of naval stores.

2343 **Young, James Richard** (b. 1853), Papers, 1916-1938, 826 items. Includes papers (1935-38) of a Wake County, North Carolina, forest, fish, and game warden. Card index.

Duke University, The University Archives

Archivist, 341 Perkins Library 27706

2344 **DeLong, Thomas Stover,** "Logging Railroads and Their History in the Coastal Plain of North Carolina," Duke University, M.F. thesis, 1947, 28 pp.

2345 **Duke University. School of Forestry,** Records. The Archives has just begun accessioning records of the School (1974).

2346 **Duke University. Botany Department.** The papers of some members pertain to the Duke Forest.

RALEIGH

North Carolina State University, D.H. Hill Library, University Archives

University Archivist 27607

2347 **Biltmore Forest School-Carl Alwin Schenck** (1868-1955), Collection, 1890-1970, 28 vertical file drawers. Forester. Correspondence, diaries (1890-1954), slides, photos, notebooks, printed material, scrapbooks, an herbarium, and other papers relating to Schenck and the Biltmore Forest School, near Asheville, North Carolina, which Schenck founded and which closed in 1913. Includes student records, reports, correspondence, biographical information on alumni, and other records of the school. Some of the papers are in German. Described in: *Listing of Biltmore Room Library Collections,* North Carolina State University, Raleigh, North Carolina, July, 1965. Processed, 27 l.

2348 **Forest Resources School, Dean's Office,** General Correspondence, 1927-1962, 4 1/10 feet. Typed list of contents, 3 l.

2349 **Forest Resources School, Wood and Paper Science Department,** Correspondence files, 1953-1968, 5 cubic feet. Unprocessed.

2350 **Forest Resources School, Wood and Paper Science Department,** Research Project Records, 1957-1965, 3 3/10 cubic feet. Includes files for projects for these years, administrative reports, records relating to various paper companies, such as Halifax, International, Riegel, Rome Kraft paper companies, Union Bag-Camp Paper Corporation, Clemson Agricultural College, Duke University, and the U.S. Forest Service.

North Carolina, Department of Cultural Resources, Division of Archives and History

Head, Archives Branch 27611. Some of the holdings are described in *Guide to Private Manuscript Collections in the North Carolina State Archives,* prepared by Beth G. Crabtree, Raleigh: State Department of Archives and History, 1964. The Historical Records Survey, North Carolina, *Guide to the Manuscript Collections in the Archives of the North Carolina Historical Commission,* Raleigh: North Carolina Historical Commission, 1942, has been largely superseded by the Crabtree work, but does describe some organizational records not in that volume.

Public Records

2351 **North Carolina. Conservation and Development Department,** Records, 1883-1973, ca. 90 cubic feet. Includes Economic and Geological Survey files (1883-1926); Board minutes and reports (1926-73); administrative reports and correspondence (1925-49); activities file (1925-49); and miscellaneous subject file (1920-45). Of particular note are Forestry-Weeks Law correspondence (1911-22), and the minutes of the board meetings.

2352 **North Carolina. Education Board,** Swamp Land Papers, ca. 1868-1939, ca. 4 cubic feet. Correspondence and survey books relating to swamplands in eastern North Carolina surveyed and sold for the support of public schools.

2353 **North Carolina. Wildlife Resources Commission,** 1927-1964, ca. 12 cubic feet. General correspondence, studies, reports, and investigations relating to game lands; marsh resources; habitat restoration; and habitat distribution of wildlife.

2354 **North Carolina National Park, Parkway and Forests Development Commission,** 1927-39, ca. 29 cubic feet. Records of the North Carolina Park Commission created to administer the acquisition of land for the Great Smoky Mountains National Park. Comprised of title file for tracts of land; cancelled checks and vouchers of the special state fund and the Rockefeller fund; opinions; deeds; and condemnation proceeding against the Champion Fibre Company, the Ravensford Lumber Company, and the Suncrest Lumber Company.

2355 **U.S. Civilian Conservation Corps,** ca. 60 cubic feet. Enrollment and discharge records arranged alphabetically by county and surname.

Private Manuscripts

2356 **Appalachian National Park Association,** 1899-1907, 1929, 1936, 147 items (including 20 letters). Correspondence relative to a convention to be held in Asheville, North Carolina, to promote a national park in the southern Appalachians, including letters to members of Congress soliciting their support, message from President Theodore Roosevelt, and addresses to the Senate by J.C. Pritchard, F.M. Simmons, and Chauncey M. Depew; newspaper clippings of conventions; bills presented, bills passed; addresses and reports of meetings; two map plats; bylaws and minutes of meetings; list of members; and list of contributions. Also included are C.P. Ambler's history of the activities of the park from 1899 and a program of the Silver Jubilee in observance of the passing of the Weeks Law.

2357 **Bayside Plantation Record,** 1860-1868, 1 item. A Louisiana plantation account of hands engaged in wood chopping, among other activities. This is a transcript of the original in the Southern Historical Collection, University of North Carolina.

2358 **Biggs, Kader,** Papers, 1787, 1814-1951, ca. 1000 items. The collection includes letters, bonds, bills, and invoices of William F.G. Pierce of Williamston and Windsor, North Carolina, concerning sale of shingles and staves and hire of slave labor.

2359 **Blount, John Gray** (1752-1833), Papers, 1706-1933, ca. 9000 items. Papers of the Blount family, of Washington, North Carolina. The bulk of the correspondence was written 1770-1800. Included is considerable material on the manufacture of tar, naval stores, and lumber.

2360 **Blount, John Gray** (1752-1833), Letterbook, 1796-1797, Microfilm, 15 feet. Includes letters of a merchant and large landowner in eastern North Carolina with commission merchants in New York and Philadelphia relative to the shipment of lumber, tar, and staves north and merchandise south. Film purchased from the New-York Historical Society.

2361 **Brevard Lumber Company** (Brevard, North Carolina), Account books, 14 vols. Ledgers and daybooks (1912-21).

2362 **Brooks, Eugene Clyde** (1871-1947), Papers, 1925-1931, 702 items. Correspondence, telegrams, speeches, reports, audits, legislative bills, newspaper clippings, and other items. The papers concern, among other activities of Brooks, material relating to the North Carolina Park Commission (1926-29) and the establishment of the Great Smoky Mountains National Park. The park commission correspondence relates to the sale of bonds to raise money for the purchase of necessary land for the park, and correspondents include Governor Angus W. McLean, Vance Rhodes, Mark Squires of the Commission, and bond attorneys Reed, Hoyt and Washburne. Negotiations with the Champion Fibre Company and litigation with the Suncrest Lumber Company appear in the papers.

2363 **Chatham, Richard Thurmond** (1896-1957), Papers, 1776-1956, ca. 57,000 items. U.S. Representative from North Carolina, 1949-1957. The collection includes material related to the Blue Ridge Parkway.

2364 **Crane, Alexander,** Manuscript, n.d., 1 item. Typescript of Crane's proposed book on North Carolina furniture of the coastal plain section, 1685-1835, including specifications for the cover, illustrations, etc.

2365 **Ehrman, Ada,** Collection, 1853-1863, 3 items. Letters addressed to B.F. Strickland, Wakefield, North Carolina (October, 1853) relative to freight and sale of turpentine.

2366 **Grist, Allen and J.R.,** Papers, 1821-1866, 5 items. Includes letters addressed to Allen Grist, Beaufort County, and J.R. Grist, New Bern, North Carolina (1866) quoting cotton and turpentine prices.

2367 **Henries Collection,** 1807-1877, 21 items. Includes statements of account for sale of turpentine.

2368 **Hines, Wait, and Leone Hines,** Collection, 1839-1898, 498 items and 3 vols. Includes papers and account books of James and W.T. Hines of Lenoir, North Carolina, including bills for lumber bought by Hines Bros., Dover, North Carolina.

2369 **Jones, Arthur,** Papers, 1793-1800, 90 items and 1 vol. Records relating to Jones' shipping business of Edenton, North Carolina. The material concerns, among other topics, purchases of lumber, shingles, staves, tar, turpentine, charges for inspecting and piling shingles, and piling and culling staves.

2370 **King, Richard Hugg** (1767-1825), Papers, ca. 25 items. Included is a manuscript diary (1819-23) with notes on sawmill affairs.

2371 **Mebane and Sutton** (Windsor, North Carolina), Accounts, 1837-1839, 21 items. Includes accounts for the sale of staves.

2372 **Newman, C.L.,** Letters, 1920-1921, 12 items. Original letters and carbons relating to efforts of Newman, Agricultural Advisor, State Prison Farm, Cary, North Carolina, to use part of the farm for experiments in reforestation. Correspondents are J.S. Holmes, State Forester, Colonel Joseph Hyde Pratt, State Geologist, and Governor Cameron Morrison.

2373 **Norden, Eric,** Collection, 1735-1957, 1967, ca. 18 cubic feet. Eric Norden was the agent for the state in the surveying and selling of swamp lands for the support of public schools. Includes maps, plats, grants, deeds, surveys, title abstracts, and descriptions of swamp and timber lands in eastern North Carolina. Largely relates to the period 1892-1949.

2374 **Pate Family,** Papers, 1815-1898, 1588 items. Correspondence and other papers of the Pate family of Northampton County, North Carolina, and related families. Includes records of lumber and railroad ties delivered, and the correspondence of John Wesley Pate (1847-96) with commission merchants in Norfolk and Petersburg, Virginia; Baltimore, Maryland; and New York; and with an agent in Portsmouth, Virginia, regarding the cutting and sale of railroad ties.

2375 **Patrick, Benjamin M.,** Collection, 1812-1903, 35 items. Includes an agreement (1821) between Eliakim Patrick and John Gray Blount relating to cutting timber.

2376 **Quinlan-Monroe Lumber Company** (Waynesville,

North Carolina), Account books, 1901-1906, 25 vols. Daybooks, ledgers, order book, deposit book, and bills receivable and payable.

2377 **Quinlan-Monroe Lumber Company** (Waynesville, North Carolina), Records, 1900-1914, 16 boxes. Letterbooks, statements, accounts, contracts, invoices, reports, promissory notes, insurance policies, and other papers, of the lumber company and an affiliated company, the Tuckaseigee Lumber Company operating in western North Carolina; together with a letterbook of McNabb and Bennett (Oct.-Dec., 1900) and some personal and business papers of Edwin E. Quinlan. The correspondence reflects various phases of timber cutting and lumber manufacturing and such lumber industry problems as transportation and employment of personnel. After the death in 1909 of E.E. Quinlan, his son Charles wrote many letters relating to his father's estate and to business matters.

2378 **Sanford Sash and Blind Company** (Sanford, North Carolina), Account book, 1912, 1 vol. Letterbook and accounts.

2379 **Sargeant and Mann** (Fayetteville, North Carolina), Account book, 1866, 1 vol. Naval stores. Daybook.

2380 **Simmons, Samuel,** Account book, 1834-1835, 1 vol. Tyrrell County, North Carolina. Lumber, wrecks, and chattels.

2381 **Vanderbilt, George W. and Louis Carr,** Timber contract, 1913, 1 item. Facsimile of a timber contract registered in Transylvania County Deed Book No. 34, p. 132 (January 25, 1913).

2382 **Wilder, Gaston Henry,** Papers, 1798 (1830-1855) 1907, ca. 175 items. The collection includes annual reports (1851 and 1852) of the Fayetteville and Western Plank Road Company.

2383 **Wilkins, Ashley,** Papers, 1884-1900, 531 items. Businessman of South Gaston, North Carolina. The collection includes correspondence relating to Wilkins' interest in the lumber business in which he was associated with Miss E.E. Dillehay. Includes inspection reports on staves.

2384 **Williamson, Hugh,** Letters, 1778-1815, Microfilm (4 feet). There are occasional references to shipbuilding in Winton, North Carolina, and descriptions of the land at the heads of the Scuppernong and Pasquotank Rivers (1795), with a discussion of the possibility of taking out timber and shingles. The papers were filmed from the Williamson Papers in the Historical Society of Pennsylvania.

WINSTON-SALEM

Wake Forest University

Director, Baptist Historical Collection, The Z. Smith Reynolds Library, Box 7777, Reynolda Station 27109

2385 **Cates, John Wesley,** Papers, 1890-1919, 6 4/5 feet. Burlington, North Carolina, lumber dealer. This collection contains correspondence with his brother and family members involved in lumber and naval stores with the Myrtlewood Lumber Company, Wakulla, Florida, and correspondence with other lumber, coal, and naval stores dealers in North and South Carolina, and West Virginia. The collection also contains price lists, shipping invoices, and bills for lumber and other forest products.

North Dakota

BOTTINEAU

*North Dakota State University, Bottineau Branch, Library

First and Simrall Ave. 58318. The following entry was listed for the North Dakota School of Forestry Library in *Forest History Sources of the United States and Canada,* p. 92.

2386 **Turtle Mountain Woodland Association,** Records. The records of the only forestry cooperative in North Dakota, including marketing reports on fence posts and fuel wood from the Turtle Mountains.

Ohio

ATHENS

Ohio University Library

Head, Special Collections Division 45701

2387 **Diller, Oliver D.,** Collection, ca. 1900-1960, 195 slides. Photographs depicting scenes from Ohio forestry.

2388 **Society of American Foresters, Ohio Chapter,** Records, 1948-1965, ca. 1½ feet.

2389 **U.S. Forest Service,** Reports and maps, 1945-50, 3 feet. Included are records of a survey of stripmining areas in Ohio, with a view toward possible reclamation.

2390 **Zaleski State Forest,** Records, 1931-1969, ca. ½ foot. Xerox copies.

BOWLING GREEN

Bowling Green State University, Northwest Ohio-Great Lakes Research Center

Archivist, 214-A Graduate Building 43403 Telephone: (419) 372-2474

2391 **Hirsch, George E., Lumber Co.** (Grand Rapids, Ohio), Records, 1888-1931, 23 folders. Account ledgers, describing lumber sold, prices, etc.

CINCINNATI

Cincinnati Historical Society

Curator of Manuscripts, Eden Park, Cincinnati 45202 Telephone: (513) 241-4622

2392 **Waldschmidt, Christian** (d. 1814), Journal, 1812-1814, 1 vol. (56 l.). Daybook of transactions for Waldschmidt in a lumber mill by John Keller. Photostatic copy made from an original owned by Keller family, Milford, Ohio (1947).

2393 **West, J.C., Lumber Co.,** Collection, ca. 1910, 16 items. Photographs portraying lumber production in West Virginia (ca. 1910, 15 items), and a letter from Dr. George Engberg to Ranald S. West of the J.C. West Lumber Co.

CLEVELAND

Western Reserve Historical Society

Librarian, 10825 East Boulevard 44106 Telephone: (216) 721-7522. The holdings are described in Kermit J. Pike, *A Guide to the Manuscripts and Archives of the Western Reserve Historical Society,* Cleveland: Western Reserve Historical Society, 1972.

2394 **Burton, Theodore Elijah** (1851-1929), Papers, 1870-1958, 80 feet. U.S. Representative, 1889-91. 1895-1909, 1921-28, and Senator, 1909-15, 1928-29, from Ohio. The papers pertain to Burton's interests in conservation, especially as a member and chairman of the House Rivers and Harbors Committee, the Inland Waterways Commission, and the National Waterways Commission. Included are material on tariff legislation, flood control, and much material relating to the proposed White Mountains and Appalachian Forest Reserves (1902-07), and other conservation issues, such as the Niagara Preservation Bill (1906). Unpublished register.

2395 **Case, Quincy A., Family,** Diaries, 1864-1871, 1903, 9 vols. Includes diaries (1868-71, 3 vols.) of Quincy A. Case (b. 1848), containing references to his work in a planing mill, Kingsville, Ashtabula County, Ohio.

2396 **Conservation Organizations** (Cleveland, Ohio), Records, 1939-1942, 3 boxes. Contains material of the National Wildlife Federation, Ohio Natural Resources Council, and Cuyahoga County Conservation Council.

2397 **Ely, Heman** (1775-1852), Papers, 1817-1849, 5 boxes. Account books, financial receipts, and business and personal letters addressed to Heman Ely, operator of a sawmill and other businesses in Elyria, Ohio. The papers include sawmill accounts (1826-31); tax records; and other business and personal materials.

2398 **Manuscripts Vertical File,** 3 file cabinets. Single items or very small collections filed in two alphabetical series. In series II, filed by organizational names, are materials of the Cleveland Paper Company and the Cleveland Board of Park Commissioners.

2399 **Ritchie, Samuel J.** (1838-1908), Papers, 1870-1920, 40 boxes, 11 vols. Correspondence, memoranda, legal documents, account books, biographical data, photographs, receipts, and other papers of Samuel J. Ritchie, who was active in the railroad, lumber, and iron industries. The bulk of this collection consists of Ritchie's business correspondence (1870-1910). Unpublished register (6 pp.).

2400 **Rust, John Franklin** (1835-1899), Papers, 1849-1889, 25 boxes. Correspondence, commercial and legal documents, and other papers relating to the activities of John F. Rust, Sr., who was concerned with the lumber and shipping businesses on the Great Lakes during the last half of the 19th century, while residing principally in Saginaw, Michigan, and Cleveland, Ohio; and to the activities of the companies with which he was associated. Rust's lumber firm in Cleveland was known as Rust, King & Company and later became Rust, King & Clint. Unpublished register (10 pp.).

COLUMBUS

Ohio Historical Society

Director, Ohio Historical Center, Interstate 71 and 17th Avenue 43211 Telephone: (614) 466-2060. The holdings are described in *A Guide to the Manuscripts at the Ohio Historical Society,* ed. by Andrea D. Lentz and Sara S. Fuller, Columbus: The Ohio Historical Society, 1972.

2401 **Bareis, George F.** (1852-1932), Papers, 1883-1921, 4½ feet. Lumber merchant, Canal Winchester, Ohio, 1881-1932. The papers mainly concern his lumber business. Included are correspondence, account books, ledgers.

2402 **Edgerton, Alfred P.** (1818-1897), Records, 1841-1876, 2 feet. Included are cashbooks of a Hicksville, Ohio, sawmill.

2403 **Friends of the Land,** Records, 1940-1962, 115 feet. Records of a national conservation and ecology organization, headquartered primarily in Columbus, Mansfield, and Zanesville, Ohio, including correspondence, reports, publications, financial records, clippings, membership lists, programs, memoranda, and notes kept by Ollie E. Fink, secretary. Topics include Malabar Farm, wildlife, nutrition, conservation legislation, National Garden Institute, National Association of Soil Conservation. Among prominent members, officers, and correspondents are Hunter C. Baker, Paul Bestor, Louis Bromfield, Edward J. Condon, Robert T. Crew, Chester C. Davis, Ollie E. Fink, Jonathan Forman, Mrs. Luis J. Franke, Russell Lord, Paul B. Sears, Victor Weybright.

2404 **Garford, Arthur L.** (1858-1933), Papers, 1877-1933, 40 feet. Industrialist, Elyria and Cleveland, Ohio. Papers contain correspondence, speeches, articles, other materials, concerning political issues, including conservation. Published inventory (1961, 20 pp.).

2405 **Herbert, Thomas J.** (b. 1894), Papers, 1947-1949, 32½ feet. Attorney general of Ohio; governor, 1947-49. The papers relate in part to conservation, and contain correspondence on the State Division of Forestry. Inventory (22 pp.).

2406 **Inskeep, John D.** (1834-1909), Diaries, 1861-1865, ¼ foot. Farmer of East Liberty and Bellefontaine, Ohio. The diaries mention lumbering in Ohio (1861).

2407 **Jacobs, Victor** (b. 1903), Papers, 1944-1953, ¾ foot (800 items). Lawyer, of Dayton, Ohio. Mostly correspondence (1946-47) of Jacobs, member of the Ohio Conservation of Natural Resources Committee, with the chairman of the committee, Arthur Ernest Morgan. Included are annotated revisions of Ohio legislative bill creating the State Department of Natural Resources (1949). Unpublished inventory.

2408 **Lee, Mary E. (Brown)** (1869-1952), Papers, 1893-1948, 2½ feet. Westerville, Ohio. The papers contain material relating to the Fish and Game Protective Association and the League of Ohio Sportsmen. Typescript inventory (4 pp.).

2409 **Mansfield, Newton D.** (1873-1957), Papers, 1933-1952, 1 foot. President of the board of directors of Muskingum Watershed Conservancy District, 1933-50. Papers contain correspondence, clippings, reports, maps, plans, photographs of the district, which constructed a system of flood control and water conservation for the Muskingum River drainage area.

2410 **Ohio Forestry Association,** Collection, ca. 1948-1958, ca. 50 feet. Correspondence, minutes, membership lists, photographs, and publications. The collection includes materials formerly held at Ohio State University, Columbus.

2411 **Smith, William Henry** (1833-1896), Papers, 1800-1896, 8 feet (ca. 180 items), and 7 rolls microfilm. Correspondence, diaries, scrapbooks, albums, and other papers relating to Ohio and national politics, conservation, and journalism. Unpublished inventory. The microfilm (1855-95) is of originals held by the Indiana Historical Society.

2412 **Society of Separatists of Zoar** (1817-1898), Records, 1816-1942, 22 feet. The records relate to various economic enterprises of Zoar, Ohio, including lumbering. Typescript inventory (16 pp.).

2413 **Stauffer, Abraham,** Records, 1835-1839, 1 vol. Lumber account journal of a firm in Canfield, Ohio.

2414 **Taylor, Albert D.** (1883-1951), Papers, 1919-1950, 11 feet. Landscape architect, Cleveland, Ohio. The papers contain correspondence and business records, articles, material relating to national parks and the Forest Service; newspaper clippings, personal accounts, notes for articles.

2415 **Turney, Benjamin,** Account book, 1851-1873, 1 vol. Included are lumber and carpentry accounts from Union County, Ohio.

2416 **Waddell Wooden Ware Work** (Greenfield, Ohio), Records, 1899-1923, 1 foot. Furniture manufacturing company, founded by John M. Waddell. Records contain business correspondence and company records.

2417 **Wilson, Milton,** Records book, 1839-1879, 1 vol. Included are lumber accounts and other records of Wilson, of Newark, Ohio.

Ohio State University, The University Archives

University Archivist, 2070 Neil Avenue 43210 Telephone: (614) 422-2409

2418 **Agriculture and Home Economics College, Department of Agricultural Economics and Rural Sociology,** Records, 1906-1961, 5 feet. Chiefly material relating to farms, but also includes case studies on forest recreation enterprises in Ohio (1963-65).

2419 **Dambach, Charles,** Papers, 1935-1966, 3 1/5 feet. Chiefly material on wildlife conservation (ca. 1935-65), but also material on tree studies, erosion control (1936-69), ecological study and material on grazing in northeastern Ohio woodlots (ca. 1938-67).

2420 **Horticulture and Forestry Department,** Records, 1938-1963, 7/10 foot. Included are administrative records, general correspondence (1955-63), executive committee correspondence (1956-57), enrollment records, records of the National Shade Tree Conference (1955-60), etc.

2421 **Muskingum Conservancy Project, Division of Research,** Records, 1913-1966, 3 feet. Records relating to the Muskingum watershed, containing annual reports to the chief engineer (1935-65), board minutes (1935-65), correspondence, a draft history, miscellaneous materials, clippings, etc.

2422 **Natural Resources School,** Records, 1963-1969, 6 feet. General administrative and academic records, including files on forestry applicants (1967-68), conservation education, general correspondence, environmental education (1969), forestry (1968-69), Nature Conservancy (1968-70), Ohio Conservation Congress (1968-69), Ohio Department of Natural Resources (1968-70), Ohio Parks and Recreation Association (1968-69), Society of American Foresters (1967-69), trainee programs, creation of the school. Inventory, typed (6 pp.).

DAYTON

Federal Records Center

Director, 2400 W. Dorothy Lane 45439. There is no Regional Archives Branch.

2423 **U.S. Forest Service** (Record Group 95), 88 cubic feet. Included are materials from U.S. Forest Service office, Bedford, Indiana; Hiawatha National Forest, Escanaba, Michigan; Huron-Manistee National Forest, Cadillac, Michigan; and the Northeastern Forest Experiment Station-Forest Insect and Disease Laboratory, Delaware, Ohio. The records chiefly consist of contracts, including timber sales, special use permits, and general correspondence files. The records of the Northeastern Forest Experiment Station contain documentation on research projects. Permission of the U.S. Forest Service is required for access.

Oklahoma

NORMAN

The University of Oklahoma, Western History Collections

Curator, Library, 401 West Brooks, Room 141 73069. Some of the holdings are described in A.M. Gibson, *A Guide to Regional Manuscript Collections in the Division of Manuscripts, University of Oklahoma Library,* Norman: University of Oklahoma Press, 1960.

2424 **Douglas, Helen Gahagan** (b. 1900), Papers, 1920-1952, 96 feet. U.S. Representative from California, 1945-51. The collection contains largely congressional papers, including material relating to the redwoods.

2425 **Kerr, Robert Samuel** (1896-1963), Papers, 1909-1950, 50 feet. U.S. Senator from Oklahoma, 1949-63. The papers contain extensive files relating to "Land, Wood, and Water," the title of his book (1960).

2426 **Thomas, John William Elmer** (b. 1876), Papers, 1900-1950, 508 feet. U.S. Representative from Oklahoma, 1922-26, serving on the Public Lands Committee; U.S. Senator, 1926-50, serving on the Agriculture and Forestry Committee. The papers contain correspondence and other papers pertaining to committee service and to legislation.

2427 **Williams, Arthur James** (b. 1877), Papers, 1910-1950, 8 feet. Geologist, University of Oklahoma. The collection contains chiefly geological reports and bulletins, historical scrapbooks, and maps; but also included are many photographs of forests and forest industries in eastern Oklahoma.

OKLAHOMA CITY

Oklahoma Department of Libraries, Archives and Records Division

Archivist, 200 Northeast 18th Street 73105 Telephone: (405) 521-2502

2428 **Oklahoma. Agriculture Department,** Records. Biennial Reports (1907-48) and Annual Reports (1949 +) include statistical information on forestry.

2429 **Oklahoma. Agriculture Department, Forestry Division.** Correspondence of the Director and Assistant Director (1966-72, 1 cubic foot); and records dealing with seed tree plantings and seedling distribution (1928 +).

2430 **Oklahoma. Attorney General,** Miscellaneous Files. Condemnation proceedings for land purchases for Lake Murray (1935) and Honey Creek deer preservation projects (1944).

2431 **Oklahoma. Governor,** Records. Records of Charles N. Haskell include a monographic account of President Theodore Roosevelt's hunting trip to the "Big Pasture" in the Indian Territory, just prior to statehood (1 folder). Administrative files of Robert L. Williams contain correspondence, survey maps, and plats dealing with the sale of Choctaw-Chickasaw land in McCurtain County to the State of Oklahoma for use as a game preserve (1917-19, 2 folders). Records of Henry S. Johnston contain correspondence regarding the Southern Forestry Congress (1927, 1 folder). Records of the Administrative Assistant to Dewey F. Bartlett include material on the Weyerhaeuser Company, with correspondence, reports, news releases, speeches, trust and lease agreements, and abstracts of proceedings of the Governor's Committee on Forest Protection (1969-70, 1 folder); also minutes of the Oklahoma Industrial Development and Park Commission, including reports and exhibits (1966-71).

2432 **Oklahoma. Secretary of State,** Miscellaneous Filings. Included are Fish and Game Commission records (1 cubic foot), with Quarterly reports (1909-52); Biennial Reports (1934-36; 1948-50); Reports of Audit (1909-15); Minutes (April 19, 1916); authorizations and administrative orders for the purchase of land for wildlife conservation purposes.

2433 **Oklahoma Water Resources Board, Planning Division,** Aerial Survey Photographs, 1940s, 127 cubic feet. Arranged by counties with photographic index.

2434 **U.S. Department of Interior, General Land Office,** Field Notes. Copy of survey notes of area later designated to be the Oklahoma Territory. Consists of handwritten records of several surveys (1873-95) which include description of the land, the soil, tree cover, general vegetation and wildlife.

Oklahoma Historical Society

Director, Indian Archives, Historical Building 73105

2435 **Indian Archives,** ca. 2,600,000 pp. and 6,000 vols. References to forest history include: Cheyenne and Arapaho Agency, timber (1870-1933, 1,030 pp.); Cheyenne and Arapaho timber cutting permits (1922-35, 1,237 pp.); Chilocco Indian School, timber (1909, 3 pp.); Kiowa Agency, timber and saw and grist mills (1869-1924, 888 pp.); Pawnee Agency, timber (1899-1924, 279 pp.); Quapaw Agency, timber (1874-91, 95 pp.); Sac and Fox Agency, and Shawnee Agency timber (1860-1924, 310 pp.); Section X, Cherokee Timber (1899, 1 p.), Chickasaw Timber (1899, 9 pp.), and Choctaw Timber (1900, 2 pp.); Cherokee timber (1886-98, 12 items); Chickasaw timber (1889-1903, 22 items); Choctaw timber (1872-1902, 284 items); Cherokee (Tahlequah) timber (1872-1903, microfilm). The Indian and Pioneer History Collection of interviews (112 vols.) includes many references to timber and sawmills in various Indian and territorial locations.

Oregon

CORVALLIS

Oregon State University, Archives Department

97331 Telephone: (503) 754-2165

2436 **Matthews, Oliver V.** (b. 1892), Papers, 1928-1969, 25 feet. Amateur botanist, of Oregon. Logs (1928-67) containing correspondence, narratives of Matthews' botanical trips, descriptions of virgin tracts of timber, receipts, maps, road information, and photos; and scrapbooks (1928-69) arranged by tree species, containing correspondence, clippings and photos. The correspondence is with herbariums, newspapers, and State and U.S. Forest Service, regarding the location and identification of trees, and old logging methods. Unpublished description.

2437 **Oregon State University, School of Forestry,** Records, 1913-1969, Microfilm, 25 reels. This collection includes materials concerning the Prospect Tract (1926-54), Pacific Northwest Regional Committee on Postwar Programs, and Elliot State Forest; records of the Office of the Dean of Forestry, containing correspondence (1910-69) and historical file (1906-67); and the George Barnes File (1947-64). Described in *Guide to the Microfilm Edition of the School of Forestry Records, 1926-1954* (RG#139), edited by William F. Schmidt and Sarah J. Wilson, Corvallis: Oregon State University Archives, 1968. Supplements, 4 vols., 1970, 1971.

CRATER LAKE

Crater Lake National Park

The Superintendent 97604

2438 **Crater Lake National Park,** Papers, 1885-1965, 10 feet, ca. 400 items. In part, transcripts (handwritten and typewritten) and photocopies of originals in various universities and federal agencies, particularly the U.S. Geological Survey, Surface Water Branch, Medford and Portland, Oregon, offices. Included are correspondence, telegrams, documents, photos, circulars, leaflets, research papers, reports, and diaries, relating to the human history, geology, botany, zoology, ecology, limnology, meteorology, climatology, visitor use, and administration of Crater Lake National Park.

EUGENE

University of Oregon

Special Collections, The Library 97403. The holdings are described in *Catalog of Manuscripts in the University of Oregon Library,* by Martin Schmitt, Eugene: University of Oregon, 1971.

2439 **Abilene & Southern Railway Company.** West Coast Lumbermen's Association vs. Abilene & Southern Railway Company, Interstate Commerce Commission docket no. 13211 (1923, 1½ feet). The complaint was against the railroad rates charged for lumber shipments from eastern Oregon. The files include correspondence, exhibits, testimony, briefs, and the decision. These are records of A.S. Titus, traffic manager of the California White and Sugar Pine Manufacturers' Association.

2440 **Ames, Fred Elijah** (1880-1957), Diaries, 1892-1919, and photographs, 24 vols. U.S. Forest Service, chief of silviculture and assistant regional forester in Oregon, 1908-42. The diaries include Ames' life at Yale Forestry School (graduated 1906), and in the U.S. Army 20th Engineers, World War I. The photographs are of his military service (1918-19). There is also a collection of speeches and reports.

2441 **Appel, Benjamin** (b. 1907), Papers, 1932-1966, 5 feet. Appel held various jobs, ranging from bank clerk to lumberjack. He writes poetry, fiction, juvenile books, short stories, biographies, current affairs, articles. Papers include manuscripts of books, galley and page proofs, copies of published books and articles, and minor correspondence. Inventory.

2442 **Arneson, Arnold** (1885-1957), Diaries, 1922-31, 1941-1947, 1 foot. Official diaries of the district ranger, Tieton District, Snoqualmie National Forest, Washington, reporting such matters as trail building, fire fighting, sales, inspecting, and office work.

2443 **Associated General Contractors of America, Inc., Oregon-Columbia Chapter,** Records, 1929-1961, 48 feet. Correspondence and documents relating to labor agreements, labor legislation, and the internal affairs of the association. Inventory.

2444 **Baker, Fentress & Co.** (Chicago, Illinois), Records, 1914-1943, 73 feet. Records, largely management correspondence, of Pacific Northwest logging and lumber companies in which Baker, Fentress had financial interest. These include Algoma Lumber Co., Kirk, Oregon (1929-41); Carlisle Lumber Co., Onalaska, Washington (1922-40); Coeur d'Alene Pine Co., Coeur d'Alene, Idaho (1924-42); Great Northern Lumber Co., Leavenworth, Washington (1914-42); and others. Inventory.

2445 **Bartrum, Smith C.** (1865-1942), Autobiography, 1 folder. U.S. Forest Service ranger, 1899-?. Much of this autobiography describes the locating of a trail from the Robert Cavit homestead to Fish Lake in Douglas County, Oregon. Typewritten.

2446 **Base Line Lumber Co.** (Portland, Oregon), Letters,

1907-1910, 1 vol. A letterpress book of a company owned by Theodore and William Nicolai.

2447 **Blakely, James** (1812-1913), Papers, 1853-1901, 1 vol. Territorial legislator, sawmill owner, and woolen mill promoter, of Brownsville, Oregon. Scrapbook containing correspondence, receipts, photos, and other mementos, assembled by Mabel McClain.

2448 **Bourne, Jonathan** (1855-1940), Business papers, 1887-1927, 24 feet. Includes material relating to the Glasgow Coal, Land and Lumber Co. Inventory.

2449 **Boutelle, Frazier Augustus** (1840-1924), Papers, 1869-1933, 3 feet (including about 800 letters). Boutelle was superintendent of Yellowstone National Park, 1889-90.

2450 **Chaney, Ralph Works** (1890-1971), Papers, 1917-1966, 27 feet (including about 12,000 letters). Paleontologist. From 1943 Chaney was a member of the U.S. National Park Service Advisory Board on National Parks, Historic Sites, Buildings and Monuments, and from 1928 a member and officer of the Save-the-Redwoods League. Inventory.

2451 **Civilian Public Service Camps,** Histories, 1944-1945, 1 folder (including 7 items). Histories (1944-45) of Camps No. 21 (Cascade Locks), No. 56 (Waldport), No. 59 (Elkton), written by members of the respective camps.

2452 **Clark, Frank Jones** (1891-1960), Papers, 1925-1955, 1½ feet. Seattle, Washington, conservationist. Correspondence with the Northwest Conservation League, the White River Recreation Association; and documents.

2453 **Cleator, Frederick William** (1883-1957), Papers, 1930-1956, 2 feet. Forester. The papers consist of diaries (1944-54), correspondence, speeches, documents, and slides relating to the recreational aspects of national forests in the Pacific Northwest and the establishment of the Cascade Skyline Trail. The collection is largely "working files."

2454 **Cronise, Ralph Weeks** (1886-1962), Papers, 1918-1961, 3 feet (ca. 1,700 letters). Newspaper publisher, Albany, Oregon. The collection includes correspondence concerning the Oregon State Parks Advisory Committee. Inventory.

2455 **Daggett, Junior F.** (1891-1960), Papers, 1936-1955, 9 feet (including about 6,000 letters). The correspondence and accompanying documents concern Daggett's career as general manager of the Alexander-Yawkey Lumber Co., Prineville, Oregon; member of the Oregon State Board of Forestry, and Eastern Oregon Lumber Survey Group, and the Western Pine Association; and president of the Midstate Lumber and Manufacturing Co.

2456 **Dana, Marshall Newport** (1885-1966), Papers, 1949-1962, 12 feet (including about 6,000 letters). Included are files concerning various aspects of conservation, such as timber as collateral, the Columbia Basin Inter-Agency Committee, and the National Reclamation Association. Inventory.

2457 **Davidson, Crowe Girard** (b. 1910), Papers, 1956-1962, 18 feet. Davidson was a member of the Natural Resources Committee of the Western States Democratic Conference, the Advisory Council on Natural Resources of the Democratic National Committee, and the Kennedy-Johnson Campaign Committee on Natural Resources. The papers relate to Davidson's service on these committees, and to Oregon and national politics generally. Inventory.

2458 **Disque, Brice Pursell** (1879-1960), Papers, 1899-1959, 16 feet. Disque was officer in charge of the Spruce Production Division of the Bureau of Aircraft Production, and president of the United States Spruce Production Corp., 1917-19. The papers contain records of the United States Spruce Production Corp., including general orders, special orders, bulletins, memoranda, specifications, daily reports, timber location maps, and official photographs. Correspondence relating to this activity is with J.J. Donovan, Frank D. Eamon, Harold C. Eustis, Felix Frankfurter, Edward W. Freeman, Alexander McAndrew, William C. Potter, J.D. Reardon, John D. Ryan, Mott Sawyer, Charles Van Way, I.D. Wolf. Miscellaneous papers include drafts of an autobiography, and the original plaque of the Loyal Legion of Loggers and Lumbermen.

2459 **Dixon, Algernon Cyrus** (1875-1962), Correspondence, 1906-1962, 2 feet (including about 3,000 letters). Lumberman of Oregon. The correspondence includes both letters received and sent and concerns the Board of Regents, University of Oregon; correspondence mainly with Robert A. Booth, president of the Ochoco Timber Co., Eugene, Oregon, concerning negotiations with the National Forest Reservation Commission for consolidation of the company's timber holdings (1934-40); and miscellaneous reports, financial records, and minor correspondence concerning the Booth-Kelly Lumber Co., Springfield, Oregon (1920-58).

2460 **Dixon, Algernon Cyrus** (1875-1962), History of the code of fair competition for the lumber and timber products industries, Eugene, Oregon, 1935, 408 pp., typewritten. With a cover letter from C.H. Crawford of the Forest Products Division, National Recovery Administration.

2461 **Dodson, William Daniel Boone** (1871-1950), Correspondence and miscellaneous papers, 1947-1949, 1 box. Some of the correspondence concerns the establishment of the Willamette Valley Wood Chemical Company, Springfield, Oregon. Inventory.

2462 **Durno, Edwin Russell** (b. 1899), Papers, 1956-1964, 15 feet. Oregon State Senator, 1958-60; U.S. Representative, 1961-63. Most of the papers are the office files of Representative Durno, in office file order, relating to legislation and projects of interest to Oregon, especially the proposed Oregon Dunes National Seashore. Other files include Oregon legislative files (1959-61), and personal and other political records. Inventory.

2463 **Ellsworth, Harris** (b. 1899), Papers, 1943-1957, 19 feet (including about 4,000 letters). U.S. Representative, 1942-57. The papers are the office files of Representative Ellsworth and deal particularly with flood control, river and harbor development, forestry, mines and mining, and hydroelectric power projects. Inventory.

2464 **Evarts, Hal George** (1887-1934), Papers, 1919-1951, 2 feet (including 2,900 letters). Evarts wrote articles on hunting and outdoor life, and fiction with a western background. For some years he was outdoor editor of *Saturday Evening Post.* His major interest was conservation of wildlife and the preservation of natural resources. The papers include minor book and article manuscripts, and correspondence with editors, agents, and conservationists, including Horace M. Albright, Will H. Dilg, Izaak Walton League, Stephen T. Mather, John R. White, and other conservationists and publishers. Inventory.

2465 **Federation of Western Outdoor Clubs,** Records, 1931-1953, 4⅔ feet. The Federation of Western Outdoor Clubs, a loosely organized combination of outing clubs, mostly in the Pacific Northwest and California, was first organized as the Western Federation of Outdoor Clubs in Portland, Oregon, in 1932. The records include organizational papers, minutes, financial and membership records, position papers, and correspondence, in various stages of completion.

2466 **Forest Grove Door and Lumbering Co.** (Forest Grove, Oregon), Minute Book of the Board of Directors, 1892-1894, 1 vol.

2467 **Forest History Society, Inc.,** Oral history interviews, 1957-1960, 9 items. Transcripts of interviews conducted by Elwood R. Maunder with Emmit Aston (1958, 41 l.), Charles S. Cowan (1957, 55 l.), George Lincoln Drake (1958, 32 l.), Francis Frink (1958, 13 pp.), Mrs. William Buckhout Greeley (1960, 21 pp.), Donald Mackenzie (1957, 18 pp.), Stuart Moir (1958, 18 pp.), Lowell Thomas Murray (1957, 60 pp.), and James Stevens (1957, 33 pp.). Subjects include the Pacific Logging Congress, forest fire control, logging equipment, William B. Greeley, logging in the Pacific Northwest and in Montana, Paul Bunyan, etc. These items are cataloged individually in the library.

2468 **Forest Industries Radio Communications,** Records, 1947-1960, 4 feet. An organization to coordinate radio communication in the lumber and forest products industries, formerly known as National Forest Industries Communications. The records include minutes of meetings, correspondence, reports, and documents. Inventory.

2469 **Fries, Almos Alfred** (1873-1963), Papers, 1903-1952, 3½ feet (including about 3,000 letters). Fries was a U.S. Army engineering officer, 1898-1929, and in 1917 he was in charge of construction of roads and bridges in Yellowstone National Park. Fries' papers include a typed autobiography (200 pp.), and correspondence, manuscripts, and published writings concerning his career.

2470 **Gardiner Mill Co.** (Gardiner, Oregon), Records, 1864-1923, 34 feet (including 130 vols.). Records of a lumber mill. Includes records of the Gardiner Mill Co., Gardiner Mill Store, and the A.M. Simpson Co. Includes accounts, ledgers, daybooks, general correspondence.

2471 **Gorman, Martin W.** (1853-1926), Papers, 1872-1926, 1 box (including 15 vols.). Botanist, Oregon and the Yukon; Curator of the Forestry Building of the Lewis and Clark Exposition, 1906-26. Included are diaries (1872-74, 1892, 1896, 1897, 1899, and 1903) which usually include plant lists; also plant lists (1899, 1922-23, and 1926); and manuscripts, among them a "Report on the Vegetation of the Washington Forest Reserve" (n.d.).

2472 **Greeley, William Buckhout** (1879-1955), Papers, 1909-1955, 7 feet (17 boxes and 2 small boxes). Chief Forester, U.S. Forest Service, 1920-28; thereafter secretary-manager of the West Coast Lumbermen's Association. The papers include diaries (1908-09) kept when Greeley was in charge of Forest Service District 1; diaries (1917-18) when he was commanding officer of the Forestry Section, 20th Engineers, A.E.F.; correspondence and documents (1931-55) concerning the West Coast Lumbermen's Association; also manuscripts of Greeley's books, articles, and addresses.

2473 **Greene, Thomas Gabbert** (1860-1944), Papers, 1894-1935, ca. 2 feet. Lawyer, of Oregon. Political and business papers, including corporation records of the Wilson-Case Lumber Company (1903-13), of Portland, Oregon.

2747 **Haefner, Henry E.,** Manuscript, 1959, 161 pp. Some reminiscences of an early forester, 1909-25, and 1930. Port land, Oregon (1959). Typewritten. Short sketches of men and events. Haefner worked in the Siskiyou National Forest.

2475 **Hanzlik, Edward John** (1886-1959), Papers, 1921-1934, 1 box. Forester. Included are field diaries (1921-34) kept while Hanzlik served with the U.S. Forest Service in the Pacific Northwest, and unrelated correspondence.

2476 **Hawks, William,** Correspondence, 1902-1903, 2 feet (2,284 items). Business correspondence of Hawks and George S. Canfield, real estate agents, of Portland, Oregon, concerning timber, mine, and ranch property in Oregon and Washington.

2477 **Hermann, Binger** (1843-1926), Papers, 1888-1920, 2 feet (including 495 letters). U.S. Representative from Oregon, 1885-97, 1903-07; Commissioner of the General Land Office, 1897-1903. The collection is largely family letters; included is a manuscript reminiscence of Hermann's days in Congress.

2478 **Horning, Walter Harold** (1892-1961), Papers, 1937-1961, 2½ feet. Professor of forestry, Iowa State College; forester in the National Park Service, 1935-38; chief forester, Oregon and California Revested Lands Administration, 1938+; Regional Administrator, Bureau of Land Management, Region I, 1947+. The papers include a report and documents on proposed Mount Olympus National Park, including a copy of Horning's official report, "The Proposed Mount Olympus National Park and Its Probable Effect Upon the National Forest and Economic Interests of the Olympic Peninsula" (1937-38); correspondence and documents concerning a proposed Department of Conservation (1937-38); reports, correspondence, and documents of the Oregon and California Revested Lands Administration, reflecting Horning's reorganization of the O. & C. Administration and the introduction of the sustained-yield principle. Major correspondence is with Harris Ellsworth, Guy Cordon, and N.O. Richardson. There is also a file of Horning's class notes at the University of Pennsylvania and Iowa State College.

2479 **Howard, James A.,** Correspondence, 1911-1918, 1 box. The collection includes material on Howard's real estate business in Albany, Oregon, which included the sale of timber tracts.

2480 **Hoyt, Vance Joseph** (1889-1967), Papers, 1913-1965, 4 feet. Hoyt was an amateur naturalist, wrote nature columns for various Los Angeles area newspapers, wrote nature novels, was author and technical director of the motion picture *Sequoia* (1935), and wrote free-lance articles for a variety of magazines. The papers include manuscripts of one book, two motion picture scripts, a series of radio scripts for the Conservation Association of Los Angeles County (1932), and a collection of published books and articles. Letters include correspondence with Hamlin Garland, Rupert Hughes, Ronald Reagan, Jim Tully, John R. White. There is a scrapbook of reviews, notices, photographs, and mementos relating to *Sequoia*. Inventory.

2481 **Ireland, Asher,** Field Diary, 1914-1920, 1 folder, Ireland was a forester in U.S. Forest Service, District 6 (Pacific Northwest).

2482 **Keep Oregon Green Association,** Records, 1948-1963, 2 feet. The association encourages carefulness with fire in Oregon forests. The files include minutes of meetings, correspondence, and promotional literature. The association was headed by Albert Weisendanger and Arthur Priaulx.

2483 **Kelsay, Robert C.** (1898-1954), Forest Service diaries, 1948-1954, 21 vols. Daily work reports of a Forest Service employee stationed at Oakridge, Oregon. Most of his work was road surveying and road building.

2484 **Kerr, Raymond Earle** (1888-1966), Papers, 1957-1966, 5 feet. The papers include correspondence, minutes, reports, and documents of various public and private conservation agencies in Oregon. Major organizations represented are Oregon Chapter, Izaak Walton League; Lane County Water Resources Advisory Committee, Oregon Committee on Natural Resources, Oregon Legislative Interim Committee on Wildlife, Oregon State Access Advisory Committee, Oregon State Water Resources Board, Western Forestry and Conservation Association, Willamette Basin Project Committee. Major correspondents are Donel J. Lane, Ivan E. Oakes. There is also a file of material on the proposed Oregon Dunes National Park. Inventory.

2485 **Kyle, William, & Sons Co.** (Florence, Oregon), Records, 1887-1940, 38 feet. Includes the records of the Florence Lumber Co.

2486 **Langille, William Alexander** (1869-1956), Papers, 1893-1955, 1½ feet. Forester, of Alaska and Brazil, and administrator, Oregon State Parks Department. Includes correspondence (482 letters), scattered diaries, a series of historical articles on Oregon state parks. There is material on forestry in Brazil, history of the Paulista Lumber Co., and history of Hood River, Oregon. Correspondents include Harold Douglas Langille (1874-1954) and Samuel H. Boardman.

2487 **Lemon Family,** Papers, 1861-1934, 3 feet (including 138 letters, 6 vols.). Includes the papers of Millard Lemon (1852-1943), Seattle engineer, who was employed for a while after 1888 in locating railroad lines for logging companies.

2488 **Linn County Fire Patrol Association,** Minutes of directors' meetings, 1911-1934, 1 folder. Xerox copies of originals held by the association in Sweet Home, Oregon. With the minutes are filed the published annual reports of the association (1913-53).

2489 **Locke, Marvin Elliott,** Collection, 1883-1933, ca. 85 items and 1 vol. Business and legal records of Dutch Flat, California, including correspondence of the Towle Lumber Company (1932-33).

2490 **Loyal Legion of Loggers and Lumbermen,** Corporate records, 1919-1936, 1 vol. and loose papers. The corporation was formed by 26 men, among them Charles T. Early, George Gerlinger, A.C. Dixon, and G.F. Hagenbush. The records include minutes of the Board of Directors, annual conference minutes, constitutions, and transcripts of meetings.

2491 **McBain, Bertram Telfar,** Papers, 1913-1936, 1½ feet. Paper mill manager and consultant. The papers include nine proposals for pulp mill sites in the Pacific Northwest, and one in Alaska; a manuscript history of the pulp and paper industry in the Pacific Northwest; and a photograph album of the Crown Pulp Mill, Oregon City (1914).

2492 **MacGregor, Dugald W.L.** (b. 1880), Letters, 1904-1905, 1 vol. MacGregor was manager of the Buenos Aires office of the Liverpool firm of Balfour, Williamson & Co. The correspondence in this letterpress book concerns commercial affairs, particularly trade in Pacific Northwest lumber.

2493 **McKay, Douglas** (1893-1959), Papers, 1925-1958, 30 feet (including ca. 25,000 letters). Governor of Oregon, 1948-52; U.S. Secretary of the Interior, 1953-56; member of the International Joint Commission on Water Resources of the United States and Canada. There are files of correspondence relating to the Oregon legislature, the governorship, and the Interior Department. They are arranged alphabetically by correspondent. There is also a trip file, appointment books, miscellaneous personal correspondence, and speeches. Overall, however there is slight evidence of controversy or decision making. Inventory.

2494 **McKinley, Charles** (1889-1970), Papers, 1930-1968, 37 feet. Professor of Political Science, Reed College, Portland. McKinley was a member of the President's Committee on Administrative Management and made several studies of the field services in the Pacific Northwest of the Department of Agriculture and the Interior. He was author of *Uncle Sam in the Pacific Northwest* (1952). The papers are the files accumulated for specific projects undertaken for government or private agencies or for personal research. Major files include: Forest Service, Northwest Forest Pest Action Committee, Oregon State Board of Forestry, and other agencies. Included are interviews with government employees and private citizens. Inventory.

2495 **McNary, Charles Linza** (1874-1944), Papers, 1921-1941, 3 feet. U.S. Senator, 1917-44. This collection deals exclusively with Oregon internal improvements, especially rivers and harbors. The bulk of the McNary office and personal files are in the Library of Congress.

2496 **Merritt, Melvin Leroy** (1879-1961), Papers, 1870-1952, 2½ feet. Forester in District 5 (Philippine Islands), 1905-09; forest assistant, Region 6 (Pacific Northwest), 1909-20; assistant regional forester (Alaska), 1921-34; Region 6, 1936-41. The papers include field diaries (1913-42), and personal diaries and account books (1905-09, 1937, 1944-49, 1952). The correspondence and documents include personal letters, lists of botanical collections, official correspondence, articles concerning forestry work in the Philippines (1905-09); special files concern the Tillamook, Oregon, subrogation case (1947-50), Save the Myrtle Woods, Inc. (1950-53), and fire control papers (1944-45). There are also papers of the Merritt family, and the field diaries of John D. Guthrie, assistant regional forester, District 6 (1923-32). A reminiscence, "Four Years in the Philippine Forestry, 1905-1909," is available on microfilm.

2497 **Mills, Abbot Low** (b. 1898), Papers, 1930-1951, 9 feet. Banker. Correspondence, mostly relating to the business of the U.S. National Bank, Portland, Oregon, including material on the Youngs Bay Lumber Company. Inventory.

2498 **Mitchell, Glenn E.** (1888-1958), Papers, 1940-1955, 1 box. Mitchell was game manager, U.S. Forest Service, Portland, Oregon. The papers consist of addresses and manuscripts of articles on game management. Included are minutes of the Big Game Committee, Multnomah Anglers and Hunters Club (1947-49, 1952, 1953).

2499 **Morfitt, James** (b. 1859), Manuscript, 1942, 26 pp. "A history of James Morfitt and his family, Nelscott, Ore." Typewritten. Author engaged in logging and ranching in Malheur County, Oregon.

2500 **Morse, Wayne Lyman** (1900-1974), Papers. U.S. Senator from Oregon, 1945-69. Personal and senatorial papers, including materials on forest access roads, timber cut allowances, Oregon and California revested lands, sustained yield, grazing policies, and forest fire control. Described in *Inventory of the Papers of Senator Wayne L. Morse, 1919-1969,* Eugene: University of Oregon Library, 1974, 257 pp.

2501 **National Hells Canyon Association,** Records of the Washington office, 1953-1958, 21 feet (including about 9,000 letters). Files include association data and federal information on Idaho Power Co. Hells Canyon projects No. 1971, 2132, 2133, and federal power policy.

2502 **Neuberger, Maurine Brown** (b. 1907), Papers, 1950-1967, 54 feet. Member of Oregon State Legislature, 1950-55; U.S. Senator, 1960-67. Most of the papers are the office files of her Senate term, maintained in the original subject classification. Major subjects include conservation projects (power and reclamation). Included are press releases, newsletters, and statements. Inventory.

2503 **Neuberger, Richard Lewis** (1912-1960), Papers, 1931-1960, 75 feet. Member of the Oregon State Legislature, 1940-42, 1948-53; U.S. Senator, 1954-60. The papers are mainly Senate office files (1955-60), retained in their original subject classification. Major subjects are conservation, reclamation, public and private power, and public lands. Also included are press releases, newsletters, tape-recorded weekly reports, phonorecords of radio appearances, etc. Personal files (1931-60) consist of general political correspondence, minor correspondence from the Oregon legislature, campaign material, and correspondence and documents concerning literary work. There is a file of published and unpublished manuscripts, most of them magazine articles, scrapbooks, and mementos. Inventory.

2504 **Neuhausen, Thomas Brues** (1872-1944), Papers, 1850-1936, 23 feet (including about 15,000 letters). Special agent, U.S. land office, Ashland, Wisconsin, 1900; land fraud investigator, The Dalles, Oregon, 1903; confidential inspector for Secretary of the Interior Ethan Allen Hitchcock, 1906, and later for Francis J. Heney, government prosecutor; broker in Portland, Oregon, in real estate, timber lands, and other investments, 1908. U.S. General Land Office papers (1900-16, 9 boxes) include major cases involving Alvin A. Muck, Pacific Lumber and Furniture Co., Oregon Lumber Co., Hyde-Benson, and William C. Bristol.

2505 **Nicolai Brothers Company** (Portland, Oregon), Records, 1870-1878, 5 vols. Daybooks (1870-71) and order books (1875-78) of a sash and door products company.

2506 **Nilsson, Adolph,** field diaries, 1913-1917, 1923, 1925-1927, 1 folder (including 40 pieces). His diaries describe his work for the U.S. Forest Service, Region 6 (Pacific Northwest).

2507 **Norblad, Albin Walter, Jr.** (1908-1964), Papers, 1951-1964, 8 feet. U.S. Representative from Oregon, 1946-64. The records include Oregon forestry material.

2508 **Onthank, Karl William** (1890-1967), Conservation Papers, 1950-1967, 24 feet (including about 7,000 letters). The papers include administrative files, correspondence, documents, and publications of conservation organizations and committees, and correspondence with public officials and conservationists. Major projects and organizations represented include: the proposed Beaver Marsh hydro-electric project on the upper McKenzie River; the Oregon Dunes National Seashore, including papers of the Committee for the Oregon Dunes, hearings in Eugene and Florence, Oregon (1959), and Eugene, Oregon (1963), and other papers; organizational papers, minutes, and publications of the Friends of the Three Sisters Wilderness, Inc., with correspondence of Ian Campbell, Paul Keyser, James A. McNab; miscellaneous correspondence, especially from the U.S. Forest Service, relating to proposed recreational development of Waldo Lake, Oregon, administrative papers of the Waldo Lake Advisory Council and the Waldo Lake Appeal Committee, and statements of interested organizations; records of the Columbia Basin Inter-Agency Committee, including minutes of the recreation subcommittee (1956-1967), and the Willamette Basin Task Force (1963-67); Federation of Western Outdoor Clubs constitution, minutes, administrative documents, and correspondence with Arthur Blake, Leo Gallagher, Emily Haig, Arthur B. Johnson, Ora Niemula, Luella K. Sawyer, Philip H. Zalesky; material relating to the Oregon state and Eugene chapters of the Izaak Walton League of America, including constitutions, minutes, and publications, and major correspondence with Dan. P. Allen and Joe Penfold; and major correspondence with Dan P. Allen and Joe Penfold; Lane County Parks and Recreation Commission minutes (1953-67), and superintendent's reports (1956-67); Nature Conservancy minutes and publications, and major correspondence with Walter S. Boardman, Jane C. Dirks-Edmunds, Paul B. Dowling, William Drake, and Huey D. Johnson; Oregon County Parks Association minutes (1963-64), and correspondence with Charles S. Collins and L.V. Epsey; Oregon Water Resources Board minutes (1956-67), policy statements, material on various projects, and major correspondence with Mark O. Hatfield, Izaak Walton League, Oregon Fish Commission, Oregon Game Commission, Oregon Sanitary Authority, Portland General Electric Co., Al Ullman, U.S. Bureau of Reclamation, U.S. Forest Service; Sierra Club minutes of the council, directors, and conservation committee (1957-67), policy statements, publications, and memoranda, and major correspondence with David R. Brower, Alfred Schmitz, and David R. Simons. There are also files of correspondence and publications of a miscellaneous group of conservation and recreation organizations, and committees, including the Desert Protective Council, Glacier Peak Wilderness Area, National Parks Association, Obsidians, Oregon Cascades Conservation Council, Oregon Roadside Council, Oregon Wildlife Federation, Wilderness Society. General correspondence is with Eunice Brandt, Campbell Church, Jr., O.K. DeWitt, Pauline Dyer, Harris Ellsworth, Virlis Fischer, Edith Green, Ruth Hopson, Harry C. James, Richard E. McArdle, Michael McCloskey, Borys Malkin, Wayne L. Morse, Maurine Neuberger, Richard L. Neuberger, Charles O. Porter, Luella Sawyer, Robert W. Sawyer, Harold Schick, J. Herbert Stone, Edgar Wayburn. Inventory.

2509 **Oregon American Lumber Co.** (Vernonia, Oregon), Records, 1926-1952, 82 feet. This company was one of the last railroad logging operations in the state. The files consist of office correspondence and reports, including daily cash statements, minutes of directors, labor relations, logs and lumber, railroad data, timber purchases and sales, and inventories. There is also correspondence with, and publications of the West Coast Lumbermen's Association, and other forest industry associations. The personal and political correspondence of Judd Greenman, president of the company, is part of the file. Index.

2510 **Oregon & California Advisory Board,** Correspondence and Minutes, 1950-1961, 3 feet. Files of Frank Sinclair Sever, member of the board, which dealt with the administration of Oregon and California revested lands.

2511 **Oregon Roadside Council,** Records, 1922-1968, ca. 2 feet. Correspondence (1922-65); minutes (1964-68); records of legislative campaigns (1951-61); and publications, from the files of Thornton T. Munger, of an organization which campaigned to protect Oregon from billboard blight and tourist litter.

2512 **Overhulse, Boyd R.** (1909-1966), Legislative papers, 1957-1964, 3 feet. Oregon state representative, 1951-55; state senator, 1961-65, Jackson County. The papers include minutes and accompanying documents of Oregon state senate committees on which Overhulse served, including the Interim Committee on Natural Resources (1961-62). Correspondence consists of letters from constituents and lobbyists (1957).

2513 **Pacific Northwest Hardwood Association,** Minutes of Meetings, 1933, 1 vol. The association was formed to encourage compliance with the lumber code of the National Recovery

Act. Officers were J.A. Irwin, B.B. Ostlind, F.C. Goodyear, A.R. DeBurgh, and M.L. Mead.

2514 **Patrick Lumber Co.** (Portland, Oregon), Sales books, 1915-1956, 25 vols. A wholesale lumber company.

2515 **Pierce, Walter Marcus** (1861-1954), Papers, 1933-1943, 98 feet. U.S. Representative from Oregon, 1933-43; a member of the House Agriculture Committee. Pierce was noted for his strong support of public power projects. The papers are the congressional office files, containing a general information file on topics of current legislative concern; correspondence, documents, clippings concerning major issues before Congress, major subjects including Civilian Conservation Corps, forests and forest policy, public lands, and others; copies of bills introduced by Pierce, with reports, hearings, and correspondence; similar material relating to public and private power development. Inventory. The Pierce papers relating to his term as governor of Oregon and certain personal files are held by the Oregon State Library, Salem.

2516 **Pitch Pine Lumber Co.** (New York), Correspondence, 1874-1909, 1 folder. Correspondence of William D. Wheelright relating to the Pitch Pine Lumber Co.; the Pacific Export Lumber Co., Portland, Oregon; etc. The file refers in part to the Elm Branch case, and to the Charles D. Ward land case.

2517 **Powell Family,** Papers, 1851-1918, 12 vols. Included are an account book (1857-1871), and daybook (1868-69), of lumber and mill work sales by Powell mill, Linn County, Oregon.

2518 **Rase, Frederick W.,** Field diaries, 1915, 1916, 15 pieces. Rase worked as a surveyor in U.S. Forest District 6 (Pacific Northwest). The diaries describe various jobs in Oregon and Washington.

2519 **Rettie, James Cardno** (1904-1969), Papers, 1924-69, 13 feet. Economist. Records of assignments from the U.S. Forest Service and Department of Interior. Major files include the North Pacific Planning Project (1942-44); Outdoor Recreation Resources Review Commission (1957-58); etc. There are several files on forest practices and forest economics, including forest credit, forestry research, forest taxation, and forest valuation. Inventory.

2520 **Ridgeway, George W.,** Field diaries, 1913, 1914, 16 pieces. Ridgeway worked as surveyor in the U.S. Forest Service District 6 (Pacific Northwest). The diaries describe jobs in Oregon and Washington.

2521 **Robins, Thomas Matthew** (1881-1965), Papers, 1913-1958, 10 feet (including about 2,000 letters). Author served in corps of engineers, 1904-45. Most of the papers concern Robins' work as private consultant after 1945, and consist of correspondence, reports, drawings, and memoranda on such projects as the Yale Dam, the Pelton hydroelectric development, and the dams of the Idaho Power Company. There are also reports of commissions, corporations, and associations, including the National Water Conservation Conference, Pacific Northwest Development Association, National Reclamation Association, and Interstate Compact Commission. Inventory.

2522 **Robinson, Elmo Arnold** (1887-1972), Papers, 1889-1969, 3 feet. Clergyman and professor of mathematics, San Jose State College, San Jose, California. Correspondence, diaries, scrapbooks, manuscripts, etc. Topics include the Sierra Club. Inventory.

2523 **Rupp, Norman Nienstedt** (1882-1942), Family and business papers, 1875-1950, 6 feet. Rupp, with his father, John J. Rupp, bought, leased, and sold timber land in Florida, Louisiana, Michigan, Minnesota, Oregon, Washington, Idaho, and California. He was associated with the Huron Timber Co.; the Mitchell, Door Real Estate Co.; the Rupp Investment Co.; and the Rose Gold Mining and Milling Co. The papers include diaries of John Jacob Rupp (1875-1927), and letter books, journals, and ledgers of the several companies with which the Rupps were concerned.

2524 **Russell, Thomas Osmonde** (1881-1961), Papers, 1908-1954, 8½ feet (including about 4,000 letters). Russell was a designer and builder of railroad lines and railroad facilities in Oregon, California, and Brazil after 1908. The papers include designs and plans for the Minarets & Western Railroad, Fresno, California, and the Sugar Pine Lumber Co., Madera, California (1921-26), among others.

2525 **Sawyer, Robert William** (1880-1959), Papers, 1913-1933, 26 feet (including about 22,000 letters). Editor, the Bend, Oregon, *Bulletin,* 1914-53; member of the Oregon State Highway Commission; member of the Hoover Commission Task Force on Water Resources and Power. His papers contain information on state and national politics, reclamation, public power, forestry, the highway billboard controversy, and Oregon state parks. Inventory.

2526 **Silver Creek Recreational Area Advisory Committee,** Records, 1937-1942, 2 folders. Correspondence, minutes, reports, and other papers, from the files of Karl William Onthank, committee member.

2527 **Smith-Powers Logging Co.** (Marshfield, Oregon), Ledger, June-December, 1881, 1 vol.

2528 **Staggs, Ira D.,** Papers, 1918-1966, ca. 2 feet. Rancher of eastern Oregon. Includes correspondence concerning Staggs' relations with state and federal committees and agencies including the U.S. Department of the Interior, Society for Range Management (Northwest Section), and the Pacific Northwest Regional Advisory Council of the Forest Service (1949-62).

2529 **Standard Appraisal Co.** (Portland, Oregon), Appraisal report on the plant of the Oregon-Kalama Lumber Co., Winlock, Washington, 1923, 258 pp., illustrated.

2530 **Standard Corporation** (Portland, Oregon), Records, 1904-1939, 8 feet. Included are the major account files, among them the Pacific Northwest Timber Co.

2531 **Stephens-Weatherford Co.** (Albany, Oregon), Journal and ledger, 1919-1945, 1 vol. Records of a timber holding company.

2532 **Stevens, Hazard** (1842-1918), Papers, 1876-1918, 14 feet. Attorney for the Northern Pacific Railway, 1871-73, active in right-of-way proceedings and the suppression of timber pirates. Inventory.

2533 **Sugar Pine Mill and Fixture Co.** (Albany, Oregon), Records, 1894-1903, 1 vol. Includes articles of incorporation and minutes of board meetings. Original incorporators were George W. Hochstedler, Edmund Zeyss, and C.W. Sears.

2534 **Thomas, Bert Carl** (1881-1962), Papers, 1911-1947, 4 feet. Includes General Land Office papers (1917-30) relating to forest history.

2535 **United States Spruce Production Corporation,** Minutes, August 21, 1918-November 12, 1946, 3 vols. Incorporated in 1919 by John E. Morley, Prescott W. Cookingham, and John P. Murphy. The minute file includes articles of incorporation, and copies of documents referred to in the minutes. This is the file of G.R. Sweetser, secretary of the corporation.

2536 **United States Spruce Production Corporation,** Records, 1918-1919, 1 foot. Records kept by supply officer Richard Stevens Eskridge.

2537 **Van Waters, George Browne** (1856-1934), Papers, 1879-1939, 1½ feet (including 653 letters). Among the miscellaneous papers is an expense ledger of George R. Vosburg on behalf of John E. DuBois lumber and timber purchases in Oregon (1899-1901). Inventory.

2538 **Weatherford, John Knox** (1850-1935), Papers of J.K. Weatherford and John Russell Wyatt, 1890-1923, 12 feet (including about 14,000 letters). The law firm of Weatherford and Wyatt was a major legal office in the upper Willamette Valley. Weatherford was personally interested in several businesses and assorted timber lands. Inventory.

2539 **Western Pine Association,** Records, 1906-1959, 26 feet. Originally incorporated as the Western Pine Manufacturers Association by A.W. Cooper, John K. Kollock, and John H. White. The records include incorporation papers, membership lists, reports of the secretary-manager, committee reports, minutes of meetings, statistical summaries, the *WPMA Barometer* (weekly report of orders, shipment, and production), price reports, legal case files (especially Interstate Commerce Commission cases), National Wooden Box Association records, circulars of the California White and Sugar Pine Manufacturers Association, and circulars of the Western Pine Association (1907-48). Inventory.

2540 **Whisnant, Archibald McNeill** (1876-1961), Manuscripts, n.d., 1½ feet. The manuscripts consist of speeches and reports to the Pacific Logging Congress, a few short stories, a novel, and a history of the lumber industry in the United States, "Mankind and the Forests."

2541 **White, John Roberts** (1879-1961), Papers, 1897-1961, 12 feet (including about 2000 letters). From 1920-41 White served as ranger, chief ranger, or park superintendent at Grand Canyon, Sequoia, and General Grant National Parks, Death Valley National Monument, and as Director of Regions 3 and 4. The papers include an autobiography, diaries and notebooks (1897, 1906-16, 1920-48, 52 vols.), and personal letters. There is considerable correspondence on national park policy and conservation, especially about Sequoia and Kings Canyon. Manuscripts of White's books, including *Sequoia and Kings Canyon National Parks,* Stanford, California: 1949, as well as correspondence with publishers and readers, are included.

2542 **Woodlawn Plywood Co.** (Hoquiam, Washington), Records, 1946-1954, 6 feet. Included are ledgers, journals, accounts receivable, and general correspondence of the Woodlawn Plywood Co. and the Northern Timber Co.

JACKSONVILLE

Southern Oregon Historical Society, Jacksonville Museum

P.O. Box 480 97530

2543 Holdings include records (1916-17) of A.L. Peachey, U.S. Forest Service employee in Jackson County, Oregon, and resident of Lakecreek, including daybooks, expense, and mileage accounts. The Museum also has photographs of the logging industry.

PORTLAND

Georgia-Pacific Historical Museum

Director, 900 S.W. Fifth Avenue 97204 Telephone: (503) 222-5561 The holdings are available to researchers by appointment.

2544 Holdings include 300 cubic feet of archival records of the Georgia-Pacific Corporation and related firms. There are all types of materials from financial records to photographs, including extensive holdings for operations in Georgia, Arkansas, Virginia, Washington, Oregon, California, and Alaska. Smaller amounts of materials represent operations in Maine, Connecticut, Michigan, Mississippi, Idaho, Pennsylvania, and Louisiana. Some material dates from the mid- to late 18th century, but the bulk is from the late 19th century to the 1970s. The more recent material is restricted.

Oregon Historical Society

Manuscripts Librarian, 1230 S.W. Park Avenue 97205 Telephone: (503) 222-1741. Holdings are described in: *Oregon Historical Society Manuscript Collections,* Oregon Historical Society Research and Bibliography Series, No. 1, Portland: Oregon Historical Society, 1971; *Supplement to the Guide to the Manuscript Collections,* Research and Bibliography Series, No. 3, 2 vols., 1973-74; and *Oregon Historical Society Microfilm Guide,* Oregon Historical Society Research and Bibliography Series, No. 4, Portland: 1973.

2545 **Abernethy, George** (1807-1877), Letters, 1840-61, ca. 275 items. Letterpress copybook (1847-50) dealing with business affairs, including the lumber trade and sawmills.

2546 **Allen, Edward Tyson** (1875-1942), Papers, 1909-31, 13 items. Papers of E.T. Allen, including some Western Forestry Conservation Association materials; other correspondence on forestry matters; a tribute to Allen by A. Whisnant (1949); and some of Allen's poems.

2547 **Associated Forest Industries of Oregon,** Records, 1947-1958, 6 feet. Non-profit organization to protect and advance the interests of Oregon forest industries. Includes correspondence of AFIO Secretary Charles E. Ogle (1947-55) relating to memberships and activities; minutes of Board meetings (1948-58).

2548 **Associations, Institutions—Miscellaneous.** Pacific Logging Congress, Papers (1945, 1 folder) include photos, clippings related to the sale of a painting of pioneer logging sponsored by Pacific Logging Congress.

2549 **Author Unknown,** Ledger, 1909-1912, 1 vol. Unidentified ledger possibly relating to a railroad construction project, including expenses for lumber and ties.

2550 **Autzen, Thomas J.** (1888-1958). "Portland Manufacturing Co.," January 29, 1953, 17 pp., typescript. Xerox copy. Refers to lumbering.

2551 **Bailey, Maida Rossiter** (d. 1973), Papers, 1901-1962, 2½ feet. Includes correspondence (1953-56) regarding logging in the Sisters, Oregon, area.

2552 **Beach, Clarke,** Papers, 1946, 8 items. Report on the U.S. Spruce Production Corporation submitted to the Associated Press; correspondence regarding the authenticity of the report, including a report on the U.S. Spruce Production Corporation by Gordon G. McNab, Portland, Oregon.

2553 **Bear Creek Logging Co.** (Portland, Oregon), Records, 1925-31, 1 box. Includes price report (1925); production cost analysis (1926); financial statements (1927-31); log sales reports (1928); miscellaneous agreements (1930); log train reports (1924-25, ca. 200 items).

2554 **Birnie, James** (1797?-1864), Papers, 1854-64, 2 feet. Includes material on lumbering, Pacific Northwest. Unpublished guide.

2555 **Blyth, Eastman, Dillon and Co.** (Portland, Oregon), Papers, 1927-1931, 1 folder and 7 vols. Includes investment opportunities notebooks relating to lumber and other industries.

2556 **Booth-Kelly Lumber Co.,** Papers, 1911, 1933-59, 2 boxes. Correspondence; financial reports; minutes of annual meetings; operation statements for Booth-Kelly and Oregon, Pacific & Eastern Railway; suit with Springfield Plywood Corp (1956); data relating to the purchase of Booth-Kelly by Georgia-Pacific Corp. (1959).

2557 **Braly, J. Fred** (b. 1889), Papers, 1934-1959, 1 box. Realtor, Albany, Oregon. Includes material relating to sawmills.

2558 **Bridal Veil Timber Company** (Bridal Veil, Oregon), Papers, 1921-35, 15 items. Correspondence, documents, clippings relating to labor organizing and conflicts between Loyal Legion of Loggers and Lumbermen and American Federation of Labor (1935); financial reports (1921-23); history of the company (n.d., 9 pp.).

2559 **Briggs Family,** Papers, 1847-1863, 5 items. Includes an agreement regarding sawmill rights between Aurelia Briggs and Seth Markham (1863).

2560 **Brooks-Scanlon, Inc.** (Bend, Oregon), Log scale analysis, November 1957, January-May 1958, 7 vols. Count for Trout Creek Butte, Dugout Lake, Brooks Inc. Company logging, Lookout Mountain.

2561 **Brown, Chandler Percy** (b. 1909), Papers, 1905-1970, 2 boxes, 16 vols. Includes correspondence with various U.S. Senators from Oregon regarding conservation. Restricted.

2562 **Brown and Brown,** Timber cruising records, ca. 1910-1960, ca. 35 feet. Material relating to Washington, Oregon, and California. Uncataloged. In storage at Sauvie Island facility.

2563 **Buehner, Philip,** Papers, 1893-1917, 6 items, 1 vol. Includes catalog for Buehner Lumber Co.; Minute Book (1910-17), Buehner Lumber Co., Portland, Oregon.

2564 **Burlington Northern,** Records, 1898-1970. Primarily the records of the traffic department of the Spokane, Portland, and Seattle Railway Company (1898-1970). Included is information on commodity rates, especially for forest products, including logs and finished and semifinished products such as pulp, paper, doors, etc. Uncataloged.

2565 **Burnham, Howard J.** "Hudson Bay Company sawmill," 4 pp., typescript. Early lumbering history in the Pacific Northwest.

2566 **Businesses—Miscellaneous.** First National Bank (Klamath Falls, Oregon), Records, (1900-23) include a contract agreement of California Northeastern Railroad (1907); incorporation documents and corporation leases; and sales contracts (1 box). Also wood products catalogs (1 box), and lumber documents (1 box).

2567 **Campbell, Hamilton,** Letter, 1840, 2 pp. Letter from Campbell to Ezekial Pilcher (September 12, 1840). Subjects include sawmills for mission. Copied from the [Springfield, Illinois] *Sangamo Journal* (May 21, 1841), by H.E. Pratt of Springfield, Illinois. Also on microfilm.

2568 **Carver-Creason Lumber Company,** Account book, 1918-19, 1 vol.

2569 **Cascade Plywood Corporation** (Portland, Oregon), Records, 1944-59, 1 box and 1 vol. Includes correspondence (1944-49); company history with profit/loss records (1945-54); biography of Max D. Tucker, President of Cascade; scrapbook of clippings and photos; miscellaneous agreements, minutes (1944) and ownership summaries (1958-59).

2570 **Cascade Plywood Corporation** (Portland, Oregon), Scrapbook, microfilm (1 vol.). The original is also in the possession of the Oregon Historical Society.

2571 **Chamberlain, George Earle** (1854-1928), Papers, ca. 1903-1922, 4 boxes and 13 vols. (partly on microfilm). Governor of Oregon, 1902-08; U.S. Senator, 1909-21; served on the Senate committees on Agriculture and Forestry and on Public Lands. The papers include correspondence, scrapbooks, speeches, articles, clippings, etc., relating to his career as governor and senator, with references to conservation and forest reserves.

2572 **Chamberlain, Levi,** Papers. General references to lumbering in correspondence.

2573 **Clark Family,** Papers, ca. 1905-1960, 20 boxes. Papers of Alfred Edward Clark (1873-1951) and Malcolm Hamilton Clark (1885-1964), attorneys of Portland, Oregon; included are correspondence and business papers concerning a suit involving the Pacific Spruce Corp. Restricted. Guide.

2574 **Columbia Gorge Preservation Committee,** Record book, May 23, 1931—July 15, 1932, 1 vol. Includes minutes (July 3, 1931-July 15, 1932), correspondence, maps, etc.

2575 **Consolidated Timber Co.** (Portland, Oregon), Records, 1934-1951, 1 vol. Includes minutes of meetings, certificate of dissolution. Restricted until 1995.

2576 **Cordon, Guy F.** (1890-1969), Papers, 1917-1943, 1 box. Roseburg, Oregon, attorney, representing Special Public Lands Committee, Association of Oregon Counties, and other organizations in litigation over Oregon and California Railroad land grants, national parks and reserves, and other land in the public domain. Correspondence, documents, literature, clippings, and other materials concerning land use (especially timber lands), roads, wildlife conservation, taxes, military installations, communities, etc.

2577 **Crosby, W.E.** "Highlights in the History of Logging Equipment and Methods," 1930, 16 pp., typescript.

2578 **Dagner, O. Edward,** Papers, 1926-1944, 1 folder. Correspondence; truck operating cost estimates, especially logging truck costs in Oregon and California; speeches, including radio broadcasts of the Office of Defense Transportation.

2579 **Dant and Russell, Inc.** (Portland, Oregon). History of the lumber brokerage business owned by Charles E. Dant and C.S. Russell, founded in 1904 (30 pp., typescript).

2580 **Diaries, Reminiscences—Miscellaneous.** Starker, T.J., Early forestry experiences, 1973, 6 pp., typescript.

2581 **Doty Lumber and Shingle Company,** Inventory of logging plant, 1911-1914, 65 pp.

2582 **Douglas Fir Export Co.** (Portland, Oregon), Records, 1913-1962, 30 boxes. Minutes of meetings (1913-53); export declarations; general correspondence (1955-62); contracts (1956-58); bookkeeping records (1955-62); records of shipments. Guide. Restricted.

2583 **Duluth-Oregon Lumber Co.** (Portland, Oregon), Stock certificate book, 1923, 1 vol.

2584 **Durham, Albert Alonzo** (1814-1901), Correspondence, 1852-1856, 6 items. Correspondence relating to shipment of lumber from Oregon to Hawaii.

2585 **Dyche, William K.,** Manuscripts, 1899-1915, 3 folders. "An Idaho Incident" (18 pp.), "Log Drive on the Clearwater"

(33 pp.), "A Wyoming Incident" (4 pp.), "Tongue River Tie Camp" (63 pp.), relating to Dyche's logging and railroad work experiences, 1899-1915.

2586 **Eastern Oregon Land Co.** (San Francisco, California), Papers, 1914-1946, 27 boxes. Correspondence, financial records, leases, land statistics relating to agricultural operations in Malheur, Grant, Sherman, Wasco, and Wheeler counties, including references to timber lands. Restricted. Guide.

2587 **Ekin, Richard H.,** Account book, 1845-57, 1 vol. Accounts of timber and mills, Oregon City vicinity.

2588 **Failing, Henry** (1834-1898), Papers, 1851-1913, 6 feet. Portland merchant. Includes correspondence, bills, and receipts, diary, etc., regarding investment interests. Guide.

2589 **Fir Tree Timber Co.,** Records, ca. 5 feet. Subsidiary of Silver Falls Lumber Company (?). In the Sauvie Island storage facility. Uncataloged.

2590 **Forest Park Committee of Fifty,** Papers, 1945-1970, 1 box. Documents regarding land use, fire protection, rock quarry business in Forest Park, Portland, Oregon.

2591 **Forestry—Miscellaneous.** Collection of pamphlets and miscellaneous documents relating to forests and forestry.

2592 **Foster, Philip** (1805-1884), Papers, 1832-88, 2 feet. Correspondence, documents, poll books, accounts, business agreements, clippings, relating to various business ventures, including the Eagle Creek lumber mill. Guide.

2593 **Furuset, Oscar** (1885-1961), Papers, ca. 1930-1950, 10 boxes. Portland, Oregon, attorney. Included are case files involving the Noble Lumber Company, Robert Dollar Company, and others. Inventory.

2594 **Garyville Land Co., Inc.** (Garyville, Louisiana), Records, 1912-49, 9 vols. Subsidiary of Lyon Lumber Co. of Chicago and Louisiana; owned timber lands in Louisiana. Included are journals (1912-49, 4 vols.); sales record books (1917-32, 2 vols.); cash books (1917-33, 2 vols.); land and timber ledger of holdings (1918-31, 1 vol.).

2595 **Gilbert, Wells Smith** (1870-1965), Papers, 1897-1965, 3 boxes. Correspondence regarding personal matters and lumber business (1904-50); lumber business records (1912-39). Guide.

2596 **Gordon Creek Tree Farms, Inc.,** Reports and prospectuses, 1908-1938, 56 pieces. Timber dealers, Oregon. Eleven cruise reports on blocks of timber mostly in western Oregon giving potential operation possibilities, and including an extensive report on the Metolius River area (1920); also prospectuses for timber bond issues (45 folders).

2597 **Haefner, Henry E.** "Some reminiscences of an early forester." Reminiscence of his service in Siskiyou National Forest, 1909-25. Reprinted in part in *Oregon Historical Quarterly,* v. 50, no. 1, March 1975. Restricted.

2598 **Hammond, E.W.** "The Forest in Relation to Horticulture," 1891, 53 pp. Pencil draft.

2599 **Hammond Lumber Company** (Portland, Oregon), Papers, 1891-1955, 21 boxes and 31 vols. The firm was incorporated in 1913; sold to Georgia-Pacific Corp. in 1956. Included are correspondence between Hammond Lumber Company officers and their attorney, John Knox Weatherford (1910-17); articles of incorporation and dissolution (1913, 1955); timber and fire maps; miscellaneous timber reports; abstracts of title; deeds and mortgages; contracts and agreements; Public Utilities Commission reports (1950-54); miscellaneous minute books (1887-1930); payroll vouchers (1913-55); tax schedules (1927-40).

2600 **Harpham, Vernon V.,** Papers, 1932-1946, 26 items. Forest Supervisor, Roseburg, Oregon, ca. 1930-46. Includes correspondence (1932), relating to employment practices; radio addresses (1944-45), relating to forest fires, grazing on national forests, Umpqua National Forest activities; retirement party program (1946).

2601 **Hawley Pulp and Paper Company,** Records, ca. 5 feet. At Sauvie Island storage facility. Uncataloged.

2602 **Hill Military Academy,** Collection. Includes Joseph W. Hill correspondence (1910-13), relating to the Oregon Land Fraud case of 1910. Uncataloged.

2603 **Hogan, Clarence Arthur** (b. 1897). "Historical Development of the Lumber Industry of the Pacific Northwest," Portland: Reed College, Senior thesis, 1921.

2604 **Holbrook, Stewart Hall** (1893-1964), Papers, 1928-1962, 2 boxes. Author, of Oregon. Business correspondence, research materials, drafts, and literary manuscripts of *The Columbia* (1956), *The Rocky Mountain Revolution* (1956), and *300 Years in the North Woods.* Correspondents include James Stevens, letters regarding the Paul Bunyan Series (1928-29, 1951-52); and John L. Tower, International Paper Company.

2605 **Hudson's Bay Company—Miscellaneous,** 2 boxes. A collection of materials relating to Hudson's Bay Company history and operations, containing correspondence, documents, articles, accounts, and inventories, including material relating to the lumber trade. Guide.

2606 **Hug, Bernal D.,** microfilm, 1 reel. History of Elgin, Oregon. Includes information on early Union County lumbering. 376 pp. Typescript.

2607 **Hunt, Harrison H.,** Accounts, 1845-1847, 1 vol. Accounts of a sawmill and store near Astoria (September 1, 1845-April 30, 1847).

2608 **Hutchinson, Francis E.,** Papers, ca. 1920-25, 1 box. Business and financial records regarding the Hutchinson Lumber Co., Oregon City, Oregon.

2609 **Inman, Poulsen & Co.** (Portland, Oregon), Account book, 1892, 1 vol. Accounts of lumber manufacturers, Robert Inman and Johan Poulsen (February-April, 1892).

2610 **Jacobsen, Louis A.** (b. 1871), Papers, ca. 1852-1960, 6 boxes. Records of various businesses owned and operated by Jacobsen, including Jacobsen Construction Co., Jacobsen-Reid Lumber Co., and Jacobsen-Jensen Co., all of Portland. Included are correspondence, ledgers, blueprints of piles and bridges, legal documents. Guide.

2611 **Jones, E.K., and Co.** (Portland, Oregon), Papers, 1896-1917, 15 items. Legal documents including suit involving Elihu K. Jones and Co., Jones Lumber Co., and John Halsey Jones Co.; corporation record book of E.K. Jones and Co. (1897-99).

2612 **Keep Oregon Green Association** (Salem, Oregon), Papers, 1941-1970, 29 items and 1 reel microfilm. Organized in 1940 to help reduce the number of forest fires. The papers include minutes of meetings (1941-43, 1 vol.); annual reports (1947-48, 1952-54, 1956-70); educational materials; ephemera. Scrapbooks (1941-42) are on microfilm.

2613 **Kicking Horse Forest Products,** Papers, 1955-1961, 25 items. Maps, diagrams and sketches of log rafting operation on upper Columbia River, British Columbia.

2614 **Kirkham, Arthur P.,** Papers, 1 box, 2 scrapbooks. Materials relating to conservation, particularly Save the Myrtle Woods, Inc. (ca. 1950s).

2615 **Kummel, Julius F.,** Papers, ca. 1915-1945, 4 vols. Stock investment book (1915-24). Kummel was a forester.

2616 **Labor—Miscellaneous.** Collection of documents, pamphlets, etc., relating to labor organizations, including material on lumber and longshoremen's unions. Also Thomas R. Cox, "Sails and Sawmills: the Pacific Lumber Trade to 1900," Eugene: University of Oregon, thesis (1969, 1 reel microfilm); and K.A. Erickson, "Morphology of Lumber Settlements in Western Oregon and Washington," Berkeley: University of California, thesis (1965, 1 reel microfilm).

2617 **Lane, Joseph** (1801-1881), Papers, 1848-1859, 5 boxes, 9 vols. Correspondence from Nathaniel Crosby to Lane (1854, 3 items), relating to lumbering; letters from Ira H. McKean to Lane (1852-53, 2 items), relating to lumbering. Guide.

2618 **Latourell Falls Wagon Road and Lumber Company,** Records, 1887-1899, 10 vols. Minute book (1887-99); daybook; journal; ledgers; secretary's book; timber estimates on parts of township 1 south, range 5 east, and township 1 north, ranges 5 and 6 east, Willamette Meridian (Multnomah County, Oregon); and other business records.

2619 **Logan, John Francis** (1868-1961), Papers, 1897-1928, 1 box. Attorney, Portland, Oregon. Records of legal practice, including records of Pacific Coast Handle and Manufacturing Co. (1908-13, 3 vols.); Columbia Timber Co. (1902-13, 1 vol.); Parks Concession Co. (1918-19, 1 vol.).

2620 **Loyal Legion of Loggers & Lumbermen.** Record of convention held in Portland (January 6, 1919). Typescript carbon.

2621 **Ludwig, Mason.** "Technology of the logging industry in the United States." Portland: 1950, 62 l., typed.

2622 **Lyon Lumber Co.,** Records, 1892-1946, 6 feet (37 vols.). Owner of timberland in Oregon and Louisiana. Known as the Lyon Cypress Lumber Company until 1916.

2623 **McGuire, Charles Alexander.** Autobiography. Typescript copy, 4 pp. Included is information on logging in Washington, and a sawmill, Westport (Oregon?).

2624 **McKay, James** (1824-1891), Daily journals, 1879-1891, 2 vols. Pioneer sawmill owner, Washington County, Oregon. Journal includes daily weather observations.

2625 **Marion County, Oregon, County Clerk,** microfilm, 1 reel. Enumeration of inhabitants and industrial products of Marion County (1895). Includes information on lumber.

2626 **Mason, David Townsend** (1883-1973), Diaries, 1906-1973, microfilm (14 reels). Diaries (1906-37) microfilmed from originals at Yale University Library; diaries (1937-73), microfilmed from originals in the possession of the Oregon Historical Society. Restricted until 1996.

2627 **Mason, David Townsend** (1883-1973), Papers, 1858-1973, 14 feet. District forester, Montana; professor of forestry, University of California, Berkeley; manager, Western Pine Association; consulting forester, Mason, Bruce & Girard (Portland, Oregon); author, lecturer, philanthropist; developed concept of sustained yield in forestry. Collection includes business correspondence (1911-73); business papers (1911-73); addresses (1909-66); publications (1915-72); trip papers (1950-73); military papers (1917-42); financial records (1920-72); family papers (1858-1973); ephemera, including awards, photos, recordings (ca. 1850s-1973); diaries (1937-73). Unpublished guide. Diaries are closed to researchers until 1996.

2628 **Mason, David Townsend** (1883-1973). "Timber Ownership and Lumber Production in the Inland Empire," Portland: Western Pine Manufacturers Association, 1920. Microfilm (1 reel).

2629 **Masten, C.C., Logging Company,** Records, ca. 1907-1910, 3 items. Ledger (1907-10, 1 vol.); certificate of decrease in capital stock (1910); township map with timber land (n.d.).

2630 **Minto, John** (1822-1915), Papers, 1875-1915, 15 folders. Correspondence, manuscripts of articles and poems written by Minto; clippings and scrapbooks; material on forestry, conservation, Jefferson State Park, Oregon State Board of Horticulture, wagon roads; diary (1894); biographical data, ephemera; and miscellaneous documents.

2631 **Minto, Lord,** Papers, microfilm, 1 reel. Correspondence with Sir Wilfred Laurier concerning Alaska boundary (1898-1905) includes reference to timber, other natural resources.

2632 **Modoc Lumber Co.,** Records, ca. 10 feet. In Sauvie Island storage facility. Uncataloged.

2633 **Moe, J.C.** (1880-1957), Memoirs, 1957, 39 pp. Reminiscences of logging and steamboating days in Kitsap County, Washington. Includes references to Puget Sound, Chinook language, and the Industrial Workers of the World. Typescript carbon.

2634 **Moir, Stuart** (1892-1961), Papers, 1922-1961, 9 boxes. Chief Forester, Western Pine Association; Forest Counsel, Western Forestry & Conservation Association; Consulting Forester, Oswego, Oregon. Papers pertain to forestry and forest management in the Pacific Northwest; forestry organizations; and related subjects such as logging, lumber and plywood industry, and sawmills. Guide.

2635 **Murnane, Francis J.** (1914-1968), Papers, 1935-68, 9 boxes, 1 scrapbook. Labor leader of Portland, Oregon. Included is International Woodworkers of America correspondence (1938-49). Guide.

2636 **Neuberger, Richard Lewis** (1912-1960), Papers, 1954-1966, 13 boxes. Office files of Richard Neuberger, U.S. Senator from Oregon, 1954-60; member of Interior and Insular Affairs Committee; and Maurine Brown Neuberger, U.S. Senator from Oregon, 1960-67; member of Agriculture and Forestry Committee; retained in their original subject classification order. Subjects include conservation (containing billboard and Oregon Dunes legislation), and public lands (including forest lands). Guide.

2637 **Noyes-Holland Logging Co.** (Portland, Oregon), 1915-1940, Records, 1890-1940, 1 box. Includes papers of predecessor firm, Portland Lumber Co. (1908-14), and affiliate businesses, Columbia County Lumber Co., Milton Creek Logging Co., and St. Helens Lumber Co. Collection includes minutes of Board of Directors' meetings (1915-40); correspondence; operation reports (1922-27); legal documents relating to mortgage, leases, land deeds, bills of sale; financial papers. Unpublished guide.

2638 **Oakridge Timber Co.** (Portland, Oregon), Records, 1944-1945, 1 vol. Merged with Cascade Plywood Corp., May, 1945. Includes articles of incorporation (December 1944); minutes of meetings (1944-45).

2639 **O'Keane, James Joseph** (b. 1871), Records, 1885-1960, 4 boxes, 8 vols. Timber assessor, surveyor, and realtor; insurance agent; receiver of U.S. Land Office at Vancouver, Washington; and trustee of Alaska United Copper Exploration Co., Seattle. Includes business correspondence (1905-06, 1909-

12, 1917, 1920-25, 1955-60); correspondence of Frank L. Huston (1909-12), and Victor H. Beckman (1918-21); Alaska United Copper Exploration Co. correspondence (1911-15), minutes and reports (1909, 1911-15), proxies, and stock certificate book; miscellaneous land cruises (1905-13); estimates (1907); deeds (ca. 1916-27); Dole Lumber Co. inventory (1913); timber cruise notebooks of W.C. Franklin and J.W. Shaw (December 1911-13); Northern Pacific Railway claims, agreements, deeds; journals (1904-14); private account book (August 1885-November 1889); and pocket memo books.

2640 **Oregon & Washington Lumbermen's Exchange of Portland,** Records, 1890-1891, 1 vol. Articles of incorporation; stock records; minutes of meetings; bylaws; membership roll.

2641 **Oregon Council for the Protection of Roadside Beauty,** Papers, 1927-1962, 1 box. Mostly newspaper clippings in scrapbooks; also petitions to the Scenic Area Commission; includes items on the Oregon Roadside Council (1927-40), state highways, timber, wildlife, conservation, and pollution.

2642 **Oregon Historical Society,** microfilm, 1 reel. Lumber resources of Klamath County, Oregon, 1840-1910. Research notes compiled by O.H.S. staff (303 pp.).

2643 **Oregon Klamath River Commission,** Records, 1951-1958, 2 feet. Correspondence, minutes, and other records of the commission created by the Oregon legislature to make a study of the land and water use in the Klamath River Basin, to be used as the basis of a compact with the State of California concerning the division of the waters of the Klamath River. Cooperating bodies and correspondents include California Klamath River Commission, California Oregon Power Company, Federal Power Commission, Klamath Indian Agency, and the U.S. Bureau of Reclamation. There is correspondence and documentation relating to the Federal Termination Act and Klamath Indian Reservation administration (1954-58); statement of Weyerhaeuser Timber Company at subcommittee hearings (1957); "Forestry Implications of the Klamath Termination Law," by Earl R. Wilcox (April 15, 1958); table of appraised stumpage analysis for tribally owned timber lands (1958?).

2644 **Oregon Roadside Council,** Papers, 1941-1968, 24 items. Correspondence and documents relating to development and maintenance of Larch Mountain Corridor.

2645 **Oregon Timber Transport Operators,** Records, 1951-1957, 102 items. Safety education programs for log truck drivers, incorporated into Willamette Valley Logging Conference (1957). Includes correspondence relating to organization and membership (1951-57); official publications, including rules and bylaws; press releases; financial records (1952-56); membership lists; clippings; ephemera.

2646 **Pacific Export Lumber Company** (Portland, Oregon), Charter parties, 1899-1930, 1 box and 1 scrapbook. Charter parties of a firm which shipped lumber to Australia, the Orient, and South America. Charter parties (1899-1917) are in folders in box; charter parties (1918-30) are pasted in a scrapbook.

2647 **Parker & Morris** (Albany, Oregon), Accounts, 1874-1876, 2 vols. Accounts of flour and shingle mill, including notice of bankruptcy (1877). Index.

2648 **Penn Timber Co.** (Warren, Pennsylvania), Records, 1912-42, 3 vols. Journal (1912-33, 1 vol.); cancelled checks (1913, 1 vol.); report on timber lands in Lane County, Oregon, compiled by Brown & Brown (1942, 1 vol.).

2649 **Portland Iron Works** (Portland, Oregon), Records, ca. 1860-1939, 1 box, 48 vols. Manufacturers of lumber, marine, railroad, and flour mill machinery. Includes correspondence, letterpress books, ledgers, journals, cashbooks, time books, inventories, and other financial records. Unpublished guide.

2650 **Portland Milling Company,** Records, 1851-1855, 7 vols. Daybooks, William S. Land, assignee (1852-55, 1 vol.); daybook (1854, 1 vol.); journals (1851-55, 2 vols.); cash book (1851-55, 1 vol.); ledger (1851-54, 1 vol.); ledger, William S. Land, assignee (1854-55, 1 vol.).

2651 **Priaulx, Arthur W.,** Papers, ca. 1940-1950, 70 items. Essays relating to timber, logging, history, education, architecture.

2652 **Rakestraw, Lawrence** (b. 1912). "A History of Forest Conservation in the Pacific Northwest, 1891-1913," Seattle: University of Washington, Ph.D. thesis, 1955. Microfilm.

2653 **Ransom, F.C.,** Papers, 1916-1921, 38 pieces. Communications with Pacific Export Lumber Co., and expense accounts as the company's representative in South America (1919-21).

2654 **Rodgers, Andrew,** Papers, 1846-1847. References to timber and lumbering in general correspondence. Originals, photostats, and typescripts.

2655 **Ryan, Thomas Gough** (1888-1933), Papers, 1915-1933, 1 box. Lawyer, of Portland, Oregon. Correspondence and legal documents, relating to legal transactions and litigations involving lumbering, among other matters. Unpublished guide.

2656 **Sailors' Union of the Pacific,** Minutes, 1907-1910, Microfilm (1 reel [4 vols.]). References to Hammond Lumber Co.

2657 **Save the Myrtle Woods, Inc.,** Papers, 1946-1964, 1 box. Corporate records and minutes (1 vol.); minutes of annual meetings; treasurers' reports; correspondence, including that of various logging companies; and personal correspondence of Thornton T. Munger.

2658 **Scenic Preservation Association of Oregon,** Papers, 1920-1923, 2 folders. Association devoted to preservation of timber and other scenic resources along Oregon roadsides; also billboard control; included are articles of incorporation, bylaws, minutes and correspondence from the files of Harold C. Jones, secretary.

2659 **Science—Miscellaneous.** F.P. Keen, notes on forest insects (ca. 1930s, 10 pp.), typescript.

2660 **Shaw, Alva C.R.,** Daybook, 1852-1858, 1 vol. Daybook, general mercantile store and sawmill, Eola (Cincinnati), Oregon (1852-58).

2661 **Shaw, J. Royal** (b. 1885). "The history and background of four generations of the John A. Shaw family as related to the lumber industry and interrelated to the growth and development of the State of Oregon." (March 21, 1967, 5 pp.). Xerox copy.

2662 **Shaw, Laurence L.,** Selected papers, 1955-1959, 3 folders. Correspondence and other material relating to the sale of Klamath Indian Reservation tribal timber lands to the U.S. Forest Service, including "Appraisal of Klamath Tribal resources for those economic units in which timber is the resource of principal value," from Vol. I, Sec. I, "Appraisal of the property of the Klamath Indians, Oregon." Transcript and Xerox copies.

2663 **Shields, Roy Franklin** (1888-1966), Papers, 1924-1966, 2 boxes. Correspondence, documents, relating to his career as an attorney, and his involvement with the lumber business. Guide.

2664 **Silver Falls Timber Co.,** Records, ca. 20 feet. Uncataloged. Stored at Sauvie Island facility.

2665 **Smith, Amedee M.** (1839-1894), Papers, ca. 1851-1945, 4 boxes. Includes forestry resolution from Portland Chamber of Commerce (1929); lumber codes from National Recovery Administration (1934-35); records of Ochoco Timber Co. (1923-36). Guide.

2666 **Spaulding, C.K., Logging Company** (Newberg, Oregon), Time books, 1901-1920, 26 vols. Names of employees, days worked, wage rate per day, and monthly wages earned.

2667 **Stamm, Edward Philip** (1892-1965), Papers, 1917-1964, 12 boxes. Forester of Portland, Oregon. Includes miscellaneous personal papers (1917-25; 1952-64); correspondence, reports, etc., concerning Pacific Lumber Co. (1924-35), and Crown Zellerbach Corp. (1933-64); American Forestry Association correspondence, rosters, accounts (1953-64); speeches (1941-64); notes, reports, etc., regarding forestry organizations and conferences (1953-64). Guide.

2668 **Stanley, L.C.,** Journal, 1909-1917, 1 vol. Journal of the estate of L.C. Stanley, including records of property and investments in Oregon and Wisconsin, including Coos, Hood River, Multnomah, and Union Counties, primarily timber land and business investments.

2669 **Stark, Benjamin** (1820-1898), Papers, 1835-1890s, 5 boxes. Portland, Oregon, merchant, land proprietor, U.S. Senator. Includes correspondence, business records, land claim papers, lawsuits. There is material on the lumber trade. Guide.

2670 **Steiwer, Frederick** (1883-1939), Records of Butte Creek Land, Livestock & Lumber Co. (Fossil, Oregon), 1891-1927. Includes journals (1891-1927, 5 vols.), daybooks (1908-26, 2 vols.), ledgers (1891-1919, 3 vols.); financial statements (1915, 1922-24, loose sheets).

2671 **Sylvester, Avery,** Manuscript, 1843-45, 18 pp. Typescript copy of the log of the brig *Pallas,* Avery Sylvester, Captain. Includes reference to sawmills. Printed in *Oregon Historical Quarterly,* XXXIV, pp. 259 ff.

2672 **Taylor, James** (1809-1894), Family letters and documents, 1837-1865, 20 items. Includes reference to lumber trade; biographical sketch.

2673 **Teal, Joseph Nathan** (1858-1929), Papers, 1879-1955, 24 feet. Attorney and partner, Teal, Winfree & McCulloch, Portland, Oregon; U.S. Shipping Commissioner, 1920-21; authority on railroad rate and transportation matters; responsible for 1% reduction in freight rates for Pacific Northwest, ca. 1920. Collection includes business correspondence (1916-25) relating to Interstate Commerce Commission and freight rate case, shipping, commerce, and river and harbor bill, local and national politics; legal papers of Teal, Winfree & McCulloch (ca. 1900-55), including cases dealing with shipping, lumber, and mercantile businesses.

2674 **Tide Water Mill Co.** (Portland, Oregon), Records, 1912-49, 1 box and 2 vols. Includes business correspondence (1924-28, 1942); minute books (1912-18, 1919-49, 2 vols.), with articles of incorporation, balance sheets, and certificate of dissolution.

2675 **Tri-County Association,** Papers, ca. 1930-1935. An association of timber owners and lumbering interests organized to combat gerrymandering by county taxing bodies. Includes file of Elmer E. Montague, cost accountant and conciliator of school board.

2676 **U.S. Forest Service.** A record concerning the Wind River Forest Experiment Station (July 1, 1913-June 30, 1924), and the Pacific Northwest Forest and Range Experiment Station (July 1, 1924-December 30, 1938). Text by June H. Wertz (1940, 122 pp.).

2677 **U.S. Attorney, Oregon** (Portland), Records, 1896-1926, 4 boxes (45 vols.). Includes bound correspondence files of John McCourt, U.S. Attorney, Oregon district (1908-11, with gaps; 21 vols.), with references to Oregon land frauds, sale of Indian lands, criminal cases; letterpress correspondence of U.S. Attorneys (1914-19, with gaps, 21 vols.) including attorneys Clarence N. Reames, Robert R. Rankin, and Bert E. Haney, with references to fishing rights, land frauds, Industrial Workers of the World, German aliens residing in Oregon, criminal cases.

2678 **U.S. Shipping Board, Emergency Fleet Corp.** (Portland, Oregon), Papers, 1918-19, 1 box. Collection contains guides to Oregon district shipbuilding plants; plans and maps of shipyard property; magazines and articles regarding shipping and shipbuilding in Portland and the U.S.

2679 **Warm Springs Lumber Company,** Records, ca. 60 feet. Office files, including payroll, accounts payable and receivable, correspondence, bank and financial statements, cancelled check records, general and log agreements, and invoices (ca. 1942-65). Legal papers, general ledgers, and appraisal ledger. Uncataloged. In storage at Sauvie Island facility.

2680 **Watzek, Aubrey R.,** Papers, ca. 1920-1950, 25 items. Papers, including materials relating to Crossett Timber Company, Crossett Western Lumber Company, Big Creek Logging Company, and Roaring River Logging Company. Crossett materials include financial papers (1913-29, 1 vol.); financial reports, notes (1920s-40s); references to Wauna Lumber Co.

2681 **Weatherford, John Knox** (1850-1919), Papers, 1880-1939, 14 boxes. Attorney of Albany, Oregon. Includes correspondence, transcripts, and dispositions of cases of his partnerships (1889-1939); personal and office papers (1900-18); and accounts (1880-90, 2 vols.). Includes 14 folders concerning lumber companies.

2682 **West, Oswald** (1873-1960), Papers, 1895-1960, 1 box. Includes correspondence with Oregon politicians and business leaders, including Guy Cordon regarding Oregon and California land grant, timber lands.

2683 **West Coast Lumberman's Association,** Papers, ca. 1910-1965, 15 boxes, 220 vols. Includes minutes and other organizational records, correspondence, publications, surveys and reports, publicity and promotional literature, labor relations material, scrapbooks. Guide.

2684 **West Coast Steamship Company,** Records, 1946-1963, 2 boxes, 15 vols. Materials relating to ships operated by the firm, including coastal lumber carriers. Guide.

2685 **Western Forestry & Conservation Association,** Records, 1903-1960, 23 boxes, 54 vols. Records of member associations in California, Canada, Idaho, Montana, Oregon, and Washington; committee records (1922-60); conference records (1910-59); correspondence (1922-55); financial papers; forest insurance records (1918-56); forest legislation records (1911-60); publicity; research reports; statistics; weather forecast reports; scrapbooks; ephemera. Guide.

2686 **Western Timber Co.** (Portland, Oregon), Stock ledger, 1904-45, 1 vol. Lists stockholders; includes trial balance.

2687 **Westport Lumber Co.** (Westport, Oregon), Records, 1909-47, 1 box and 1 vol. Includes minutes (1909-46); certificate of dissolution (December 31, 1947); receipts (1923-47); miscellaneous stock subscriptions, proxies (1946-47), and stockholder lists (1945).

2688 **Whitman, Marcus,** Papers, 1834-1847. Correspondence, journals, miscellaneous documents of Marcus and Narcissa Prentiss Whitman. There are general references to lumbering and forestry. Partly on microfilm.

2689 **Whitney Company,** Records, 1889-1927, 1 vol. Abstract of timber lands in Tillamook County.

2690 **Willamette Iron and Steel Co.** (Portland, Oregon), Records, ca. 1902-1944, 8 file cabinets, 77 boxes. Work orders, engineering drawings and blueprints, miscellaneous papers. Includes plans for machinery used in the logging industry.

2691 **Young, Ewing** (d. 1841), Account book, 1838-1844, 1 vol. Account book kept by Ewing Young and administrators of his estate. Included are sawmill accounts (1838). Printed in *Oregon Historical Quarterly,* XXI, pp. 244-70.

Reed College Library

3203 S.E. Woodstock 97202

2692 **Thesis Collection.** Senior theses relating to forest or forest products history on file in the Reed College Library include Paul Abramson, "The Industrial Workers of the World in the Northwest Lumber Industry" (1952, 138 pp.); Dorothy L. Aseman, "Oregon Timber Land Frauds" (1937, 153 pp.); John F. Carter, "Governmental Policies in Relation to Wheat and Lumber Export Trade in the Pacific Northwest since 1929" (1939, 108 pp.); Rayburn Clifton, "Prosperity and Depression in the Douglas Fir Industry" (1929, 90 pp.); Myles E. Flint, "Private Industrial Forest Associations in the Pacific Northwest: Their Activities and Influence on Public Policy: Three Case Studies" (1958, 141 pp.); William F. Graham, "The Problem of the Radical Laborer in the Logging Industry of the Northwest" (1924, 27 pp.); Allen Hodgson, "The Development of the Pulp and Paper Industry in Oregon and Washington" (1936, 84 pp.); Clarence A. Hogan, "The Historical Development of the Lumber Industry of the Pacific Northwest" (1921, 92 pp.); Roger Johnston, "Recreation and Forest Service Multiple-Use Policy in the Oregon Cascades: Can Smokey Bear Change his Personality through Self-analysis or Group Therapy?" (1971, 131 pp.); Jerry L. Kelley, "The Tariff and the Lumber Industry of the Pacific Northwest" (1944, 67 pp.); David F. Lent, "Federal Land and Forest Policy in Connection with the Rise of Conservation, 1850-1910" (1953, 140 pp.); Donald McKinley, "Wilderness Politics of the North Cascades" (1967, 86 pp.); Richard Newton, "Chemistry and Economics of Wood Waste in Relation to the Manufacture of Paper with Particular Reference to the Portland Industrial Region" (1928, 100 pp.); Robert M. Noel, "The Economic Significance of the Small Sawmill in the Douglas Fir Region" (1948, 76 pp.); Tara O'Toole, "On the Social Benefits and Costs of a Redwood National Park" (1967, 79 pp.); Samuel K. Polland, "European Methods of Forestry as Applied in the United States" (1917, 58 pp.); Roger Randall, "Labor Relations in the Pulp and Paper Industry of the Pacific Northwest" (1942, 166 pp.). Howard M. Rondthaler, "Some Developmental Steps in the History of a Sustained Yield Program in the Pacific Northwest Forests" (1955, 112 pp.); Fred C. Shorter, "Forest Management Policies of the Federal Government in the Pacific Northwest" (1944, 76 pp.); William M. Sutherland, "The Douglas Fir Plywood Industry of the Pacific Northwest" (1941, 68 pp.); Louise Tanner, "Industrial Relations in the Logging Camp of the Northwest from 1900 to 1932" (1938, 135 pp.); Helen E. Watt and C.E. Zollinger, "The Lumber Industry in Oregon" (1921, 238 pp.); Jacob Joseph Weinstein, "The Jurisdictional Dispute in the Northwest Lumber Industry with Particular Reference to Portland, Oregon, 1937-38" (1939, 178 pp.); Elliott White, "The Association of O&C Counties: Pressure Politics and Forest Administration" (1958, 186 pp.).

SALEM

Archives Division, Office of the Secretary of State

Archivist, 1005 Broadway, N.E. 97310. Some of the holdings are described in *Guide to Legislative Records in the Oregon State Archives,* Oregon State Archives Bulletin No. 8, Publication No. 29, March 1968; and *Guide to Legislative Records Received 1968-1970,* Oregon State Archives, Bulletin No. 8 Supplement, Publication No. 33, January, 1971.

2693 **Oregon. Attorney General,** Records, Correspondence (1906-55, 45 cubic feet) is indexed for legal problems relating to forests. Case records (1904-41, 96 cubic feet) include forestry cases.

2694 **Oregon. Control Board,** Map of Columbia County, 1920, 1 roll. The map shows the holdings of lumber companies in Columbia County, Oregon.

2695 **Oregon. County Records,** 1857-1925. Tax rolls for Grant County (1869-98), Marion County (1857-1925), and Wasco County (1866-72) include evaluations of forest lands and industries, with tax rates. Brand records for Grant County (1866-1915), Polk County (1849-90), and Wasco County (1857-92) include logging brands.

2696 **Oregon. Defense Council,** Papers, July, 1942-June, 1944. Correspondence and issuances relative to the Forest Fire Fighters Service (1 folder).

2697 **Oregon. Governors Committee on Forest Resources,** Records, 1970-1971. Minutes (10 folders); exhibits (4 folders); tape recordings (9 reels).

2698 **Oregon. Governors Committee on Natural Resources,** 1949-1967, 20 boxes. Includes files on forestry and U.S. Forest Service programs.

2699 **Oregon. Industrial Accident Commission,** Records. Ledgers (1919-20, 1922-23, 1925-28, 3 vols.) include receipts, disbursements, gains and losses for various industries, including forest and forest products industries. Closed case files (1914-62, 70 cartons) are indexed (2 vols.). Minutes (1913-65, 40 vols.) also have supporting documents (1947-65, 13 cartons).

2700 **Oregon. Legislature,** Records, 1843 +. Includes minutes of the Senate Standing Committees on Fish and Game (1961-69), with exhibits and tapes; Natural Resources (1951, 1955-69), including some exhibits and tapes (partly on microfilm); Forest and Forest Products (1949), exhibits only; the House Standing Committees on Fish and Game (1955-69), chiefly including exhibits and tapes; Forestry (1951), with exhibits; Forestry and Mining (1953-59), with exhibits (partly on microfilm); Game (1951, microfilm); Natural Resources (1961-69), with exhibits and tapes; minutes of the Interim Committees on Natural Resources (1959-63), with exhibits and tapes; Wildlife (1963), with exhibits and tapes; Fish and Game [Fish and Wildlife] (1957-59), with exhibits; Public Lands (1965-69), with exhibits and tapes; and original bills, memorials, and resolutions (1856-76, 1931-69, incomplete).

2701 **Oregon. Planning Board,** Data and Research Studies 1935-1939. The studies contain a section on forest taxation and legislation. 76 items are listed in the index under "Forestry," including special items on forest products relating to economic problems, highways, land, mapping, Pacific Northwest Regional Planning Commission, state planning, water resources, and wildlife. Unprocessed.

2702 **Post War Readjustment and Development Committee,** Papers, 62 folders. Correspondence, reports, and other documents. The papers contain a file on forests and forest products (1943-49).

2703 **U.S. Agriculture Department.** Forest Service hearings relating to the proposed establishment of Three Sisters Wilderness Area, Mt. Washington and Diamond Peak Wild Areas, held at Eugene, Oregon, February 16 & 17, 1955. Microfilm (1 reel, 16 mm.).

Oregon State Library

State Librarian, State Library Building, Summer and Court Streets 97310

2704 **Capital Lumber Company,** Records, 1853-1934. Daybooks, ledgers, correspondence, and other business papers of a Salem, Oregon, lumber company and its predecessor firms.

2705 **Kent, F.C.,** "Wood Manufacturing Industries of Oregon," 1928. Paper compiled for the Statistical Bureau of the Industries Department of the Portland Chamber of Commerce.

2706 **Lloyd, W.W.,** "History of the Early Lumbering Industry in Pine Valley," 2 pp.

2707 **U.S. Census,** 10th, 1880. Original schedules for Oregon. Included are special schedules for lumber and sawmills, containing data on capital investment, motive power, number of employees, production, wages and hours of labor, months in operation, saws, mill products, remanufactures, and region from which logs were procured. Also available on microfilm.

2708 **Watts, Lyle F.** (1890-1962), "Planning for Future Forests." Copy of a speech given before the Eugene Chamber of Commerce, February 23, 1940. 13 pp., typescript.

2709 **Work Projects Administration. Oregon Writer's Project,** Papers, 147 feet. The papers contain unpublished articles, prepared for the *Dictionary of Oregon History,* on the flora and fauna of the state; clippings on flora, fauna, and lumbering; photographs of lumber mills, pioneer and modern logging operations, rolleo contests, and Douglas-fir plywood manufacturing processes. Notes relating to lumbering are found in the industry files of various counties, and include abstracts from newspapers, periodicals, and county records. There are also references to lumbering in the biographical files; and, in the folklore files, there are tall tales of Paul Bunyan, and various lumberjack songs.

Pennsylvania

AMBRIDGE

Old Economy Village, Pennsylvania Historical and Museum Commission

Curator, 14th and Church Streets 15003. Serious researchers are welcome, but should inquire in advance.

2710 **Harmony Society,** Records, ca. 1790-1905 (with portions to 1950), ca. 350 feet. Included are letters, legal documents, ledgers, letterbooks, etc. The Harmony Society was involved in the manufacture of barrels, lumber, and shingles, for internal use after 1804, and exported lumber after ca. 1825, although large transactions are not reflected in the records until the 1840s. Beginning in the 1850s, the society purchased large tracts of timber lands in northwest Pennsylvania and later in Michigan, operated a sawmill in Tidioute, Pennsylvania (1860s-80s), and later a sash and door factory in Economy (1880s-90s). Records are scanty in the early period, but become more complete later. There are no finding aids.

BETHLEHEM

Archives of the Moravian Church

Archivist, Main Street at Elizabeth Avenue 18018

2711 **General Economy** (Northampton County, Pennsylvania), Records, 1750-1771, 80 feet. Accounts of the farms, forests, and other holdings, including, among the occupations represented, that of sawmill operator. There are records of lumber produced, and some maps showing types of timber growth in 1757. The materials are largely written in the German language.

CARLISLE

Dickinson College, Special Collections

Special Collections, Spahr Library 17013. There is a printed guide to the holdings.

2712 **Maulsby, Samuel,** Papers, 1790-1840, 1 foot. Montgomery County, Pennsylvania. Includes an account book (1803-20) dealing in part with lumber (mostly receipts for boards); and business papers (1790-1840), also relating in part to lumber.

2713 **Slifer-Dill Family,** Papers, 1850-1866, 6 feet (also on microfilm [6 reels]). Correspondence and related papers of Eli Slifer (1818-88) contain occasional references to the lumber industry in Pennsylvania. Included are a list of quantities and types of timber shipped from Reading, Pennsylvania (1846), and a letter (October 8, 1866) of Henry R. Walton relating to plans to elect political candidates favorable to lumber interests in Union and Snyder Counties. There is also material on the building of canal boats.

DOYLESTOWN

*Bucks County Historical Society

Pine and Ashland Streets 18901. The information in the following entries was taken from Neiderheiser's *Forest History Sources* and from the Eleutherian Mills Historical Library's "Union Catalog of Business and Economic History Manuscripts."

2714 **Coryell, L.S., and Company** (New Hope, Pennsylvania), Account book, 1819-1821. Record of an account with William Watts who conducted a branch lumber business in Doylestown for Coryell and Company.

2715 **Fackenthal, B.F.,** Papers. The papers (1778-89) contain numerous mentions of wood, especially cord wood, wood cutting, and wages of wood cutters. Papers relating to islands in the Delaware River contain incidental mentions of rafting, sawmills, and the sawmill industry. A bound manuscript by Fackenthal, "Durham Iron Works and Other Historical Papers," contains references to rafting pine logs on the Delaware, price of logs, and expenses of rafting.

2716 **Gillingham, J.,** Daybook, 1837-1846, 1 vol. The daybook shows the cost of sawed wood, prices of various woods, such as oak, hickory, gum; and the charges for hauling wood by boat on the Delaware Division Canal for a sawmill, lumberyard, and country store located on the Delaware River in east-central Bucks County, Pennsylvania.

2717 **Kendrick, F.,** Accounts, 1 vol. Bucks County, Pennsylvania; surveyor, also engaged in sawing, 1850-59.

2718 **Kulp, Enos,** Daybook, 1873-1885, 1 vol. The daybook of a sawyer for Tohickon Mills in north-central Bucks County.

2719 **Ledgers,** 1834-1838 and 1859-1866, 2 vols. Records of a small lumberyard operated in connection with a mercantile business (1834-38); accounts of a lumber business, principally charge accounts for sawing lumber, of a mill in central Bucks County (1859-66).

2720 **Rowland, William H.,** Papers. The papers contain letters (1842-43, 4 items), written by George Green, Belvidere, New Jersey; Joseph D. Murray, New Hope, Pennsylvania; and G.S. Condin, Morristown, New Jersey; to William Rowland of Dublin, Pennsylvania, regarding a double sawmill plant, shipping lumber by canal, and rafts on the Delaware River.

2721 **Tinicum Saw Mills,** Daybook, 1805-1835, 1 vol. Contains a record of lumber sales of a sawmill in northern Bucks County.

2722 **Tohickon Mills (Enos Kulp & Co.),** Account, 1871-1874, 4 vols. Grist and sawmill.

ELVERSON

Hopewell Village National Historic Site

R.D. #1, Box 345 19520

2723 **Clingan, A.H.,** Accounts, 1890-1909, 20 vols. Birdsboro, Pennsylvania. Included are a woodcutter's ledger (1896-98, 1 vol. plus loose papers), containing accounts for cordwood, rails, posts, railroad ties, charcoal, etc.

2724 **Hopewell Furnace** (Hopewell Village, Pennsylvania), Accounts, 1784-1896 (with gaps), 83 vols. (partly on microfilm). Accounts of a charcoal iron furnace. Brooke and Buckley, proprietors of Hopewell, owned a large acreage of woodland, and also purchased wood from others. The records include only one wood account book (1819-22), but each journal contains a spring wood purchase and wood cutting settlement, and later payment to the master colliers. This data has been extracted on a card file index. One timebook also contains wood accounts. Journals and ledgers after the 1830s carry entries for the Union Canal Coaling Account, shipments of charcoal to Hopewell from a tract of woodland in Lebanon County, Pennsylvania.

2725 **Joanna Furnace,** Account books, 1791-1901 (7 rolls microfilm). Microfilm of originals at the Historical Society of Berks County. Included are wood books (1831-52; 1887-1901; 3 vols.).

2726 **Long Family,** Ledger, 1843-1854, 1 vol. Ledger of a saw and grist mill at Birdsboro, Pennsylvania.

HARRISBURG

Department of Environmental Resources, Forest Advisory Services, Library

Chief, Forest Advisory Services, P.O. Box 1467 17120. The library is basically for the use of departmental personnel, but public access is permitted.

2727 Holdings include correspondence, reports, manuals, minutes, newsletters, and miscellaneous other records of the State Bureau of Forestry and predecessor agencies, among them a photograph album of windfall damage in the Cook Forest Cathedral Area, Pennsylvania (1956); Pennsylvania Department of Forests and Waters, Bureau of Forestry, lists of private forest tree planters in the state by counties (1910-61), also, nursery and planting reports (1920-61); Pennsylvania Forestry Reservation Commission minutes (1899-1945, 20 vols.); Pennsylvania Department of Forestry, General Forestry Correspondence, onionskin copies (1899-1901, 8 vols.); miscellaneous letters, memos, reports, and pamphlets relating to the Mont Alto Academy state forestry school (1902-25, 2 vols.); Pennsylvania Department of Forests and Waters, forest management plans (1950-84), and four-year summary reports (1935-38); etc. There is also one filing cabinet of records (ca. 1922-74) of the Society of American Foresters, Allegheny Section, containing correspondence, memos, reports, and other records, in chronological order.

Pennsylvania Historical and Museum Commission, Division of Archives and Manuscripts

State Archivist, William Penn Memorial Museum and Archives Building, P.O. Box 1026 17108. The State Archives has in its possession ca. 17,000 cubic feet of permanently valuable state records from practically every department, board, and commission of the executive branch, as well as records from the legislative and judicial branches. In addition, the archives hold over 5,000 cubic feet of historical manuscripts, which include the records or papers of prominent individuals, business concerns, and social or professional organizations. The commission planned to have published *A Guide to Historical Manuscripts* late in 1975.

State Records

2728 **Department of Forests and Waters,** Records (Record Group 6), 1727-1971, 235 cubic feet. Records of the Office of the Secretary (1923-35, 3 vols., 3 folders, and 1½ cartons), contain largely reports, appointments, and newsletters. Records of the Department of Forestry contain records of the Office of the Commissioner, with reports (1902-18, 7 cartons), letterpress books (1902-12, 75 vols.), general correspondence of the Foresters' and Rangers' Service (1902-21, 9 cartons), and an account book (1874-1904, 1 vol., including Commissioner of Forestry, Department of Agriculture, 1897-1901). Records of the Bureau of Parks contain minutes (1928-38, 1½ cartons), reports (1924-41, 1½ cartons), general correspondence (1923-41, including that of the Department of Forestry, 1922-23, 9 cartons), and addresses (1935-36, 1 folder). Records of the Bureau of Forest Management (1 carton) include general correspondence of the Bureau (1923-30), and of the Department of Forestry (1920-23).

2729 **Special Commissions,** Records (Record Group 25). Records of the Chestnut Tree Blight Commission (1911-14, 4½ cubic feet) contain minutes (1911-14, 5 folders); proceedings of the Governor's Conference (1912, 1 folder); reports (1912-13), including general report and reports of districts and counties; special studies; general correspondence (1911-14, 1 folder); accounts (1911-14, 13 folders); and photographs (774 items).

Historical Manuscripts

2730 **Barclay Family,** Papers, 1906-1934, 8 cubic feet. Account books (15 vols.) and correspondence files, including bills and invoices, of W.L. Barclay and his sons, S.D. and George S., pertaining to family business and financial interests. Prominently mentioned are the Lacquin Lumber Company, and the Northwest Lumber Company (Seattle, Washington). Unarranged.

2731 **Bining, Arthur C.,** Collection, ca. 1898-1955, 2½ cubic feet. Notes, photographs, clippings, and bibliographical references used by Arthur C. Bining in the preparation of books and articles relating to the development of the iron and steel industry. Included also are manuscripts of some of his writings and letters. There are references and photographs relating to charcoal iron furnaces.

2732 **Business Records Collection,** 1681-1963, 40 cubic feet. Business records of Pennsylvania turnpike, canal, and railroad companies, iron forges and furnaces, etc., including cordwood entries in account books and other record books of the Berkshire Furnace and Charming Forge (1681-1902, 45 vols.); Elizabeth Furnace and Speedwell Forge (1764-1839, 5 vols.); Laurel Furnace (1804-12, 1 vol.); Octoraro Forge and Conowingo Furnace (1833-60, 1 vol.); Roxberry Furnace (1760-62); Stanhope Furnace (1847-60, 2 vols.); and similar records of other furnaces and forges. There is also a notebook on the Juniata Forge and Mary Ann Furnace (1850, 1 vol.), containing a paragraph on the consumption of cords of wood; and a ledger (1845-69, 1 vol.) and daybooks (1845-50, 2 vols.) for the Bucks Mills (May Lumber Company).

2733 **Cornwall Furnace** (Cornwall, Pennsylvania), Collection, 1768-1940, 2 cubic feet. Consists primarily of account books (1768-1892, 27 vols.) of or pertaining to Cornwall Furnace, owned by the Coleman family after 1798, and several other furnaces and forges. There is a charcoal book (1768, 1 vol.) for Hopewell Forge, containing credits to colliers; workmen's accounts (1806-16, 2 vols.) for cutting, coaling, and hauling cordwood; Cornwall Furnace accounts (1840-43, 1 vol.) for cutting, hauling, and coaling wood; general accounts of a forge, probably Hopewell (1845-46, 1 vol.), including entries for construction of hauling roads from the charcoal pits to the forge.

2734 **Curtin Iron Works** (Bellefonte, Pennsylvania), Records, 1810-1941, 6 cubic feet. Records consist primarily of account books (53 vols.) which make references to the purchase of cordwood by the Eagle Iron Works, Centre County, Pennsylvania, a firm owned and operated by the Curtin family of Bellefonte; there is also correspondence (½ cubic foot).

2735 **Diaries and Journals Collection,** 1763-1938, 2½ cubic feet. Includes diaries (1855-84, with gaps, 4 vols.) of William T. Langdon (b. ca. 1828), lumberman, rafter, carpenter, etc., in the Clearfield County area, with a few brief references to lumbering.

2736 **Dock Family,** Papers, 1865-1951, 5½ cubic feet. Primarily papers of Mira Lloyd Dock (1853-1945), prominent after 1900 in the promotion of forestry in Pennsylvania and the U.S. She served on the State Forestry Reservation Commission, 1901-13. Her papers include correspondence (1878-1943, 33 folders) featuring numerous letters with the American Forestry Association, the State Forestry Academy at Mont Alto, the Women's School of Horticulture at Ambler, and the General Federation of Women's Clubs. There are also State Forestry Reservation Commission minutes (1913); reports and speeches (1902-13); school notebooks and professional and personal diaries (1869-1918, 32 vols.).

2737 **Emporium Lumber Company,** Records, ca. 1883-1972, ca. 85 cubic feet. Records (mostly 1892-1920) contain correspondence, accounts, legal papers, maps, blueprints, and related materials of the Emporium Lumber Company of Potter County, Pennsylvania, incorporated in 1892 by William L. Sykes and William Caflisch and controlled by members of the Sykes family until its dissolution in 1972. During part of this time the firm was the largest hardwood company in Pennsylvania, operating mills at Keating Summit, Galeton, and Austin, Pennsylvania; and Danby, Vermont. Included is a great deal of correspondence relating to the leasing of gas and oil rights, Sykes family holdings in the Embreeville Timber Company (Tennessee), and other family interests; also correspondence with the subsidiary Emporium Forestry Company, which operated in New York State from 1912-50. There is an account book (1883-97) of Sykes and Caflisch, a predecessor firm. Unprocessed.

2738 **Fall Brook Railroad and Coal Company** (Bath, New York), Records, 1768-1938, 345 cubic feet. Records of the business and financial interests of the Magee family of Bath, New York, including John Magee (1794-1868), who was involved in land speculation, lumbering, and other activities in north central Pennsylvania and south central New York, and owned large tracts of timber land in Michigan and Wisconsin. This collection contains only one account book (1854-61, 1 vol.) pertaining to lumbering.

2739 **Gross Family,** Papers, 1805-1918, 4/25 cubic feet. Legal papers, accounts, correspondence, etc., pertaining chiefly to the various business activities of David Gross of Winfield, Union County, Pennsylvania. Includes a statement of lumber taken from the Smith and Gross tract, 1880-82.

2740 **Lebanon County Historical Society Manuscript Collections,** 1757-1940, 207 cubic feet. This group includes 34 separate collections pertaining mainly to the history of Lebanon County, Pennsylvania. Most prominent, both in size and importance, is the Coleman Collection (1757-1940, 174 cubic feet), which consists primarily of business records of the iron furnaces and ore hills operated by the family of Robert Coleman (1748-1825) and his descendants. Included is a charcoal book (1812-13), and numerous cordwood entries in the account books of various companies and furnaces controlled by the Colemans.

2741 **McFarland, John Horace** (1859-1948), Papers, 1859-1866, 1880-1951, 20 cubic feet. Editor of the "Beautiful America" department of *Ladies Home Journal*; lecturer of wide influence; president of the American Civic Association, 1904-24; and member of the National Park Trust Fund Board after 1935, McFarland campaigned widely for the preservation of Niagara Falls, the preservation of national parks, and roadside development. The collection includes his private and professional papers. Files relating to the American Civic Association (1901-50, 17 cubic feet), contain correspondence, articles, news releases, and clippings concerning Hetch Hetchy (1908-18, 1934, 1946), and similar materials concerning national parks (1908-50). Files relating to the American Planning and Civic Association (1920-51, ½ cubic foot) include correspondence, minutes, reports, budgets of the Council on National Parks, Forests and Wildlife, formerly the National Parks Committee (1920-51); correspondence, clippings, addresses, articles, etc., on Cumberland Falls (1926-29); Everglades National Park (1930-48); Grandfather Mountain (1944-46); and the Roadside Development Committee (1932-34).

2742 **Map Collection,** 1681-1973, 48 cubic feet. More than 1,000 historical maps. Several show such information as the location of sawmills and charcoal furnaces. The Collection is described in the "Descriptive List of the Map Collection of the Pennsylvania State Archives," by Martha L. Simonetti, revised annually.

2743 **Pennsylvania, Governors,** Papers. The state archives hold papers of ten governors which contain varying small amounts of materials relating to forestry (between one folder and one carton), generally consisting of reports and correspondence pertaining to the Department of Forests and Waters. These collections are catalogued individually in the repository. Among them are papers of William A. Stone, governor from 1899-1903; Samuel W. Pennypacker, governor, 1903-07; John

S. Fisher, governor 1927-31; Arthur H. James, governor 1939-43; James H. Duff, governor 1947-51; and David L. Lawrence, governor 1959-63. The archives also contain collections, which are presently restricted, for governors John S. Fine, George M. Leader, William W. Scranton, and Raymond P. Shafer.

2744 **Philadelphia Commercial Museum,** Collection, ca. 1906-1954, 12 cubic feet. Photographic files, including pictures of charcoal kilns, the lumber industry, etc.

2745 **Phillips, John M.** (1861-1953), Papers, 1891-1966, 7 cubic feet (11 boxes and 2 vols.). Member and president of the Board of Game Commissioners of Pennsylvania, 1905-24; trustee of the American Wildlife Institute; and member of the National Executive Board of the Boy Scouts of America. Papers, including correspondence, addresses, articles, speeches, tributes and awards, publications, pictures, accounts, and scrapbooks, reflect the prominence of Phillips in the field of conservation.

2746 **Photograph Collections,** ca. 1853-1974, 10 cubic feet. Collections of photographs, generally arranged by counties, include items pertaining to Pennsylvania's forges and furnaces, sawmills, etc.

2747 **Pine Grove Furnace** (Cumberland County, Pennsylvania), Collection, 1785-1914, 60 cubic feet. Records of the early operations of the Pine Grove Furnace under various owners consist entirely of account books, with references to cordwood. Over half of the collection consists of letterpress books, correspondence, and account books, including charcoal books (1879-81, 1883, 1885-87) of the South Mountain Mining and Iron Company, which purchased the furnace in 1877, and the Fuller Brick and Slate Company, Ltd., which purchased several tracts of Pine Grove land in Adams and Cumberland Counties, Pennsylvania, in 1891. The collection also includes account books of the Cumberland Furnace, Laurel Forge, and other businesses.

2748 **Stoey, William M.** (1846-1925), Collection, 1893-1925, 11 cubic feet. Photograph collection, including pictures of forges and furnaces, sawmills, etc.

2749 **Thompson, Edward Shippen,** Collection, 1684-1941, 4½ cubic feet. Chiefly papers of succeeding generations of the Thompson family, which founded Thompsontown, on the lower Juniata River, Pennsylvania. The papers of William Thompson, Jr. (1785-1834) contain an account book pertaining to sawing (1817-18, 1 vol.).

2750 **Tonkin, Vincent,** Papers, 1867-1899, ½ cubic foot. Business correspondence, legal papers, and accounts of a lumber and real estate dealer of Cherry Tree, Indiana County, Pennsylvania. Much of the correspondence pertains to dealings with lumber companies.

2751 **Welles Family,** Papers, 1805-1898, 1½ cubic feet. Correspondence, accounts, etc., of the family of Charles F. Welles (1789-1866), of Wyalusing, Bradford County, Pennsylvania. Personal correspondence of Charles Welles discusses lumbering, bridge construction, and other business concerns. Business correspondence of Charles Welles and his son, George (1818-88), partly relates to the use of timber.

2752 **Wheeler and Dusenbury Lumber Company** (Newtown, Pennsylvania), Records, 1874-1951, 28 cubic feet. Cashbooks, journals, ledgers, order books, payroll books, sales books, and various other account books (101 vols.) of the Wheeler and Dusenbury Lumber Company (1874-1951) and the Hickory Valley Railroad Company (1914-46). One of the largest lumber companies in Pennsylvania in the 19th century, Wheeler and Dusenbury owned at one time up to 55,000 acres of pine, hemlock, and hardwood timber. In 1922 it sold two-thirds of its land to the United States government as part of the Allegheny National Forest.

2753 **Wirt, George Hermann** (1880-1961), Papers, 1878-1959, 2½ cubic feet. State Forester of Pennsylvania; director of the State Forest Academy at Mont Alto, 1903-10; chief forest fire warden in the Pennsylvania Department of Forests and Waters, 1915-45. The collection includes correspondence, notes, printed matter, clippings, transcripts of letters, reports, etc. Most of the material relates to the life and career of Joseph T. Rothrock (1839-1922), and includes biographical articles by Wirt; the Pennsylvania Conservation Education Laboratory for Teachers; and the development of forestry in Pennsylvania, including the policies and programs of the Department of Forests and Waters, the State ForestCommission, etc. There are a number of letters on Gifford Pinchot's criticisms of lumbering in Pennsylvania state forests in 1942. Some correspondence referring to living persons is restricted.

LANCASTER

Armstrong Cork Company, Archives

Coordinator, Management Reference Service, Liberty and Charlotte Streets 17604 Telephone: (714)397-0611, ext. 2258

2753a Holdings (1860 +, 650 feet) consist of photographs, annual reports, house organs, press clippings, training brochures, memorabilia, minutes, articles of incorporation, historical bulletins, price lists, and a card file. Researchers should make appointments in advance.

MECHANICSBURG

Federal Records Center

Center Director 17055. This center does not have a regional archives branch.

2754 **U.S. Forest Service,** Records, 204 cubic feet. Records of the Allegheny National Forest office, Warren, Pennsylvania, include general correspondence (1958-71, 95 cubic feet), financial records (1963-70, 40 cubic feet), contracts and purchase orders (1967-70, 37 cubic feet); commercial and timber sales (1954-71, 28 cubic feet), miscellaneous diaries of ranger employees (1957-60, 4 cubic feet). The records remain under the jurisdiction of the Forest Service.

PENNSBURG

Schwenkfelder Library

Administrator 18073

2755 **Bauer, Schultz & Schall** (Hereford Township, Berks County, Pennsylvania), Accounts, 1869-1872, 4 vols. Coal and lumber dealers.

2756 **Cope, Edward,** Accounts, 1837-1839, 1 vol. Marlborough Township, Montgomery County, Pennsylvania. Flour and grist mills, sawmill.

2757 **Heebner & Krauss** (Lansdale Borough, Montgomery County, Pennsylvania), Records, 1 vol., 1856-1865. Coal and lumber dealers.

2758 **Krauss, Edwin B.,** Accounts, 1876-1909, 6 vols. Upper Hanover Township, Montgomery County, Pennsylvania. Sawmill.

2759 **Krauss, George S.,** Accounts, 1841-1882, 7 vols. Upper Hanover Township, Montgomery County, Pennsylvania. Sawmills.

2760 **Schantz, Milton B.,** Journals, 1870-1883, 3 vols. Sawmill, Lower Milford Township, Lehigh County, Pennsylvania.

2761 **Schultz, Adam,** Cashbook, 1842-1844, 1 vol. Sawmill, Hereford Township, Berks County, Pennsylvania.

2762 **Schultz, Enoch,** Ledger, 1852-1868, 1 vol. Sawmill, Upper Hanover Township, Montgomery County, Pennsylvania. In German language.

2763 **Stauffer, Daniel G.,** Ledger, 1861, 1 vol. Lower Milford Township, Lehigh County, Pennsylvania. Sawmill.

2764 **Stauffer, Samuel,** Ledger, 1861-1872, 1 vol. Upper Milford Township, Lehigh County, Pennsylvania. Sawmill.

2765 **Sugar Valley Lumbering Company** (North Wales Borough, Montgomery County, Pennsylvania), Journal, 1871-1888, 1 vol. Coal and lumber dealers.

PHILADELPHIA

American Philosophical Society, Library

105 South Fifth Street 19106 Telephone: (215) 925-9545. Holdings are described in the *Guide to the Archives and Manuscript Collections of the American Philosophical Society,* compiled by Whitfield J. Bell, Jr., and Murphy D. Smith, Philadelphia: American Philosophical Society, 1966.

2766 **Bartram, John** (1699-1777), Collection, 1738-1777, ca. 150 items. Naturalist. This collection contains several groups cataloged separately in the repository, including papers of the John Bartram Association (1929-32, 1 box) relating to the bicentennial of the founding of the first botanical garden in the American Colonies; typescripts (719 pp.) of Bartram's correspondence in the Historical Society of Pennsylvania and other repositories, collected by Edward E. Wildman and Francis D. West; Bartram's journal (1765, microfilm) of a trip to South Carolina, Georgia, and Florida (original at the Pennsylvania Historical Society); and other papers (microfilm of originals at the College of Physicians of Philadelphia and New-York Historical Society).

2767 **Burd, James** (1726-1793), Business records and accounts, 1747-1768, 7 vols. Includes an account book, Shippensburg, Pennsylvania (1752-53), and Philadelphia (1750-56, 1 vol.), giving sales of wood, barrels, and other goods.

2768 **Coates, Samuel** (1748-1830), Account and memorandum books, 1785-1830, 5 vols. and 1 reel microfilm. Philadelphia merchant. The collection includes a receipt book (1803-30, 1 vol.) containing signed receipts for the purchases of hickory wood, ships' stores, and other goods.

2769 **Du Hamel Du Monceau, Henri Louis** (1700-1782), Papers, 1716-1789, 17 boxes, 25 folders. French agronomist and botanist. Principally on botany and agriculture, this collection includes manuscripts on trees, shrubs, and plants, copies of botanical essays by others. Included are some materials on American trees. In French language.

2770 **Girard, Stephen** (1750-1831), Papers, 1769-1831, microfilm (ca. 600 reels). The collection includes correspondence, with translations of letters in French, bank records, account books, ledgers, cashbooks, journals, etc., and other papers. Girard purchased much wood for fuel, barrel staves, shipbuilding, etc.

2771 **Lewis, Meriwether** (1744-1809), Journal, 1803, 1 vol. Journal of the winter trip from Pittsburgh to the winter camp of the Lewis and Clark expedition. There are many comments on trees across the United States. Printed in Milo M. Quaife, ed., *The Journals of Captain Meriwether Lewis and Sergeant John Ordway, Kept on the Expedition of Western Exploration, 1803-1806,* State Historical Society of Wisconsin, *Collections,* XXII, 1916.

2772 **Lewis, Meriwether** (1744-1809), and **William Clark** (1770-1838). Manuscript journal of travels to the source of the Missouri River and across the American continent to the Pacific Ocean (1804-06, 18 bound codices and 12 loose-leaved codices). Contains many comments on trees across the United States. Arranged, annotated, and indexed by Elliott Coues; interlineations throughout by Nicholas Biddle. Printed in Reuben G. Thwaites, ed. *Original Journals of the Lewis and Clark Expedition, 1804-1806,* New York: 1904-05.

2773 **Michaux, André** (1746-1802), Collection. Included are Michaux' botanical journal in North America (1787-97, 8 vols.) containing comments on American trees (printed in American Philosophical Society *Proceedings,* XXVI, 1889); and letters and papers (1783-1885, 6 pieces), including letters to his son, François André, and an act of New Jersey authorizing Michaux to purchase lands in the state to establish a botanical garden. There are comments on American trees. Largely in the French language.

2774 **Michaux, François André** (1770-1855), Collection. Included are papers (1802-1911, ca. 300 pieces) relating to the Michaux bequest to the American Philosophical Society to further silviculture in the United States, the Society's developing interest in American forests, the acquisition of books on silviculture by the Society's library, and the placing and care of the Michaux grove of oaks in Fairmount Park, Philadelphia. There are also microfilm copies of essays (1820, 1849, 2 items) on the trees of North America, and on a project for a nursery of foreign trees and plants in the vicinity of Bayonne, France, with accompanying letters. Originals are in the Muséum d'Histoire Naturelle, Paris.

2775 **Muhlenberg, Gotthilf Heinrich Ernst** (1753-1815), Collection. Included are Muhlenberg's journals (1777-1815, 2 vols.); letters (1798-1811, 19 pieces) from Christian Frederick Heinrich Dencke on botanical matters; and Muhlenberg's writings on botany and natural history (1784-1813, 24 vols.) including many comments on Pennsylvania trees. In Latin or German languages.

2776 **Murphy, Grace Emeline Barstow,** Papers, 1835-1973, ca. 2000 items. Included is professional correspondence demonstrating her interest in conservation, and articles and speeches relating to this interest. Table of contents.

2777 **Murphy, Robert Cushman** (1887-1973), Papers, ca. 1925-1973, ca. 10,000 items. Zoologist, ornithologist. Full and detailed record of his activities as a speaker, conservationist, naturalist, traveller, etc., augmenting his journals (38 vols. in library). Much material on his activities on Long Island on community education, conservation, etc. Table of contents.

2778 **Lyon, John** (d. 1814), Botanical journal, 1799-1814, 1 vol. Pertains to his travels in eastern part of the United States, and concerns U.S. trees. Edited by Joseph and Nesta Ewan and printed in *American Philosophical Society Transactions* LIII, 2 (1963).

2779 **Rothrock, Joseph Trimble** (1839-1922), Letters to Eli Kirk Price, 1878-1884, 9 pieces. Physician, botanist, forester. The correspondence relates to botanizing expeditions in the Chesapeake Bay and Virginia area, to the University of Pennsylvania, the American Philosophical Society, etc.

2780 **Royal Society of London,** Letters and communications from Americans, 1662-1900, microfilm (10 reels). Letters about America selected from the Society's manuscripts and collections of private papers. Includes an item, "A Description of the Artifice and Making of Tarr & Pitch in New England" (1662). Table of contents and index to the collection (107 pp.).

Federal Archives and Records Center

Chief, Archives Branch, 5000 Wissahickon Ave 19144
Telephone: (215)438-5591

2781 **Forest Service** (Record Group 95), 1571 cubic feet. Not accessioned into the archives branch are records of the Northeastern Forest Experiment Station, Upper Darby, Pennsylvania. They are not open to research without approval of the originating agency.

The Historical Society of Pennsylvania

Director, 1300 Locust Street 19107. The collections are described in the Historical Records Survey, *Guide to the Manuscript Collections of the Historical Society of Pennsylvania,* 2nd ed.; Philadelphia: 1949. As the collections are not cataloged by subject, there may be other material pertinent to forest history. There are restrictions on the use of the collections.

2782 **Burd, Edward Shippen,** Collection, 1790-1859, 12 vols. Philadelphia. Personal, business, legal records including account book of lumber and sundries for Pine Street, Philadelphia, building; receipts for wages, taxes, etc.

2783 **Business Account Books,** 1845-1876, 6 vols. Includes the journal (1851-70) of William Pollock, lumber merchant; his account book (1845-53); and the ledger of a Pottsville, Pennsylvania, lumber and coal merchant (1853-69).

2784 **Business, Professional, and Personal Account Books,** Collection, 1676-1904, ca. 500 vols. Records of miscellaneous business enterprises. Those containing information on the cost of wood and lumber include: James Bird, Philadelphia cord and wood dealer, account book (1777-81, 1 vol.); Boyer, Brooke, and Wilson, Philadelphia shipbuilders, account book (1788-90, 1 vol.); Josiah Bunting, Philadelphia lumber merchant, ledger (1812-13, 1 vol.); David Evans, Philadelphia carpenter and maker of coffins and venetian blinds, daybooks (1774-1812, 3 vols.); Robert Jordan and Moses Lancaster, Philadelphia carpenters and builders, receipt book (1790-1833, 1 vol.); Moses Lancaster and John Lancaster, Philadelphia carpenters and builders, receipt books, daybooks, and account books (1809-44, 8 vols.); Samuel Leedom, Haverford, Pennsylvania, lumber merchant, ledger (1837-78, 1 vol.); William R. McBay, Philadelphia coal, wood, and iron merchant, account book (1778-81, 1 vol.); S.J. Marshall, carriage maker, account book (1853-66; 1 vol.); Marshall and Wier, carriage makers, account book (1865-69, 1 vol.); Gregory Marlow, Philadelphia shipbuilder, account book (1676-1703, 1 vol.); Jonathan Meredith, Philadelphia tanner, account books with bark accounts (1784-1800, 34 vols.); Joel Mount, Philadelphia cabinetmaker, ledger (1829-65, 1 vol.) and daybook (1837-49, 1 vol.); Samuel Randolph and Company, Philadelphia carpenters and builders, receipt book (1836-46, 1 vol.); Robeson and Paul, iron, lumber, and general merchandise, letterbook (1807-13, 1 vol.); Charles C. Robinson, Philadelphia chair maker, daybook (1809-25, 1 vol.); Thomas Savery, Philadelphia carpenter and builder, account book (1782, 1 vol.); Joseph and Jesse Sharpless, Philadelphia merchants, lumber sellers, etc., receipt book (1794-1819, 1 vol.); Joseph Trotter, carpenter, account book (1829-30, 1 vol.); Issac H. Whyte, Philadelphia carpenter and builder, receipt book (1824-57, 1 vol.); also several other account books of boat builders, cabinet and chair makers, shipbuilders, etc.

2785 **Coryell, Lewis S.,** Correspondence, 1806-1867, ca. 700 items. Pennsylvania lumber dealer and politician. Included are letters concerning Coryell's lumber business at New Hope, Pennsylvania.

2786 **Cummins, John, and Company,** Correspondence, 1800-1812, 90 items. Letters addressed to John Cummins and Co., merchants of Duck Creek and Smyrna, Delaware, dealing partially in lumber. The letters are for two brief periods, April, 1800, and September, 1812.

2787 **DeWitt, Peter,** Letterbook, 1794-1820, 1 vol. Letterbook of a Philadelphia merchant relating to his trade in lumber and other commodities, and including cash accounts and records of land transactions.

2788 **Gilpin Papers,** 1720-1859, 18 vols., 2 boxes. Wilmington, Delaware; paper and textile mills on Brandywine. Correspondence, miscellaneous.

2789 **Griffith Manuscripts,** 1837-1853, 150 items. Correspondence (1837-53) addressed to Joseph D. Murray, postmaster at New Hope, Pennsylvania, and to his son, Thomas. The letters deal with business affairs, particularly the lumber trade.

2790 **Hale, John Mulhallan** (1815-1869), Correspondence, 1835-1869, ca. 4500 items. Papers relating to Pennsylvania timber, and Hale's many other business interests.

2791 **Hollingsworth Family,** Papers, 1740-1909. Merchant family, papers relating to trade, including lumber, in West Indies, Europe, America. Papers of Luke W. Morris contain material on lumber, land, railroads.

2792 **Humphreys, Joshua,** Papers, 1682-1835, 20 vols. The business records of a shipbuilding enterprise, including letterbooks (1793-1835, 3 vols.); account books (1784-1813, 2 vols.); ledgers (1766-77, 1784-1805, 3 vols.); roll call book (1794-99, 1 vol.); daybook (1791-1823, 1 vol.); navy yard mast book (1797-1806, 1 vol.); records of the building of the ship *United States* (1798-1801), and notes on the plans and construction of other ships.

2793 **Huntington, Joshua** (1751-1821), Papers, 1776-1786, ca. 525 items. Shipbuilder, Norwich, Connecticut. Accounts, payrolls, reports, and other documents relating to the construction of the frigate *Confederacy,* launched in 1779. The records present many details of the construction, provisioning, and outfitting of the vessel, and include information on the cost of lumber.

2794 **Irvine Family,** Newbold-Irvine Papers, 1760-1924, 15,000 items. Includes correspondence and account books of William A. Irvine (1803-86), businessman of Irvine, Warren County, Pennsylvania, relating to his operating a sawmill and other concerns.

2795 **Jessup, William** (1797-1868), Account books of William Jessup and William Huntting Jessup, 1814-1896, 166 vols. Daybooks, ledgers, journals, and other account books and legal papers from the law offices of Jessup and his son, of Montrose, Pennsylvania. Includes accounts of various clients in Montrose and Susquehanna County, Pennsylvania, involving timber and other concerns.

2796 **Jones and Clarke Papers,** 1784-1816, ca. 500 items. Letters, accounts, and other business records relating to the commercial enterprises of William Jones and Samuel Clarke of Philadelphia, Pennsylvania, and Charleston, South Carolina. They engaged in trade with the West Indies and Europe in turpentine. The papers also contain information on shipbuilding.

2797 **Knowles, Gustavus W.,** Account books, 1870-99, 9 vols. Included are accounts and correspondence relating to railroad ties.

2798 **Leaming, Aaron,** Diaries, 1750-1751, 1761, 1775-1776, 4 vols. The diaries contain information relating to land transactions, surveys, early settlers, farming, trade in timber and other commodities in the Cape May and adjacent area of New Jersey.

2799 **Lightfoot, Benjamin,** Records, 1770-1772, 2 vols., Photostat. Journals of a tour from Reading, Pennsylvania, to Tankhannink Creek, and of surveys of a large tract of land there containing pine timber suitable for masts.

2800 **Logan Family,** Business papers, 1808-1836, 23 vols. The papers include Charles F. Logan's Stenton Mill daybooks (1816-23, 4 vols.); ledger (1818); letterbook (1818-19); sawmill journal (1820-23); cashbook (1822-23); daybook (1823-25); journal (1823-25); the Loganville Mill journals (1819-24); daybook (1822-24); and book of accounts of team hauling (1820-23).

2801 **Meredith, Joseph D.,** Papers, 1840-1859, 2 boxes. Pottsville, Pennsylvania. Accounts, correspondence, miscellaneous, of a land agent involved in railroads, coal companies, and lumber.

2802 **Philadelphia. Lumbermen's Exchange,** Papers, 1886-1905, 5 vols. Minute books (1887-94) and a scrapbook of social events (1886-1905) of a Philadelphia business association, together with the minute book (1893-94) of the Retail Lumber Merchants' Protective Association, an organization formed to combat price cutting in the lumber trade.

2803 **Potts, William McCleery,** Isabella Furnace Papers, 1880-1929, 161 boxes and 66 vols. One of the last producing charcoal blast furnaces in the U.S. The papers include accounts, correspondence, and miscellaneous, including information on the cost of wood and charcoal.

2804 **Richards, Samuel,** Papers, 1787-1845, ca. 400 items. The papers include accounts of sales of land, timber, and shipping.

2805 **Righter Lumber Company,** Records, 1847-1881, 36 vols. Business records of the firm founded by Washington Righter, lumber dealer of Philadelphia and of Columbia, Pennsylvania.

2806 **Walker, Juliet C.,** Collection, 1845-1876, 6 vols. Journal and account book of William Pollock, lumber merchant; ledger of a lumber and coal merchant; account books (1870s, 3 vols.).

2807 **Welsh, Herbert** (1851-1937), Correspondence, 1858-1934, ca. 50,000 items. Contains material on the Society for the Protection of Forests (1890-1929, 1 box).

2808 **Wistar, Isaac Jones,** Journals, 1892, 2 vols. The journals describe incidents in Wistar's life including enterprises in lumbering and other activities in the Midwest, California, Oregon, Alaska, and Mexico.

Philadelphia Department of Records, City Archives

City Archivist, 160 City Hall 19107. The holdings of the Philadelphia City Archives are described in the *Descriptive Inventory of the Archives of the City and County of Philadelphia,* by John Daly, Philadelphia: 1970.

2809 **City Forester,** Report, 1898, 1 vol.

2810 **Fairmount Park Commission,** Records, ca. 1867-1967, ca. 81 vols. and ca. 31 cubic feet. Included are minutes of the board of commissioners, appendix to the minutes, correspondence, deeds, journals, ordinances, committee minutes, maps, and other records relating to city park policies, improvements, surveys and acquisition of lands, tree planting for ornamental purposes, administration, protection of water purity, structures, and memorials.

Philadelphia Maritime Museum

Director, 321 Chestnut Street 19106

2811 Holdings include records relating to the use of lumber by the shipbuilding industry.

Temple University, Samuel Paley Library, Special Collections Department

Head, Special Collections Department 19122

2812 Holdings include a ledger (1840-57, 1 vol.) of an unidentified firm, containing lumber accounts; and a ledger (1865-71, 1 vol.) of the Oil Lake Mining and Lumber Company (Meadville, Crawford County, Pennsylvania), containing accounts, correspondence, and other records of a saw and shingle mill owned by a Philadelphia group and operating in Meadville.

Wagner Free Institute of Science

Director, Montgomery Avenue and 17th Street 19121

2813 **Wagner, William** (1796-1885), Papers, 1819-1884, 24 feet. Includes the letters, bills, and statements of Wagner's commission lumber business between North Carolina and Philadelphia along the Atlantic Coast (1819-28), in which Wagner operated a sawmill and general store in Lennoxville, Beaufort County, North Carolina, while his partner sold the lumber in Philadelphia. Later, and prior to 1840, Wagner participated in the operation of steam rail wagons which hauled lumber from Columbia, on the Susquehanna River, to Philadelphia. Some of the letters between Wagner and his partner contain incidental descriptions of mill operations, and the condition of the sawed lumber trade in Columbia.

PITTSBURGH

Carnegie-Mellon University, Hunt Botanical Library

Archivist 15213 Telephone: (412) 621-4619. Holdings include ca. 2,000 letters, chiefly from the 18th and 19th centuries, cataloged for author and recipient, but not for subject matter. There are also more than 11,000 photographs and other likenesses of plant scientists and a file containing over 150,000 citations to biographical information, both published and unpublished, concerning plant scientists of various countries and times.

2814 **Ball, Charleton Roy** (1873-1958), Papers, 1881-1957, 11 boxes. Correspondence and manuscripts relating largely to Ball's work on the taxonomy of willows, their range, nomenclature, usefulness and ecology, with references to the work of various other botanists.

2815 **Boott, Francis** (1792-1863), Notebooks, 1814-1819 and n.d., 3 vols. The notebooks concern botanical excursions in New England; volume three contains notes from lectures on botany and natural history, some of them concerning the principal U.S. forest trees. Undated.

2816 **Braun, Emma Lucy** (1889-1971), Forestry research notes, ca. 1919-1944, ca. 2½ feet. Professor of Botany, University of Cincinnati. The notes consist of her field notebooks compiled during studies of eastern forests, which culminated in the publication of *Deciduous Forests of Eastern North America* (1950, 1967), card files of deciduous forest literature, and a file giving the percentage composition of tree species in the areas she studied.

2817 **Friesner, Ray Charles** (1894-1952), Papers, 1919, 1937-1948, 1 foot. Professor of Botany, Butler University, Indianapolis, Indiana. Included is correspondence with John Shepard Wright, Frederick Edward Clements, and N.H. Moore on the study of tree growth and the use of dendrometers.

2818 **Leiberg, John Bernhard** (1853-1913), Notes on Oregon specimens, ca. 1895-1896, 1 vol. A numbered list of plant and tree specimens collected in Oregon, with notes as to date, locations, and altitudes, probably compiled during 1895-96, when Leiberg worked in Oregon as field agent for the Botanical Division, U.S. Department of Agriculture. Bound with string.

2819 **Oral History Interviews.** Interview with Nicholas T. Mirov (b. 1893) by Lois Stone of the University of California, Berkeley, concerning Mirov's work for the U.S. Forest Service and his interest in the taxonomy, physiology, geography, and chemistry of pines (1970, 17 pp. transcript and tape). Also interviews by William L. Stern, University of Maryland, with Bohumil F. Kukachka (b. 1915), Louis O. Williams (b. 1908), John R. Millar (b. 1899), Carl de Zeeuw (b. 1912), Ralph H. Wetmore (b. 1892), and Oswald Tippo (b. 1911), concerning the wood collections at the Forest Products Laboratory, Madison, Wisconsin; Field Museum of Natural History, Chicago; Harvard University; Syracuse University; and the Smithsonian Institution; and elsewhere; and discussing the question of a national wood collection (tapes). Restricted.

2820 **Slate, George Lewis** (b. 1899), Arnold Arboretum papers, ca. 1954-1967, 1 foot. Correspondence concerning a lawsuit between the Association for the Arnold Arboretum, Inc., and the Harvard Corporation, concerning the location of the arboretum. Also contains court reports, briefs, opinions, etc., and items concerning the Association.

The Historical Society of Western Pennsylvania

4338 Bigelow Boulevard 15213 Telephone: (412) 681-5533. Holdings are described in "The Manuscript and Miscellaneous Collections of the Historical Society of Western Pennsylvania," *Western Pennsylvania Historical Magazine,* 49-52 (1966-69).

2821 **Alliance Furnace,** Correspondence, 1788-1816, 27 items. Records of the first iron furnace west of the Allegheny Mountains. Included are scattered references to the use of logs and coal [charcoal?] (1790-1813).

2822 **Hogg, George E.,** Sawmill book, 1850-1851, 1 vol. Brownsville, Pennsylvania? Records kept by Hogg and S.P. Patterson.

POTTSVILLE

The Historical Society of Schuylkill County

Curator, 14 North 3rd Street 17901

2823 **Yost, Daniel,** Collection, 400 items. Miscellaneous items dealing with land, lumbering, and other topics.

READING

Historical Society of Berks County

Executive Director, 940 Center Ave. 19601 Telephone: (215) 375-4375

2824 **Charming Forge,** Records, 20 vols. Records of a charcoal iron forge, Berks County, Pennsylvania.

2825 **Joanna Furnace,** Records, ca. 1791-1901. Account books of a charcoal iron furnace, including wood books (3 vols.). A microfilm copy of this collection is held by the Hopewell Village National Historical Site, Elverson, Pennsylvania.

2826 **Schlegel, Daniel,** Ledger, 1842-1878, 1 vol. Sawmill, near New Jerusalem, Rockland Township, Pennsylvania.

SWARTHMORE

Swarthmore College, Friends Historical Library

19081

2827 **Account Books Collection.** Includes ledger of Abraham Sharpless (1793-1853, 1 vol.) for a grist and sawmill in Concord Township, Delaware County, Pennsylvania; and a cashbook of H.C. Williams (1824-27, 1 vol.) for a mahogany and lumber yard, Philadelphia.

UNIVERSITY PARK

Pennsylvania State University, Fred Lewis Pattee Library, Pennsylvania Historical Collections and Labor Archives

Archivist 16802. Some of the holdings are described in *Materials Relating to Forest History: A Guide,* Pennsylvania State University: 1973, 3 pp., processed.

2828 **Aaronsburg General Store and Mill,** Records, 1855-1894, 1 vol. Sawmill account book.

2829 **Beaver, James Addams** (1837-1914), Papers, 1855-1914, 20,000 items. Governor of Pennsylvania, 1887; President of the American Forestry Association, 1888-90. The collection includes business and personal correspondence and miscellaneous printed materials and photographs, chiefly relating to his term as governor. Correspondence series 5 and 6 (1887-1911) contain some letters on forest conservation, the Pennsylvania Forestry Association, and the Southern States Forestry Association. Unpublished guide.

2830 **Brush Valley Saw Mill,** Records, 1884-1900, 2 vols. Account books.

2831 **Centre Furnace,** Records, 1813-1880, 63 vols. Furnace ledgers and daybooks, store ledgers and daybooks, gristmill and pig iron account books, pay book, time books, furnace and store inventory, ore and coal books. Coal books (1836-51, 2 vols.) contain records of charcoal received at the furnace, and a diary-like account of the making of 'coal,' its transportation and consumption at the furnace.

2832 **Clepper, Henry Edward,** Papers, 1936-1957, 1½ feet. Forester of Pennsylvania. Correspondence relating to professional forestry education in Pennsylvania and to his book of that title. Also materials relating to the Pennsylvania Forestry Academy/School located at Mont Alto (1903-29) and its alumni.

2833 **Clepper, Henry Edward,** Additional papers, 1896-1972, 3 feet. Correspondence, booklets, pamphlets, reports, maps, research notes relating to the Penn State Mont Alto Alumni Society, the publication of a book entitled *Professional Forestry in the United States* (1972), and the 6th and the 7th World Forestry Congresses (1966, 1972).

2834 **Dock, Mira,** Papers, 1898-1947, 1½ feet. Botanist and Commissioner of Forestry on the Pennsylvania State Forestry Reservation Commission (1901-13). Correspondence relating to her activities in forestry conservation, as chairman of the

Forestry Committee of the Pennsylvania Federation of Women's Clubs, and to her role in the establishment of the College of Home Economics at Penn State.

2835 **Drake, George,** Oral history transcript, 1958, 1 vol. Typescript (carbon copy) of an oral history interview with Drake by Elwood R. Maunder of the Forest History Foundation, Inc. The interview concerns the Pacific Logging Conference.

2836 **Edwards, William,** Collection, ca. 1914-1944, 21 feet. Professor of lumbering. Photographic materials of various aspects of forestry including approximately 850 glass slides of views of early lumbering and forestry industry, forestry in other countries, and conservation.

2837 **Ferguson, John Arden,** Collection, ca. 1912-1952, 5 items. Professor of Forestry and head of the Department of Forestry at Penn State. Photographs of campus and buildings at Mont Alto. Glass slide of Penn State forestry students at Mix Run summer camp (1912). Typescript of paper, "Forestry Education at Penn State."

2838 **Forest History—Miscellaneous.** Included are "Forestry, 1900-1927: a detailed report on forestry in Pennsylvania"; clippings files; Pennsylvania Forests and Waters Department publications (1950-58); William B. Greeley, excerpts from diary, "A Forester At War" (1917-19); "Dr. C. A. Schenck, German Pioneer in American Forestry," manuscript; material on National Christmas Tree Growers Association Information Center, Pattee Library (1965); State Foresters' Conference, Harrisburg (1920), proceedings; Cherry Tree Centennial (1922); Film: "The Last Raft" (1938), plus clippings; Photographs of log-rafting on the Susquehanna and of R. Dudley Tonkin, pioneer in Christmas tree industry; *Birch Bark Log,* issued by the Penn State Forest School, Mont Alto, 1924-27; and printed items relating to the history of forestry in Pennsylvania.

2839 **Martin, Edward** (1879-1967), Papers, 1947-1958, 37 feet. Governor and U.S. Senator from Pennsylvania, 1947-58. Legislative Files on Natural Resources and Public Works, General (1957-58) contain correspondence with constituents on the wilderness bill; Interior Department file (1957-58) contains a letter from a forest management consultant.

2840 **Reynoldsville Table Works,** Records, 1930, 1½ feet. General business correspondence and bankruptcy accounts illustrative of how the Depression ruined a thriving business.

2841 **Swanson, David C.,** Photograph collection, 1924-26, ca. 60 items. 2¾ x 4½″ negatives of scenes at the Pennsylvania State Forestry School, Mont Alto, Pennsylvania.

2842 **Tyrone Planing Mill,** Records, 1881-1927, 24 items. Ledgers, daybooks, account books, and order books.

2843 **Wirt, George Hermann** (1880-1961), Papers, 1899-1951, 1½ feet. Forester, of Pennsylvania. Business and personal correspondence; diaries (1901, 1902); forestry legislation reports and proposals; notebooks (1913-42) containing lecture notes, speeches, radio addresses, reports, and journal articles; and clippings relating to various aspects of forestry, forestry education and protection in Pennsylvania, and to Wirt's work as chief fire-warden in the Pennsylvania Department of Forests and Waters.

WARREN

Warren County Historical Society

Secretary, P.O. Box 427 16365 County Court House Annex

2844 **Chase, Theodore,** Collection, 1832-1882, ca. 3 cubic feet. Includes rafting diaries, business records, and other papers of an individual whose career included lumbering.

2845 **Clemons, Thomas,** Collection, 1825-ca. 1890, ca. 9 cubic feet. Business and personal papers of a Warren lumberman.

2846 **Irvine-Wynn Papers,** 1832-1865, ca. 1/5 cubic foot. Lumbering papers and personal correspondence.

2847 **Miscellaneous Volumes.** Ledger of Joseph E. Fay, with lumbering accounts (1889-98, 1 vol.) in Farmington Township, Warren County, Pennsylvania; daybooks of the Allen Family (1849-57, 2 vols.) with lumbering accounts; minute book of the Allegheny Lumber Co. (1899-1925, 1 vol.) with subscription list, incorporation papers and bylaws, relating to a firm in Warren, Pennsylvania, with operations in Washington and Oregon.

2848 **"Rafting, Lumbering,"** 1 vol. A large ring binder of typescripts of reminiscences, diaries, letters, clippings, etc.

2849 **"Recreation, Forestry, Conservation,"** 1 vol. A large ring binder containing clippings and articles.

2850 **Wetmore Family,** Collection. Papers of a family active in lumbering in Pennsylvania and on the West Coast. The bulk of the collection consists of correspondence and records of their various lumber companies. Unprocessed.

WEST CHESTER

Chester County Historical Society

Executive Director, 225 North High Street 19380

2851 **Ash, William R.,** Correspondence, 1860-69. Coatesville, Chester County, Pennsylvania; lumber dealer.

2852 **Barnard, Worth, & Co.** (West Chester Borough, Pennsylvania), Bills and vendue list, 1860-69. Lumber dealer.

2853 **Brinton Mill** (Birmingham Township, Chester County, Pennsylvania), Accounts, 1870-1889, 1 vol. Sawmill.

2854 **Hawley Saw Mill** (West Bradford Township, Chester County, Pennsylvania), Accounts, 1850-69, 1 vol.

2855 **Hemphill, E. Dallett,** Papers, 1890-99, 1 vol. West Chester, Pennsylvania. Lumber dealer.

2856 **Hoopes, David,** Accounts, 1800-1849, 1 vol. West Goshen Township, Chester County, Pennsylvania. Sawmill.

2857 **Hooper Brothers & Darlington Wheel Works** (West Chester, Pennsylvania), Records, 9 feet. Account books and newspaper clippings for manufacturer of wheels and spokes.

2858 **James, Arthur E.,** "The Paper Mills of Chester County, Pennsylvania, 1779-1967." Manuscript history of papermaking in Chester County, treating operations of paper mills at 65 locations, including the manufacture of pulp from poplar wood by mechanical process at the Pleasant Garden Wood Pulp Works on the Big Elk Creek, New London Township, 1867-69, and chemical manufacture by the soda process of pulp from poplar and spruce wood at the American Wood and Paper Company, on the Schuylkill River, Springville (Spring City), 1863-1890s. Restricted.

2859 **McMullen & Land** (Kennett Square Borough, Chester County, Pennsylvania), Accounts, 1864-68, 1 vol. Lumber, coal, and grain dealer.

2860 **Potts Saw Mill** (Highland Township, Chester County, Pennsylvania), Accounts, 1888-1924, 1 vol.

2861 **Pyle, George, Saw Mill** (West Bradford, Chester County, Pennsylvania), Accounts, 1855-1859, 1 vol.

2862 **Rothrock, Joseph T.,** Papers, ca. 1860s-1890s. The papers comprise a few miscellaneous letters, and 4 large scrapbooks of clippings and articles and printed programs, etc., illustrating his lifetime forestry work, and appear to include all of his public lectures on the subject.

2863 **Scott, David J.,** Account book, 1903-1904, 1 vol. Lumber dealer, West Chester, Chester County, Pennsylvania.

2864 **Sheeder Farm and Sawmill** (West Vincent Township, Chester County, Pennsylvania), Accounts, 1793-1897, 17 vols.

2865 **Sheeder Saw Mill** (East Vincent Township, Chester County, Pennsylvania), Accounts, ledger.

2866 **Taylor, Abiah,** Accounts, 1794-1852, 3 vols. East Bradford Township, Chester County, Pennsylvania; flour mill and sawmill.

2867 **Taylor, Lownes,** Accounts, 1819-1852, 2 vols. West Goshen Township, Chester County, Pennsylvania; sawmill.

2868 **Taylor, Mordecai,** Accounts, 1850-1862, 1 vol. New Garden Township, Chester County, Pennsylvania; sawmill.

2869 **Thompson, James,** Accounts, 1796-1813, 1 vol. West Fallowfield Township, Chester County, Pennsylvania; sawmill.

2870 **Valley Paper Mill** (Lower Oxford Township, Chester County, Pennsylvania), Letters, 1868-70.

2871 **Walton, Joseph,** Daybook, 1834-45, 1 vol. London Grove Township, Chester County, Pennsylvania; sawmill.

2872 **Worth Lumber Yard** (West Chester, Chester County, Pennsylvania), Records, 1860-84, 28 vols. Accounts, letterbooks.

WILKES-BARRE

Wyoming Historical and Geological Society

Executive Director, 49 South Franklin Street 18701
Telephone: (717) 823-6422

2873 **Lewis, Albert, Company** (Bear Creek, Luzerne County, Pennsylvania), Records (partly on microfilm). Microfilm (30 reels) covers account books (ca. 30 vols.), some relating to Lewis Lumber Company, some to the Lewis Ice Company. There are lumber sales books and paybooks, both ranging from about 1880-1910. Originals, also at Wyoming Historical and Geological Society, were damaged by flood in 1972.

2874 **Piolett,** Papers, 1807-1903, ca. 2¼ cubic feet. Business papers dealing in part with lumber and land transactions along the Susquehanna River in the region of Wilkes-Barre, Pennsylvania.

South Carolina

COLUMBIA

South Carolina Department of Archives and History

Director, P.O. Box 11,669, Capitol Station 29211 1430 Senate Street. Holdings are described in Marion C. Chandler, *Colonial and State Records in the South Carolina Archives: A Temporary Summary Guide,* Columbia: South Carolina Department of Archives and History, 1973.

2875 **British Public Record Office,** Records relating to South Carolina, 1663-1782, microfilm 12 rolls. These records contain lists of ships which cleared from Charleston, South Carolina; lists of imports and exports; and manifests of cargo. Included are CO 5/508-CO 5/511, covering 1716-1719, 1721-1767 (with gaps). Indexed. This film is part of the British Manuscript Project microfilm, available at the Library of Congress.

2876 **General Assembly,** Records, 1800 cubic feet. Journals of the Commons House of Assembly (1692-1775, 79 vols. and 2 flat files) include petitions and acts concerning the production of potash and the marketing of tar, pitch, rosin, turpentine, shingles, and barrel staves, and make reference to the kinds of trees used in shipbuilding. The Journals have been published, in part, as *The Colonial Records of South Carolina, Series 1: Journals of the Commons House of Assembly, November 10, 1736-March 19, 1750,* edited by J.H. Easterby and Ruth S. Green, 9 volumes, Columbia: University of South Carolina Press, 1951-1962.

2877 **Surveyor General,** Records, 1680-1932, 150 cubic feet. Colonial and state plats provide information on the type of forests standing when the lands were surveyed. Included are duplicate and recorded Colonial Plats (1731-76), and duplicate and recorded State Plats (1784-1932), each in several series.

University of South Carolina, South Caroliniana Library

Manuscripts Division 29208

2878 **Atlantic Coast Lumber Corporation** (Georgetown, South Carolina), Papers, 1716-1946, 19 boxes (ca. 16,000 items). Included are deeds for land and timber rights in the counties of Berkeley, Florence, Georgetown, Marion, and Williamsburg (15 boxes); miscellaneous deeds and other land records (1716 +) chiefly of the Theodore L. Gourdin, D.W. Alderman, and Rhem family estates, including plats and conveyances to the Allston, Pawley, Peyre, Sinkler, Porcher, Richardson, Serre', Lenud, Delavillette, Chardon, Mazyck, Gaillard, Childs, Jeanneret, Kinloch families; general correspondence and legal records (1899-1942), minutes and other records (1902-15) of the stockholders and directors meetings of the Georgetown and Western Railroad Co., abstract of title to lands owned by Edward Beers (1936, 1 vol.) and Raymond S. Farr (1941, 1 vol.) in Georgetown County; stock certificates (1902-46, 3 vols.) of the Atlantic Coast Lumber Corporation and the United Timber Corporation; cashbooks, journals, and other financial records (1922-46, 5 vols.); record of timber and timberland, Georgetown County (1899-1943); records of real estate in Georgetown and Andrews, South Carolina (1899-1943); minutes of board of directors and miscellaneous correspondence chiefly relating to trustees' board meetings, proxies, shares, and taxes (1903-16, 3 vols.). There are also maps and charts.

2879 **Black, John,** Papers, 1790-1848, 354 items and 79 vols. Chiefly business papers of a merchant and planter, Laurens district, South Carolina. The bound volumes, chiefly accounts, include timber records.

2880 **Brown Family,** Papers, 1878-1927, 139 items. Prosperity, Newberry County, South Carolina, and various places. Family correspondence, orders for lumber from Brown & Moseley Lumber Mills, and other business records.

2881 **Brunson, Joel E.** (1847-1913), Letterbooks, 1882-1911, 13 vols. Kingstree and Sumter, South Carolina.

2882 **Clark, Henry** (fl. 1870), Diary, 1870, 1899-1900, 1 vol. Descriptions of Florida, with comments on timber cutting.

2883 **Frost, J.D.,** Record book, 1850-1853, 1 vol. Record

book of lumber cut for the Greenville and Columbia Railroad Co.

2884 **Guignard Family,** Papers, 1761-1952, 2992 items. Correspondence, letterbooks, account books, land grants and titles, etc., of this Columbia, South Carolina, family, showing details of plantation management, the antebellum lumber industry, etc. One of the family's holdings was a mill making cypress shingles in Georgetown, South Carolina. The papers also refer to timber cutting at Mars Bluff, Florence County, South Carolina.

2885 **Hammond, James Henry** (1807-1864), Papers, 1823-1920, 3000 items and 106 vols. Governor of South Carolina and U.S. Senator. Chiefly personal and plantation papers. There are various account books including lumber records, etc. Processing incomplete.

2886 **Jefferies Family,** Papers, 1771-1936, 547 items. Chiefly land and papers of John R. Jefferies, including business papers, land and timber records, Union and Cherokee Counties, South Carolina, with bills for sale of lumber from Jefferies' sawmill.

2887 **Lukens Lumber Company** (Sumter County, South Carolina), Daybook, 1903, 1 vol. Accounts with individuals and businesses in Sumter County.

2888 **Martin, Joseph Barry.** "The Lumber Industry in South Carolina," Columbia: University of South Carolina, M.A. thesis (Business Administration), 1964, ca. 120 pp.

2889 **Seeley, Fred R.,** Letters to Henry Savage, Jr., May, 1953, 2 items. Giving information about the rivers of the Santee, Seeley's experience in the lumber business, history of the Santee River Cypress Lumber Co., methods and changes in the industry, living conditions and social customs of employees, malaria, wildlife, vegetation, and topography.

2890 **Wade, E.W.,** Account book, 1860-1865, 1 vol. Lumber account book.

South Dakota

MADISON

Dakota State College, The Karl E. Mundt Library

Director 57042

2891 **Mundt, Karl Earl** (b. 1900), Papers, ca. 1,500,000 items. U.S. Representative from South Dakota, 1939-48; U.S. Senator, 1948-73; served on Senate Committee on Agriculture and Forestry. There are files relating to the U.S. Forest Service (4 boxes), National Park Service (2 boxes); extensive materials relating to the Black Hills National Forest filed under Interior, Agriculture, etc.; and various legislative and project files and a speech file contain material relating to forest history. Processing incomplete (1974).

MITCHELL

Dakota Wesleyan University

Librarian 57301 Telephone: (605)996-6511

2892 **Case, Francis Higbee** (1896-1962), Papers. U.S. Representative from South Dakota, 1937-51; U.S. Senator, 1951-1962; member, National Forest Reservation Commission. The papers include a manuscript, "Forestry in the Black Hills of South Dakota and Wyoming," by Guy Bjorge, general manager of the Homestake Mine (10 pp.); and U.S. Forest Service correspondence, mostly of a routine nature, during the 76th, 77th, 79th, 80th, 81st, 85th, 86th, and 87th Congresses (ca. 1939-43, 1945-51, 1957-63). Guide.

PIERRE

South Dakota Department of Education and Cultural Affairs, Historical Resource Center

Director, Memorial Building 57501 Telephone: (605)224-3615

2893 **Pringle, W.H.** (d. ca. 1969), Papers, 1,500 items (2 boxes). Pierre, South Dakota. Pringle was national president of the Izaak Walton League of America, 1956-58. Includes copies of incoming correspondence and a folder of outgoing letters concerning the League.

SIOUX FALLS

Pettigrew Museum

131 N. Duluth Ave 57104 Telephone: (605)336-6272

2894 **Pettigrew, Richard Franklin** (1848-1926), Papers, 1887-1904, 56 vols. and 215 folders. U.S. Senator from South Dakota, 1889-1901; member of the Public Lands Committee; sponsor of the forest management legislation of 1897. The papers include letterpress books (42 vols. [20,419 items]) of congressional correspondence; scrapbooks (14 vols.); and correspondence, clippings, and photographs relating to Pettigrew's personal and business affairs (215 folders). There is much information concerning the Black Hills forest.

VERMILLION

University of South Dakota, I.D. Weeks Library

Reference Librarian 57069

2895 **Norbeck, Peter** (1870-1936), Papers, 1900-1936, 90 feet. Governor of South Dakota, 1917-21; U.S. Senator, 1921-36; member of the committees on Agriculture and Forestry, Public Lands and Surveys, and the Special Committee on Conservation of Wildlife Resources. Correspondence and papers (18 boxes) relating to parks probably chiefly concern the Black Hills National Monument, South Dakota, and Norbeck's other interests in national park legislation. There is some material on Civilian Conservation Corps work in Harney National Forest, South Dakota. Similar material on conservation (5 boxes) relates largely to wildlife.

Tennessee

GATLINBURG

Great Smoky Mountains National Park

Chief of Park Interpretation 37738

2896 **Lumbering Records.** The collection, mostly relating to logging operations conducted within the park area, 1895-1930, includes: William Whitmer and Sons, West Virginia, minutes (1895-June, 1924); William Whitmer and Sons, Delaware, minutes (June, 1909-June, 1918); Parsons Pulp and Lumber Company, Straight Fork, Oconaluftee River, Swain County, North Carolina, Records (1909-23), including certificate of incorporation, bylaws, minutes (1900-23), accounts (December, 1918); William Whitmer and Sons and Parsons Pulp and Lumber Company, report by Coverdale and Colpitts, Consulting Engineers, New York; Whitmer-Parsons Pulp and Lumber

Company, Records, (1923-27), including certificate of incorporation, bylaws, minutes (December, 1923-October, 1927), accounts receivable (April, 1924-September, 1927), miscellaneous audits, balance sheets, estimates; Champion Lumber Company, Crestmont, North Carolina, records (1911, 1917-18), a few miscellaneous notebooks and reports, Suncrest Lumber Company, Sunburst and Waynesville, North Carolina, records (1917-30), including correspondence (1920-28), annual audits (1919-24), account ledger (1917), account of lumber sold (1924-30), correspondence and estimates relating to condemnation proceedings of the North Carolina Park Commission; timber type map of park holdings (1929); Ravensford Lumber Company, Ravensford, North Carolina, appraisal report (1931); Holstein Lumber Company, Wilmington, Delaware, minutes (1908-10), important as a holding of the Whitmer interests; logging records (n.d.), logging by crews, company not named; transcripts of the proceedings by which the park lands were condemned (1930s).

MEMPHIS

Memphis State University, John Brister Library

Curator of Special Collections 38152

2897 **Meeman, Edward J.,** Papers, ca. 2 feet. Editor of the Memphis *Press-Scimitar,* 1931-63. The papers include material on conservation, state parks, and forests in general.

2898 **Oral History Reserach Office.** Interviews with J.S.F. Wilson on the hardwood industry in Memphis (1 reel); Stanley F. Deas (3 reels) and H.C. Berckes (1 reel) on the Southern Pine Association; Harry Kirk, Sam Carey, Dick Boren, and Frank Riddick on lumber and the lumber industry in Memphis (1 reel).

NASHVILLE

Tennessee State Library and Archives, Division of Archives and Manuscripts

Archivist 37219 State Library and Archives Building, Ground Floor. Holdings include the collections of the Tennessee Historical Society, described in *Guide to the Processed Manuscripts of the Tennessee Historical Society,* ed. by Harriet Chappel Crosley, Nashville, 1969.

Manuscripts Section

2899 **Adams, Paul Jay** (b. 1901), Papers, 1918-1962, ca. 50 items and 19 vols. Park guide and nurseryman, of Crab Orchard, Tennessee. Diaries, correspondence, scrapbooks (2 vols.), and few other papers. The diaries reflect Adams' great love of nature and contain lists of birds, trees, and flowers, maps, photos, field notes, and other nature observations made during his many trips into the Great Smoky Mountains and other places. Also included are accounts of the Smoky Mountain Conservation Association. Unpublished register.

2900 **American Camping Association, Tennessee Valley Section,** Records, 1946-1969, ca. 3 feet (ca. 1200 items). Minutes of the board, reports, convention proceedings, bylaws and constitution, membership and officer lists, news notes, programs, publications, training guides, directories of camps, and other material, dealing with the history of the Tennessee Valley Section and its subdivisions, including Cumberland Valley District (Nashville), Moccasin Bend District (Chattanooga), and Smoky Mountain District (Knoxville). Subjects include campcrafter and leadership training, Negroes, Tennessee Valley Authority projects, and the work of Henry G. Hart, first president of the section. Unpublished register.

2901 **Bills Family,** Papers, ca. 9,000 items and 29 vols. Includes the correspondence of the John V. Wright Lumber Mill, Bolivar, Tennessee (1914-20). Unprocessed.

2902 **Henry, Joseph Milton** (b. 1913), Collection, 1650-1969, ca. 4 feet (2,000 items). Chiefly transcripts (typewritten) and photocopies of originals in possession of various owners. Henry was professor of history at Austin Peay State University, Clarksville, Tennessee. Correspondence, business, school, cemetary, and church records, genealogical data, government and legal documents, personal interviews, and clippings, primarily concerning Stewart County, Tennessee, compiled by Henry, director of the project, for writing a history of the Land Between the Lakes area (Tennessee and Kentucky) for the Tennessee Valley Authority. Includes many letters (1966-69) from Henry to officials of the TVA; information on the timber and other industries. Unpublished register.

2903 **Porter, Mary Rhoda (Montague),** Collection, 1822-1963, ca. 50 items and 1 vol. In part, photocopies of originals owned by various persons. Includes an article, "Why Invest in Forest Land?" and information about foresters in the Chattanooga area. Unpublished register.

2904 **Primm Lumber Company** (Nashville, Tennessee), Records, 1915-1937, 3 vols. Cash and sales books.

2905 **Rugby Papers,** 1872-1942, ca. 8 feet (ca. 5,000 items, 25 vols., and 12 reels of microfilm). Chiefly photocopies and microfilm largely from originals owned by the Rugby Restoration Association, and housed in the Hughes Library, Rugby, Tennessee. Papers concerning the founding and organization of Rugby, an English colony established in Fentress, Morgan, and Scott Counties, Tennessee. Subjects include timber lands. Unpublished register.

2906 **Schulman, Steven A.** "Logging in the Upper Cumberland River Valley: A Folk Industry," Bowling Green: Western Kentucky University, M.S. Thesis (Geography), 60 pp.

2907 **Seward, W.H.,** Correspondence, 1900, 13 items. Twelve letters by Seward to F.N. Moore concerning 6,000 acres of timberland near Pikeville, Bledsoe County, Tennessee, terminus of the Sequatchie Valley. Also included is a map of part of Polk County, Tennessee, showing lands surveyed in 1881.

Archives Section

2908 **Tennessee. Office of the Attorney General and Reporter,** Records. Included is material relating to land acquisition for the Great Smoky Mountain National Park.

SEWANEE

University of the South, Jessie Ball DuPont Library, Archives

37375

2909 **Wiggins, Benjamin Lawton,** Papers. Vice Chancellor of the University of the South, 1893-1909. His files contain letters from Gifford Pinchot, C.A. Schenck, and John Foley, and others (1898-1907, ca. 60 items), relating to the forest lands owned by the University, to proposed plans of forest management, formation of the Society of American Foresters, formation of the Tennessee Valley Authority, etc.

Texas

AUSTIN

Lyndon Baines Johnson Library

78705

2910 **Johnson, Lyndon Baines** (1908-1973), White House Central Files. U.S. President, 1963-69. Files on natural resources contain material on forests, with annual reports on wildlife preservation, executive orders dealing with the transfer of land to parks, etc. (4½ inches); and material dealing with land preservation, including a special report "Surface Mining and Our Environment" (4½ inches). Files on parks and monuments contain material concerned with national parks, including executive orders for the creation of new parks, correspondence with conservation groups, and proposals for trails and parks (11½ inches); also material on park reservations including a memo dealing with the development and management of the North Cascades (2 folders); legislation files contain material dealing with natural resources conservation, such as memoranda, bills, interim reports, and a list of beautification and conservation measures enacted by the 89th and 90th Congresses (3 inches); a report on wilderness areas in forests (1 folder); and legislation for new parks, including Redwood National Park (3 inches). Files on acquisition contain material dealing with wilderness preservation and national and state park movements (9 inches). Files on federal government agencies include material pertaining to federal loans to the National Park Service, a status report on the growth of the Service, etc. (1 inch); material dealing with the sale of timber by the Forest Service (1 inch); materials of the President's Council on Recreation and Natural Beauty dealing with conservation and parks, and including progress reports on the Council (3 inches); and material of the Public Land Law Review Commission relating to its study of the laws and administration of public lands (1 inch). Also dealing with forestry or conservation are files of the Outside Task Force on National Resources (1964), Outside Task Force on Preservation of Natural Beauty (1964), Interagency Task Force on Resources-Recreation (1966), Interagency Task Force on Natural Resources (1966), and the Interagency Task Force on Quality of the environment (1966, 1967, and 1968).

University of Texas, The Library, Barker Texas History Center

Librarian-Archivist 78712. Holdings are described in *The University of Texas Archives: A Guide to the Historical Manuscripts Collections in the University of Texas Library,* comp. and ed. by Chester V. Kielman, Austin: University of Texas Press, 1967.

2911 **Bryant, Charles Granderson** (1803-1850), Papers, 1842-1867, 83 items. Soldier, farmer, artisan, and merchant. Includes materials concerning Bryant's financial reverses in cedar-getting ventures, among other matters. In part, transcripts (typewritten).

2912 **Burges, Richard Fenner** (1873-1945), Papers, 1897-1941, ca. 4 feet. Lawyer, soldier, and politician, of Texas. The papers include material relating to his activities as president of the Texas Forestry Association, 1921-23. Correspondents include Gifford Pinchot.

2913 **Gilmer, Alexander** (1829-1906), Papers, 1872-1929, ca. 102 feet. Shipyard owner and lumberman. Correspondence, order books, stockbooks, salesbooks, payroll books, bills, reports, catalogs, reference books, time books, land plats, maps, blueprints, promissory notes, timber books, business account books, yard books, mill log books, log-cut books, receipts, cashbooks, invoice books, clippings, photos, and other papers concerning Gilmer's career and the operations of his sawmill and lumber business, from acquisition of timber land to the shipping of finished lumber to yards in Texas, elsewhere in the United States, and abroad. Includes material on local and national lumbermen's associations. In part, photocopies.

2914 **Haydon, Charles E.,** Papers, 1888-1898, 246 items. The collection contains material on the Woodworth Lumber Company in Louisiana.

2915 **Jones, M.T., Lumber Company,** Records, 1883-1914, ca. 32 feet. Correspondence, minutes, constitution and bylaws, promissory notes, bonds, contracts and franchises, deeds, legal papers, tax records, journals, ledgers, financial papers, price lists, invoices, logging reports, speech, clippings, photo, map, and other papers of Martin Tilford Jones (d. 1898), lumberman, banker, and financier. Subjects include logging; the lumber industry and trade in Texas; timberland leasing, timber cutting, rail transport, and litigation; wholesale and retail lumber business; Jones' membership in the Southern Cypress Shingle Association, the Lumbermen's Association of Texas, the Lumber Dealer's Association of Texas, the Texas and Louisiana Lumber Manufacturers' Association; Jones' personal affairs; the Phoenix Lumber Company (Houston); the Bel-Bunker Lumber Company; the Emporia Lumber Company (Houston); D.R. Wingate Lumber Company; the Orange Lumber Co.; and business interests of Jesse Holman Jones in lumber and other concerns. Places associated with the records include Lake Charles, Louisiana; and Angelina County, Beaumont, Dallas, Ennis, Jasper County, La Porte, Mexia, Nacogdoches County, and Waxahachie, Texas. Persons associated include Robert Holmes Baker, Thomas Henry Ball, David E. Bryant, William Waldo Cameron, Samuel Fain Carter, Eber Worthington Cave, William D. Cleveland, Charles Dillingham, Henry Exall, Royal Andrew Ferris, Lafayette Lumpkin Foster, Alexander Gilmer, Malcolm K. Graham, Edward Howland Robinson Green, William Smith Herndon, Thomas William House, Jr., Joseph Chappell Hutcheson, John T. Jones, William E. Jones, John Henry Kirby, Henry Dickinson Lindsley, Cesar Maurice Lombardi, Robert Scott Lovett, Henry Jacob Lutcher, Henry Frederick McGregor, Jones Shearn Rice, James M. Rockwell, Henry B. Sanborn, George Washington Smyth, Jr., William H. Stark, John Grant Tod, Jr., and Benjamin Franklin Yoakum.

2916 **Kleberg, Robert Justus** (1803-1888), Papers, 1842-1921, ca. 1 foot. Personal and business papers concerning Kleberg's various interests, including timber.

2917 **Rockwell Brothers,** Records, 1883-1947, ca. 22 feet. Correspondence, postcards, telegrams, journals, ledger, register, passport, announcements, certificates, account papers, lists, musical scores, circulars, programs, brochure, pamphlet, books, clippings, and photographs of James Morton Rockwell (1863-ca. 1932), lumberman. Records relate to M.T. Jones Lumber Company (1894-1901); Rockwell Bros. & Co. (Houston); retail lumber yards in Texas; James M. Rockwell's work as president of the Waxahachie Lumber Company, auditor for M.T. Jones Lumber Company, and director and vice president, and as delegate to the 1929 Annual Convention of the Lumbermen's Association of Texas in Waco; political career of Jesse Holman Jones, and other matters. Places associated with the collection include Albany, Amarillo, and Hereford, Texas. Persons include James Addison Baker, William Waldo Cameron, Oscar F. Holcombe, Thomas William House, Jr., Martin Tilford Jones, Cecil Rockwell, Henry M. Rockwell, James Wade Rockwell, Lillian Rockwell, and James Henri Tallichet.

FORT WORTH

*Federal Archives and Records Center

P.O. Box 6216 76115 4900 Hemphill Street
Telephone: (817) 334-5515

2918 **Forest Service** (Record Group 95), ca. 1910-1970, 2,501 cubic feet. Records not yet accessioned into the Archives Branch, and accessible only with agency permission, consist chiefly of records of the Southern Forest Experiment Station, New Orleans (2,225 cubic feet), and small amounts of records from Forest Service offices in Louisiana, Arkansas, and Texas, chiefly administrative service files, research program and publication files, and records of timber sales.

2919 **National Park Service** (Record Group 79), 104 cubic feet.

HOUSTON

University of Houston, University Libraries

Cullen Boulevard 77004

2920 **Kirby, John Henry,** Papers, 1886-1930, 260 boxes. Lumberman of east Texas. The collection contains records of his lumber and other business investments. Included are general correspondence, containing materials on numerous lumber companies and associations, lumbermen, land companies, turpentine companies, railroads, etc. There are files relating to the Southern Lumber Operators Association (1911-13, 1 box), mostly concerning labor relations; the Southern Tariff Association (1920-29, 1 box); turpentine (1909-14, 1 box); lands; correspondence with banks; and files concerning various lumber products, such as railroad ties, etc. Typed container list, 34 l.

2921 **Rockwell, James,** Papers. These papers deal with the Rockwell Brothers Lumber business (1880s-1914). Closed to research (1974).

2922 **Warner, Charles A.,** Papers, ca. 1899-1926, 225 boxes + loose items. Largely relating to the Houston Oil Company and the Kirby Lumber Company. Included are files relating to transfer of timberland from HOCO to Southwestern Settlement and Development Company (1916); Maryland Trust Co. vs. Kirby Lumber Company, et al., in the U.S. Circuit Court, Southern District of Texas (1905-11); East Texas Pulp and Paper Co., certificate of incorporation and a memorandum on closing; stamp taxes on sale of assets (timber) (1953-56); settlement between HOCO and the Kirby Lumber Co.; chronological history of HOCO and the Kirby Lumber Co. (7 vols. + index); Bancroft Lumber Co., title papers for Calcasieu Parish, Louisiana; Louisiana Lumber Co. and Beaumont Lumber Co., statements and miscellaneous title papers; Texas Pine Land Association, miscellaneous papers; documents relating to timber land agreement between HOCO and J.H. Kirby (1902); court records relating to HOCO vs. Kirby Lumber Co.; White, Weld, & Co., appraisals of Houston Pine Lumber Co.; East Texas Pulp Co. minute book; original Kirby contract (1920); East Texas Pulp Co.; Kirby Lumber Co. audit (1909); litigation between HOCO and the Kirby Lumber Co. (1904-05). Typed container list, 27 pp.

LUBBOCK

Texas Tech University, Southwest Collection

P.O. Box 4090 79409

2923 **Murphy, John Joseph** (1879-1957), Papers, ca. 1881-ca. 1953, 16 feet (ca. 14,570 items). Includes records (1909-46) of the J.J. Murphy Lumber Co. (Crosbyton and Lubbock, Texas), consisting primarily of letters received, invoices, deeds, legal papers, invitations, membership cards, clippings, magazines, photos (1890-1945), and other papers, including material on lumber business, lumber yards, taxes, the chamber of commerce, and the Murphy family. Inventory.

2924 **Willson, James M.,** Papers. Largely materials dealing with the philanthropic activities of Willson, owner of a number of lumberyards in west Texas.

NACOGDOCHES

Stephen F. Austin State University, Library

Special Collections Librarian 75961

2925 **Cronister Lumber Company,** Records, 1898-1945. The records consist of minute books (1898-1945, 2 vols.), scattered correspondence, and a few maps of lumber holdings.

2926 **Frost Lumber Industries,** Records. These records of Hayward and Frost, and Frost-Johnson, consist of a few maps of timber holdings, deeds, and an agreement with the Kirby Lumber Company.

2927 **Jones, W. Goodrich,** Papers, 1881-1907, ca. 1000 items. Lumberman and conservationist of Trinity, Texas. Personal and business correspondence.

2928 **Kirby Lumber Corporation,** Papers, 1886-1967, ca. 100,000 items. Correspondence, ledgers, and other business papers of the firm organized in 1901 by John Henry Kirby. Calendar.

2929 **Kurth Collection,** 1887-1930, 150 feet. Correspondence, personal and business papers, ledgers, time books, plats of timber holdings, contracts with logging railroads, and commissary records of Joseph H. Kurth and Ernest L. Kurth and their company, the Angelina Lumber Company of Keltys, Texas. There are also records of the San Augustine Lumber Company, the Henderson and Kurth Lumber Company, and the Newton County Lumber Company.

2930 **Lutcher and Moore Lumber Company,** Records, 1882-1964, 44,000+ items. Founded in Orange, Texas, in 1876; absorbed into the Boise Southern Company, 1970. Included is material relating to lumber exports. Calendar.

2931 **Nacogdoches & Southeastern Railroad Company,** Records. Articles of incorporation, first meeting of the stockholders, agreement between the railroad and the Hayward Lumber Company, and a minute book (1905-30).

2932 **Temple Industries, Inc.,** Records, 1908-1957, ca. 70,000 items. Correspondence, invoices, receipts, and other business papers. Originally the Southern Pine Lumber Company (Diboll, Texas), 1894. Also files relating to the Southland Paper Mills. Calendar.

Utah

LOGAN

*Utah State University, Merrill Library and Learning Resources Program

Special Collections and Archives 84332. The holdings are described in *An Annotated Bibliography of Western Manuscripts in the Merrill Library at Utah State University, Logan, Utah,* compiled by Mary Washington, Logan: Utah State University Press, 1971, from which the following entries were taken.

2933 **Anderson and Sons Company,** Ledgers, 1904-1906, 1909-1910, 1919, 3 vols. Accounts of lumber yards in Trenton, Wellsville, Logan, Smithfield, and Lewiston, Utah; and in Preston, Downey, Bancroft, Dayton, Grace, Idaho Falls, and Blackfoot, Idaho.

2934 **Belnap, Hyrum,** Interview, Ogden, Utah, 1938, 7 l. Belnap worked for the Eccles Lumber Company.

2935 **Card, Charles O.** (1839-1906), Journals, 1886-1903, microfilm (21 vols.). Card describes leading a group to Alberta, Canada, where he established a lumber business.

2936 **Card, Charles O.** (1839-1906), Memoranda books and journals, 1871-1883, microfilm (18 vols.). These journals cover Card's visit to several northern lumber factories in 1871, and his business affairs in Logan, Utah. Card was involved with the United Order Manufacturing and Building Company in Logan.

2937 **Cliff, Edward Parley** (b. 1909), Papers, 7 vols. Chief, U.S. Forest Service, 1962-72. Duplicate sets, in 3-inch loose-leaf binders, of articles, speeches, photographs, and other papers have been given to Utah State University, the Forest History Society, Inc., the Denver Public Library, the University of Wyoming, and the Washington, D.C., office of the U.S. Forest Service. Cliff has designated Utah State University as the eventual repository for the original drafts of his published papers and articles, speeches, clippings, photographs, etc. (ca. 30 3-inch loose-leaf binders).

2938 **Cole, Gilbert W., and George A. Bell,** "Beginning of the Cache Valley Lumber Industry," Manuscript, 6 l. Describes the United Order Manufacturing Company, organized in 1868, to its consolidation with Cole Brothers Lumber Company in 1890; the operation of a mill located on Steam Mill Flat, Steam Mill Canyon, in 1867; and floating logs down the Logan River to a mill.

2939 **Eccles Lumber Company** (Ogden, Utah), Journal, 1887-1888, 1 vol. (600 pp.). Contains accounts of the Ogden Lumber Company and receipts from the Hood River, Oregon, Lumber Company.

2940 **Greenwell, Mrs. Fannie E.,** Interview, 1939, 9 pp. Mrs. Greenwell's father, Edmund Ellsworth, ran an early sawmill on Mill Creek Canyon.

2941 **Hanks, Ebeneezer** (b. 1814), Biographical note, ca. 1936, 3 pp. Written by his daughter. Hanks operated a sawmill in Parowan, Utah.

2942 **Hansen, Lorenzo E.,** "History of Early Preston, Idaho," 4 l. Dictation. Includes account of Hansen's affairs as a lumber salesman in Preston after 1885.

2943 **Hayball, Edith,** Manuscript, "Hyrum Hayball," 2 l. Typescript. Hayball (1852-1936) was superintendent of the United Order Lumber Mill in Cache Valley, Idaho, after 1868.

2944 **Huntsman, Orson Welcome,** Journal, 1891-1926, 3 vols. Typescript from original in possession of Mrs. O.W. Huntsman, Enterprise, Utah. The journal mentions conflicts between cattlemen and forest rangers.

2945 **Monson, C.H.,** Account books, 1885-1898, microfilm (2 vols.). Lists accounts for a lumber company. Original in Relic Hall, Franklin, Idaho.

2946 **Moore, David** (b. 1851), Interview, 1938, 3 pp. Moore's father worked in the first sawmill in Ogden, Utah.

2947 **Potter, Albert F.,** Journal, July-November, 1902. Potter was an associate chief of the U.S. Forest Service in Utah, and in this journal he records the condition of land prior to its becoming a forest reserve, observed on a field trip around Logan, Bear Lake, Wasatch County, American Fork, and Provo; indicates the crops growing, types of trees and their locations, and the use of land by animals, including a list of the numbers of sheep in Sanpete County (1902). Original in the possession of Forest Experiment Station, Ogden, Utah.

2948 **Shaw, Myrtillo,** Interview, 1939, 4 pp. Describes hauling wood from Ogden Canyon, near Ogden, Utah.

2949 **Shurtliff, Francis Marion** (b. 1851), Autobiography, 11 pp. Shurtliff worked at Eccles Lumber Mill in Oregon.

2950 **U.S. Work Projects Administration, Federal Writers' Project,** Papers, 13 boxes. Materials collected under the supervision of George F. Willison, 1940-41, for a projected history of grazing in seventeen western states: Arizona, California, Colorado, Idaho, Kansas, Montana, Nebraska, Nevada, New Mexico, North Dakota, Oklahoma, Oregon, South Dakota, Texas, Utah, Washington, and Wyoming. The collection includes Willison's draft, his correspondence with field workers in the various states, and some of the chapter drafts for each state. Box 3 includes a file on the national forests and the field offices of the United States Forest Service, and in Box 12 are files on the conservation activities of the U.S. Forest Service and U.S. National Park Service.

2951 **Wilcox, James David** (1827-1916), Journal, 1856-1861, microfilm (2 items). The journal includes details of work at a lumber mill in Farmington, Utah.

2952 **Winsor, Luther M.** (1884-1968), Papers, 13 boxes. Papers (2 boxes) and photographs (11 boxes). Papers include 13 articles on such topics as "Control of floods in mountain streams," "The barrier system of flood control," and "Canal management and use of water." Box II contains a summary of cooperative irrigation studies in Utah from 1922 to 1924, and a report on dike construction at Bear River Bay Migratory Bird Refuge, among others. The photographs, indexed, depict land use or water conservation and were taken throughout the western states.

2953 **Wright, Angus Taylor** (1856-1928), Autobiography, 1856-1928, 219 pp. Typed. Describes work in a sawmill, Richmond, Utah, after 1860.

PROVO

Brigham Young University, Clark Library

Manuscript Librarian, Clark Library, Box 6 84602
Telephone: (801) 374-1211

2954 **Smoot, Reed** (1862-1941), Papers. U.S. Senator from Utah, 1903-33; served on Public Lands Committee. The collection includes material on the Hetch Hetchy Reservoir controversy (1 folder); and material on the Bridal Veil Falls, Utah (1 folder).

2955 **Watkins, Arthur Vivian** (b. 1886), Papers. U.S. Senator from Utah, 1947-1959; served on the Interior and Insular Affairs Committee, Public Works Committee, the Irrigation and Reclamation Subcommittee and the Indian Subcommittee of the Interior and Insular Affairs Committee. The papers contain substantial materials documenting his activities in all of these areas, especially his concern with the Colorado River Storage Project bill and other reclamation projects, the controversy over the proposed Echo Park dam in Dinosaur National Monument, Rainbow Bridge National Monument, Hells Canyon Dam, mining, the U.S. Bureau of Reclamation, etc. Papers include correspondence, bills, drafts of bills, newspaper clippings, photographs, and other types of records. Portions of the collection were severely weeded by Senator Watkins and his staff while engaged in work on his book, *Enough Rope*.

SALT LAKE CITY

Department of Development Services, Division of State History

Director, 603 East South Temple 84102 Telephone: (801) 328-5755

2956 **Anderson, William,** "Reminiscences of the Forest Service in Utah," 14 pp., typescript. Experiences in the Forest Service, 1897-1914, related by a supervisor of Ashley National Forest, Utah.

2957 **Antrei, Albert C.,** "A Geographical Interpretation of Timber Production in Utah," Salt Lake City: University of Utah, M.A. thesis (Geography), 1951, 133 pp.

2958 **Baker, F.S., and A.G. Hague,** "Report on Tie Operation, Standard Timber Company, Uinta National Forest, 1912-1913," 52 pp., maps, illustrations. Photocopy of typed manuscript.

2959 **DeMoisy, Charles,** "Some early history of the Uinta National Forest," 11 pp., typescript. DeMoisy was supervisor, Uinta National Forest, Utah, 1925-38.

2960 **Guild, Charles F.,** "History of Old Piedmont, Wyoming," Tape recording, 1964. Piedmont in Uinta County, Wyoming, was the center of a lumber industry, cutting timbers for the Union Pacific Railroad.

2961 **Humphrey, J.W.,** Oral history interview. Supervisor, Manti-LaSal National Forest. The interview covers the years 1906-41.

2962 **Jonas, Frank H.,** Papers, microfilm. Correspondence and papers concerning efforts to establish a Utah State Department of Conservation in 1948. Utah State Archives also has the originals.

2963 **Nixon, James William,** "Autobiography," 106 pp. Typed photocopy. Included is material on the lumber industry in Utah.

2964 **Savage, Levi Mathers,** Journal, July 4, 1873-November 18, 1873, 52 pp. Typed photocopy. A lumberman in southern Utah.

2965 **U.S. Forest Service. Dixie National Forest,** Collection, 1890-1926, 4 items (34 l.). Chokecherry Point, Dixie National Forest, visitors' notes (1890-1926), plus other items. Photocopies of original notes and transcript copies.

2966 **U.S. Forest Service. Manti-LaSal National Forest,** Records, 1904-1970, microfilm, 15 rolls. Forest inspection reports and supervisors' letterbooks.

2967 **U.S. Forest Service, Uinta National Forest,** Records, 1910-1969, microfilm, 13 rolls. Forest inspection reports; supervisors' letterbooks; photographs.

2968 **Varner, I.M.,** Interview with Michael Barclay, 4 l. Photocopy of typewritten report (June, 1952). Mr. Barclay relates his experiences in the sheep business in Utah and Idaho, especially in the Uinta National Forest.

2969 **Wood, Rhoda M.,** "Zion Cable," 11 pp., typescript (2 copies). Based on diary of William Flannigan. Concerning the building of a cable in Zion Canyon by Dave Flannigan and his brothers to haul lumber from the plateau above to the settlements. Accompanied by correspondence.

Vermont

BURLINGTON

University of Vermont, Guy W. Bailey Library

Manuscripts Librarian 05401

2970 **Allen, Ira** (1751-1814), Papers, 1730-1906, 10 feet (ca. 1050 items). Allen was an author, politician, landowner, and pioneer of Vermont. The papers are originals and transcripts and photocopies of originals in various repositories in the eastern United States. The collection contains the papers of Ira's brothers, Ethan, Heman, Heber, and Levi, and contains many references to land purchase, lumbering, log rafting, and milling in Vermont. It is described in the Historical Records Survey, *Calendar of the Ira Allen Papers in the Wilbur Library of the University of Vermont,* Montpelier, Vermont: 1939, 149 pp.

2971 **Austin, Warren R.,** Wilbur Timber Case Files, 1924-1945, 1 foot, 7 inches. These files are the record of Austin's role as agent of the University of Vermont in handling a bequest from James R. Wilbur of stock in the Wilbur Timber Company Ltd. (Vancouver, British Columbia); they contain correspondence (1936-46) dealing with timber licenses, indentures, agreements concerning land in Vancouver, correspondence with the Bank of Montreal concerning timber rights, stumpage, and land sales; also Timber Holder's Association of British Columbia reports (1934-36); Wilbur Timber Company correspondence (1933-40), records and resolutions (1936); University Executive Committee resolutions concerning the sale of land and bonds (1936-40); a small diary kept by Austin during a trip to Vancouver (1938), together with clippings from British Columbia newspapers concerning lumbering; and a valuation of San Mateo County, California, forest lands (1924-25).

2971a **Booth, John Rudolphus** (1827-1925), Scrapbook, ca. 1910-1935, 1 vol. Chiefly newspaper clippings relating to the J.R. Booth Lumber Co. of Ottawa, Ontario, and the E.J. Booth Lumber Co. of Burlington, Vermont.

2972 **Congdon, Herbert Wheaton** (1876-1965), Photographs and papers, 1910-1972, ca. 5 inches. Congdon, of Vermont, was a Green Mountain Club member and professional photographer. The papers are originals and transcripts relating to the early history of the Green Mountain Club (founded 1910), especially regarding the Long Trail in Vermont; they include maps, blueprints, photos, slides, and other materials describing the breaking of trails, lodges, and points of interest along the Long Trail.

2973 **Coolidge, Philip T.,** Papers, 1924, 1927-1931, 2 folders (45 items). Coolidge was a timber cruiser, Bennington County, Vermont. The papers include manuscripts, letters, and printed maps. There are reports to clients on timber stands, prices, etc.; a report on the Emporium Forestry Co. timberland in Mount Tabor, Vermont (1927); and a report on lots in Searsburg, Vermont (1924).

2974 **Cowles, Clarence P.** (1875-1963), Papers, 1930-1963, 2 inches. Papers and manuscripts of unpublished articles on the Green Mountain Club of Vermont, on Will Monroe, and on various aspects of the Long Trail, and side trails. Unpublished inventory.

2975 **Emporium Lumber Co.,** Records, ca. 1900-1950, ca. 8 feet. The collection relates to operations in Pennsylvania, the Adirondack region (St. Lawrence County, New York), and Danby, Vermont. It contains financial records, maps of forests, photographs of logging, etc. Unprocessed.

2975a **Marsh, George Perkins** (1801-1882), Papers, 1812-1900, 20 feet. Marsh was a scholar, writer, diplomat, lawyer, farmer, manufacturer, U.S. Representative from Vermont, 1843-1849, and pioneer thinker in the conservation movement. The papers contain incoming correspondence, diaries, notebooks, drafts of writings, and other records relating to most of his varied interests, including material concerning his *Man and Nature: Or Physical Geography as Modified by Human Action*

(1864). Portions of the collection have been published in *Life and Letters of George Perkins Marsh,* by Caroline Crane Marsh (New York: 1888). The manuscript of the unpublished second volume of this work is among the holdings of the Guy W. Bailey Library. The collection is described in "The George Perkins Marsh Papers," by T.D. Seymour Bassett, *Dartmouth College Library Bulletin*, X (November, 1969), pp. 9-14.

2976 **Merrill, Perry** (b. 1894), Papers, ca. 1900-1960, 2 inches (ca. 200 pp.). Merrill was Commissioner of the Vermont State Department of Forests and Parks, 1929-66. The papers include manuscripts of an autobiography and articles on Vermont life, and a typescript of an interview by T.D.S Bassett. The articles deal with the Green Mountains, lumber business, municipal forests, agriculture, skiing and mountain recreation, and the history of forestry in Vermont. Partly transcripts and photocopies of originals still in the possession of Mr. Merrill.

2977 **Pike, Robert E.,** Photographs, 1890-1960, 133 items. Prints of logging operations in northern New England, especially the Upper Connecticut River, including pictures of men and machinery. Many of the photos were published in Robert E. Pike, *Tall Trees, Tough Men* (New York: 1967).

2978 **St. Albans Lumber Co.** (St. Albans, Vermont), Accounts, 1912, 1 vol. (125 pp.). Customer accounts for lumber and cutting lumber.

2979 **Stockbridge Mill** (Stockbridge, Vermont), Photographs, ca. 1900, 8 items. Photographic prints of the mill and surroundings, accessioned with the records of the Eagle Square Company (Shaftsbury, Vermont), which purchased the Stockbridge Mill in 1946.

2980 **Stone and Webster Corporation,** Report, 1914, 1 vol. (150 pp.). Carbon copy of a report on "Storage Investigations: Connecticut River Headwaters," made for the Turner Falls Co., including blueprints, photographs, and a written outline describing log-driving methods, run-off studies, storage studies, and undeveloped water-power studies.

2981 **Taber, Edward Martin** (1863-1896), Diary, 1882-1913, 2 vols. (500 pp.). Kept during visits to Stowe, Vermont, the manuscripts act as a diary of Taber's contacts with the mountains, containing day-to-day reflections, memorabilia, clippings, verses, etc. Published as *Stowe Notes, Letters, and Verses,* by E.M. Taber, edited by Florence T. Hold (Boston and New York: 1913).

2982 **Taft, Elihu B.,** Papers, 1905-1926, 1 folder (ca. 30 items). The collection includes correspondence, reports, manuscripts of articles, clippings, printed matter, and photographs relating to the Green Mountain Club of Vermont, the Long Trail, and the buildings of the Taft lodge. Inventory and calendar of correspondence, 4 pp.

2983 **White and Whittemore Sawmill** (Eden, Vermont), Accounts, 1870-1879, 1 vol. (90 pp.). Records of wood cut and delivered to customers.

MONTPELIER

Vermont Historical Society

Librarian 05602

2984 **Brainerd, Lawrence Robbins** (1819-1863), Letterbook, December 10, 1850-March 7, 1853, 1 vol. (428 pp.). The book contains press copies of letters written by Brainerd of St. Albans, Vermont, while engaged in the wholesale lumber business in northern New York, Canada, and Vermont.

2985 **Burgess, Lyman,** Papers, ca. 1820-1862. Several hundred general merchandise bills, lumber records, lumber expenses, estate accounts, and gristmill costs (1820-30s) in Milton, Vermont; deeds (ca. 200 items) covering the same period; several account books and diaries, including an account book (1842) covering spruce logs, and miscellaneous bills and notes.

2986 **Howland, Fred Arthur** (1864-1953), Correspondence, 1911-1928. Correspondence with Henry S. Wardner and Frank S. Greene pertaining to Green Mountain National Forest.

2987 **Sumner, David Hubbard** (1776-1867), Papers, 1791-1888, ca. 2,700 pieces. Scattered among other business records are bills and accounts (1804-71, ca. 300 items) for general merchandise, construction of mills, purchase of potash, of timber, and of land in Lewis and Ferdinand townships, Vermont, and also Dalton township, New Hampshire; lumber agreements; Hartford lumber records with information about log runs on the Connecticut River (1820s-70s); various lumber bills (1830s-40s); receipt for toll paid on 65 tons of lumber; letters about the lumber business (1825-82, 94 items).

2988 **Vermont-New York Boundary Line Survey** (1903-1904), October, 1904, 1 item. Record of trees cut on farms traversed in running the preliminary lines by E.H. Randall, Acting Engineer for Vermont.

Virginia

CHARLOTTESVILLE

University of Virginia, Alderman Library, Manuscripts Department

22901 Telephone: (804) 924-3025

2989 **Albemarle Ledgers,** 1818-1902, 44 vols. The collection includes a journal of Seamonds & Marshall, Albemarle County, Virginia, lumber dealers.

2990 **Atkinson, Homer,** Letterpress book, 1886-1890, 1 vol. (ca. 400 letters). Relates to Atkinson's Petersburg, Virginia, wood retailing business.

2991 **Baxter, Thomas and William H.,** Records, 1833-1915, 1500 pieces and 70 vols. The records are largely those of the Blandford (lumber) Mill Company, which was organized in 1833, and include bound account books, bank books, and check stub books.

2992 **Buford, Edward Price** (1865-1931), Papers, 1894-1931, 41 feet (ca. 7520 items). Lawyer of Lawrenceville, Brunswick County, Virginia. The records include material on the Lawrenceville Land and Timber Co., of which Buford was an officer.

2993 **Cleveland, Grover,** Document, February 7, 1888. Grant of land in Kansas, subject to a law encouraging the growth of timber.

2994 **Conservation Council of Virginia,** Records, 1969-1973, 2000 items.

2995 **Cox-McPherson,** Ledgers, 1884-1917, 4 vols. The ledgers include the accounts of C.W. Cox, a carriage firm of Albemarle County, Virginia.

2996 **Davis, William W.,** Papers, ca. 1840-1907, 1000 items. Part III of these manuscripts, letters and financial papers relating to the Horn or Horne family of Rockbridge County, Virginia, contains papers having to do with John Horn, a lumber dealer.

2997 **Drummond Brothers** (Wiseville, Virginia), Records 1817-1879, 3 feet. Ledgers, daybooks, cashbooks, and journals of general merchandise stores located in Accomack County, Virginia, at Grape-on-Hunting Creek, Centreville, and Wiseville, Virginia, relating in part to the lumber trade. Includes the firms of Drummond Bros., James Dix, John Y. Bagwell, Segar & Goffigon, and accounts for business in Baltimore, Maryland, and Washington, D.C. Indexed in part.

2998 **Easley,** Ledgers, 1837-1904, 115 vols. Business records of the Easleys of Halifax, Virginia, and their associates in a merchandise business there which dealt in lumber, general merchandise, mill products, and Western lands. The records comprise ledgers, daybooks, journals, letterbooks, land books, cashbooks, invoice books, and a map book. The ledgers and daybooks contain accounts of Easley, Logan, & Co.; Easley and Hurt; Easley, Carrington & Co.; Easley, Holt & Co.

2999 **Graham-Robinson,** Records, 1818-1906, 77 vols. The business records of David Graham, Graham & Son, and Graham & Robinson contain accounts for iron, mill products, merchandise, labor, sawmill, and boardinghouse in connection with the iron foundry, general store, and grist mill, Graham's Forge, Wythe County, Virginia.

3000 **Heidelbach and Penn** (Danville, Virginia), Records, 1897-1908, 9 vols. Ledgers, daybooks, journals, and cashbook, containing records of lumber transactions in Danville and Pittsylvania County, Virginia. Indexed in part.

3001 **Hotchkiss, Jedediah** (1845-1905), ca. 5,000 items. Chiefly business papers including surveys (1883-99) relating to land development in Virginia and West Virginia, especially mineral resources and timber in the Shenandoah Valley.

3002 **Hume,** Ledgers, 1795-1860, 3 vols. An account book (1823-24) contains accounts for grain, lumber, whiskey, and farm products, Fauquier County, Virginia.

3003 **Irvine-Saunders,** Journals, 1865-1882, 6 vols. Records of a blacksmith, and a lumber and general merchandise business of the Saunders Family of Evington, Campbell County, Virginia. A cashbook (1865-66) contains lumber accounts.

3004 **Jefferson, Thomas,** Papers. There are a number of letters relating to lumber and the cutting of fuelwood on Jefferson's lands in Albemarle County, Virginia. Abstracted in *The Jefferson Papers of the University of Virginia,* Charlottesville: University Press of Virginia, 1973.

3005 **Johnson-Drummond Merchandise Ledgers,** 1818-1879, 24 vols. Ledger (1838-53, 1 vol.) deals with lumber cutting and selling.

3006 **Nelson, Wilbur Armistead** (b. 1889), Papers, 1913-1937, 1 foot (ca. 1000 items). State geologist of Tennessee (1918-25) and of Virginia (1925-28) and professor of geology at the University of Virginia. Subjects which his papers concern include the proposed Appalachian National Park.

3007 **Parker, John Crafford,** Papers, 1884-1946, 3650 items, 10 vols. Some of the papers deal with the political implications of lumber, such as the tariff.

3008 **Shackelford, George Scott** (1856-1917), Papers, 1910-1948, 6 feet (ca. 3500 items). Political and business correspondence of Shackelford and his son, Virginius Randolph Shackelford (b. 1885), lawyer of Orange, Virginia, relating to political and other matters, including material on the Shenandoah Park Association.

3009 **State Bank of Pamplin,** Records, 1905-1931, 80 vols. The records include a daybook (1926-29) containing lumber sales of a store apparently owned by the bank.

3010 **Thomas, Waverly,** Ledgers, 1814-1824, 2 vols. The ledgers contain accounts for farm and mill products, lumber, and other services at Smithfield, Isle of Wright County, Virginia.

3011 **Warren County, Virginia,** Records, 1857-65, 2 vols. Journals of W.E. Sumers, a carpenter of Front Royal, Virginia (1857-61), and of Lewis A. Smith (1857-65).

3012 **Winiker-Heidelbach & Penn,** Ledgers, 1897-1906, 9 vols. Ledgers (1897-1904), daybooks (1901-08), journals (1897-1906), and a cashbook (1903-04), of Heidelbach & Penn, Danville, Virginia, lumber dealers.

RICHMOND

Virginia State Library, Archives Branch

Head, Archives Branch 23219

3013 **Allason, William,** Papers. The papers contain records (2 vols.) of the Pine Forest Sawmill; a daybook (June, 1787-September, 1793) and a record of planks sawed (June, 1787-June, 1789).

3014 **Virginia. Department of Conservation and Development,** Records, 1927-1950, 71,000 items.

Virginia Historical Society

Curator of Manuscripts, P.O. Box 7311 23221 (Boulevard and Kensington Avenue)

3015 **Adams, William C.,** Account book, 1830-1892, 1 vol. (468 pp.). Concerns his lumber and carpentry business in Scottsville, Virginia.

3016 **Allmand Family,** Papers, 1796-1891, 573 items. Includes materials concerning the procurement of live oak timber in St. Mary Parish, Louisiana, to fulfill the contract of John Driver Allmand (1799-1851), William H. Ivey, and Daniel Sanford with the U.S. Navy as follows: accounts (1835-42, 30 items) of Allmand; correspondence of Allmand (1836-43, 60 items); of Ivey (1842, 2 items); of Sidney A. Sanford (1838-40, 8 items); and of others (1841, 3 items); also agreements (1835-41, 13 items); and materials concerning related lawsuits in the Circuit Superior Court of Law and Chancery of Norfolk, Virginia (1840-43, 43 items), including affidavits, answers, complaints, summons, decrees, and miscellaneous notes; accounts of Ivey (1836-42, 134 items); accounts of Daniel Sanford (1840-44, 7 items); checks drawn by Daniel Sanford, Allmand, and Ivey upon the Bank of Virginia, Norfolk (1841-43, 8 items); and accounts of Sidney A. Sanford (1838-41, 73 items), kept in St. Mary Parish.

3017 **Aylett Family,** Papers, 1776-1945, 2,848 items. Papers (1906-24, 18 items) kept by Philip Aylett concerning the administration of the estate of William Roane Aylett, at "Montville," King William County, Virginia, include specifications, contract, and accounts of the sale of timber to W.D. Rouzie.

3018 **Burwell, George Harrison** (1799-1873), Account book, 1826-1827, 1 vol. (30 pp.). Concerns delivery of wood and agricultural activities, Frederick (now Clarke) County, Virginia.

3019 **Commonwealth Lumber Corporation** (Richmond, Virginia), Records, 1911-1917, 80 items. This firm was previously known as the Garrett Lumber Company (Richmond, Virginia). The records include account books (1911-17, 2 vols.); unbound accounts; estimates; etc.

3020 **Douthat Family,** Papers, 1795-1922, 1,432 items. Includes instructions for preservation of wood (1 item?).

3021 **Edmundson Family,** Papers, 1781-1949, 1,402 items. Includes an agreement (1946) of Anne Beale Edmundson, Granville Eskridge Edmundson, and Maria Antoinette Edmundson with J. Gilbert Cox concerning timber at "Fotheringay," Montgomery County, Virginia.

3022 **Edrington Family,** Papers, 1766-1967, 503 items. Includes an agreement (1825) of Samuel Seldon Brooke, John Catesby Edrington, and Maxey & Mercer concerning locust trees; and a deed (1912) of Angelina Selden Edrington to W.S. Embrey, Inc. (Fredericksburg, Virginia) for timber in Stafford County, Virginia.

3023 **Hopkins & Bro.** (Onancock, Virginia), Papers, 1838-1964, 373 items. Includes correspondence, accounts, daybooks, ledgers, cashbooks, invoice books, inventory books, money letter books, and freight books concerning the general merchandise operations, including lumber, of Hopkins & Bro.; Stephen Hopkins & Son (Onancock); and T.S. Hopkins & Company (Tasley, Virginia); also barrel books (1879-1921) and daybook (1904-07) of the Eastern Shore Barrel Company (Onancock); and other accounts. Guide.

3024 **Horner, Henry,** Letterbook, 1888-1901, 1 vol. (497 l.). Letterpress book concerning sawmilling operations at Horner's, Westmoreland County, Virginia. Indexed.

3025 **Lee, Robert Edward** (1807-1870), Letter, n.d., 1 item. Letter to Lorenzo Sitgreaves concerning the sale of Smith's Island, Northampton County, Virginia, and trees. Xerox copy from the original in the possession of the United Daughters of the Confederacy Library, Richmond, Virginia.

3026 **Lee Family Papers,** 1824-1918. Robert Edward Lee (1807-70) letterbook (1842-60, p. 191) includes a letter (June 21, 1845) to Colyer & Dugard Steam Saw Mill, North River, New York, concerning the delivery of lumber to Fort Hamilton, New York. There is also a letter of Robert E. Lee to William Henry Fitzhugh Lee (March 31, 1846) concerning the devotion of young Harry to his father while engaged in wood cutting operations in New Hampshire.

3027 **Ruffin, Edmund** (1794-1865), Papers, 1818-1865, 826 items. Includes correspondence (1859) of Ruffin relating to timber.

3028 **Sanford, John C.,** Letter, n.d., 1 p. Letter from Richmond, Virginia, to Richard Borden, Fall River, Massachusetts, concerning loading coal and wood on board a ship for Massachusetts.

3029 **Sinton, John C.,** Account book, 1846-1866, 1 vol. (128 pp.). Includes accounts of James R. Watson, Albemarle County, Virginia, concerning the sale of lumber (1862-63). Indexed.

WILLIAMSBURG

The College of William and Mary in Virginia, Earl Gregg Swem Library

Curator of Manuscripts 23185 Telephone: (804) 229-3000

3030 **Account Books,** Collection. Included are an unidentified account book (1853-99) containing a record of lumber sawed at Valley Mills, Shenandoah County, Virginia; an unidentified account book (1885-87) containing records of a sawmill operator, Buckingham County, Virginia; records (1845-65, 2 vols.) of John L. Beard, Augusta County, Virginia, cabinet maker, including a daybook (1845-49) and a ledger (1863-65); a ledger (1841-50) of Samuel O. Beard, Augusta County cabinet maker; account book (1837-61) of Henry Neff, Buckingham County cabinet maker; account book (1792-1824, 84 pp.) of Robert Piper, carpenter; and a ledger (1840-57) of Jacob B. Smith, Augusta County carpenter and miller.

3031 **Ferrell Family,** Papers, 1866-1873, 70 items. Includes business correspondence of P.W. Ferrell & Company, lumber merchants, of Danville, Virginia, principally with C.T. Sutherlin & Company of Brooklin, Halifax County, Virginia, and Thomas King, of Wolf Trap, Halifax County, Virginia.

Washington

BELLINGHAM

Whatcom Museum of History and Art

121 Prospect Street 98225 Telephone: (206) 734-5791

3032 **Mount Baker Club,** Scrapbook, 1911-1958, 1 vol. Scrapbook on the history of Mount Baker compiled by Charles Finley Easton, club historian. Includes articles, partly manuscript or transcripts; maps, mostly hand-drawn; and photographs. Indexed. Largely pertains to early mountaineering expeditions, Indians, settlement, natural history, etc. Original manuscripts by Easton relate to geographical names (4 pp.); proposed Mount Baker National Park (5 pp.), House Public Lands Committee hearings on the proposed park, 1916 (4 pp.); and the Mount Baker National Forest (7 pp.).

OLYMPIA

Department of General Administration, State Archives

218 General Administration Building 98504. Records are described in *General Guide to the Washington State Archives,* n.p., n.d., processed.

3033 **Governor's Office,** Records. A detailed listing of specific series can be found in the individual indexes of the governors, which specify forestry-related records for practically every governor from Isaac Stevens to the present. General correspondence and subject files of Daniel J. Evans (1965-71) contain material (ca. 2 cubic feet) relating to the Governor's committee to develop a position on the North Cascades National Park proposals, and to Snoqualmie River dam proposals.

3034 **Interagency Committee for Outdoor Recreation,** Records. Administrative general correspondence relates to natural rivers and national parks (1965-67, 1 cubic foot).

3035 **Legislative Council,** Records, 1947-1963, 2 cubic feet. Research studies and reports include material on state parks (1957-59).

3036 **Natural Resources Department,** Records. Included are records of the former Department of Conservation. Records of the Natural Resources Accounting Section contain files of the Forestry Trust Fund (1937-41), forest patrol assessment collections (1942-45), forest assessment collections by county (1931-40), forest assessment fund (1948), forest assessment (1958-59), scale sale records, forest product brand register indexes (1926-69, on microfilm), expired log brand registrations (1950-64). Department of Public Lands records include ledgers (1890s+), pertaining to leases, inspections, appraisement and sale of public lands for forest industries usage; and Board of State Land Commissioners proceedings (1893+). There are fire patrol records (1950s-70) including fire reports (on microfilm), forest patrol assessment lists by counties, slash clearance records, and loggers fire equipment inspections. There are Forest Board minutes (1923-62); also capital land grants involving forest lands (1910+); and Department of Public Lands correspondence files (1915+).

3037 **Parks and Recreation Commission,** Records, 34 cubic feet, 200 feet microfilm. The State Parks Commission was created in 1921 and superseded by the State Parks and Recreation Commission in 1947. The records include minutes (1954-65, microfilm) and general correspondence (1963-67) of the commission; administrative correspondence, minutes, reports, and other files; records of the Governor's Mt. Rainier Study Commission (1962), and of the Governor's Conference on Outdoor Recreation (1962-63), park engineering records, and correspondence with cities and counties.

3038 **Planning Council,** Records, 1934-1945, 42 cubic feet. The records contain files relating to the proposed Cascade Ridge National Park, forestry, Olympic Peninsula, including Mt. Olympus National Park; parks and playgrounds, including Ginkgo Petrified Forest, North Cascades, Olympic Highway, Mt. Rainier and roadside development; Columbia River Gorge recreation.

PORT ANGELES

Olympic National Park

600 East Park Avenue 98362

3039 Holdings include notes, clippings, and copies of correspondence relative to legislation pertaining to the establishment of Olympic National Park (1 vol., looseleaf).

PULLMAN

Washington State University, The Library

Chief, Manuscripts-Archives Division 99163. Major holdings are described in *Selected Manuscript Resources in the Washington State University Library,* Pullman, 1974, 94 pp.

3040 **Babb, Veva V.** Personal narrative of M.P. Bogle, Spokane, in an interview with Veva V. Babb and G.H. Lathrop (1937, 3 pp.), typed carbon. Relates to lumber trade in the Inland Empire.

3041 **Babbitt, Ray.** Olympic National Forest, 1930, 11 pp. Holograph signed. Prepared for the Works Progress Administration.

3042 **Baird-Naundorff Lumber Company** (Spokane, Washington), Records, 1924-1961, 24 feet. Ledgers, accounts, personnel records, and other papers. Container list.

3043 **Bayle, G.J.,** Papers, 1930-1940, 15 items. Timber cruiser and logger for the Consolidated Mining and Smelting Company, Trail, British Columbia. The collection includes letters, maps, and published briefs relating to the Trail smelter fume case; also 1938 decision of the Trail Smelter Arbitral Tribunal.

3044 **Black Lumber Company** (Orchards?, Washington), 1908-1909, ca. 60 pp. Accounts and loose pages from account book.

3045 **Bockmier, Ralph H.** (b. 1893). "The Inland Empire Lumber Industry, 1900-1965." Typescript (1967, 34 pp.).

3046 **Bockmier, Ralph H.** (b. 1893). "Mr. Lumber (In the Days When Lumber Was a Thriving Industry), An Autobiography of Action, War, Romance, Progress and Transition, As It Actually Happened in the Inland Empire and Northwest Over a Period of 60 Years, 1905-1965, From the Personal Memoirs of a 'Palouser' " (1967, ca. 300 pp.), typed, photographs. Relates to the history of the lumber trade in the Inland Empire (of northern Idaho and eastern Washington).

3047 **Carroll, Robert Wesley** (b. 1948), Papers, 1971-1972, ca. 400 items. Correspondence, notes, photocopies, drafts, tape recordings of interviews, and other papers written and collected while researching the Civilian Conservation Corps in the state of Washington for a master's thesis.

3048 **Carty, William Edward** (1894-1962), Papers, 1898-1962, 22 feet (14,500 items). Carty was a representative in the Washington State Legislature, 1933-41, 1947-51, 1955-59. His papers largely relate to legislation and contain material relating to forests and forestry. Described in *William Edward Carty: An Indexed Register of His Papers, 1898-1963, in the Washington State University Library,* Pullman: 1967, 42 pp.

3049 **Cox, Earl Blake, Family,** Papers, 1759-1970, ca. 50 items. Includes diaries of Nathaniel P. Millard (1856-59), Louisiana sawmill operator.

3050 **Craig Mountain Lumber Company,** Statement, 1911, ca. 20 pp. Relates to lumber trade, finance.

3051 **Crosby, W.E.** "Highlights in the History of Logging Equipment and Methods" (n.d., ca. 17 pp.). Photocopy from original in the Seattle Public Library.

3052 **Exchange National Bank,** Spokane, Washington, Records, 1880-1930, ca. 12 feet (ca. 7500 items). Includes records of bank subsidiaries and lumber and other corporations of the Inland Empire, of which the bank assumed partial management, especially the affairs of lumberman Fred Herrick (b. 1853). Container list.

3053 **Fall, Albert Bacon** (1861-1944), Papers, 1912-1941, 2 feet (1222 items). U.S. Senator from New Mexico, 1913-21, and Secretary of the Interior, 1921-23. Official and private correspondence, copies of government contracts, proposed and final congressional legislation, memoranda, press releases, congressional resolutions, clippings, executive orders, and interoffice communications, relating to Fall's service as Secretary of the Interior, his trial for conspiracy, and personal matters. Photocopies from originals acquired by Dr. David H. Stratton from Mrs. Alexina Fall Chase. Present location of the originals is not known. Container list (4 pp.). Restricted.

3054 **Fisher, Vardis** (1895-1968), Papers, 1948-1957, 194 items. Author, of Hagerman, Idaho. The papers include correspondence discussing conservation and restoration, among other topics. Unpublished guide. Additions to the collection are anticipated.

3055 **Garber, C.Y.** "Just Reminiscing" (January 20, 1965, 9 pp.), typescript (photocopy). Relates to forest fire prevention and control, the Pine Creek, Idaho, fire, 1924.

3056 **Gleason, Jay Mark** (1881-ca. 1955), Papers, 1933-1942, 8 feet (ca. 2000 items). Correspondence, orders, official manuscripts, notes and photos, relating to Gleason's duties as chaplain, superintendent, and welfare and public relations officer, Fort George Wright District, Civilian Conservation Corps. Includes ca. 900 photos for District yearbooks, and CCC yearbooks and ephemeral publications. Container list.

3057 **Gyppo Tools,** Account book, 1928-1929, 1 vol. Relates to lumber machinery.

3058 **Horan, Walter Franklin** (1898-1967), Papers, 1943-1967, ca. 250 feet (360,800 items). U.S. Representative from Washington, 1943-65; member of the House Committee on Agriculture, and the Agriculture Subcommittee of the House Appropriations Committee. The Horan papers consist of correspondence, printed matter, working papers, speeches, etc., representing his office files as congressman; they include material relating to the Forest Service (1943-64), National Park Service, Wilderness Areas (1951-64). Described in: *Walt Horan: A Register of his Papers, 1943-1965, in the Washington*

State University Library, Pullman: Washington State University, 1965, 20 pp.

3059 **Hult, Ruby El** (Mrs. S.J. Sether) (b. 1912), Papers, 1924-1971, 12 feet. Correspondence, notes, clippings, taped interviews, and photographs gathered during her research on *The Untamed Olympics, Lost Mines and Treasures,* and *Steamboats in the Timber,* a study of steamboating on Lake Coeur d'Alene at the turn of the century. The collection contains ledgers of the Red Collar Steamship Line for the run between St. Maries and Harrison, Idaho (1921); photographs of logging in the Pacific Northwest; and pamphlets of scenic attractions of various areas and labor organizations such as the Industrial Workers of the World. Container list.

3060 **Inland Empire Paper Company,** Papers, 1929-1934, 3 items. Brief outline of operating history (1934, 18 pp.); appraisal report on water power rights (December 14, 1929); report on examination of the Company (1933).

3061 **Jessuph, Ward,** Papers, 1926-1959, ca. 200 items. Correspondence, notices, drafts, charts, maps, clippings, announcements relating to the Washington State Good Roads Association and the Sunset-Stevens Pass Highway Association.

3062 **Jewett, George Frederick** (1896-1956). "Thoughts on Forest Ownership," (May 14, 1938, 12 pp.), typescript.

3063 **Lathrop, G.H.** "A Personal Narrative by W.G. Leonard, Carlyle Hotel, Spokane, based on his experience and that of his father, the late Frederick Charles Leonard (1853-1932), in the Timber Business" (January 20, 1937, 5 pp.), typescript (carbon), incomplete. Relates to the Weyerhaeuser Company.

3064 **Lewis River Log and Boom Company,** Accounts, 1902-1903, ca. 90 pp.

3065 **Lucia Mill** (Vancouver, Washington). "Account of Lumber Cut by the Lucia Mill and Sold to H.R. Duniway & Co., and Others, 1881-1891" (334 pp.). Manuscript relates to lumber industry, prices.

3066 **Martinson, Arthur David** (b. 1933), Papers, 1961, 11 items. Correspondence, clippings, documents, and notes relating to the Longmire Family and the Mount Rainier National Park.

3067 **Mason, David Townsend** (b. 1883). "Putting the Brakes on Lumber Production." 1927. 28 pp., graphs.

3068 **May, Catherine Dean Barnes** (b. 1914), Papers, 1958-1971, 210 feet (400,000 items). U.S. Representative from Washington, 1959-71. The May papers are Congressional office files. They relate to political affairs of the Fourth Congressional District, Republican Party politics, and agricultural policy and politics, especially relating to her membership on the House Committee on Agriculture. Described in: *Catherine May: An Indexed Register of Her Congressional Papers, 1959-1970, in the Washington State University Library,* Pullman, Washington: Washington State University, 1972, 32 pp.

3069 **Miller, Francis G.** "The Lumber Industry of Washington." n.d. 25 pp. Transcript.

3070 **Nelson, Arnt.** "Log Scale and Tie Tally, 1908-1913." Ca. 30 pp. Manuscript.

3071 **Oakdale Lumber Company** (La Center, Washington), Papers, 1906-1913, 52 l. Articles of incorporation, minutes, and legal papers.

3072 **Plummer, Fred G.** "Field Notes of Examination of Cascade Range, Washington, 1900." 72 pp. Typescript, maps. Forestry survey of the Cascade Range.

3073 **Plummer, George H.** "Early Loggers, ca. 1905." 4 l., manuscript. Relates to lumbering and lumbermen in Washington and Idaho.

3074 **Pratsch, Charles Robert** (1857-1937), Photographs, 1888-1913, ca. 1000 items. Photographer, Aberdeen, Washington. Glass negatives, lantern slides, prints by Pratsch and Colin S. McKenzie of woods work, milling, lumber transportation, shipping, fishing, sealing, and other maritime activities. Locations include towns, harbors, and woods in the Grays Harbor area, Washington. Unpublished register. Some of these photographs have been published in Robert A. Weinstein, "Grays Harbor County, 1880-1920: Maritime Scenes," *The Record, 1973,* Pullman, Washington: Friends of the Library, Washington State University, 1973, 36-49.

3075 **Reinhard, L.F., & Co.** Audit of firm of A.C. White (September 20, 1917, 52 pp.), typescript. Relates to accounting in the lumber trade.

3076 **Richardson, Elmo Race** (b. 1930), Papers, 1962-1968, ca. 2,400 items. Historian. Correspondence, notes, photographs, and other papers collected during research on the Civilian Conservation Corps, including diaries and reminiscences of corpsmen, correspondence with congressmen, officials, and corpsmen, notes on archival and manuscript material and a subject index to *Happy Days* (1932-36), a corps newspaper. Container list.

3077 **Robinson and Peterson,** Account book, 1924-1927, 1 vol. Logging firm in the Inland Empire.

3078 **Russell, Carl Parcher** (1894-1967), Papers, 1920-1967, 37 feet. U.S. National Park Service official. Correspondence, diaries, journals, reminiscences, notes, book reviews, accounts, minutes, business records, reports, documents, newspaper articles, photographs, motion pictures, microfilm, drawings, published and unpublished writings of Russell and others, and other papers, chiefly collected, arranged, edited, and indexed by Russell. The papers relate to Russell's interest in the American West, the National Park Service, the preservation of sites, park history, etc. Described in *Carl Parcher Russell: An Indexed Register of his Scholarly and Professional Papers, 1920-1967, in the Washington State University Library,* 1970, 149 pp.

3079 **Simmons and Cole,** Account book, 1903-1908, 1 vol. An Inland Empire logging firm.

3080 **Soundview Pulp Company** (Everett, Washington), Minute books, 1932-1949, 5 vols. Minutes, reports, and notices of the trustees, directors, officers, and shareholder meetings. Relates to the wood pulp industry.

3081 **Spinning, George T.,** Papers, 1935-1938, ca. 500 items. Educational advisor, Civilian Conservation Corps Camp F-188, Emida, Idaho. Correspondence, reports, memoranda, announcements, bulletins, accounts, photographs, employee rosters, brochures and other papers relating to educational work in the camp.

3082 **Stevens, James F.** (b. 1892), Papers, 1909-1965, 12 items. Author. Correspondence, clippings, unpublished manuscripts, "The Green Glory" (523 pp.); transcript of oral interview by Warren L. Clare (1963); articles by and about Stevens. Contains material on forest management and the lumber trade. Container list.

3083 **Washington State University. Department of Forestry and Range Management,** Papers, 1912-1914, 8 items. Reports on timber cruising of the W.S.U. land grants.

3084 **Wikstrom, Charles Erik,** Papers, 1873-1942, 225 items.

Miner, Jackson County, Oregon. Among the papers are deeds, indentures, mortgages, letters, and other papers relating to mineral and timber lands in Jackson County, Oregon.

3085 **Wolf, R.B.** "The Function of the Chart Room" (n.d., 5 pp.), typescript. Relates to the function of the chart room of the Weyerhaeuser Timber Company.

SEATTLE

Federal Archives and Records Center

Chief, Archives Branch, 6125 Sand Point Way, N.E. 98115. Records not accessioned into the Archives Branch are not available for research without agency approval.

Archives Branch

3086 **Alaskan Territorial Government** (Record Group 348), 1884-1958, 297 cubic feet. The general correspondence file (1909-58, 176 linear feet) is arranged according to 3 separate file schemes. For the period 1908-18 there are files on forests, Forest Service, land matters other than homestead, Sitka National Monument, Juneau Land Office, pulp industry, Mt. McKinley National Park. For 1919-33 there are files on development of Alaska and its resources; U.S. Forest Service; lumber and timber, pulp, paper industries; national parks and monuments. Unpublished preliminary inventory (1968, 15 pp.).

3087 **Chief of Engineers Office** (Record Group 77), 425 feet. Records of the Northern Pacific Division, Seattle District, and Portland District. Records of the Portland District include 4 volumes and unbound materials relating to Crater Lake National Park roads (1911-19, 2 feet; indexed). In the records of the Seattle District are miscellaneous materials with correspondence relating to supplies for the Spruce Production Division, Bureau of Aircraft Production (1918). There are unpublished preliminary inventories for each subgroup.

3088 **District Courts of the United States** (Record Group 21), 1,300 feet. Records of the Districts of Idaho, Oregon, and Eastern Washington. There is material on land titles that affected forest boundaries. There is an unpublished preliminary inventory for the District of Oregon, shelf lists for records of the other districts.

3089 **Indian Affairs Bureau** (Record Group 95), 5039 feet. Included are records of Indian agencies, schools, irrigation projects, etc., in the Pacific Northwest. There are files for some agencies concerning forestry and lumbering on reservations. There are unpublished preliminary inventories available for the Colville Agency records (1865-1943), and for the Warm Springs Agency records (1861-1925). There are also shelflists.

Federal Records Center

3090 **Forest Service** (Record Group 95), 13,998 cubic feet. Records include files of the Northern Region headquarters, Missoula, Montana; Alaska Region headquarters, Juneau, Alaska; and the Pacific Northwest Region headquarters, Portland, Oregon. Each group contains general records, records documenting executive direction, timber management, recreation, watersheds, wildlife management, cooperation with state and private forest owners, tree planting, forest fire research, forest fire prevention and control, research on insects, lands, engineering, etc. There are also small lots of records for forest and range experiment stations in the Pacific Northwest, Intermountain, and Alaska regions. Shelf lists.

3091 **Land Management Bureau** (Record Group 49), 8398 cubic feet. Included are records relating to forest management. Shelf list.

University of Washington Libraries

Curator of Manuscripts 98195. Some holdings are described in: Richard C. Berner, "Sources for Research in Forest History: The University of Washington Manuscripts Collection," *Business History Review,* XXXV, Autumn, 1961, 420-425; *The Manuscript Collection of the University of Washington Libraries,* Seattle: University of Washington Libraries, Library Leaflet No. 1, revised, November, 1967, 33 pp., processed; and M. Gary Bettis, *Recent Accessions to the Archives and Manuscript Division of the University of Washington Library, 1967-69,* n.p., 6 l., processed. *North Cascades Archival Resources in Washington State Repositories,* prepared by Harry M. Majors and the staff of the Manuscripts Section, University of Washington Library, and edited by Karyl Winn, Seattle: University of Washington, 1974, 17 l., processed, describes regional materials in the University of Washington Library and other repositories.

3092 **Aldwell, Thomas Theobald** (b. 1868), Papers, 1890-1950, 1 foot. Founder, vice president, and general manager of the Olympic Power Company, Port Angeles, Washington, and president of the Port Angeles Port Commission. Correspondence, documents, and clippings used by Mr. Aldwell for his autobiography, *Conquering the Last Frontier* (1950), and relating to his activities with the Olympic Power Company, its financing and construction of Elwha River Dam, the resultant attraction of the forest products industry to Port Angeles and the Olympic Peninsula, his activities with the Port Angeles Port Commission in developing its harbor, his role in the creation of the Olympic National Park, and subsequent disputes about the size of and logging in the park. Correspondents include George A. Glines; Crown Zellerbach, Inc.; Peabody, Houghteling, and Company, Chicago; Rayonier, Inc.; the Seattle Lighting Department; and L.L. Summers and Company, Chicago.

3093 **Aloha Lumber Corporation,** Records, 1909-1951, 6 feet. Correspondence concerning lumber for airplanes, financial materials, accident reports for the early 1920s, and War Department contracts. Correspondents include Harold J. Barrett; O.W. Brown Timber and Supply Company, Seattle; William E. Campbell; the Lumbermen's Indemnity Exchange, Seattle; John G. McIntosh and Company, Seattle; Mackie and Barnes, Inc., Hoquiam, Washington; Moore Dry Kiln Company, Portland, Oregon; Nestos Pole Company, Seattle; Northwestern Lumber Company, Hoquiam, Washington; Pacific Northwest Loggers Association; the Field Service of the U.S. Bureau of Indian Affairs; the Spruce Production Division of the U.S. War Department; the auditor of Grays Harbor Co., Washington; the Washington Department of Labor and Industries; and the Washington Commission on Public Lands. Inventory.

3094 **Alpine Lakes Protection Society,** Records, 1969-71, ca. 17 inches and 4 reels of tape. Articles of incorporation; minutes of general meetings, directors' minutes, trustees' minutes, research and resources committee minutes; some correspondence; notices; membership lists; miscellaneous; one tape recording of an interview for a slide show; ephemera and business records. Restricted. Inventory.

3095 **Ames, Edwin Gardner,** Papers, 1888-1935, ca. 50,000 items. Northwest manager of Pope and Talbot, 1907-31. The records relate primarily to the Puget Mill Company, constructed at Port Gamble, Washington, in 1853 by Pope and Talbot. There is business and personal correspondence of Ames, and of his predecessor, Cyrus Walker; also correspondence files of the Douglas Fir Company, Douglas Fir Exploitation and Export Company, Pacific Lumber Inspection Bureau, Admiralty Logging Company, Puget Sound Tug Boat Company, Puget Sound Stevedoring Company, the Washington Forest Fire Association, Western Forestry and Conservation Association, West Coast Lumberman's Association, and others.

3096 **Andrews, Ralph** (b. 1897), Papers, 1956-1965, ca. 1 foot. Manuscripts, including "Glory Days of Logging," "This Was Sawmilling," "Redwood Classic," etc., and related correspondence. In part relating to forest industries in the Pacific Northwest.

3097 **Association of Western Pulp and Paper Workers,** Records, ca. 1960-1970, 14 feet. Court papers and exhibits, relating to litigation following formation of the union and prior to certification by the National Labor Relations Board, 1964.

3098 **Bagley Family Papers,** 1859-1932. Papers of Clarence Booth Bagley (1843-1932), Daniel Bagley, and other members of the family. Correspondence and other papers concerned mainly with politics, early Washington business and history. There is material of Henry L. Yesler relating to the Henry Yesler Company and to Yesler and Denny Company (1863-71), the Puget Mill Company of Port Gamble, Washington (1903, 1 item); the Washington Mill Company of Hadlock, Washington (1894, 1 item); and the Pacific Pine and Lumber Company. Described in *The Bagley Family Papers, 1859-1932, University of Washington Libraries Manuscript Series, Number 4,* n.p.: 1966, 33 pp.

3099 **Ballaine, John Edmund** (1868-1941), Papers, 1889-1940, 6 feet. Journalist; private secretary to John R. Rogers, governor of Washington; and founder of the Alaska Central Railroad. Correspondence, business papers, speeches, documents, financial records, court papers, memorabilia, and writings by Ballaine. Business papers relate, in part, to the Alaska Pulp, Paper, and Hardwood Company. Inventory.

3100 **Ballinger, Richard Achilles** (1858-1922), Papers, 1907-1920 (also on microfilm, 13 reels). Commissioner, U.S. General Land Office, 1907-09; Secretary of the Interior, 1909-11. The collection includes correspondence (1907-08, 800 items) relating to the General Land Office and other matters; correspondence, clippings, scrapbooks, memoranda on the U.S. Forest Service, records of hearings at which Ballinger was a principal witness, speeches and writings, and ephemera relating to his career as Secretary of the Interior (1909-11, ca. 6,500 items); correspondence, bills, and receipts (1915-20, ca. 200 items) relating to the Lake Ballinger Land Comnpany, Seattle; and letters of condolence and clippings relating to Ballinger's death (ca. 200 items). The collection is described in National Historical Publications Commission, Microfilm Publication Program, *Richard A. Ballinger Papers, 1907-1920, in the University of Washington Libraries,* project director, Richard A. Berner, 1965, 26 pp., which contains a complete list of correspondents.

3101 **Beck, Broussais Coman** (1886-1937), Papers, 1919-1920, 1 foot. Department store manager, of Seattle, Washington. Papers relating to industrial espionage, including reports from agents about union and strike meetings both during and shortly after the Seattle general strike, 1920, and copies of newspapers such as *The Industrial Worker* and *Lumber Workers Bulletin.* Includes material relating to Eugene Debs, the Industrial Workers of the World, and the Seattle Cooperative movement. Inventory.

3102 **Benson, Naomi Achenbach,** Papers, 1898-1961, 21 feet. Teacher and Democratic Party worker in Snohomish County, Washington. Contains material reflecting her interest in politics and other matters, including conservation. Inventory.

3103 **Bloedel-Donovan Lumber Mills** (Bellingham, Washington), Records, ca. 2 inches. Payroll records and labor agreements from Beaver Camp, Beaver, Washington.

3104 **Brainerd, Erastus** (1855-1922), Papers, 1880-1919, ca. 3000 letters. Newspaper editor, Seattle. Friend of Richard A. Ballinger; member of Western Forestry Association; park commissioner. The collection is described in *The Erastus Brainerd Papers, 1880-1919,* Manuscript Series Number Two, Seattle: University of Washington Library, Reference Division, 1959, 13 pp.

3105 **Brant, Irving Newton** (b. 1885), Papers, 1926-1961, microfilm, 5 reels. Material relating to Brant's activities in conservation, including correspondence and related speeches, writings, and legislation, in chronological order. Typed list of correspondents, 8 pp. Filmed from originals at the Library of Congress.

3106 **Brinkley, Joseph Arthur,** Biographical manuscript, 1969, 78 pp. Forest ranger. Xerox copy.

3107 **Brockman, Christian Frank** (b. 1902), Papers, 1946-1970, 13 feet. Chief Park Naturalist, Mt. Rainier National Park, 1928-41; Yosemite National Park, 1941-46; Professor of Forest Resources, College of Forestry, University of Washington, 1946-67. Manuscripts of *Recreational Use of Wild Lands, Flora of Mount Rainier National Park, Trees of North America,* related correspondence, galley proofs, reviews, etc.; reports and printed matter relating to national parks, especially Mt. Rainier, North Cascades, and Olympic; materials on dendrology; miscellaneous correspondence and speeches.

3108 **Burke, Thomas** (1849-1925), Papers, 1875-1925, 26 feet. Burke was a central figure in the political and economic life of Seattle and Washington Territory and State. The collection includes incoming and outgoing correspondence, legal documents, financial documents, speeches, drafts of articles, and letterpress correspondence and other materials relating to numerous law firms with which Burke was affiliated. Authors of correspondence include, among many others, Richard Achilles Ballinger (1893-1910, 5 items); Fowler-Boyer Lumber Company, Centralia, Washington (1900-01, 4 items); George H. Megguire (1893, 4 items); Compact Shingle Package Company, Anacortes, Washington (1893-94, 23 letters); National Parks Association (1923-25, 17 letters, 3 enclosures); Natural Parks Association of Washington (1919-22, 7 letters); Olmsted Brothers (1907-08, 5 letters); Gifford Pinchot (1897-1916, 7 letters); Port Blakely Mill Company (1900, 3 letters); Puget Mill Company (1890-91, 5 letters); Rainier National Park Company (1916-24, 39 letters, 6 enclosures); Rocky Mountain Club (1910-17, 9 letters); Seattle Cedar Lumber Manufacturing Company (1900-17, 5 letters); Seattle-Tacoma Rainier National Park Committee (1915, 2 letters); Stetson & Post Mill Company, Seattle (1890-93, 3 letters); William Howard Taft (1911-25, 5 letters); U.S. Department of the Interior (1901-16, 4 letters); Washington Board of State Land Commissioners (1899-1901, 11 letters); Washington Commissioner of Public Land (1898-1924, 15 letters); Washington Conservation Association (1910-14, 3 letters); Weyerhaeuser Timber Company (1915-23, 2 letters). The collection is described in *The*

Thomas Burke Papers, 1875-1925, Manuscript Series Number 3, Seattle: University of Washington Library, Reference Division, 1960, 36 pp.

3109 **Burke Millwork Company** (Seattle), Records, 35 feet. Manufacturing and remanufacturing firm. General correspondence (mainly 1940-54); files of the Puget Supply Co., their jobber.

3110 **Clark, Donald Hathaway** (1890-1965), Papers, 1880-1964, 60 feet. Secretary-manager of the Northwest Hardwood Association and of the Red Cedar Shingle Association, owner of the Cascade Cedar Company, sales manager of the Colonial Cedar Company, research associate in forest products at the University of Washington College of Forestry, director of the Institute of Forest Products, and author. Correspondence, subject files, manuscripts and printed copies of writings, speeches, published articles, clippings, and other papers. Subgroups of papers relate to the Bellingham Bay and Eastern Railroad, the Blue Canyon Coal Mining Company, J.J. Donovan, the Forest Club of the University of Washington, the Concatenated Order of Hoo-Hoo, the Loggers Information Association, the Pacific Northwest International Writers Conference, the Pennsylvania Civic League, the Red Cedar Shingle Bureau, the Rite Grade Shingle Bureau, the Institute of Forest Products, the Stark Stained Shingle Company, the Stark Manufacturing Company, the Totem Shake Company, the U.S. Bureau of Indian Affairs, the U.S. Forest Service and the World Forestry Congress. Correspondents include Guy Allison, American Forest Products Industries, the American Forestry Association, *American Forests, American Heritage, American Magazine,* Deane C. Bartley, Harry C. Bauer, Charles T. Conover, Don W. Donaldson, Joel E. Ferris, June Wetherell Frame, Thomas R. Garth, Arthur Garton, J. H. Georgregan, Bror Grondal, Dean Guie, Harry Hamilton, Robert Hitchman, Lloyd Hougland, Aline Howell, Roland Huff, Fred Lockley, Peter Long, Olga E. Pattison, Herbert M. Peet, John Pietz, Lancaster Pollard, Charles E. Putnam, Click Relander, Arthur K. Roberts, Frank I. Sefrit, Chris Siegel, Walter W. Smith, James F. Stevens, Leon L. Stock, F.E. ("Jess") Willard, and Hugo Winkenwerder. Inventory.

3111 **Clark, Irving M.** (1883-1960), Papers, 1934-1960, 3 feet. Lawyer and conservation leader of Seattle, Washington. Correspondence, financial miscellany, legal documents, maps, minutes, reports, writings, photos, and printed miscellany, relating mainly to Clark's activities in forestry, conservation of wildlife and natural resources, and national parks and reserves in Washington. Includes some political correspondence and ephemera. Represented in the collection are materials on Arthur Watts Clark, Elam M. Hack, Henry Martin Jackson (b. 1912), Robert Marshall, the Robert Marshall Wilderness Fund, George Marshall, National Parks Association, Richard Lewis Neuberger, North Cascades Conservation Council, Olympic National Park Associates, Inc., Sierra Club, and Wilderness Society. Index and inventory.

3112 **Coe, Earl Sylvester** (1892-1964), Papers, 1939-1963, 5 feet. Member of the Washington House of Representatives and Senate, Secretary of State, and Director of Conservation for Washington. Correspondence, reports, news releases, speeches, campaign material, photos, and memorabilia, relating to Coe's political career and business affairs. Includes papers relating to Earl S. Coe & Associates, Earl S. Coe & Company, Nordby Lumber and Box Company, and politics, government, conservation, and the forest industry in the state of Washington. Correspondents include Willis Tryon Batcheller, Ken Billington, Henry P. Carstensen, Clarence Cleveland Dill, James A. Farley, Warren G. Magnuson, Hugh B. Mitchell, and Richard L. Neuberger. Inventory.

3113 **Columbia Basin Inter-Agency Committee,** Records, 1946-1966, 48 feet. Correspondence, agenda, minutes, reports, and miscellaneous items from the committee and its many subcommittees. Includes material on the natural resources of the Columbia River Basin. Inventory.

3114 **Curtis, Asahel** (1874-1941), Papers, 1898-1941, ca. 8 feet. Photographer, conservationist, and civic leader, of Seattle. Correspondence relating to Curtis' photographic activities, his work as founder of the Mountaineers, and chairman of its Outing Committee, the creation and development of Olympic and Mt. Rainier National Parks, his work as chairman of the Rainier National Park Advisory Board, as a founder of the Washington State Good Roads Association, as a member of the Seattle Chamber of Commerce's Highway Committee, and as president of the Washington Irrigation Institute, regional publicity and resource development activities as chairman of the Northwest Development Committee of the Seattle Chamber of Commerce, and activities as executive of the Marsh-Curtis "Know Your State" Bureau. Includes diary (1898) kept at Dawson, Yukon Territory, Canada; ledgers from Curtis' studio; and scrapbooks of correspondence, photos, clippings relating to the Seattle-Tacoma Mt. Rainier National Park Committee. Correspondents include Homer T. Bone, Clarence Cleveland Dill, Lindley Hoag Hadley, Knute Hill, Stephen B. Hill, Wesley Livsey Jones, Clarence Daniel Martin, Edmond Stephen Meany, Lewis Baxter Schwellenbach, Monrad C. Wallgren, Marion Anthony Zioncheck, the U.S. Department of Agriculture, U.S. Department of the Interior, and U.S. Forest Service. Inventory. The scrapbooks are also available on microfilm (2 reels). A portion of Curtis' photographs are in Library's photography collections.

3115 **Darrington Pioneers' Exchange** (Darrington, Washington), Records, 1935-1937, 144 items. Correspondence, minutes, constitution, bylaws, financial miscellany, and ephemera of the exchange, a nonprofit cooperative association formed, in part, to sell lumber products. Inventory.

3116 **De Shaw, William** (ca. 1830-1900), Papers, 1852-1898, 10 vols. and loose material. Business record books of Point Agate Store. Includes materials about relations with the Port Madison Mill Co.

3117 **Difford, Wallace Ellsworth,** Papers, 1952-1968, 5 inches. Chief executive, Douglas Fir Plywood Association. Correspondence, reports, speeches and writings, and ephemera.

3118 **Disque, Brice P.** (1879-1960), Collection, 1917-1960, 2 feet (ca. 3000 items). Commander of Spruce Production Division during World War I, president of the U.S. Spruce Production Corporation, founder of the Loyal Legion of Loggers and Lumbermen, and warden of Michigan State Prison. Personal papers include correspondence, autobiographical writings, and a logbook; U.S. War Department Spruce Production Division material includes correspondence, memoranda, reports, minutes, history, articles of incorporation, photos, and ephemera; and Loyal Legion of Loggers and Lumbermen papers include correspondence, minutes of conventions, bulletins, constitution, and bylaws, speeches, and writings. Names represented include Clarence F. Lea, George Meany, Leonard Wood, American Federation of Labor, and the U.S. Department of Labor. These papers were collected by Dr. Harold Hyman for his book *Soldiers and Spruce,* 1963.

3119 **Evans, Brock,** Papers, 1961-1972, 36 feet. Northwest Representative of the Sierra Club and Federation of Western

Outdoor Clubs. Correspondence and subject files relating to environmental preservation issues primarily in Washington, Oregon, Idaho, Montana, with some material on those in Alaska, California, British Columbia, and Wyoming. Also includes records of two previous Northwest Representatives, J. Michael McCloskey (1959-65) and Roger W. Pegues (1965-67).

3120 **Everett Prisoners' Defense Committee,** Papers, 1916, 4 items. Publicity material, resolution, speakers' data, relating to Industrial Workers of the World strikes.

3121 **Federation of Western Outdoor Clubs,** Records, 1965-1969, 5 feet. Correspondence and subject files of Betty Hughes as an officer and president (1972-73) of the FWOC. Inventory and list of correspondents. Restricted.

3122 **Force, Horton** (b. 1878), Letters, 1860-1861, 5 items. Relate to the forest industry in Washington state.

3123 **Forest History Society,** Oral History Interviews, 1957-1960, 2 inches. Includes interviews with Emmit Aston, Charles S. Cowan, George Drake, Francis Frink, Mrs. William B. Greeley, Stuart Moir, L.T. Murray, Sr., James Stevens.

3124 **Gilmore, Hugh P.,** Papers, 1901-1937, 14 vols. Captain on Puget Sound tugboats from ca. 1900-38. Logbooks (13 vols.) and reminiscences (1 vol.), partly relating to the towing of logs.

3125 **Griggs, Chauncey L.,** Scrapbooks, 1961-1966, 5 vols. Scrapbooks containing clippings, letters, press releases on log exports. List of major authors of letters.

3126 **Haig, Emily Huddart (Mrs. Neil Haig)** (b. 1890), Papers, ca. 1934-1967, 27 feet. Correspondence and related items regarding her work as officer in Sierra Club, Mountaineers, North Cascades Conservation Council, Federation of Western Outdoor Clubs, and other organizations.

3127 **Hansen, Julia Butler** (b. 1907), Papers, 1961-1974, 163 feet; Microfilm, 1950-1970, 3 rolls. Member of Washington State House of Representatives, 1939-60. U.S. Representative from Washington, 1960-74; served on the House Appropriations Committee, and the Committee on Interior and Insular Affairs. Correspondence, reports, legislation, speeches, budget requests, clippings, brochures, photographs, legal documents and ephemera. Relevant subject files include the Bureau of Land Management, Bureau of Indian Affairs, Bureau of Outdoor Recreation, Land and Water Conservation Fund, National Park Service. Julia B. Hansen was a board member of the Forest History Society. Preliminary inventory. Preliminary inventory.

3128 **Hartsuck, (Mrs.) Ben,** Reminiscences, 1905-1919, 35 pp. Relates to the forest products industry in Washington and Oregon, and to the Industrial Workers of the World in Washington.

3129 **Heermans, Harry Clay** (b. 1852), Papers, 1889-1930, ca. 21 feet. Corporation executive, real estate investor, and civil engineer of Corning, New York, and of Hoquiam, Raymond, and Grays Harbor, Washington. Correspondence and business records relate, in part, to the State Lumber and Box Company. Inventory.

3130 **Henry, Horace C.,** Collection, 1890-1959, 45 feet. Includes a diary (1891-1924, incomplete, 1 foot); correspondence; records of the Pacific Creosoting Company (ca. 1906-59, 4 feet); West Coast Wood Preserving Company (1930-43, 1½ feet); Henry & Larson Company (ca. 1903-55, 5 feet, 2 vols., and maps); Russeltown Timber Company (1909-55, 1 foot); Ozette Timber Company (1940-55, 1 foot); and other business firms. Largely financial records, but also minutes, correspondence, etc., with some photographs. Typed inventory, 7 pp.

3131 **Hobi, Frank** (b. 1894), Autobiography, 193 pp. Xerox copy of original in the possession of Hobi; relates to the forest industry in Washington state, including the Hobi lumber interests.

3132 **Hodges, Leander L.,** Letters, 1854-1855, 2 items. One letter, written at Ashburnham, Massachusetts (March 16, 1855) relates to forest industries in Massachusetts.

3133 **Holbrook, Stewart Hall** (1893-1964), Papers, ca. 1904-1965, 3 feet. Portland, Oregon, and Seattle, Washington, author, journalist, historian. Correspondence, literary manuscripts, diaries, scrapbooks, speeches, writings, clippings, notes, and research material. Correspondents include the Forest History Society, Ralph Chaplin, and James Floyd Stevens. Inventory. Access to the diary is restricted.

3134 **Ihrig, Herbert G.,** Collection, 1932-1949, 2½ feet. Correspondence, clippings, articles, reports, photographs, lists, and ephemera.

3135 **Industrial Workers of the World, Seattle Branch,** Records, ca. 1905-1950, 6 feet. Correspondence, minutes, photographs, reports, ephemera.

3136 **Iverson, Peter** (b. 1861), Papers, 1898-1936, 2 feet. Newspaper publisher and editor, and Washington state senator. Correspondence (chiefly 1915-19), financial miscellany, legal documents, legislative bills, newspapers and clippings, petitions, reports, speeches, and writings, and legislative scrapbooks containing bills submitted to the state senate. The correspondence is concerned primarily with bills before the legislature, which deal with logged-off lands, among other subjects. Index and inventory.

3137 **Jacobson, Norman F.** (1887-1960), Papers, 1903-1965, ca. 10 feet. Forester. Speeches, notebooks, legal records, maps, photos, reports, mainly on the St. Paul and Tacoma Lumber Company and the Potlatch Lumber Company, and ephemera (6 feet) relating to the timber industry, related associations and organizations, the U.S. Forest Service, and the U.S. Department of Agriculture. Correspondents include the St. Paul and Tacoma Lumber Company, E.T. Allen, and William B. Greeley. Inventory.

3138 **Jones, Wesley L.** (1863-1932), Papers, 1899-1932, 147 feet. U.S. Representative from Washington, 1899-1909, and U.S. Senator, 1909-32. He served on the Senate Committees on Conservation of Natural Resources and Reclamation of Arid Lands. General correspondence, mostly relating to legislation, includes material relating to the tariff on lumber and on shingles (1929-30), national parks, the U.S. Forest Service, etc. Inventory and name index.

3139 **Keep Washington Green,** Records, 1940-1963, 18 feet. Correspondence, financial records, promotional material.

3140 **King County Assessor's Office,** Records, 1877-1967, 324 feet. Includes timber cruise records, forest inventories of the county, land examination records, timber assessment worksheets, statistics, correspondence, and aerial topographic maps.

3141 **Lacey, James D., & Co.** (Chicago and New Orleans), Papers, 1930-1932, 4 vols. and 1 folder. Court papers relating to the case of Lacey & Co. vs. T. H. McCarthy, et al., in Clallam County, Washington.

3142 **Langlie, Arthur Bernard** (b. 1900), Papers, 1936-1956, 54 feet. Lawyer, politician, and business executive of

Seattle. Correspondence and other papers, most of which date from his terms, 1941-45 and 1949-57, as governor of Washington. Subjects mentioned include the Washington Forestry Advisory Committee. Inventory.

3143 **Luark, Michael Fleeman** (1818-1901), Papers, 1846-1899, 1 foot. Farmer and lumberman. Diaries (25 vols.) relate in part to the daily routine of a pioneer farmer-logger in western Washington Territory. There is also a journal of the Sylva Mill Company, Montesano, Washington, which Luark operated from 1869 to 1885.

3144 **McCormick, Charles R., Lumber Company,** Records, 1925-1949, ca. 27 feet. Miscellaneous personnel, production, and sales records; financial records, chiefly bound in volumes. Inventory.

3145 **McEwan, (Mrs.) Alexander F.,** Papers, ca. 1929-1946, about 3,000 items. Papers relating to her conservationist activities, mainly with the Washington State Conservation Society, of which she was a president.

3146 **McGillicuddy, Blaine H.,** Papers, ca. 1902-1965, 2 feet. Forester. Correspondence, timber cruise books, journals and photographs concerning the Corkery Logging Company.

3147 **Magnuson, Donald H.** (b. 1911), Papers, 30 feet. U.S. Representative from Washington, 1953-63. Correspondence, reports, financial materials, ephemera. Subject files of correspondence include U.S. Forest Service (1961-62); forest ski areas at Crystal Mountain, near Chinook Pass Highway, and Mt. Rainier (1959-60); Lewis River Road, Gifford Pinchot National Forest (1955-60); wilderness preservation (1956-60). Legislative files, arranged by congressional session, include correspondence on forestry (1961-62); wilderness bill (1961-62); North Cascades National Park (1957-61); proposed office of Assistant Secretary of Agriculture for Forestry Resources (1962); and First World Conference on National Parks (1961-62). Inventory.

3148 **Mahaffay, Robert E.** (1908-1967), Papers, 1932-1967, 1½ feet. Writer, publicist. Correspondence, writings, clippings, and related items, concerning his work mainly with the West Coast Lumberman's Association and the Western Wood Products Association.

3149 **Mallory, Virgil W.,** Papers, ca. 1950-1962, 8 inches. Correspondence and related items regarding conservation activities and his public relations consulting business, the West Coast Reporting Service. Inventory.

3150 **Manning, Harvey M.,** Papers, 1953-1973, ca. 30 feet. Correspondence and manuscripts of an officer of the Mountaineers and the North Cascades Conservation Council, and editor for Friends of the Earth. Includes files relating to Mountaineers publications and the Sierra Club. Restricted.

3151 **Marckworth, Gordon D.,** Papers, 1934-1964, 10 feet. Includes files relating to the Washington State Board of Natural Resources (1957-64); Society of American Foresters (1941-63); Governor Langlie's Forest Advisory Committee (1942-43); Western Forestry and Conservation Association; West Coast Forestry Procedures Committee; American Forest Congress (1946); and the University of Washington College of Forestry. Inventory.

3152 **Marshall, Louise B.,** Papers, 1965-1971, 13 inches. Correspondence, ephemera, and notes relating to the Washington Environmental Council, the Alpine Lakes Protection Society, and other outdoor organizations and government agencies; also writings relating to outdoor recreation in Washington state.

3153 **Mason County Central Railroad Co.,** Journal, 1891, 1 vol. Relates to Simpson Timber Company.

3154 **Meany, Edmond Stephen** (1862-1935), Papers, 1883-1935, 58 feet. Professor of history at the University of Washington, state legislator, editor, and author. Correspondence, speeches, writings, minutes, clippings, note cards, and pamphlets. Correspondents include the American Forestry Association, Frank Pierrepont Graves, David Starr Jordan, Ezra Meeker, Charles V. Piper, Edwin C. Starks, Hazard Stevens, the Forestry Division of the U.S. Department of Agriculture, and Young Naturalists. Inventory.

3155 **Merrill and Ring Lumber Company,** Records, 1902-1942, 93 feet, ca. 155,000 items. Correspondence, legal documents, business documents, maps and plats, and other records, arranged in annual series and relating primarily to the business, legal and fiscal activities of the company which operated in Seattle, Washington, the Olympic Peninsula, and British Columbia. Material reflects its labor and political relations, including dealings with affiliated firms, accountants, bankers, lawyers, real estate and insurance companies, architects, lumbermen's and trade associations, federal agencies (chiefly the Forest Service and the Spruce Production Division of the War Department), state agencies, politicians, and others. Includes annual financial statements (1923-31), indexed in compilations of cost, assets, and land holdings, with comparative tabulations of various companies; and a reference file (1903-42) containing estimates of timber stands and timber cruises. Correspondents include Thomas T. Aldwell, William Edward Boeing, Homer Truett Bone, Clarence Cleveland Dill, Lindley Hoag Hadley, Ralph Ashley Horr, Guy C. Myers, Miles Poindexter, Lewis Baxter Schwellenbach, Martin Fernard Smith, and George Corydon Wagner. Executives include Thomas D. Merrill, Sr., Clark Lombard Ring, William D. Chisholm, Timothy Jerome, Thomas D. Merrill, Jr., Richard Dwight Merrill. Affiliated companies include Crescent Logging Company, Ozette Timber Company, Polson Logging Company, Pysht River Boom Company, Pysht River Lumber Company, and Sol Duc Investment Company. Index and inventory. Restricted.

3156 **Mitchell, Hugh Burton** (b. 1907), Papers, 1945-1952, 72½ feet. U.S. Senator from Washington, 1945-46; U.S. Representative, 1949-53. His congressional papers include correspondence, clippings, and ephemera on forest and timber problems, Hells Canyon, the Forest Service, the Bonneville Power Administration, export and import of lumber, the National Park Service, the lumber industry, Mt. Rainier and Olympic National Parks, the pulp and paper board industry, tariff on cedar shingles, tariff on plywood, export control of lumber, and the Washington State Park and Recreation Commission. Inventory.

3157 **Morris, William W.,** Letters, 1909-1911, ca. 140 pp. Letters of Morris during his career with the U.S. Forest Service. Xerox copies of carbons, relating to the Coeur d'Alene National Forest.

3158 **Mountaineers** (Seattle, Washington), Records, 1906-1970, 21½ feet (microfilm, 6 reels). Outdoor recreation and conservation organization. Minutes; reports; records of committees, including Literary Fund Committee (1956-70); Climbing Committee (1950-70); Outdoor Division (1958-70); Alpine Scramblers Committee (1961-70); summit registers; and slides from annual outings (1906-32, 2000 items).

3159 **Nordstrom, Anna E.,** Papers, 3½ feet. Reference files relating to editorial work with the West Coast Lumbermen's Association. Subject files relate to many aspects of Pacific

Northwest lumber industry and lumber companies. Inventory, typed, 2 pp.

3160 **North Cascades Conservation Council,** Records, 1958-1969, 7 inches. Correspondence and ephemera. Restricted.

3161 **North Cascades History Project,** Oral History Collection. Includes tape recordings relating to the history of the mountainous region between Snoqualmie Pass and the Canadian border. Topics include mountain climbing, pioneer photography, A.H. Sylvester (Supervisor of the Wenatchee National Forest), the U.S. Forest Service in the Skykomish district, development of skiing in Snoqualmie Pass, trout planting in mountain lakes, early cartography of the region.

3162 **Northern Pacific Railway Company,** Records, ca. 1879-1959, ca. 100 items. Letters, reports, scrapbooks (1879-80), and proposals. Microfilm copies of records in their files regarding Wilkeson Coal and Coke Company and the St. Paul and Tacoma Lumber Company, filmed from the originals loaned for copying by the Seattle office of the Northern Pacific.

3163 **Northwestern Lumber Company** (Hoquiam), Records, 1887-1934, 27 feet, microfilm, 18 rolls. Articles of incorporation, bylaws, minutes (1914-26), correspondence (1891-1934), including letterbooks (1891-1912) of George H. Emerson, financial and business records (1883-1934). George H. Emerson supervised the construction of the original mill at Hoquiam, Washington; and became manager when the mill began operating in late 1882; later vice-president of the Northwest Lumber Company and general manager of the mill.

3164 **Olympia Veneer Company,** Records, 1921-1954, 2 feet. Minutes and stock books of the company which was consolidated into Associated Plywood Mills in 1946 and sold to U.S. Plywood in 1955.

3165 **Overly, Fred J.** (1907-1973), Papers, 1935-1971, 2 feet. Employee in National Park Service; Director of Bureau of Outdoor Recreation, Pacific Northwest Region. Correspondence (1961-71) relating to work with Olympic National Park, Great Smoky Mountains National Park, and Bureau of Outdoor Recreation; correspondence, photographs, news clippings, reports, speeches and writings, and legislation relating to North Cascades National Park, Alaska Pipeline, and boundary changes in Olympic National Park. Inventory.

3166 **Pacific Coast Oyster Growers' Association,** Records, 1930-1964, 26 feet. Correspondence, committee files, legislative materials, reports and clippings. Includes the activities of the pollution committee, and the Hemlock looper spraying project, Willapa Harbor, Washington; also the report on the Industrial Waste Conference, Corvallis, Oregon, 1963. Inventory.

3167 **Pacific Lumber Inspection Bureau,** Records, 1903-1954, 12 feet. Correspondence and financial records.

3168 **Pacific Northwest Loggers Association,** Records, ca. 1917-1962, 150 feet. Articles of incorporation and bylaws, minutes, correspondence, weekly sales reports, log inventories, logging cost reports, bulletins, ephemera. Inventory.

3169 **Parker, Edwin S.,** Papers, 1893-1955, 5 inches. Manuscripts, Xeroxed notes from the book *Timber* (1893-1916), and historical material on Seattle City Light (1917-55).

3170 **Pearce, J. Kenneth,** Papers, 1921-1965, 7 feet. Professor of Logging Engineering, University of Washington College of Forest Resources. Major correspondents include the Society of American Foresters. Typed description, 5 pp.

3171 **Plumb, Herbert L.** (b. 1889), Papers, 1918-1970, 2 feet; Microfilm, 1 reel. Diaries, letters, ephemera, relating to his career in the U.S. Forest Service, 1918-50. The diaries are on microfilm. Places include the Snoqualmie National Forest and Olympic National Park. Additional material relates to his work as a consulting forester after 1950. 2 pp. typed description.

3172 **Plummer, George,** Correspondence, 1863-1865, 13 items. Relates to forest industry in Washington state.

3173 **Pope and Talbot,** Records, 1875-1951, 137 feet. Correspondence, financial records, reports, journals, ledgers, minutes, and related material, concerning a lumber industry firm based in San Francisco, with many operations in the Pacific Northwest; and records of some of its subsidiaries, including the Admiralty Logging Company, the Admiralty Tug Boat Company, the Broadmoor Maintenance Commission (a municipal corporation of Seattle, Washington), Brown's Bay Logging Company, Hall Brothers Marine Railroad and Shipping Company, Hayden Company (loan and investment company, Seattle), Industrial Company (investment company, Seattle), McCormick Lumber Company, Puget Mill Company, Puget Sound Commercial Company, Puget Sound Mercantile Company, Puget Sound Towage Company, Puget Sound Tugboat Company, Union River Logging Railroad Company, and the Walker Estate Company (real estate firm, Seattle). Inventory.

3174 **Port Blakely Mill Company** (Port Blakely, Washington), Papers, 1875-1923, 84 feet. Records of a sawmill founded by William Renton and John A. and James Campbell, 1864. Included are incoming letters (ca. 40,000 items), letterpress books (49 vols.), and unbound outgoing letters (ca. 2,000 items); documents and business records (36 feet); papers of affiliated companies (6 feet); and private correspondence of the executives (1 carton). The records are most complete for the period 1876-1902, when the Campbells sold out to a group in San Francisco headed by David E. Skinner and John and James Eddy. Inventory.

3175 **Port Townsend Cargoes,** Record book, 1864-1865, 1 vol. Journal kept by James S. Woodman, including entries for shipping of forest products, Washington state.

3176 **Pracna, Arthur B.,** Papers, 1904-1921, 5 inches. Correspondence, photographs, documents, journal, of a lumber mill designer of Everett, Washington.

3177 **Price, William Howard,** Papers, 1955-1969, 18 inches. Correspondence, reports, pamphlets, photographs, from Price's forestry consulting work (ca. 1955-59). Inventory.

3178 **Pulsipher, Catherine Savage,** Letters, 1960. Letters reminiscing about the logging activities of her father (George Savage) along the Skagit River, Washington.

3179 **Redfern, Donald** (1916-1965), Papers, 1944-1965, 14 feet. Correspondence and related items regarding Century 21, Seattle Area Industrial Council, State Air Pollution Control Board, American-Marietta Company, and other organizational work. Contains material relating to the forest products industry.

3180 **Reed, Mark Edward** (1866-1933), Papers, 1918-1942, 5 feet. Lumberman, financier, politician, and Washington state representative. Chiefly correspondence, much of which relates to the operation of Simpson Logging Company (of which Reed was president), the Phoenix Logging Company, the Reed Mill Company, and banks which he directed. Correspondents include Charles E. Clise, James J. Donovan, Marion E. Hay, Alex Polson, Edgar A. Sims, and the American Legion Post 31 (Shelton, Washington), the Columbia Basin Irrigation League, and the Republican Party. Inventory.

3181 **Roberts, Arthur Kitchel** (b. 1895), Papers, 1945-1962, 3½ feet. Promoter, advertiser, and researcher in lumber, of Oregon and Washington. Subject files. Names represented include American Forest Products Industries, Crown Zellerbach Corporation, the Forest History Foundation, Stewart Hall Holbrook, the Industrial Forestry Association, the Izaak Walton League, the National Lumber Manufacturers Association, Pacific Northwest Forest Industries, the Society of American Foresters, the West Coast Lumbermen's Association, the Western Pine Association, and Weyerhaeuser Company. Inventory.

3182 **Robinson, Dwight,** Papers, 1954, 7 inches. Correspondence, minutes, briefs, exhibits, notes, and ephemera relating to the Pacific Northwest Lumber Strike, 1954.

3183 **Ruegnitz, William C.,** Correspondence, 1917-1944, 1½ feet. Correspondence relating to two associations which consist primarily of employers concerned with labor relations in the forest industries of the Pacific Northwest: the Loyal Legion of Loggers and Lumbermen, of which Ruegnitz was president, and the Columbia Basin Loggers, Portland, Oregon, of which he was manager. Chief correspondents are the Weyerhaeuser Timber Company, the St. Paul and Tacoma Lumber Company, and other firms in the industry. Inventory.

3184 **Rust, William Ross** (1850-1928), Papers, 1879-1938, 6 feet. Correspondence, financial records, legal documents, notebooks, and subject files. Among the subject files are materials on the Shaffer Pulp Company. Inventory.

3185 **St. Paul and Tacoma Lumber Company,** Records, 1876-1958, 504 feet. In part, microfilm (2 reels). Correspondence, financial records, reports, journals, ledgers, printed material, and related papers of the firm and its subsidiaries including Cascade Timber Company, Chehalis & Pacific Land Company, Consolidated Lumber Company (Los Angeles), Griggs and Company (grocers, St. Paul, Minnesota), Griggs and Foster (investment firm, St. Paul), Griggs and Johnson (real estate and loans, St. Paul), Interlaken Water Company, Natches Pass Railway Company (Tacoma), Pacific Coal and Lumber Company (Tacoma), Pacific Meat Company (Tacoma), Puget Sound Dry Dock and and Machinery Company, Riverside Land Company (Tacoma), Tacoma Bituminous Paving Company, Tacoma Land and Improvement Company, Union Stockyards Company (Tacoma), and Wilkeson Coal and Coke Company (Pierce County, Washington). Inventory.

3186 **Satsop Railroad Co.,** Journal, 1884-1888. Relates to forest products industry in Washington state; the Simpson Timber Co.

3187 **Savage, George Milton** (b. 1844), Reminiscences, 1916, 5 folders. Pencilled notebooks of an early settler in Skagit Valley, relating to the forest products industry.

3188 **Schafer Bros. Logging Company,** Film, 1 reel. Film, "The Schafer Story," narrated by Stewart Holbrook.

3189 **Schmitz, Henry** (1892-1965), Papers, 1925-1965, 11 feet. Papers include personal correspondence; official papers as Dean of School of Agriculture, Forestry and Home Economics at the University of Minnesota; official papers as President of the University of Washington; files relating to Fifth World Forestry Conference, Seattle, 1960; American Forestry Association, American Forest History Foundation, Society of American Foresters, Forest History Society, and University of Washington College of Forestry.

3190 **Schmoe, Floyd W.** (b. 1895), Papers, ca. 1942-1964, 7 inches. Correspondence, notes, ephemera, clippings, manuscripts, including manuscripts of his *Year in Paradise*, Harper, 1959, and *For Love of Some Islands*, Harper, 1964. *Year in Paradise* describes Schmoe's experiences as a ranger in Mount Rainier National Park.

3191 **Seattle Lighting Department,** Records, 1899-1963, 322 feet. Papers of the Seattle City Light, including the superintendency of James Delmage Ross, and material on natural resources conservation, flood control, automobile mileage control, and multiple use of lands. Index and inventory. Restricted.

3192 **Semple, Eugene** (1840-1908), Papers, 1865-1907. Oregon and Washington attorney, farmer, lumberman, editor, and politician. The papers document his career, including his governorship of Washington Territory, 1887-89. Among the incoming correspondence are letters (1895-1900, 24 items) from the Lucia Mill Company of Vancouver, Washington, a lumber firm which Semple owned. There is also a group of Lucia Mill Company records, including agreements (1883-85), correspondence from Brigham (A.J.) Huston Co. of Vancouver (1893-99, 50 items), a letterbook (1884, 300 pp.), and a ledger (1884, 218 pp.). The collection is described in *The Eugene Semple Papers,* University of Washington Libraries, Manuscript Series Number 5, n.p.: 1967, 18 pp.

3193 **Slipper, John H.,** Letters, 1908-1911, 1 vol. Letterpress copybook of letters and accounts. Slipper was the owner of the Eagle Shingle Company.

3194 **Society of American Foresters, Puget Sound Section,** Records, 1935-1967, 13 feet. Correspondence, committee reports, minutes, ephemera. Inventory, typed, 4 pp.

3195 **Spada, John,** Papers, 1919-1963, ca. 500 items and microfilm (1 reel). Correspondence and related material, in part pertaining to the Washington Forest Products Cooperative (1947, 1951, 1962). Typed inventory, 33 pp.

3196 **Stevens, James Floyd** (1892-1971), Papers, 1916-1966, 20 feet. Author and public relations counsel. Correspondence, photos, clippings and scrapbooks, notes, and other papers relating to Stevens' activities as a free-lance writer, as public relations counsel of the West Coast Lumbermen's Association, as a member of the Pacific Northwest Writers' Conference, and as trustee of the Washington State Forestry Conference. Includes files of literary manuscripts, files of syndicated newspaper column, "Out of the Woods," notes for a projected poem, "American Glory," news releases, articles, and scripts for the Mason County, Washington, Forest Festival and for the National Broadcasting Company's National Farm and Home Hour. Correspondents include John Main Coffee, Stewart Holbrook, Hubert Humphrey, Henry M. Jackson, Alfred A. Knopf, Blanche Knopf, Sinclair Lewis, Charles L. McNary, Donald Hammer Magnuson, Warren G. Magnuson, James Marshall, H.L. Mencken, Richard L. Neuberger, Carl Sandburg, Harry Leon Wilson, *American Forests,* Crown Zellerbach, Inc., Doubleday and Company, Forest History Society, Simpson Logging Company, Society of American Foresters, West Coast Lumbermen's Association, and the Weyerhaeuser Company. Inventory.

3197 **Stimson, Thomas Douglas** (1828-1898), Records, 1871-1913, 50 feet. Correspondence, financial records, including records of the T.D. Stimson Company in Michigan, family business and correspondence, and some records of Stimson Land Company, Stimson and Clark Manufacturing Company, and Stimson, Fay and Company.

3198 **Stimson, Thomas Douglas, Jr.** (b. 1884), Manuscript, 1910-1929, 3 folders. "A record of the family, life, and activities of Charles Douglas Stimson, compiled by Thomas Douglas Stimson, Jr., ca. 1910-1929." Photocopy of the original held by Mrs. Thomas D. Stimson.

3199 **Stimson Mill Company** (Seattle), Records, 1880-1957, 33 feet. Correspondence, financial records, reports on timber and labor activities, freight records, specifications for buildings and equipment, legal documents, minutes of meetings, ledgers, and other papers, relating to the lumber industry, railroads and banking in Washington. Includes papers of and relating to Charles D., Fred S., and William H. Stimson, the Charles D. Stimson family, Stimson and Fay Company, Ballard Logging Company, Clear Lake Shingle Mill, General Insurance Company, Ives Investment Company, J.F. Ives family, Manufacturers Water Company of Ballard, Marysville and Northern Railway Company, Metropolitan Building Company, Second Holding Corporation, and the West Coast Wood Agency. Inventory.

3200 **Suzzallo, Henry** (1875-1933), Papers, 1903-1940, 33 feet (ca. 20,000 items). Educator. Correspondence and other material dealing in part with his work toward adoption of better working and living conditions for loggers in the lumber industry. Represented in the collection are James Hutchings Baker, Newton Diehl Baker, William Edward Boeing, William Edgar Borah, James H. Brady, Erastus Brainerd, Robert Somers Brookings, Nicholas Murray Butler, Samuel Paul Capen, Lotus Delta Coffman, Elwood Patterson Cubberly, Clarence Cleveland Dill, Brice P. Disque, Felix Frankfurter, Samuel Gompers, Frank Pierrepont Graves, Lindley Hoag Hadley, Louis F. Hart, Roland Hill Hartley, Howard Heinz, Samuel Hill, Ernest O. Holland, Herbert Clark Hoover, Hiram Warren Johnson, Wesley Livsey Jones, Charles Hubbard Judd, William Henry Langdon, Ernest Lister, Percival Lowell, Loyal Legion of Loggers and Lumbermen, Ruth Karr McKee, John Franklin Miller, Winlock William Miller, Robert Moran, John Flesher Newsom, Charleton Hubbell Parker, Miles Poindexter, Henry Smith Pritchett, Mark Edward Reed, Theodore Roosevelt, Lewis Baxter Schwellenbach, Rexford Guy Tugwell, the Washington State Federation of Labor, Benjamin Ide Wheeler, Ray Lyman Wilbur, and Park Weed Willis. Inventory and index. Restricted.

3201 **Sylvester, Albert H.,** Papers, 1897-1944, 5 inches. Includes photographs from his work as a topographer with the U.S. Geological Survey, 1895-1905, in the Sierra Nevada and Cascades; photographs from his career as supervisor of the Wenatchee National Forest; an account of the origin of place names in the Icicle Creek drainage; and a manuscript narrative of his surveying experiences in the North Cascades during 1897-1902.

3202 **U.S. Civilian Conservation Corps, Company 936** (Humptulips, Washington), Papers, 1933-1941, 2 inches. *C.C.C. Scarifier* (1934-35).

3203 **U.S. Forest Service,** 1966-1967, 2 inches. Construction progress reports of the Berg Construction Co. relating to the building of the Goose Cove Road on the Tongass National Forest, Alaska. Includes daily and weekly reports giving location, weather, a list of equipment and men used on the job, expenditures, and status of work in progress.

3204 **Wagner, George Corydon** (b. 1895), Papers, 1895-1959, 60 feet. Lumber executive, of the Pacific Northwest. Correspondence, notices, minutes, financial records, reports, notes, news releases, speeches, photos, clippings and ephemera. Subgroups of papers include American Forest Products Industries, Inc., Chamber of Commerce of the U.S., Forest History Foundation, Inc., Forest Industries Council, Joint Committee on Forest Conservation, Merrill and Ring Lumber Company, National Lumber Manufacturers Association, the Republican Party, the St. Paul and Tacoma Lumber Company, Timber Engineering Company (Washington, D.C.), West Coast Lumbermen's Association, Western Forestry and Conservation Association, the Winthrop Hotel (Tacoma, Washington), and Yale University. Correspondents include John W. Blodgett, Jr., Prentice Bloedel, Homer T. Bone, Harry P. Cain, John M. Coffee, Arthur B. Comfort, Arthur B. Langlie, Warren G. Magnuson, Thomas Pelly, Albert D. Rosellini, Mark Tobey, and Jack Westland. Inventory.

3205 **Warth, John,** Papers, 1955-1971, microfilm (2 reels). Personal papers concerned primarily with the North Cascades, including correspondence with the U.S. Forest Service, National Park Service, Sierra Club, Mountaineers, Federation of Western Outdoor Clubs, etc.; also notes, reports, writings, ephemera; material relating to Salmon La Sac, Baker Lake, Packwood Lake, watersheds. Originals in Warth's possession. Inventory, 6 pp., typed.

3206 **Washington Environmental Council,** Records, 1969-1973, 9 feet. Primarily subject files and legislative drafts from office of the comprehensive environmental organization in Washington state.

3207 **Washington Forest Protection Association,** Papers, 1917-1967, 18 feet. Fire reports.

3208 **Washington Mill Company** (Seabeck, Washington), Records, 1857-1890, 6 feet. Correspondence, bills, business records, documents, private papers of the executives of the company, scrapbook, and records of companies affiliated with the Washington Mill Co. The papers reflect the economic conditions, and political and moral climate of Washington Territory. Persons and organizations mentioned include William J. Adams; W.J. Adams, Blinn, and Co., San Francisco; William E. Barnard; Marshall Blinn; Samuel Blinn; Thomas Burke; Charles Miner Bradshaw; Francis A. Chenoweth; Wa Chong & Co., Seattle; James Manning Colman; B.F. Dennison; Arthur Armstrong Denny; James Douglas; Elwood Evans; D.B. Finch; Henry C. Hale; Granville Owen Haller; Cornelius Holgate Hanford; L.B. Hastings; Richard Holyoke; Dexter Horton; Paul K. Hubbs; Daniel Bachelder Jackson; Orange Jacobs; William Jameson; Henry Landes; John Jay McGilvra; Angus MacKintosh; John McReavy; Ezra M. Meeker; George Anson Meigs; Thomas T. Minor; Puget Mill Co., Teekalet (Port Gamble), Washington; F.C. Purdy; D.C.H. Rothchild & Co., Port Townsend, Washington; William P. Sayward; James Seavey; William B. Sinclair; U.S. Indian Agent, Neah Bay; U.S. Indian Agent, Skokomish; Cyrus Walker; Waterman & Katz, Port Townsend, Washington; and E.A. Wilson & & Co., Union City, Washington. The collection is described in *Washington Mill Company Papers, 1857-1888, at the University of Washington Libraries*, pamphlet accompanying National Historical Publications Commission Microfilm Publication, n.p.: n.d., 4 pp.

3209 **Washington Natural Resources Association,** Minutes, 1927-1929.

3210 **Washington State Conservation Society,** Records, 1925-1959, 3½ feet. Correspondence, speeches, writings, financial records, organizational records, minutes, reports, miscellaneous, photographs, clippings, ephemera and other printed material, relating to a variety of conservation problems in Washington state. Inventory, typed, 5 pp., with list of correspondents.

3211 **Washington State Forestry Conference,** Minute book, 1929-1954, 1 vol.

2312 **Washington, University. Arboretum,** Records, 1945-1968, 25½ feet. Correspondence, reports, bulletins, proposals, maps, and blueprints.

3213 **Washington, University. College of Forest Resources,** Records, 1908-70, ca. 52 feet. Correspondence, pertaining to forest research in Washington state; subject files pertaining to forest resources and the forest products industry; subject files relating to management of Charles Lathrop Pack Demonstration Forest; records of State Forestry Conference (1921-38); Forest Products Commission Biennial Reports; annual reports of the College (1933-70); scrapbook of newspaper clippings pertaining to the College (1916-29).

3214 **Washington, University. College of Forest Resources, Institute of Forest Products,** Records, 1960-66, 5 feet. Institute correspondence; committee reports; financial records.

3215 **Westland, Alfred John** (b. 1904), Papers, 1952-1964, 6 feet. U.S. Representative from Washington, 1953-65; served on the House Interior and Insular Affairs Committee and the U.S. National Forest Reservation Commission. Subject files include material on Bonneville Power, Hells Canyon, lumbering, national parks and forest legislation. Inventory, typed, 6 pp., and list of correspondents.

3216 **Weyerhaeuser Timber Company,** Oral History Transcript, 1954, 98 pp. Interview conducted by Ralph W. Hidy and used for *Timber and Men,* by Hidy, Frank Hill, and Allan Nevins, New York: Macmillan, 1963. Subjects include Clyde S. Martin, C.D. Weyerhaeuser, and E.F. Heacox.

3217 **Williams, Theodore O.,** Letter, 1904, 1 item. Typescript of letter to Albert Wist, Seattle (March 13, 1904). Relates to the forest products industry of Washington state.

3218 **Wimmer, Thomas O.,** Papers, 1947-1970, 9 feet. Seattle conservationist; president of the Washington State Sportsmen's Council, Washington Environmental Council, and Seattle Audubon Society; member, Washington Pollution Control Commission. Correspondence, subject files, legislative materials related to wildlife, water pollution, and other environmental issues in Washington. Includes legislative materials on the Washington Forest Practices Act, 1973.

3219 **Winkenwerder, Hugo,** Papers, 1934-1942, 1 foot. Dean of the College of Forestry, University of Washington. The papers relate to his work with the Society of American Foresters.

3220 **Yesler, Denny and Company** (Seattle, Washington), Papers, 1861-1876. Records relating to a sawmill, including letters from Henry L. Yesler to his partner, Captain George Plummer in San Francisco (41 items), and from Plummer to Yesler (2 items); merchandise lists (24 items); and lottery tickets (1876) offering his mill as grand prize.

SPOKANE

Eastern Washington State Historical Society

Archivist, West 2316 First Avenue 99204 Telephone: (509) 456-3931. In addition to the items cited below, there are scattered references to logging and lumbering in a number of reminiscences and autobiographies.

3221 **Borgh, Avis Larson.** "With Swedish Accent: Family History of Borgh and Larson Families," 15 pp., mimeographed. Relates to logging camp conditions in Seattle area, 1910-25.

3222 **Hay, Marion E.** (1865-1933), Papers, 1909-1916, 15 feet. Governor of Washington. The collection includes material relating to the Ballinger-Pinchot controversy; and to the Washington State Good Roads Association. Index.

3223 **LaSota, Felix,** "Autobiographical Sketch," 7 pp., Xerox copy of typed manuscript. Includes brief statements of experiences in Wisconsin logging camps in the 1890s, and in eastern Washington in 1908-10.

TACOMA

Tacoma Public Library

1102 Tacoma Avenue South 98402 Telephone: (206) 383-1574

3224 **Boland, Marvin D.,** Collection, ca. 1911-1950, ca. 83,000 items. Photographs and photographic negatives of ships and shipping, forests and forestry, railroads and railroading, mining, commercial work, etc., in the Tacoma area.

3225 **Heritage, Clark C.** (1891-1972), Papers, 1891-1972, 30 feet. Materials cover Heritage's career as a consulting chemical engineer in the Pacific Northwest, dealing with wood and wood products. Includes letter books, diaries, correspondence, and reports of investigations. Materials relating to his work for the Weyerhaeuser Company have been removed from the collection. Heritage was concerned with developing new uses for wood products.

3226 **Northern Pacific Railway, Tacoma Division,** Papers, 1910-1964, 900 feet. Includes material relating to operations in southwestern Washington state, relating to expenses, crew reports, spur laying and abandonment, bridge tenders reports, investigations, building and track laying, etc.

3227 **Reynolds, Clinton S.** (1892-1972), Papers, 1921-1972, 10 feet. Memoranda, letters, documents, publications, correspondence, etc., relating to Reynolds' interests in Washington State Automobile Association, Washington State Good Roads Association, Washington State Highway Commission, etc.

3228 **Stapleton, Margaret** (b. 1903), Bibliography of Mount Rainier National Park, 1928, 17 pp., typescript. An attempt to list all material of importance relating to the subject in Tacoma and Seattle libraries, or noted in earlier bibliographies or in magazine and document indexes. Not annotated. This is supplemented by a more up-to-date slip file on Mount Rainier bibliography compiled by the library staff.

3229 **Tacoma Land and Improvement Company,** Records, 1873-1923, 3 feet. Includes minutes, stock certificates, ledgers, journals, of the Tacoma Land and Improvement Company of Washington, the Tacoma Land Company, and the Tacoma Land and Improvement Company of New Jersey.

3230 **Wilhelm, Honor** (1870-1957), Collection, 3 feet. Among the more than 1000 photographs, negatives, and postcards, many of which were used to illustrate the *Coast Magazine* of Seattle, which Wilhelm edited, are scenes of lumbering.

Washington State Historical Society

Archivist, 315 North Stadium Way 98403

3231 **Chambers, Thomas McCutcheon** (1795-1876), Papers, 1838-1876, ca. 450 items. Lumber and flour mill owner, and local official, Lewis County, Oregon Territory. The papers include correspondence, legal papers, receipts, bills, and circulars relating to the flour and lumber business. Described in the *Inventory of the Thomas M. Chambers Collection* (1972).

3232 **Chaplin, Ralph** (1887-1961), Papers, 1875-1960, ca. 1750 items and 28 vols. Publicist for the Industrial Workers of the World, editor of the union newspaper, *Solidarity.* The collection includes correspondence, manuscripts of writings, including an autobiography, *Wobbly: The Rough and Tumble Story of an American Radical* (1948); also proofs, photos, printed matter, cartoons, and ephemera, relating to the labor movement in the Pacific Northwest. *Inventory of the Ralph Chaplin Collection* (1967), 30 pp.

3233 **Chittenden, Hiram Martin** (1858-1917), Papers, 1876-1962, 316 items. Army officer, engineer, and historian. The papers include correspondence, diaries, journals, scrapbooks, pamphlets, photos, sketches, clippings, diplomas, commissions, and other material used in the publication of books by and about Chittenden. Relates chiefly to Chittenden's engineering achievements in Yellowstone National Park, early efforts toward flood control in the Missouri Valley, and the building of the Port of Seattle. Also includes material on Chittenden's pioneer work in the history of the fur trade. Inventory.

3234 **Hanson, Howard** (1876-1957), Papers, 1899-1957, ca. 4000 items and 22 vols. Lawyer and public official of Seattle and King County, Washington, and state legislator. Correspondence, reports, minutes, speeches, articles, printed material, photos, maps, clippings, drafts, and other documents, relating to Hanson's activities in the development of Seattle and his advocacy of conservation and flood control in Washington. *Inventory of the Howard Hanson Collection* (1970).

3235 **Logging, Lumber, and Forest Photographs,** ca. 1870-present, 5 cubic feet. Photographs and photo albums (compiled by both John Cress and Asahel Curtis) covering aspects of logging and the lumber industry from oxen logging to use of the helicopter; mills; railroads; general forest scenes; reforestation; forest fires. Cataloged and indexed.

3236 **Mentzer Brothers Lumber Company,** ca. 1889-1930s. 10 cubic feet. Earlier known as the Western Lumber Company, and later as Coenen and Mentzer. There were mills located in both Tacoma and Tenino. Included are books, catalogs, and pamphlets associated with the lumber industry; business records and correspondence, real estate transactions, and legal proceedings concerned with the Mentzer vs. Mentzer case. Inventory.

3237 **Slater, Harold** (1894-1959), Papers, 1937-1959, ca. 1000 items. Labor official, a founder of the International Woodworkers of America, and Secretary-treasurer of the Washington State Labor Council. Correspondence, legal papers, minutes, proceedings, reports, press releases, financial statements, manuscripts, printed material, and organizational papers, relating to the activities of organized labor in Washington state in the 1940s and 1950s, especially to the attempts of the CIO to organize the lumber industry and influence political action in the state. Correspondents include George Adams, Francis Artz, Claud Ballard, Daniel Bandmann, George Brown, J.M. Bullock, Chester Dusten, E.J. Flanagan, Adolph Germer, Peter Giovine, Bernard Griffin, Alan Haywood, Eric Johnston, Ray Kroeger, Worth Lowery, Michael McMahon, Philip Murray, Robert Ollinger, Walter Reuther, Agnew Lumber Company, Castle Rock Lumber Company, Hart Mill Company, Llewellyn Logging Company, Mayfield Winston Company, National Labor Relations Board, National War Labor Board, Nettleton Timber Company, U.S. Office of Price Administration, Packwood Lumber Company, Simpson Logging Company, United States Plywood Corporation, Weyerhaeuser Timber Company. Inventory.

3238 **Tacoma Mill Company,** Records, 1855-1892. Account and shipping records (1855, 1869, 1885, 1892); ledgers (1873-75, 1877-79, 1882).

3239 **Thomas, Norman,** Papers, 1920-1940, 5 cubic feet. Materials gathered by Thomas when compiling information on the Weyerhaeuser Timber Company. Uninventoried.

Weyerhaeuser Company, Historical Archives

Director, Historical Archives 98401 Telephone: (206) 924-2345. Records are open to research by bona fide scholars.

3240 **Records,** 1900-1970, 530 cubic feet. Included are correspondence and other records of the Weyerhaeuser Company and subsidiaries, and some articles, artifacts, photographs, and oral history interviews. Additions are made continually. Restrictions prevail on some specific material.

West Virginia

HARPERS FERRY

Appalachian Trail Conference

Special Projects, P.O. Box 236 25425
Telephone: (304) 535-6331

3241 **Appalachian Trail Conference Archives,** 1921-1975, ca. 140 boxes. Holdings include printed newsletters, guidebooks; proceedings of annual conferences; correspondence relating to National Trails System legislation; scrapbooks; Trailway Agreements of 1938; photographs and slides. There is material relating to the U.S. Forest Service, National Park Service, multiple-use, trail maintenance and rights-of-way. Correspondents include Myron H. Avery and Stanley A. Murray, chairmen of the conference; Daniel Hoch, U.S. Representative from Pennsylvania; and Gaylord Nelson, U.S. Senator from Wisconsin.

National Park Service Harpers Ferry Center, National Park Service Archives

Staff Curator 25425

3242 Holdings include ca. 400 taped oral history interviews (1962-74) with present and past employees of the National Park Service, about half of them transcribed, all dealing with aspects of conservation and forest recreation, 1872 to the present. Subjects include the National Park Service, the Civilian Conservation Corps and other relief projects, historic preservation, scenic parkways, interpretative planning and techniques, and other aspects of national park programs. Among major interviewees are Horace M. Albright, Conrad L. Wirth, Freeman Tilden, Louis Cramton, and Ronald F. Lee. Restricted: may be quoted only with permission of the Director, Harpers Ferry Center. There is also a large accompanying collection of printed and processed National Park Service documents and ephemeral materials. Copies of some of the oral history transcripts have been deposited at the Conservation Library of the Denver Public Library.

MORGANTOWN

West Virginia University Library, West Virginia Collection

26505. Major holdings are described in James W. Hess, *Guide to Manuscripts and Archives in the West Virginia Collection,* Morgantown: West Virginia University Library, 1974.

3243 **Account books,** 1795-1956, 3 feet and 106 vols. Includes account book of John N. Waters (1811-92), a farmer and operator of a gristmill and sawmill at Jamestown, Monongalia County (1841-64, 1 vol.).

3244 **Bishop, Charles Mortimer** (1827-1896), Papers, 1857-1897, 2 boxes. Preston County, West Virginia; farmer, merchant, preacher, and politician. Papers include material relating to the Rowlesburg Lumber and Iron Company; stave making; and other matters.

3245 **Brown, David Dare,** Notebooks, ca. 1900-1957, 10 vols. Notebooks contain correspondence, photographs, and manuscript histories of lumbering operations and sawmills in the forest regions of central West Virginia, compiled by Brown for a projected study on lumbering in West Virginia.

3246 **Camden, Gideon D.** (1805-1891), Papers, 1785-1958, 38 feet, 4 folders, and 13 items. Correspondence, and business and legal papers of a Harrison County, West Virginia, lawyer and politician. Includes material on his extensive interests in land, timber, coal, and oil exploitations. The collection also includes the papers of Camden's grandson, Wilson Lee Camden (1870-1956), which deal with the Camden estate and its vast coal, timber, land, railroad, and oil interests in West Virginia and Pennsylvania.

3247 **Camden, Johnson Newlon** (1828-1908), Papers, 1845-1908, 51 feet. U.S. Senator from West Virginia, 1881-87, 1893-95. Correspondence, maps, business records and other papers, of Camden, promoter of the oil industry, railroads, and coal and timber resources of West Virginia.

3248 **Collins, Justus** (1856-1934), Papers, 1896-1934, 56 boxes and 2 reels microfilm. Collins' business activities included speculation in timber lands in West Virginia; the collection includes material on the lumber business.

3249 **Courtney Family,** Papers, 1804-1920, ca. 200 items. Papers of a Monongalia County farm family residing near Maidsville include diary of Ulysses J. Courtney (1878-83, 7 vols.), pertaining in part to lumbering; a record of lumbering operations (1878); and other records.

3250 **Crogan, Patrick J.** (1856-1949), Papers, ca. 1900-1930, 34 feet. Preston County, West Virginia. Includes material on cases tried by attorney Crogan for lumber companies.

3251 **Davis, Henry Gassaway** (1823-1916), Papers, 1865-1916, 86 feet. Businessman; politician; U.S. Senator, 1871-83. There is material on Davis' interests in lumbering.

3252 **Deakins Family,** Papers, 1778-1925, 2 feet. There is material on the Preston Lumber and Coal Company, and correspondence between George S. Deakins and the Hancock Cooperage Company, Hancock, Maryland, relative to working Deakins' timber in Preston County.

3253 **Elkins, Stephen B.** (1841-1911), Papers, 1861-1946, 8 feet. U.S. Senator, 1895-1911. Included are papers relating to railroads, coal and silver mining, and lumbering in West Virginia, New Mexico, Texas, and California.

3254 **Fayette County (West Virginia),** Archives, 1787-1923, 19 feet. The archives include the accounts of the Gauley Tie and Lumber Company (1887-92).

3255 **Fleming Family,** Papers, 1810-1943, 6 feet. Business and personal papers and other records relating to various business enterprises carried on by the family in Fairmont, including hardwood sales.

3256 **Flynn Lumber Company,** Map and photographs, 4 items, 1945. The map is a section of a U.S. Geological Survey map showing a portion of the holdings of the Flynn Lumber Company in Nicholas County, West Virginia. The photos show railroad cars loaded with coal mined from holdings of the Flynn Lumber Company.

3257 **Frankenberry, Allen D.,** Diaries and manuscript, 1861-1870, 1 folder and 1 reel microfilm. The diaries include material concerning cutting, sawing, and selling timber in and around Point Marion, Pennsylvania. Filmed from originals in the possession of Mrs. I.L. Voorhis, Maidsville, West Virginia.

3258 **Garrison, M.J., and Company** (Wadestown, West Virginia), Records, 1858-1922, 18 vols. and 42 items in 5 boxes. These are other papers relating to lumber buying and selling.

3259 **Guseman, Jacob** (1786-1878), Records, 1836-1866, 1930, 8 vols. Ledgers and daybooks of a sawmill and other manufactories at Muddy Creek, Preston County, West Virginia.

3260 **Hall, William,** Account book and papers, 1850-1853, 1861, 3 items. These are entries for wages paid to loggers, Egypt, West Virginia (1850-53).

3261 **Hansford, Felix G.,** Papers, 1790-1876, 1 foot. Correspondence, business, legal, and land papers; subjects include Hansford's sawmill, Kanawha County.

3262 **Hart and Woolworth** (Martinsburg, West Virginia), Ledger, 1810-1820, 1 vol. Lumber mill business owned originally by Stephen Hart and Reuben Woolworth.

3263 **Hendricks (West Virginia),** Business records, 1897-1944, 25 feet. Correspondence, business papers and account books of several companies in the town of Hendricks, including Hendricks Lumber Co.

3264 **Hoffman, John Stringer** (1821-1877), Papers, 1856-1892, 1 foot. Lawyer and judge, Clarksburg, West Virginia. Legal correspondence relating to land title litigation includes material on timber lands on Shaver's Mountain.

3265 **Holt-Keyes Families,** Papers, 1878-1944, 2 feet. Barbour County, West Virginia. Included are business records of the Keyes Lumber Company, and records relating to building supplies, mine post sales, etc.

3266 **Hopkins, Andrew Delmar** (1857-1948), Papers, 1873-1943, 2 feet. Vice-director of the West Virginia Agricultural Experiment Station and professor of entomology at West Virginia University, 1890-1902; forest entomologist with the United States Department of Agriculture, 1902-1923; etc. The papers include correspondence, reports, clippings, project notes, and other materials.

3267 **Jackson, William A.,** Papers, 1892-1930, 2 reels microfilm. Papers and records of various businesses in Jane Lew, West Virginia, including the sale of lumber by W.W. Smith & Co.

3268 **Johnson, George W.** (1837-1902), Papers, 1860-1960, 5 feet. Diaries, correspondence, notes, receipts, and newspaper clippings of a lumberman and stockman from Morgantown, West Virginia. Subjects include Johnson's extensive lumber dealings in Monongalia County, and the rafting of logs to Pittsburgh and intermediate points on the Monongahela River.

3269 **Jones, Clement Ross** (1871-1939), Papers, ca. 1890-1958, 7 feet. Papers of a genealogist and former dean of the College of Engineering, West Virginia University, 1911-32, including family correspondence, university records, business and legal papers, clippings, and genealogical studies and correspondence. Subjects include the Athens Lumber Company (1921-32).

3270 **Kanawha County,** Archives, 1779-1933, 152 feet. Includes, among other records, the records of a sawmill (1879).

3271 **Lewis Family,** Papers, 1825-1936, 8 feet. Personal and business papers of the Lewis family, mainly of John D. Lewis (1800-82), and Charles C., Sr. (b. 1839), and Jr. (b. 1865), of Kanawha County, West Virginia. Includes papers relating to lumbering.

3272 **Longacre, Mrs. Glenn V.** (Collector), Papers, 1814-1915, 27 items. Includes two photographs of lumbering activities near Holly River, West Virginia; a short history of the Webster Hardwood Lumber Company at Dixie, West Virginia.

3273 **Lumber Corporations, Tucker County (West Virginia),** Minutes, 1884-1910, 3 vols. Minute books of the J.L. Rumbarger Lumber Company (1884-1910), the Condon-Lane Boom and Lumber Company (1892-1910), and the Dry Fork Lumber Company (1904-10). These three companies had their central office in Philadelphia and were under the financial control of R.F. Whitmer. The collection also includes typescripts pertaining to the Dry Fork Lumber Company, William Whitmer & Sons, Parsons Pulp & Paper Company, J.L. Rumbarger Lumber Company, Condon-Lane Boom and Lumber Company, and the Holstein Lumber Company.

3274 **Lynch, John R.,** Papers, 1908-1924, 11 items. Includes a notebook containing measurements of lumber, a memo book of logs sold, and a tally book of rafts floated down the Little Kanawha River.

3275 **McGraw, John Thomas** (1856-1920), Papers, 1842-1948, 4 feet. Includes records relating to the purchase, sale, and development of timber lands.

3276 **Marshall, Jacob Williamson** (b. 1830), Papers, 1852-1899, 1 foot. The papers of a livestock broker, farmer, and merchant of Mingo Flats, who was associated with John T. McGraw in the development of Marlintown and the purchase and sale of land, coal, and timber in Pocahontas County, West Virginia.

3277 **Maxwell Family,** Papers, ca. 1845-1950, 9 feet. Papers of Hu Maxwell (1860-1927), historian, editor, and author of several county histories of West Virginia, along with the papers and records of other family members. There are manuscripts of fiction, verse, and local history written by Maxwell, as well as a number of his manuscripts and printed compilations dealing with wood uses and forestry prepared while he was a member of the U.S. Forest Service. Included is a diary (1901-19).

3278 **Miller, George W.,** Papers, 1848-1971, 10 feet. Correspondence and financial records of a Gilmer County, West Virginia, businessman; his investments included timber. The collection includes material relating to timber sales.

3279 **Miscellany,** Papers, 1774-1960, 1 foot. Includes the articles of incorporation of the Graham-Yeager Lumber Company.

3280 **Monongalia County,** Archives, 1774-1936, 198 feet. Includes, among other records, account books for lumber firms.

3281 **Monongalia County. Clinton District,** Papers, 1796-1964, 5 items. Includes a photocopy of a manuscript, "History of Ridgedale School District," by C.G. Howell, 12 pp., which mentions grist and sawmills in the area.

3282 **Moreland, James Rogers** (1879-1955), Papers, 1809-1948, 65 feet. Includes a memorandum book (1858-65) and an account book (1849-50) of John Rogers, Morgantown, West Virginia, sawmill operator.

3283 **Morrison, Gross and Company** (Preston and Randolph Counties), Records, ca. 1917-1950, 21 feet. Includes records of the Woodford Lumber Company.

3284 **Nuttall, Lawrence William** (1857-1933), Papers, 1946-1952, 1971, 13 items. Includes information on the estate of John Nuttall (1817-97), describing the coal and timber lands in Fayette and Nicholas Counties, West Virginia, and other holdings.

3285 **Orton, Clayton Robert** (b. 1885), Papers, 1938-1955, 35 items. Office files of a former dean of the College of Agriculture, West Virginia University, including papers of the Resources Committee of the West Virginia State Planning Board (1938-45); Potomac River Basin (1945); Soil Conservation Society, West Virginia Chapter (1953-55); West Virginia Forest Council; and the Watershed Development Conference (1949).

3286 **Pardee and Curtin Lumber Company,** Records, 1889-1938, 99 items. Account books (56 vols.) of the Pardee and Curtin Lumber Company for the Cherry Run Mill, Hominy Mill, Palmer Mill and Sutton Mill. From company offices located in Curtin, Sutton, and Clarksburg, and logging operations in Nicholas, Braxton, and Webster Counties. Also photographs (43 items) of railroads, mills, logging, and coal mining operations.

3287 **Parsons, Job W.,** Diaries, 1874, 1875, 1879-1884, 1886-1888, 1893, 1894, Microfilm (2 reels [13 vols.]). Randolph County, West Virginia; farmer and lumberman. Films of pocket diaries (originals in the possession of Dickson W. Parsons and W.E. Parsons [1954]) with daily entries centering on routine farm chores, weather conditions, prices, and wages.

3288 **Pierce, Carleton Custer** (1877-1958), Papers, ca. 1840-1955, 3 boxes and 2 vols. Papers of a Kingwood, West Virginia, resident. Includes material on his timber and other business activities.

3289 **Preston County (West Virginia),** Papers, 1836-1966, ca. 95 items. Miscellaneous papers including an early map of Coopers Rock State Forest.

3290 **Preston County (West Virginia),** Papers, 1788 (1802-1916) 1943, 200 items. Account book, and legal, business, and personal papers of some early settlers at Aurora, Preston County, West Virginia, including the records of a sawmill.

3291 **Price, Calvin W.,** Tape recording and transcripts, 1956. Tape recording with two typed transcripts, of an interview with Dr. Calvin W. Price (March 12, 1956), in Marlintown, West Virginia. Subjects covered include Price family history, Pocahontas County local history, lumbering, wildlife and conservation, agriculture, and education.

3292 **Ranwood Lumber Company** (Pickens, West Virginia), Photographs, ca. 1920s, 46 items. Postcards and pictures of lumbering activities; most of the scenes were taken in Randolph and Webster Counties, West Virginia.

3293 **Rogers, John** (1776-1864), Papers, 1777-1857, 3 feet. Business and personal papers of John Rogers, merchant, land agent, and owner of grist, carding, fulling, and sawmills in Monongalia County, West Virginia.

3294 **Ruffner, Henry** (1790-1861) **and William Henry** (1824-1908), Papers, 1829-1913, 1 reel microfilm. Microfilm of the original held by the Historical Foundation of the Presbyterian and Reformed Churches, Montreat, North Carolina. Subjects of the correspondence include iron and timber.

3295 **Savage, Jesse Spillman** (1836-1917), **and William Allen** (1841-1929), Manuscript, ca. 1925, 1 item. A one-page biographical sketch typed from an original manuscript of the Savage brothers, who migrated from Maine to West Virginia in 1871 and engaged in the cattle, iron manufacturing, and lumber business.

3296 **Scott Lumber Company,** Ledger, 1911-1929, microfilm, 1 reel. Ledger containing financial statements and reports of the company's operations in Morgantown and Follansbee, and records of a subsidiary, the Cecil Lumber and Hardwood Company of Elm Grove.

3297 **Shahan, James B.,** Letter, 1861, 1 item. Letter (September 12, 1861) by Dr. E. Mead, concerning the cutting, burning, and marketing of pine timber in the Grafton, West Virginia, area.

3298 **Thompson, George Benjamin** (1870-1957), Tape recording, 1956, 2 items. Recorded interview and transcript, concerning family genealogy, Tucker County history, and lumbering operations.

3299 **Trendley, Frederick,** Records, 1841-1951, 30 items. The records include a ledger for a paper manufacturing mill at Newton Fall, Ohio, an estimate of paper production costs, a method for coloring paper, and a lease for part of the mill.

3300 **Tucker County (West Virginia),** Records, 1909-1922, 51 items. Includes volumes of business records from Red Creek of the Elkins Lumber Company and the J.C. Myers Lumber Company.

3301 **Webster County (West Virginia),** Archives, 1851-1951, 90 feet and 12 reels microfilm. Includes articles of incorporation of boom and lumber companies.

3302 **West Virginia Fibre Company,** Papers, 1884, 3 items. Bids and specifications for paper-making machinery furnished by Pusey & Jones Company, Wilmington; the Logan Iron Works, Brooklyn; and the Harris-Corliss Steam Engines, Providence.

3303 **West Virginia Pulp and Paper Co.** (Cass, West Virginia), Records, 1892-1939, microfilm, 19 reels. Subjects treated are the Blackwater Lumber Company; Condon-Lane Boom and Lumber Co.; Chesapeake and Ohio Railroad; Greenbrier River; logging; lumber agents; Martin-Lane and Company; railroads; J.L. Rumbarger Lumber Company; R.M. Sutton Company; Stearns Manufacturing Company; Saint Lawrence Boom and Manufacturing Company; S.E. Slaymaker and Company; timber lands; West Virginia Pulp and Paper Company; West Virginia Spruce Lumber Company; William Whiter and Sons, Inc. Places mentioned include Greenbrier County; Green Bank, West Virginia; Horton, West Virginia; Cass, West Virginia; Covington, Virginia; Piedmont, West Virginia; Pocahontas County; Randolph County; and Ronceverte, West Virginia. Correspondents include C.T. Boon, William A. Bratton, Joseph K. Cass, William H. Cobb, T.B. Davis, H.G. Davis, S.B. Elkins, James Gibson, H.F. Harrison, F.A. Hauck, David C. Luke, John G. Luke, Thomas Luke, William A. Luke, C.F. Moore, George L. Miller, L.M. McClintic, William Musgrave, P.I. Reed, Shelton L. Reger, H.C. Savidge, S.E. Slaymaker, E.P. Shaffer, A. Thompson, E.D. Talbott, W.S. Tolbard, R.S. Turk, Thomas M. Williamson, George F. Willis, James N. White, James A. Whiting, H. Yocum, and Harry Zinger. Originals held by Kyle Neighbors, Box 134, Cass, West Virginia. Restricted: may be consulted only with written permission from Mr. Neighbors.

3304 **West Virginia Pulp and Paper Company,** Reports, 1920, Microfilm, 1 reel. Two typescript reports on the company's timber lands in Pocahontas, Randolph, and Webster counties, including two maps and four photographs on forestry and lumbering.

3305 **Whitmer, William, and Sons, Inc.,** Minute book, 1895-1924, microfilm, 1 reel (1 vol.). Minute book of a West Virginia lumber corporation with mills and timber lands in Tucker County area. Original in possession of Fred J. Overly (1960).

3306 **Withers and VanDevender** (Parkersburg, West Virginia), Records, 1899-1951, 5 feet. Correspondence, ledger books, account and cashbooks of a firm (later known as Wiant and VanDevender), specializing in timber cutting, but also with other concerns. Subjects include timber cutting methods, costs, timber shipping, floods, freezes, droughts, logjams, timberland locations, timber purchases and sales, salaries of timber workers, etc. Includes material relating to the Nixolette Lumber Company, Parkersburg Mill Company, and other firms.

Wisconsin

ASHLAND

Northland College, Area Research Center

Curator, Dexter Library, 1411 Ellis Avenue 54806
Telephone: (715) 682-4531 (ext. 205). Holdings of Area Research Centers are described under the State Historical Society, Madison.

EAU CLAIRE

Eau Claire Public Library

Reference Librarian, 217 S. Farwell Street 54701
Telephone: (715) 832-8341. Collections are housed in a disorganized condition in a vault. Those described below are as cited in Clodaugh Neiderheiser's *Forest History Sources of the United States and Canada* (St. Paul: Forest History Foundation, Inc., 1956), pp. 113-114.

3307 **Bartlett Collection,** 1854-1938. The collection includes letters, clippings, maps, pictures, business contracts, federal and state patents, log marks, stock certificates, and records of the following companies: Alexander Stewart Lumber Company, Wausau, Wisconsin (1891-98); Badger State Lumber Company, Badger Mills, Wisconsin (1876); Bradstreet Brothers, Gardiner, Maine (1868-78); Cruikshank Lumber & Coal Company, Hannibal, Missouri (1870-1901); S.C. & S. Carter Company, Keokuk, Iowa (1890-1900); Daniel Shaw Lumber Company, Eau Claire, Wisconsin (1856-1922); Del Norte Company, California (1902-09); Dole, Ingram & Kennedy, Eau Claire, Wisconsin (1860-66); J.L. Gates Land Company, Milwaukee, Wisconsin (1899-1906); Half Moon Lake Canal Company, Eau Claire, Wisconsin (1855-1912); Horton & Hamilton, Winona, Minnesota (1866-77); Lake City Lumber Company, Lake City, Minnesota (1874-78); Montreal River Lumber Company (1891-1901); Northwestern Lumber Company, Eau Claire, Wisconsin (1869-91); Porter & Moon Lumber Company, Eau Claire, Wisconsin (1869-71); Valley Navigation Company, Wabasha, Minnesota (1873-1900); Mississippi River Logging Company, Chippewa Falls, Wisconsin (1873-1909). There are also several typewritten papers: William Warren Bartlett, "Logging Days" (n.d., 103 pp.); W.W. Bartlett, "Lumbering in the Lake States of the Northwest" (1922, 11 pp.); W.W. Bartlett, "Lumbering Story" (1932, 42 pp.); W.W. Bartlett, "Story of the Early Days in the Lumber Game" (n.d., 12 pp.); Orrin Henry Ingram, "Lumbering in Wisconsin" (n.d., 5 pp.); John C. Storlie, "Eau Claire's First Industry" (1953, 20 pp.); Laura Sutherland, "Adin Randall, Pioneer and Builder" (1950, 9 pp.); and Bruno Vinette, "Early Lumbering on the Chippewa" (n.d., 8 pp.).

3308 **Phoenix Manufacturing Company** (Eau Claire, Wisconsin), Records, 1900+, 2 boxes. Correspondence, orders, and bills of a firm which manufactured logging and lumbering machinery.

3309 **Wisconsin Lumber Inspectors,** Record books, 1861-97.

University of Wisconsin, Area Research Center

Curator, William D. McIntyre Library, 105 Garfield Avenue 54701 Telephone: (715) 836-2739. Holdings of Area Research Centers are described under the State Historical Society of Wisconsin, Madison.

GREEN BAY

University of Wisconsin, Area Research Center

Librarian, Library-Learning Center, 110 South University Circle Drive 54305 Telephone: (414) 465-2539. Holdings are described under the State Historical Society of Wisconsin, Madison.

KENOSHA

University of Wisconsin, Parkside, Area Research Center

University Archivist, Parkside Library 53140
Telephone: (414)443-2411. Holdings are described under the State Historical Society of Wisconsin, Madison.

LA CROSSE

University of Wisconsin, Area Research Center

Curator, Eugene W. Murphy Library, 1631 Pine Street 54601 Telephone: (608) 784-6050 (ext. 237). Holdings are described under the State Historical Society of Wisconsin, Madison.

MADISON

The State Historical Society of Wisconsin, Division of Archives and Manuscripts

Director, 816 State Street 53706. Some of the holdings of the State Historical Society of Wisconsin and of its various Area Research Centers are described in the following publications: *Guide to the Manuscripts of the Wisconsin Historical Society,* edited by Alice E. Smith, Madison: 1944; *Supplement Number One,* edited by Josephine L. Harper and Sharon C. Smith, 1957; *Supplement Number Two,* by Josephine L. Harper, 1966; *Labor Manuscripts in the State Historical Society of Wisconsin,* compiled by F. Gerald Ham, Madison: 1967; and *Guide to the Wisconsin State Archives,* compiled by David J. Delgado, Madison: 1966. The listing below includes the Society's relevant holdings in Madison, and in its various area research centers.

Manuscripts

3310 **American Forest History Foundation,** Oral History Interviews, 1955. Typewritten copies (87 pp.). Transcripts of interviews with Mrs. Maud Carlgren, Mrs. Hope Garlick Mineau, Wirt Mineau, Mrs. Maggie Orr O'Neill, and Hugo Schlenk discussing early lumbering activities in northern Minnesota and Wisconsin, particularly in the St. Croix Valley. Interviews were made, transcribed, and edited by Helen McCann White. This organization is now the Forest History Society, Inc., Santa Cruz, California.

3311 **American Fur Company,** Papers, 1803-1866, 19 vols. and 42 reels of microfilm. The papers include a combination ledger and sales book (1837-47), probably kept at Prairie du Chien, Wisconsin, of the accounts of the company's sawmill at Chippewa Falls, Wisconsin, with a record of sales, and a few additional records of lumber shipped by raft and steamboat.

3312 **Ames, Jesse H.** (b. 1875), Autobiography, 1 item. Typewritten copy (59 pp.). Reminiscences of the president of the Wisconsin State College at River Falls, 1917-46, describing the settlement of his parents and grandparents in Wisconsin, pioneer life and lumbering experiences in the vicinity of Shiocton and Shawano, and his professional career.

3313 **Andrews, James A., and John Comstock** (1812-1890), Papers, 1837-1934, 2 boxes. Comstock was a state legislator from Hudson, Wisconsin, 1861. The collection includes some material on railroads, lumbering, and politics in northeastern Wisconsin. Correspondence with William Wilson, of Knapp, Stout, and Company, Menomonie, Wisconsin, and with Joseph G. Thorp of the Eau Claire Lumber Company, discusses the legislative measures needed by the lumber industry in the 1860s. The collection is in the River Falls Area Research Center.

3314 **Arpin, Edmund P., Jr.** (b. 1894), Papers, 1920-1921, 1956, 1 item, and microfilm, 1 reel. The collection includes a letter (1956) containing Arpin's recollections of "Indian Jeff," an old-time lumberman on the Wisconsin River. Portions of this letter were published in the *Wisconsin Magazine of History,* XL, Autumn, 1956, p. 30.

3315 **Babbitt, T.T.** "Comparison of the lumber industry of Wisconsin with that of the Northwest." Typewritten copy.

3316 **Babcock, Joseph Weeks** (1850-1909), Papers, 1864-1916, 1 box. Lumberman and U.S. Representative from Wisconsin, 1893-1907. Mainly family correspondence. Includes letters written by Babcock while traveling in Iowa for the Ingram, Kennedy, and Day Lumber Company.

3317 **Bailey, Thomas J.,** Papers, 1842-1845, 1852-1918, 3 boxes (including 14 vols.). Ledgers (2 vols.) and diaries (12 vols.) of Bailey and his son, Henry J. Bailey, who operated machine and woodturning shops at Fort Howard, Green Bay, and De Pere, Wisconsin. Inventory. Green Bay Area Research Center.

3318 **Baird, Henry Samuel** (1800-1875), Papers, 1798-1890, 1930, 7 boxes and 1 vol. In part, photocopies. Lawyer of Green Bay, Wisconsin, and territorial political figure. Mrs. Baird's correspondence includes letters concerning relief for sufferers in the forest fires in 1871.

3319 **Baker, Harry D.,** Manuscript, 1 folder (46 pp.). "Letters to Clarence: Early Days in St. Croix Falls." Gives details of life in St. Croix Falls, Wisconsin, 1875 to the early 1900s. River Falls Area Research Center.

3320 **Ballard, Anson** (1821-1873) **and Nathaniel Edwards** (b. 1837), Papers, 1838-1874, 3 boxes (including 2 vols.) and 4 additional vols. Papers of the first and second husbands of Harriet M. Story of Appleton. Includes correspondence relating to the operations of a planing mill owned by Edwards and two partners until 1872. Green Bay Area Research Center.

3321 **Barker, Herbert C., Family,** Papers, 1872-1912, microfilm, 1 reel. Included are correspondence and legal and financial papers concerning family timber land holdings in Wisconsin, some of which were owned in partnership with R.M. Forsman and Co., Williamsport, Pennsylvania. Originals are in the possession of John Klinger, Chippewa Falls, Wisconsin. Eau Claire Area Research Center.

3322 **Barker and Stewart Lumber Company,** Record books, 1899-1915, 29 vols. Journals, ledgers, cashbooks, sales books, letterbooks, and other records of the saw and planing mill owned by Christopher C. Barker and Hiram C. Stewart at Wausau, Wisconsin. Stevens Point Area Research Center.

3323 **Bartlett, W.W.,** Papers, January, 1921. Typewritten copies of letters written from camps of the Park Falls Lumber Company and the Edward Hines Lumber Company.

3324 **Bayfield Transfer Railway Company,** Papers, 1836-1913, 30 boxes (including 36 vols.) and 20 additional vols. Papers of William F. Dalrymple (1825-1901), a native of

Sugar Grove, Warren County, Pennsylvania, dealing with his investments in mills, railroads, farmlands, and city real estate in Pennsylvania, North Dakota, Minnesota, and Wisconsin. The collection consists primarily of incoming letters, but includes occasional drafts and copies of Dalrymple's replies. There is correspondence of Andrew Tate and R.D. Pike after 1886 containing brief observations on lumbering, land sales, road building, and residents of Bayfield, Wisconsin, and numerous letters exchanged after 1891 between Dalrymple and Herbert C. Hale, general manager of the railroad at Bayfield, concerning the building, financing, and operation of the short line, which hauled passengers and freight in addition to its main cargo of lumber. The unbound papers are supplemented by letterbooks (1892-1916, 8 vols.), annual reports (8 vols.), and trial balances (2 vols.) of the Bayfield Transfer Railway Company.

3325 **Bender, Peter,** Papers, 1828-1914, 1 box (including 1 vol.), and 5 additional vols. The papers of a Milwaukee County justice of the peace, commission merchant, and operator of a sawmill including an account book (1854-57) for merchandise and sawing lumber, and papers concerning the building and management of the Milwaukee and Green Bay Plank Road.

3326 **Bentley, Stephen R.** An account of Wisconsin River lumber rafting days. September 30, 1911.

3327 **Bergstrom Paper Company** (Neenah, Wisconsin), Records, 1904-1919, 44 pp. Articles of organization and corporate minutes. Xerox copies of originals held by the Bergstrom Paper Company. Oshkosh Area Research Center.

3328 **Biron, Joseph.** Biography of Joseph Biron, and history of his family, particularly his eldest son, Francis Xavier Biron, an early lumberman (written by Capt. Joseph Cotey?), 5 pp., typewritten.

3329 **Black River Falls, Wisconsin,** Letters, 1913, 1926, 1 folder (12 items). Pertain to the early settlement and lumber industry of Black River Falls, as recalled by Amos Elliott, early lumberman of the area.

3330 **Black River Log Driving Company** (La Crosse, Wisconsin), Papers, 1856-1858, 1 folder. Unbound daybook (1856-57) kept by W.W. Crosby, clerk; minutes of meetings (February 5, 1856; October 11 and December 1, 1858); and miscellaneous bills and receipts. La Crosse Area Research Center.

3331 **Blaine, John J.** (1875-1934), Papers, 1873, 1894-1934, 69 boxes (including 1 vol.) and 8 additional vols. Governor of Wisconsin, 1921-27; U.S. Senator, 1927-33. The collection consists mainly of unbound correspondence, and speeches (8 vols.), and relates primarily to the years of his governorship. The gubernatorial papers' major topic throughout is the issue of state taxation and finances in various forms, including Blaine's opposition to the creation of a Northern Lakes state park.

3332 **Blair, Emma Helen** (1851-1911), Papers, 1809-1910, 1 box (including 6 vols.) and 1 additional vol. Letters from Thomas Bartlett and his son, George, to his wife, Lucy Bartlett, in Orono and Old Town, Maine, and Milwaukee, Wisconsin, containing references to the Aroostook boundary dispute, and to lumber business on the Canadian boundary.

3333 **Blanding, William Martin** (1829-1901), Papers, 1847-1958, 24 vols. and 13 boxes. Businessman, land speculator, and local politician, of St. Croix Falls, Wisconsin. Business and political correspondence (1847-1901), financial papers, and land records comprise the bulk of the collection. There are a few letters and records relating to transportation and logging on the St. Croix River. River Falls Area Research Center.

3334 **Blodgett, Benjamin** (b. 1830), Correspondence, 1889-1915, microfilm, 1 reel. Letters exchanged by members of the Benjamin Blodgett family, who left Jefferson County to engage in mining, lumbering, trapping, and trading in California, Oregon, Alaska, and the Yukon.

3335 **Borth, Henry F.** (b. 1883), Manuscript, microfilm, 1 reel. Biography of Frank Borth, written by Henry Borth, entitled "Frank Borth: The story of my father's life," concerning sixty years of Wisconsin logging and lumbering. Original manuscript owned by Henry F. Borth (1957).

3336 **Bridgman, Charles** (d. 1962?), Papers, 1797-1961, microfilm, 1 reel. Papers of Charles Bridgman, Wautoma, Wisconsin, historian and genealogist, dealing with early families and events of the region. They include reminiscences and articles about early farming, lumbering, circuses, roads, and bridges in Waushara County written or collected by Bridgman.

3337 **Broughton, Charles Elmer** (1873-1956), Papers, 1916-1953, 14 boxes, 14 vols., and 2 packages. Sheboygan, Wisconsin; editor. Correspondence (1916-53); articles and speeches (1924-52); and clippings (1925-51). Some of the material reflects Broughton's interest in conservation. Milwaukee Area Research Center.

3338 **Brown, Charles Edward** (1872-1946), Papers, 1906-1940, 12 boxes. Curator of the museum of the State Historical Society of Wisconsin and Secretary of the Wisconsin Archaeological Society. Correspondence and notes on various subjects, including lumbering.

3339 **Brown, Neal** (1856-1917), Papers, 1913-1923, 1955, 1 folder. Relating to Neal Brown, Wausau, Wisconsin, an organizer of the Marathon Paper Mills Co., containing his views on Democratic politics in the state, and including a few letters (1920-23) from C.S. Gilbert, also an organizer of the paper company, relating to timber land in northern Wisconsin; with one clipping (1955) on Gilbert's death. Stevens Point Area Research Center.

3340 **Brown Brothers Lumber Company** (Rhinelander, Wisconsin), Papers, 1854-1959, 1 box, 28 vols., and microfilm, 1 reel. Scattered correspondence (1854-1959), letterbooks of company correspondence (1897-99, 1904-08, 2 vols.), and assorted financial files, land inventories, and other records (1890-1943). A diary kept by Anderson W. Brown (1874) records the trip on which the Brown brothers based their decision to acquire land in the Pelican Rapids vicinity. The collection also contains several contracts and agreements, an inventory (1934), a few clippings, and a history of the Rhinelander Paper Company, which the brothers formed early in the twentieth Century as the lumber supply became depleted. The collection is in the Stevens Point Area Research Center. Microfilm available in State Historical Society of Wisconsin, Madison, and in Stevens Point Area Research Center. The film includes Brown's diary (1874) and scattered correspondence (1854-59), and reminiscences.

3341 **Buchen, Gustave William** (1886-1951), Papers, 1915-1951, 4 boxes. Wisconsin state senator, 1941-51. Correspondence (1941-51); an autobiography; clippings (1943-47); speeches (1915-49); committee reports; and an unfinished manuscript about the Kettle Moraine State Forest. The correspondence, which contains comments on Buchen's sponsorship of several bills, reveals his interest in conservation.

3342 **Burke, Fred C.,** Papers, 1949-1953, 3 items. Mainly recollections of early medical practice in a Wisconsin sawmill village, including data on group medical insurance in the lumber industry.

3343 **Burmeister, Charles** (1855-1891), Papers, 1864-1891, 1918, 2 boxes (including 34 vols.). The papers include the correspondence (1879-80) of Hannah, Lay, and Co. (Traverse City, Michigan), a lumber firm. Milwaukee Area Research Center.

3344 **Burns, Mathew J.** (b. 1887), Papers, 1939-1963, 1 box. Correspondence and a few speeches of the editor of the *Pl lp Workers Journal* and president of the International Brotherhood of Paper Makers. The papers relate to Burns' career in the labor movement and the papermakers union, his work as a labor specialist for the Department of Commerce, his commentary on the "Carey injunction," and his philosophy of unionism. Included is a resume of Jeremiah T. Carey's union record.

3345 **Butler, James Davie** (1815-1905), Papers, 1706, 1776-1905, 6 vols. and 14 boxes. Includes papers relating to a telepathic meeting between John Muir and Butler in the Yosemite Valley, California. There are two letters by Muir, 1869 and 1888; Butler's correspondence with John E. Woodhead, editor of *Mind in Nature*; and typewritten copies of two articles written by Butler for that periodical.

3346 **Carpenters District Council of Milwaukee, Waukesha, Washington, and Ozaukee Counties** (Milwaukee, Wisconsin), Minute books, 1887-1921, 1923-1926, 3 boxes (including 11 vols.). An affiliate of the United Brotherhood of Carpenters and Joiners of America. Container list. Milwaukee Area Research Center.

3347 **Carson and Eaton,** Records, 1840-1852, 1 box (including 4 vols.). Records of William Carson and Henry Eaton, kept in connection with the Carson and Eaton lumber business on the Eau Galle River in west central Wisconsin, including a ledger (1843-45) kept by Eaton and Richardson, and a journal and ledger (1845-49) kept by Bradley and Richardson, jobbers for Carson and Eaton. Inventory. Stout Area Research Center.

3348 **Chido, George** (1908-1964), Journals, 1934-1939, 1 box (including 4 vols.). Ranger for the Wisconsin Department of Conservation; included are day-to-day notations of Chido's work with the Civilian Conservation Corps. Superior Area Research Center.

3349 **Christensen, P.N.,** Manuscript, 1917, 6 pp. "The Danish Settlement in Nasonville, Wood County, Wisconsin," by a descendant of settlers there, describing living conditions, farming, lumbering, and Indians, and including biographical sketches of some of the early settlers. Xerox copy. In the River Falls Area Research Center.

3350 **Clark, Alexander** (?), Record book, 1850-1872, 1 package (including 1 vol.). Kept by a cabinetmaker near Delavan, in the Town of Richmond, Walworth County, Wisconsin. In the Whitewater Area Research Center.

3351 **Cole, Charles D.** (1806-1867), Papers, 1831-1847, microfilm, 1 reel. Correspondence of Cole, Sheboygan County, Wisconsin, merchant and lumberman, with his family in New York state, concerning business, agriculture, family affairs, and the activities of the Fourierites near Sheboygan Falls.

3352 **Collins, Cornelius C.** (1863-1950), Manuscript, 1 item. Reminiscences, typed copy (14 pp.). An account of experiences in logging camps and lumber mills in Michigan and northwestern Wisconsin, Collins' later investments in Wisconsin timberlands, and his establishment of C.C. Collins & Son, Inc., a Madison, Wisconsin, lumber company.

3353 **Colman, Henry Root** (1800-1895), Papers, 1817-1894, 2 boxes. Includes copies of Colman's diaries (1854-57) describing the beginnings of the Colman Lumber Co., La Crosse, Wisconsin.

3354 **Comstock, Elizabeth** (b. 1875), Papers, 1779-1952, 10 boxes (including 79 vols.). Included are some financial records of the 1860s and 1870s relating to the land speculations and business operations of Dr. Comstock's father, Noah D. Comstock (1832-90), a lumber dealer and a flour and feed mill owner in Arcadia, Wisconsin. La Crosse Area Research Center.

3355 **Connor, William D.** (1864-1944), Scrapbooks, 1904-1924, microfilm, 1 reel. Scrapbooks relating to W.D. Connor of Marshfield, Wisconsin, a lumberman and politician.

3356 **Conover, Edith W.,** Papers, 1843-1924, 3 boxes, including 3 vols. The papers include one letter describing the calling of the militia in Madison, Wisconsin, for use in the strike of mill and lumber workers at Eau Claire in 1881.

3357 **Coogan, Mrs. Grace** (d. ca. 1946), Papers, 1918-1946, 1 box. Papers of a resident of Lodi, Wisconsin, concerning sales of timberland in Vilas, Forest, and Marinette counties, Wisconsin, mainly to the U.S. Forest Service for reforestation projects.

3358 **Cornell Western Lands Collection,** 1859-1925, microfilm, 36 reels. Selections pertaining to Ezra Cornell's land business in Wisconsin. Correspondence (21 reels) of Eastern speculators and Cornell University officials, including Ezra Cornell, Jeremiah W. Dwight, Henry W. Sage, W.A. Woodward, and J.W. Williams, and of the Cornell land agent in Wisconsin, Henry C. Putnam; also related tax registers, contracts, survey notes, plat books, and other legal and financial records. Originals are in the possession of the Cornell University Archives and Collection of Regional History, Ithaca, New York.

3359 **Dodge, Joseph T.** (1823-1904), Papers, 1845-1899, 5 boxes (including 13 vols.) and 11 additional vols. Professional and personal papers of a Wisconsin civil engineer relating to railroad building. A few personal letters are scattered through the volumes, including some to his brother, A. Clarke Dodge, in regard to a planing mill at Monroe, Wisconsin.

3360 **Donald, John Sweet** (1869-1934), Papers, 1811-1948, 34 boxes (including 72 vols.). Mainly correspondence, articles, diaries, and notebooks of a Dane County, Wisconsin, farmer, agricultural expert, and politician. Some letters indicate Donald's continued interest in conservation and highway improvement. Also included are records (2 boxes) of the Friends of Our Native Landscape (1916-48), including correspondence, minutes of meetings, and other papers concerned with landscape preservation.

3361 **Dousman, Hercules Louis** (1800-1868), Papers, 1835-1912, 62 items and 54 boxes. Fur trader, lumberman, land speculator, and businessman of Wisconsin. His papers include material on the Chippewa Falls Lumbering Company; land speculation; and contain statements of Dousman's mill.

3362 **Dumke, Lorenz A.** (b. 1902), Reminiscences, 1967, 2 pp. Dumke's family built sawmills in Wisconsin at Theresa in Dodge County, and at Caroline and Tilleda in Shawano County, in the middle 1900s. Typewritten. In the Green Bay Area Research Center.

3363 **Dyer, Harry C.,** Manuscript, 1 item. "Historical Account of Log Rafting on the Mississippi," 16 pp. This account of rafting on the Mississippi and St. Croix Rivers after 1839 contains descriptions of logging techniques, riverboats, and shipbuilding companies.

3364 **Eau Claire Lumber Company** (Eau Claire, Wisconsin), Papers, 1846-1888, 1 box (including 3 vols.). Copies from the surveyor general's records of timber contracts between Nelson C. Chapman and Joseph G. Thorp and various Wisconsin individuals and firms (1861-65, 1 vol.); a stock book dating from the incorporation of the company in 1866; descriptions and maps of lands owned by the company (1 vol.); and miscellaneous maps, deeds, and contracts. In the Eau Claire Area Research Center.

3365 **Edwards, John,** Papers, 1846-1868, 3 boxes (including 1 vol.). The papers contain letters of John Edwards, Jr., relating to his lumber enterprises on the Wisconsin River which later developed into the Nekoosa-Edwards Paper Company at Port Edwards, Wisconsin.

3366 **Emerson, Hugh R.,** Manuscripts, ca. 1966, 2 items (12 pp.). An essay on the short-lived logging community of Emerson, Iron County, Wisconsin; and a glossary of technical terms and expressions used in logging camps. Typewritten.

3367 **Esch, John Jacob** (1861-1941), Additional papers, 1880-1941, 3 boxes. U.S. Representative from Wisconsin; sponsor of the Federal Water Power Act, 1920. The bulk of this collection is composed of letters saved by Esch primarily because of their autograph value. Among the writers represented are Robert U. Johnson, John Muir, Isaac Stephenson, Charles R. Van Hise, and others.

3368 **Everest, David Clark,** Papers, ca. 1919-1955, 127 boxes. Personal and general correspondence, speeches, reports, and miscellaneous materials of a leader of the pulp and paper industry and a founder of the Marathon Corporation and other firms in the forest products industries. Topics and major correspondents include the Marathon Corp., Ontonagon Fiber Corp. (1925-44), Ward Paper Co., (1942), Wausau Paper Mills Co. Preliminary inventory.

3369 **Fairchild, Lucius** (1831-1896), Papers, 1819, 1830-1923, 85 boxes (including 26 vols.) and 116 additional vols. This extensive collection contains a small amount of information on early Madison, Wisconsin, buildings, boatbuilding at Necedah, Wisconsin, and the Fairchild lumbering interests.

3370 **Farrish, Roy A.,** Letter, March 12, 1952. To William F. Groves, Lodi, Wisconsin, recalling life in a lumber camp in Wisconsin when Farrish was boss of a small lumber and logging crew, 1900.

3371 **Fifield Central Lumber Company,** Timebook, June-September 1910, 1 vol. in package.

3372 **Fisk, W.D., and Company** (Green Bay, Wisconsin), Records, 1853-1908, 2 boxes (including 6 vols.). Contractors of ties, poles, and other lumber to the Chicago and Northwestern Railroad. The records consist of account books and letterpress copy books of correspondence relating to the company and to earlier business ventures of William J. Fisk and Joel S. Fisk.

3373 **Follstad Brothers** (Elcho, Wisconsin). Descriptions and scenes from Follstad Brothers Logging Operations near Elcho, Wisconsin, 1890. Typewritten copy, 2 pp. Four related photographs are filed in the picture collection.

3374 **Folsom, William Henry Carman** (1817-1900), Papers, 1836-1900, 1922, 1 box (including 1 vol.) and 2 additional vols. Papers of an Indian trader, lumberman, land speculator, Minnesota politician, and regional historian, including correspondence (1841, 1851-1900); account books (1841-44); and papers relating to Folsom's activities in the Mississippi-St. Croix Valley. In the River Falls Area Research Center.

3375 **Foster, Jacob** (b. 1782) and **Jacob T.** (1827-1906), Reminiscences, 1846, 1900, microfilm, 2 reels. Memoirs of early settlers in the Sheboygan, Wisconsin, area. Jacob T. Foster's memoirs, compiled in 1900, cover the years 1827-94, and include discussion of his career as a sawmill operator.

3376 **Foster, Nathaniel C.** (1834-1923), Papers, 1830-1936, 1 box (including 1 vol.) and 1 vol. Wisconsin lumberman, railroad builder, and businessman. Included are political correspondence (1921-22), miscellaneous business papers (1830-1920) concerning land interests, and a journal (1908-36) of the N.C. Foster Company, Fairchild, Wisconsin. Inventory. In the Eau Claire Area Research Center.

3377 **Four County Development Group** (Ashland, Wisconsin), Papers, 1957-1962, 1 folder. An organization promoting the economic development of Ashland, Bayfield, Iron and Price counties. Included are minutes, financial reports, president's letters to members, and occasional reports.

3378 **Frear, James Archibald** (1861-1939), Papers, 1906, 1919-1940, 3 boxes and 6 vols. Wisconsin state legislator, 1902-06; U.S. Representative from Wisconsin, 1913-35. His congressional correspondence contains some information on his views on conservation.

3379 **Gebhardt, Herman J.** (d. 1954), Diaries, 1898-1954, 3 boxes (including 44 vols.). Diaries kept by a Black River Falls socialist, cranberry grower, and exponent of reforestation. Includes material on his conservation interests. In the La Crosse Area Research Center.

3380 **Germer, Adolph** (b. 1881), Papers, 1898-1958, 26 boxes (including 28 vols.). Correspondence (1901-58), diaries (1930-32, 1935-58), speeches, reports, and other papers. Records of the International Woodworkers of America (IWA) comprise the largest segment of the collection. Germer was a director of the union during most of the 1940s and early 1950s. Many papers relate to controversies between communist and right-wing factions within the IWA, and between the IWA and the AFL carpenters union. Other papers concern IWA efforts to organize in the Pacific Northwest. A few letters (1943-44) describe organizing activities in the South and the civil rights problems encountered by organizers in this region.

3381 **Giese, Gustave A.,** Manuscript, 1947, 7 pp. "The rafting and running of lumber down the Wisconsin and Mississippi Rivers to the southern lumber market in bygone days." Hectographed copy of typewritten original.

3382 **Green Bay and Prairie Du Chien Papers,** 1774-1895, 123 vols. The collection includes letters from Robert M. Eberts and Andrew J. Vieau to John Lawe of Green Bay, Wisconsin, regarding operation of a sawmill and gristmill at Two Rivers, Wisconsin, after 1836.

3383 **Grimmer, George** (1827-1907), Papers, ca. 1859-1892. Correspondence and an account book of a Kewaunee lumber and general merchant. Unprocessed.

3384 **Hall, Richard L.** (1833-1892), Manuscript, 1876, 1 item. "The Centennial History of Oconto County." Typewritten. Contains considerable detail on the development of the lumber industry. In the Green Bay Area Research Center.

3385 **Hamilton, Alfred K.** (1840-1918), Papers, 1848-1900, 1 box (including 1 vol.). Papers of Hamilton, concerning a lumber firm operated at Fond du Lac under various firm names, and other business enterprises. Includes a volume of accounts and an inventory of lumber camp property (1878-80).

3386 **Harrington, Michael M.** (1880-1961), Papers, 1853-1955, 3 boxes (including 24 vols.), and 4 additional vols. Papers of a railroader and collector of railroad memorabilia. There is considerable material relating to hauling lumber and supplies for the John Arpin Lumber Company of Grand Rapids, Wisconsin.

3387 **Harris, Joseph, Sr.** (1813-1889), Papers, 1844-1894, 1 box. Sturgeon Bay, Wisconsin; newspaper publisher. Information on canal building, railroad land grants, lumbering, and quarrying in Door County, Wisconsin.

3388 **Hawn, E.L.,** Account book, 1884-1897, 1 vol. Account book kept at Hawn's sawmill at Olivet, Pierce County, including notations on the kinds of wood sawed. River Falls Area Research Center.

3389 **Heath Indian Manuscripts,** 1809-1940, 2 boxes. Miscellaneous papers relating to the Stockbridge Indians of New Stockbridge Community in Wisconsin (1809-84); the Chippewa Indians of Leech Lake Reservation in Minnesota (1890-1917); and other United States Reservation Indians (1915-18, 1926-40). Included are timber contracts.

3390 **Henrickson, Henry E.,** Letter, August 22, 1937, 1 item. Letter describing rivalry between crews of lumbermen on the Flambeau River during the 1880s.

3391 **Heuston, Benjamin Franklin** (1823-1894), Papers, 1849-1894, 2 boxes (including 32 vols.). Lumberman, land speculator, and local official in Trempealeau County, Wisconsin. Included are articles on local history. In the La Crosse Area Research Center.

3392 **Hiestand, William D.** (b. 1861), Miscellaneous papers, 1900-1925, 1 folder. Included are papers concerning timber land owned by Hiestand and Ransome A. Moore in Florence County, Wisconsin.

3393 **Hixon and Company** (La Crosse, Wisconsin), Papers, 1856-1928, 100 boxes (including 157 vols.) and 87 additional vols. Records of the lumber business and other investments of Gideon C. Hixon (1826-92) and his son, Frank P. Hixon (1862-1931). The elder Hixon operated a sawmill in the La Crosse region after 1856, purchased the Wisconsin timber holdings of the T.B. Scott Company, and entered the lumber business in Wisconsin, Missouri, and Kansas. In 1900 Frank P. Hixon and his four brothers incorporated Hixon and Company, expanded its investments in timber operations into Minnesota, the Pacific Northwest, the South, Ontario, and British Columbia. G.C. Hixon's Wisconsin business ventures after 1856 are represented by assorted journals, ledgers, payroll records, log and raft records, camp-store financial accounts, and memoranda books. The major part of the collection is composed of letterbooks and unbound correspondence of G.C. Hixon, Frank P. Hixon, and Hixon and Company (1869-1922). The correspondence and related records pertain especially to northwestern Wisconsin and deal with transportation and freight rates, lumber grading and pricing, the use of water and water rights, taxation, logmarks prior to 1900, labor relations, union activity, and the Industrial Workers of the World (1917 to 1920). A few letters (1919) discuss the preservation of stands of virgin timber through the possible creation of national parks. The collection also contains some correspondence and many financial records and legal papers pertaining to the subsidiary lumber companies in other states and to other business investments of the Hixons. La Crosse Area Research Center.

3394 **Holt Lumber Company** (Oconto, Wisconsin), Records, 1839-1943, 6 boxes (including 17 vols.) and 230 vols. Records of one of the largest lumber companies operating in northeastern Wisconsin, which had its headquarters in Chicago, Illinois, and at Oconto, Wisconsin. The collection includes extensive series of letterbooks, journals, ledgers, cashbooks, purchase records, cruisers' reports, log drive records, log books, price lists, time books, payrolls, and other papers. The majority of the volumes cover the period from 1865 to 1888, when Devillo R. Holt (1823-99) and Uri Balcom (1815-93) were in partnership. These two natives of New York state founded a company which engaged in the lumbering of white pine in Oconto and Marinette counties. Its operations included cargo vessels on the Great Lakes, a sawmill, feedmill, general store, several farms, and a boardinghouse in the Oconto area. Despite gaps in several areas the records contain data on almost every aspect of the business from logging to marketing—the structure of the company, its operations, its varying financial condition, and its relations with other companies. Later records (1890-1943) describe the formation and operation of the Holt Lumber Company, successor to Holt and Balcom. This organization was operated by William Holt, son of D.R. Holt, with the assistance of other members of the family until the dissolution of the company in 1938. Records contain detailed information on the type and volume of business done and indicate the changes resulting from the shift from pines to hardwoods and hemlock, from oxen to motorized equipment, from lake carriers to railroads. Scattered records of several subsidiary organizations are included in the collection: the Oconto River Improvement Company (1893-1901); the American Lumber Company (1901-04); the Sever Anderson Logging Company (1925-26); and the Oconto Electric Company (1911-24). In the Green Bay Area Research Center.

3395 **Howard, James C.** (d. 1880), Papers, 1813-1881, 8 boxes. Correspondence of the Howard family, touching on such topics as wood conservation.

3396 **Hoxie and Mellor Company** (Antigo, Wisconsin), Records, 1884-1891, 1 box, 39 vols. Records of a lumber company at Antigo, with branches at Bryant, Wisconsin, and Bessemer, Marenisco, and Ironwood, Michigan, including volumes of invoices, statements and estimates, assets and liabilities, court testimony, journals, ledgers, cashbooks, salesbooks, and order books; and a box of correspondence. The company was founded by John C. Hoxie and Edward N. Mellor. The collection includes information on a legal controversy during which the company's property was assigned to Charles V. Bardeen. In the Stevens Point Area Research Center.

3397 **Hummel, William,** Reminiscences, 1959, 6 pp. Logger and riverman on the Wolf River in Wisconsin. Recorded by Mrs. Franklin Neuschafer. Xerox copy of original handwritten manuscript owned by Mrs. Neuschafer, Fremont, Wisconsin (1964).

3398 **Husting, Paul Oscar** (1866-1917), Papers, 1909-1918, 25 boxes and 27 vols. Wisconsin state senator, 1907-13; U.S. Senator, 1915-17; served on the Public Lands Committee. The papers include some correspondence, data, and drafts of bills (1909-13) relating to the conservation movement in Wisconsin, including forest reserve lands, navigable waters, riparian rights, and control of water power; and two boxes of reports of Wisconsin state legislative committee hearings on water power, forestry, and drainage (1909-10).

3399 **Inabit, Robert E.,** Papers, 1 box. Articles and folk tales written and collected by Inabit, mainly concerning the history of logging and railroading in western Wisconsin; also including clippings, photographs, and maps of Eau Claire and Chippewa Falls, Wisconsin. In the River Falls Area Research Center.

3400 **Ingram, Hulius G.,** Letters, 1886, 1889, 36 items (ca. 95 pp.). President of the Woodville Lumber Co. at Woodville, St. Croix County, Wisconsin. Letters to his wife, relating to his daily activities and family life. In the River Falls Area Research Center.

3401 **Ingram, Orrin H.** (1830-1918), Papers, 1857-1904, 63 boxes (including 46 vols.) and 4 vols. The business papers of a Wisconsin lumberman who operated in the Chippewa Valley, with headquarters in Eau Claire, Wisconsin, and sold manufactured lumber through subsidiary wholesale companies along the Mississippi River. Ingram was associated with A.M. Dole and Donald Kennedy in 1857. In 1881 the company was reorganized as the Empire Lumber Company. The correspondence emphasizes the sales aspect of the business and pertains to business plans, policies, and practices; filling of orders; market prospects; prices, types, and qualities of lumber; credit and collection matters; details of rafting on the Chippewa and Mississippi Rivers; acquisition of timberlands in Wisconsin and other states; and organization of the Pacific Empire Lumber Company of South Bend, Washington. Some letters deal with the coming of the railroads and competition with the Chicago market, and the sale of Cornell University lands in Wisconsin. The letterbooks (1873-1904) contain correspondence of Ingram and of Clarence A. Chamberlin, assistant secretary of the Empire Lumber Company. There are scattered record books of various lumbering operations, and of the Half Moon Lake Canal Company. In the Eau Claire Area Research Center.

3402 **International Brotherhood of Pulp, Sulphite and Paper Mill Workers,** Papers, 1906-1957, microfilm, 284 reels. Mainly correspondence of the union with its locals, its organizers, the AFL-CIO, and other unions, companies, and government agencies covering many facets of the union's activities. Interspersed in the correspondence files are speeches delivered by union officers at conventions and conferences, and a few financial records. Originals in the possession of the Union headquarters at Fort Edward, New York.

3403 **Irvine, William** (1851-1927), Papers, 1896-1926, 1954, 1 folder (including 43 letters, 1 biography, and 1 speech). Manager of the Chippewa Lumber and Boom Company, Chippewa Falls. Letters from Frederick Weyerhaeuser and his son, Charles A., including reminiscences by William Irvine concerning the Chippewa River lumber days, and a biography of Irvine. In the Eau Claire Area Research Center.

3404 **Izaak Walton League, Wisconsin Branch,** Papers, ca. 1938-1945. Unprocessed.

3405 **Jones, Granville D.** (1856-1924), Papers, 1886-1962, 6 items. Papers by or relating to a Wausau, Wisconsin, lawyer, lumberman, and member of the University of Wisconsin Board of Regents. Included are Jones' address of acceptance delivered at the unveiling of a bust of John Muir on December 6, 1916, and reminiscences of Jones by his daughter.

3406 **Judkins, Henry** (1823-1850), Correspondence, 1848-1863, 4 items. Three letters from a timber worker in Michigan and Wisconsin to his family in Maine, including information on the lumbering operations of Judkins' employer, Daniel Wells, Jr.; and a letter to Judkins from Wells, enclosing a poem by Wells.

3407 **Kaiser, John H., Lumber Company** (Eau Claire, Wisconsin), Papers, 1905-1957, 5 boxes (including 12 vols.). Includes correspondence (1925-56), board of directors' minutes (1905-57), record of stock issues (1909-48), and miscellaneous legal and business records. Inventory. In the Eau Claire Area Research Center.

3408 **Kanneberg, Adolph,** Document, 1944, 84 pp. "Statutory Provisions and the Common Law of Wisconsin pertaining to log driving and the rafting of lumber in Wisconsin," a survey prepared by Kanneberg, Counsel for the Public Service Commission of Wisconsin. Mimeographed. Includes appendix.

3409 **Keenan, Walter E.** (b. 1910), Letters, 1925-1952, 1 box. Includes letters to his parents from Keenan as a lieutenant in Civilian Conservation Corps camps in Illinois, Michigan and Wisconsin (1934-37).

3410 **Kiel Manufacturing Company** (Kiel, Wisconsin), Records, 1 box (including 3 vols.). Makers of furniture. Included are minutes (1903-19); stock registers (1892-1930); and stock certificates (November 1892 and September 1897). In the Green Bay Area Research Center.

3411 **Kimberly-Clark Corporation** (Neenah, Wisconsin), Papers, 1880-1906, 1940-1952, microfilm, 2 reels. One reel contains a volume of minutes of the board of directors and stockholders' meetings of the company (1880-1906, 1 reel [1 vol.]) relating to incorporation and the formative years, and to the activities of the leaders in the Wisconsin water power and paper industries. There are also materials used in preparing the company's 75th anniversary history (1947, 1 reel). Included are letters from retired employees; some copy for publicity and news releases; individual departmental histories and reports; Melvin E. Bartz, "Origins and Development of the Paper Industry in the Fox River Valley (Wisconsin)," State University of Iowa, M.A. thesis (History); and an unpublished 80th anniversary history. In the State Historical Society, Madison. Microfilm also available at the Oshkosh Area Research Center.

3412 **Kirkendall, Paul** (1884-1953), Papers, 1 folder. Recollections by Kirkendall of logging activities for the Knapp, Stout Company in Polk, St. Croix, Barron, and Washburn counties, Wisconsin, in the 1890s; with an introduction and related newspaper articles about Birchwood, Wisconsin, by Ethel Elliott Chapelle. Partially Xerox copies. In the River Falls Area Research Center.

3413 **Kleinpell, Henry** (1869-1951), Papers, 1904, 1912, 1918-1951, 4 boxes (including 10 vols.), and 1 additional vol. Gassville and Prairie du Chien, Wisconsin; physician. Includes business papers (1926-45) consisting mainly of financial statements of the Flint (Michigan) Lumber Company, and other records.

3414 **La Budde, Wilhelmine D.** (1880-1955), Papers, 1924-1956, 7 feet. Papers relating to Mrs. La Budde's activities as state conservation chairman for the Milwaukee County Federation of Women's Clubs, her campaigns for conservation education, for the restoration of Horicon Marsh in Wisconsin, for a resident fishing license bill, for reforestation, and for preservation of tracts of virgin timber. Correspondents include John J. Blaine, Charles E. Broughton, Charles E. Brown, George Washington Carver, F. Ryan Duffy, Glenn Frank, Frank Grass, Edgar A. Guest, Julius P. Heil, Daniel W. Hoan, Louis M. Howe, Merlin Hull, Walter J. Kohler, Sr., Walter J. Kohler, Jr., Philip F. La Follette, Robert M. La Follette, Jr., Orland S. Loomis, Joseph R. McCarthy, William Mauthe, William Morris, Haskell Noyes, Mary Pickford, Louis Radke, Albert G. Schmedeman, Robert F. Wagner, Henry A. Wallace, Alexander Wiley, and Fred A. Zimmerman. Inventory. In the Milwaukee Area Research Center.

3415 **La Follette, Philip F.** (1897-1965), Papers, 1876-1969, 166 boxes, 24 vols., and 1 package. Governor of Wisconsin, 1931-33 and 1935-39. General correspondence (1921-39) includes the official papers of the various state officials, boards, and commissions operative during La Follette's governorship.

3416 **La Follette, Robert M., Sr.** (1885-1925), Papers, 1879-1910, 230 boxes, including 63 vols. U.S. Representative from Wisconsin, 1885-91; governor of Wisconsin, 1901-05; U.S. Senator, 1906-5. The collection covers La Follette's Wisconsin period as an attorney and governor, and includes chiefly political and business papers. Letters (1905) discuss the establishment of the Wisconsin State Board of Forestry. Also on microfilm, 166 reels.

3417 **Landt, Sophronius S.** (1842-1926), Reminiscences, 1926, 7 vols. Reminiscences of pionner life at Lake Mills, Wisconsin, in the 1840s, and later at Big Springs, Adams County, Wisconsin, including accounts of claim jumping, and logging and lumber rafting on the upper Wisconsin River.

3418 **Langley and Alderson,** Papers, 1889-1909, 1 box. Business papers of a wholesale and retail logging concern run by George Langley and Nathaniel L. Alderson, of Merrill, Wisconsin, involving dealings with retail lumber dealers in the upper Wisconsin Valley, owners of timberland, employoment agencies in Minneapolis, concern handling camp supplies, and railroad companies. In the Stevens Point Area Research Center.

3419 **Lapham, Increase A.** (1811-1875), Papers, 1811-1876, 34 boxes (including 61 vols.), 5 vols., and 3 folios. Papers of an early exponent of soil, fish, and forest conservation in Wisconsin.

3420 **Lincoln, Ceylon Childs.** Lumbering in early days in Wisconsin. Also a short autobiography.

3421 **Lincoln Lumber Company,** Record books, 1881-1892, 1 box (including 11 vols.). Records of a company incorporated in 1881 by Thomas F. Mathews and others of Merrill, Wisconsin, with articles of incorporation, minutes of meetings, a stock certificate book, records of bills payable, and log scale books of several camps. In the Stevens Point Area Research Center.

3422 **Loy, David M.** (1816-1873), **and Peter S.** (1846-1931), Papers, 1839-1911, 9 boxes (including 14 vols.) and 10 additional vols. Business and family records of a father and son of De Pere, Wisconsin, including information on sawmilling.

3423 **Ludington, Harrison** (1812-1891), Papers, 1848, 1855-1890, 5 boxes. Governor of Wisconsin, 1876-78. The collection relates only to Ludington's business activities, particularly to his partnership with Daniel Wells, Jr., of Milwaukee and Anthony Van Schaick of Chicago in the lumber industry. There are many letters from Van Schaick concerning the land investments and lumber production of the Ludington, Wells & Van Schaick Company which operated near Menominee, Michigan, largely concentrated in the years 1885-86 and 1890. In the Milwaukee Area Research Center.

3424 **McCann, Henry,** Reminiscences, 1938, 6 pp. Concerns experiences as a lumberjack in Wisconsin and Minnesota, the rivers on which he was a driver, names of men with whom he worked, and brief sketches of the life of a river man. In the River Falls Area Research Center.

3425 **McCorison, Joseph Lyle, Jr.** (b. 1900), Letter, 1971, 4 pp. Reminiscence about logging and railroading along the Flambeau River at Ladysmith, Wisconsin, ca. 1910 (Xerox); including a postcard showing logging on the Connecticut River.

3426 **McDonald, Arthur,** Recollections, 13 pp. Logging experiences in the early twentieth century around Bruce, Rusk County, Wisconsin, as told to and recorded by John Doyle. Typewritten. In the River Falls Area Research Center.

3427 **McMillan, B.F., Lumber Company,** Papers, 1884-1937, 1 box (including 4 vols.) and 4 additional vols. Records of a Marathon County, Wisconsin, timber and milling company, including bills receivable, sales records, ledgers, journals, and correspondence. Letters pertain to tax assessments, contracts, and the settlement of the estate of one partner. In the Stevens Point Area Research Center.

3428 **McNutt, James O.** (1848-1931), Reminiscences, 1930, microfilm, 1 reel. Describing, among other things, logging near Black River Falls, Wisconsin.

3429 **Macomber, William Harvey,** Papers, 1856-1863, microfilm, 1 reel. Mainly letters (1856) by Macomber to his wife, describing Fond du Lac and his work as a mill hand in Winnebago County.

3430 **Marathon Paper Mill Company** (Rothschild, Wisconsin), Papers, 1918-1945, microfilm, 1 reel. Includes correspondence of David C. Everest and other company executives relating to comparative wage schedules of pulp and paper companies. The originals are in the possession of the company. Copies are held at the State Historical Society, Madison, and at Stevens Point Area Research Center.

3431 **Martinson, Martha,** Manuscript, 7 pp. Essay on Porter's Mills, a lumbering community from ca. 1873-1903, in Eau Claire County, Wisconsin. Xerox copy. In the Eau Claire Area Research Center.

3432 **Mason, Jeremiah** (1836-1913), Papers, 1861-1876, 1 box (including 14 vols.). Scattered pocket diaries, some personal accounts, and few personal letters kept by a Fort Atkinson farmer who also operated a lumber business.

3433 **Menasha Wooden Ware Corporation,** Papers, 1857-1946, 19 boxes and 202 vols. A manufacturer of wooden washtubs, churns, pails, etc.; then barrels and kegs; and, more recently, corrugated containers. Contains early business records, correspondence, inventories, etc. In the Oshkosh Area Research Center.

3434 **Menominee Bay Shore Lumber Company** (Soperton, Wisconsin), Records, 1881-1931, 7 feet (3 boxes [including 14 vols.] and 19 vols.). Records (chiefly 1888-1924, with a few items to 1931), including business correspondence, journals, ledgers, trial-balance books, a cashbook, sales records, individual and payroll records, and some bills; records of the North Branch Pine Improvement Company (1890-1904), including minute books, ledgers, and journals which cover the activities of the company with its relationship with the Menominee Bay Shore Lumber Company; and a short history of the Menominee Company, photos, and clippings. In the Stevens Point Area Research Center.

3425 **Menomonie, Wisconsin—The Knapp, Stout & Co., Company,** Papers, 1841-1962, 4 boxes (including 44 vols.) and 43 additional vols. Papers of The Knapp-Stout & Co., Company and its predecessor, Knapp-Stout & Co., including diaries and records of its first president, John H. Knapp, and his son, Henry E. Knapp; various balance sheets; inventories; records of logging camps and company boats; maps and descriptions of company-owned lands; historical sketches; and miscellaneous letterbooks. The additional vols. include cashbooks (1848-68); ledgers (1846-1962); contracts (1867-1902); etc. Inventory. In the Stout Area Research Center.

3436 **Metcalf, John** (1788-1864), Business records, 1831-1861, 1 box (including 2 vols.). Includes accounts and correspondence concerning Metcalf's operation of a sawmill at Baraboo.

3437 **Meyer Lumber Company,** Records, ca. 1917-1957, 5 cartons. Unprocessed.

3438 **Michigan Pine Lands Association,** Papers, 1863-1886, microfilm, 1 reel. Mainly correspondence and records of Cyrus Woodman, agent for the association. Includes survey maps, land deeds, tax papers, and timber estimates for the association, and papers relating to the St. Mary's Ship Canal Company. Originals in the possession of the Burton Historical Collection, Detroit Public Library, Detroit, Michigan.

3439 **Millar, James D.** (1869-1948), Papers, 1913, 1927-1929, microfilm, 1 reel. State Assemblyman from Menomonie, Wisconsin, 1911-15 and 1923-32. Included are correspondence, and reminiscences of lumbering in Wisconsin.

3440 **Millard, Burton** (d. 1862), Records, 1857-1860, 2 vols. Ledger (1857-59) showing household expenses, and journal (1857-60) recording lumber sawed, for whom, and expenses for mill equipment. Millard was a Wausau, Wisconsin, millwright. Inventory. In the Stevens Point Area Research Center.

3441 **Morris, William W.** (b. 1880), Papers, 1957, 1 item (120 pp.). Typescript of comments gleaned from the general notes of the original survey of Wisconsin by counties and townships (1833-62), including data on original land cover, streams, geology, agriculture, fishing, forestry, settlements, and archaeology.

3442 **Muench, Virgil J.** (b. 1904), Papers, 1940-1955, 1 box. Papers of a Green Bay attorney, including correspondence relating to his tenure as executive secretary of the Green Bay Trade Independent Association, 1943-44, and as president of the Brown County Chapter and the Wisconsin Division of the Izaak Walton League of America, 1948-50. The collection also includes articles and speeches by Muench on free enterprise and water pollution problems.

3443 **Muir, John** (1838-1914), Papers, 1861-1914, 31 items. Letters from Muir to members of the Edward Pelton family of Prairie du Chien (mostly 1861-72), containing observations on his family home in Wisconsin, his employment in Madison, and his experiences as a school teacher, impressions of the Civil War, accounts of geological and botanical expeditions into the Mississippi and to Yosemite, San Francisco, and Martinez, California, and several poems. Typewritten copies are also included.

3444 **Muir, John** (1838-1914), Correspondence, 1913-1917, 1954, 10 items. Two letters written by Muir to James Whitehead concerning a statement in Muir's book, *The Story of My Boyhood and Youth,* about the practice of child beating. These are accompanied by letters to Whitehead written by William Frederic Badè, Muir's literary executor, and Muir's brother David and sister Mary.

3445 **Munson, Israel,** Papers. The papers include letters (1854, 4 items) written by Israel M. Hill describing his lumber mill operations on Robinson Creek, Jackson County, Wisconsin, near Black River Falls.

3446 **Nekoosa-Edwards Paper Company,** Papers, 1907-1920, 1 folder. Material relating to operations of the company and including miscellaneous accounts of a subsidiary, the Mellen Lumber Company, and its Shanagolden logging operations to 1920. In the Stevens Point Area Research Center.

3447 **Nekoosa Lumbering Company,** Records, 1858-1961, 1 box (including 2 vols.). Daybook and ledger for merchandise and other expenses while building a dam on the Wisconsin River at Nekoosa. In the Stevens Point Area Research Center.

3448 **Nohl, William George** (b. 1861), Reminiscences, 23 pp. Timber cruiser and outdoorsman in the Ashland, Wisconsin, area. Includes 1 clipping. Xeroxed copy of original owned by Mr. Ted Nohl, Ashland, Wisconsin (1964).

3449 **North Western Lumber Company** (Eau Claire, Wisconsin), Papers, 1881-1944, 5 boxes (including 15 vols.) and 24 additional vols. Records of a firm which engaged in lumbering in northwestern Wisconsin from 1873 until the depletion of timber supplies halted operations about 1920. The company was not legally dissolved until the mid-1940s. Records include journals, ledgers, cashbooks, and tax records covering the company's operations (1894+); annotated plat books and land inventories show property holdings, and appraisals (2 vols.) describe the plant and equipment at Stanley, Wisconsin (1909); correspondence relates primarily to the management and gradual liquidation of the company's assets (1920-1940s). Most of the letters were written by George W. Hipke of Stanley, an officer of the company, and many pertain to his career in the state legislature (1930s). The collection also contains scattered records of related or subsidiary firms: The North Western Mercantile Company (1919-32); the Stanley, Merrill, and Phillips Railway (1902-26), both in Wisconsin; and the Del Norte Company, a lumber company operating in northern California. In the Eau Claire Area Research Center.

3450 **North Wisconsin Lumber Company** (Hayward, Wisconsin), Correspondence, 1883-88, 1897-1901, microfilm, 1 reel. Incoming letters of the North Wisconsin Lumber Company, a supplier of logs and lumber, with information on their sales operations throughout the Midwest and on their relationship with the Laird, Norton Company, Winona, Minnesota, and with Frederick Weyerhaeuser. Originals owned by Keith Schwab. In the Eau Claire Area Research Center.

3451 **Noyes, Haskell** (b. 1886), Papers, 1924-1943, 1 box. Papers of a Milwaukee conservationist, president of the Wisconsin division of the Izaak Walton League of America in 1925, and member of the State Conservation Commission 1928-33. Some correspondence and reports relate to his work with the Izaak Walton League. Numerous congratulatory letters concern his appointments to the Conservation Commission in 1928, to its chairmanship in 1931, and to the federal advisory committee on the Migratory Bird Act in 1932. Among his correspondents were other Wisconsin leaders of the conservation movement, including William Aberg, Charles Broughton, Halbert L. Hoard, Wilhelmine La Budde, and William Mauthe. In the Milwaukee Area Research Center.

3452 **Owen, John S., Lumber Company,** Papers, 1875-1955, 95 boxes and 565 vols. Business records covering lumbering and logging operations in northwestern Wisconsin, and related enterprises in lumber, real estate, railroads, and cotton in the Pacific Northwest, South Dakota, Missouri, Arkansas, and Louisiana. Owen engaged in the lumber business in Wisconsin after 1873; with Aloney J. Rust and Ralph E. Rust, he organized in 1882 the Rust-Owen Lumber Company with offices in Eau Claire and Drummond. In 1893 Owen organized the John S. Owen Lumber Company with offices in Eau Claire and Owen. His sons, John G., Ralph W., and Frank G., took active parts in both of these firms. Subsidiary and related companies operating mainly in other states included the Oregon Lumber Company; the California and Oregon Lumber Company; Del Norte Company, Limited; Three States Lumber Company; Gilbert Lumber Company; Owen Box and Crating Company; Pierre Lumber Company; Owen Falls Mining Company; the

Blytheville, Burdette and Mississippi River Railroad Company; and the Drummond and Southwestern Railway Company. The records contain letterpress books (1875-1951, 246 vols.), and incoming correspondence (1920-51, 40 boxes) relating primarily to the Wisconsin operations of the Rust-Owen Lumber Company and the John S. Owen Lumber Company. Additional letterbooks (37 vols.) and incoming letters (22 boxes) comprise correspondence files for some of the other related Owen firms. The correspondence files are supplemented by series of incorporation papers, minutes, ledgers, journals, cashbooks, payroll balances, operating statements, stock books, land inventories, and contracts. Many of the series of minutes and financial records are fragmentary, particularly for the companies operating outside of Wisconsin. In the Eau Claire Area Research Center.

3453 **Paine Family,** Papers, 1822-1888, microfilm, 1 reel. Notebook containing typewritten copies of papers of the family of Edward Lathrop Paine, Oshkosh lumber manufacturer. Originals in the possession of the Paine Art Center, Oshkosh, Wisconsin.

3454 **Paine Lumber Company** (Oshkosh, Wisconsin), Papers, 1866-1915, 35 feet (155 vols.). Financial records of lumber milling, and allied business operations, including daybooks, cashbooks, customers' journals, accounts receivable, ledgers, and indexes. The collection also includes financial records, and inventories for the company's office and lumber operations at Merrillan, Wisconsin, and for the Minnesota Percheron Horse Company, which may have been part of the Paine enterprises; lumber order books for yards at Hastings, Nebraska, Grand Forks, Dakota Territory, and Arcadia, Whitehall, and Independence, Wisconsin; pine-tax books with legal descriptions of pineries throughout Wisconsin and the Midwest; and log records which describe the type and amount of wood shipped.

3455 **Philipp, Emanuel L.** (1861-1925), Papers, 1914-1920, 20 boxes (including 8 vols.), 2 packages, and 15 vols. Governor of Wisconsin, 1915-20. Includes correspondence on Wisconsin's part in the relief work for northern Minnesota fire victims in 1919.

3456 **Pike, Robinson D.** (1838-1905), Papers, 1 folder. Biographical material, including a military record and newspaper clippings, concerning Robinson D. Pike, a lumber merchant and founder of Bayfield, Wisconsin. In the Ashland Area Research Center.

3457 **Plumbe, John** (1809-1857), Diary, 1838-1839, 1 vol. Diary kept by the owner and promoter of the boom town of Sinipee on the Mississippi River. Some entries touch on lumber rafting. In the Platteville Area Research Center.

3458 **Polleys, William H.** (1824-1906), Papers, 1853-1895, 3 boxes (including 21 vols.). Correspondence (1853-95); miscellaneous business papers; and various logging camp account books (21 vols.). Inventory. In the La Crosse Area Research Center.

3459 **Prentice, Jackson L.** (1827-1902), Papers, 1852-1922, 2 boxes (including 18 vols.). Papers of a Stevens Point, Wisconsin, surveyor, including fragmentary correspondence (1857-1907), mostly from business associates concerning timber sales; timber survey maps (1850s-60s) for Adams, Juneau, Marathon, and Portage counties, Wisconsin; survey notebooks (1852-89, 15 vols.) containing descriptions of land conditions, plant cover, and soil; and employee time records (1881, 1916-22, 2 vols.). In the Stevens Point Area Research Center.

3460 **Reinhard, David G.** (b. 1867), Manuscript, 1944 and 1945, microfilm, 1 reel. Recollections of early lumbering and pioneering experiences at Omro, near Horicon Marsh, and in Shawano County, Wisconsin.

3461 **Rich Brothers Manufacturing Company** (Ontonagon, Michigan), Records, 1846-1895, 6 vols. Mainly records of business enterprises engaged in by three brothers, A.J., H.H., and Martin Rich, one-time residents of Horicon, Wisconsin. Letterbooks (1874-82, 3 vols.) relate to the manufacture of lumber and shingles, and trade in general merchandise in Wisconsin and Michigan. There are also a few sales records of an unidentified merchant dealing in iron, lumber, and other products in Southport (Kenosha), Wisconsin (1846-47, 1 vol.). In the Oshkosh Area Research Center.

3462 **Ritchie, James,** Letter, December 3, 1884, 1 item. Letter describing wages and conditions in a lumber camp at Weyerhauser, Wisconsin.

3463 **Roese, Alfred E.** (1862-1920), Correspondence, 1904-1910, 69 items. Osceola, Wisconsin; editor. Chiefly letters by Roese, offering assistance to John F. Dietz of Winter, Wisconsin, during the latter's controversy with the Chippewa Lumber and Boom Company over the company's practice of shipping logs past Dietz's dam without payment. Includes a few of Dietz's replies. In the Eau Claire Area Research Center.

3464 **Rusk, Jeremiah McLain** (1830-1893), Correspondence, 1862-1898, 14 vols. and 5 boxes. Governor of Wisconsin and U.S. Secretary of Agriculture, 1889-93. Correspondence and letterbooks concerning state and national politics, the St. Croix and Lake Superior Railroad land grant. Material relating to the Department of Agriculture seems to relate primarily to meat and livestock marketing and inspection of agricultural products. Part of a letterbook contains letters of Norman J. Colman, the last federal Commissioner of Agriculture, relating almost wholly to personnel.

3465 **Sawyer, Philetus** (1816-1900), Papers, 1866, 1883, photostatic copies, 3 items. Includes a letter from William D. Goodman to Edgar P. Sawyer concerning the affairs of the Sawyer-Goodman Lumber Company. Originals were in the Oshkosh Public Museum.

3466 **Scholfield (Schofield), William B.** (d. 1863), Papers, 1820, 1845-1857, 8 items and 2 vols. Includes business papers (1845-57) at Wausau, consisting mainly of contracts to sell lumber.

3467 **Sherman, Simon Augustus** (1824-1906), Papers, 1848-1906, 3 boxes (including 45 vols.). Pioneer settler and lumberman of the upper Wisconsin River. Original diaries, containing genealogical data, financial and business records, information on early Portage County, and entries relating to a number of inventions made by Sherman. Also typed transcripts of a few excerpts from the diaries.

3468 **Sherman, Simon A.** (1824-1906), Reminiscences, 1866, microfilm, 1 reel. Describes Sherman's experiences lumber rafting on the Wisconsin River, near Plover, Portage County, Wisconsin.

3469 **Sibley, Henry H.** (1811-1891), Papers, 1826-1848, microfilm, 1 reel. Mendota, Minnesota; fur trader. This collection consists of letters and documents relating to Wisconsin (ca. 325 items), selected from the papers of Sibley at the Minnesota Historical Society. They contain some information on timber speculation. Originals are in the Minnesota Historical Society. Microfilm copies are available at the State Historical Society of Wisconsin, Madison, and in the River Falls and Ashland Area Research Centers.

3470 **Smith, Andrew J.** (1832-1929), Autobiography, 1926, 5 pp. Mimeographed. Includes a Xerox copy of a newspaper article describing his life. Smith cruised timber in northeastern Wisconsin for 29 years. In the Green Bay Area Research Center.

3471 **Sorenson, Gordon,** Report, ca. 1969. Report by Sorenson on the history of logging and logging camps in the Washburn district of the Chequamegon National Forest, Wisconsin.

3472 **Spaulding, Dudley J.** (b. 1834), Papers, 1849-1899, 5 boxes (including 4 vols.). Black River Falls, Wisconsin; lumberman. The collection consists largely of letters received (1856-75), containing information on Spaulding's lumber business in Iowa, Texas, Arkansas, Louisiana, and Wisconsin. Other papers pertain to his operation of a sawmill, a sash, door, and blind factory, and other businesses in Wisconsin. In the La Crosse Area Research Center.

3473 **Sprague, Joseph D.** (1825-1873), Correspondence, 1801-1873, 19 items. One letter (1867) gives an account of the I.W. Woodruff Company's sawmill operations near Green Bay, Wisconsin.

3474 **Starr, William J.** (1861-1921), Papers, 1886, 1902-38, 5 boxes (including 7 vols.). Eau Claire, Wisconsin; lumberman. Most of the papers relate to the period 1919-27. One group pertains to the Davis & Starr Lumber Company of Eau Claire, including correspondence concerning the company's timber and real estate holdings in Wisconsin and the development of the town of Frederic, incorporation papers, and letterbooks (1920-24). Other papers relate to Starr's investments in real estate and timberlands in Shasta County, California. Among Starr's business associates represented in this correspondence are A.L. Arpin, C. T. Bundy, W.L. Davis, Burt E. De Yo, J.D.R. Steven, J.B. Trasker, the Red River Lumber Company (representing the California timber interests of T.B. Walker), and the California Forest Protective Association. In the Eau Claire Area Research Center.

3475 **Stout, James Huff** (1848-1910), Papers, 19 pp. Biography (5 pp.) of Wisconsin state senator, James Huff Stout, of Knapp, Stout & Co., Company, and founder of Stout Institute, Menomonie, Wisconsin; and the genealogy (14 pp.) of his wife, Angeline Wilson Stout, descendant of Capt. William Wilson, one of the first settlers of Menomonie and member of the lumber company. Xerox copies. In the Stout Area Research Center.

3476 **Strange, John** (1853-1923), Papers, 1902-1935, microfilm, 3 reels. Papers of the Menasha, Wisconsin, paper manufacturer and lieutenant governor of Wisconsin, 1909-11, including an autobiography (ca. 1922) relating to business and political activities, and correspondence (1909-23), with political leaders; also a minute book (1902-35), for the meetings of the directors and stockholders of the John Strange Paper Company. Copies in the State Historical Society, Madison, and the Oshkosh Area Research Center.

3477 **Strong, Moses McCure** (1810-1894), Papers, 1774-1894, 41 boxes (including 50 vols.), and 34 additional vols. Mineral Point, Wisconsin. The collection includes lumber papers dealing with operations around Stevens Point and Nekoosa, Wisconsin. They touch on many details of the business, such as methods, wages, personnel, and organization, stressing problems of capital, and the special difficulties connected with lumbering on the Wisconsin River.

3478 **Stuntz, Albert C.** (1825-1914), Diaries, 1858, 1863-1865, 1867-1869, 1882, 1 box (including 8 vols.) and 1 typescript vol. Original diaries, with typed copies, in part relating to Stuntz's work as timber agent for school and university lands in the upper St. Croix Valley. In the State Historical Society, Madison. A Xerox copy of the typescript volume is in the River Falls Area Research Center.

3479 **Superior Fire Sufferers' Fund,** Papers, 1894-1895, 1 box. A citizens' relief committee organized to aid victims of the 1894 forest fires in northern Wisconsin; included are proceedings, correspondence, financial records, statements by investigators of individual losses and injuries, and clippings. Inventory. In the Superior Area Research Center.

3480 **Synnott, J.J.,** Papers, 1896-1916, 13 pieces. Contains a letter concerning the sale of hospital insurance in Wisconsin lumber camps, and hospital tickets for hospitals in Wausau and Merrill, Wisconsin.

3481 **Taylor, William R.** (1820-1909), Papers, 1848-1919, 4 boxes. Includes some items concerning lumber rafting on the Wisconsin River.

3482 **Timlin, William H.** (1852-1916), Reminiscences, 1915, 1 item, typewritten copy (69 pp). Describes experiences as a lumber worker, educator, and attorney; also describes the forest fires of 1871, and the changes brought about in Anglo-Saxon communities by the influx of immigrants. In the State Historical Society, Madison. Xerox copy available at the Milwaukee Area Research Center.

3483 **United Brotherhood of Carpenters & Joiners of America, Local 91** (Racine, Wisconsin), Records, 1899-1947, 6 boxes (including 35 vols.). Minute books (1905-47), daybooks (1912-36), dues books (1899-1920, 1934), and a book listing members in arrears on dues (1930-34). In the Parkside Area Research Center.

3484 **United Brotherhood of Carpenters & Joiners of America, Local 161** (Kenosha, Wisconsin), Minute books, 1897-1921, 2 boxes (7 vols.). In the Parkside Area Research Center.

3485 **United Brotherhood of Carpenters and Joiners of America, Local 454** (West Superior, Wisconsin), Financial book, 1888-1897, 1 box (including 1 vol.). In the Superior Area Research Center.

3486 **United Brotherhood of Carpenters and Joiners of America, Local 630** (Neenah, Wisconsin), Statement, 1946, 13 pp. Statement (June 1, 1946) by the Hardwood Products Corporation, Neenah, Wisconsin, detailing their position during wage arbitration. Typewritten. In the Oshkosh Area Research Center.

3487 **United Brotherhood of Carpenters and Joiners of America, Local 755** (Superior, Wisconsin), Records, 1901-1946, 6 boxes (including 22 vols.). Minute books (1901-44), dues books (1901-26, 1939-46), and a business agent's record book (1905-07). In the Superior Area Research Center.

3488 **United Brotherhood of Carpenters and Joiners of America, Local 1143** (La Crosse, Wisconsin), Records, 1894-1938, 2 boxes (including 11 vols.). Minute book (1926-36, 1 vol.); daybooks (1932-36); and dues books (1894-1906, 1915-19, 1930-32, 1934-38). In the La Crosse Area Research Center.

3489 **United Brotherhood of Carpenters and Joiners of America, Local 1246** (Marinette, Wisconsin, and Menominee, Michigan), Records, 1896-1949, 4 boxes (including 28 vols. and 1 package). Minutes of meetings (1913-46); daybooks (1902-49); ledgers (1902-46); treasurers' cashbooks (1896-1939); and miscellaneous correspondence, receipts and other papers. Container list. In the Green Bay Area Research Center.

3490 **U.S. Department of the Interior. Office of Indian Affairs,** Records, 1857, 1909-1939, microfilm, 10 reels. Includes intermittent reports by federal inspectors, with reference to lumbering within the area under the jurisdiction of the reporting officer.

3491 **Van Ostrand, Edwin H.** (1860-1928), Papers (1738, 1805) 1829-1911, 2 boxes (including 2 vols.). Dealer in timber lands in the Neenah and Antigo areas of Wisconsin, including business correspondence (1899-1901, 1909-11); some family letters and genealogical notes; agreements and deeds; and two diaries (1881, 1884). In the Oshkosh Area Research Center. Inventory.

3492 **Vilas, William Freeman** (1840-1908), Papers, 1827-1959, 83 boxes (including 29 vols.) and 16 vols. U.S. Senator from Wisconsin, 1891-97; served on Public Lands Committee; U.S. Secretary of the Interior, 1888-89. The collection includes correspondence, diary, legal and business papers and miscellaneous papers. The bulk of the material relates to state and national politics for two decades beginning 1880. Includes material on the U.S. General Land Office in the 1890s. Additional papers contain memoranda and statistics compiled for reports; notations on Land Office cases; and other papers relating to Vilas' position as Secretary of the Interior. Partial inventory.

3493 **Warner, Jared** (ca. 1811-1880), Papers, 1834-1902, 1 box (including 9 vols.) and 15 vols. Business and personal papers of a Grant County, Wisconsin, pioneer, containing daybooks (1836-80), a ledger (1849-59), and a number of small memorandum and account books relating to his sawmill, lumberyard, and general mercantile activities at Millville and Patch Grove, Wisconsin. In the Platteville Area Research Center.

3494 **Waterston, Healy A.** (1862-1955?). Notes concerning Healy A. Waterston, a lumber camp cookee and logger above Chippewa Falls.

3495 **Wheeler, J. Russell** (1875?-1966), Papers, 1848-1927, 16 boxes, including 12 vols. Correspondence after 1900 pertains mainly to the business enterprises of Wheeler's father, John Russell Wheeler, a stockholder and director of the Rib Lake Lumber Company; these letters describe hardwood lumbering operations in northern Wisconsin and farmland promotion, financing, and acquisition in Jackson and Clark counties.

3496 **Wheelock, Mary Bascom.** A history of Clifton Hollow, a settlement in northwestern Pierce County, Wisconsin, 1849-1898, mentioning lumber and grist mills. 16 pp. In the River Falls Area Research Center.

3497 **Wisconsin Audubon Society,** Records, 1898-1917, 1 box. Papers concerning effort to arouse public sentiment and secure legislation for the protection of wild birds.

3498 **Wisconsin Colonization Company,** Papers, 1916-1938, 1 box and 1 vol. Papers collected by Benjamin Faast, president of the company relating to the organization's sale of "ready-made" farms in southern Sawyer County and the development of the village of Ojibwa. Includes historical data on the upper Chippewa River area and the lumber industry. In the Eau Claire Area Research Center.

3499 **Wisconsin State Federation of Women's Clubs,** Papers, 1899-1954, 5 boxes and 14 vols. Correspondence, minutes of board meetings and annual conventions (1914-54), audit reports (1926-53), financial reports (1900-52), a history of the organization, scrapbooks, and other materials concerning the federation's work in conservation, and other activities.

3500 **Woodman, Cyrus** (1814-1889), Papers, 1833-1889, 2 boxes (including 9 vols.) and 206 additional vols. Correspondence and accounts (1840-44, 6 vols.), dealing with Woodman's services in northern Illinois for the Boston and Western Land Company. Correspondence after 1844 deals with the transactions of the real estate office of Woodman and Cadwallader C. Washburn at Mineral Point, Wisconsin, which included entry of public lands for settlers, acting as agent between eastern landholders and western purchasers, and investing in and managing agricultural, lumber, and mineral lands in Wisconsin, Minnesota, Iowa, Nebraska, and other states. Woodman's allied operations included lumbering on the Saginaw, the upper Wisconsin, the Black, and other rivers, and managing the St. Mary's Ship Canal Company's grant of pinelands in lower Michigan, 1862-64.

3501 **Wright, J.C.** "The Old Wiscons." A Wisconsin River lumbering song recited to Dr. C.F. Rudolf by Dr. J.C. Wright. Typewritten copy.

3502 **Wright, William W.** (1819-1903), Papers, 1841, 143 pp. Typed transcripts of reminiscences, describing the construction of a sawmill and gristmill at Brotherton, Wisconsin.

3503 **Yawkey Lumber Company** (Wausau, Wisconsin), Records, 27 cartons. Unprocessed.

Wisconsin State Archives

3504 **Agriculture Department,** Records. Records of the Information Division contain notes and working papers for 1937 and 1941 land cover surveys (1928-41, 93 boxes). Records of the Plant Industry Division include laboratory reports, survey reports, correspondence, and related material concerning detection of Dutch elm disease (1956-63, 3 cubic feet [Unorganized]).

3505 **Conservation Department,** Records. Included are minutes of the State Board of Forestry (1904-15, 1 envelope); minutes of the State Park Board (1909-15, 1 envelope); minutes of the Conservation Commission (1927-32, 1941-64, 8 boxes and 2 vols.); correspondence of E.M. Dahlberg, secretary of the Conservation Commission (1928-31, 1 box); report to the Conservation Commission concerning field position salaries and working conditions in Wisconsin, as compared with other states (1952, 1 vol.); subject file (1922-49, 90 boxes and 7 vols.) including minutes, correspondence, memoranda, reports, and other materials originating throughout the department, pertaining to fish and game matters, lumbering, and enforcement of conservation laws; animal, fish, and tree population control; prevention, control, and repair of game, fire, and disease damage; parks and forests; control over navigable waters; coordination with other public and private agencies; projects planned; education, training, and publicity; research; legislation; administration of the department; federal aid; and other activities. Also additional unorganized material (1927-57, 116 cubic feet).

3506 **Defense Councils,** Records. State Council of Defense, World War I, Records (1917-19, 28 boxes and 1 vol.) include records of Non-War Construction Committee.

3507 **Executive Department,** Records. Papers of individual governors include those of Oscar Rennebohm (1947-51, 71 boxes), with correspondence, departmental, appointment, and subject files; Walter J. Kohler, Jr. (1951-56, 195 boxes), with correspondence, reports, and memoranda concerning legislation, and correspondence with various federal and state departments. Records concerning settlement and internal improvements include correspondence (3 boxes) pertaining to the appointment, supervision and functions of timber agents

(1858-90) and lumber inspectors (1861-1905). Records concerning relief activities contain correspondence and reports (1871-1915, 1 cubic foot) largely relating to relief efforts following the Peshtigo fire, 1871; and bound ledger sheets (1871, 1 vol.) showing contributions received and amounts paid out for relief of fire sufferers.

3508 **Lumber Inspectors,** 1864-1927. Six state lumber inspection districts, based on river systems, were established in 1864; new districts were added until 1919 when the 17 old districts were replaced by 4 new districts covering the entire state. The inspectors existed to protect the interests of owners of logs floated downstream to market; their two main functions were to prepare statements of numbers of logs and numbers of board feet in shipments of logs; and to register log marks. The system was abolished in 1927. The records include: miscellaneous records of District 6 lumber inspector, Chippewa Falls (1864-1927, 1 box and 17 vols.); weekly record of lumber scaled at logging camps in District 6 (1881-82, 1 vol.); record of amount of timber cut, listed by area in District 6 (undated, 1 vol.); miscellaneous records of District 11 lumber inspector (Ashland, 1875-1927, 2 vols.); and record of marks registered for use on logs in Districts 6 and 14 (1864-1911, 3 vols.).

3509 **Public Land Commissioners,** Records. Chief Timber Cruiser's correspondence (1934-43, 2 boxes).

3510 **Secretary of State,** Records. Records of the Division of Elections and Records contain materials of the Commissioners of the Public Lands, including approvals for the sale of lands to the U.S. for national forests (1928-36, 1 box); also deeds and abstracts of title to land purchased by the state for forestry purposes (1884-1957, 46 boxes), sometimes accompanied by attorney general's opinions and governors' approvals of purchase, with an alphabetical card index. Records of the Corporations Division include annual reports of finances and operations of log driving and booming companies (1893-97, 1 box).

3511 **Treasurer,** Records. Records concerning railroad, telegraph, and boom companies include materials related to railroad lands (1877-89, 2 boxes and 2 vols.), containing record of payments made by persons occupying railroad lands to obtain title, lists relating to taxation of railroad lands; annual reports (12 boxes) of railroads (1852-1905), telegraph companies (1856-98), and boom companies (1891-1917). Records relating to various subjects contain abstracts of reports received from state timber agents (1862-64, 1 vol.) concerning trespass and illegal lumbering on state lands; and list of forest crop lands under county forest reserves (1930-35, 1 vol.).

University of Wisconsin, Division of Archives

Director, Memorial Library 53706. Records in the Archives may be used only with permission of the donor or office of origin.

3512 **Forestry Department,** Records. Correspondence, newsletters, and publications. Materials of particular interest include a "Fifty-Year History of Wisconsin Forestry" (1 folder); Wisconsin wood resources (1 box); and material on the Shelterbelt Project.

3513 **Leopold, Aldo** (1886-1948), Papers, 1903-1948, 25 cubic feet (75 boxes). Leopold served with the U.S. Forest Service, Southwestern Division (District 3), 1909-24; Forest Products Laboratory, Madison, Wisconsin, 1924-28; professor of wildlife management, University of Wisconsin, 1933-48. The records are largely dated since 1928. The collection chiefly pertains to wildlife management, but some parts relate to the preservation of forest lands. Among the many conservation organizations and agencies represented by correspondence, those concerned with forests include the American Forestry Association (1 folder, 1928-47), relating to publications; Friends of the Land (1940-41, 1 folder); Isle Royale (1946-47, 1 folder), relating to a proposed wildlife study; Izaak Walton League of America (1937-48, 1 folder) and of Wisconsin (1925-48, 1 folder); National Land Use Planning Committee (1932-33, 1 folder); National Park Service (1934-48, 2 folders), chiefly relating to wildlife, particularly elk in Yellowstone; National Research Council, Aircraft Committee (1946-47, 1 folder), concerning control of airplanes over wilderness areas; and the Committee on Aircraft vs. Wilderness (1945-47, 1 folder); Quetico-Superior Council (1 folder, 1942-47); Society of American Foresters (1934-38, 1 folder); U.S. Forest Service (ca. 1932-35, 4 folders), relating to game activities in Wisconsin, Emergency Conservation Work in Wisconsin and the Southwest, etc.; Wilderness Society (1934-48, 4 folders), relating to Leopold's work as a Council member and a trustee of the Robert Marshall Wilderness Fund. Subject files include a file on wilderness areas (1922-48, 48 items), including correspondence, memos, maps, manuscripts, clippings, etc., relating to the Gila Wilderness Area and wilderness policy in general. Research Projects files include material relating to Leopold's work as director of animal research on the University of Wisconsin Arboretum (1933-48, 1 box). Leopold's publications files contain voluminous material, including correspondence, manuscripts, reprints, etc. There are U.S. Forest Service diaries (1909-16, 4 vols. and loose material, incomplete), and other diaries and journals of Leopold. Forest Service records in the collection include a file of the *Carson Pine Cone,* staff publication of the Carson National Forest, New Mexico, with related material (1911-14); mimeographed U.S. Forest Service District 3 handbooks compiled by Leopold on fish and game (1915) and watersheds (1923); copies of miscellaneous Leopold documents from the Apache and Carson National Forests, and Region 3 office, Albuquerque, New Mexico, including correspondence, memoranda, and reports by Leopold or relating to his work (1909-23, 7 folders and 1 folio, Xerox copies of originals in Forest Service offices, in the Federal Records Center, Denver, the Carhart Papers in the Denver Public Library, and in Grand Canyon National Park). Also among Forest Service records in the collection are Leopold's official federal service personnel record (1909-39, 1 reel microfilm of originals in the St. Louis Federal Records Center), and selected records relating to Leopold's work for the Forest Service (ca. 1910-23, 1 reel microfilm of originals in the National Archives). There are also course notes taken by Leopold as a student at Yale Forest School (1905-09). Guide, including cross-references to Leopold materials located elsewhere, bibliographies of published and unpublished writings, etc., 179 l., typed.

3514 **Van Hise, Charles R.,** Papers, 1875-1918, 9 boxes. Incoming letters and copies of outgoing letters, telegrams, and mimeographed and printed items. Those subjects in this material relating to forest history and related topics include conservation of natural resources, U.S. Geological Survey, forestry conferences, and proposed soils and minerals report.

MENOMONIE

University of Wisconsin, Stout, Area Research Center

Curator, Robert L. Pierce Library 54751 Telephone: (715) 232-1225. Holdings are described under the State Historical Society of Wisconsin.

MILWAUKEE

University of Wisconsin, Area Research Center

University Archivist, Library, 2311 East Hartford Avenue 53201 Telephone: (414) 963-5402. Holdings are described under the State Historical Society of Wisconsin, Madison.

NEENAH

Kimberly-Clark Corporation, Archives

North Lake Street 54956 Telephone: (414) 231-2010

3514a Holdings include material dating from 1872 to the present, and consist of photographs, annual reports, financial records, house organs, press clippings, training brochures, pamphlets, minutes, articles of incorporation, correspondence, and other records. Restricted: open to scholars upon request and approval.

OSHKOSH

Oshkosh Public Museum

1331 Agoma Boulevard 54901 Telephone: (414) 231-2010

3515 **Chippewa Lumber Industry,** Collection, 1859-1927, 30 items. Copies of letters from Donald Kennedy to O.H. Ingram (1859-60, 4 items, originals in the State Historical Society, Madison); typescripts of letters by Joseph Viles, S.T. McKnight, and others (1869-94, 8 items, apparently published in the Eau Claire *Leader* ca. 1920, and copied therefrom); typescripts of newspaper articles (memoirs, etc.).

3516 **Fox-Wolf River Valleys, Lumbering Collection.** The collection contains papers, clippings, and booklets relating to the following: Paine Lumber Company; American Excelsior Company, formerly Oshkosh Bottle Wrapper; Pluswood, formerly Paine Lumber Company's veneer mill; Morgan Lumber Company; Buckstaff Company; McMillans; Foster-Lotham; Badger Lumber Company, formerly W.J. Campbell; Robert Brand & Son; Sawyer and Son; Bandrob and Chase Furniture; Gould Manufacturing; Diamond, formerly Star, Match Company; the Oshkosh sash and door industry; and carriage and furniture manufacture. There are papers on water transportation which include material concerning the steamboat in the growth of the local lumber industry. Biographical files are available on Leander Choate, James L. Clark, E.N. Conlee, Carleton Foster, James Larson, Daniel Libbey, the Morgans, the Paines, and the Radford and Sawyer families. Another section of the files comprising stories of the early settlers contains much information on early lumbering. There are also log marks from Oshkosh and vicinity (1870s, 3 vols.). They contain about 400 log marks, of which some 230 are identified as to name of owner, quantity of logs, etc.

3517 **Lake Poygan Boom-Bay,** Collection, ca. 1941-1958, 5 items. Manuscripts: W.J. Campbell, "Bay Boom Days in Oshkosh" (1941, 11 pp.); Elaine Neumann, "History of Boom Bay" (4 pp.); Royal Richter, "Logging on Boom Bay" (6 pp.); clippings.

University of Wisconsin, Area Research Center

Curator, Forrest R. Polk Library, 800 Algoma Boulevard 54901 Telephone: (414) 424-3320. Holdings are described under the State Historical Society of Wisconsin, Madison.

PLATTEVILLE

University of Wisconsin, Area Research Center

Curator, Elton E. Karrman Library. 725 West Main Street 53818 Telephone: (608) 342-1643 (a.m.), 342-1757 (p.m.). Holdings are described under the State Historical Society, Madison.

RIVER FALLS

University of Wisconsin, Area Research Center

Curator, Chalmer Davee Library, 120 Cascade Avenue 54022 Telephone: (715) 425-3567. Holdings are described under the State Historical Society, Madison.

STEVENS POINT

University of Wisconsin, Area Research Center

Curator, Learning Resources Center, 2100 Main Street 54481 Telephone: (715) 346-3726. Holdings are described under the State Historical Society of Wisconsin, Madison.

SUPERIOR

Douglas County Historical Museum

Director, 906 East 2nd Street 54880

3518 Holdings include unpublished manuscript local histories, relating, in whole or in part, to logging and lumbering in Wisconsin (ca. 1940, 6 items). They include: Norris C. Dickey, "Connors Point, Brief History: Its Rise and Decline"; Benjamin Finch, "Early Logging in the Northwest," "History of George B. Stuntz"; Otto Wieland, "Lumber Industry at the Head of the Lakes"; Achile H. Bertrand, "Recollections of Old Superior"; John A. Bardon, "Recollections of Early Days in Superior" (2 vols.). There are also photographs (275 items) of logging, lumbering, sawmills, and wood products factories in northern Wisconsin and northeastern Minnesota. A clipping file contains newspaper clippings and transcripts of newspaper articles on logging, lumber, and wood products.

University of Wisconsin, Area Research Center

Curator, Jim Dan Hill Library, 18th and Grand Avenue 54880 Telephone: (715) 392-8101 (ext. 235 or 333). Holdings are described under the State Historical Society of Wisconsin, Madison.

WHITEWATER

University of Wisconsin, Area Research Center

University Archivist, Harold Anderson Library, West Main Street 53190 Telephone: (414) 472-1025. Holdings are described under the State Historical Society of Wisconsin, Madison.

Wyoming

CHEYENNE

Wyoming State Archives and Historical Department, Historical Research and Publications Division

State Office Building 82002

3519 Holdings include several manuscript histories: Paul L. Armstrong, "History of the Medicine Bow National Forest" (ca. 1935, 28 pp.); James F. Conner, "History of the Bighorn National Forest and Vicinity" (1940, 123 pp.); Neal L. Blair, scrapbook of correspondence, manuscripts, diaries, notes, concerning Bridger National Forest (ca. 1940s-60s, 1 reel microfilm); anonymous untitled collection of notes, correspondence, and memoranda relating to the Medicine Bow National Forest (ca. 1922-28, 20 pp., typescript); Hazel Hall, oral history interview relating to the tie-cutting operations of the Clear Creek Tie and Timber Company, Buffalo, Wyoming (10 minutes tape, 6 pp. typescript); and U.S. Forest Service records (ca. 1906-58) relating to Bighorn National Forest, including diaries (26 vols.), notebooks (2 vols.), and correspondence and documents (14 folders).

LARAMIE

University of Wyoming, Conservation History and Research Center

Research Historian, Box 3334, The Library 82070

3520 **American National Cattlemen's Association,** Records, 1898 +, 55 feet. Includes files on public lands and grazing. There are publications, reports, proceedings and minutes (1900-59), resolutions (1917-55), state resolutions (1953-57), programs of national conventions, etc.

3521 **Baker, Joseph,** Collection, 50 items, 1906. Photographs of Baker and his bride on their honeymoon in Yellowstone National Park.

3522 **Barrett, Frank A.** (1892-1962), Papers, 1942-1959, 60 feet. U.S. Representative from Wyoming, 1943-50; member of House committees on Irrigation and Reclamation, and Public Lands; Governor of Wyoming, 1951-53; U.S. Senator, 1953-59; member of Senate Interior and Insular Affairs Committee. The collection contains his congressional files of correspondence, reports, legislation, and miscellaneous items, arranged by subject. Many of the papers deal with western land policies and resource development.

3523 **Cliff, Edward P.,** Collection, 10 inches. Includes 2 color filmstrips, 2 slide sets, and narrative guides relating to national forests; reports on timber management policies on the Bitterroot, Monongahela, Teton, Bridger, Shoshone, and Bighorn National Forests (1970-71); speeches, articles, statements, and press reports of Cliff (1936-72, 3 vols. looseleaf notebooks). Inventory, 18 pp.

3524 **Coe, Isaac** (1816-1899) **and Levi Carter** (1830-1904), Business papers, ca. 1861-1934, 3 feet (ca. 700 items). Records of a partnership, ca. 1859-89, holding diverse interests in mining, railroad, real estate, lumber, freighting, and cattle industries in California, Nebraska, Idaho, Texas, Utah, Colorado, Oregon, Wyoming. In 1866 the firm contracted to furnish hay and wood to Fort Philip Kearny, Wyoming, and other western posts. In 1867 it contracted to furnish ties for the Union Pacific Railroad, and maintained tie camps at Pine Bluffs, Wyoming, and later at Sherman Station, 2½ miles north of Fort Senders. The collection comprises personal notebooks of Isaac Coe (4 vols.); Coe-Carter bankbook and cashbook of Isaac's son, Frank Coe (1909-15). There are individual business papers (1861-1934, 691 items), including deeds, abstracts, bills of sale, tax receipts, maps, mortgages, memoranda, and sales notes.

3525 **Demaray, Arthur E.,** Papers, ca. 13½ feet. National Park Service employee after 1917, director in 1951. The collection includes correspondence (1902-58); diaries (scattered, 1937-55); notebooks concerning park administration; notes of Demaray's travels to various regions (1935-51); printed materials; photographs (18 albums and ca. 1,800 items) concentrating on the western parks, especially Yellowstone and Yosemite.

3526 **Drury, Newton B.,** Collection, 1938-1950, 5 inches. Xerox copies of materials pertaining to Jackson Hole National Monument; includes articles, documents, correspondence (1943-50), lists of conservation organizations and witnesses, memoranda (1943-50), notes, clippings, reports, statements, statistics, miscellaneous items. Correspondents and authors of memoranda include Harold P. Fabian, Valentine Webb, Joseph C. O'Mahoney, Kenneth Chorley, John C. Hazen, Ray Lyman Wilbur, Hillory A. Tolson, Harold L. Ickes, Arthur E. Demaray, Leslie A. Miller, Oscar L. Chapman, Maud Noble, James E. Shelton, Lester C. Hunt, Charles C. Moore, Laurence S. Rockefeller, Arthur H. Carhart, Struthers Burt, Horace H. Albright, Olaus J. Murie, Waldo G. Leland, Frederick J. Lawton, Dale E. Doty, Roger W. Jones, Ethel L. Larson, E.E. Byerrum, J.W. Penfold, John S. McLaughlin, Lawrence C. Merriam, Michael W. Straus, Charles W. Porter, Ben H. Thompson, Herbert J. Slaughter, Conrad L. Wirth, Mr. Price, Mr. Richey, Superintendent Grand Teton National Park, Mr. Breeze, and other National Park Service officials. Inventory, 17 pp.

3527 **Fryxell, Fritiof M.,** Papers, 10 feet. Largely research notes and artifacts gathered for a biography of Ferdinand Vandeveer Hayden. There is also correspondence; geological writings of Fryxell; photographs (ca. 300 items), including prints of Yellowstone area by William Henry Jackson; manuscripts of western U.S. history by Adam N. Keith; extracts from Fryxell's diary on Grand Teton National Park.

3528 **Georgia Forest Research Council,** Reports and research papers, 1956-72, 10 inches. Reports 1-30 (1956-72), and research papers, 1-71, (1961-71).

3529 **Handley, Rolland B.,** Papers, ca. 1½ feet. Professional papers, arranged by subject, including manuscript and printed material. Topics include forestry and forest game management.

3530 **Hunt, Lester C.,** Papers, 93 feet. Governor of Wyoming, 1943-48; U.S. Senator, 1949-54. The collection includes congressional papers, personal correspondence, scrapbooks, speeches. There is considerable material on western resource development and planning.

3531 **Izaak Walton League, Wyoming Division,** Records, 1925-62, 4 feet. Minutes and proceedings of the annual conventions, constitution and bylaws, convention programs, minutes of the board of directors (1942-51), minutes and reports of committees, scrapbooks (2 vols.), correspondence (1 box), and miscellaneous records of League activities.

3532 **Medicine Bow National Forest,** Collection, 12½ feet. Historical scrapbook; correspondence, clippings, memoranda, reports, and other material concerning artifacts; biographical resumes of foresters; lists of forest personnel (1903-51); information on early residents of Albany and Carbon counties; reports, letters, memoranda, other documents including maps concerning Forest Service telephone lines (1909-64); development and improvements; Sand Lake Road; fire control records; forest administration; fiscal and financial matters; place names and other geographical information; grazing; Hayden National Forest; historical data; forest inspections (1951-60); land classification; land ownership and administrative sites; minerals management; range management; reclamation; recreation; safety; timber management; transportation; timber trespass; watersheds; weather; wildlife; maps; notebook containing "History of the Cheyenne National Forest," by J.H. Mullison and P.S. Lovejoy; statistics from the Pole Mountain Nursery (1929-39); photographs; printed materials; and miscellaneous items. Inventory, 25 pp.

3533 **Murie, Olaus J. and Margaret,** Papers, 10 inches. Items concerning the establishment of and opposition to the Jackson Hole National Monument, including congressional bills, court cases, and miscellaneous bulletins, circulars, debates, fliers, reports, press releases, and newspaper reports. There are also included papers of the Jackson Hole Preserve, Inc. (12 items), including minutes of annual meetings (1940-45).

3534 **O'Mahoney, Joseph C.** (1884-1961), Papers, 1920s-1962, 930 feet. U.S. Senator, 1934-53; 1954-61. Committee assignments included Public Lands and Surveys, Irrigation and Reclamation; Interior and Insular Affairs (chairman); Select Committee on Government Organization. Congressional files are arranged alphabetically by subject within years (310 boxes); also scrapbooks (57 vols.).

3535 **Ordway, Samuel Hanson, Jr.** (b. 1900), Papers, 8 feet. Correspondence (1938-71, 300 items) including letters to and from various civil service reform and conservation organizations, especially the Conservation Foundation, of which Ordway was an officer and trustee; annual reports of the Conservation Foundation; miscellaneous manuscript and published materials; notebooks (4 items); speeches and articles for the Conservation Foundation (25 items); and manuscripts of Ordway's published writings.

3536 **Rodgers, Andrew Denny, III,** Collection, 5 feet. Published materials, interviews, and research notes for Rodgers' writings; also scrapbooks.

3537 **Simpson, Milward Lee** (b. 1897), Papers, 31¼ feet. Personal papers, gubernatorial and senatorial files, administrative and organizational files, historical files. Simpson was a member of the Wyoming Game and Fish Commission in the 1930s and state legislator; governor, 1955-59; and U.S. Senator, 1963-67. There is material on legislation during Simpson's term in the Senate (73 boxes), in which he was active in conservation and reclamation affairs. There is also correspondence, minutes, and convention material for the Wyoming Izaak Walton League (1935-39), and correspondence of the Wyoming Fish and Game Commission (1931-40).

3538 **Society for Range Management** (Denver, Colorado), Records, 61 feet. Material relating to the organization of the Society (1947); directors' minutes; correspondence; records of sections; printed material. Additional accessions are anticipated.

3539 **Spring, Agnes Wright,** Papers, ca. 13½ feet. State historian, Colorado, and State Historian, Wyoming. This collection includes research notes, files, manuscripts, clippings, and scrapbooks used in preparing her various writings. There is material on range management, forestry, and historical materials for the Shoshone and Medicine Bow National Forests, Wyoming, and the U.S. Forest Service office at Lander, Wyoming.

3540 **Steinke, Harold A.,** Reports, 1947-1960, 10 inches. Reports on Wisconsin wildlife, containing forest habitat surveys (4 items).

3541 **Teton National Forest,** Collection, 2 feet. Teton National Forest (2 vols., looseleaf); correspondence of J. Neilson Barry; recreation master plan (1939); Walter Sheppard complaints (1913-24); Jackson Hole National Monument material; grazing reports and range allowances (1910-38); material on the Jackson Hole elk herd (1909-58); material on the insect and disease control project (1958-68); miscellaneous historical and other materials.

3542 **Towell, William Earnest** (b. 1916), Speeches, 26 items. Copies of speeches by Towell, executive vice president of the American Forestry Association (1967+), on miscellaneous forestry, environmental, and conservation topics.

3543 **Warne, William E.,** Papers, 22 feet. Reclamation engineer; Assistant Secretary of the Interior, 1947-51; California Director of Water Resources and Administrator, California Resources Agency, 1961-67; holder of other positions in water and power administration. The bulk of the collection consists of material, printed and manuscript, collected and used by Warne in his many administrative positions; also correspondence (1936-72, ca. 5,000 items).

3544 **Warren, Francis E.** (1844-1929), Papers, 1873-1929, 99 feet. Governor of Wyoming Territory; first governor of Wyoming; U.S. Senator, 1891-93, 1895-1929. Committee assignments included Agriculture and Forestry and Forest Reservations and the Protection of Game. The collection includes materials relating to public lands controversies, especially in the early 20th century.

3545 **Williamson, C.D.,** Papers, 4½ feet. Laramie, Wyoming, and Carbon City, Wyoming, banker and businessman. The collection includes correspondence relating to the Wyoming Timber Company (1917-55, 728 items). There are bound records for the Carbon and Wyoming Timber Companies, including account books (1936-58, 5 vols.); deposit record (1928-43); production records (1916-50); stock books (1916, 1926-41, 3 vols.); ledger (1900-14, 1 vol.); letterpress book (1902-03, 1 vol.); commissary cashbooks (1950, 2 vols.). There are loose financial records for the Carbon Timber Company (1914, 1920), and the Wyoming Timber Company (1921-59). For the Wyoming Timber Company there are Union Pacific tie contracts (23 items) and maps (16 items); tax records (1919-50), and papers relating to tax suits by the U.S. Internal Revenue Service (1919-21). There are also, relating to the Wyoming Timber Company, legal papers and minutes of the board of directors meetings (1916-30). Some of these records may be intermingled with records of Williamson's banking interests.

3546 **Wing, Leonard D.,** Papers, 33 feet. Papers relating to Wing's teaching and research concerning the study of wildlife and geophysical cycles. The collection includes correspondence (1925-68); research notes (4 feet), including material on tree rings; and published material. Much of the collection concerns forestry, forest ecology, and forest management.

3547 **Wright, A.T.,** Papers, 24 items. Correspondence (1968-73, 7 items); statements by Wright before Izaak Walton League chapters and U.S. congressional committees, concerning national parks and recreation, wilderness proposals, and other conservation matters.

3548 **Wyoming Stock Growers Association,** Papers, 1881-1922, 131 feet. Letterbooks (62 vols.), correspondence (71 boxes), proceedings of annual conventions, brand books, books, publications, manuscripts (8 vertical file drawers), prints by Charles M. Russell (65 items). The collection relates to aspects of the cattle business, including range management, public lands, irrigation, etc.

3549 **Wyoming Wool Growers Association,** Records, 24 feet. Most of this collection relates to the period 1918-27, the presidency of J.M. Wilson. Subject files, alphabetical, include abstract of testimony in U.S. Forest Service vs. Moroni A. and Ray Smith, material relating to public land commissioners, congressional relations, the Forest Service Committee (1943), grazing, etc.

MOOSE

Grand Teton National Park, Library

83012

3550 Holdings of papers relating to Grand Teton National Park (1872-1968, 3 feet [ca. 50 items]) include sketches by Thomas Moran; photographs made by William H. Jackson (ca. 1872); research reports on park elk and moose ecology; and Moran's diary (1879, 1 reel microfilm, original located at park headquarters).

YELLOWSTONE NATIONAL PARK

Yellowstone National Park Museum

82190

3551 **Yellowstone National Park,** Collection, 1824-1951. Material relating to the park and the surrounding region. Items covering the Hayden Survey include a copy of a diary (1872) of William Blackmore; copies of letters (1872) of Sidford F. Hamp; excerpts from diaries (1872, 1878) kept by William Henry Holmes, artist with the survey; and a diary of Albert C. Peale (1871-72), mineralogist. Other papers relating to Yellowstone Park include some of Nathaniel P. Langford (1872 and later, 2 vols. and 1 piece); and letters and a journal kept by other park officials (1903-21); letters by Stephen Tyng Mather (1917-19); and letters and legal papers of John W. Meldrum, acting governor of Wyoming Territory and U.S. Commissioner of Yellowstone National Park (1894-1933, 480 pieces). There are also official diaries of park scouts (1872-1912, 46 vols.); a record of burials at Fort Yellowstone (1886-1914, 2 vols.); and scattered letters and diaries of travelers to and through the park region (1824-1908). Included are letters of Horace Albright (1926); letters and maps of J. Neilson Barry (1935); miscellaneous material of Hiram M. Chittenden (1892-94); Xerox copy of report of Gustavus C. Doane (1870; original in Montana State University Library, Bozeman); typescript of David Folsom diary (1869); F. Jay Haynes diary (1915-17).

CANADA

CANADA

Alberta

BANFF

Archives of the Canadian Rockies

Director, Box 160, 111 Bear Street TOL OCO

3552 **Sanson, Norman B.** (1862-1949), Collection, 1890-1944, 2 feet and 2 inches. Notebooks and correspondence of a meteorological observer and museum curator at Banff. They relate to his extensive plant collections, and include notes on flora of Banff National Park. Preliminary Inventory.

3553 **Van Kirk, Sylvia,** "The Development of National Park Policy in Canada's Mountain National Parks, 1885 to 1930," University of Alberta, M.A. thesis (History), 1969.

3554 **Warren, Mary Schaffer** (d. 1939), Collection, 1908-1913, 10 inches. Diaries, articles, and other written manuscripts of an amateur botanist and collaborator with Stewardson Brown in *Alpine Flora of the Canadian Rocky Mountains,* including diaries of explorations north of Banff, and notes and articles written afterwards. The diaries and some of the articles relate to the flora of the region. Preliminary Inventory.

CALGARY

Glenbow-Alberta Institute, Archives Division

Archivist, 902 Eleventh Avenue S.W. T2R OE7. All of the collections noted have finding aids.

3555 **Banff National Park,** Banff, Alberta, Register, 1922-23, 1 inch. Register of pack parties going into wilderness areas of the park.

3556 **Bilsland, William W.** "A History of Revelstoke and the Big Bend," Vancouver: University of British Columbia, thesis?, 1955. Microfilm, 1 reel. Includes discussion of lumbering.

3557 **Bow River Saw and Planing Mills** (Calgary, Alberta), 1884-1903, 1 vol. Letterbook of lumber and contracting business conducted by James Walker. Consists of tenders, accounts, and estimates of various people who were building in the Bow River district.

3558 **Byrne, Anthony Roger.** "Men and Landscape Change in the Banff National Park Area before 1911," Calgary: University of Calgary, M.A. thesis, 1964. Microfilm, 1 reel.

3559 **Calgary Brewing and Malting Company** (Calgary, Alberta), Papers, 1892-1937, 127 boxes. Correspondence of Alfred Ernest Cross contains scattered items of interest, such as a letter from the Minister of Interior relating to conservation of big game in Banff National Park (ca. 1898-99); report on tree planting campaign of Canadian Forestry Association (1924); article by William Pearce, on history of parks in Rocky Mountains; correspondence from Canadian National Parks Commission relating to protection of birds (1925). Cross' ranch files include correspondence with the Department of the Interior relating to timber and grazing leases (1925), including copy of letter from R.H. Campbell, Director of Forestry, relating to use of reserves. Restricted. Typed list of contents, ca. 100 pp.

3560 **Canadian Pacific Railway,** Papers, 1886-1958, 150 feet. These papers deal mainly with colonization work of the Canadian Pacific Railway and subsidiary organizations; land, coal, and irrigation activities are included. The area involved is chiefly Alberta with some extension into British Columbia and Saskatchewan. The coal files consist of reports, maps, and correspondence containing incidental information concerning timber on coal lands in British Columbia and Alberta (ca. 1904-10). British Columbia Land Department files (1898-1926) contain correspondence, maps, plats, permits, licenses, applications, agreements, leases, contracts, blueprints, reports, statistics, etc., relating to such subjects as railroad construction, railroad land grants, colonization in British Columbia; timber cutting, permits, berths, cruises, examinations, dues, rights, leases, reservations, limits, revenues, rangers, surveys, licenses, sales; tie reserves; tie contracts; sawmill sites; trespass on CPR timber lands; forest insects; lumber; lumber prices; treatment of structural and tie timber; fires; telephone poles; purchase and sale of timber lands; stumpage; taxes; salaries of fire wardens; flumes; lumber for settlers; fire fighters' wages; grazing. This series pertains to many individual localities in British Columbia, chiefly in the Kootenay and Yale districts. Among the lumber companies and other firms included are the Baker Lumber Co.; B.C. Southern Railway; Breckenbridge & Lund Co.; Canadian Pacific Timber Co.; Columbia River Lumber Co.; East Kootenay Lumber Company; John Lineham Lumber Co.; King Lumber Mills; Kootenay Lumber and Exploration Co.; McNab Lumber Co.; Wardrop Bros.; Moyie Timber Co.; Pugh and Livingston Lumber Co., Ltd.; and the Standard Lumber Co. Typed inventory.

3561 **Daggett, Frank Austen** (1867-1936), Diaries, 1891-1913, microfilm, 1 reel. Includes material on work in lumbering in Washington state during the 1890s. Original in possession of Mrs. J. Weld, Davisburg, Alberta.

3562 **Eastern Rockies Forest Conservation Board,** Records, 1948-1971, 15 inches. This collection consists of a sampling of files relating to timber berths, applications and permits, press releases concerning the activities of the board, and guides for a number of Forest Conservation Units established by the board. These guides contain information on the history, physical description, watershed hydrology, wildland management, land use classifications, watershed damage and condition, and management problems of the unit.

3563 **Eau Claire and Bow River Lumber Company** (Calgary, Alberta), Records, 1883-1954, 4 boxes and 1 oversize. Correspondence, legal documents, tax forms, financial statements, maps, three bound volumes, and an oversize plan of a timber berth, relating to logging operations along the Bow and Kananaskis Rivers and sawmill operations at Calgary, ca. 1886-1945, by an American-Canadian firm.

3564 **Getty, Ian.** "A History of Human Settlement at Waterton Lakes National Park, 1800-1937," 1971, 272 1. Photocopy. Prepared for National and Historic Parks Branch. Restricted.

3565 **Kerr, Isaac Kendall,** Papers, 1883-ca. 1910, ½ inch. Diary of a trip from Calgary into Kananaskis to select timber limits (1883); license to cut timber on Bow River and tributaries (1884 and 1886); article and clippings on history and accomplishments of the Eau Claire and Bow River Lumber Company (Calgary).

3566 **Luxton, Norman Kenny** (b. 1876), Papers, 1902-1926, 6 boxes and 2 oversize. Includes business correspondence (1902-26); manuscript by H.C. Stovel, Rocky Mountains Park (1915); regulations, tariffs and petitions for new livery regulations and the site of a new campground, bird game licenses (1912-14), and a building permit (1914) relating to Rocky Mountains Park (1 folder); Eau Claire and Bow River Lumber Co. price list. Typed container list, 4 pp.

3567 **McLaren Lumber Co** (Mountain Mill, Northwest Territories), Ledger book, 1888-1891, microfilm (16 feet).

3568 **Pearce, William** (1848-1930), Papers, 1892-1925, 28 1. Includes a manuscript history of the establishment of the Banff, Jasper, Kootenay, and Waterton Lakes National Parks, read at the Historical Society of Calgary meeting, December 16, 1924. Photocopies.

3569 **Prince, Peter Anthony,** Papers, 1875-1926, 2 inches. Includes agreement between Eau Claire and Bow River Lumber Co. and Prince (1885); letters relating to mill and light company (1885-90); agreements relating to the mill (1889-1913); log patent (1875).

3570 **Revelstoke Companies, Ltd.** (Revelstoke, British Columbia), Records, 1901-1955, 10 inches and 47 vols. Minute books and account books of fourteen predecessor firms to the Revelstoke Sawmill Co., Ltd., and Revelstoke Companies, Ltd., which had offices in Winnipeg, Manitoba; Calgary and Edmonton, Alberta; and Revelstoke, British Columbia, and were involved directly or indirectly in the lumber trade of Western Canada.

3571 **Rump, Paul Charles.** "The Recreational Land Use of the Bow, Kananaskis, and Spray Lakes Valleys," Calgary: University of Calgary, M.A. thesis, 1967, microfilm, 1 reel.

3572 **Scace, Robert C.** "Bibliography of Waterton Lakes National Park and Waterton-Glacier International Peace Park, Alberta," 1972, 1 vol. (249 1.). Photocopy. Commissioned by National and Historic Parks Branch, Department of Indian Affairs.

3573 **Scace, Robert C.** "Banff National Park Area Annotated Bibliography," 1970, 229 1. Photocopy. Prepared for the National and Historic Parks Branch, Department of Indian Affairs and Northern Development.

3574 **Soole, Mrs. David M.,** Papers, 1937-1953, 1 inch. Includes a history of Banff National Park, and correspondence relating to proposed exclusive federal jurisdiction over Banff National Park.

3575 **Strom, Theodore,** Reminiscences, 1886-1912, 23 pages. Engineer; Calgary, Alberta. Transcripts of reminiscences of the establishment and operation of the Eau Claire Lumber Co., Calgary. Published in Part in *Alberta Historical Review* (Summer, 1964).

3576 **Tennant, Charles Irvain** (1864-1957), Papers, 1898-1954, 3 feet, 4 inches. Dawson, Yukon Territory; woodcutter. Personal correspondence (1900-49); papers relating to fraternal societies; mining, land, and legal papers; business accounts; miscellaneous items.

3577 **Wilson, Thomas,** 1882-1929, 81 pp. Banff, Alberta; packer and outfitter. Autobiography and notes on locating Canadian Pacific Railway route through Rocky Mountains, discovery of Lake Louise, and early events in Banff National Park.

3578 **Young, Hilton,** Papers, ca. 1960s, 1¼ inches. Articles and notes relating to personalities and places of the Kootenay district, British Columbia. Handwritten and photocopies. Articles relate to lumbering near Crows Nest Pass, Great Northern Railway, settlement of the area, etc.

EDMONTON

Provincial Archives of Alberta

12845 102nd Ave. T5N OM6 Telephone: (403)452-2150

Archives

3579 **Alberta. Environment Department,** Records. Specimens of Surface Reclamation Council Files (1963-72).

3580 **Alberta. Executive Council,** Records. Records of committees and commissions deposited by R. Crevolin, Clerk of the Legislative Assembly. Includes material concerning the transfer of natural resources from federal to provincial jurisdiction (6 items).

3581 **Alberta. Lands and Forests Department,** Records. Annual reports; licenses, permits, tags, and seals; tree growth survey master plans (1944-55); surveyors' reports on districts in Alberta prepared for settlement (1922-24); maps of forest reserves in Alberta (1936-72); speeches and scrapbooks of N.A. Willmore, Minister of Lands and Forests (1955-65); mill returns, grazing and timber permits, and mill owner registers (1909-46); timber case files (ca. 1920-37); Xerox copies of notes on Lake Saskatoon City; Report by W.T. Aiken on Provincial Parks of Alberta for the Provincial Parks Board (1935); photocopies relating to the Hinton district, supplied by the Forestry Technology School, Hinton.

3582 **Alberta. Premier's Office,** Records, 1921-1945. Files of H. Greenfield, J.E. Brownlee, R.G. Reid, W. Aberhart, and E.C. Manning, including subject files relating to the lumber trade, national parks and reserves, provincial parks, forest fires, etc. Inventory.

3583 **Eastern Rockies Forest Conservation Board,** Records. Annual reports (1948-65, 1966-69, 1970-71); annual reports of the East Slopes (Alberta) Watershed research program (1963, 1965, 1966); management reports nos. 1 and 2; "Watershed Management of the East Slopes of the Rocky Mountains in Alberta," by W.L. Hanson and G. Tunstele, n.d.; conservation unit guides, case studies, etc. (2 boxes); aerial photographs of the area of the Eastern Rockies Forest Conservation Board (ca. 1944-69).

3584 **Environmental Conservation Authority,** Reports, 1973, 2 items. "The Resources of the Foothills: A Choice of Land Use Alternatives," 1973; Third Annual Report, 1973.

Manuscript and Photograph Collections

3585 **Athabasca Delta Region,** Aerial photographs, 1927.

3586 **Cooking Lake Forest Reserve.** Photographs.

3587 **Dominion Forest Supervisors, Alberta and British Columbia, Annual Convention,** Photograph, 1926, 1 item.

3588 **Douglas Family,** Papers. Includes a small amount of material on Howard Douglas, Superintendent of Banff National Park, 1896, and Commissioner of National Parks in Western Canada, 1911.

3589 **Edmonton Botany Club,** Records, 1947-1966. Business records and other items relating to the study of natural history. Organization later known as the Natural History Club.

3590 **Grande Prairie Forest Division, Bold Mountain**

3591 **Guest, R.** "Wilderness Areas in the Peace River Country." Manuscript.

3592 **Hanson, W.R.** "Conserving a Watershed." Manuscript.

3593 **Jasper National Park,** Notes relating to the park. Includes photographs of Pocahontas area.

3594 **National Parks.** Noncurrent federal records on National Parks in Alberta (1912-58). Inventories.

3595 **Oral History Interviews.** Tapes and transcripts of interviews with the following persons: Henry Stelfox (b. 1883), naturalist, Rocky Mountain House area, on the role of beaver and fire in forest ecology (1958, tape, 2 hours; also transcript); Charles O. Saunders (b. 1878), on the importance of the lumber mill to the economy of Okotoks, Alberta (1957, tape, 95 minutes); Carl C. Ranche, lumberman, and timber inspector for the provincial Department of Lands and Forests (1973, tape, 2 hours); Harold William Parnall (b. 1889), forester, Edson and Peace River vicinities (1970, tape, 1 hour and 40 minutes); William Norman Morrison (b. 1875), on the Lacombe Experimental Station (tape, 10 minutes, with transcript); William John Kirby (b. 1866), on the development of the lumber industry near Rocky Mountain House, Alberta, ca. 1911-58 (1958, tape, 15 minutes, and transcript); Willis G. Cole (b. 1875), on lumber and construction, Edmonton vicinity, ca. 1890s-1900s (1957, tape, 90 minutes); Colin Bray, on reforestation (tape, 1 reel). Interviewers were E.S. Bryant and Naomi Redford.

3596 **Pearce, W.** "History of the Establishment of the Chief Parks along the Main Line of the Canadian Pacific Railway." Paper read to the Historical Society of Calgary, 1926.

3597 **Poplar Lake Timber Camp.** Photograph, n.d.

3598 **Scouler, R.A.** "History of the Alberta Natural History Society." Manuscript.

3599 **Willmore, Norman Alfred** (b. 1909), Papers. Minister of Lands and Forests, 1955-65. Speeches, photographs, certificates.

The University of Alberta, University Archives

Archivist, Rutherford Library (South).

3600 **Pearce, William,** Papers. Pearce served as Vice President for Alberta of the Canadian Forestry Association (1900-10); Director (1911-30). The collection contains correspondence relating to American Forestry Association membership (1901-02, 10 pp.); annual reports, correspondence, pamphlets, etc., of the Canadian Forestry Association (1900-29, 10 inches), including an article, "Some Remarks Regarding the Protection and the Promotion of the Growth of the Canadian Forests," by Pearce (14 pp.); and correspondence requesting information relating to tonnage of sawmill products shipped by rail (1916, 15 pp.).

British Columbia

KAMLOOPS

Kamloops Museum Association

Curator, 207 Seymour Street. In addition to the item below, the Museum has an index to local newspapers (1880-1945), containing all references to lumbering in the Thompson River drainage basin. There are also copies of historical articles relating to the Adams River Lumber Company, the lumber town of Chase, and biographical details of lumbermen.

3601 **Shuswap Milling Company** (Kamloops, British Columbia), Ledger, 1889-93, 1½ inches. Ledger relating to flour and lumber milling.

VANCOUVER

British Columbia Forest Products Limited, Archives

Archivist and Historian, 1050 West Pender Street V6E 2X3
Telephone: (604) 665-3821

3602 **Records,** ca. 20 feet. General business correspondence and records relating to logging, sawmilling, plywood manufacture, and pulp and paper manufacture by British Columbia Forest Products (after 1946) and its predecessors; also photographs of logging and milling operations on the coast and Vancouver Island, including some dating from ca. 1900; and tape recorded interviews, accompanied by transcripts, with retired and long-time company employees. Additions to the oral history collection are made at frequent intervals. Copies of some tapes are available at the Provincial Archives of British Columbia, Aural History Division, Victoria, British Columbia.

University of British Columbia, Library

Head, Special Collections Division

3603 **Anderson, P.B.,** Collection, 1885-?, 2 inches. Pioneer lumberman of the Pacific Northwest. The collection includes reminiscences, clippings, and several albums of photographs.

3604 **Hastings Saw Mill** (Vancouver, British Columbia), Papers, 1865-1890, 5 inches. Correspondence, maps, photographs.

3605 **Lougheed, Nelson Seymour,** Papers, 1910-1954, 7 inches. Correspondence, financial records, photographs, journal and diaries relating to lumbering in British Columbia.

3606 **Orchard, C.D.,** Collection, 7 feet. British Columbia Deputy Minister of Forests. The collection contains correspondence, reports, printed material, and recordings relating to forests in British Columbia.

3607 **Pircock, R.H.,** Reminiscences, n.d., 120 pp. Reminiscences of R.H. Pircock, Comox, British Columbia, sawmill proprietor, who came to British Columbia in 1862.

3608 **Victoria Lumber and Manufacturing Company** (Chemainus, British Columbia), Annual reports, 1906-1950, 7 inches.

3609 **Wilkinson Family,** Papers, ca. 1920s, 3 feet. Photographs, documents, correspondence, and printed material pertaining to family interests and the lumber industry in British Columbia.

---

*Vancouver City Archives

3610 Russell G. Hann, et al., *Primary Sources in Canadian Working Class History, 1860-1930,* Kitchener, Ontario: 1973, credits the Vancouver City Archives with holding correspondence and records (1876-89) of James McIntosh, Vancouver carpenter and sawmill manager, and a letterbook (1870-74) of the Hastings Saw Mill Co.

VICTORIA

Provincial Archives

Provincial Archivist, Parliament Buildings V8V 1X4

3611 **Bloedel, Welch & Stewart Corp.,** Papers, 1914-1927, 3½ inches. Survey of the timber resources in the Sayward District, British Columbia (1914); appraisals of company property (1927).

3612 **British Columbia and Vancouver Island Spar Lumber and Sawmill Company, Limited.** Letter to Governor Seymour (July, 1966, 2 1.) regarding a flagpole for the Crystal Palace.

3613 **British Columbia. Speaker and Legislative Assembly,** Petition of holders of special licenses asking for annual renewal (1903, ¼ inch).

3614 **British Columbia. Lands Department,** Records, 1858-1938, 50 feet. Records of the Department of Lands and its predecessors, Department of Lands and Works, Department of Lands and Forests, and the various subsidiaries, Water Rights Branch, Lands Branch, and Forest Branch, and the Land Officer of the Colonies of British Columbia and Vancouver Island. Included are a number of early mining and road records.

3615 **Canadian Western Lumber Co., Ltd.,** Historical sketch, 1958, 4 1., typescript.

3616 **Canadian Western Timber Co., Ltd.** Copies of correspondence with the British Columbia Department of Lands regarding a request for Crown timber reserve in the vicinity of Duncan Bay to enable the establishment of a new pulp mill (1947, ½ inch).

3617 **Costello Sawmill and Mining Company** (Lightning Creek, Cariboo District, British Columbia), Minute book, 1873-78, 1 inch.

3618 **DeBeck, George Ward,** Letter to ? Keary, relating to the early history of Brunette Sawmill, Vancouver (July 16, 1931, 3 1.).

3619 **Dominion Sawmill Company** (New Westminster, British Columbia), Records, 1883-84, 1½ inches (2 items). An undated register of shareholders, and a register of shares, no. 1-500 (1883-84).

3620 **Grainger, Martin Allerdale** (1873-1941), Papers, 1876-1938, 2½ inches. Chief forester of British Columbia. Xerox copies of miscellaneous items, including correspondence (1876-1938), principally with Eve Smith and a Mr. Denny; printed material (1905 and n.d.); articles (1901-32); deed (1938).

3621 **Insurance Policies.** Fraser River Sawmills, Ltd. (1908, 2 items); Canadian Pacific Lumber Co., Ltd. (1909, 2 items); E.H. Heaps and Company (1909-10, 3 items); Rat Portage Lumber Company (1909, 3 items); W.F. Hunting Lumber Co., Ltd. (1908, 1 item).

3622 **Kinghorn, Hayward K.,** Report on Timber Reconnaissance of Peace River Company, 1923, typescript, 64 1.

3623 **Lawrence, Joseph C.** "Markets and Capital: A History of the Lumber Industry of British Columbia, 1778-1952," Vancouver: University of British Columbia, unpublished M.A. thesis (History), 1957.

3624 **MacMillan, H.R.,** Notebooks, 3 vols. Notebooks, Yale Forest School.

3625 **Mercer, William M.** "Growth of Ghost Towns: the Decline of Forest Activity in the East Kootenay District and the Effect of the Growth of Ghost Towns on the Distributing Centres of Cranbrook and Fernie." Prepared by W.M. Mercer, Bureau of Economics and Statistics, Board of Trade and Industry, for the Royal Commission on Forestry, 1944, typescript, 23 1.

3626 **Provincial Archives of British Columbia.** "A Preliminary Checklist of Sawmills in the Area Covered by the Interior Lumber Manufacturer's Association, 1880-1920," prepared at the Provincial Archives of British Columbia, typescript, 45 1.

3627 **Royal Botanical Gardens** (Kew, England), Letters, 1825-34, ½ inch. Letters (1825-34, 27 items) from David Douglas, apparently to Sir William Jackson Hooker, Director of Kew Gardens, containing accounts of Douglas' botanical expeditions to the Pacific Coast of North America. Typescripts of originals in the Royal Botanical Gardens, Kew, England.

3628 **Royal City Planing Mills Co., Ltd.,** Petition to the Legislative Assembly for the Purpose of Floating Logs to the Nicomekl River, 1884, 2 pp., sketch map.

3629 **Society for the Preservation of Native Plants of British Columbia,** Records, 1936-48, 1½ inches. Minutes (1936-48), plus miscellaneous printed material.

3630 **Stewart, John,** "Early Days at Fraser Mills British Columbia from 1889 to 1912," 1955. Typescript, 29 1. Memoir by general store manager, Canadian Western Lumber Co., Ltd.

3631 **Vancouver Island Steam Sawmill Co.,** Prospectus, 1851, 3 pp. Prospectus of a steam sawmill to be located on Cordova Bay, Vancouver Island.

3632 **Vancouver Island Steam Saw Mill Co.,** Articles of Agreement, 1847, 9 pp.

3633 **Victoria Coal and Lumber Company** (Victoria, British Columbia), Collection book, 1864-65, 1 inch. Kept by James Frain.

3634 **Weir, Adam** (1842-1897), Papers, 1872-74, ¼ inch. Lumberman. Diary (1874), with a few notes, accounts, etc. (1872-73).

3635 **Young, Hilton** (1886?-1969), "Early Days of Logging on the Crow," 1964, typescript, 46 1. Memoirs of a logger in the Fernie-Wardner area, British Columbia.

Manitoba

WINNIPEG

Provincial Archives of Manitoba

200 Vaughan Street R3C OP8

3636 **Cairnes, William,** Papers, 1918-1942, 2 vols. Day and account book of a lumber merchant (1919-42); daybook of a lumber merchant (1918-30).

3637 **Canada. Interior Department, Dominion Lands Branch,** Records, 1891-1901, 1½ inches. Crown lumber office, Winnipeg, letterbook (November 13, 1891-January 29, 1901, 1 vol.).

3638 **Gill, C.B.,** Manuscript, ca. 1962, 26 pp. Photostat. Manitoba forest history.

3639 **Healy, William James,** Paper, 16 pp. Historical notes relating to the development of the lumber industry in Manitoba, 1870-1900.

3640 **Keewatin Lumber and Manufacturing Co. Ltd.** (Keewatin, Ontario), Letterbooks, 1885-87, 2 inches.

3641 **McArthur, Alexander** (1842-1887), Papers, 1868-87, 4 inches. Lumber merchant, banker, and writer; Winnipeg, Manitoba. Correspondence (1868-87); manuscripts, notes, and papers (1872-87); diaries (February 2-March 29, 1887).

3642 **McArthur, Peter,** Recollections, 1934-1935, 1 vol. (22 pp.). These recollections of a pioneer lumberman of Manitoba describe the Riel disturbances, riverboats, and logging operations (1869-82).

3643 **Manitoba. Department of Mines and Natural Resources,** Reports, 1965-66, 2 inches. Transcripts of Renewable Resources Section quarterly progress reports.

3644 **Manitoba. Department of Mines and Natural Resources,** Records, 1913-1958, 42 feet. Deputy Minister's correspondence regarding transfer of natural resources (1930-48); Flin Flon townsite (1930-49); Pasquia Reclamation and River Dam (1953-55); Rowell-Sirois Commission submission (1937); Canada/Manitoba/Ontario waters agreement (1922); power, water, flood and timber reports (1927-31); Churchill (1913-32). Correspondence concerning Slave Falls, Seven Sisters hydroelectric sites. Deputy Minister's Office reports. Minister's files concerning transfer of natural resources to province.

3645 **Manitoba. Department of Mines and Natural Resources, Deputy Minister,** Records. Records (1927-31, 5 inches) include miscellaneous reports on power development, water resources, flood prevention, and timber surveys (file list). Records (1930-58, 40 feet) include correspondence, reports, and surveys of the Game, Fish, Forest, Mines and Land Branch.

3646 **Manitoba. Department of Mines and Natural Resources, Forestry Branch,** Records. Records (1875-1931, 6 feet) contain timber berth registers, licenses, and ledgers. Records (1923-69, 3 feet) contain miscellaneous files (file list). Papers (1933-36, 10 inches) include monthly diaries of forest ranger G.A.W. Lockhart, Rice Creek, Manitoba. Records (1931-51, 2 feet) are diaries of A. Bainbridge (1931, 1934, 1939-46); files on postwar planning (1943-45); forest fire research, control, legislation, and fighting (1945-51); Winter Transportation Committee (1947-49); information, equipment, and supplies (1945-50). Records (1963-67, 2 feet) contain rangers' and conservation officers' diaries. Other records (8 feet) consist of personnel files (1908-1966); diaries (1929-1964); files relating to timber sales, clearances, supervision, management, and improvements (1921-1967). There are also correspondence, reports, maps, and articles of C.B. Gill, provincial forester and forest consultant (1916-1970, 2½ feet).

3647 **Manitoba. Department of Mines and Natural Resources, Renewable Resources Section,** Records, 1961-1967, 6 inches. Correspondence relating to legislation, including the Canada Forestry Act; drafts of new act and regulations.

3648 **Manitoba. Department of Mines and Natural Resources, Resource Operations and Administration,** Records, 1968-1969. Monthly progress reports of the Forestry Branch.

3649 **Mathieu, James Arthur,** Oral history interview transcript, 1957, 10 pp. Interview by Bruce C. Harding. Relates to lumber industry in Minnesota. Original at the Forest History Society, Santa Cruz, California.

3650 **Minnedosa, Manitoba, Council,** Correspondence, 1927, ½ inch. Town of Minnedosa Council correspondence concerning the establishment of Riding Mountain National Park (June 6-November 24, 1927).

3651 **Somers, J.C.,** Manuscript, ca. 1962, 3 pp. Photostat. History of forest legislation relating to timber disposal in Manitoba.

3652 **Winkler, Valentine,** Records, 1885-1894, 1899, 4 vols. Lumber sales books (1885-91, 2 vols.); daybook and account book of lumber sales (1899, 2 vols.); and abstracts of lumber orders (1887-94, 1 vol.).

New Brunswick

FREDERICTON

Legislative Library, New Brunswick

3653 **Hamilton, James Francis.** "The Pulp and Paper Industry of New Brunswick, Canada," Bloomington: Indiana University, M.A. thesis, 1950. 93 l.

3654 **Smith, Kenneth George.** "Impact of the Sawmilling Industry on the Economy of New Brunswick," Fredericton: University of New Brunswick, M.Sc.F. thesis, 1970. 94 l. Microfilm.

Provincial Archives of New Brunswick

Box 39, Centennial Building, P.O. Box 6000 E3B 5H1

Archives

3655 **New Brunswick. Executive Council,** Records, 1784-1880, ca. 23 feet. Includes land and timber records (1852-57).

3656 **New Brunswick. Natural Resources Department,** Records, ca. 1900-1950. Several feet of unsorted material dealing with forest management by the government of New Brunswick, including rangers' reports, ministers' correspondence, etc.

Manuscripts

3657 **Burchill Family,** Papers, 1824-1966, 260 feet. The lumbering papers of Percy Burchill, South Nelson, Northumberland County, New Brunswick, including correspondence books, accounts, correspondence on operations, bills, employee records, and related papers; also miscellaneous church records, family papers, legal agreements, and artifacts.

3658 **Clark, R. Corry, Company** (Newcastle, New Brunswick), Records, 1899-1955, 9 feet. The business papers of the R. Corry Clark Spoolwood Manufacturing Company, including account books (1908-39); bank books (1921-36); shipping contracts, orders, settlements, specifications, wood storage papers, insurance papers, accounts of shipments (1900-50); correspondence (1933-53); and miscellaneous plans, statistics, test results, and printed forms (1898-1928).

3659 **Cushing Sulphite Fiber Company** (Saint John, New Brunswick), Records, 1878-1909, ca. 36 feet. Minute books, stock ledgers and financial ledgers, mill reports, and correspondence relating to the sale of pulp and by-products.

3660 **Elkin, D.E.,** Papers, ca. 1840-1970, 1 foot. Research notes and articles concerning lumbering and local history of Newcastle-Chatham, New Brunswick, area; also the papers of Monkhouse and Company, lumbering firm (1946-58).

3661 **Loggie, W.S., Company,** Papers, ca. 1873-1931, 78 feet. Contains accounting ledgers, orders, letterbooks, and subsidiary company papers of this Miramichi River lumbering firm. Unprocessed. Restricted.

3662 **Merchandizing, Samuel, Family,** Papers, ca. 1838-1860, 5 feet. Correspondence, agreements, accounts, invoices, orders, receipts, bills of lading and miscellaneous papers, including material on the white pine lumber trade in small New Brunswick towns.

3663 **Nepisiquit Lumber Company** (Bathurst, New Brunswick), Records, 1907-1912, 27 feet. Accounts, correspondence, legal papers, minutes and bylaws, financial statements and reports, investors' lists and shares, stock certificates, property ledgers and cashbooks. Included is information concerning partnerships in the company, acquisition of property, and liquidation.

3664 **Raymond, Joseph, and Joseph Cunard,** Papers, 1843-1850, microfilm, 50 feet. Letters to Joseph Cunard (1844-47); diary of Joseph Cunard (1843-50); letters (1847); largely relating to lumbering, Newcastle, New Brunswick.

3665 **Todd Family** (St. Stephen, New Brunswick), Papers, 1870-1930, 42 feet. The bulk of the collection comprises ledgers, correspondence, and papers, especially relating to lumbering (1870-1900) and to William Todd's political career.

The University of New Brunswick, Harriet Irving Library

Manager, Archives and Special Collections, P.O. Box 7500 E3B 5H5

Manuscript Collections

3666 **Gillespie, Thomas F.,** Papers, 1867-93, 1 foot and 4 inches. Manufacturer, Chatham, New Brunswick. Daybooks (1880-93); timebooks (1867-69); insurance books (1877-84); ledgers (1873-93); belonging to Mr. Gillespie's manufacturing, sawmill and foundry interests.

3667 **Gilmour & Rankine** (Douglastown [Miramichi], New Brunswick), Records, 1751-1893, 5 inches. Lumber merchants and ship builders. Correspondence, business and personal; bills of lading; statement of cash paid and received (1815-22); lists of ships; travel journals.

3668 **Winslow Family,** Papers, 1695-1815, 8 feet. Family correspondence and letters, letterbooks, diaries, land grants, deeds, accounts, commissions and memorials. There are numerous references to lumbering and economic conditions relating to lumbering, such as letters of William Pagan to Dr. William Paine (May 2, 1784) and Peter Clinch to Edward Winslow (December 19, 1784), describing the timber resources on the St. Croix River and Passamaquoddy Bay, and discussing opportunities for development of the lumber industry and the export of lumber to the West Indies. There is also material relating to the Maine-New Brunswick boundary arbitration (1796-98). Name index. The collection is also available on microfilm (6 reels), and parts have been published as *Winslow Papers, A.D. 1776-1826,* ed. by William Odber Raymond, St. John, New Brunswick: The Sun Printing Company, under the auspices of the New Brunswick Historical Society, 1901, 732 pp.

Theses

3669 **Arnold, J.E.M.** "An Analysis of Forest Drain and Growing Stock Relationships in Eastern New Brunswick between 1944 and 1955," Fredericton: University of New Brunswick, M.Sc. thesis (Forestry), 1957.

3670 **Elliott, George P.** "Physiography and Economic Development of the St. Maurice Valley, and their Influence on the Pulpwood Operations in the Region," Fredericton: University of New Brunswick, M.Sc. thesis (Forestry), 1959.

3671 **McGinley, Eugene Gregory.** "The Lumber Industry of the Atlantic Region, 1945-1955," Fredericton: University of New Brunswick, M.A. thesis (Economics), 1958.

3672 **Smith, Kenneth George.** "Impact of the Sawmilling Industry on the Economy of New Brunswick," University of New Brunswick, M.Sc. thesis (Forestry), 1970.

3673 **Vincent, Arleigh Burton.** "The Development of Balsam Fir Stands in the Green River Watershed Following the Spruce Budworm Outbreak of 1913-1919," Fredericton: University of New Brunswick, M.Sc. thesis (Forestry), 1954.

3674 **Webb, Horace P.** "The Development of Forest Law in New Brunswick," Fredericton: University of New Brunswick, M.Sc. thesis (Forestry), 1923.

MONCTON

Université de Moncton, Centre d'Etudes Acadiennes

Director

3675 **Carbonneau, Hector,** Papers. Writings on Canadian forests, deforestation, protection, forests in the colonial period, glossary of trees and forests, timber in the fishing industry. In French language. Some of the manuscripts have been published. Unarranged.

SAINT JOHN

The New Brunswick Museum

Archivist, 277 Douglas Avenue Telephone: (506)693-1196

3676 **Baird, George Thomas** (1847-1914), Collection, 1858-1914, ca. 3 feet. Perth, New Brunswick, politician and merchant. Includes family, business, and political papers and correspondence. Material relating to forest history (ca. 10 inches) includes papers pertaining to lumbering on the Tobique Indian Reservation, New Brunswick; accounts for lumber sold to New Brunswick dealers; contracts for lumber and sales, Victoria County, New Brunswick (1878-1908); Crown timber licenses, Victoria County (1894-1906); survey accounts; etc.

3677 **Charlotte County, New Brunswick,** Records, 1784-1867. Includes a list of the number of mills and the quantity of lumber manufactured in Charlotte County (1831).

3678 **Clark, R. Corry,** Account books, 1920-1931, 4 vols. Account books for the sawing and export of spool wood from Newcastle and vicinity in New Brunswick, principally to the firm of J. & P. Coats, Ltd., Manchester, England.

3679 **Crane & Allison** (Sackville, New Brunswick), Papers, 1834-52, 1 inch. Correspondence relating to the firm's overseas business in Liverpool, Boston, New York, and Philadelphia, the correspondence to England relating mainly to the timber trade (1834); also correspondence to Hon. William Crane relating to timber, vessels, and overseas business (1851-52).

3680 **Dalhousie, New Brunswick,** Account book, 1857, 1 inch. Account book of an unnamed general merchant and lumber dealer, with names of many people in the district.

3681 **Emmerson, John,** Account books, 1843-1909, 3 feet. Emmerson operated a trading company at Little Falls (Edmundston, New Brunswick), supplying J. & S. Glasier and other lumber dealers.

3682 **Ganong, William Francis** (1864-1941), Collection, 1686-1941, 7 feet, 9 inches. Includes a scrapbook (1900-10) containing newspaper clippings, pamphlets, and correspondence concerning New Brunswick forests and their conservation.

3683 **Glasier, Thomas S.,** Reminiscences. Glasier spent several periods early in the 19th century with his uncle, Senator John Glasier, known as "The Maine John," in the lumber woods at the head of the St. John River, where it enters New Brunswick from Maine.

3684 **Gregory, R.S.,** Account books, 1856-1919, 34 vols. Account books of a prominent lumberman, mill owner, and tugboat owner operating in the Saint John area.

3685 **Hazen, Sir John Douglas** (1860-1937), Collection, 1745-1870, 1 foot, 4 inches. Premier of New Brunswick and federal cabinet minister. Includes John Hazen Jr.'s instructions to seize timber for masts (1819).

3686 **Hilyard Family,** Papers, 1765-1955, 1 foot, 4½ inches. Lumber manufacturers and shipbuilders, Saint John, New Brunswick. The collection includes foreshore rights in Portland, bills and correspondence (1765-1901); deeds relating to Portland (Saint John) (1788-1877); deeds, mortgages and bonds (1825-69); A. McL. Seely, receipts and mercantile insurance papers (1848-79); papers relating to Capt. John Thain (1849); sawmill papers, miscellaneous items (1854-1901); ship's contracts and agreements (1856-69); log and deal contracts (1860-88); timber lands leased and owned by Hilyard Bros. (1865-1911); Tapley papers relating to lumber (1868); marine railway bills, correspondence, and plans (1870-72); and other items.

3687 **Johnston, R.,** Account book, 1855-1872, 1 vol. Account book of a sawmill at Chatham, New Brunswick.

3688 **McAllister, Captain George,** Journal, 1831-1832, 1 vol. Daily journal of the captain of a brig engaged in the lumber trade from St. Andrews, New Brunswick, to the West Indies. Transcript of an original in the possession of Evans Hill, St. Stephen, New Brunswick (1953).

3689 **New Brunswick, Timber Surveyors' Bonds,** 1836-1842, 60 pp. Bonds filed in pursuance of "An Act to Regulate the Exportation of Lumber" in Westmorland County, New Brunswick. Part of the New Brunswick Historical Society Collection.

3690 **New Brunswick Shipping Registers,** 1826-1928, microfilm, 24 rolls. These shipping registers give a very complete picture of the New Brunswick shipping industry. Originals are in the Public Archives of Canada, Ottawa, R.G. 12.

3691 **Parker Family,** Papers, 1823-86, 1 foot, 6 inches. Shipbuilders. Account books for labor, ships, sawmills, and a record of all ships built by Robert Ellis of Tynemouth Creek, New Brunswick, from the establishment of the yard in 1823.

3692 **Reynolds Family,** Papers, 1865-81, ½ inch. Saint John, New Brunswick; builders. Papers of William Kilby Reynolds, Sr., including plans (12 items) showing timber surveys and license in the Lepreau, New Brunswick, area.

3693 **Saint John, New Brunswick, Account Books,** 1787-1912, 5 inches (10 items). One account book (1815-16), includes accounts for a Saint John sawmill; another relates to lumber sold (1877-78).

3694 **Saunders, Allan A.M.,** Account books, 1872-1913, 2 inches (2 vols). Miramichi, New Brunswick, general merchant. Includes account for labor done by lumberjacks in the Miramichi area.

3695 **Simonds, Hazen, and White,** Records, 1764-1810, 4 feet. Correspondence and other papers of a firm engaged in fur trading, lumbering, fishing, and lime burning on the Saint John River. An interesting series of letters (1780-82) was written from a camp where masts were being cut for the Royal Navy.

3696 **Ward, John, and Son,** Papers, 1820-1840, 11 feet. Correspondence and accounts of a Saint John, New Brunswick, firm. One of its activities was the export of lumber, staves, shingles, etc., to the West Indies and Great Britain. About one-third of the collection relates to lumbering and shipbuilding (chiefly 1830s).

3697 **Wetmore, David** (1803-1882), Papers, 1855-79, ½ inch. Lumber merchant; Clifton, New Brunswick. Ledger containing references to surveyors and activities in running a logging camp.

Nova Scotia

HALIFAX

*Dalhousie University Library

Telephone: (902) 429-3601

3698 The following holdings are described in Russell G. Hann, et al., *Primary Sources in Canadian Working Class History, 1860-1930,* Kitchener, Ontario: 1973. Records (1865-1935) of Colin and G.D. Campbell and Campbell Lumber Company (Weymouth, Nova Scotia); papers (1860-1929) of Alfred Dickie, lumber merchant of Stewiacke, Nova Scotia, and Halifax; papers (1866-1972) of Rufus Edward Dickie, Stewiacke lumber merchant; records (1860 +) of the Dominion Chair Co. (Bass River, Nova Scotia), furniture manufacturerers; records (1882-1914) of S. St. C. and H. Jones, lumber merchants; papers (1853-91) of Archibald Woodbury McLelan, Londonderry, Nova Scotia, lumber merchant and shipbuilder; records (1881-1903) of James P. Mitchell & Co. (Mill Village and Lawrencetown, Nova Scotia) lumber merchant; records (1924-72) of the Scott Paper Co. (Sheet Harbour, Nova Scotia), and related firms; and records (ca. 1913-19) of the Waterman Tanning Co., Ltd. (Bridgewater, Nova Scotia). Dalhousie University also holds papers of the Weymouth family, merchants, lumberers, and marine insurers in the Maritimes (described in Alan Wilson, "Maritime Business History," *Business History Review,* XLVII (1973), p. 276).

Public Archives of Nova Scotia

Provincial Archivist, Coburg Road B3H 1Z9

3699 **Acadia Charcoal Iron Company** (Londonderry, Nova Scotia), Records, 1857-73, 10 inches. Ledgers, cashbooks, etc.

3700 **Cunard, Joseph,** Papers, 1837-1841, 1 inch. Halifax, Nova Scotia, and Chatham, New Brunswick. Shipbuilder and merchant. The collection contains timber agreements.

3701 **Davison, Edward Doran, & Sons** (Bridgewater, Nova Scotia), Records, ca. 1889-1915, 3 feet, 10 inches. Business papers of Doran, E.D. Davison & Sons, pulp and paper company.

3702 **Fraser, W.J.,** Papers, 1856-60, 1 inch. Chatham, New Brunswick. Agreements for supply of timber and lumber.

3703 **Nova Scotia. Governor,** Records, 1713-1914, 20 feet, Papers of the Governor's Office. Vol. 137 (1784-91, 112 pp.) is a letterbook of Governor John Parr and of Richard Bulkeley, Secretary of Nova Scotia, containing routine correspondence, including an order to a collector of customs prohibiting the importation of lumber from the United States; vols. 49-54 (1783-1807, 1 foot) are letterbooks of Governor John Wentworth with the Secretary of State and various British officials and others while he was Surveyor-General of the King's Woods and Forests and Governor of Nova Scotia; vols. 55-57 (1767-78, 5 inches) are letterbooks of Wentworth while Governor of New Hampshire and also in his capacity as Surveyor-General of the King's Woods and Forests.

3704 **Nova Scotia. Land Department.** Records, 1732-1904, 70 feet. Petitions, grants, instructions, warrants to survey, deeds, accounts of quit rents, poll taxes, and assessment rolls and other papers relative to settlement and establishment including Minutes of Council relative to land grants (1772-1827), Cape Breton grants, plans and warrants to survey (1787-1871), and land grant docket books (1773-75, 1784-1935), together with the land papers of various counties, and those on boundaries (1819-26). These records pertain in part to timberlands.

3705 **Smith, Titus** (1768-1850), Papers, 1801-02, 3½ inches (179 pp.). Survey of eastern and northern parts of province in 1801-02, with general observations thereon; also survey of lands between Sackville (Bedford) and Shubenacadie; and observations on western parts of province, with lists of trees, shrubs, grasses, and plants with observations on nature and the uses of the trees. Transcript from a copy of the original then in the possession of James Irons, 1857. Largely in journal form. Smith reported to the provincial government on the suitability of lands in the interior for the settlement of emigrants. Original and transcript.

Ontario

CHATHAM

Chatham-Kent Museum

Secretary, 59 William Street North N7M 4L3

3706 **Ferguson, John,** Accounts, 1868-87, 2 inches. Merchant, Thamesville, Ontario. Account book (1868-81) records sales of different kinds of wood; ledger (1883-86) lists sales of cordwood to the Grand Trunk Railway.

KINGSTON

Queen's University at Kingston, Queen's University Archives

Archivist, Douglas Library K7L 5C4 Telephone: (613) 547-5950

3707 **Calvin Company** (Garden Island, Ontario), Records, 1836-1923, 100 feet. Correspondence, timber records, transportation records, financial records, and personal records, relating to rafting, shipbuilding, and salvaging.

3708 **Kingston Naval Dockyard** (Point Frederick, Ontario), Account book, 1813, 1 vol. (11 pp.). Photocopy; original in private possession. Account of timber delivered.

3709 **Mair Family** (Lanark, Ontario), Papers, 1841-1885, 3 feet. Lumbering business.

3710 **St. Maurice Lumber Co.** (Trois Rivières, Quebec), Records, 1856-57, 2 vols. Letterbook (1857); daybook (1856-57).

3711 **Tett Family** (Newboro, Ontario), Papers, 1820-1968, 38 feet. Lumber and general merchants. Daybooks, journals, cashbooks, ledgers, letterbooks.

LONDON

*University of Western Ontario Library, Regional History Collection

Telephone: (519)679-3165

3712 Holdings listed in Russell G. Hann, et al., *Primary Sources in Canadian Working Class History, 1860-1930,* Kitchener, Ontario: 1973, include the following: business papers (1873-1927) of C. Beck Co., Ltd. (London), containing material on a lumber company, planing mill and box factory, and pail and tub factory; account books (1866-1913) of Dennis Daly and Son (London), coal and wood dealers; account book (1879-84) of James Dawson, lumber merchant of Sombra, Ontario; account books (1871-78) of the Huron Planing Mill (Seaforth, Ontario); and account book and papers (1867-71) of John Ross and William Machan, lumber merchants of London.

MORRISBURG

The St. Lawrence Parks Commission, Upper Canada Village

Librarian, Box 340 Telephone: (613)543-2911

3713 **Baker Family,** Papers, 1796-1861, 1874, 4½ inches. Osnabruck, Ontario. Papers of members of the Baker family, principally of Adam Baker, including timber contracts (1807-26, 12 items), relating to masts, bowsprits, squared timber, staves, boards, etc.; types of timber include white pine, white oak, white ash.

3714 **Rankin Family,** Collection, 1798-1883, 8 inches. Inclues papers of Anthony McQuin, merchant and industrialist of Kingston, Ontario, containing a few papers of a sawmill. Two notes request boards from the sawmill. There are also two short lists of timbers for ships.

NAPANEE

Lennox and Addington Historical Society

Curator, 176 Water Street K7R 1W4. Major holdings are described in Public Archives of Canada, *Collections of the Lennox and Addington Historical Society: Preliminary Inventory,* Ottawa: 1959.

3715 **Bell, William** (1760-1833), Papers, 1779-1836. Included are correspondence of Thomas Merritt relating to the cutting of timber (1810, pp. 284-89); and of the Mohawk Indians, Council of Chiefs, with a letter from Chief J. Claus relating to the cutting of timber on reservations (1828, pp. 290-300).

3716 **Benson Family,** Papers, 1794-1899. Correspondence of John Benson (1819-99) of Napanee includes letters of C.A. McConnell (1851, pp. 3005-06) relating to timber, and H.B. Rathbun (1866-80, pp. 3308-36) relating to lumbering. Correspondence of Samuel Manson Benson (1818-82) of the Midland District includes a letter of Chester Hoskins relating to the theft of lumber (1859, pp. 8489-90).

3717 **Carscallen, John A.,** Papers, 1858-1904. Included are a notice of the Department of Crown Lands concerning lumbering (1874, 1 p.); a license to cut timber (1875, 3 pp.); and an agreement of the Rathbun Company to deliver logs (1896, 7 pp.).

3718 **Casey, Stephen N.,** Letter from David Roblin & Company, 1853, 2 pp. Offer to purchase logs.

3719 **Draper, W. George,** Papers, 1861-1863, 7 pp. Includes licenses to cut timber (1861-63).

3720 **Gibbs, T.F.,** Account and receipts for lumber, 1875, 1 p.

3721 **Jaynes, John,** Rental agreement with John Ellis for a sawmill, 1865-1866, 2 pp.

3722 **Napanee River Improvement Company,** Papers. Uncataloged.

3723 **Newburgh & Napanee Paper Mills,** Records, 1835-1957, ca. 2 inches. Sources of information; memoranda; notes on the Thompson family; correspondence on the mills and the family.

3724 **Rathbun Lumber Co.** (Deseronto, Ontario), Documents, 1897-1898, 3 items (55 pp.). Petitions to Crown Lands Department, to the Attorney General, and to Sir Wilfred Laurier, relating to the lumber business, fire hazards, exports to the United States, and United States competition.

3725 **Roblin Family,** Papers, 1818-1873. David Roblin engaged in the lumber business in Napanee, Ontario, after ca. 1832; his business correspondence (1847-63, ca. 100 items [892 pp.]) contains letters relating to land, lumber, and the contracting business; his business papers contain licenses to cut timber (1851-61, 17 pp.) and other lumber trade records (1850-62, 53 pp.), such as contracts for cutting timber, sales of lumber, delivery and price specifications. Papers of David Allan Roblin include a contract (1862, 2 pp.) with H.B. Rathbun to cut and deliver lumber, and a letter (1863, 2 pp.) of C.S. Ross concerning lumbering. Correspondence of Marshall Perry Roblin includes a letter (1857, 3 pp.) of John Geddes concerning log rafting.

3726 **Stevenson, John** (1812-1884), Papers, 1833-1884. Lumberman, sawyer, miller, contractor, and merchant of Napanee, Ontario, after 1850. General correspondence includes timber licenses of Hon. M.C. Cameron (1871, 3 pp.). Business correspondence (1834-78) contains letters relating to real estate, lumber milling, lumber sales, etc., including letters (1854-59, 98 pp.) of the Crown Timber Office (James F. Way). Business papers contain files relating to lumbering and milling (1849-80, 211 pp.) and accounts of lumber sales (1849-71, 49 pp.).

OTTAWA

Public Archives of Canada

Dominion Archivist, 395 Wellington Street K1A 0N3. Holdings of the Manuscripts Division are described in *General Inventory: Manuscripts,* Ottawa: Information Canada, 4 vols. +, in progress. The following volumes have been published: Vol. I (1971): M.G. 1-10; vol. III (1974): M.G. 17-21; vol. IV (1972): M.G. 22-25; vol. V (1972): M.G. 26-27. The Public Records Division has in preparation a series of inventories for the Record Groups. Earlier preliminary inventories are now out of print. The *General Inventories* cited above contain reference to many more detailed finding aids for the manuscript groups. Outdated, but still of some use in revealing subject content of the holdings which they cover, are two volumes by David W. Parker, *Guide to the Materials for United States History in Canadian Archives,* Washington: Carnegie Institution of Washington, 1913; and *A Guide to the Documents in the Manuscript Room at the Public Archives of Canada,* Ottawa: Government Printing Bureau, 1914.

Public Records Division

3727 **R.G. 1. Executive Council,** 1759-1874, 612 feet. State Records of the Executive Council contain Minute Books (1764-1867), submissions to council (1841-67), Orders-in-Council (1841-67), blue books (1824-64), and other records including files relative to timber licenses and the supply of masts to the Royal Navy, cutting of timber on Crown and clergy reserves and related matters. There are also accounts relative to timber dues and license revenues prior to confederation. Land records, including minute books (1787-1867); petitions, Upper and Lower Canada (1791-1867); land board records, Upper Canada (1764-1804); and other records; contain reports of surveys, with descriptions of the forest cover.

3728 **R.G. 4. Civil and Provincial Secretaries (Canada East).** Records of the Office of the Civil Secretary, Lower Canada, and its predecessor of Quebec (1760-1846, 128 feet) contain correspondence (indexed) which may include information relative to timber licenses and dues, trade and commerce, establishment of sawmills, etc. Records of the Office of Provincial Secretary, Canada East, and its predecessors of Lower Canada and Quebec (1763-1867, 382 feet) contain lumber cullers' bonds and applications for commissions, Lower Canada (1808-34); and accounts relative to timber dues and license revenues. In the correspondence of the Provincial Secretary is material similar to that noted in correspondence of the Civil Secretary above.

3729 **R.G. 5. Civil and Provincial Secretaries (Canada West).** Official records, sometimes known as the Upper Canada Sundries, of the Civil Secretary's Office, Upper Canada and Canada West (1788-1867, 66 feet) include correspondence (1766-1841, indexed), which may contain information relative to timber licenses and dues, trade and commerce, establishment of sawmills, etc. Correspondence of the Provincial Secretary of Canada West and Upper Canada (1788-1869, 261 feet) contains bonds of officials, which may include lumber cullers' and Crown Lands Agents' bonds, etc.; accounts relative to timber dues and license revenues; and correspondence (1821-67, indexed) with information similar to that noted in correspondence of the Civil Secretary, above.

3730 **R.G. 7. Governor General's Office,** 1771-1953, 696 feet. Records of the Governor General's Office and its predecessors of the United Province of Canada, Upper Canada, Lower Canada, and Quebec. The records consist largely of dispatches from the Colonial Office plus records of various departments of the Governor General's Office, dispatches from Lieutenant Governors (1820-69), etc. Among the subjects mentioned are timber duties, proposed timber duties, and repeal of duties; proposed timber channel at Chaudière Falls, Upper Canada; use of the Ottawa River by the lumber trade; timber cutting on the Maine-New Brunswick boundary; decline of imports of turpentine and rosin by Great Britain during the U.S. Civil War, 1864, etc. *Preliminary Inventory: Record Group 7, Governor General's Office,* Ottawa: 1953, 20 pp.

3731 **R.G. 15. Interior Department,** 1821-1934, 425 feet. Records of the Dominion Lands Bureau, Timber and Grazing Branch, include incoming correspondence together with memoranda and returns relating to lease and permit applications, selected to document the branch's activities on both a geographical and chronological basis. Vol. 106 is a "History of Timber Regulations," by G.W. Payton, Chief Forester. Policy files, also retained, relate to pulpwood, grazing, and timber regulations, and other subjects. Published preliminary inventory (1957).

3732 **R.G. 21. Department of Mines and Technical Surveys,** 1885-1961, 184 feet, 8 inches. Includes records of the Dominion Forest Service (1899-1945), and its predecessor, the Department of the Interior, Forestry Branch.

3733 **R.G. 22. Department of Northern Affairs and National Resources,** 1890-1962, 570 feet. Records of the Department and its various branches, including National Parks, and Forestry Branch (Dominion Forest Service) records pertaining to the National Forestry Programme (1936-47).

3734 **R.G. 30. Canadian National Railways,** Records, 1848-1950, 520 feet. Records consisting of stock transfer books, minute books, correspondence, letterbooks, etc., of over three hundred railway, land and steamship companies which at one time or another were absorbed by the C.N.R.

3735 **R.G. 33. Royal Commissions.** Records of the Pulpwood Commission (1923-24, 2 feet, 4 inches) contain administrative files of the commission as kept by the secretary, E.H. Finlayson. Records of the Commission on Rentals Charged for Lots in the Banff and Jasper National Parks (1950, 9 pp.) consist of Report of the Commissioner, Mr. H.O. Patriquin. Records of the Natural Resources of Saskatchewan Commission (1933-34, 7 feet) contain correspondence, exhibits, reports of hearings, and related papers. Records of the Natural Resources of Alberta Commission (1934-35, 3 feet, 8 inches) contain correspondence, exhibits, briefs, and reports of hearings. Records of the Manitoba Resources Commission (1928-29, 3 feet, 8 inches) include correspondence, document files, transcripts of proceedings, and related papers. Records of the Newsprint Commission (1917-20, 4 feet) contain correspondence, transcripts of hearings, and other files of the commission established to investigate the cost of production of pulp and paper and to assure the publishers of Canadian newspapers of adequate supply of newsprint at reasonable prices.

3736 **R.G. 36. Boards, Offices, Commissions (Non-Continuing).** Records of the St. John River Commission (1909-16, 3 feet) include two minute books, one each for meetings held in 1909 and 1916, respectively; the last three of six volumes of hearings; working papers including various analyses, statistics, and field books. Records of the Eastern Rockies Forest Conservation Board (1947-55, 5 feet) consist of financial statements, reports, contracts, and subject files created by the board. Records of the Forest Insect Control Board (2 feet, 4 inches) relate to the administrative and operational aspects of the Board, survey work, scientific experiments, and reports of various kinds.

3737 **R.G. 39. Department of Forestry,** 1899-1953, 25 feet. Records of the Department and its predecessors, the Forestry Branch, Department of the Interior, and the Dominion Forest Service, Department of Mines and Resources. The records consist of correspondence, memoranda, subject files, etc.

3738 **R.G. 45. Geological Survey of Canada.** Vols. 123-77 are field survey notebooks, which contain information on vegetation cover. Available on microfilm.

3739 **R.G. 84. National Parks Branch, Department of Northern Affairs and National Resources,** ca. 1900-1971, ca. 130 feet. All the records which relate to national parks, historic sites, and the Wildlife Service which are in the Public Archives are located in this Record group. Other material which indirectly relates to the parks may be found in Record Group 22, Department of Northern Affairs and National Resources; R.G. 86, Mines Branch; R.G. 88, Surveys and Mapping Branch. Records in R.G. 84 relate to all the various aspects of and functions of the National Parks Branch, including files started by predecessor agencies. Included are files relating to overall administration and individual files for various subjects on each park. Much of the material relates to opening up of the parks, depression projects, deportation of Japanese-Canadians from the Pacific Coast during World War II, and other subjects relating to parks and their administration. Also correspondence between R.B. Bennett and various officials of the National Parks Branch, Department of the Interior (1912-21, 8 inches), and manuscript project reports of the Historic Sites Service (1964-71, ca. 82 vols.).

Manuscripts Division

3740 **M.G. 1. Archives des Colonies—Paris.** Copies of originals in the *Archives Nationales,* Paris. In the French language. More detailed finding aids for these records are cited in the *General Inventory,* vol. I. Series A, *Actes du Pouvoir Souverain* (1670-1782, transcripts, 1 foot) contains documents of the same nature as those in Series B and F^3, described below. Series B, *Lettres envoyées* (1663-1774, microfilm, 194 reels; 1663-1789, transcripts, 46 feet) contains copies of letter books, including dispatches, memoranda, and other papers sent by the king and the minister to officials, ecclesiastics and private persons in the colonies, with letters relating to shipbuilding, potash manufacture, and naval stores production in Canada. Series C^{11}A, *Correspondence Générale, Canada* (1458-1784, microfilm, 128 reels; transcripts, 31 feet) contains correspondence to the ministre de la marine from the civil and military officers, ecclesiastics, and other persons in the colony, including many references to the production of naval stores, shipbuilding, and the preservation of woods and forests. Series C^{11}B, *Correspondence Générale, Ile Royale* (1712-62, microfilm, 39 reels; transcripts, 7½ feet) contains similar documents, including references to timber and shipbuilding. Series C^{11}C, *Amerique du Nord* (1661-1898, microfilm, 18 reels; 1661-1861, transcripts, 5 feet), contains similar documents concerning Terre-Neuve, l'Ile Royale, l'Ile Saint Jean, l'Ile Madame, les Iles de la Madelein et Gaspé. Series C^{11}D, *Correspondence générale, Acadie* (1603-1788, 1814, microfilm, 9 reels; transcripts, 3 feet), contains similar items. Series C^{11}G, *Correspondence Raudot-Pontchartrain, domaine d'occident et Ile Royale* (1677-1758, microfilm, 12 reels; transcripts, 3 feet) complements series A, C^{11}A, C^{11}B, and C^{11}C, and includes the same type of documents, with references to shipbuilding, the production of naval stores, the timber trade, and manufacture of tar. Series C^{13}A, *Correspondence générale, Louisiane* (1678-1781, microfilm, 36 reels; 1700, transcripts, 13 pp.), includes correspondence and reports to the ministre de la marine from civil, military, and ecclesiastical officials in the colony. There is supplementary material in Series $C^{13}B$, *Correspondence générale, Louisiane* (1699-1803, microfilm, 1 reel). Series F^3, *Collection Moreau de Saint-Méry* (1540-1806, microfilm, 21 reels; 1603-1804, transcripts, 11 feet) consists of material relating to the French colonies collected by Moreau de Saint-Méry, including references to the manufacture of potash and shipbuilding.

M.G. 9. Provincial, Local, and Territorial Records

3741 *New Brunswick. Executive Council, Records, 1784-1867, 42 reels microfilm.* Records accumulated by the Provincial Secretary and the clerk of the Executive Council. Copied from originals in the Provincial Archives of New Brunswick. Finding Aid No. 121. Subject files include Crown Lands and Forests (1784-1851, 4 reels), with papers on timber and lumber duties, seized timber, mill reserves, licenses, disputes, conservation, trespass, the Upper St. John and disputed boundary, and Joseph Cunard's timber concerns on the North West Branch of the Miramichi and on the Nepisiquit Rivers; Boundaries (1803-59, 1 reel), with papers relating to disputes between Maine and Canada; Indians (1809-59, 1 reel), with correspondence, petitions, and reports relating to the sale of Indian reserve lands and timber rights on the reserves; Companies (1836-56, 1 reel), including documents relating to several boom, mill, and land companies.

3742 *Ontario. Department of Lands and Forests, Records, 1784-1848, 2½ feet; 35 reels microfilm, 1780-1958.* Includes land and survey papers from the Crown Lands Department of the Province of Canada, the Ontario Crown Lands Department, and surveys from the Ontario Department of Lands and Forests.

3743 *Ontario Local Records.* Lennox and Addington Counties, collections of the Lennox and Addington Historical Society (1768-1921, 24 reels microfilm). Finding Aid No. 297. See description under Lennox and Addington Historical Society, Napanee, Ontario. L'Original, Papers (1828-45, 2 vols.) include a printed notice forbidding the cutting of timber on certain lands in Longueuil Township by Charles Treadwell at L'Original, Ontario, August 13, 1845.

3744 *British Columbia. Royal Commission on Forest Resources, Records, 1944-1945, 2 feet, 4 inches.* Incomplete copies of the hearings before Chief Justice Gordon Sloan concerning matters affecting forest resources and allied industries, containing testimonies from which Sloan prepared his report in 1945.

3745 **M.G. 11. Colonial Office Papers (London).** Governor's dispatches to the Colonial Office and its predecessors, military and naval correspondence, survey notes, government gazettes, legislation, etc., copied from originals in the Public Record Office, London. These records contain material pertaining to the problem of naval stores for the Royal Navy and efforts to identify and preserve for the Crown appropriate timberlands. In most cases, the Public Archives of Canada holds complete copies of the correspondence classes for each colony in North America. For the remaining four classes (acts, sessional papers, government gazettes, miscellanea), holdings are mostly selections, often supplementing sources in the public archives library. The originals are described in the *Guide to the Contents of the Public Record Office,* 3 vols., London: Her Majesty's Stationery Office, 1963-68. Many of these documents, 1574-1738, have been printed or abstracted in *Calendar of State Papers, Colonial Series, America and West Indies, Preserved in the Public Record Office,* 44 vols., London: Her Majesty's Stationery Office, 1860-1969. Holdings of the Public Archives of Canada are described in *Preliminary Inventory: Manuscript Group 11, Public Record Office, London: Colonial Office Papers,* Ottawa, 1961, 81 pp. C.O. 1, Colonial Papers, General Series (transcripts, 1597-1697, 15 inches) contain material relating to American and West Indian colonies. C.O. 5, America and West Indies (1689-1819; transcripts, 21 feet; microfilm, 27 reels) consists of correspondence and entry books of the Board of Trade and the Secretary of State, together with Acts, Sessional Papers, and miscellaneous records arranged by colony or province, and containing frequent reference to naval stores, pitch and tar trade, timber cutting licenses, etc., in British North American and the American colonies before the revolution. C.O. 6, British North America, Correspondence General (1816-68, microfilm, 27 reels) includes letters, journals, etc., of various geological and other explorations which may yield reports on forests. C.O. 42, Canada, Correspondence (1700-1922; transcripts, 5 feet; microfilm, 723 reels) contains letters from governors, lieutenant governors, and administrators of Upper and Lower Canada, and includes materials relating to the procurement of masts, yards, etc., in America for the Royal Navy, the importation of potash into Quebec, and the import of potash and timber from the U.S. into Great Britain. C.O. 42, Canada, Correspondence "Q Series" (1760-1841, 145 feet), which parallels a portion of C.O. 42, but also contains additional material, includes many references to such topics as commerce in potash, lumber, lumber trade between British North America and the United States, sawmills in Canada, timber licenses, revenues from sales of timber from Crown lands, improvement of timber navigation of the Ottawa River, the office of Surveyor-General of Woods and Forests, lumber duties, proposals for timber-slides at Chats Falls and the Chaudière Falls. For Canada there are also selections from C.O. 43, entry books of correspondence (1763-1872; transcripts, 13 feet; microfilm, 32 reels); C.O. 44, Acts (1764-1841, microfilm, 30 reels); C.O. 45, sessional papers (1809-10, transcripts, 4 inches); C.O. 47, Miscellanea (1786-1865; transcripts, 2½ inches; microfilm, 10 reels). cluding shipping returns, blue books of statistics, surveyor's reports, etc. For British Columbia there are selections from C.O. 60, correspondence (1858-71, microfilm, 35 reels); C.O. 61, Acts (1858-64, microfilm, 65 feet); C.O. 63, Government Gazettes (1863-71, microfilm, 3 reels); C.O. 64, Miscellanea (1860-70, microfilm, 2 reels), including blue books of statistics; and C.O. 398, entry books of correspondence (1858-71, microfilm, 3 reels). Holdings on New Brunswick include the largest part of C.O. 188, correspondence (1784-1867, microfilm, 85 reels; 1784-1840, transcripts, 17 feet), with references to despoliation of timber, lumber trade, and timber licenses; C.O. 189, entry books of correspondence (1784-1867, microfilm, 5 reels); Acts (1786-1853, microfilm, 16 reels); sessional papers (1822-30, microfilm, 2 reels); Government Gazettes (1842-68, microfilm, 10 reels); and miscellanea (1786-1865; transcripts, 2½ inches; microfilm, 10 reels). Relating to Newfoundland are selections from C.O. 194, correspondence (1696-1922, microfilm, 212 reels); C.O. 195, entry books of outgoing correspondence (1632-1872, microfilm, 7 reels); C.O. 197, sessional papers (1833-55, 1860-1901, microfilm, 9 reels); and C.O. 199, miscellanea (1677-1903, microfilm, 18 reels), including blue books of statistics. Relating to Nova Scotia and Cape Breton are the largest part of C.O. 217, original correspondence (1603-1867, microfilm, 112 reels); C.O. 217, correspondence "Nova Scotia A" (1603-1840; transcripts, 30 feet; microfilm, 8 feet) including references to sawmills, the lumber trade, trees found in the province, timber exports; C.O. 217 and 220, composite series, Nova Scotia Executive Council minutes (1720-85, transcripts, 3 feet), Legislative Council Journals (1758-1807, transcripts,

3 feet), and Assembly Journals (1758-1805, transcripts, 6 feet); C.O. 218, Nova Scotia and Cape Breton, entry books of correspondence of the Board of Trade and the Secretary of State (1710-1867; photocopies, 11 inches; microfilm, 8 reels); C.O. 219, Nova Scotia and New Breton, Acts (1749-1899, microfilm, 6 reels); C.O. 220, Nova Scotia and New Breton, sessional papers (1767-1869; photocopies, 48 pp.; microfilm, 2 reels), and Cape Breton Executive Council minutes (1785-1807, transcripts, 2 feet); and C.O. 221, Nova Scotia and Cape Breton, miscellanea (1730-1866; transcripts, 1⅔ feet; microfilm, 14 reels). Holdings for Prince Edward Island include C.O. 226, correspondence (1769-1873, microfilm, 61 reels; 1763-1840, transcripts, 14 feet); C.O. 226 and 229, Executive Council minutes (1770-98, 1805-19, transcripts, 18 inches), Legislative Council Journals (1773-1803, 1806, transcripts, 8 inches), and Assembly Journals (1776-1806, 1818-19, transcripts, 18 inches); C.O. 227, entry books (1769-1872, microfilm, 3 reels); C.O. 228, Acts (1770-1864; transcripts, 1 inch; microfilm, 7 reels); C.O. 229, sessional papers (1770-1858, microfilm, 8 reels); C.O. 230, Government Gazettes (1832-75, microfilm, 9 reels); and C.O. 231, miscellanea (1807-71; photocopies, 52 pp.; and microfilm, 11 reels).

3746 **M.G. 12. Admiralty and War Office (London).** Copies of originals in the Public Record Office, London, and described in *Guide to the Contents of the Public Record Office,* 3 vols., London: Her Majesty's Stationery Office, 1963-68. Included is military and naval correspondence containing materials pertaining to the problem of supplying naval stores for the Royal Navy and efforts to identify and preserve to the Crown appropriate forest areas. Adm. 1, Secretary's Department In-Letters (1697-1864; transcripts, 10 feet; microfilm, 128 reels), contains communications from naval officers concerning the preservation of mast trees in the colonies, production of naval stores, surveyors of the woods, etc. Adm. 49, Accountant General's Department Miscellanea (1740-1851, microfilm, 4 reels), includes material on such topics as surveys of naval timber in Nova Scotia by John Wentworth (1783-94); prices and production of naval stores and ship timber in the colonies (1712-20); abstracts of correspondence (1804-05) relating to Quebec oak timber supplied by Beatson and Company; and abstracts relating to the supply of masts and timber by Henry Caldwell (1804-05); etc. Adm. 106, Navy Board records (1763-64, 1790-1882, microfilm, 10 reels), includes information on naval stores. Adm. 128, Station records, North America and West Indies (1810-1912, microfilm, 56 reels) may also contain references of interest.

3747 **M.G. 14. Audit Office and Treasury.** Copies of originals in the Public Record Office, London. M.G. 14B, Treasury, includes scattered references to such topics as masts exported (1771), licenses to cut timber (1811-17), delivery of timber at Kingston, Ontario (1814), etc.

3748 **M.G. 16. Customs and Plantations (London).** 1766-1857, Microfilm, 68 reels. Official papers sent to the Commissioners of Customs from officers in Quebec, Montreal, Newfoundland, New Brunswick, Nova Scotia, Cape Breton, and Prince Edward Island relating chiefly to customs establishments, revenue, legislation, vessel registry, and customs enforcement including seizures. Originals are in the Kings Beam House, London.

3749 **M.G. 20. Hudson's Bay Company Archives.** Vancouver's Island Steam Saw Mill Company, Correspondence (1852-1856, 1 reel microfilm [1 vol.]) is the records of a firm formed in 1851 to provide shipbuilding timber on the coast of British Columbia and Vancouver Island. Included are invoices, minutes, accounts, and other papers. A list of the documents precedes the volume.

M.G. 21. Transcripts from Papers in the British Museum

3750 *Egerton Manuscripts.* 929. Transcripts (n.d. and 1705-09, 1747, 25 pp.). Finding Aid No. 599. From miscellaneous papers (1692-1761), mainly addressed to Charles Montagu, Earl of Halifax. Included are anonymous and undated memoranda (3 items), sent to Halifax apparently by John Chamberlayne (1666-1723), urging the ejection of the French from Canada and Nova Scotia, and proposing that colonists from Scotland and the Palatinate be brought in to settle these areas and the Kennebec region, and to produce naval stores. Also a similar proposal set forth in a memorial and letter (1705) from colonial agent Jeremiah Dummer. Copies of originals in the British Museum, London.

3751 *Sloane and Additional Manuscripts.* Add. Mss. 8133 B (1783, transcript, 1 p.), from papers relating to the revenue of the customs, contains accounts of spruce essence imported from Canada each year, from Christmas, 1778, to Christmas, 1782. Add. Mss. 14035 (ca. 1761-79, transcripts, 16 pp.), from papers of the Board of Trade and Plantations relating to Britain's trade with Europe and the colonies, includes references to timber, etc. Add. Mss. 15485 (ca. 1769, transcript, 45 pp.) contains accounts of exports and imports of the British North American colonies, 1768-69, including statistics relating to the quantities and value of particular goods, their sources and destinations, and the number and tonnage of vessels engaged in the trade. Add. Mss. 40467 (1841-42, microfilm, 1 reel) contains correspondence between Sir Robert Peel and the Secretary for the Colonies, Lord Stanley, later Earl of Derby, discussing timber duties.

3752 *Stowe Manuscripts.* 246. Transcripts (1711-19, 8 pp.) from letters chiefly to James Craggs, the younger (1686-1721), relating principally to state affairs. Included is a letter of William Popple, secretary to the Board of Trade, to Martin Bladen, a commissioner of the Board (1719) reporting that William Lowndes, secretary to the treasury, had changed his mind and now opposed lifting duties on American lumber and timber.

M.G. 23. Late Eighteenth Century Papers

3753 *Nova Scotia.* Diaries (1766-1812, transcripts, 1 foot; 1777-1812, microfilm, 2 reels) of Simeon Perkins (1735-1812), Liverpool, Nova Scotia, lumberman, cover his business affairs. The originals are in the possession of the Town of Liverpool. Microfilm (reels M-71 and M-72) may be obtained on inter-institutional loan.

3754 *New Brunswick.* Papers (1751-1844, 24 feet, and 1783-1839, 3 reels microfilm) of Ward Chipman, Sr. (1754-1824) and Jr. (1787-1851), Series I, "Lawrence Collection," include records of Thomas Roy, Deputy Surveyor General of the Woods (1809-11 [vol. 9, pp. 106-24]). Finding Aid No. 92. Papers (1695-1877, microfilm, 7 reels) of Edward Winslow, Surveyor General of Woods for New Brunswick, are copies of the originals held by the University of New Brunswick, Fredericton. Finding Aid No. 210 provides a shelf list of the Winslow papers in the order of their filming, and includes a subject and name index. Some items have been printed in William O. Raymond, *Winslow Papers* (Saint John, 1901). Microfilm (reels M-145 to M-151) is available on inter-institutional loan.

3755 *Quebec and Lower Canada.* Among collections relating to political figures are papers (1759-84, transcripts, 32 pp.) of Christopher Chapman Bird, containing correspondence

(1783-84) and memoranda (1782-83) relating to his office as Surveyor General of Woods in Canada, in particular the difficulties he had in obtaining his salary. Included are several letters from William Van Felson concerning the suitability of pine for naval use. Originals are in the Register House, Edinburgh, Scotland. Among collections relating to merchants and settlers are Detroit, Michigan, garrison and land records (1770-84, 119 pp.), containing a list of suppliers of firewood to the garrison (1776).

M.G. 24. Nineteenth Century Pre-Confederation Papers

3756 *British Officials and Political Figures.* All records of governors may contain material relative to timber licenses and dues, particularly in New Brunswick and Upper Canada. Papers (1746-1839, transcripts, 7 feet; 1789-1839, microfilm, 15 reels) of George Ramsey, Ninth Earl of Dalhousie (1746-1839), include subject files with material relating to forestry (1815-28). The originals are in the Scottish Record Office, Edinburgh. Calendared in the Public Archives of Canada, *Report for 1938,* pp. 1-175; microfilm is described in Finding Aid No. 399. In the papers (1825-41, 2 feet, 2 inches; transcripts, 1813-41, 2 inches) of Sir John Harvey (1778-1852) are letterbooks (Series I, vol. I, 1839-41) with transcripts of many letters relating to the Maine-New Brunswick boundary dispute. Finding Aid No. 306. Papers (1841-48, 2 inches) of Sir William MacLean George Colebrooke (1787-1870) also contain letters and dispatches relating to the Maine-New Brunswick boundary dispute.

3757 *North American Political Figures and Events.* Papers (1666-1912, 37 feet) of the Neilson Family of Quebec include records of the Admiralty Lake Service with correspondence (1814-15, 1817) of the Naval Storekeeper at Montreal; also newspaper records relative to the publication of trade notices (sales of timber, etc.) and government notices relative to dues and licenses. Papers (1778-1850, photocopies, 256 pp.) of Charles Wright include letters relating to the business and land interests of John Cambridge, a Bristol merchant with very extensive interests in the timber trade and shipbuilding on Prince Edward Island. Originals are owned by Mrs. J.T. McIntyre of Calgary, Alberta. A manuscript (1865, 15 pp.) by James Skead (1817-84), a lumber merchant of Ottawa and member of the Legislative Council of the Province of Canada, is a speech on reciprocity entitled "The Lumber Trade of Canada," read before the Detroit Commercial Convention, July, 1865.

3758 *Industry, Commerce, and Finance.* Papers (1742-1910, 1⅔ feet) of the Hayes Family of Toronto, Ontario, contain personal and business correspondence relating to shipbuilding, Crown lands and timber administration, etc. Finding Aid No. 81. Papers (1803-88, 2 inches) of George Hamilton (1781-1839) include correspondence and business records relating to the lumber business operated by Hamilton and his son John (1827-88); mills were near the present town of Hawkesbury, Ontario, and both square timber and deals were shipped to Quebec and Liverpool. The firm was sold to the Hawkesbury Lumber Company in 1888. Finding Aid No. 169. The Wright Family Papers include the personal and business records (1792-1864, 45 feet) of Philemon Wright (1761-1839), his family, and the firm of Philemon Wright and Sons, Hull, Lower Canada, relating to various enterprises, including lumbering; there are various financial records and raft books (1806-47); and a diary (1834-35, photocopy, 74 pp.) recording Philemon Wright's trips conveying lumber to the Quebec market. Original of the diary is in the possession of Mrs. T.G. Maybury, Hull, Quebec. The correspondence is described in Finding Aid No. 542. There is also correspondence and other papers (1816-62, microfilm, 1 reel) kept by Ruggles Wright relating to the lumbering and land business of Philemon Wright and Sons. Originals are in the possession of Bruce Wright, Fredericton, New Brunswick. Finding Aid No. 86. Newfoundland Papers (1813, 2 items) include a warrant authorizing the importation of turpentine, pitch and other goods into Newfoundland, issued by the Prince Regent and sent to Sir Thomas Duckworth, Governor. Accounts (1834-48, 2 inches) of Theophilus Yale relate to the estate of this lumber merchant of La Chute, Lower Canada. Records (1866-1920, photocopies, 32 pp.) of the James Maclaren Company (Buckingham, Quebec), include a biographical sketch of James Maclaren (1818-92), together with abstracts of titles, copies of deeds and extracts of letters relating to the Maclaren milling interests on Green Island (Ottawa) and at Buckingham, Quebec. Originals are in the possession of the James Maclaren Company. Lester-Garland Papers (1761-1834, microfilm, 1 reel) contain extracts relating to Canada from the diaries and letterbooks of Benjamin Lester, George Garland, and John Bingley Garland, of the firm of Benjamin Lester and Co., fish and timber merchants of Poole, England, and Trinity, Newfoundland. Originals are in the possession of G.H. Lester-Garland, Bath, England. Finding Aid No. 96. Letter (1809, 1 item) of N.G. Graham to a Mr. Rose (George Rose, Vice President of the Board of Trade?) concerns the fur and lumber trades in British North America, and the general state of trade between those colonies and the West Indies. Papers (1829-31, photocopies, 18 pp.; also microfilm) of William and George Harper, timber merchants in Moncton, New Brunswick, contain letters (1830-31) of William and timber accounts, tables of deals, and an invoice of goods wanted (1829-31) prepared by George Harper. Originals are at Fort Beauséjour Historic Park. The Archibald M. Campbell Collection includes business records (1828-50, 2 vols.) of William and John Bell, general merchants at Perth, Lanark County, Ontario, and commission agents for the sale of potash, lumber, and real estate. Address (1866, photocopy, 1 p.) to Daniel McLachlin (1810-72) by the residents of Arnprior, Ontario, and vicinity, recognizes his service to the community and to the lumber industry of the Ottawa Valley. Original is in the possession of Donald McLachlin. Papers (1838-51, 1⅔ feet) of James Stevenson (1815-51), Crown Timber Agent at Bytown, Upper Canada, include letters with routine enumerations of timber from agents on the Upper Ottawa River and at Québec City, accounts of fees forwarded to the Commissioner of Crown Lands, and information concerning disputed timber berths, as well as private matters. P. Musgill and Company (Québec), bill of lading (1844, 1 p.) relates to lumber to be shipped on the *Mersy* from Quebec to Menai Bridge. Papers (1808-1907, 8 inches) of William Loch (ca. 1786-1856), Newcastle, New Brunswick, lumber merchant, include correspondence and related records.

3759 *Immigration, Land, and Settlement.* Reminiscences (2 pp.) of Robert Charles Wilkins (1782-1866), timber merchant of Belleville, Canada West, concern transportation on the St. Lawrence River system and particularly at Carrying Place, Upper Canada, during the War of 1812. Papers (1792-1844, ca. 14 feet) of Alexander Hamilton (d. 1839) include material relating to a milling and lumbering business established by Hamilton and Alexander Askin at Canboro, Upper Canada, which failed in 1817. Finding Aid No. 211. Journal (1819-22, 1 inch) of a Mr. Johnson includes an account of Johnson's work as a logger in Canada. Letter (1828, 3 pp.) to William Allen from Robert Dickson concerns the operation of a sawmill at Trollhattan, Upper Canada. Papers (1859-60, microfilm, 1 reel) relating to Gatineau River, Quebec, consist of a diary and notebook attributed to John Mather, woods manager of the Gilmour Company. Originals are in the possession of Mrs. Arthur Davidson, Ottawa. Letters (1826-27, photocopies, 8 pp.) to William Fussey from William Wallace describe trips as a timber buyer for Longley and Dyke of Quebec through Upper and Lower Canada. Originals are in the possession of T.B. Higginson, Sharbot Lake, Ontario. Papers of Pierre-Gustave Joly (1798-1865) relating to the Seigneurie de Lotbinière (1821-65, microfilm, 2 reels) include material relating to the lumber trade. In the French language. Finding Aid No. 177. Originals are in the archives of the Seigneurie de Lotinbière, Quebec. Papers (1834-86, 1902, 35 pp.) of William Farmer relate to land and timber transactions near Gatineau Falls, Quebec. Originals are in the possession of Dr. G.R.D. Farmer, Rockcliffe, Ottawa. Papers (1808-1923, ca. 2 feet) of Thomas J. Barrow (1832-1912) include lumber accounts of John Chesser, Terrebonne, Quebec (1808-09). Papers (4 inches) of Robert Nelles (1761-1842) relate to his lumbering business at Grimsby, Upper Canada, and personal affairs. Correspondence (1846-47, 14 pp.) of Cyprien Blanchet, Crown Lands Agent, is with R.M. Harrison, Arthur Ross, and H. Breakey, concerning timber licenses in Beauce County, Quebec.

3760 *Miscellaneous.* Delancy-Robinson Collection (1789-1876, ca. 1 foot) is a collection of miscellaneous documents relating to the history of New Brunswick, including material concerning the New Brunswick fires of 1825. Finding Aid No. 298.

M.G. 27. Political Figures, 1867-1950

3761 *1867-1896.* Papers (1877-99, 1 1/6 feet) of Edgar Dewdney (1835-1916), Canadian Minister of the Interior, 1888-92, relating to the Department of the Interior and Indian Affairs (1 vol.), include correspondence, memoranda, and reports on Banff Park development (1877-92). Finding Aid No. 56. Letters (1883, photocopies, 4 pp.) of William Little are 2 items from Sir John A. MacDonald regarding the duty on American lumber, and possible renewal of reciprocity negotiations. Originals are held by the Bonar Law-Bennett Library, University of New Brunswick.

3762 *1896-1921.* Papers (1875-76, 1 inch) of Richard Reid Dobett (1837-1902), include a record of lumber purchased, wintered, and sold by R.R. Dobett and Co. (Quebec), with information on prices, methods of purchase and sale, types of lumber, composition of shiploads, and other details of the operation of the lumber industry. A letter (1885, 4 pp.) to Sir John Carling (1828-1911) from J. Beaufort Hurlbert concerns the Canadian Wood Display at the 1862 World's Exhibition. Papers (1882-1939, 8 feet) of Francis Robert Latchford (1854-1939) include records of the proceedings of the Timber Commission Inquiry (1920-27) conducted by Latchford for the government of Ontario concerning the administration of Crown timber lands. Finding Aid No. 31.

M.G. 28. Records of Post-Confederation Corporate Bodies

3763 *Boyd, Mossom, & Company (Bobcaygeon, Ontario), Records, 1839-1941, 120 feet.* Correspondence, reports, memoranda, account books, letterbooks, etc., being the records of the lumbering and sawmill operations, land transactions, and personal papers of Boyd, Mossom, & Company, including the Big Island Stock Farm, Lindsay, Bobcaygeon and Pontypool Railway, and Trent Valley Navigation Company.

3764 *Gilmour & Hughson, Ltd., Records, 1845-1926, 125 feet.* Business records of a lumbering and shipbuilding firm. Correspondence (1847-1924), letterbooks (1877-1921), account books (1873-1925), accounts (1853-1925), miscellaneous record books (1857-1926), and other papers.

3765 *Kerry & Chace, Ltd. (Toronto, Ontario), Records, 1887-1938, 30 feet.* Company records of an engineering firm that was concerned with large-scale public utility, mining, and forest resource development. Included are incoming reports (1887, 1898-1938), incoming specifications (1887-1926), estimates (1910-38), appropriations (1910-15), specifications and contracts (1909-29), rates for electric power (1907-18), photographs (1915-27).

M.G. 29. Nineteenth Century Post-Confederation Manuscripts

3766 *Exploration and Travel.* Papers (1854-1919, 17 feet) of Robert Bell (1841-1917) include correspondence, memoranda, clippings, and notebooks of Bell and his family, including material relating to the Canadian Geological Survey (1857-1908), forestry in Canada, etc.

3767 *Economic Development and Social Life.* Reminiscences (ca. 1911, 1 inch) of W.A. Robertson (b. 1832) include an account of mining and lumbering in British Columbia, 1865-1911. A manuscript (9 pp.) of Andrew Sibbald mentions lumbering at Morley, Northwest Territories, ca. 1875-90. A manuscript of George L. Graham is a copy of supply lists for timber shanties at Brennan Lake, Quebec, 1897-1903.

3768 **M.G. 30. Manuscripts of the first half of the twentieth century.** Papers (1916-19, 4 inches) of William Fishbourne McConnell (1884-1940) relate chiefly to the Canadian Forestry Corps and his duties as a chaplain therein. Reminiscences (1937, 6 pp.) of William Adair (b. 1847) are the memoirs of a Middlesex County, Ontario, charcoal maker. Recorded by Dr. Edwin Seaborn.

PETERBOROUGH

Peterborough Public Library

Librarian K9H 3R8

3769 **Crowe, Joseph,** Reminiscences, ca. 1870-1919, 3 items. Peterborough, Ontario. Includes a memoir (9 pp., typed) relating to life in a lumber camp of the Ludgate and McDougall Lumber Company in Digby Township, Victoria County, Ontario, in 1875.

---

*Trent University, Thomas J. Bata Library

Telephone: (705) 748-1011

3770 Russell G. Hann, et al., *Primary Sources in Canadian Working Class History, 1860-1930,* Kitchener, Ontario: 1973, describes papers (1838-1935) of the Fowlds family, lumbermen and merchant-millers of Hastings, Ontario.

TORONTO

Archives of Ontario

Archivist, 77 Grenville Street, Queen's Park M7A 1C7

Archives

Department of Lands and Forests (Ministry of Natural Resources)

3771 *Survey field notes, and diaries, 1783-1944, 9 feet.* Along with technical data and other observations, species of trees are noted. Most of these records are still in custody of the Surveys and Mapping Branch of the Ministry of Natural Resources.

3772 *Timber Branch of the Department of Lands and Forests and the Woods and Forests Branch of the Crown Lands Department, Records, 1800-1970, 305 feet.* They are comprised of timber license files, returns of operations, reports and memoranda, correspondence files and accounting records. Restricted: permission from the Ministry is needed for access to records less than 30 years old.

3773 *Forest Management Branch, Records, 1905-1970, 20 feet.* The records consist of Deputy Minister of Forestry correspondence, private land forestry correspondence, forestry management reports, reforestation records, Ontario Forestry Board minutes and files of the Advisory Committee to the Minister of Lands and Forests. Restricted: for access to files less than 30 years old permission from the Ministry is needed.

3774 *Local Timber Agencies (Ottawa, Belleville, Windsor), of the Department of Lands and Forests, Records, 1830-1929, 33 feet.* Bound volumes consisting of license registers, statements and returns of operations, correspondence registers, letterbooks and account ledgers.

3775 **Department of Municipal Affairs,** Financial returns by township clerks and audit reports, 1879-1961, 750 feet. In addition to financial information these returns include general statistics, such as acreages of swamp, woodland, cultivated land, etc.

3776 **Commissions and Committees.** Ontario Royal Commission on Forestry records (1947, 15 feet) consist of the final printed report, the verbatim proceedings, briefs and submissions from interested parties and exhibits of the inquiry into conservation, management, and development of forest resources of Ontario, conducted by Howard Kennedy. Royal Commission on Forest Reservation and National Park records (1893, ½ inch) consist of the final report and appendices which were published as an Ontario Sessional Paper No. 31 in 1893, and which led to the creation of Algonquin Provincial Park. Royal Commission on Forestry Protection in Ontario records (1897, 1899, ½ inch) consist of the preliminary and final reports of the commission appointed to study the reforestation of burnt-out or cut-over lands. Records (1922, 1 foot) of the Commission to inquire into charges of negligence on the part of the Attorney-General's Department and others in the investigation into the death of Captain Orville Huston at Fort Frances consist of the report, proceedings, and exhibits, and include lengthy testimony by E.W. Backus on his lumber business dealings with Huston in Fort Frances and northern Manitoba. Timber Commission records (1922, 8 feet) consist of the printed interim and final reports, press clippings, proceedings, and exhibits of an inquiry by Justices William R. Riddell and Francis Latchford into Ontario's timber industry, concerning timber licenses, the government's relations with companies, cutting procedures, depletion of natural resources, etc. Northern Ontario Fire Investigation records (1922, 1 foot) consist of the report of the commission headed by E.P. Heaton to inquire into the forest fire which ravaged the Temiskaming District in northern Ontario, 1922, and include appendices, proceedings, and photographic exhibits, and information on land usage with regard to lumbering and farming, fire prevention legislation, firefighting, etc. Select Committee in Relation to the Supervisor of Cullers' Office records (1865, ¼ inch) consist of the printed report of a committee, chaired by W.F. Powell, named to investigate the Supervisor of Cullers' office, an agency with responsibility for overseeing the measurement and cutting of wood in Upper Canada. Select Committee on the Department of Lands and Forests records (1941, 2 feet) consist of reports and proceedings of a committee named by the legislature to investigate the administration, licensing, sale, supervision, and conservation of natural resources by the department, and to make recommendations for the reorganization of the department. Select Committee on Conservation Authorities records (1967, 5 feet) consist of interim and final reports of a committee under Dr. Arthur Evans appointed to review the provisions of the Conservation Authorities Act and to examine the methods of financing, land appropriation powers, etc., of these agencies. Select Committee on Conservation records (1950, 2½ feet) consist of briefs, proceedings, and reports of a committee under F.S. Thomas appointed by the legislature to study all aspects of conservation including soil depletion, drainage, flood control, reforestation, soil analysis, etc. The report was published as Sessional Paper No. 43, 1950.

Manuscripts

3777 **Anonymous.** Miscellaneous items of uncertain origin include an account book (1828-38, 10 pp.) of ashes received, Brockville, Ontario; history of the White River District (1953, 1959, 10 pp.) describing lumbering there from fur trade days to 1959; scrapbook (ca. 1874-92, ½ inch) containing clippings regarding the lumber industry on Ontario Crown lands, timber licenses, and export duties; account book (1875-80, 2 inches) of a lumber company recording logs accepted, lumber cut, daily expenditures, lumber and lath cut, Monteith area, Ontario.

3778 **Burton, Rev. W.W.,** Manuscript, ca. 1910, 1 p. Handwritten list of lumbermen killed in lumber camps and rivers in North Hastings, Ontario, 1848-92.

3779 **Campbell, Jack, and Marshall Dobson,** Typescript of oral history interview, 1963, 20 pp. Lumbermen in Parry Sound area. Interview concerns early days of lumbering in northern Ontario. Taped by John Macfie in 1963.

3780 **Cameron, Douglas,** Collection, 1826-1918, 2 inches. Photocopies. Alex, John, and Dougald Cameron engaged in logging, lumbering and farming in Glengarry and western Québec. Papers collected by Douglas Cameron including correspondence relating to the operation and conditions of the logging and lumber industry and farming and family matters, etc.

3781 **Carpenter, William H.,** Letterbooks, 1874-1891, microfilm, 4 reels. Accounts and correspondence related to work as supplier for the Canadian Pacific Railway in the Thunder Bay area, to Carpenter's lumbering operations and to his activities as Sheriff of Rainy River District, and a few personal letters. Microfilm catalog introduction and cards.

3782 **Dawson Family,** Manuscripts, 1842-1895, 2⅔ feet. Included are papers of William McDonell Dawson, Crown Timber Agent, Superintendent of Woods and Forests Branch, Crown Lands Office, Member of Parliament from Three Rivers, 1858-62, and Ottawa County, 1862-63, and papers of other family members. There is correspondence and papers relating to procedures of the Crown Timber Office, the lumber business, land and taxes, family matters, exploration of the Red River area. Calendar.

3783 **Douglas, William** (1838-1902), Papers, 1826-1914, 2½ feet. Correspondence, incoming and outgoing, of a personal, business and legal nature of the Douglas family of Chatham, Ontario, together with several diaries, manuscripts of a speech and poetry and documents dealing with lumbering and mining, and mineral and timber exploration in Ontario and Québec in the late 19th century. Inventory and catalog cards.

3784 **Ford Family Papers.** Correspondence of David Ford and Nathan Ford (1795-1810, 10 items) concerns the cutting, sale, and transport of timber, Morristown, Upper Canada. Calendar.

3785 **Guelph Lumber Company,** Account book, 1878-1880, 2 inches. Lists expenditures for lumber, payments by mill to lumbering companies.

3786 **Hamilton, George** (1781-1839), Letterbooks, 1804-1815, 3 vols. Vols. 1 & 2 (1804-10) are also on microfilm. The letterbooks contain copies of letters written at Québec by George Hamilton as Canadian agent for Robert and George Hamilton of Liverpool, England, and also as a partner in the firm of George and William Hamilton of Québec. The letters are addressed to his brother Robert in Liverpool, and to various individuals and firms in the United Kingdom, Newfoundland, the United States, the British West Indies, and Canada. They are concerned with arranging shipments of wheat, flour, potash and lumber from Québec, and the sale in Canada of consignments of imported goods. Catalog cards.

3787 **Hawkesbury Lumber Company,** Records, 1889-1939, 11½ feet. Messrs. Blackburn, Egan, Robinson, and Thisle purchased Hamilton Brothers lumber company in 1889 and formed a joint stock company which operated the firm under the name of the Hawkesbury Lumber Company, until 1940. The papers consist of correspondence, accounts, employment contracts, and sketches. Inventory and catalog cards.

3788 **Heenan, Peter** (1874-1948), Papers, 1919-1948, 1¼ feet. Engineer on Canadian Pacific Railway line at Kenora, Ontario, elected representative for Kenora-Rainy River; M.P.P. in Farmer-Labour Government, 1919-25, Minister of Labour in Mackenzie King's cabinet, 1926-30, Minister of Lands and Forests and Northern Development, 1934-41, and Minister of Labor, 1941-43 in Mitchell Hepburn's government. There is a small amount of personal correspondence, debates, speeches, newspaper clippings, photographs, scrapbooks containing miscellany connected with social activities. Subjects include collective bargaining, Winnipeg Strike 1919, unemployment 1930, Lake of the Woods Control Board, production of pulpwood and other forest products. Inventory and catalog cards.

3789 **Henry, George S.** (1871-1958), Papers, 15½ feet. Henry held various political offices, and served as Prime Minister, 1930-34. Included is material relating to paper and lumber companies. Preliminary inventory. Partially restricted: closed if less than 40 years old. Memoirs closed.

3790 **Inderwick, Mrs. Cyril,** Papers, 1829-1904, 2 inches. Also on microfilm. Includes letters of Robert Gemmell to his fiancé, Joanna Lees, containing information on lumbering on the Ottawa River.

3791 **Johnson, Edward Ellsworth** (1890-1953), Papers, 1938-1953, 15 feet. Johnson founded the Pigeon Lumber Co. in 1925, served as special assistant to President of Great Lakes Paper Co., E.W. Backus. In 1940 he built a sawmill at Fort William for his firm, Great Lakes Lumber Co. Its name changed to Great Lakes Lumber and Shipping in 1942. In 1950 it shut down for lack of sawlogs caused by competition with pulp and paper companies. Correspondence, speeches, publications, news clippings relating to the lumbering business, to Johnson's activities as a pioneer in forest management, furranching, animal husbandry, produce-growing in northwestern Ontario, and to his campaign against the pulpwood monopolists, and the Ontario government. Correspondents include various lumber and paper companies, M. Hepburn, George Drew, Leslie Frost, C.D. Howe. Restricted: donor's permission required for publication.

3792 **Langton Family,** Collection, 1821-1949, 1 foot. Included are papers of John Langton, an investor in sawmills, Peterborough, Ontario, 1851-55, and later auditor-general of Canada. Family correspondence includes descriptions of lumbering with Mossom Boyd at Fenelon Falls, 1849. Inventory and catalog cards.

3793 **Lumbermen's Association of Ontario,** Records, 1887-1908, 4 inches. Business correspondence (1900-08); constitution; list of lumber companies; miscellaneous memoranda; minute book (1887-1908), with inserted newspaper clippings and correspondence; other correspondence (1900-08).

3794 **McLachlin Family,** Records, 1834-1941, 1 foot. Merchants; Ottawa Valley, Ontario. Business records relating chiefly to lumbering in Bytown (1834-51), and to the milling establishment at Arnprior (1851-1900). Correspondence, invoices, accounts, receipts, supply lists, contracts, timber licenses. Photographs of logging operations of the McLachlin Brothers Lumber Co. taken by Charles McNamara (ca. 1900).

3795 **Macpherson, A.F.,** Letterbook, 1855-1864, 2 inches. Crown Land Agent (Lennox, Addington, Frontenac), 1844-60, Kingston lumber merchant. Copies of outgoing correspondence concerning price and sale of timber and the lumber business, activities as land agent including woodlot inspections and disputes, licenses, cutting rights on Canada Co. Land, property sales and other personal matters. Correspondents include: Hon. John A. Macdonald, Hon. W.B. Robinson, Frederick Widder. Also correspondence (1861-64) of W. Macpherson of the Government Emigrant Office, Kingston, regarding the arrival and placement of immigrants. Catalog cards.

3796 **Madawaska Improvement Co., Ltd.,** Records, 1887-1905, 2¾ feet. Company formed in 1887 by lumber firms using Madawaska River to construct and maintain improvements to facilitate the passage of logs down the river. J.R. Booth, lumber manufacturer of Ottawa was President and G.B. Greene, Secretary-Treasurer. Company liquidated in 1904 when stocks of timber in Madawaska area depleted. Corporate records of minutes of directors' meetings, the sale of stock, the transfer of shares, costs, profits, losses, letters patent and bylaws of the company; routine business correspondence; operating records including detailed accounts and receipts; maps; and agreements with Francis W. Dunn for damages to his farm. Inventory and catalog cards.

3797 **Mickle, Charles** (d. 1881), Papers, 1832-1915, 2⅓ feet. Lumber merchant of Guelph, Ontario. Personal and family correspondence, papers, including two cashbooks, relating to his lumber business, real estate, loans, patents, his estate and the estate of his sister, Mrs. Anne Smith. Catalog cards.

3798 **Northern Ontario Fire Relief Committee,** 1916-1922, 5 inches. Established to extend relief to sufferers in forest fires of July and August 1916 in Hearst, Cochrane, Iroquois Falls, and New Liskeard areas and of October 1922 in Cobalt and Haileybury areas. Reports containing accounts, 1916 and 1922, scrapbook containing stories of fires, clippings, pamphlet, "Memoirs of the Great Fire of 1922," photographs.

3799 **Morrisburg Collection.** Cashbook (1849-52, ½ inch) of Elisha Owen contains records of lumbering operations, listing expenses for various commodities, value and quality of logs and lumber.

3800 **Quetico Provincial Park,** Ontario, Diaries, 1917-19, 2 inches (8 items). Diaries of rangers containing daily reports of activities.

3801 **Ross, Alexander Herbert Douglas,** Papers, 1878-1949, 1⅔ feet. Manuscripts, postcards, photographs, and clippings relating to historical writings by Ross, in charge of the Dominion Forest Survey in Manitoba and consulting forester with the Canadian Pacific Railway. Included is a draft history of the Canadian Society of Forest Engineers (1945).

3802 **Smith, Elias,** Letterbook, 1799-1800, photostats, 2 inches. Smith conducted a milling, lumber and general trading business near Port Hope, Ontario. Copies of outgoing correspondence concerning supply and sale of goods, prices and payments, movement of trading vessels and other routine business; and concerning the development of Hope Township, Durham County. Letters to David Smith, his son and partner in New York, John Crysler, Peter Hunter, William Jarvis, Joseph Papineau, Thomas Ridout, John G. Simcoe, Joel Stone.

3803 **Stewart, William,** Letterbooks, 1834-1846, 3 inches. Bytown lumber merchant. Copies of letters to Quebec and Montreal timber merchants, other business associates, and friends regarding lumbering in the Ottawa Valley, financial matters, and politics. Correspondents include Malcolm Cameron, Hon. W.H. Draper, James Stevenson, Jr. Significant subjects include the culler's bill, amending of lumber act, fear of a war (1837-38) with Americans. Catalog cards.

3804 **Tett, Benjamin,** Papers, 1847-1924, 2 inches. J.P. Tett & Brother, merchants, lumbermen, and forwarders at Bedford Mills, Frontenac County. Receipts, inquiries, orders, memoranda regarding lumbering, leases, letters, agreements, deed relating to the management of the Benjamin Tett estate and to the leasing of a mica mine on Devil Lake. Catalog cards.

3805 **Tucker, Stephen** (1798-1884), Papers, 1862-1906, ½ inch and 2 reels of microfilm. Included are photostats of business papers and microfilm copies of letterbook and diary. Lumberman on Ottawa River at Papineauville, Québec, until 1871 and at Clarence, Ontario, until 1875. Ledger sheets of expenses incurred in transporting lumber by raft. Information on the composition of each raft, varieties of timber shipped, the construction and loading of a crib, and instructions for the care and maintenance of cribs and rafts. Letters concerning lumbering operations, financial arrangements for purchasing logs, and details of transporting them. Personal letters referring to activities in the community and politics. Diary containing entries relating to wood lot inspections and to other timber operations, to social events, and to personal activities.

3806 **Wallis, James** (ca. 1815-1893), Papers, 1830-75, 1¼ feet. Sawmill and land owner, Fenelon Falls, Ontario. Land documents (1833-74), accounts, and items concerning financial affairs of local churches.

Metropolitan Toronto Central Library, Canadiana and Manuscripts Section

Head, Canadiana and Manuscripts Section, 214 College Street M5T 1R3 Telephone: (416) 924-9511

3807 **Bracebridge (Ontario) Tanneries,** Business records, 1862-1906, 11 vols. The volumes are chiefly account books of various tanneries in Bracebridge (1862-93). There is one journal recording shipments of bark (1883).

3808 **Matheson, Alexander,** Papers, 1861-1904, 18 vols. and 569 pieces. The collection includes correspondence concerning investments in various mining and lumbering operations, including the Keewatin Lumbering and Manufacturing Company, Lake of the Woods Mining and Milling Company, and Sabaskong Mining and Lumbering Company (1882-1901, 375 pieces).

3809 **Powell, William Dummer** (1755-1834), Papers, 1735-1847, ca. 5¾ feet. The collection includes the papers of Hay and Glenny, a stave and lumber business at Quebec City, Quebec (1771-87, 1 vol.). They consist of accounts and letters between the partners, used as evidence in a legal dispute.

*United Church of Canada Central Archives

Victoria University, Queen's Park

3810 Russell G. Hann, et al., *Primary Sources in Canadian Working Class History, 1860-1930,* Kitchener, Ontario: 1973, describes papers (1911-15) of James Allen, General Secretary, Home Department, Missionary Society of the Methodist Church, containing correspondence on work in logging camps on the Pacific Coast.

University of Toronto Library, University Archives

University Archivist M5S 1A5

3811 **University of Toronto. Faculty of Forestry,** Records, ca. 170 feet. Administrative records of the Faculty; papers of the Deans of the Faculty and staff members, including Bernhard Eduard Fernow, G.G. Gosens, J.W.B. Sisam, and Edmund John Zavitz; logging reports (ca. 1907-60), with some photographs, resulting from field trips to observe lumbering operations; minutes of the advisory committee to the Ontario Minister of Lands and Forests; records of the Forestry Alumni Association; records of the University of Toronto Foresters Club; etc.

Québec

QUÉBEC

Archives Nationales

Parc des Champs de Bataille G1A 1A3. Holdings are described in *L'Etat Général des Archives Publiques et Privées du Québec,* Québec: 1968. Most of the following collections are in the French language.

Public Archives

3812 **Québec. Terres et Forêts.** 1859-1930, 4 feet. Correspondence, reports, and other papers of the Ministère des Terres et des Forêts.

3813 **Québec et Bas-Canada. Arpentage.** 1765-1881, 4 feet. Included is correspondence, instructions, reports, etc. (1827-63, 4 inches), of John Davidson, Surveyor of Crown Woods.

3814 **Québec et Bas-Canada. Terres de la Couronne.** 1793-1839, 9 feet. Records relating to permits, revenues, etc., for lumbering in districts (1828-41, 8 inches); correspondence relating to lands and forests (ca. 1808-63, 4 feet); woodcutting licenses (1838-48, 4 inches).

3815 **Nouvelle-France. Collection de Pièces Judiciaires et Notariales.** 1638-1759, 41 feet. Documents of August 29, 1669 (no. 64), April 15, 1710 (no. 440½), January 21, 1722 (no. 626), November 9, 1754 (no. 1731), and of 1746 (no. 2701) concern forests. Described in P.-G. Roy, *Inventaire d'une collection de pièces judiciaires et notariales, etc.,* Beauceville: L'Eclaireur, 1917, 2 vols.

3816 **Nouvelle-France. Ordonnances des Intendants.** 1666-1760, 2½ feet. Included are orders relating to forests (1706-47, ca. 100 items). Described in P.-G. Roy, *Inventaire des Ordonnances des intendants de la Nouvelle-France,* Beauceville: L'Eclaireur, 1919, 4 vols.

Private Papers

3817 **Chouinard, Flavien.** 1877-1944, 18 feet. Account books of logging camps (1893-1944); correspondence relating to logging (1918-43, 6 inches); miscellaneous documents concerning the sale of timber (1919-29, 2 feet).

3818 **Hart, Famille.** 1760-1865, microfilm, 15 reels. Correspondence of Moses Hart with D. Thomas, agent of the corporation of clergy reserves, on logging by Moses Hart in the Nicolet region (1832-33); documents concerning sale of wood in the canton of Shipton (1831-33); documents concerning mills, forestry and lumber industry operations (ca. 1820-50). Originals are in the Archives of the Seminary of Trois-Rivières. Finding aid.

3819 **Honorat, J.B.** 1845, 9 pp. Memoranda probably to the Commissioner of Colonization, describing Saguenay. Manuscript and typed copies.

3820 **Joly de Lotbinière, Famille.** 1672-1923, 8⅔ feet, and microfilm, 6 reels. In the papers of Pierre-Gustave Joly (1821-65, 2 feet) and the papers of Henry-Gustave Joly de Lotbinière (1844-1908, 5⅔ feet) there are documents relating to the lumber trade. The microfilm includes account books of sawmills. Finding aids.

3821 **Laurier, Wilfried.** 1847-1957, 2½ feet. Included is a book of statistics on the commerce of Canada (1855-83, 2 inches).

3822 **Marchmont, Edwin.** 1870-1879, 4 inches. Licenses for timber cutting issued by Levison Sewell for the estate of Edwin Marchmont in the Trois-Pistoles River area (1875-78, 2 inches); accounts, timber estimates, deeds, etc. (1870-79, 1 inch).

3823 **Mazuret, Famille.** 1816-1831, 42 pp. Record of a timber sale by Amable Mazuret to Louis and Antoine Mazuret (1816, 4 pp.).

3824 **Menier, Famille.** 1896-1928, 3 1/6 feet. Records of daily activities of Georges Marting-Zédée at Anticosti Island (1909-14, 6 items).

3825 **Ross, Famille.** 1738-1967, 11 1/6 feet. Wood accounts, logging permits and contracts, and related correspondence (1851-1938, 4 inches).

3826 **Silsby Lumber Co.** 1908-1955, 6 pp. List of logging contracts of B.C. Howard with the Silsby Lumber Co.

3827 **Trigge, T.** 1840-1862, 1 inch. Merchant; Quebec. Account book mentioning a mill and a sale of wood.

Université Laval, Bibliotèque Générale. Division des Archives

3828 **Dunn Family,** Papers, 1821-1939, 43 cubic feet. Correspondence, daily journals, cashbooks, daybooks, accounts, etc., of Dunn's companies or associations and other businesses.

3829 **Laval University. Faculty of Forestry and Geodesy,** Records. The thesis collection includes about 700 theses and student papers on forestry and geodesy. Archival records of the Faculty are scheduled for early accession under the University's records management program (1974).

Université Laval, Québec Seminary Archives

3830 **Québec Seminary Corporation,** Papers, 1680-1955. References to sawmill operations and rental of timberlands are scattered through the collection.

Saskatchewan

REGINA

Saskatchewan Archives Office

Provincial Archivist, Saskatchewan Archives Office, Library Building, University of Regina

3831 **Bain, John,** Manuscript, 1918, 133 pp. Typescript. Report to the government of Saskatchewan relative to the alienation by the Dominion government of the lands and natural resources situated within the province.

3832 **Barr, George Herbert** (1878-1960), Papers, 1913-58, 3 feet. Regina, Saskatchewan; lawyer. Correspondence, briefs, addresses, etc. Topics include natural resources.

3833 **Saskatchewan. Forestry Commission,** Papers, 1946, 10 inches. Proceedings at sittings (January 29-May 16, 1946).

Saskatchewan. Natural Resources Department

3834 *Records of the Survey Branch, 1883-1958, 76 feet.* Files concerning surveys and development plans for villages and towns, Saskatchewan-Northwest Territories boundary surveys, surveys of roads, parks, reserves, etc., equipment and supplies, aerial photography, maps and mapping, place names, employees, budget and expenses, etc. (1933-58); Department of the Interior township plans, township files, and surveyors' diaries (1883-1930).

3835 *Records of the Deputy Minister's Office, 1933-52, 15 inches.* Included are files concerning the Royal commission on Natural Resources of Saskatchewan; notes and correspondence with regard to transfer of resources, and exhibits (December, 1933-March, 1934).

3836 *Records of the Research and Planning Branch, 1953-61, 7 feet.* Files of the Director concerning such matters as departmental administration and policies, research studies and projects, industrial development, northern development, Crown corporations, forestry, wildlife, etc. (1956-61); Provincial Secretary's Parks Branch files (1953-57); files of the Parks Branch, Department of Travel and Information (1956-60).

3837 **Western Canada Society for Horticulture,** Papers, 1955-56, 5 inches. Files relating to collection of information on historical development of horticulture on the prairies including correspondence and reports on contributions, including forestry stations.

SASKATOON

University of Saskatchewan, Saskatchewan Archives Office

Provincial Archivist S7N 0W0

3838 **Canada. Interior Department, Dominion Lands Branch,** Records, 1864-1907, microfilm, 15 reels. Orders in Council (1864-1905); departmental files concerning York Farmers' Colonization Co., Temperance Colonization Society (Saskatoon); Barr Colony, Regina Townsite. Originals in the Department of Northern Affairs and National Resources, Ottawa.

3839 **Central Saskatchewan Game Protective Association,** Saskatoon, Saskatchewan, Records, 1922-24, 1 inch. Minutes.

3840 **Cowan, N.,** Reminiscences, 1953, microfilm (5 feet). Lumberjack and pioneer, northeastern Saskatchewan. Reminiscences, 1903-07.

3841 **Kerr, William Franklin** (b. 1876), Papers, 1935-44, 7 feet. papers relating to Mr. Kerr's duties as Minister of Natural Resources, Saskatchewan. Included are correspondence files, memoranda, clippings, reports and pamphlets.

3842 **Phelps, Joseph Lee** (b. 1899), Papers, 1944-48, 17 feet. Ministerial papers of Hon. J.L. Phelps as Minister of Natural Resources and Industrial Development, Saskatchewan.

3843 **Prince Albert, Saskatchewan,** Papers, 1905-23, microfilm, 3 reels. Includes records concerning the Great West Iron, Wood, and Chemical Works, Ltd. (1912-19). Filmed from originals in the Prince Albert Historical Society.

3844 **Saskatchewan Lake and Forest Products Corporation,** Records, 1945-59, 147 feet. Records of the Saskatchewan Lake and Forest Products Corporation (1945-59), a Crown corporation, and its successors, Saskatchewan Marketing Services (1949-59), and Saskatchewan Forest Products (1949-57).

3845 **Saskatchewan. Natural Resources Department,** Records, 1886-1958, 390 feet. Correspondence, ledgers and records: Accounts Division (1945-50); Assistant Deputy Minister (1942-55); Deputy Minister (1930-58); Games Branch (1918-57); Provincial Lands Branch (1886-1948); Water Rights Branch (1937-44); Forestry Branch (1891-1950); Mines Branch (1889-1944); Coal Administration Branch (1931-38); Fisheries Branch; Industrial Development Board.

BIBLIOGRAPHY

BIBLIOGRAPHY

Armstrong, Robert D., comp. and ed. *A Preliminary Union Catalog of Nevada Manuscripts.* Reno, Nevada: University of Nevada Library, 1967. 218 pp.

Bell, Whitfield J., Jr., and Murphy D. Smith, comps. *Guide to the Archives and Manuscript Collections of the American Philosophical Society.* Memoirs of the American Philosophical Society, Vol. LXVI. Philadelphia: American Philosophical Society, 1966. 182 pp.

Berner, Richard C. "Sources for Research in Forest History: The University of Washington Manuscripts Collection." *Business History Review,* XXXV (Autumn, 1961), 420-25.

Bettis, M. Gary. *Recent Accessions to the Archives and Manuscripts Division of the University of Washington Library 1967-69.* Processed. n.d. 6 l.

Billington, Ray Allen, comp. "Guides to American History Manuscript Collections in the Libraries of the United States." *Mississippi Valley Historical Review,* XXXVIII (December, 1951), 467-96.

Blosser, Susan Sokol, and Clyde Norman Wilson, Jr. *The Southern Historical Collection: A Guide to the Manuscripts.* Chapel Hill: University of North Carolina Library, 1970. 317 pp.

Born, Lester K. *British Manuscript Project: A Checklist of the Microfilms Prepared in England and Wales for the American Council of Learned Societies, 1941-1945.* Washington, Library of Congress, 1955.

Breton, Arthur J. *A Guide to the Manuscript Collections of the New York Historical Society.* 2 vols. Westport, Connecticut: Greenwood Press, 1972.

Bridges, Roger D., comp. "Illinois Manuscript and Archival Collections." *Journal of the Illinois State Historical Society,* LXVI (Winter, 1973), 412-27.

Brinks, Herbert, ed. *Guide to the Dutch-American Historical Collections of Western Michigan.* Grand Rapids and Holland, Michigan: Dutch-American Historical Commission, 1967. 52 pp.

California State Archives. *Records of the California Legislature.* Compiled by David L. Snyder. California State Archives Inventory No. 2. Processed. Sacramento: 1971.

Carman, Harry J., and Arthur W. Thompson. *A Guide to the Principal Sources for American Civilization, 1800-1900, in the City of New York: Manuscripts.* New York: Columbia University Press, 1960. 435 pp.

Chandler, Marion C. *Colonial and State Records in the South Carolina Archives: A Temporary Summary Guide.* Columbia: South Carolina Department of Archives and History, 1973. 52 pp.

Cockhill, Brian, and Dale L. Johnson, comps. and eds. *Guide to Manuscripts in Montana Repositories.* Missoula: University of Montana Library, 1973.

Cocks, J. Fraser, III. *A Bibliography of Manuscript Resources Relating to Natural Resources and Conservation in the Michigan Historical Collections of the University of Michigan.* Processed. n.p.: 1970. 14 l.

Colley, Charles O., comp. *Documents of Southwestern History: A Guide to the Manuscript Collections of the Arizona Historical Society.* Tucson: Arizona Historical Society, 1972. 233 pp.

Columbia University Libraries. *Manuscript Collections in the Columbia University Libraries: A Descriptive List.* New York: 1959. 104 pp.

Coombs, William H., Director; Anthony Zito, comp. *A Guide to the Historical Collections of Michigan State University.* East Lansing, Michigan: 1969. 110 pp.

Cornell University Libraries. *Collection of Regional History and the University Archives: Report of the Curator and Archivist, 1950-1954.* Ithaca: 1955. 77 pp.

———. *Collection of Regional History and the University Archives: Report of the Curator and Archivist, 1954-1958.* Ithaca: 1959. 152 pp.

———. *Collection of Regional History and the University Archives: Report of the Curator and Archivist, 1958-1962.* Ithaca: 1963. 141 pp.

———. *Collection of Regional History and the University Archives: Report of the Curator and Archivist, 1962-1966.* Ed. by Kathleen Jacklin. Ithaca: 1974. 258 pp.

Daly, John. *Descriptive Inventory of the Archives of the City and County of Philadelphia.* Philadelphia: City of Philadelphia, Department of Records, 1970. 545 pp.

Delgado, David J., comp. and ed. *Guide to the Wisconsin State Archives.* Madison: State Historical Society of Wisconsin, 1966. 262 pp.

Diaz, Albert James. *Manuscripts and Records in the University of New Mexico Library.* Albuquerque: University of New Mexico Library, 1957. 55 pp.

Eisenhower, Dwight D., Library. *Historical Materials in the Dwight D. Eisenhower Library.* Abilene, Kansas: 1972. 43 pp.

Florida. Division of Archives, History and Records Management. *Preliminary Catalog No. 2: Accessions of the Florida State Archives.* Processed. Tallahassee: 1974. 65 l.

Fogerty, James E., comp. *Preliminary Guide to the Holdings of the Minnesota Regional Research Centers.* St. Paul: Minnesota Historical Society, 1975. 20 pp.

Gibson, A.M. *A Guide to Regional Manuscript Collections in the Division of Manuscripts, University of Oklahoma Library.* Norman: University of Oklahoma Press, 1960. 222 pp.

Goddard, Jeanne M., and Charles Kritzler, comps. *A Catalogue of the Frederick W. and Carrie S. Beinecke Collection of Western Americana.* Vol. I: *Manuscripts.* Ed. by Archibald Hanna. New Haven: Yale University Press, 1965. 114 pp.

Greene, Evarts Boutell, and Richard Brandon Morris. *A Guide to the Principal Sources for Early American History (1600-1800) in the City of New York.* 2nd ed. rev. by Richard B. Morris. New York: Columbia University Press, 1953. 400 pp.

Griffin, Grace Gardner. *Guide to Manuscripts Relating to American History in British Depositories, Reproduced for the Division of Manuscripts of the Library of Congress.* Washington: Library of Congress, 1946. 313 pp.

Hale, Richard W., Jr. *Guide to Photocopied Historical Materials in the United States and Canada.* Ithaca: Cornell University Press, published for the American Historical Association, 1961. 241 pp.

Ham, F. Gerald, comp. *Labor Manuscripts in the State Historical Society of Wisconsin.* Madison: The State Historical Society of Wisconsin, 1967. 48 pp.

Hamer, Philip M., ed. *A Guide to Archives and Manuscripts in the United States.* Compiled for the National Historical Publications Commission. New Haven: Yale University Press, 1961. 775 pp.

Hann, Russell G., Gregory S. Kealey, Linda Kealey, Peter Warrian. *Primary Sources in Canadian Working Class History, 1860-1930.* Kitchener, Ontario: Dumont Press, 1973. 169 pp. + index.

Harper, Josephine L. *Guide to the Manuscripts of the State Historical Society of Wisconsin. Supplement Number Two.* Madison: State Historical Society of Wisconsin, 1966. 275 pp.

Harper, Josephine L., and Sharon C. Smith. *Guide to the Manuscripts of the State Historical Society of Wisconsin. Supplement Number One.* Madison: State Historical Society of Wisconsin, 1957. 222 pp.

Hess, James W. *Guide to Manuscripts and Archives in the West Virginia Collection.* Morgantown: West Virginia University Library, 1974. 317 pp.

Historical Records Survey. California. *Guide to Depositories of Manuscript Collections in the United States: California.* Processed. Los Angeles: 1941. 76 l.

———. Florida. *Guide to Depositories of Manuscript Collections in the United States: Florida.* Processed. Jacksonville: 1940. 27 pp.

———. Iowa. *Guide to Depositories of Manuscript Collections in the United States: Iowa.* Processed. Des Moines: 1940. 47 pp.

———. Louisiana. *Guide to Depositories of Manuscript Collections in Louisiana.* 2nd edition. Processed. University, Louisiana: Department of Archives, Louisiana State University, 1941. 50 pp.

———. Massachusetts. *Guide to Depositories of Manuscript Collections in Massachusetts.* Processed. Boston: 1939. ca. 160 pp.

———. ———. *Guide to the Manuscripts Collection in the Worcester Historical Society.* Processed. Boston: 1941. 54 l.

———. Michigan. *Guide to Manuscript Depositories in the United States: Michigan.* Processed. Detroit: 1940. 78 l.

———. ———. *Guide to the Manuscript Collections in Michigan, Vol. II: University of Michigan Collections (Exclusive of those in the Michigan Historical Collections).* Processed. Detroit: 1941.

———. Minnesota. *Guide to Depositories of Manuscript Collections in the United States: Minnesota.* Processed. Saint Paul: 1941. 84 l.

———. Montana. *Bibliography of Graduate Theses, University of Montana, Montana State University, Montana State College, Montana School of Mines.* Processed. Bozeman: 1942. 71 pp.

———. New Hampshire. *Guide to Depositories of Manuscript Collections in the United States: New Hampshire (Preliminary Edition).* Processed. Manchester, N.H.: 1940. 44 pp.

———. New York (State). *Guide to Depositories of Manuscript Collections in New York State (Exclusive of New York City).* Processed. Albany: 1941. ca. 500 pp.

——— ———. *Supplement.* Cooperstown, N.Y.: [New York State Historical Association], 1944. 23 pp.

———. North Carolina. *Guide to the Manuscript Collections in the Archives of the North Carolina Historical Commission.* Raleigh: The North Carolina Historical Commission, 1942. 216 pp.

———. Oregon-Washington. *Guide to Depositories of Manuscript Collections in the United States: Oregon-Washington.* Processed. Portland: 1940. 42 pp.

———. Pennsylvania. *Guide to Depositories of Manuscript Collections in Pennsylvania.* Ed. by Margaret Sherburne Eliot and Sylvester K. Stevens. Harrisburg: Pennsylvania Historical Commission, 1939. 126 pp.

———. ———. *Guide to the Manuscript Collections of the Historical Society of Pennsylvania.* 2nd ed. Philadelphia: Historical Society of Pennsylvania, 1949.

———. Wisconsin. *Guide to Manuscript Depositories in Wisconsin.* Processed. Madison: 1941. 36 pp.

Hoover, Herbert, Presidential Library. *Historical Materials in the Herbert Hoover Presidential Library.* West Branch, Iowa: 1973. v + 26 pp.

Hughes, Katherine W., comp. *Forestry These Accepted by Colleges and Universities in the United States, 1900-1952.* Bibliographic Series Number 3. Corvallis: Oregon State University. 1953. 140 pp.

———. *Forestry Theses Accepted by Colleges and Universities in the United States, 1953-1955.* Bibliographic Series Number 5. Corvallis: 1956.

———. *Forestry Theses Accepted by Colleges and Universities in the United States, 1956-1969.* Published annually in *Forest Science,* 1959-1970.

Ireland, Florence. *The Northeast Archives of Folklore and Oral History: A Brief Description and A Catalog of Its Holdings, 1958-1972.* Orono, Maine: Northeast Folklore Society under the auspices of the Department of Anthropology, University of Maine, 1973. [*Northeast Folklore,* XIII (1972), pp. 1-86].

Kane, Lucile M. *Guide to the Public Affairs Collection of the Minnesota Historical Society.* St. Paul: Minnesota Historical Society, 1968. 46 pp.

———, and Kathryn A. Johnson. *Manuscripts Collections of the Minnesota Historical Society, Guide No. 2.* Saint Paul: Minnesota Historical Society, 1955. 212 pp.

Kielman, Chester V., comp. and ed. *The University of Texas Archives: A Guide to the Historical Manuscripts Collections in the University of Texas Library.* Austin: University of Texas Press, 1967. 594 pp.

Larson, David R., ed. *Guide to Manuscripts Collections & Institutional Records in Ohio.* [Columbus?]: Society of Ohio Archivists, 1974. 315 pp.

Larson, Henrietta M. *Guide to Business History.* 1948. Reprint. Boston: J.S. Canner & Co., 1964. 1181 pp.

Lentz, Andrea D., and Sara S. Fuller, eds. *A Guide to Manuscripts at the Ohio Historical Society.* Columbus: The Ohio Historical Society, 1972. 284 pp.

Library of Congress. *Handbook of Manuscripts in the Library of Congress.* Washington: Government Printing Office, 1918. 750 pp.

———. *Manuscript Sources in the Library of Congress for Research on the American Revolution,* compiled by John R. Sellers, Gerard W. Gawalt, Paul H. Smith, and Patricia Molen van Ee. Washington: 1975. 372 pp.

———. *The National Union Catalog of Manuscript Collections, 1959-1961.* Ann Arbor, Mich.: J.W. Edwards, Inc., 1962.

———. *The National Union Catalog of Manuscript Collections, 1962.* 2 vols. Hamden, Conn.: The Shoe String Press, Inc., 1964.

———. *The National Union Catalog of Manuscript Collections, 1963-1972.* 9 vols. Washington: 1965-1974.

———. Manuscript Division. *Manuscripts on Microfilm: A Checklist of the Holdings in the Manuscript Division.* Compiled by Richard B. Bickel. Washington: 1975. 82 pp.

Lovett, Robert W., and Eleanor C. Bishop. *List of Business Manuscripts in Baker Library.* 3rd ed. Boston: Baker Library, 1969. 334 pp.

Majors, Harry M. *North Cascades Archival Resources in Washington State Repositories.* Edited by Karyl Winn. Processed. Seattle: University of Washington, 1974. 17 l.

Mason, Elizabeth B., and Louis M. Starr. *The Oral History Collection of Columbia University.* Rev. ed. New York: Oral History Research Office, 1973. 460 pp.

Massachusetts Historical Society. *Catalog of Manuscripts of the Massachusetts Historical Society.* 7 vols. Boston: G.K. Hall & Co., 1969.

Maxwell, Robert S. "Manuscript Collections at Stephen F. Austin State College." *American Archivist,* XXVIII (July, 1965), 421-26.

Mississippi Department of Archives and History. *Guide to Official Records in the Mississippi Department of Archives and History.* Compiled by Thomas W. Henderson and Ronald E. Tomlin. Jackson: 1975. 115 pp.

Morgan, Dale L., and George P. Hammond, eds. *A Guide to the Manuscript Collections of the Bancroft Library.* Vol. I: *Pacific and Western Manuscripts (Except California).* Berkeley and Los Angeles: University of California Press, 1963. 379 pp.

National Library of Canada. *Canadian Theses.* Otawa: 1953-1972.

Neiderheiser, Clodaugh M., comp. *Forest History Sources of the United States and Canada: A Compilation of the Manuscript Sources of Forestry, Forest Industry, and Conservation History.* Processed. Saint Paul, Minn.: Forest History Foundation, Inc., 1956. 140 pp.

New York Botanical Garden. Library. *Catalog of the Manuscript and Archival Collections, and Index to the Correspondence of John Torrey.* Compiled by Sara Lenley, et al. Boston: G.K. Hall & Co., 1973. 473 pp.

New York Public Library. Research Libraries. *Dictionary Catalog of the Manuscript Division.* 2 vols. Boston: G.K. Hall & Co., 1967.

North Carolina. State Department of Archives and History. *Guide to Private Manuscript Collections in the North Carolina State Archives.* Prepared by Beth G. Crabtree. Raleigh: 1964. 492 pp.

Nute, Grace Lee, and Gertrude W. Ackermann. *Guide to the Personal Papers in the Manuscript Collections of the Minnesota Historical Society.* Saint Paul: Minnesota Historical Society, 1935. 146 pp.

Oregon Historical Society. *Oregon Historical Society Manuscript Collections.* Oregon Historical Society Research and Bibliography Series, No. 1. [Portland: 1971]. 264 pp.

———. *Oregon Historical Society Microfilm Guide.* Oregon Historical Society Research and Bibliography Series, No. 4. Portland: 1973. 162 pp.

———. *Supplement to the Guide to the Manuscript Collections.* Research and Bibliography Series, No. 3. Processed. 2 vols. Portland: 1973-1974.

Oregon State Archives. *Guide to Legislative Records in the Oregon State Archives.* Bulletin No. 8 (Publication No. 29). Salem: 1968. 13 pp.

———. *Guide to Microfilms in the Oregon State Archives.* Bulletin No. 9 (Publication No. 32). Salem: 1968.

Osborn, Katherine H., and Elizabeth W. Finch, comps. *Forestry Theses Accepted by Colleges and Universities in the United States. Supplement, July 1969-June 1970.* Processed. Corvallis: Oregon State University Library, n.d.

Owsley, Harriet Chappell, ed. *Guide to the Processed Manuscripts of the Tennessee Historical Society.* Nashville: Tennessee State Library and Archives, 1969. 70 pp.

Parker, David W. *A Guide to the Documents in the Manuscript Room at the Public Archives of Canada.* Publications of the Archives of Canada, No. 10. Ottawa: Government Printing Bureau, 1914.

———. *Guide to the Materials for United States History in Canadian Archives.* 1913. Reprint. New York: Kraus Reprint Corp., 1965. 339 pp.

Peckham, Howard H., comp. *Guide to Manuscript Collections in the William L. Clements Library.* 1st ed. Ann Arbor: University of Michigan Press, 1942.

———, and William S. Ewing. *Guide to Manuscript Collections in the William L. Clements Library.* 2nd ed. Ann Arbor: University of Michigan Press, 1953.

Pedley, Avril J.M. *The Manuscript Collections of the Maryland Historical Society.* Baltimore: 1968. 390 pp.

Phillips, Audrey E., comp. and ed. *Guide to Special Collections, University of California, Berkeley, Library.* Metuchen, N.J.: The Scarecrow Press, Inc., 1973. 151 pp.

Pike, Kermit J. *A Guide to the Manuscripts and Archives of the Western Reserve Historical Society.* Cleveland: Western Reserve Historical Society, 1972. 425 pp.

Public Archives of Canada. *Collections of the Lennox and Addington Historical Society: Preliminary Inventory.* Ottawa: 1959. 127 pp.

———. *Union List of Manuscripts in Canadian Repositories.* Edited by Robert S. Gordon. Ottawa: 1968. 734 pp.

———. Manuscript Division. *General Inventory: Manuscripts.* Vol. I: *MG1-MG10.* Ottawa: Information Canada, 1971. 272 pp.

———. ———. *General Inventory, Manuscripts.* Vol. III: *MG17-MG25.* Ottawa: 1974. 407 pp.

———. ———. *General Inventory: Manuscripts.* Vol. IV: *MG22-MG25.* Ottawa: 1972. 449 pp.

———. ———. *General Inventory, Manuscripts.* Vol. V: *MG26-MG27.* Ottawa: 1972. 242 pp.

———. ———. *Preliminary Inventory. Manuscript Group 29: Nineteenth Century Post-Confederation Manuscripts, 1867-1900.* [Ottawa]: 1962. 48 pp.

———. ———. *Preliminary Inventory. Record Groups No. 14: Records of Parliament, 1775-1915; No. 15: Department of the Interior; No. 16: Department of National Revenue.* [Ottawa]: 1957. 30 pp.

Québec. Archives Nationales. *L'État général des archives publiques et privées du Québec.* Québec: Ministère des Affaires Culturelles, 1968. 312 pp.

Richman, Irwin, comp. *Historical Manuscript Depositories in Pennsylvania.* Harrisburg: Pennsylvania Historical and Museum Commission, 1965. 73 pp.

Riggs, John Beverly. *A Guide to the Manuscripts in the Eleutherian Mills Historical Library: Accessions Through the Year 1965.* Greenville, Del.: Eleutherian Mills Historical Library, 1970. 1205 pp.

Ring, Elizabeth, ed. *A Reference List of Manuscripts Relating to the History of Maine.* University of Maine Studies, 2nd Series, No. 45 *(Maine Bulletin,* XLI [August, 1938]; XLII [August, 1939]; XLIII [February 20, 1941]). 3 vols. Orono, Maine: 1938-1941.

Roosevelt, Franklin D., Library. *Historical Materials in the Franklin D. Roosevelt Library.* Hyde Park: [1973]. 16 pp.

Schmitt, Martin, comp. *Catalogue of Manuscripts in the University of Oregon Library.* Eugene: University of Oregon, 1971. 355 pp.

Smith, Alice E., ed. *Guide to the Manuscripts of the Wisconsin Historical Society.* Madison: State Historical Society of Wisconsin, 1944. 290 pp.

Smith, Charles Wesley. *A Union List of Manuscripts in Libraries of the Pacific Northwest.* Seattle: University of Washington, 1931. 57 pp.

Smith, David C., comp. *Lumbering and the Maine Woods: A Bibliographical Guide.* Portland, Maine: Maine Historical Society, 1971. 35 pp.

Smith, Herbert F., comp. *A Guide to the Manuscript Collections of the Rutgers University Library.* New Brunswick, New Jersey: Rutgers University Library, 1964. 179 pp.

Smithsonian Institution. *Preliminary Guide to the Smithsonian Archives.* Washington: 1971. 72 pp.

Society of American Archivists. Business Archives Committee. *Directory of Business Archives in the United States and Canada.* Chicago: 1975. 38 pp.

Steen, Judith A., comp. *A Guide to Unpublished Sources for A History of the United States Forest Service.* Processed. [Santa Cruz, California]: Forest History Society, Inc., 1973. 67 pp.

Stewart, Bruce W., and Hans Mayer, comps. *A Guide to the Manuscript Collection, Morristown National Historical Park.* Morristown, New Jersey: n.d. 142 pp.

Tilley, Nannie M., and Noma Lee Goodwin. *Guide to the Manuscript Collections in the Duke University Library.* Durham, North Carolina: Duke University Press, 1947. 362 pp.

Truman, Harry S, Library. *Historical Materials in the Harry S Truman Library: An Introduction to their Contents and Use.* Independence, Missouri: 1971. 29 pp.

U.S. National Archives and Records Service. *Catalog of National Archives Microfilm Publications.* Washington: National Archives Trust Fund Board, 1974. 184 pp.

———. *Guide to the National Archives of the United States.* Washington: Government Printing Office, 1974. 884 pp.

———. *Guide to the Records in the National Archives.* Washington: Government Printing Office, 1948. 684 pp.

———. *Inventory of the Records of the War Manpower Commission, Record Group 211.* Compiled by Charles Zaid. National Archives and Records Service Inventory Series No. 6. Washington: 1973. 75 pp.

———. *Preliminary Inventories.*

No. 4: War Labor Policies Board Records. Compiled by Mary Walton Livingston and Leo Pascal. Washington: 1943. 22 pp.

No. 5: Records of the National War Labor Board. Compiled by Herbert Fine. Washington: 1943. 16 pp.

No. 7: Records of the Federal Trade Commission. Compiled by Estelle Rebec. Washington: 1948. 7 pp.

No. 10: Records of the Bureau of Yards and Docks. Compiled by Richard G. Wood. Washington: 1948. 28 pp.

No. 11: Records of the Civilian Conservation Corps. Compiled by Harold T. Pinkett. Washington: 1948. 16 pp.

No. 15: Records of the War Production Board. Compiled by Fred G. Halley and Josef C. James. Washington: 1948. 59 pp.

No. 18 (Revised): Records of the Forest Service. Compiled by Harold T. Pinkett; revised by Terry W. Good. Washington: 1969. 23 pp.

No. 22: Land-Entry Papers of the General Land Office. Compiled by Harry P. Yoshpe and Philip P. Brower. Washington: 1949. 77 pp.

No. 23: Records of the United States Senate. Compiled by Harold E. Hufford and Watson G. Claudill. Washington: 1950. 284 pp.

No. 25: Records of the Office of War Mobilization and Reconversion. Compiled by Homer L. Calkin. Washington: 1951. 156 pp.

No. 26: Records of the Bureau of Aeronautics. Washington: 1951. 9 pp.

No. 32: Records of the Accounting Department of the Office of Price Administration. Compiled by Meyer H. Fishbein and Elaine E. Bennett. Washington: 1951. 108 pp.

No. 44: Records of the National Recovery Administration. Compiled by Homer L. Calkin, Meyer H. Fishbein, and Leo Pascal. Washington: 1952. 226 pp.

No. 50: Central Office Records of the National Resources Planning Board. Compiled by Virgil E. Baugh. Washington: 1953. 66 pp.

No. 58: Records of the United States Court of Claims. Compiled by Gaiselle Kerner. Washington: 1953.

No. 64: Records of the Regional Offices of the National Resources Planning Board. Compiled by Virgil E. Baugh. Washington: 1954. 47 pp.

No. 66: Records of the Bureau of Plant Industry, Soils, and Agricultural Engineering. Compiled by Harold T. Pinkett. Washington: 1954. 49 pp.

No. 78: Records of the National War Labor Board (World War II). Compiled by Estelle Rebec. Washington: 1955. 188 pp.

No. 81: Cartographic Records of the Office of the Secretary of the Interior. Compiled by Laura E. Kelsay. Washington: 1955. 11 pp.

No. 83: Records of the Extension Service. Compiled by Virgil E. Baugh. Washington: 1955. 37 pp.

No. 95: Records of the Price Department of the Office of Price Administration. Compiled by Meyer H. Fishbein, Walter W. Weinstein, and Albert W. Winthrop. Washington: 1956. 272 pp.

No. 97: Records of the United States Shipping Board. Compiled by Forrest R. Holdcamper. Washington: 1956. 165 pp.

No. 102: Records of the Rationing Department of the Office of Price Administration. Compiled by Meyer H. Fishbein, Martha Chandler, Walter W. Weinstein, and Albert W. Winthrop. Washington: 1958. 175 pp.

No. 103: Cartographic Records of the Bureau of the Census. Compiled by James Berton Rhoads and Charlotte M. Ashby. Washington: 1958. 108 pp.

No. 109: Records of the Bureau of Reclamation. Compiled by Edward E. Hill. Washington: 1958. 27 pp.

No. 113: Records of the United States House of Representatives, 1789-1946. Compiled by Buford Rowland, Handy B. Faut, and Harold E. Hufford. 2 vols. Washington: 1959.

No. 116: Records of the United States District Court for the Southern District of New York. Compiled by Henry T. Ulasek and Marion Johnson. Washington: 1959. 68 pp.

No. 118: Records of the Farmers Home Administration. Compiled by Stanley W. Brown and Virgil E. Baugh. Washington: 1959. 62 pp.

No. 120: Records of the Enforcement Department of the Office of Price Administration. Compiled by Meyer H. Fishbein and Betty R. Bucher. Washington: 1959. 65 pp.

No. 124: Records of the United States District Court for the Eastern District of Pennsylvania. Compiled by Marion M. Johnson, Mary Jo Grotenrath, and Henry T. Ulasek. Washington: 1960. 44 pp.

No. 133: Records of the Bureau of Ships. Compiled by Elizabeth Bethel, Ellmore A. Champie, Mabel E. Deutrich, Robert W. Krauskopf, and Mark N. Schutz. Washington: 1961. 241 pp.

No. 134: Records of the Bureau of Public Roads. Compiled by Truman R. Strobridge. Washington: 1962. 34 pp.

No. 149: Records of the Bureau of Agricultural and Industrial Chemistry. Compiled by Helen T. Finneran. Washington: 1962. 35 pp.

No. 154: Records of the Office of Territories. Compiled by Richard S. Maxwell and Evans Walker. Washington: 1963. 117 pp.

No. 160: Records of the Smaller War Plants Corporation. Compiled by Katherine H. Davidson, Washington: 1964. 87 pp.

No. 161: Records of the Bureau of the Census. Compiled by Katherine H. Davidson and Charlotte M. Ashby. Washington: 1964. 141 pp.

No. 163: Records of the Bureau of Indian Affairs. Compiled by Edward E. Hill. 2 vols. Washington: 1965.

No. 166: Records of the National Park Service. Compiled by Edward E. Hill. Washington: 1966. 52 pp.

No. 167: Cartographic Records of the Forest Service. Compiled by Carlotte M. Ashby. Washington: 1967. 71 pp.

No. 171: Records of the Solicitor of the Treasury. Compiled by George S. Ulibarri. Washington: 1968. 35 pp.

No. 173: Records of the Reconstruction Finance Corporation, 1932-1964. Compiled by Charles Zaid. Washington: 1973. 90 pp.

No. 177: Records Relating to International Claims. Compiled by George S. Ulibarri. Washington: 1974. 73 pp.

———. *Records of the Public Land Law Review Commission: Inventory of Record Group 409.* Compiled by Richard C. Crawford. National Archives Inventory Series No. 8. Washington: 1973. 12 pp.

———. *Reference Information Paper No. 71: Cartographic Records in the National Archives of the United States Relating to American Indians.* Compiled by Laura E. Kelsay. Washington: 1974. 35 pp.

———. *Special Lists.*

No. 10: Lists of Wage Stabilization Cases Acted on by the Headquarters Office of the National War Labor Board, 1942-45. Compiled by Estelle Rebec, Arthur Hecht, and Paul Flynn. Washington: 1953, 162 pp.

No. 12: Select List of Documents in the Records of the National Recovery Administration. Compiled by Homer C. Calkin and Meyer H. Fishbein. Washington: 1954. 190 pp.

No. 13: List of Cartographic Records of the Bureau of Indian Affairs. Compiled by Laura E. Kelsay. Washington: 1954. 127 pp.

No. 18: Index to Appropriation Ledgers in the Records of the Office of the Secretary of the Interior, Division of Finance, 1853-1923. Compiled by Catherine M. Rowland. Washington: 1963. 127 pp.

No. 19: List of Cartographic Records of the General Land Office. Compiled by Laura E. Kelsay. Washington: 1964. 202 pp.

No. 35: Printed Hearings of the House of Representatives Found Among its Committee Records in the National Archives of the United States, 1824-1958. Washington: 1974. 197 pp.

———. Federal Records Center, East Point, Georgia. Archives Branch. *Research Opportunities.* East Point: 1972. [5 pp.].

———. ———, Los Angeles. *A Guide to Research Records.* Processed. 1967. 21 l.

———. Region 9. *Preliminary Inventory of the Records of the Bureau of Land Management.* Compiled by Gilbert Dorame. Processed. Los Angeles: 1966. 83 l.

———. Federal Records Center, San Francisco. Regional Archives Branch. *Preliminary Inventory of the Records of the United States District Court, Northern District of California, San Francisco, 1851-1950.* Compiled by Arthur R. Abel. Processed. [San Francisco]: 1964. 40 pp.

———. ———. ———. *Opportunities for Research in Federal Records for California, Nevada, and the Pacific Ocean Area: Special List of Research Records.* Processed. San Francisco: 1970. 16 pp.

———. ———. ———. *Preliminary Inventory of the Records of the Bureau of Indian Affairs, Northern California and Nevada Agencies.* Compiled by Thomas W. Wadlow and Arthur R. Abel; revised by Arthur R. Abel. Processed. San Francisco: 1966. 29 l.

———. ———. ———. *Preliminary Inventory of the Records of the Bureau of Land Management, Northern California and Nevada.* Compiled by John P. Heard. Processed. San Francisco: 1970. 60 pp.

University of California. *Dictionary Catalog of the Water Resources Center Archives, University of California, Berkeley.* 5 vols. Boston: G.K. Hall & Co., 1970. *Supplement.* 2 vols. 1973.

University of California Library at Los Angeles. *Guide to Special Collections in the Library of the University of California at Los Angeles.* U.C.L.A. Occasional Paper Number 7. Processed. Los Angeles: 1958. 76 pp. + index.

University of Illinois at Congress Circle, Chicago. Library. *Guide to the Manuscript Collections in the Department of Special Collections, the Library.* Processed. [Chicago]: 1974.

University of Iowa Libraries, Special Collections Department. *Alphabetical Index to Manuscript Collections.* 3rd revision. Processed. [Iowa City]: 1969. 19 l.

University of Missouri, Western Historical Manuscripts Collection. *Guide to the Western Historical Manuscripts Collection.* University of Missouri Bulletin, Vol. 53, No. 33 (Library Series No. 22). Columbia: 1952. 125 pp.

———. *Guide to the Western Historical Manuscripts Collection.* By John A. Galloway. The University of Missouri Bulletin, Vol. 58, No. 13 (Library Series No. 24). Columbia: 1956. 53 pp.

University of Southwestern Louisiana, University Libraries. *Guide to Southwestern Archives and Manuscripts Collection.* Lafayette: 1970. 29 l.

University of Washington Library. *The Manuscript Collection of the University of Washington Libraries.* University of Washington Libraries, Library Leaflet, New Series, No. 1. Processed. N.p.: 1967. 33 pp.

Warner, Robert M., and Ida C. Brown. *Guide to Manuscripts in the Michigan Historical Collections of the University of Michigan.* Ann Arbor: 1963. 315 pp.

Washington, Mary, comp. *An Annotated Bibliography of Western Manuscripts in the Merrill Library at Utah State University, Logan, Utah.* Logan: Utah State University Press, 1971. 157 pp.

Washington State University. *Selected Manuscript Resources in the Washington State University Library.* Pullman: Washington State University, 1974. 94 pp.

Wilson, Alan. "Maritime Business History: A Reconnaissance of Records, Sources, and Prospects." *Business History Review,* XLVII (Summer, 1973), 260-76.

Withington, Mary C., comp. *A Catalogue of Manuscripts in the Collection of Western Americana Founded by William Robertson Coe, Yale University Library.* New Haven: Yale University Press, 1952. 398 pp.

INDEX

This index refers the user to entry numbers, not to pages. Numbers in boldface type indicate persons or organizations cited in titles or associated with major subdivisions of papers or records. Additional information on the use of the index will be found in the introduction to the front of this volume.

INDEX

New York. 1982, 2036, 2040, 2071, 2084
North Carolina. 2227, 2232, 2233
North Dakota. 1821
Ohio. 2403-2405, 2411
Ontario. 3776
Oregon. 2456, 2484, 2561, 2598, 2612, 2614, 2630, 2641
Pacific Northwest. 2071, 2652
Pennsylvania. 593, 2834, 2849
South Dakota. 1821, 2895
Southern states. 990
Tennessee. 2897
Texas. 2927
United States. 126, 188, 411, 419, 589, 920, 2017, 2021, 2110, 3542, 3547
(1800—1900). 347, 348, 360, 509, 593, 595, 603, 612, 615, 617, 901, 902, 1268, 1280, 1893, 1903, 2102, 2105, 2187, 2394, 2411, 2692, 2745, 2975a, 3242, 3514
(1900—1920). 45, 178, 282, 347, 348, 360, 440, 504, 509, 593, 595, 602, 603, 612, 615, 617, 618, 621, 622, 679, 900, 901, 902, 910, 912, 1268, 1280, 1728, 1893, 1942, 2016, 2027, 2051, 2089, 2101, 2102, 2105, 2187, 2328, 2394, 2571, 2692, 2745, 2776, 3378, 3514
(1920—1940). 45, 178, 282, 347, 348, 360, 362, 382, 385, 391, 590, 591, 593, 595, 600, 602, 612, 617, 621, 679, 682, 835, 895, 900, 901, 909, 910, 912, 921, 922, 923, 1280, 1302, 1602, 1675, 1745, 1779, 1787, 1812, 1893, 2016, 2023, 2027, 2051, 2062, 2064, 2079, 2102, 2105, 2187, 2328, 2394, 2464, 2571, 2745, 2776, 2777, 3105, 3378
(1940—1953). 282, 347, 348, 360, 362, 363, 385, 391, 600, 612, 617, 679, 682, 806, 835, 895, 909, 910, 912, 913, 923, 1280, 1302, 1588, 1590, 1602, 1611, 1621, 1675, 1745, 1779, 1812, 1815, 1818, 2016, 2023, 2027, 2051, 2062, 2064, 2079, 2187, 2328, 2403, 2464, 2745, 2776, 2777, 3054, 3105
(1953—1974). 282, 347, 348, 360, 362, 363, 391, 617, 682, 806, 835, 909, 910, 913, 923, 932, 933, 935, 937, 943, 944, 1249, 1251, 1252, 1280, 1302, 1588, 1590, 1602, 1611, 1621, 1630, 1675, 1745, 1779, 1786, 1815, 1861, 2016, 2027, 2051, 2064, 2079, 2187, 2403, 2503, 2636, 2777, 2910, 3054, 3105, 3537
Utah. 1858
Washington. 3102, 3111, 3112, 3114, 3145, 3149, 3191, 3210, 3218, 3234
West Virginia. 2084, 3291
Wisconsin. 3337, 3341, 3360, 3379, 3398, 3451, 3499, 3505
CONSERVATION and Administration of the Public Domain, Committee on, see U.S. Conservation and the Administration of the Public Domain, Committee
CONSERVATION Associates. **114**
CONSERVATION Association of Los Angeles County. 2480
CONSERVATION Association of Southern California. 286
CONSERVATION camps see Labor camps
CONSERVATION conferences. 600, 1812
International. 601
Missouri. 1787
CONSERVATION Council of Colorado. 427
CONSERVATION Council of Virginia. **2994**
CONSERVATION education
See also Environmental education; Forestry education
California. 273, 286, 344
Michigan. 1365
Minnesota. 1615
New York. 2777
Pennsylvania. 2753
United States. 346, 407
Wisconsin. 3414
CONSERVATION Education Association. 344
CONSERVATION Foundation. 427, 1280, 3535
CONSERVATION organizations
See also American Forestry Association; Butano Forest Associates; Conservation Foundation; Emergency Conservation Committee; Federation of Western Outdoor Clubs; Forestry organizations; Izaak Walton League; Jackson Hole Preserve, Inc.; National Audubon Society; National Parks and Conservation Association; National Wildlife Foundation; Nature Conservancy; Quetico-Superior Council; Sierra Club; Soil Conservation Society of America; Wilderness Society; etc.
Alaska. 3119
Arizona. 37
British Columbia. 3119, 3629
California. 71, 104, 110, 114, 125, 132, 179, 213-215, 220, 257, 286, 342, 344, 3119
Colorado. 412
Connecticut. 499, 500
Idaho. 3119
Illinois. 824
Iowa. 900, 903
Louisiana. 1001
Maryland. 1145
Minnesota. 1559, 1592, 1675, 1690, 1691
Missouri. 1793
Montana. 1880, 3119
Nebraska. 1980
New Hampshire. 498
New York. 409, 418, 1983, 1988
Ohio. 2396, 2403
Oregon. 2482, 2484, 2508, 2511, 2574, 2590, 2614, 2641, 2644, 2657, 2658, 3119
Pacific Northwest. 2501, 2807
Tennessee. 2899
United States. 205, 220, 344, 346, 347, 351, 363, 385, 407, 408, 414, 418, 427, 431, 498, 499, 606, 2062, 2501, 2508, 2910, 3111, 3150, 3513, 3535
Virginia. 2994, 3008
Washington. 2452, 3094, 3095, 3108, 3111, 3119, 3121, 3126, 3139, 3145, 3150, 3152, 3158, 3160, 3181, 3205, 3206, 3209, 3210
Wisconsin. 3360, 3404, 3451, 3497
Wyoming. 1871, 3119, 3526, 3531, 3537
CONSOLIDATED Lumber Company. 3185
CONSOLIDATED Mining and Smelting Company. 3043
CONSOLIDATED River Improvement Company. 1425
CONSOLIDATED Timber Company. **2575**
CONSTRUCTION industry
See also Bridge construction; Building materials industry and trade; Building specialties trade; Carpentry; Lumber industry and trade; Plastering mills; Railroad construction
Alberta. 3557, 3595
British Columbia. 139
California. 144, 241
Colorado. 141, 149, 226
Indiana. 879
Kentucky. 953
Maine. 1012, 1108
Michigan. 1409
Minnesota. 1623
Montana. 1845
New Brunswick. 3692
New Jersey. 1971
New York. 2049, 2098
North Carolina. 2313
Ontario. 3725
Oregon. 2443, 2610
Pennsylvania. 2782, 2784
United States. 703, 704, 719, 721, 746-748
Utah. 2936
Wisconsin. 3369, 5506
CONTRA COSTA Hills Fire Protection Committee. **115**
CONTRACTORS see Construction industry
CONVERTED paper products
United States. 721
CONWAY, James. 400
CONZET, Grover M. 1525, 1693
COOGAN, Mrs. Grace (d. ca. 1946). **3357**
COOK, C.W. 292
COOK, Marshall L. **1274**
COOK, William Randolph. **1274**
COOK & Jones Company. **1466**
COOKE, Morris Llewellyn (1872—1960). **2015**
COOKE and Hobe Company. 1608
COOKINGHAM, Prescott W. 2535

EHRMAN, Ada. 2365
EICHHOLZ, Duane. 1502
EICHORN, Noel D. 427
EISENHOWER, Dwight David (1890–1969). **932, 933**
EKIN, Richard H. 2587
EL DORADO Lumber Copmpany. 31
EL DORADO Wood and Flume Company. 248, **1923**
ELDREDGE, Inman F.
  Interviews. 375, 596, 2110
ELDRIDGE, Edward. 123
ELECTRICAL power see Hydroelectric power
ELEMENTAL wood products see Forest products
EL HULT, Ruby see Hult, Ruby El
ELIOT, Charles N. 397
ELIOT, Samuel. 1171
ELK River Mill and Lumber Company. **124**
ELKIN, D.E. **3660**
ELKINS, Stephen B. (1841–1911). **3253,** 3303
ELKINS Lumber Company. 3300
ELLIOTT, Amos. 3329
ELLIOTT, George P. 3670
ELLIOTT, Henry W. 292
ELLIOTT, John (U.S. Interior Department). 1812
ELLIOTT, John Bennett (New Harmony, Indiana, ca. 1870). **854**
ELLIOTT, Louise. 1992
ELLIS, Harold D. 2040
ELLIS, John. 3721
ELLIS, Robert. 3691
ELLIS, William (1821–1905). 519
ELLIS and Leikem. 1737
ELLISON, Smith. **1583**
ELLSBREE Family. **2039**
ELLSWORTH, Edmund. 2940
ELLSWORTH, Harris (b. 1899). **2463**
  Correspondence with Walter H. Horning, 2478; Karl Onthank, 2508
ELLSWORTH, Rodney Sydes. 125, 253
ELSE, John Hubert (1907–68). **934**
ELY, Heman (1775–1852). **2397**
ELY, Northcutt. **386**
ELY, Richard T. 2091
EMBREE, Lucius C. **878**
EMBREEVILLE Timber Company. 2737
EMBREY, W.S., Inc. 3022
EMERGENCY Conservation Committee. 414, 600
EMERGENCY Conservation Work see U.S. Emergency Conservation Work
EMERGENCY Rubber Project see U.S. Forest Service, Emergency Rubber Project
EMERSON, George H. 3163
EMERSON, Hugh R. 3366
EMERSON and Stevens Axe Factory. **1016**
EMERY, H.A. 2075
EMERY, Moses. 1088
EMERY & Stetson. 1206
EMERY, Stetson & Company. 1206
EMIGRATION see Immigration and emigration
EMMERSON, John. **3681**
EMMET, Thomas Addis. 2118, **2160**
EMMONS, Francis Whitefield. **855**
EMPARAN, Madie Brown. **396**
EMPIRE Lumber Company. 373, 3401
EMPIRE Millwork Company. 916
EMPIRE State Forest Products Association. **2040,** 2071, 2092, **2182**
EMPLOYERS
  United States
    Collective bargaining. 689
EMPLOYERS' associations
  Oregon. 3183
  Pacific Northwest. 2490, 3183
  Washington. 3118
EMPLOYMENT
  See also Unemployment
  Minnesota. 3418
EMPORIA Lumber Company (Houston). 2915
EMPORIUM Forestry Company. **1993,** 2737, 2973
EMPORIUM Lumber Company. **2737,** 2975
ENGBERG, George. 2303
ENGELMANN, George (1809–84). **1826,** 2000
ENGINEERING see Forest engineering
ENGLISH, W. Francis. 1782
ENGLISHMEN
  New Brunswick. 1077
ENTOMOLOGY
  See also Insect control; Insect research; Insects; U.S. Forest Service, research
  Michigan. 1336, 1375
  West Virginia. 3266
ENVIRONMENTAL education. 2189
  North Carolina. 2255
  Ohio. 2422
ENVIRONMENTAL laws
  See also Forest laws; Pollution; etc.
  Connecticut. 1243
  Maine. 1243
  Massachusetts. 1243
  New Hampshire. 1243
  Rhode Island. 1243
  Vermont. 1243
ENVIRONMENTAL Pollution see Pollution; Pulp and paper industry; Smelter fumes; Turpentine industry
ENVIRONMENAL quality
  United States. 2910
EPSEY, L.V. **2508**
ERICKSON, John Edward (1863–1946). **620**
ERICKSON, K.A. 2616
EROSION see Soil
ERVIN, Edwin J. 2161
ERVIN, Thomas. 2161
ERVIN, William Ethelbert (1809–60). 2207
ESCH, John Jacob (1861–1941). **3367**
ESKRIDGE, Richard Stevens. 2536
ETHNIC groups see Finns; Frenchmen; Germans; Negroes; Swedes; etc.
EUBANKS, Wallace B. 377
EUCALYPTUS trees. 181
  California. 189, 276a, 304
  Florida. 656
EUREKA Steam Quartz Mill. 1924
EUSTIS, Benjamin. 1217
EUSTIS, Harold C. 2458
EVANS, Arthur. 3776
EVANS, Brock. **3119**
EVANS, Daniel J. **3033**
EVANS, David. 2784
EVANS, Elwood. 3208
EVANS, Joseph R. **2273**
EVANS, Oscar (b. 1878). 1288
EVARTS, Hal George (1887–1934). **2464**
EVEREST, David Clark. **3368,** 3430
EVERETT Prisoner's Defense Committee. **3120**
EVESMITH, Hansen. 1584
EVISON, Herbert. **415**
  Interview. 205
  Interviews by. 233
EWAN, Joseph. 2778
EWAN, Nesta. 2778
EXALL, Henry. 2915
EXPERIMENT (sloop). 2126
EXTENSION forestry
  Alabama. 1
  California. 205, 378
  Illinois. 832
  New York. 351, 2093
  United States. 645, 1838
EYSTER, Weiser Company. **543**
FAAST, Benjamin. 3498
FABIAN, Harold P. 2112, 3526

Prices (19th century). 1279, 1409, 1428, 2033
Middle West
(19th century). 2808, 3374, 3401, 3450
Interviews. 373
Minnesota. 1502, 1506, 1725, 1761
See also Lumber industry, St. Croix Valley; Lumber trade, Great Lakes
(1800—1850). 1568, 1607, 1642, 1709, 1724, 1734, 2057, 3374
(1850—1900). 1507, 1530, 1531, 1533, 1535, 1536, 1540, 1548, 1562, 1653, 1568, 1572, 1573, 1578, 1589, 1593, 1600, 1603, 1607-1609, 1628, 1629, 1642, 1653, 1654, 1664, 1667, 1669, 1673, 1681, 1684, 1702, 1708, 1709, 1716, 1719, 1731, 1735, 1737, 1755, 1757, 1758, 1821, 257, 3185, 3307, 3374
Lumber trade. 1617, 1635, 1638, 1712, 1724
(1900—1935). 1505, 1530, 1531, 1533, 1535, 1540, 1547, 1548, 1562, 1572, 1574, 1584, 1600, 1603, 1607-1609, 1613, 1614, 1620, 1626, 1628, 1629, 1638, 1642, 1652, 1653, 1664, 1667, 1668, 1669, 1673, 1676, 1681, 1684, 1685, 1702, 1704, 1708, 1712, 1719, 1731-1733, 1741, 1755, 1758, 1821, 3185, 3393
Legislation. 1525
Supplies. 1705
(1935—1970). 1562, 1584, 1600, 1608, 1614, 1638, 1664, 1667, 1668, 1704, 1731, 1732, 1741, 1821, 3185
Interviews. 373, 374, 1517, 2111, 2114, 3310, 3649
Prices (1891—1914). 1530
Waste materials. 1673
Mississippi
(1813—1865). 959, 972, 993, 1766, 1770, 2225
(1865—1900). 972, 993, 1003, 1764, 1767, 1770, 1772, 2544
(1900—1925). 972, 993, 1767, 1771, 1772, 2544
(1925—1972). 209, 974, 1767, 1771, 1773, 1774, 2544
Photographs. 1762
Prices (1841—1842). 959
Missouri. 1790
(1850—1900). 1778, 1781, 1783, 1791, 1792, 1797, 1802, 3307, 3393, 3452
(1900—1955). 842, 843, 1782, 1783, 1792, 1797, 1801, 1802, 3452
Interview. 1784
Prices (1869—1894). 1781, 1791
Montana
(1867—1955). 93, 620, 1600, 1836, 1851, 1856, 1866, 1867, 1869, 1872, 1873, 1882, 1884
Interview. 1847
Photographs. 1883
Wages (1942—1946). 710, 711
Nebraska
(1850—1890). 818, 919, 1903, 3524
(1890—1920). 919
Netherlands. 1141
Nevada. 86, 134, 1913
See also Railoads, Nevada
(1862—1900). 109, 190, 248, 330, 340, 1909, 1911, 1914, 1919, 1921, 1922, 1923, 1925, 1927, 1928
(1900—1950). 248, 264, 340, 1915, 1917, 1922, 1923, 1928
Interview. 1916
New Brunswick. 1077, 3672
(1695—1800). 3667, 3668, 3686, 3695, 3741, 3745
(1800—1856). 495, 3657, 3660, 3662, 3664, 3667, 3668, 3677, 3681, 3683, 3686, 3688, 3695, 3696, 3741, 3745, 3758
(1856—1900). 3660-3662, 3665, 3667, 3676, 3680, 3681, 3684, 3686, 3693, 3702, 3745, 3757, 3758
(1900—1920). 3657, 3660, 3661, 3663, 3676, 3681, 3684, 3686
(1920—1970). 3657, 3660, 3661
Interviews. 1076
Laws regulating exports. 3689
New England
(1800—1807). 529
(1941—1946). 720
New Hampshire
(1636—1767). 1095
(1767—1800). 1095, 1243, 1937, 2257, 2987
(1800—1869). 1095, 1130, 1175, 1205, 1243, 1390, 1933, 1943, 2185, 2987
(1870—1920). 1076, 1208, 1243, 1933, 2050, 1987
(1920—1952). 1243
Prices (1818—1833). 2106
New Jersey
(1749—1776). 1957, 2798
(1822—1875). 1947, 1960, 1966, 1967, 1969
Prices (1759—1769). 2138
New Mexico. 337, 584, 1973, 1975, 3253
Bankruptcies. 446
Interview. 468
New York. 1989, 1992, 1998, 2100
(1695—1750). 2121, 2140, 2149, 2150
(1750—1800). 485, 1141, 1980, 1990, 2129, 2130, 2141, 2142, 2150, 2152, 2163, 2360, 3802
(1800—1850). 485, 753, 1141, 1499, 1980, 1984, 1990, 2002, 2007, 2011, 2028, 2031, 2053, 2056, 2061, 2065, 2082, 2085, 2099, 2118, 2123, 2128, 2129, 2130, 2133, 2142, 2153, 2163, 2171, 2177, 2178, 2186, 2273, 2275, 3679
(1850—1900). 1499, 1536, 1980, 1984, 1995, 2002, 2006, 2007, 2008, 2011, 2028, 2031, 2037, 2038, 2039, 2041, 2053, 2055, 2056, 2059, 2075, 2077, 2085, 2099, 2177, 2178, 2186, 2275, 2738, 2984, 3679
(1900—1970). 1499, 1987, 1993, 1994, 1995, 2028, 2041, 2049, 2055, 2068, 2075, 2077, 2178, 2186, 2737, 2975
Interviews. 371
Prices (1759—1769). 2138
Newfoundland. 3758
Northwest Territory
(1875—1891). 3567, 3767
North Carolina. 786
See also Lumber industry and trade, Atlantic Coast
(1717—1800). 1141, 2119, 2265, 2360
(1800—1865). 535, 1141, 2208, 2213, 2228, 2265, 2283, 2337, 2339
(1865—1900). 2213, 2237, 2238, 2265, 2283, 2313, 2337, 2385
(1900—1960). 500, 2237, 2265, 2283, 2299, 2330, 2385
Eastern North Carolina
(1700—1800). 2221, 2240, 2275, 2326, 2327, 2359, 2369
(1800—1865). 2221, 2240, 2244, 2258, 2275, 2282, 2286, 2302. 2318, 2326, 2327, 2336, 2359, 2368, 2374, 2380, 2813
(1865—1900). 2221, 2239, 2240, 2258, 2275, 2286, 2318, 2326, 2327, 2336, 2359, 2368, 2374, 2383
(1900—1962). 2217, 2221, 2222, 2239, 2240, 2286, 2324, 2326, 2327, 2359
Western North Carolina
(1796—1800). 2215
(1800—1865). 2198, 2205, 2215, 2231
(1865—1900). 2205, 2215, 2252, 2281
(1900—1960). 2205, 2248, 2252, 2281, 2361, 2376, 2377
North Dakota. 1821
Nova Scotia
(1600—1700). 3740, 3745
(1700—1800). 1116, 1937, 3740, 3745, 3753
Prohibition of United States imports. 3703
(1800—1900). 373, 1209, 3698, 3745, 3753
(1900—1930). 3698
Interviews. 373, 1077
Ohio
(1800—1850). 2400, 2412, 2413, 2417
(1850—1900). 2391, 2399, 2400, 2401, 2406, 2412, 2415, 2417
(1900—1942). 2391, 2399, 2401, 2412
Ontario. 3777, 3779
(1750—1800). 3729, 3730, 3802
(1800—1850). 1382, 3707, 3709, 3711, 3713, 3725, 3726, 3730, 3747, 3758, 3759, 3763, 3770, 3780, 3782, 3783, 3790, 3792, 3794, 3797, 3799, 3803, 3804
(1850—1900). 2058, 2075, 3640, 3707, 3709, 3711, 3712, 3716-3718, 3720, 3724-3726, 3730, 3758, 3763, 3770, 3777, 3780-3783, 3785, 3787, 3790, 3793-3795, 3797, 3799, 3804, 3805
Competition with U.S. 3724
(1900—1930). 2075, 2971a, 3393, 3707, 3711, 3712, 3730, 3763, 3770, 3776, 3780, 3783, 3787, 3791, 3793, 3794, 3797, 3804, 3805, 3811
(1930—1968). 3711, 3730, 3763, 3770, 3787, 3791, 3794, 3811
Prices (1855—1864). 3795
Oregon. 2572, 2589, 2664, 2706, 2709
(1834—1850). 2545, 2654, 2669, 2672, 2688

New York. 2092, 2174, 2181
See also New York (geographical names), Adirondack Forest Reserve
Ohio. 2398
Oregon. 2590
Pennsylvania. 593, 2774, 2810
Saskatchewan. 3834
South Dakota. 2895
United States
(19th century). 593, 609, 610
(1900—1930). 609, 610, 836, 899
(1930—1950). 593, 609, 610, 836, 1775, 1780, 2316
(post 1950). 836, 1780, 2910
Preservation of (1932-33). 2316
Washington. 3038, 3104
PARKS Concession Company. 2619
PARMELE, Norman L. **510**
PARNALL, Harold William. 3595
PARR, John. 3703
PARRATT, Lloyd S. 293
PARSONS, Edward T. **152, 195**
PARSONS, Ivan. **1422**
PARSONS, Job W. **3287**
PARSONS, Marion R. **195**
PARSONS, Nathaniel. 1129
PARSONS, W.B. 662
PARSONS Engineering Company. 556
PARSONS Family. **195**
PARSONS Pulp and Lumber Company. 2896
PARSONS Pulp & Paper Company. 3273
PATENTS. 2046, 3307, 3797
Bark mill equipment. 544
Bark mills. 1230
Fire-proofing timber. 218
Log rafts. 1079
Lumber industry. 1576
Naval stores. 625
Nitrocellulose. 541
Paper making. 555
Paper making machinery. 556
Sawmill machinery. 1472, 2218
Wood preservation. 218
PARTRIDGE, William. 1936
PASSMORE Paper Company. 1856
PAST, John Comly. **1905**
PAST and Marsh Steam Saw Mill and Lumber Yards (Beatrice, Nebraska). 1905
PATCH, Edith. 2027
PATE, John Wesley. 2374
PATE Family. **2374**
PATRICK, Benjamin M. **2375**
PATRICK, Eliakim. 2375
PATRICK, W.S. 2075
PATRICK Lumber Company. **2514**
PATRIQUIN, H.O. 3735
PATTEE, Dora (Jewett). 1677
PATTEE, Edward Sidney. 1677
PATTEN, Amos, & Associates. **1200**
PATTERSON, Arthur H. 1885
PATTERSON, James N. **2228**
PATTERSON, Rufus T. 2311
PATTERSON, Samuel Finley. **2311**
PATTERSON, S.P. 2822
PATTERSON, Walter. 1379
PATTERSON, William Davis (b. 1858). **1115**
PATTERSON and Fries. 2215
PATTERSON Family. 2215
PATTISON, Olga E. 3110
PAUL, Helen Longyear. 1449
PAUL, M. Rea. 734
PAULISTA Lumber Company. 2486
PAW PAW Lumber Company. **2312**
PAWLEY Family. 2878
PAYETTE Improvement and Boom Company. 791
PAYETTE Lumber & Manufacturing Company. 791
PAYTON, G.W. 3731
PEABODY, Houghteling, and Company, Chicago. 3092
PEABODY, Selim H. 841
PEACE River Company. 3622
PEACHEY, A.L. 2543
PEALE, Albert C. 3551
PEARCE, J. Kenneth. **3170**
PEARCE, William. 3559, **3568,** 3696, 3600
PEARSON Charles Gottfried. **1678**
PEARSON, C.H., Hardwood Company. **2068**
PEARSON, Eben. 1238
PEARSON, Gustav Adolph. 2089
PEARSON, Mary S. 32
PEARSON, Tappan. 1238
PEASE, Robert G. 1913
PEASLEE, Nathaniel. 1238
PEATTIE, Donald Culross. **345**
PECK, S.J. 2057
PEDERSON, Thomas. 1679
PEEL, Robert. 3751
PEET, Herbert M. 3110
PEGUES, Rodger W. 3119
PEARCE, Joseph Harrison. **46**
PEIRCE, Earl S. 205, 379
PEIRCE, Ebenezer. **1201**
PEIRCE, Hayford. **1049**
PEIRCE, Job. **1201**
PEIRCE, Waldo T. **1049**
PEIRCE Family. **1049**
PENCE, Kingsley A. **463**
PENDILL, James P. 1450
PENDILL'S Sawmill (Tahquamenon River, Michigan). 1452
PENDLETON, Nathaniel. 2136
PENFOLD, Joseph W. 2508, 3526
PENINSULAR Land Company. 1396
PENN Timber Company. **2648**
PENNELL, Henry. 1119
PENNSYLVANIA (geographical names)
Aaronsburg. 2828
Adams County. 2747
Allegheny National Forest. 2752, **2754**
Alliance Furnace. 2821
Ambler. 2736
Ambridge see Pennsylvania (geographical names), Economy
Austin. 2737
Bartville, Lancaster County. 550
Bear Creek, Luzerne County. 2873
Bellefonte. 2734
Berks County. 2725, 2824
Berkshire Furnace. 2732
Bib Elk Creek, New London Township. 2858
Birdsboro. 2723, 2726
Birmingham Township, Chester County. 2853
Brandywine Springs. 540
Brownsville. 2822
Brush Valley. 2830
Bucks County. 552, 2716-19, 2721
Bucks Mill. 2732
Centre County. 1121, 2734
Centre Furnace. **2831**
Charming Forge. 2732, **2824**
Cherry Tree, Indiana County. 2750
Chester County. 2858
Clearfield County. 1121, 2735
Coatesville, Chester County. 2851
Columbia. 2115, 2805, 2813
Concord Township, Delaware County. 2827
Conowingo Furnace. 2732
Cook Forest
Photographs. 2727
Cornwall Furnace. **2733**
Cumberland County. 2747
Cumberland Furnace. 2747
Delaware Division Canal. 2716

Alabama
(19th century). 9, 1713, 2202, 2206, 2251, 2277
Photographs (1880s). 763
Alaska. 668
Alberta. 218, 3563, 3569, 3581
Interviews. 3595
British Columbia. 3618
(1850s). 3631, 3632, 3749
(1860s—1890s). 3560, 3601, 3604, 3607, 3610, 3617, 3619, 3626
(20th century). 139, 3560, 3602, 3626
British Honduras. 985
California. 73, 80, 85, 343
(1840—1850). 74, 79, 88, 89, 184, 222
(1850—1900). 124, 142, 165, 185, 207, 218, 221, 247, 299, 302, 309, 321, 323-325, 335
(1900—1970). 124, 185, 264
Photographs. 307
Canada. 139, 557
Colorado (1863—1886). 121, 404, 438, 461
Connecticut. 510, 2010
Delaware
(1827—1870). 549, 554, 558, 576
(1880—1919). 527, 562
Delaware River. 2715
Florida
(19th century). 753, 2206, 2210
(20th century). 780, 782
Photographs (1880s). 763
Georgia. 2210
Idaho. 791, 1539, 1872
Interviews. 794
Illinois. 834, 2081
Indiana. 864
(1830—1850). 861, 870, 892
(1850—1884). 851, 854, 858, 866, 885, 896
Kentucky. 948
Lake states. 809
Louisiana. 999
(1741—1800). 963
(1800—1870). 963, 976, 984, 992, 993, 995, 3049
(1870—1910). 970, 976, 988
Maine. 1236
(1631—1700). 1107, 1131, 1222, 1234
(1700—1780). 1088, 1092, 1106, 1131
(1780—1835). 1015, 1020, 1023, 1024, 1093, 1106, 1109, 1117, 1120, 1131, 1237, 1953
(1835—1882). 1008, 1013, 1014, 1020, 1024, 1044, 1059, 1082, 1106, 1111, 1117, 1120, 1128, 1953
(1882—1938). 1020, 1059
Interviews. 1077
Maryland
(1712—1800). 1153
(1800—1900). 540, 1153, 2115
(1900—1944). 1153
Massachusetts
(1770—1800). 1201, 1228, 1230, 1238, 1254, 1258
(1800—1860). 1181, 1183, 1201, 1210, 1228, 1230, 1238, 1254, 1256, 1258
(1860—1907). 1210
Michigan. 1380, 1397, 1451, 1452, 3352
(1820s—1840s). 1290, 1293, 1325, 1353, 1389, 1411, 1427
(1850s). 1264, 1279, 1284, 1293, 1305, 1325, 1389, 1427, 1449, 1450, 1501
(1860s). 1186, 1270, 1279, 1293, 1325, 1344, 1389, 1501
(1870s). 1293, 1325, 1344, 1389, 1445, 1501
(1880s—1890s). 1270, 1293, 1344, 1501
(20th century). 1259, 1293, 1344, 1501
Interviews. 1431
Minnesota. 1502
(1850—1900). 1508, 1530, 1553, 1560, 1568, 1586, 1591, 1598, 1603, 1623, 1642, 1651, 1667, 1681, 1701, 1716, 1718, 1731, 1821, 3324
(1900—1960). 1530, 1532, 1560, 1586, 1598, 1613, 1626, 1651, 1667, 1678, 1701, 1821, 3324
Photographs. 3518
Mississippi
(1824—1903). 972, 1713, 2207, 2301
Photographs (1905—1906). 1762
Missouri
(1837—1860). 1759, 1806, 1807
(1860—1950). 1795
Montana. 1832, 1847, 1857, 1860, 1872, 1885
Interviews. 373, 1831, 1859
Nebraska. 1905
Nevada. 264
New England. 2143
New Brunswick. 3654, 3666, 3672, 3677, 3684, 3686, 3687, 3691, 3693, 3741
New Hampshire (19th century). 1174, 1553, 1929, 1935, 2115, 2987
New Jersey. 1970
(1768—1832). 1948, 1954, 1955, 2146
(1832—1900). 1948, 1952, 1956, 1959, 1964, 1965, 2115
(1900—1924). 1952
New Mexico. 1972
New York
(1685—1700). 2131
(1700—1750). 2131, 2140, 2150, 2151, 2162
(1750—1800). 2010, 2124, 2125, 2131, 2137, 2141, 2150, 2151
(1800—1850). 614, 2001, 2010, 2012, 2043, 2053, 2120, 2131, 2157, 3026
(1850—1900). 2001, 2010, 2029, 2047, 2053, 2058, 2070, 2115, 2131, 2157
(1900—1939). 2054, 2070
Interviews. 2042
North Carolina
(1753—1800). 2236, 2250
(1800—1850). 2201, 2208, 2210, 2214, 2219, 2236, 2243, 2250, 2276, 2306, 2370, 2813
(1850—1900). 2199, 2201, 2210, 2214, 2219, 2236, 2243, 2250, 2272, 2276, 2230
(1900—1962). 2199, 2214, 2217
North Dakota. 1821, 3324
Nova Scotia. 3745
Ohio. 2392, 2397, 2402
Oklahoma. 2435
Ontario
(18th and 19th centuries). 2085, 3714, 3721, 3726, 3729, 3745, 3758, 3763, 3770, 3792, 3806
(20th century). 1532, 3758, 3763, 3770, 3791
Oregon. 2592, 2623, 2949
(1820—1850). 67, 2545, 2567, 2587, 2607, 2671, 2691
(1850—1900). 67, 524, 2447, 2470, 2545, 2624, 2650, 2660, 2707
(20th century). 245, 732, 2447, 2557, 2674
Photographs. 2709
Pacific Northwest. 374, 2565, 2634, 2692, 3096
Pennsylvania. 561, 2865
(18th century). 2711, 2864, 2866, 2869
(1800—1850). 552, 565, 580, 2170, 2716, 2720, 2721, 2726, 2756, 2759, 2761, 2800, 2826, 2827, 2856, 2864, 2866, 2867, 2869, 2871, 3324
(1850—1880). 557, 565, 580, 2115, 2170, 2710, 2718, 2719, 2722, 2726, 2758-2760, 2762-2764, 2812, 2822, 2826, 2828, 2853, 2854, 2861, 2864, 2866-2868, 3324
(1880—1924). 550, 2710, 2718, 2737, 2758-2760, 2828, 2830, 2853, 2860, 2864, 3324
Maps. 2742
Photographs. 2746, 2748
Prince Edward Island. 1391
Quebec. 3728, 3818, 3820, 3827, 3830
South Carolina. 2160, 2210, 2242, 2274, 2278, 2317, 2880, 2886
South Dakota. 1821
Southern states. 515
Tennessee. 2204, 2220, 2901
Texas. 154, 191, 961, 993, 2913
United States
(19th century). 557, 2080
(20th century). 670, 709, 719, 734
Utah. 2938, 2940, 2941, 2946, 2951, 2953
Vermont. 1176, 1179, 2120, 2970, 2983

TAYLOR, James (1809—1894). **2672**
TAYLOR, John (d. 1777). 2145
TAYLOR, Knowles. **1399**
TAYLOR, Lownes. 2867
TAYLOR, Mordecai. 2868
TAYLOR, Norman (1883—1967). 2000
TAYLOR, Will L. 267
TAYLOR, William E. 2057
TAYLOR, William R. (1820—1909). 3481
TAYLOR and Fowler. 993
TAYLOR Family. **993**
TAYLOR Grazing Act. 439
TEAL, Joseph Nathan (1858—1929). **2673**
TEALE, Edwin Way. 2027
TEBBE, Charles L. 2092
TEBBETTS, Reliance (1786—1856). 1127
TELA, Winfree & McCulloch. 2673
TELEGRAPH
  Wisconsin. 3511
TELEPHONE poles
  British Columbia. 3560
  Canada. 1578
  Maryland. 1156
  Minnesota. 1578
TELFAIR, Edward. **2334**
TELFORD, Clarence J. 839
TELLER, Henry Moore (1830—1914). **445**
TEMPERANCE Colonization Society. 3838
TEMPLE, John. 1214
TEMPLE Family. **1214**
TEMPLE Industries, Inc. **2932**
TENNANT, Charles Irvain (1864—1957). **3576**
TENNESSEE (geographical names)
  Bolivar. 2901
  Chattanooga. 2903
  Clarksville see Tennessee (state), Austin Peay State University
  Cleveland see U.S. Forest Service, Cleveland
  Corbandale. 2204
  Crab Orchard. 2899
  Fentress County. 2905
  Great Smoky Mountains. 2899
  Great Smoky Mountains National Park. 3165
    See also North Carolina (geographical names), Great Smoky Mountains National Park; Smoky Mountains Conservation Association
    Interviews. 375
    Land acquisition. 2908
    Logging. 2896
  Lincoln County. 2220
  Memphis. 2897, 2898
    See also Tennessee (state), Memphis State University, Oral History Research Office
  Morgan County. 2905
  Nashville. 2904
  Palmyra. 2204
  Pikeville. 2907
  Polk County. 2907
  Rugby. 2905
  Scott County. 2905
  Sequatchie Valley. 2907
  Stewart County. 2902
  Tennessee Valley. 2902
  Upper Cumberland River Valley. 2906
TENNESSEE (state)
  Attorney General and Reporter's Office. 2908
  Austin Peay State University, Clarksville. 2902
  Geologist. 3006
  Memphis State University, Oral History Research Office. **2898**
TENNESSEE Forestry Association. 2909
TENNESSEE Valley Authority see U.S. Tennessee Valley Authority
TERMINATION Act. 2643
TERMINOLOGY see Logging terminology
TETT, Benjamin. **3804**
TETT, J.P., & Brother. 3804
TETT Family. **3711**
TEXAS
  Albany. 2917
  Amarillo. 2917
    See also U.S. Soil Conservation Service, Amarillo
  Angelina County. 2915
  Bastrop County. 191
  Beaumont. 2915
  Crosbyton. 2923
  Dallas. 2915
  Diboll. 375, 2932
  Ennis. 2915
  Hereford. 2917
  Houston. 2915, 2917
  Jasper County. 2915
  Keltys. 2929
  La Porte. 2915
  Lubbock. 2923
  Mexia. 2915
  Nacogdoches County. 2915
  Orange. 2930
  Texarkana. 161
  Waco. 2917
  Washington County. 961
  Waxahachie. 2915
TEXAS and Louisiana Lumber Manufacturers' Association. 2915
TEXAS Forestry Association. 2912
TEXAS Pine Land Association. 2922
TEXTILE manufacturing
  Massachusetts. 1210
  Delaware. 2788
THAIN, John. 3686
THACHER, George. 1128, 1171
THAXTER, Roland. 1223
THAYER, H.H., Jr. 546, 547
THE PAS Lumber Company. 1748
THISTLE (Mr.). 3787
THOMAS, Bert Carl **2534**
THOMAS, Charles Spaulding. 467, **469**
THOMAS, D. 3818
THOMAS, F.S. 3776
THOMAS, G. Harris. 2114
THOMAS, John Warwick. **2243**
THOMAS, John William Elmer. **2426**
THOMAS, Jonathan. 572
THOMAS, Norman. **3239**
THOMAS, Waverly. **3010**
THOMAS and Beard. 2208
THOMAS and Maulsby. 572
THOMPSON, Samuel, & Son. 2118
THOMPSON, A. 3303
THOMPSON, A.B. **1919**
THOMPSON, Andrew. **561**
THOMPSON, Ben H. 3526
THOMPSON, Charles Donald (1873—1956). **1400**
THOMPSON, Charles E. 1400, **1401**
THOMPSON, Clyde. 375
THOMPSON, D.C. 1129
THOMPSON, Edward Shippen. **2749**
THOMPSON, George Benjamin. **3298**
THOMPSON, James. **1242**, 2869
THOMPSON, Joseph. 1943
THOMPSON, Lewis (Bertie County, North Carolina; l. 1808—1867). **2244**
THOMPSON, Lewis G. (Indiana, fl. 1842). **892**
THOMPSON, William, Jr. (1785—1834). 2749
THOMPSON & McClure. 1767
THOMPSON Family. 3723
THOMPSON McDonald Lumber Company. 1638
THOMPSON Yards (Minnesota). 1741
THOREAU, Henry David. **341**
THOREAU, John. 341
THOREAU, Sophia. 341
THORP, Joseph G. 3313, 3364
THRALL, William H. **342**
THREE Runs Lumber Company. 2246

WILCOX, James David (1827—1916). 2951
WILD rivers
Minnesota. 1703
WILDE, Margaret F. 1075
WILDE, Willard H. 239
WILDER, A.H. 830
WILDER, Charles T. 1212
WILDER, Gaston Henry. **2382**
WILDER, Herbert A. 1212
WILDER, Oshea. 1399
WILDER & Company. **1212**
WILDERNESS
See also Primitive areas; U.S. Forest Service, wilderness; Wilderness Act; Wildlands
Alberta. 3555, 3591
California. 255
Colorado. 412, 427
Idaho. 1858
Maine. 1068
Minnesota. 373, 427, 1510, 1592, 1611, 1615, 1630, 1640, 1672, 1690, 1691, 1704, 1726, 1761, 3513
See also Wild rivers
Montana. 1858
New Jersey. 427
New Mexico. 3513
Ontario. 1690
Oregon. 2703
Pacific Northwest. 374
United States. 274, 373, 385, 724, 933, 1168, 1611, 1630, 1675, 1942, 2023, 2910, 3058, 3513, 3547
National Conference on Outdoor Recreation. 717
National parks. 415
Utah. 1858
Washington. 2692, 3147
Wyoming. 443
WILDERNESS Act. 584, 1592, 1611, 2839, 3147
WILDERNESS areas see Wilderness
WILDERNESS Bill see Wilderness Act
WILDERNESS preservation see Wilderness
WILDERNESS Research Foundation. 1761
WILDERNESS Society. 344, 427, 1592, 1675, 3111
Correspondence with Aldo Leopold, 3513; Karl Onthank, 2508
WILDLANDS
See also U.S. Forest Service Wildland Study (1937)
Alberta. 3562
California. 34
Maine. 1044
New York. 1981
United States. 3107
WILDLIFE
See also Bird protection; Game and fish; Wildlife conservation; Wildlife management; Wildlife research; Wildlife sanctuaries
Alberta. 3559
Arizona. 40
Colorado. 447
Illinois. 1974
Kansas. 447
Maine. 1007
Minnesota. 1502, 1511, 1601, 1615, 1691, 1696, 1821
Missouri. 1783, 1798
Montana. 1874
Nebraska. 447
New England. 1974, 2143
New Mexico. 1974
New York. 1983
North Carolina. 2353
North Dakota. 1821
Oklahoma. 447, 2434
Ohio. 2403
Oregon. 2576, 2641, 2700, 2701
Pennsylvania. 1974
Saskatchewan. 3836
South Carolina. 2889
South Dakota. 447, 1821, 2895
United States. 422, 429, 3546
(19th century). 434, 501, 902
(20th century). 369, 416, 434, 501, 613, 1621, 2023, 2325, 2464, 2910
National forests. 653
Washington. 3111
West Virginia. 3291
Wisconsin. 3540
Wyoming. 447, 1841, 3513, 3532, 3541, 3550
WILDLIFE conservation
See also Game conservation
California. 392
Michigan. 1327
Ohio. 2419
Oklahoma. 2432
United States. 414, 912
WILDLIFE management
See also Game management
Alaska. 3090
Arizona. 49
California. 314
Colorado. 427
Northern Region. 3090
Pacific Northwest. 3090
United States. 419
See also U.S. Forest Service, wildlife management
Wisconsin. 3513
WILDLIFE Management Institute. 27
WILDLIFE refuges see Wildlife sanctuaries
WILDLIFE research
Illinois. 840
United States. 3546
WILDLIFE sanctuaries
Delaware. 644
Maryland. 644
Middle Atlantic states. 644
New England. 644
New Jersey. 427
Oklahoma. 385
Pennsylvania. 414
United States. 643, 694, 1611
WILDMAN, Edward E. 2766
WILEY. Alexander. 3414
WILEY, Farida. 2027
WILHELM, Honor (1870—1957).
Photograph collection. **3230**
WILHELM, S.S. 1376
WILKESON Coal and Coke Company. 3162, 3185
WILKIN, Alexander. **1746**
WILKIN, Westcott. **1746**
WILKINS, Ashley. **2383**
WILKINS, Robert Charles (1782—1866). 3759
WILKINSON, Charles D. 673
WILKINSON, E.J. 1502
WILKINSON Family. **3609**
WILL, Thomas E. 756
WILLAMETTE Iron and Steel Company. **2690**
WILLAMETTE Steam Mill Lumber and Manufacturing Company (Los Angeles). **291**
WILLAMETTE Valley Logging Conference. 2645
WILLAMETTE Valley Wood Chemical Company. 2461
WILLARD, F.E. ("Jess"). 3110
WILLCOX, Robert N. **335**
WILLIAM JONES (schooner). 489
WILLIAMS, Amos (1797—1857). 831
WILLIAMS, Arthur James (b. 1877). **2427**
WILLIAMS, Benjamin F. 2249
WILLIAMS, C.P. 2085
WILLIAMS,.C.P., & Company. **2099**
WILLIAMS, Donald A. 1252
WILLIAMS, Edwards C. (1820—1913). 242
WILLIAMS, Gerhard Mennen (b. 1911). **1371**
WILLIAMS, H.C. 2827
WILLIAMS, H.G. 2183
WILLIAMS, Henry F. (1828—1911). 241

WILLIAMS, J.W. 3358
WILLIAMS, John. 2010
WILLIAMS, Josiah B. **2085**
WILLIAMS, Louis O. (b. 1908). 2819
WILLIAMS, Mary Floyd (1866—1959). **242**
WILLIAMS, Nathan Witter. 519
WILLIAMS, Robert L. **2431**
WILLIAMS, Sarah Frances (Hicks). **2249**
WILLIAMS, T.R. 371
WILLIAMS, Theodore O. 3217
WILLIAMSON, C.D. **3545**
WILLIAMSON, Hugh. **2384**
WILLIAMSON, Maxwell W. 2114
WILLIAMSON, Thomas M. 3303
WILLIS (Mr.) (Maine, sawmill owner, 1835). 1128
WILLIS, George F. 3303
WILLIS, Park Weed. 3200
WILLIS, Raymond E. (1875—1956). **895**
WILLISON, George F. 2950
WILLMORE, Norman Alfred (b. 1909). 3581, **3599**
WILLOW trees
  Delaware. 538
  United States. 626, 2814
WILLOW wood
  Delaware. 540, 541
  Maryland. 540
WILLSON, James M. **2924**
WILSON, A.H., & Company. 764
WILSON, B.F. (b. 1832). 243
WILSON, Bethany (Mrs. Hugh E. Wilson). 1372
WILSON, Charles Scoon. 2089
WILSON, Chester S. 1690
WILSON, E.A., & Company. 3208
WILSON, Etta Smith Wolfe (1857—1936). **1406**
WILSON, Harry Leon. 3196
WILSON, J.M. 3549
WILSON, J.S.F. 2898
WILSON, James (1742—1798). 2125
WILSON, James S.R. (member of Ludlow Yellowstone expedition, 1875). 526
WILSON, Milburn Lincoln (1885—1969). **1838**
WILSON, Milton. 2417
WILSON, P. St. J. 1166
WILSON, R.C. 264
WILSON, Robert (b. 1789). **2250**
WILSON, Thomas. **3577**
WILSON, William (Minnesota). 3313, 3475
WILSON, William (New York). 2153
WILSON, Woodrow (1856—1924). **622**
  Correspondence with James R. Garfield, 597; C.L. Pack, 45
WILSON & Brother. **326**
WILSON-Case Lumber Company. 2473
WILSON Family. 1021
WILY, Hugh. **240**
WIMBUSH, Solomon Mitchell. 1000
WIMMER, Thomas O. **3218**
WINANT & Degraw. 2118
WINCHESTER, Jonas (1810—87). **308**
WINDBREAKS
  Illinois. 834
WIND-DRIVEN sawmills
  New York. 2012
WINDFALLS
  New England. 205, 651
  Pennsylvania
    Photographs. 2727
WINE trade
  British West Indies. 2279
  France. 2279
  Great Britain. 2279
  South Carolina. 2279
WINER, Herbert Isaac (b. 1921). **515**
WING, Leonard D. **3546**
WING, Nelson. **1407**
WINGATE, D.R., Lumber Company. 2915
WINIKER-Heidelbach & Penn. **3012**
WINKENWERDER, Hugo. 3110, **3219**
WINKLER, Valentine. **3652**
WINN, Frederick. **49**
WINN, John. 1239
WINNIPISEOGEE Paper Company. 1945
WINSLOW, Edward. 3668, 3754
WINSLOW, Isaac. 1104
WINSLOW, Jonathan. 1109
WINSLOW Family. **3668**
WINSOR, Luther M. **2952**
WINSTON, Frederick S. 1690, 1691
WINSTON, P.H. 2216
WINTON, Charles J., Family. 1748
WINTON, David J. (b. 1897). 1747
WINTON Lumber Company. **1748**
WINTON-Oregon Timber Company. 1748
WINTON Timber Company. 1748
WIREBOUND box industry
  United States. 720, 721
WIRT, George Hermann (1880—1961). 371, **2753, 2843**
WIRTH, Conrad L. 1278, 2112, 3242, 3526
WISCONSIN (geographical names)
  Adams County. 3459
  Alma. 1537
  Antigo
    Lumber industry. 3396
    Timber lands. 3491
  Arcadia. 3454
  Ashland. 3377, 3448
    Lumber inspectors. 3508
    U.S. Land Office see U.S. General Land Office, Wisconsin
  Ashland County
    Cutover lands. 1698
    Lands. 1729
  Badger Mills. 3307
  Baraboo. 3436
  Barron County. 3412
    Lands. 1729
  Barronett. 1530
  Bayfield. 3324, 3456
  Bayfield County. 3377
  Beef Slough. 1537
  Big Springs, Adams County. 3417
  Birchwood. 3412
  Black River. 3330
  Black River Falls, Jackson County. 3329, 3379, 3445, 3472
    Logging. 3428
  Brantwood. 1608
  Brothertown. 3502
  Bruce, Rusk County. 3426
  Bryant. 3396
  Burnett County. 1729
  Caroline, Shawano County. 3362
  Chequamegon National Forest, Washburn District. 3471
  Chippewa Falls. 1659, 1727, 1740, 3307, 3311, 3399
    Cutover lands. 1526
    Logging. 3494
    Lumber industry. 1555, 1730, 2026, 3403
    Lumber inspectors. 3508
  Chippewa River. 2090, 3498
    Booms. 1543
    Improvement. 1537
    Log driving. 1556
    Lumber industry. 3307, 3403, **3515**
  Chippewa Valley. 3401
  Clark County. 3495
  Clifton Hollow, Pierce County. 3496
  Conners Point. 3518
  Dane County. 3360
  De Pere. 3317
    Sawmills. 3422
  Delavan, Town of Richmond, Walworth County. 3350
  Door County. 3387

North American Forest History: A Guide to Archives was compiled by Richard C. Davis, who was also in charge of copy editing. Proofing was done by Judyl Mudfoot. Text design was by Shelly Lowenkopf. Composition: Camera-ready Composition, Santa Barbara, Calif., using Baskerville type faces for text and display. Printing and binding by Edwards Brothers, Inc., Ann Arbor, Mich., using Kivar 9 cover over standard binder's boards, and a 45-pound Educator's Coated paper. The cover art was designed and prepared by Jack Swartz.